Engineering Mathematics

In memory of Elizabeth

Engineering Mathematics
Second Edition

J. O. Bird, BSc(Hons) CMath, FIMA, CEng, MIEE, FCollP, FIIE

Newnes

OXFORD   AUCKLAND   BOSTON   JOHANNESBURG   MELBOURNE   NEW DELHI

Newnes
An imprint of Butterworth-Heinemann
Linacre House, Jordan Hill, Oxford OX2 8DP
225 Wildwood Avenue, Woburn, MA 01801-2041
A division of Reed Educational and Professional Publishing Ltd

R A member of the Reed Elsevier plc group

First published 1989
Second edition 1996
Reprinted 1998 (twice), 1999

British Library Cataloguing in Publication Data
A catalogue record for this book is available from the British Library

ISBN 0 7506 3121 X

Typeset by Laser Words, Madras, India
Printed and bound in Great Britain

Contents

Preface

The second edition of *Engineering Mathematics* covers the mathematics syllabuses for the Advanced GNVQ in Engineering, also providing a suitable course for BTEC National students.

The aim of this textbook is to develop the student's ability to use mathematics which is appropriate for entry to an engineering degree course; it is widely recognized that such a facility is a key element in determining subsequent success. First year undergraduates who need some remedial mathematics will also find the book meets their needs. In *Engineering Mathematics*, theory is introduced in each chapter by a simple outline of essential definitions, formulae, laws, procedures, etc. The theory is kept to a minimum, for problem solving is extensively used to establish and exemplify the theory. It is intended that readers will gain real understanding through seeing problems solved and then through solving similar problems themselves.

For clarity, the book is divided into four parts, each covering one mathematics syllabus appropriate to the Business and Technology Education Council and City and Guilds of London published units.

Part 1, Mathematics for Engineering, covers the mandatory Advanced GNVQ unit of mathematics and contains material on number and algebra, trigonometry, graphs and differentiation.

Part 2, Further Mathematics for Engineering, extends the mathematics contained in Part 1, and provides mathematical underpinning for students wishing to progress their education further. This part contains material on calculus and its applications, numerical techniques and series.

Part 3, Additional Mathematics for Engineering, extends further the mathematics of the mandatory unit and contains material on differential equations, vectors and phasors and some further trigonometry.

Part 4, Extended Mathematics for Engineering, contains material on further calculus, complex numbers and matrices and completes the suite of mathematics units for the Advanced GNVQ in Engineering.

The aim of the whole suite of four units is to provide the essential mathematics for entry to a wide range of honours degrees in engineering and physics.

Each topic considered in the text is presented in a way that assumes in the reader only the knowledge attained at Intermediate GNVQ in Engineering or GCSE Mathematics (grades A to C).

A set of 20 multiple choice questions has been included at the end of each of the four main parts of the book. This is intended to provide the reader with practise for GNVQ end-of-unit tests.

This fully revised second edition of this practical textbook contains some 360 illustrations and **800 detailed worked problems**, followed by some **1450 further problems (with answers)**.

Engineering Mathematics is a follow-on from *Early Engineering Mathematics* and leads into *Higher Engineering Mathematics*.

I would like to express my appreciation for the friendly co-operation and helpful advice given by the publishers, and to Elaine Woolley for the excellent typing of the manuscript.

John O. Bird

Part 1
Mathematics for Engineering

1

Indices, standard form and binary numbers

1.1 Indices

The lowest factors of 2000 are $2 \times 2 \times 2 \times 2 \times 5 \times 5 \times 5$. These factors are written as $2^4 \times 5^3$, where 2 and 5 are called **bases** and the numbers 4 and 3 are called **indices**.

When an index is an integer it is called a **power**. Thus, 2^4 is called 'two to the power of four', and has a base of 2 and an index of 4. Similarly, 5^3 is called 'five to the power of three' and has a base of 5 and an index of 3. Special names may be used when the indices are 2 and 3, these being called 'squared' and 'cubed', respectively. Thus 7^2 is called 'seven squared' and 9^3 is called 'nine cubed'. When no index is shown, the power is 1, i.e. 2 means 2^1.

Reciprocal

The **reciprocal** of a number is when the index is -1 and its value is given by 1 divided by the base. Thus the reciprocal of 2 is 2^{-1} and its value is $\frac{1}{2}$ or 0.5. Similarly, the reciprocal of 5 is 5^{-1} which means $\frac{1}{5}$ or 0.2.

Square root

The **square root** of a number is when the index is $\frac{1}{2}$, and the square root of 2 is written as $2^{(1/2)}$ or $\sqrt{2}$. The value of a square root is the value of the base which when multiplied by itself gives the number. Since $3 \times 3 = 9$, then $\sqrt{9} = 3$. However, $(-3) \times (-3) = 9$, so $\sqrt{9} = -3$. There are always two answers when finding the square root of a number and this is shown by putting both a $+$ and a $-$ sign in front of the answer to a square root problem. Thus $\sqrt{9} = \pm 3$ and $4^{(1/2)} = \sqrt{4} = \pm 2$, and so on.

Laws of indices

When simplifying calculations involving indices, certain basic rules or laws can be applied, called the **laws of indices**. These are given below.

(i) When multiplying two or more numbers having the same base, the indices are added. Thus $3^2 \times 3^4 = 3^{2+4} = 3^6$

(ii) When a number is divided by a number having the same base, the indices are subtracted. Thus $3^5/3^2 = 3^{5-2} = 3^3$

(iii) When a number which is raised to a power is raised to a further power, the indices are multiplied. Thus $(3^5)^2 = 3^{5 \times 2} = 3^{10}$

(iv) When a number has an index of 0, its value is 1. Thus $3^0 = 1$

(v) A number raised to a negative power is the reciprocal of that number raised to a positive power. Thus $3^{-4} = 1/3^4$. Similarly, $1/2^{-3} = 2^3$

(vi) When a number is raised to a fractional power the denominator of the fraction is the root of the number and the numerator is the power. Thus $8^{(2/3)} = \sqrt[3]{8^2} = (2)^2 = 4$ and $25^{(1/2)} = \sqrt{25^1} = \pm 5$

> **Problem 1.** Evaluate: (a) $5^2 \times 5^3$, (b) $3^2 \times 3^4 \times 3$ and (c) $2 \times 2^2 \times 2^5$

From law (i):

(a) $5^2 \times 5^3 = 5^{(2+3)} = 5^5$
$$= 5 \times 5 \times 5 \times 5 \times 5 = \mathbf{3125}$$

(b) $3^2 \times 3^4 \times 3 = 3^{(2+4+1)} = 3^7 = 3 \times 3 \times \cdots$ to 7 terms $= \mathbf{2187}$

(c) $2 \times 2^2 \times 2^5 = 2^{(1+2+5)} = 2^8 = \mathbf{256}$

> **Problem 2.** Find the value of:
> (a) $\dfrac{7^5}{7^3}$ and (b) $\dfrac{5^7}{5^4}$

From law (ii):

(a) $\dfrac{7^5}{7^3} = 7^{(5-3)} = 7^2 = \mathbf{49}$

(b) $\dfrac{5^7}{5^4} = 5^{(7-4)} = 5^3 = \mathbf{125}$

Problem 3. Evaluate: (a) $5^2 \times 5^3 \div 5^4$ and (b) $(3 \times 3^5) \div (3^2 \times 3^3)$

From laws (i) and (ii):

(a) $5^2 \times 5^3 \div 5^4 = \dfrac{5^2 \times 5^3}{5^4} = \dfrac{5^{(2+3)}}{5^4}$

$= \dfrac{5^5}{5^4} = 5^{(5-4)}$

$= 5^1 = \mathbf{5}$

(b) $(3 \times 3^5) \div (3^2 \times 3^3) = \dfrac{3 \times 3^5}{3^2 \times 3^3} = \dfrac{3^{(1+5)}}{3^{(2+3)}}$

$= \dfrac{3^6}{3^5} = 3^{6-5}$

$= 3^1 = \mathbf{3}$

Problem 4. Simplify: (a) $(2^3)^4$ and (b) $(3^2)^5$, expressing the answers in index form

---

From law (iii):

(a) $(2^3)^4 = 2^{3 \times 4} = \mathbf{2^{12}}$

(b) $(3^2)^5 = 3^{2 \times 5} = \mathbf{3^{10}}$

Problem 5. Evaluate: $\dfrac{(10^2)^3}{10^4 \times 10^2}$

From the laws of indices:

$\dfrac{(10^2)^3}{10^4 \times 10^2} = \dfrac{10^{(2 \times 3)}}{10^{(4+2)}} = \dfrac{10^6}{10^6} = 10^{6-6}$

$= 10^0 = \mathbf{1}$

Problem 6. Find the value of

(a) $\dfrac{2^3 \times 2^4}{2^7 \times 2^5}$ and (b) $\dfrac{(3^2)^3}{3 \times 3^9}$

From the laws of indices:

(a) $\dfrac{2^3 \times 2^4}{2^7 \times 2^5} = \dfrac{2^{(3+4)}}{2^{(7+5)}} = \dfrac{2^7}{2^{12}}$

$= 2^{7-12} = 2^{-5} = \dfrac{1}{2^5} = \dfrac{1}{\mathbf{32}}$

(b) $\dfrac{(3^2)^3}{3 \times 3^9} = \dfrac{3^{2 \times 3}}{3^{1+9}} = \dfrac{3^6}{3^{10}}$

$= 3^{6-10} = 3^{-4} = \dfrac{1}{3^4} = \dfrac{1}{\mathbf{81}}$

Problem 7. Evaluate (a) $4^{1/2}$ (b) $16^{3/4}$ (c) $27^{2/3}$ (d) $9^{-1/2}$

(a) $4^{1/2} = \sqrt{4} = \mathbf{\pm 2}$

(b) $16^{3/4} = \sqrt[4]{16^3} = (2)^3 = \mathbf{8}$

(Note that it does not matter whether the 4th root of 16 is found first or whether 16 cubed is found first–the same answer will result.)

(c) $27^{2/3} = \sqrt[3]{27^2} = (3)^2 = \mathbf{9}$

(d) $9^{-1/2} = \dfrac{1}{9^{1/2}} = \dfrac{1}{\sqrt{9}} = \dfrac{1}{\pm 3} = \mathbf{\pm \dfrac{1}{3}}$

Problem 8. Evaluate $\dfrac{3^3 \times 5^7}{5^3 \times 3^4}$

The laws of indices only apply to terms **having the same base**. Grouping terms having the same base, and then applying the laws of indices to each of the groups independently gives:

$\dfrac{3^3 \times 5^7}{5^3 \times 3^4} = \dfrac{3^3}{3^4} \times \dfrac{5^7}{5^3}$

$= 3^{(3-4)} \times 5^{(7-3)} = 3^{-1} \times 5^4$

$= \dfrac{5^4}{3^1} = \dfrac{625}{3} = \mathbf{208\dfrac{1}{3}}$

Problem 9.   Find the value of

$$\frac{2^3 \times 3^5 \times (7^2)^2}{7^4 \times 2^4 \times 3^3}$$

$$\frac{2^3 \times 3^5 \times (7^2)^2}{7^4 \times 2^4 \times 3^3} = 2^{3-4} \times 3^{5-3} \times 7^{2 \times 2 - 4}$$

$$= 2^{-1} \times 3^2 \times 7^0$$

$$= \frac{1}{2} \times 3^2 \times 1$$

$$= \frac{9}{2} = 4\frac{1}{2}$$

Problem 10.   Evaluate: $\dfrac{4^{1.5} \times 8^{1/3}}{2^2 \times 32^{-2/5}}$

$$4^{1.5} = 4^{3/2} = \sqrt{4^3} = 2^3 = 8,$$

$$8^{1/3} = \sqrt[3]{8} = 2, \ 2^2 = 4$$

$$32^{-2/5} = \frac{1}{32^{2/5}} = \frac{1}{\sqrt[5]{32^2}} = \frac{1}{2^2} = \frac{1}{4}$$

Hence $\dfrac{4^{1.5} \times 8^{1/3}}{2^2 \times 32^{-2/5}} = \dfrac{8 \times 2}{4 \times \frac{1}{4}} = \dfrac{16}{1} = \mathbf{16}$

Alternatively,

$$\frac{4^{1.5} \times 8^{1/3}}{2^2 \times 32^{-2/5}} = \frac{[(2)^2]^{3/2} \times (2^3)^{1/3}}{2^2 \times (2^5)^{-2/5}}$$

$$= \frac{2^3 \times 2^1}{2^2 \times 2^{-2}}$$

$$= 2^{3+1-2-(-2)}$$

$$= 2^4 = \mathbf{16}$$

Problem 11.   Evaluate: $\dfrac{3^2 \times 5^5 + 3^3 \times 5^3}{3^4 \times 5^4}$

Dividing each term by the HCF (i.e. highest common factor) of the three terms, i.e. $3^2 \times 5^3$, gives:

$$\frac{3^2 \times 5^5 + 3^3 \times 5^3}{3^4 \times 5^4} = \frac{\dfrac{3^2 \times 5^5}{3^2 \times 5^3} + \dfrac{3^3 \times 5^3}{3^2 \times 5^3}}{\dfrac{3^4 \times 5^4}{3^2 \times 5^3}}$$

$$= \frac{3^{(2-2)} \times 5^{(5-3)} + 3^{(3-2)} \times 5^{(3-3)}}{3^{(4-2)} \times 5^{(4-3)}}$$

$$= \frac{3^0 \times 5^2 + 3^1 \times 5^0}{3^2 \times 5^1}$$

$$= \frac{1 \times 25 + 3 \times 1}{9 \times 5} = \frac{\mathbf{28}}{\mathbf{45}}$$

Problem 12.   Find the value of

$$\frac{3^2 \times 5^5}{3^4 \times 5^4 + 3^3 \times 5^3}$$

To simplify the arithmetic, each term is divided by the HCF of all the terms, i.e. $3^2 \times 5^3$. Thus

$$\frac{3^2 \times 5^5}{3^4 \times 5^4 + 3^3 \times 5^3} = \frac{\dfrac{3^2 \times 5^5}{3^2 \times 5^3}}{\dfrac{3^4 \times 5^4}{3^2 \times 5^3} + \dfrac{3^3 \times 5^3}{3^2 \times 5^3}}$$

$$= \frac{3^{(2-2)} \times 5^{(5-3)}}{3^{(4-2)} \times 5^{(4-3)} + 3^{(3-2)} \times 5^{(3-3)}}$$

$$= \frac{3^0 \times 5^2}{3^2 \times 5^1 + 3^1 \times 5^0}$$

$$= \frac{1 \times 5^2}{3^2 \times 5 + 3 \times 1}$$

$$= \frac{25}{45 + 3} = \frac{\mathbf{25}}{\mathbf{48}}$$

Problem 13.   Simplify $\dfrac{7^{-3} \times 3^4}{3^{-2} \times 7^5 \times 5^{-2}}$,

expressing the answer in index form with positive indices

Since $7^{-3} = \dfrac{1}{7^3}$, $\dfrac{1}{3^{-2}} = 3^2$ and $\dfrac{1}{5^{-2}} = 5^2$ then

$$\frac{7^{-3} \times 3^4}{3^{-2} \times 7^5 \times 5^{-2}} = \frac{3^4 \times 3^2 \times 5^2}{7^3 \times 7^5} = \frac{3^{(4+2)} \times 5^2}{7^{(3+5)}}$$

$$= \frac{\mathbf{3^6 \times 5^2}}{\mathbf{7^8}}$$

Problem 14. Simplify $\dfrac{16^2 \times 9^{-2}}{4 \times 3^3 - 2^{-3} \times 8^2}$ expressing the answer in index form with positive indices

---

Expressing the numbers in terms of their lowest prime numbers gives:

$$\frac{16^2 \times 9^{-2}}{4 \times 3^3 - 2^{-3} \times 8^2} = \frac{(2^4)^2 \times (3^2)^{-2}}{2^2 \times 3^3 - 2^{-3} \times (2^3)^2}$$

$$= \frac{2^8 \times 3^{-4}}{2^2 \times 3^3 - 2^{-3} \times 2^6}$$

$$= \frac{2^8 \times 3^{-4}}{2^2 \times 3^3 - 2^3}$$

Dividing each term by the HCF (i.e. 2^2) gives:

$$\frac{2^6 \times 3^{-4}}{3^3 - 2} = \frac{2^6}{3^4(3^3 - 2)}$$

Problem 15. Simplify $\left(\frac{4}{3}\right)^3 \times \left(\frac{3}{5}\right)^{-2} \bigg/ \left(\frac{2}{5}\right)^{-3}$ giving the answer with positive indices

---

A fraction raised to a power means that both the numerator and the denominator of the fraction are raised to that power,

i.e. $\left(\dfrac{4}{3}\right)^3 = \dfrac{4^3}{3^3}$

A fraction raised to a negative power has the same value as the inverse of the fraction raised to a positive power.

Thus, $\left(\dfrac{3}{5}\right)^{-2} = \dfrac{1}{\left(\dfrac{3}{5}\right)^2} = \dfrac{1}{\dfrac{3^2}{5^2}} = 1 \times \dfrac{5^2}{3^2} = \dfrac{5^2}{3^2}$

$$= \left(\frac{5}{3}\right)^2$$

Similarly, $\left(\dfrac{2}{5}\right)^{-3} = \left(\dfrac{5}{2}\right)^3$

Thus, $\dfrac{\left(\dfrac{4}{3}\right)^3 \times \left(\dfrac{3}{5}\right)^{-2}}{\left(\dfrac{2}{5}\right)^{-3}} = \dfrac{\dfrac{4^3}{3^3} \times \dfrac{5^2}{3^2}}{\dfrac{5^3}{2^3}}$

$$= \frac{4^3}{3^3} \times \frac{5^2}{3^2} \times \frac{2^3}{5^3}$$

$$= \frac{(2^2)^3 \times 2^3}{3^{(3+2)} \times 5^{(3-2)}}$$

$$= \frac{2^9}{3^5 \times 5}$$

Further problems on indices may be found in Section 1.7, Problems 1 to 22, pages 11 and 12.

1.2 Standard form

A number written with one digit to the left of the decimal point and multiplied by 10 raised to some power is said to be written in **standard form**. Thus: 5837 is written as 5.837×10^3 in standard form, and 0.0451 is written as 4.15×10^{-2} in standard form.

When a number is written in standard form, the first factor is called the **mantissa** and the second factor is called the **exponent**. Thus the number 5.8×10^3 has a mantissa of 5.8 and an exponent of 10^3.

(i) Numbers having the same exponent can be added or subtracted in standard form by adding or subtracting the mantissae and keeping the exponent the same. Thus:

$$2.3 \times 10^4 + 3.7 \times 10^4 = (2.3 + 3.7) \times 10^4$$
$$= 6.0 \times 10^4, \text{ and}$$

$$5.9 \times 10^{-2} - 4.6 \times 10^{-2} = (5.9 - 4.6) \times 10^{-2}$$
$$= 1.3 \times 10^{-2}$$

When the numbers have different exponents, one way of adding or subtracting the numbers is to express one of the numbers in non-standard form, so that both numbers have the same exponent. Thus:

$$2.3 \times 10^4 + 3.7 \times 10^3$$
$$= 2.3 \times 10^4 + 0.37 \times 10^4$$
$$= (2.3 + 0.37) \times 10^4$$
$$= 2.67 \times 10^4$$

Alternatively, $2.3 \times 10^4 + 3.7 \times 10^3$
$$= 23\,000 + 3700 = 26\,700 = 2.67 \times 10^4$$

(ii) The laws of indices are used when multiplying or dividing numbers given in standard form.

For example,

$$(2.5 \times 10^3) \times (5 \times 10^2)$$

$$= (2.5 \times 5) \times (10^{3+2})$$

$$= 12.5 \times 10^5 \text{ or } 1.25 \times 10^6$$

Similarly,

$$\frac{6 \times 10^4}{1.5 \times 10^2} = \frac{6}{1.5} \times (10^{4-2}) = 4 \times 10^2$$

Problem 16.   Express in standard form:
(a) 38.71 (b) 3746 (c) 0.0124

For a number to be in standard form, it is expressed with only one digit to the left of the decimal point. Thus:

(a)   38.71 must be divided by 10 to achieve one digit to the left of the decimal point and it must also be multiplied by 10 to maintain the equality, i.e.

$$38.71 = \frac{38.71}{10} \times 10 = \mathbf{3.871 \times 10} \text{ in standard}$$

form

(b) $3746 = \dfrac{3746}{1000} \times 1000 = \mathbf{3.746 \times 10^3}$ in standard form

(c) $0.0124 = 0.0124 \times \dfrac{100}{100} = \dfrac{1.24}{100}$

$$= \mathbf{1.24 \times 10^{-2}} \text{ in standard form}$$

Problem 17.   Express the following numbers, which are in standard form, as decimal numbers:
(a) 1.725×10^{-2} (b) 5.491×10^4
(c) 9.84×10^0

(a) $1.725 \times 10^{-2} = \dfrac{1.725}{100} = \mathbf{0.01725}$

(b) $5.491 \times 10^4 = 5.491 \times 10\,000 = \mathbf{54\,910}$

(c) $9.84 \times 10^0 = 9.84 \times 1 = \mathbf{9.84}$ (since $10^0 = 1$)

Problem 18.   Express in standard form, correct to 3 significant figures:

(a) $\dfrac{3}{8}$ (b) $19\dfrac{2}{3}$ (c) $741\dfrac{9}{16}$

(a) $\frac{3}{8} = 0.375$, and expressing it in standard form gives: $0.375 = \mathbf{3.75 \times 10^{-1}}$

(b) $19\frac{2}{3} = 19.\dot{6} = \mathbf{1.97 \times 10}$ in standard form, correct to 3 significant figures

(c) $741\frac{9}{16} = 741.5625 = \mathbf{7.42 \times 10^2}$ in standard form, correct to 3 significant figures

---

Problem 19.   Express the following numbers, given in standard form, as fractions or mixed numbers:
(a) 2.5×10^{-1} (b) 6.25×10^{-2}
(c) 1.354×10^2

(a) $2.5 \times 10^{-1} = \dfrac{2.5}{10} = \dfrac{25}{100} = \mathbf{\dfrac{1}{4}}$

(b) $6.25 \times 10^{-2} = \dfrac{6.25}{100} = \dfrac{625}{10\,000} = \mathbf{\dfrac{1}{16}}$

(c) $1.354 \times 10^2 = 135.4 = 135\dfrac{4}{10} = \mathbf{135\dfrac{2}{5}}$

Problem 20.   Find the value of
(a) $7.9 \times 10^{-2} - 5.4 \times 10^{-2}$
(b) $8.3 \times 10^3 + 5.415 \times 10^3$ and
(c) $9.293 \times 10^2 + 1.3 \times 10^3$
expressing the answers in standard form

Numbers having the same exponent can be added or subtracted by adding or subtracting the mantissae and keeping the exponent the same. Thus:

(a) $7.9 \times 10^{-2} - 5.4 \times 10^{-2} = (7.9 - 5.4) \times 10^{-2}$

$$= \mathbf{2.5 \times 10^{-2}}$$

(b) $8.3 \times 10^3 + 5.415 \times 10^3$

$$= (8.3 + 5.415) \times 10^3$$

$$= 13.715 \times 10^3$$

$$= \mathbf{1.3715 \times 10^4} \text{ in standard form}$$

(c)   Since only numbers having the same exponents can be added by straight addition of the mantissae, the numbers are converted to this form before adding. Thus:

$$9.293 \times 10^2 + 1.3 \times 10^3$$

$$= 9.293 \times 10^2 + 13 \times 10^2$$

$$= (9.293 + 13) \times 10^2$$

$$= 22.293 \times 10^2$$

$$= \mathbf{2.2293 \times 10^3} \text{ in standard form}$$

Alternatively, the numbers can be expressed as decimal fractions, giving

$$9.293 \times 10^2 + 1.3 \times 10^3$$

$$= 929.3 + 1300$$

$$= 2229.3$$

$$= \mathbf{2.2293 \times 10^3} \text{ in standard form}$$

as obtained previously. This method is often the 'safest' way of doing this type of problem.

Problem 21.   Evaluate
(a) $(3.75 \times 10^3)(6 \times 10^4)$ and

(b) $\dfrac{3.5 \times 10^5}{7 \times 10^2}$ expressing answers in standard form

---

(a) $(3.75 \times 10^3)(6 \times 10^4) = (3.75 \times 6)(10^{3+4})$

$$= 22.50 \times 10^7$$

$$= \mathbf{2.25 \times 10^8}$$

(b) $\dfrac{3.5 \times 10^5}{7 \times 10^2} = \dfrac{3.5}{7} \times 10^{5-2} = 0.5 \times 10^3$

$$= \mathbf{5 \times 10^2}$$

Further problems on standard form may be found in Section 1.7, Problems 23 to 34, pages 12 and 13.

1.3   Binary numbers

The system of numbers in everyday use is the **denary** or **decimal** system of numbers, using the digits 0 to 9. It has ten different digits (0, 1, 2, 3, 4, 5, 6, 7, 8 and 9) and is said to have a **radix** or **base** of 10.

The binary system of numbers has a radix of 2 and uses only the digits 0 and 1.

1.4   Conversion of binary to denary

The denary number 234.5 is equivalent to

$$2 \times 10^2 + 3 \times 10^1 + 4 \times 10^0 + 5 \times 10^{-1}$$

i.e. it is the sum of terms comprising: (a digit) multiplied by (the base raised to some power).

In the binary system of numbers, the base is 2, so 1101.1 is equivalent to:

$$1 \times 2^3 + 1 \times 2^2 + 0 \times 2^1 + 1 \times 2^0 + 1 \times 2^{-1}$$

Thus the denary number equivalent to the binary number 1101.1 is $8 + 4 + 0 + 1 + \frac{1}{2}$, that is 13.5, i.e. $1101.1_2 = 13.5_{10}$, the suffixes 2 and 10 denoting binary and denary systems of numbers respectively.

Problem 22.   Convert 11011_2 to a denary number

From above:

$$11011_2 = 1 \times 2^4 + 1 \times 2^3 + 0 \times 2^2 + 1 \times 2^1$$

$$+ 1 \times 2^0$$

$$= 16 + 8 + 0 + 2 + 1$$

$$= \mathbf{27_{10}}$$

Problem 23.   Convert 0.1011_2 to a denary fraction

$$0.1011_2 = 1 \times 2^{-1} + 0 \times 2^{-2} + 1 \times 2^{-3} + 1 \times 2^{-4}$$

$$= 1 \times \frac{1}{2} + 0 \times \frac{1}{2^2} + 1 \times \frac{1}{2^3} + 1 \times \frac{1}{2^4}$$

$$= \frac{1}{2} + \frac{1}{8} + \frac{1}{16}$$

$$= 0.5 + 0.125 + 0.0625$$

$$= \mathbf{0.6875_{10}}$$

Problem 24.   Convert 101.0101_2 to a denary number

$$101.0101_2 = 1 \times 2^2 + 0 \times 2^1 + 1 \times 2^0 + 0 \times 2^{-1}$$

$$+ 1 \times 2^{-2} + 0 \times 2^{-3} + 1 \times 2^{-4}$$

$$= 4 + 0 + 1 + 0 + 0.25 + 0 + 0.0625$$

$$= \mathbf{5.3125_{10}}$$

1.5   Conversion of denary to binary

An integer denary number can be converted to a corresponding binary number by repeatedly dividing by 2 and noting the remainder at each stage, as shown below for 39_{10}.

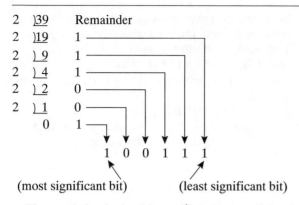

(most significant bit)   (least significant bit)

The result is obtained by writing the top digit of the remainder as the least significant bit (a bit is a **binary digit** and the least significant bit is the one on the right). The bottom bit of the remainder is the most significant bit, i.e. the bit on the left. **Thus $39_{10} = 100111_2$.**

The fractional part of a denary number can be converted to a binary number by repeatedly multiplying by 2, as shown below for the fraction 0.625.

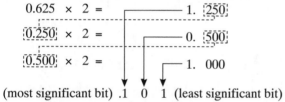

(most significant bit) .1   0   1 (least significant bit)

For fractions, the most significant bit of the result is the top bit obtained from the integer part of multiplication by 2. The least significant bit of the result is the bottom bit obtained from the integer part of multiplication by 2. **Thus $0.625_{10} = 0.101_2$.**

Problem 25.   Convert 47_{10} to a binary number

From above, repeatedly dividing by 2 and noting the remainder gives:

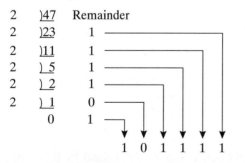

Thus $47_{10} = 101111_2$.

Problem 26.   Convert 0.40625_{10} to a binary number

From above, repeatedly multiplying by 2 gives:

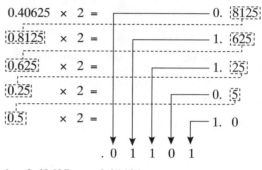

i.e. **$0.40625_{10} = 0.01101_2$.**

Problem 27.   Convert 58.3125_{10} to a binary number

The integer part is repeatedly divided by 2, giving:

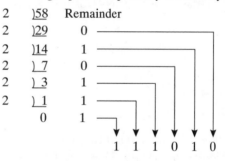

The fractional part is repeatedly multiplied by 2 giving:

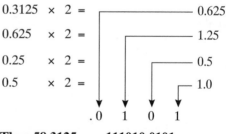

Thus $58.3125_{10} = 111010.0101_2$.

1.6   Conversion of denary to binary via octal

For denary integers containing several digits, repeatedly dividing by 2 can be a lengthy process. In this case, it is usually easier to convert a denary number to a binary number via the octal system of numbers. This system has a radix of 8, using the digits 0, 1, 2, 3, 4, 5, 6 and 7. The denary number equivalent to the octal number 4317_8 is

$$4 \times 8^3 + 3 \times 8^2 + 1 \times 8^1 + 7 \times 8^0$$

i.e. $4 \times 512 + 3 \times 64 + 1 \times 8 + 7 \times 1$ or 2255_{10}

An integer denary number can be converted to a corresponding octal number by repeatedly dividing by 8 and noting the remainder at each stage, as shown below for 493_{10}.

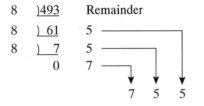

Thus $493_{10} = 755_8$.

The fractional part of a denary number can be converted to an octal number by repeatedly multiplying by 8, as shown below for the fraction 0.4375_{10}.

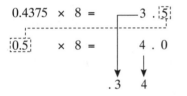

For fractions, the most significant bit is the top integer obtained by multiplication of the denary fraction by 8, thus

$$0.4375_{10} = 0.34_8$$

The natural binary code for digits 0 to 7 is shown in Table 1.1, and an octal number can be converted

Table 1.1

Octal digit	Natural binary number
0	000
1	001
2	010
3	011
4	100
5	101
6	110
7	111

to a binary number by writing down the three bits corresponding to the octal digit.
Thus $437_8 = 100\ 011\ 111_2$
and $26.35_8 = 010\ 110.011\ 101_2$.
The '0' on the extreme left does not signify anything, thus
$26.35_8 = 10\ 110.011\ 101_2$

Conversion of denary to binary via octal is demonstrated in the following worked problems.

Problem 28.   Convert 3714_{10} to a binary number, via octal

---

Dividing repeatedly by 8, noting the remainder gives:

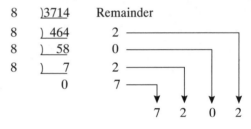

From Table 1.1, $7202_8 = 111\ 010\ 000\ 010_2$

i.e.   **$3714_{10} = 111\ 010\ 000\ 010_2$**

Problem 29.   Convert 0.59375_{10} to a binary number, via octal

---

Multiplying repeatedly by 8, noting the integer values, gives:

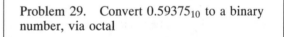

Thus $0.59375_{10} = 0.46_8$.
From Table 1.1, $0.46_8 = 0.100\ 110_2$
i.e. **$0.59375_{10} = 0.100\ 11_2$**

Problem 30.   Convert 5613.90625_{10} to a binary number, via octal

---

The integer part is repeatedly divided by 8, noting the remainder, giving:

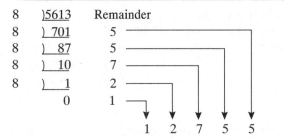

This octal number is converted to a binary number (see Table 1.1)

$$12\ 755_8 = 001\ 010\ 111\ 101\ 101_2$$

i.e. $5613_{10} = 1\ 010\ 111\ 101\ 101_2$

The fractional part is repeatedly multiplied by 8, noting the integer part, giving:

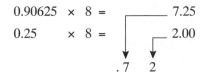

This octal fraction is converted to a binary number (see Table 1.1)

$$0.72_8 = 0.111\ 010_2$$

i.e. $0.90625_{10} = 0.111\ 01_2$

Thus, $5613.90625_{10} = 1\ 010\ 111\ 101\ 101.111\ 01_2$.

Problem 31. Convert $11\ 110\ 011.100\ 01_2$ to a denary number, via octal

---

Grouping the binary number in threes from the binary point gives: $011\ 110\ 011.100\ 010_2$

Using Table 1.1 to convert this binary number to an octal number gives:

$$363.42_8 \text{ and}$$

$$363.42_8 = 3 \times 8^2 + 6 \times 8^1 + 3 \times 8^0$$
$$+ 4 \times 8^{-1} + 2 \times 8^{-2}$$
$$= 192 + 48 + 3 + 0.5 + 0.03125$$
$$= 243.53125_{10}$$

Further problems on binary numbers may be found in the following Section 1.7, Problems 35 to 46, page 13.

1.7 Further problems on indices, standard form and binary numbers

Indices

In Problems 1 to 14, simplify the expressions given, expressing the answers in index form and with positive indices:

1 (a) $3^3 \times 3^4$ (b) $4^2 \times 4^3 \times 4^4$

$$[\text{(a)}\ 3^7\ \text{(b)}\ 4^9]$$

2 (a) $2^3 \times 2 \times 2^2$ (b) $7^2 \times 7^4 \times 7 \times 7^3$

$$[\text{(a)}\ 2^6\ \text{(b)}\ 7^{10}]$$

3 (a) $\dfrac{2^4}{2^3}$ (b) $\dfrac{3^7}{3^2}$ $[\text{(a)}\ 2\ \text{(b)}\ 3^5]$

4 (a) $5^6 \div 5^3$ (b) $7^{13}/7^{10}$ $[\text{(a)}\ 5^3\ \text{(b)}\ 7^3]$

5 (a) $(7^2)^3$ (b) $(3^3)^2$ $[\text{(a)}\ 7^6\ \text{(b)}\ 3^6]$

6 (a) $(15^3)^5$ (b) $(17^2)^4$ $[\text{(a)}\ 15^{15}\ \text{(b)}\ 17^8]$

7 (a) $\dfrac{2^2 \times 2^3}{2^4}$ (b) $\dfrac{3^7 \times 3^4}{3^5}$ $[\text{(a)}\ 2\ \text{(b)}\ 3^6]$

8 (a) $\dfrac{5^7}{5^2 \times 5^3}$ (b) $\dfrac{13^5}{13 \times 13^2}$ $[\text{(a)}\ 5^2\ \text{(b)}\ 13^2]$

9 (a) $\dfrac{(9 \times 3^2)^3}{(3 \times 27)^2}$ (b) $\dfrac{(16 \times 4)^2}{(2 \times 8)^3}$ $[\text{(a)}\ 3^4\ \text{(b)}\ 1]$

10 (a) $\dfrac{5^{-2}}{5^{-4}}$ (b) $\dfrac{3^2 \times 3^{-4}}{3^3}$ $\left[\text{(a)}\ 5^2\ \text{(b)}\ \dfrac{1}{3^5}\right]$

11 (a) $\dfrac{7^2 \times 7^{-3}}{7 \times 7^{-4}}$ (b) $\dfrac{2^3 \times 2^{-4} \times 2^5}{2 \times 2^{-2} \times 2^6}$

$$\left[\text{(a)}\ 7^2\ \text{(b)}\ \dfrac{1}{2}\right]$$

12 (a) $13 \times 13^{-2} \times 13^4 \times 13^{-3}$ (b) $\dfrac{5^{-7} \times 5^2}{5^{-8} \times 5^3}$

$$[\text{(a)}\ 1\ \text{(b)}\ 1]$$

13 (a) $\dfrac{3^3 \times 5^2}{5^4 \times 3^4}$ (b) $\dfrac{7^{-2} \times 3^{-2}}{3^5 \times 7^4 \times 7^{-3}}$

$$\left[\text{(a)}\ \dfrac{1}{3 \times 5^2}\ \text{(b)}\ \dfrac{1}{7^3 \times 3^7}\right]$$

14 (a) $\dfrac{4^2 \times 9^3}{8^3 \times 3^4}$ (b) $\dfrac{8^{-2} \times 5^2 \times 3^{-4}}{25^2 \times 2^4 \times 9^{-2}}$

$$\left[\text{(a)}\ \dfrac{3^2}{2^5}\ \text{(b)}\ \dfrac{1}{2^{10} \times 5^2}\right]$$

15   Evaluate (a) $\left(\dfrac{1}{3^2}\right)^{-1}$ (b) $81^{0.25}$ (c) $16^{(-1/4)}$

   (d) $\left(\dfrac{4}{9}\right)^{(1/2)}$ $\left[\text{(a) 9 (b) } \pm 3 \text{ (c) } \pm\dfrac{1}{2} \text{ (d) } \pm\dfrac{2}{3}\right]$

In Problems 16 to 22, evaluate the expressions given:

16   $\dfrac{9^2 \times 7^4}{3^4 \times 7^4 + 3^3 \times 7^2}$ $\left[\dfrac{147}{148}\right]$

17   $\dfrac{3^3 \times 5^2}{2^3 \times 3^2 - 8^2 \times 9}$ $\left[-1\dfrac{19}{56}\right]$

18   $\dfrac{3^3 \times 7^2 - 5^2 \times 7^3}{3^2 \times 5 \times 7^2}$ $\left[-3\dfrac{13}{45}\right]$

19   $\dfrac{(2^4)^2 - 3^{-2} \times 4^4}{2^3 \times 16^2}$ $\left[\dfrac{1}{9}\right]$

20   $\dfrac{\left(\dfrac{1}{2}\right)^3 - \left(\dfrac{2}{3}\right)^{-2}}{\left(\dfrac{3}{5}\right)^2}$ $\left[-5\dfrac{65}{72}\right]$

21   $\left(\dfrac{4}{3}\right)^4 \Big/ \left(\dfrac{2}{9}\right)^2$ $[64]$

22   $\dfrac{(3^2)^{3/2} \times (8^{1/3})^2}{(3)^2 \times (4^3)^{1/2} \times (9)^{-1/2}}$ $\left[4\dfrac{1}{2}\right]$

Standard form

In Problems 23 to 27, express in standard form:

23   (a) 73.9 (b) 28.4 (c) 197.62

   $\left[\begin{array}{l}\text{(a) } 7.39 \times 10 \text{ (b) } 2.84 \times 10 \\ \text{(c) } 1.9762 \times 10^2\end{array}\right]$

24   (a) 2748 (b) 33 170 (c) 274 218

   $\left[\begin{array}{l}\text{(a) } 2.748 \times 10^3 \\ \text{(b) } 3.317 \times 10^4 \\ \text{(c) } 2.74218 \times 10^5\end{array}\right]$

25   (a) 0.2401 (b) 0.0174 (c) 0.00923

   $\left[\begin{array}{l}\text{(a) } 2.401 \times 10^{-1} \\ \text{(b) } 1.74 \times 10^{-2} \\ \text{(c) } 9.23 \times 10^{-3}\end{array}\right]$

26   (a) 1702.3 (b) 10.04 (c) 0.0109

   $\left[\begin{array}{l}\text{(a) } 1.7023 \times 10^3 \\ \text{(b) } 1.004 \times 10 \\ \text{(c) } 1.09 \times 10^{-2}\end{array}\right]$

27   (a) $\dfrac{1}{2}$ (b) $11\dfrac{7}{8}$ (c) $130\dfrac{3}{5}$ (d) $\dfrac{1}{32}$

   $\left[\begin{array}{l}\text{(a) } 5 \times 10^{-1} \\ \text{(b) } 1.1875 \times 10 \\ \text{(c) } 1.306 \times 10^2 \\ \text{(d) } 3.125 \times 10^{-2}\end{array}\right]$

In Problems 28 and 29, express the numbers given as integers or decimal fractions:

28   (a) 1.01×10^3 (b) 9.327×10^2 (c) 5.41×10^4
   (d) 7×10^0

   [(a) 1010 (b) 932.7 (c) 54 100 (d) 7]

29   (a) 3.89×10^{-2} (b) 6.741×10^{-1} (c) 8×10^{-3}

   [(a) 0.0389 (b) 0.6741 (c) 0.008]

In Problems 30 to 33, find values of the expressions given, stating the answers in standard form:

30   (a) $3.7 \times 10^2 + 9.81 \times 10^2$
   (b) $1.431 \times 10^{-1} + 7.3 \times 10^{-1}$
   (c) $2.68 \times 10^{-2} - 8.414 \times 10^{-2}$

   $\left[\begin{array}{l}\text{(a) } 1.351 \times 10^3 \\ \text{(b) } 8.731 \times 10^{-1} \\ \text{(c) } -5.734 \times 10^{-2}\end{array}\right]$

31   (a) $4.831 \times 10^2 + 1.24 \times 10^3$
   (b) $3.24 \times 10^{-3} - 1.11 \times 10^{-4}$
   (c) $1.81 \times 10^2 + 3.417 \times 10^2 - 5.972 \times 10^2$

   $\left[\begin{array}{l}\text{(a) } 1.7231 \times 10^3 \\ \text{(b) } 3.129 \times 10^{-3} \\ \text{(c) } -7.45 \times 10\end{array}\right]$

32   (a) $(4.5 \times 10^{-2})(3 \times 10^3)$ (b) $2 \times (5.5 \times 10^4)$

   [(a) 1.35×10^2 (b) 1.1×10^5]

33   (a) $\dfrac{6 \times 10^{-3}}{3 \times 10^{-5}}$ (b) $\dfrac{(2.4 \times 10^3)(3 \times 10^{-2})}{(4.8 \times 10^4)}$

   [(a) 2×10^2 (b) 1.5×10^{-3}]

34   Write the following statements in standard form.
   (a)   The density of aluminium is 2710 kg m^{-3}
   (b)   Poisson's ratio for gold is 0.44
   (c)   The impedance of free space is 376.73 Ω
   (d)   The electron rest energy is 0.511 MeV
   (e)   Proton charge–mass ratio is 95 789 700 C kg^{-1}

(f) The normal volume of a perfect gas is 0.02241 m^3 mol^{-1}

$$\begin{bmatrix} \text{(a) } 2.71 \times 10^3 \text{ kg m}^{-3} \\ \text{(b) } 4.4 \times 10^{-1} \\ \text{(c) } 3.7673 \times 10^2 \text{ } \Omega \\ \text{(d) } 5.11 \times 10^{-1} \text{ MeV} \\ \text{(e) } 9.57897 \times 10^7 \text{ C kg}^{-1} \\ \text{(f) } 2.241 \times 10^{-2} \text{ m}^3 \text{ mol}^{-1} \end{bmatrix}$$

Binary numbers

In Problems 35 to 38, convert the binary numbers given to denary numbers:

35 (a) 110 (b) 1011 (c) 1110 (d) 1001

$$[\text{(a) } 6_{10} \text{ (b) } 11_{10} \text{ (c) } 14_{10} \text{ (d) } 9_{10}]$$

36 (a) 10 101 (b) 11 001 (c) 101 101 (d) 110 011

$$[\text{(a) } 21_{10} \text{ (b) } 25_{10} \text{ (c) } 45_{10} \text{ (d) } 51_{10}]$$

37 (a) 0.1101 (b) 0.11001 (c) 0.00111 (d) 0.01011

$$\begin{bmatrix} \text{(a) } 0.8125_{10} \text{ (b) } 0.78125_{10} \\ \text{(c) } 0.21875_{10} \text{ (d) } 0.34375_{10} \end{bmatrix}$$

38 (a) 11010.11 (b) 10111.011 (c) 110101.0111 (d) 11010101.10111

$$\begin{bmatrix} \text{(a) } 26.75_{10} \text{ (b) } 23.375_{10} \\ \text{(c) } 53.4375_{10} \text{ (d) } 213.71875_{10} \end{bmatrix}$$

In Problems 39 to 42, convert the denary numbers given to binary numbers:

39 (a) 5 (b) 15 (c) 19 (d) 29

$$[\text{(a) } 101_2 \text{ (b) } 1111_2 \text{ (c) } 10 \text{ } 011_2 \text{ (d) } 11 \text{ } 101_2]$$

40 (a) 31 (b) 42 (c) 57 (d) 63

$$\begin{bmatrix} \text{(a) } 11 \text{ } 111_2 \text{ (b) } 101 \text{ } 010_2 \\ \text{(c) } 111 \text{ } 001_2 \text{ (d) } 111 \text{ } 111_2 \end{bmatrix}$$

41 (a) 0.25 (b) 0.21875 (c) 0.28125 (d) 0.59375

$$\begin{bmatrix} \text{(a) } 0.01_2 \text{ (b) } 0.001 \text{ } 11_2 \\ \text{(c) } 0.010 \text{ } 01_2 \text{ (d) } 0.100 \text{ } 11_2 \end{bmatrix}$$

42 (a) 47.40625 (b) 30.8125 (c) 53.90625 (d) 61.65625

$$\begin{bmatrix} \text{(a) } 101 \text{ } 111.011 \text{ } 01_2 \\ \text{(b) } 11 \text{ } 110.110 \text{ } 1_2 \\ \text{(c) } 110 \text{ } 101.111 \text{ } 01_2 \\ \text{(d) } 111 \text{ } 101.101 \text{ } 01_2 \end{bmatrix}$$

In Problems 43 to 45, convert the denary numbers given to binary numbers, via octal:

43 (a) 343 (b) 572 (c) 1265

$$\begin{bmatrix} \text{(a) } 101 \text{ } 010 \text{ } 111_2 \\ \text{(b) } 1 \text{ } 000 \text{ } 111 \text{ } 100_2 \\ \text{(c) } 10 \text{ } 011 \text{ } 110 \text{ } 001_2 \end{bmatrix}$$

44 (a) 0.46875 (b) 0.6875 (c) 0.71875

$$\begin{bmatrix} \text{(a) } 0.011 \text{ } 11_2 \\ \text{(b) } 0.101 \text{ } 1_2 \\ \text{(c) } 0.101 \text{ } 11_2 \end{bmatrix}$$

45 (a) 247.09375 (b) 514.4375 (c) 1716.78125

$$\begin{bmatrix} \text{(a) } 11 \text{ } 110 \text{ } 111.000 \text{ } 11_2 \\ \text{(b) } 1 \text{ } 000 \text{ } 000 \text{ } 010.011 \text{ } 1_2 \\ \text{(c) } 11 \text{ } 010 \text{ } 110 \text{ } 100.110 \text{ } 01_2 \end{bmatrix}$$

46 Convert the binary numbers given to denary numbers, via octal:

(a) 111.011 1 (b) 101 001.01 (c) 1 110 011 011 010.001 1

$$\begin{bmatrix} \text{(a) } 7.4375_{10} \\ \text{(b) } 41.25_{10} \\ \text{(c) } 7386.1875_{10} \end{bmatrix}$$

2

Errors, calculations and evaluation of formulae

2.1 Errors

(i) In all problems in which the measurement of distance, time, mass or other quantities occurs, an exact answer cannot be given; only an answer which is correct to a stated degree of accuracy can be given. To take account of this an **error due to measurement** is said to exist.

(ii) To take account of measurement errors it is usual to limit answers so that the result given is **not more than one significant figure greater than the least accurate number given in the data**.

(iii) **Rounding-off errors** can exist with decimal fractions. For example, to state that $\pi = 3.142$ is not strictly correct, but '$\pi = 3.142$ correct to 4 significant figures' is a true statement. (Actually, $\pi = 3.141\ 592\ 65\ldots$)

(iv) It is possible, through an incorrect procedure, to obtain the wrong answer to a calculation. This type of error is known as a **blunder**.

(v) **An order of magnitude error** is said to exist if incorrect positioning of the decimal point occurs after a calculation has been completed.

(vi) Blunders and order of magnitude errors can be reduced by determining **approximate values of calculations**. Answers which do not seem feasible must be checked and the calculation must be repeated as necessary.

Problem 1. The area A of a triangle is given by $A = \frac{1}{2}bh$. The base b when measured is found to be 3.26 cm, and the perpendicular height h is 7.5 cm. Determine the area of the triangle

---

Area of triangle $= \frac{1}{2}bh = \frac{1}{2} \times 3.26 \times 7.5$

$= 12.225 \text{ cm}^2$ (by calculator)

The approximate value is $\frac{1}{2} \times 3 \times 8 = 12 \text{ cm}^2$, so there are no obvious blunders or magnitude errors. However, it is not usual in a measurement-type problem to state the answer to an accuracy greater than 1 significant figure more than the least accurate number in the data: this is 7.5 cm, so the result should not have more than 3 significant figures. Thus **area of triangle $= 12.2 \text{ cm}^2$**.

Problem 2. State which type of error has been made in the following statements:
(a) $72 \times 31.429 = 2262.9$
(b) $16 \times 0.08 \times 7 = 89.6$
(c) $11.714 \times 0.0088 = 0.3247$ correct to 4 decimal places
(d) $\dfrac{29.74 \times 0.0512}{11.89} = 0.12$, correct to 2 significant figures

---

(a) $72 \times 31.429 = 2262.888$ (by calculator), hence a **rounding-off error** has occurred. The answer should have stated:
$72 \times 31.429 = 2262.9$, correct to 5 significant figures

(b) $16 \times 0.08 \times 7 = \cancel{16}^{4} \times \dfrac{8}{\cancel{100}_{25}} \times 7 = \dfrac{32 \times 7}{25}$

$= \dfrac{224}{25} = 8\dfrac{24}{25} = 8.96$

Hence an **order of magnitude** error has occurred

(c) 11.714×0.0088 is approximately equal to $12 \times 9 \times 10^{-3}$, i.e. about 108×10^{-3} or 0.108. Thus a **blunder** has been made

(d) $\dfrac{29.74 \times 0.0512}{11.89} \approx \dfrac{30 \times 5 \times 10^{-2}}{12} = \dfrac{150}{12 \times 10^2}$

$= \dfrac{15}{120} = \dfrac{1}{8}$ or 0.125

hence no order of magnitude error has occurred. However, $(29.74 \times 0.0512)/11.89 = 0.128$ correct to 3 significant figures, which equals 0.13 correct to 2 significant figures. Hence a **rounding-off error** has occurred

---

Further problems on errors may be found in Section 2.5, Problems 1 to 5, page 19.

2.2   Use of calculator

The most modern aid to calculations is the pocket-sized electronic calculator. With one of these, calculations can be quickly and accurately performed, correct to about 9 significant figures. The scientific type of calculator has made the use of tables and logarithms largely redundant.

To help you to become competent at using your calculator check that you agree with the answers to the following problems:

> **Problem 3.**   Evaluate the following, correct to 4 significant figures:
>
> (a) $4.7826 + 0.02713$ (b) $17.6941 - 11.8762$
> (c) 21.93×0.012981

(a) $4.7826 + 0.02713 = 4.80973 = \mathbf{4.810}$, correct to 4 significant figures

(b) $17.6941 - 11.8762 = 5.8179 = \mathbf{5.818}$, correct to 4 significant figures

(c) $21.93 \times 0.012981 = 0.2846733\ldots = \mathbf{0.2847}$, correct to 4 significant figures

> **Problem 4.**   Evaluate the following, correct to 4 decimal places:
>
> (a) $46.32 \times 97.17 \times 0.01258$ (b) $\dfrac{4.621}{23.76}$
>
> (c) $\dfrac{1}{2}(62.49 \times 0.0172)$

(a) $46.32 \times 97.17 \times 0.01258 = 56.6215031\ldots$

$$= \mathbf{56.6215},$$

correct to 4 decimal places

(b) $\dfrac{4.621}{23.76} = 0.19448653\ldots = \mathbf{0.1945}$, correct to 4 decimal places

(c) $\frac{1}{2}(62.49 \times 0.0172) = 0.537414 = \mathbf{0.5374}$, correct to 4 decimal places

> **Problem 5.**   Evaluate the following, correct to 3 decimal places:
>
> (a) $\dfrac{1}{52.73}$ (b) $\dfrac{1}{0.0275}$ (c) $\dfrac{1}{4.92} + \dfrac{1}{1.97}$

(a) $\dfrac{1}{52.73} = 0.01896453\ldots = \mathbf{0.019}$, correct to 3 decimal places

(b) $\dfrac{1}{0.0275} = 36.3636363\ldots = \mathbf{36.364}$, correct to 3 decimal places

(c) $\dfrac{1}{4.92} + \dfrac{1}{1.97} = 0.71086624\ldots = \mathbf{0.711}$, correct to 3 decimal places

> **Problem 6.**   Evaluate the following, expressing the answers in standard form, correct to 4 significant figures:
>
> (a) $(0.00451)^2$
> (b) $631.7 - (6.21 + 2.95)^2$
> (c) $46.27^2 - 31.79^2$

(a) $(0.00451)^2 = 2.03401 \times 10^{-5}$

$$= \mathbf{2.034 \times 10^{-5}},$$

correct to 4 significant figures

(b) $631.7 - (6.21 + 2.95)^2 = 547.7944$

$$= 5.477944 \times 10^2$$

$$= \mathbf{5.478 \times 10^2},$$

correct to 4 significant figures

(c) $46.27^2 - 31.79^2 = 1130.3088 = \mathbf{1.130 \times 10^3}$, correct to 4 significant figures

> **Problem 7.**   Evaluate the following, correct to 3 decimal places:
>
> (a) $\dfrac{(2.37)^2}{0.0526}$
>
> (b) $\left(\dfrac{3.60}{1.92}\right)^2 + \left(\dfrac{5.40}{2.45}\right)^2$
>
> (c) $\dfrac{15}{7.6^2 - 4.8^2}$

(a) $\dfrac{(2.37)^2}{0.0526} = 106.785171\ldots = \mathbf{106.785}$, correct to 3 decimal places

(b) $\left(\dfrac{3.60}{1.92}\right)^2 + \left(\dfrac{5.40}{2.45}\right)^2 = 8.37360084\ldots = \mathbf{8.374}$, correct to 3 decimal places

(c) $\dfrac{15}{7.6^2 - 4.8^2} = 0.43202764\ldots = \mathbf{0.432}$, correct to 3 decimal places

---

Problem 8. Evaluate the following, correct to 4 significant figures:
(a) $\sqrt{5.462}$ (b) $\sqrt{54.62}$ (c) $\sqrt{546.2}$

(a) $\sqrt{5.462} = 2.3370922\ldots = \mathbf{2.337}$, correct to 4 significant figures

(b) $\sqrt{54.62} = 7.39053448\ldots = \mathbf{7.391}$, correct to 4 significant figures

(c) $\sqrt{546.2} = 23.370922\ldots = \mathbf{23.37}$, correct to 4 significant figures

---

Problem 9. Evaluate the following, correct to 3 decimal places:
(a) $\sqrt{0.007328}$ (b) $\sqrt{52.91} - \sqrt{31.76}$
(c) $\sqrt{(1.6291 \times 10^4)}$

(a) $\sqrt{0.007328} = 0.08560373 = \mathbf{0.086}$, correct to 3 decimal places

(b) $\sqrt{52.91} - \sqrt{31.76} = 1.63832491\ldots = \mathbf{1.638}$, correct to 3 decimal places

(c) $\sqrt{(1.6291 \times 10^4)} = \sqrt{(16\,291)}$
$$= 127.636201\ldots$$
$$= \mathbf{127.636},$$
correct to 3 decimal places

Problem 10. Evaluate the following, correct to 4 significant figures:
(a) 4.72^3 (b) $(0.8316)^4$ (c) $\sqrt{(76.21^2 - 29.10^2)}$

(a) $4.72^3 = 105.15404\ldots = \mathbf{105.2}$, correct to 4 significant figures

(b) $(0.8316)^4 = 0.47825324\ldots = \mathbf{0.4783}$, correct to 4 significant figures

(c) $\sqrt{(76.21^2 - 29.10^2)} = 70.4354605\ldots$
$$= \mathbf{70.44},$$
correct to 4 significant figures

Problem 11. Evaluate the following, correct to 3 significant figures:

(a) $\sqrt{\left[\dfrac{(6.09)^2}{25.2 \times \sqrt{7}}\right]}$ (b) $\sqrt[3]{(47.291)}$

(c) $\sqrt{(7.213^2 + 6.418^3 + 3.291^4)}$

(a) $\sqrt{\left[\dfrac{(6.09)^2}{25.2 \times \sqrt{7}}\right]} = 0.74583457\ldots = \mathbf{0.746}$, correct to 3 significant figures

(b) $\sqrt[3]{(47.291)} = 3.61625876\ldots = \mathbf{3.62}$, correct to 3 significant figures

(c) $\sqrt{(7.213^2 + 6.418^3 + 3.291^4)}$
$= 20.8252991\ldots = \mathbf{20.8}$, correct to 3 significant figures

---

Problem 12. Evaluate the following, expressing the answers in standard form, correct to 4 decimal places:
(a) $(5.176 \times 10^{-3})^2$

(b) $\left(\dfrac{1.974 \times 10^1 \times 8.61 \times 10^{-2}}{3.462}\right)^4$

(c) $\sqrt{(1.792 \times 10^{-4})}$

(a) $(5.176 \times 10^{-3})^2 = 2.679097\ldots \times 10^{-5} = \mathbf{2.6791 \times 10^{-5}}$, correct to 4 decimal places

(b) $\left(\dfrac{1.974 \times 10^1 \times 8.61 \times 10^{-2}}{3.462}\right)^4$
$= 0.05808887\ldots = \mathbf{5.8089 \times 10^{-2}}$, correct to 4 decimal places

(c) $\sqrt{(1.792 \times 10^{-4})} = 0.0133865\ldots$
$$= \mathbf{1.3387 \times 10^{-2}},$$
correct to 4 decimal places

Further problems on the use of a calculator may be found in Section 2.5, Problems 6 to 33, pages 19 and 20.

2.3 Conversion tables and charts

It is often necessary to make calculations from various conversion tables and charts. Examples include currency exchange rates, imperial to metric unit conversions, train or bus timetables, production schedules and so on.

> Problem 13.   Currency exchange rates for five countries are shown in Table 2.1.
>
> **Table 2.1**
>
> | France | £1 = 7.50 francs (f) |
> |---|---|
> | Italy | £1 = 2300 lira (l) |
> | Spain | £1 = 185 pesetas (pes) |
> | Germany | £1 = 2.25 Deutschmarks (Dm) |
> | U.S.A. | £1 = 1.46 dollars ($) |
>
> Calculate:
> (a)   how many French francs £27.90 will buy
> (b)   the number of German Deutschmarks which can be bought for £75
> (c)   the pounds sterling which can be exchanged for 63 250 lira
> (d)   the number of American dollars which can be purchased for £92.50, and
> (e)   the pounds sterling which can be exchanged for 2664 pesetas

(a)   £1 = 7.50 francs, hence
£27.90 = 27.90 × 7.50 francs = **209.25 f**

(b)   £1 = 2.25 Deutschmarks, hence
£75 = 75 × 2.25 Dm = **168.75 Dm**

(c)   £1 = 2300 lira, hence
63 250 lira = £$\frac{63\,250}{2300}$ = **£27.50**

(d)   £1 = 1.46 dollars, hence
£92.50 = 92.50 × 1.46 dollars = **$135.05**

(e)   £1 = 185 pesetas, hence
2664 pesetas = £$\frac{2664}{185}$ = **£14.40**

> Problem 14.   Some approximate imperial to metric conversions are shown in Table 2.2.
>
> **Table 2.2**
>
> | length | 1 inch = 2.54 cm |
> |---|---|
> | | 1 mile = 1.61 km |
> | weight | 2.2 lb = 1 kg |
> | | (1 lb = 16 oz) |
> | capacity | 1.76 pints = 1 litre |
> | | (8 pints = 1 gallon) |
>
> Use the table to determine:
> (a)   the number of millimetres in 9.5 inches
> (b)   a speed of 50 miles per hour in kilometres per hour
> (c)   the number of miles in 300 km
> (d)   the number of kilograms in 30 pounds weight
> (e)   the number of pounds and ounces in 42 kilograms (correct to the nearest ounce)
> (f)   the number of litres in 15 gallons, and
> (g)   the number of gallons in 40 litres

(a)   9.5 inches = 9.5 × 2.54 cm = 24.13 cm
24.13 cm = 24.13 × 10 mm = **241.3 mm**

(b)   50 m.p.h. = 50 × 1.61 km/h = **80.5 km/h**

(c)   300 km = (300/1.61) miles = **186.3 miles**

(d)   30 lb = (30/2.2) kg = **13.64 kg**

(e)   42 kg = 42 × 2.2 lb = 92.4 lb
0.4 lb = 0.4 × 16 oz = 6.4 oz = 6 oz, correct to the nearest ounce.
Thus 42 kg = **92 lb 6 oz**, correct to the nearest ounce

(f)   15 gallons = 15 × 8 pints = 120 pints
120 pints = (120/1.76) litres = **68.18 litres**

(g)   40 litres = 40 × 1.76 pints = 70.4 pints
70.4 pints = (70.4/8) gallons = **8.8 gallons**

Further problems on conversion tables and charts may be found in Section 2.5, Problems 34 to 36, pages 20 to 22.

2.4   Evaluation of formulae

The statement $v = u + at$ is said to be a **formula** for v in terms of u, a and t.

v, u, a and t are called **symbols**.

The single term on the left-hand side of the equation, v, is called the **subject of the formulae**.

Provided values are given for all the symbols in a formula except one, the remaining symbol can be made the subject of the formula and may be evaluated by using a calculator.

Problem 15.   In an electrical circuit the voltage V is given by Ohm's law, i.e. $V = IR$. Find, correct to 4 significant figures, the voltage when $I = 5.36$ A and $R = 14.76$ Ω

$$V = IR = (5.36)(14.76)$$

Hence voltage $V = 79.11$ V, correct to 4 significant figures.

Problem 16.   The surface area A of a hollow cone is given by $A = \pi r l$. Determine the surface area when $r = 3.0$ cm, $l = 8.5$ cm and $\pi = 3.14$

$$A = \pi r l = (3.14)(3.0)(8.5) \text{ cm}^2$$

Hence surface area $A = 80.07$ cm^2.

Problem 17.   Velocity v is given by $v = u + at$. If $u = 9.86$ m/s, $a = 4.25$ m/s^2 and $t = 6.84$ s, find v, correct to 3 significant figures

---

$$v = u + at = 9.86 + (4.25)(6.84)$$

$$= 9.86 + 29.07$$

$$= 38.93$$

Hence velocity $v = 38.9$ m/s, correct to 3 significant figures.

Problem 18.   The area, A, of a circle is given by $A = \pi r^2$. Determine the area correct to 2 decimal places, given $\pi = 3.142$ and $r = 5.23$ m

---

$$A = \pi r^2 = (3.142)(5.23)^2$$

$$= (3.142)(27.3529)$$

Hence area $A = 85.94$ m^2, correct to 2 decimal places.

Problem 19.   The power P watts dissipated in an electrical circuit may be expressed by the formula $P = V^2/R$. Evaluate the power, correct to 3 significant figures, given that $V = 17.48$ V and $R = 36.12$ Ω

$$P = \frac{V^2}{R} = \frac{(17.48)^2}{36.12} = \frac{305.6}{36.12}$$

Hence power $P = 8.46$ W, correct to 3 significant figures.

Problem 20.   The volume V cm^3 of a right circular cone is given by $V = \frac{1}{3}\pi r^2 h$. Given that $r = 4.321$ cm, $h = 18.35$ cm and $\pi = 3.142$, find the volume correct to 4 significant figures

---

$$V = \frac{1}{3}\pi r^2 h = \frac{1}{3}(3.142)(4.321)^2(18.35)$$

$$= \frac{1}{3}(3.142)(18.671)(18.35)$$

Hence volume $V = 358.8$ cm^3, correct to 4 significant figures.

Problem 21.   Force F newtons is given by the formula $F = (Gm_1 m_2)/d^2$, where m_1 and m_2 are masses, d their distance apart and G is a constant. Find the value of the force given that $G = 6.67 \times 10^{-11}$, $m_1 = 7.36$, $m_2 = 15.5$ and $d = 22.6$. Express the answer in standard form, correct to 3 significant figures

---

$$F = \frac{Gm_1 m_2}{d^2} = \frac{(6.67 \times 10^{-11})(7.36)(15.5)}{(22.6)^2}$$

$$= \frac{(6.67)(7.36)(15.5)}{(10^{11})(510.8)} = \frac{1.490}{10^{11}}$$

Hence force $F = 1.49 \times 10^{-11}$ newtons, correct to 3 significant figures.

Problem 22.   The time of swing t seconds of a simple pendulum is given by $t = 2\pi\sqrt{(l/g)}$. Determine the time, correct to 3 decimal places, given that $\pi = 3.142$, $l = 12.0$ and $g = 9.81$

$$t = 2\pi\sqrt{\left(\frac{l}{g}\right)} = (2)(3.142)\sqrt{\left(\frac{12.0}{9.81}\right)}$$

$$= (2)(3.142)\sqrt{(1.223)}$$

$$= (2)(3.142)(1.106)$$

Hence time $t = 6.950$ seconds, correct to 3 decimal places.

Problem 23.   Resistance, R Ω, varies with temperature according to the formula $R = R_0(1 + \alpha t)$. Evaluate R, correct to 3 significant figures, given $R_0 = 14.59$, $\alpha = 0.0043$ and $t = 80$

$$R = R_0(1 + \alpha t) = 14.59[1 + (0.0043)(80)]$$

$$= 14.59(1 + 0.344)$$

$$= 14.59(1.344)$$

Hence resistance $R = 19.6$ Ω, correct to 3 significant figures.

Further problems on evaluating formulae may be found in the following Section 2.5, Problems 37 to 48, page 22.

2.5   Further problems on errors, calculations and evaluation of formulae

Errors

In Problems 1 to 5 state which type of error, or errors, have been made:

1   $25 \times 0.06 \times 1.4 = 0.21$

[order of magnitude error]

2   $137 \times 6.842 = 937.4$

$\left[\begin{array}{l}\text{rounding-off error–should add} \\ \text{'correct to 4 significant figures'}\end{array}\right]$

3   $\dfrac{24 \times 0.008}{12.6} = 10.42$ [blunder]

4   For a gas $pV = c$. When pressure $p = 103\,400$ Pa and $V = 0.54$ m^3 then $c = 55\,836$ Pa m^3.

[measured values, hence $c = 55\,800$ Pa m^3]

5   $\dfrac{4.6 \times 0.07}{52.3 \times 0.274} = 0.225$

$\left[\begin{array}{l}\text{order of magnitude error and rounding-} \\ \text{off error–should be 0.0225 correct to 3} \\ \text{significant figures}\end{array}\right]$

Use of calculator

In Problems 6 to 9, use a calculator to evaluate the quantities shown correct to 4 significant figures:

6   (a) 3.249^2 (b) 73.78^2 (c) 311.4^2 (d) 0.0639^2

$\left[\begin{array}{l}\text{(a) 10.56 (b) 5443} \\ \text{(c) 96970 (d) 0.004083}\end{array}\right]$

7   (a) $\sqrt{4.735}$ (b) $\sqrt{35.46}$ (c) $\sqrt{73\,280}$ (d) $\sqrt{0.0256}$

$\left[\begin{array}{l}\text{(a) 2.176 (b) 5.955} \\ \text{(c) 270.7 (d) 0.1600}\end{array}\right]$

8   (a) $\dfrac{1}{7.768}$ (b) $\dfrac{1}{48.46}$ (c) $\dfrac{1}{0.0816}$ (d) $\dfrac{1}{1.118}$

$\left[\begin{array}{l}\text{(a) 0.1287 (b) 0.02064} \\ \text{(c) 12.25 (d) 0.8945}\end{array}\right]$

9   (a) lg 3.764 (b) lg 241.8 (c) lg 1.0 (d) lg 0.07632

[(a) 0.5756 (b) 2.383 (c) 0 (d) -1.117]

10   Evaluate correct to 3 decimal places:

(a) ln 41.62 (b) ln 0.0179 (c) $\dfrac{\lg 5.29}{\ln 5.29}$

[(a) 3.729 (b) -4.023 (c) 0.434]

In Problems 11 to 18, use a calculator to evaluate correct to 4 significant figures:

11   (a) 43.27×12.91 (b) 54.31×0.5724

[(a) 558.6 (b) 31.09]

12   (a) $127.8 \times 0.0431 \times 19.8$ (b) $15.76 \div 4.329$

[(a) 109.1 (b) 3.641]

13   (a) $\dfrac{137.6}{552.9}$ (b) $\dfrac{11.82 \times 1.736}{0.041}$

[(a) 0.2489 (b) 500.5]

14   (a) $\dfrac{1}{17.31}$ (b) $\dfrac{1}{0.0346}$ (c) $\dfrac{1}{147.9}$

[(a) 0.05777 (b) 28.90 (c) 0.006761]

15   (a) 13.6^3 (b) 3.476^4 (c) 0.124^5

[(a) 2515 (b) 146.0 (c) 0.00002932]

16   (a) $\sqrt{347.1}$ (b) $\sqrt{7632}$
     (c) $\sqrt{0.027}$ (d) $\sqrt{0.004168}$
          [(a) 18.63 (b) 87.36 (c) 0.1643 (d) 0.06456]

17   (a) $\left(\dfrac{24.68 \times 0.0532}{7.412}\right)^3$

     (b) $\left(\dfrac{0.2681 \times 41.2^2}{32.6 \times 11.89}\right)^4$

                        [(a) 0.005559 (b) 1.900]

18   (a) $\dfrac{14.32^3}{21.68^2}$ (b) $\dfrac{4.821^3}{17.33^2 - 15.86 \times 11.6}$

                        [(a) 6.248 (b) 0.9630]

19   Evaluate correct to 3 decimal places:

     (a) $\dfrac{29.12}{(5.81)^2 - (2.96)^2}$ (b) $\sqrt{53.98} - \sqrt{21.78}$

                        [(a) 1.165 (b) 2.680]

20   Evaluate correct to 4 significant figures:

     (a) $\sqrt{\left[\dfrac{(15.62)^2}{29.21 \times \sqrt{10.52}}\right]}$

     (b) $\sqrt{(6.921^2 + 4.816^3 - 2.161^4)}$

                        [(a) 1.605 (b) 11.74]

21   Evaluate the following, expressing the answers
     in standard form, correct to 3 decimal places:
     (a) $(8.291 \times 10^{-2})^2$ (b) $\sqrt{(7.623 \times 10^{-3})}$

                  [(a) 6.874×10^{-3} (b) 8.731×10^{-2}]

22   The area A of a rectangle is given by $A = lb$. The length l when measured is found to
     be 23.1 mm and the breadth b is 7.8 mm.
     Determine the area of the rectangle.
                                    [180 mm^2]

23   The velocity of a body is given by $v = u + at$.
     The initial velocity u is measured when time t
     is 15 seconds and found to be 12 m/s. If the
     acceleration a is 9.81 m/s^2 calculate the final
     velocity v.   [159 m/s]

24   Calculate the current I in an electrical circuit,
     where $I = V/R$ amperes when the voltage V
     is measured and found to be 7.2 V and the
     resistance R is 17.7 Ω.   [0.407 A]

25   Find the distance s, given that $s = \frac{1}{2}gt^2$. Time
     $t = 0.032$ seconds and acceleration due to
     gravity $g = 9.81$ m/s^2.
                        [0.00502 m or 5.02 mm]

26   The energy stored in a capacitor is given by
     $E = \frac{1}{2}CV^2$ joules. Determine the energy when
     capacitance $C = 5 \times 10^{-6}$ farads and voltage
     $V = 240$ V.   [0.1440 J]

27   Find the area A of a triangle, given $A = \frac{1}{2}bh$,
     when the base length b is 23.42 m and the
     height h is 53.7 m.   [629.0 m^2]

28   Resistance R_2 is given by $R_2 = R_1(1 + \alpha t)$.
     Find R_2, correct to 4 significant figures, when
     $R_1 = 220, \alpha = 0.00027$ and $t = 75.6$.
                                    [224.5]

29   Density = mass/volume. Find the density
     when the mass is 2.462 kg and the volume is
     173 cm^3. Give the answer in units of kg/m^3.
                              [14 230 kg/m^3]

30   Velocity = frequency × wavelength. Find the
     velocity when the frequency is 1825 Hz and
     the wavelength is 0.154 m.   [281.1 m/s]

31   Evaluate resistance R_T, given $(1/R_T) = (1/R_1) + (1/R_2) + (1/R_3)$, when $R_1 = 5.5$ Ω,
     $R_2 = 7.42$ Ω and $R_3 = 12.6$ Ω.   [2.526 Ω]

32   Find the total cost of 37 calculators cost-
     ing £12.65 each and 19 drawing sets costing
     £6.38 each.   [£589.27]

33   Power = (force × distance)/time. Find the
     power when a force of 3760 N raises an object
     a distance of 4.73 m in 35 s.   [508.1 W]

Conversion tables and charts

34   Currency exchange rates listed in a newspaper
     included the following:

     France   £1 = 7.50 francs

     Japan   £1 = 157.50 yen

     Germany £1 = 2.25 Deutschmarks

     U.S.A.   £1 = $1.50

     Spain   £1 = 190 pesetas
     Calculate (a) how many French francs £32.50
     will buy, (b) the number of American dol-
     lars that can be purchased for £74.80, (c) the
     pounds sterling which can be exchanged for
     14 000 yen, (d) the pounds sterling which can
     be exchanged for 1750 pesetas, and (e) the
     German Deutschmarks which can be bought
     for £55.
            $\begin{bmatrix} \text{(a) 243.75 f (b) \$112.20 (c) £88.89} \\ \text{(d) £9.21 (e) 123.75 Dm} \end{bmatrix}$

35   Below is a table of some metric to imperial
     conversions:
     Length   2.54 cm = 1 inch

                1.61 km = 1 mile

     Weight   1 kg = 2.2 lb (1 lb = 16 ounces)

Capacity 1 litre = 1.76 pints

(8 pints = 1 gallon)

Use the table to determine (a) the number of millimetres in 15 inches, (b) a speed of 35 mph in km/h, (c) the number of kilometres in 235 miles, (d) the number of pounds and ounces in 24 kg (correct to the nearest ounce), (e) the number of kilograms in 15 lb, (f) the number of litres in 12 gallons and (g) the number of gallons in 25 litres.

$$\left[\begin{array}{l} \text{(a) 381 mm (b) 56.35 km/h} \\ \text{(c) 378.35 km (d) 52 lb 13 oz} \\ \text{(e) 6.82 kg (f) 54.55 l} \\ \text{(g) 5.5 gallons} \end{array}\right]$$

36 Deduce the following information from the BR train timetable shown in Table 2.3:

(a) At what time should a man catch a train at Mossley Hill to enable him to be in Manchester Piccadilly by 8.15 a.m.?

(b) A girl leaves Hunts Cross at 8.17 a.m. and travels to Manchester Oxford Road. How long does the journey take? What is the average speed of the journey?

(c) A man living at Edge Hill has to be at work at Trafford Park by 8.45 a.m. It takes him 10 minutes to walk to his work from Trafford Park station. What time

Table 2.3 Liverpool, Hunt's Cross and Warrington → Manchester

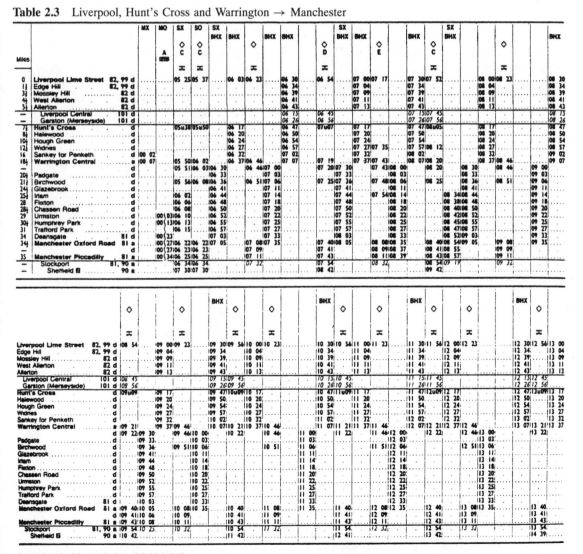

Reproduced with permission of British Rail

train should he catch from Edge Hill?

$$\begin{bmatrix} \text{(a) 7.09 a.m.} \\ \text{(b) 51 minutes, 32 m.p.h.} \\ \text{(c) 7.04 a.m.} \end{bmatrix}$$

Evaluation of formulae

37   The area A of a rectangle is given by the formula $A = lb$. Evaluate the area when $l = 12.4$ cm and $b = 5.37$ cm.
$$[A = 66.59 \text{ cm}^2]$$

38   The circumference C of a circle is given by the formula $C = 2\pi r$. Determine the circumference given $\pi = 3.14$ and $r = 8.40$ mm.
$$[C = 52.75 \text{ mm}]$$

39   A formula used in connection with gases is $R = (PV)/T$. Evaluate R when $P = 1500$, $V = 5$ and $T = 200$.
$$[R = 37.5]$$

40   The potential difference, V volts, available at battery terminals is given by $V = E - Ir$. Evaluate V when $E = 5.62$, $I = 0.70$ and $R = 4.30$.
$$[V = 2.61 \text{ V}]$$

41   Given force $F = \frac{1}{2}m(v^2 - u^2)$, find F when $m = 18.3$, $v = 12.7$ and $u = 8.24$.
$$[F = 854.5]$$

42   The current I amperes flowing in a number of cells is given by $I = (nE)/(R + nr)$. Evaluate the current when $n = 36$, $E = 2.20$, $R = 2.80$ and $r = 0.50$.
$$[I = 3.81 \text{ A}]$$

43   The time, t seconds, of oscillation for a simple pendulum is given by $t = 2\pi\sqrt{(l/g)}$. Determine the time when $\pi = 3.142$, $l = 54.32$ and $g = 9.81$.
$$[t = 14.79 \text{ s}]$$

44   Energy, E joules, is given by the formula $E = \frac{1}{2}LI^2$. Evaluate the energy when $L = 5.5$ and $I = 1.2$.
$$[E = 3.96 \text{ J}]$$

45   The current I amperes in an a.c. circuit is given by $I = V/\sqrt{(R^2 + X^2)}$. Evaluate the current when $V = 250$, $R = 11.0$ and $X = 16.2$.
$$[I = 12.77 \text{ A}]$$

46   Distance s metres is given by the formula $s = ut + \frac{1}{2}at^2$. If $u = 9.50$, $t = 4.60$ and $a = -2.50$, evaluate the distance.
$$[s = 17.25 \text{ m}]$$

47   The area, A, of any triangle is given by $A = \sqrt{[s(s - a)(s - b)(s - c)]}$, where $s = (a + b + c)/2$.
Evaluate the area given $a = 3.60$ cm, $b = 4.00$ cm and $c = 5.20$ cm.
$$[A = 7.184 \text{ cm}^2]$$

48   Given that $a = 0.290$, $b = 14.86$, $c = 0.042$, $d = 31.8$ and $e = 0.650$, evaluate v given that $v = \sqrt{(ab/c - d/e)}$.
$$[v = 7.327]$$

3

Algebra

3.1 Basic operations

Algebra is that part of mathematics in which the relations and properties of numbers are investigated by means of general symbols. For example, the area of a rectangle is found by multiplying the length by the breadth; this is expressed algebraically as $A = l \times b$, where A represents the area, l the length and b the breadth.

The basic laws introduced in arithmetic are generalized in algebra.

Let a, b, c and d represent any four numbers. Then:

(i) $a + (b + c) = (a + b) + c$
(ii) $a(bc) = (ab)c$
(iii) $a + b = b + a$
(iv) $ab = ba$
(v) $a(b + c) = ab + ac$
(vi) $\dfrac{a+b}{c} = \dfrac{a}{c} + \dfrac{b}{c}$
(vii) $(a + b)(c + d) = ac + ad + bc + bd$

> **Problem 1.** Evaluate $3ab - 2bc + abc$ when $a = 1$, $b = 3$ and $c = 5$

Replacing a, b and c with their numerical values gives:

$3ab - 2bc + abc$

$= 3 \times 1 \times 3 - 2 \times 3 \times 5 + 1 \times 3 \times 5$

$= 9 - 30 + 15 = \mathbf{-6}$

> **Problem 2.** Find the value of $4p^2qr^3$, given that $p = 2$, $q = \frac{1}{2}$ and $r = 1\frac{1}{2}$

Replacing p, q and r with their numerical values gives:

$$4p^2qr^3 = 4(2)^2 \left(\frac{1}{2}\right) \left(\frac{3}{2}\right)^3$$

$$= 4 \times 2 \times 2 \times \frac{1}{2} \times \frac{3}{2} \times \frac{3}{2} \times \frac{3}{2} = \mathbf{27}$$

> **Problem 3.** Find the sum of $3x$, $2x$, $-x$ and $-7x$

The sum of the positive terms is $3x + 2x = 5x$.
The sum of the negative terms is $x + 7x = 8x$.
Taking the sum of the negative terms from the sum of the positive terms gives:

$$5x - 8x = \mathbf{-3x}$$

Alternatively

$$3x + 2x + (-x) + (-7x) = 3x + 2x - x - 7x$$
$$= \mathbf{-3x}$$

> **Problem 4.** Find the sum of $4a$, $3b$, c, $-2a$, $-5b$ and $6c$

Each symbol must be dealt with individually:

for the 'a' terms: $+4a - 2a = 2a$
for the 'b' terms: $+3b - 5b = -2b$
for the 'c' terms: $+c + 6c = 7c$

Thus $4a + 3b + c + (-2a) + (-5b) + 6c$

$= 4a + 3b + c - 2a - 5b + 6c$

$= \mathbf{2a - 2b + 7c}$

> **Problem 5.** Find the sum of $5a - 2b$, $2a + c$, $4b - 5d$ and $b - a + 3d - 4c$

The algebraic expressions may be tabulated as shown below, forming columns for the as, bs, cs and ds. Thus:

$$\begin{array}{rrrr}
+5a & - 2b & & \\
+2a & & + c & \\
& + 4b & & - 5d \\
-a & + b & - 4c & + 3d \\
\hline
\mathbf{6a} & \mathbf{+ 3b} & \mathbf{- 3c} & \mathbf{- 2d}
\end{array}$$

Adding gives: $\mathbf{6a + 3b - 3c - 2d}$

Problem 6.   Subtract $2x + 3y - 4z$ from $x - 2y + 5z$

$$
\begin{array}{rrr}
x & -\,2y & +\,5z \\
2x & +\,3y & -\,4z
\end{array}
$$

Subtracting gives: $-x - 5y + 9z$

(Note that $+5z - -4z = +5z + 4z = 9z$)

An alternative method of subtracting algebraic expressions is to 'change the signs of the bottom line and add'. Hence:

$$
\begin{array}{rrr}
x & -\,2y & +\,5z \\
-2x & -\,3y & +\,4z
\end{array}
$$

Adding gives: $-x - 5y + 9z$

Problem 7.   Multiply $2a + 3b$ by $a + b$

Each term in the first expression is multiplied by a, then each term in the first expression is multiplied by b, and the two results are added. The usual layout is shown below.

$$
\begin{array}{rr}
2a & +\,3b \\
a & +\,b
\end{array}
$$

Multiplying by $a \rightarrow$ $2a^2 + 3ab$

Multiplying by $b \rightarrow$ $\quad\ + 2ab + 3b^2$

Adding gives: $2a^2 + 5ab + 3b^2$

Problem 8.   Multiply $3x - 2y^2 + 4xy$ by $2x - 5y$

$$
\begin{array}{rrr}
3x & -\ 2y^2 & +\ 4xy \\
2x & -\ 5y &
\end{array}
$$

$6x^2 - 4xy^2 + 8x^2y$	(1)
$\quad\ - 20xy^2 \qquad\quad - 15xy + 10y^3$	(2)
$6x^2 - 24xy^2 + 8x^2y - 15xy + 10y^3$	(3)

Multiplying by $2x \rightarrow$ (1)
Multiplying by $-5y \rightarrow$ (2)
Adding gives: (3)

Problem 9.   Simplify $2p \div 8pq$

$2p \div 8pq$ means $\dfrac{2p}{8pq}$

This can be reduced by cancelling as in arithmetic. Thus:

$$
\frac{2p}{8pq} = \frac{2^1 \times p^1}{8_4 \times p_1 \times q} = \frac{1}{4q}
$$

Problem 10.   Divide $2x^2 + x - 3$ by $x - 1$

$2x^2 + x - 3$ is called the **dividend** and $x - 1$ the **divisor**. The usual layout is shown below with the dividend and divisor both arranged in descending powers of the symbols.

$$
\begin{array}{r}
2x + 3 \\
x - 1 \,\overline{\big)\ 2x^2 + \ x - 3} \\
2x^2 - 2x \\
\hline
3x - 3 \\
3x - 3 \\
\hline
\cdot \qquad \cdot
\end{array}
$$

Dividing the first term of the dividend by the first term of the divisor, i.e. $(2x^2/x)$ gives $2x$, which is put above the first term of the dividend as shown. The divisor is then multiplied by $2x$, i.e. $2x(x-1) = 2x^2 - 2x$, which is placed under the dividend as shown. Subtracting gives $3x - 3$. The process is then repeated, i.e. the first term of the divisor is divided into $3x$, giving 3, which is placed above the dividend as shown.

Then $3(x - 1) = 3x - 3$, which is placed under the $3x - 3$. The remainder, on subtraction, is zero, which completes the process.

Thus $(2x^2 + x - 3) \div (x - 1) = (2x + 3)$.

(A check can be made on this answer by multiplying $(2x + 3)$ by $(x - 1)$, which should equal $2x^2 + x - 3$.)

Problem 11.   Simplify $\dfrac{x^3 + y^3}{x + y}$

$$\begin{array}{r}
\qquad(1)\quad(4)\quad(7)\\
x^2\ -\ xy +\ y^2 \qquad\qquad(1)\\
x+y\ \overline{)\ x^3\ +\ 0\ +\ 0\ +\ y^3}\\
x^3\ +x^2y \qquad\qquad(2)\\
\hline
\qquad\qquad\qquad\qquad(3)\\
-\ x^2y \qquad\quad +\ y^3\quad(4)\\
-\ x^2y - xy^2 \qquad\quad(5)\\
\qquad\qquad\qquad\qquad(6)\\
\hline
xy^2 + y^3\quad(7)\\
xy^2 + y^3\quad(8)\\
\hline
\cdot\qquad\cdot\qquad(9)\\
\hline
\end{array}$$

(1) x into x^3 goes x^2. Put x^2 above x^3
(2) $x^2(x+y) = x^3 + x^2y$
(3) Subtract
(4) x into $-x^2y$ goes $-xy$. Put $-xy$ above dividend
(5) $-xy(x+y) = -x^2y - xy^2$
(6) Subtract
(7) x into xy^2 goes y^2. Put y^2 above dividend
(8) $y^2(x+y) = xy^2 + y^3$
(9) Subtract

Thus $\dfrac{x^3+y^3}{x+y} = x^2 - xy + y^2.$

Zeros are not normally shown in the dividend, but are included to clarify the subtraction process and to keep similar terms in their respective columns.

Problem 12.   Divide $4a^3 - 6a^2b + 5b^3$ by $2a - b$

$$\begin{array}{r}
2a^2 - 2ab - b^2\qquad\qquad\\
2a-b\ \overline{)\ 4a^3 - 6a^2b\qquad\quad + 5b^3}\\
4a^3 - 2a^2b\qquad\qquad\qquad\\
\hline
-4a^2b\qquad\quad + 5b^3\\
-4a^2b + 2ab^2\qquad\qquad\\
\hline
-2ab^2 + 5b^3\\
-2ab^2 +\ b^3\\
\hline
4b^3\\
\hline
\end{array}$$

Thus $\dfrac{4a^3 - 6a^2b + 5b^3}{2a - b} = 2a^2 - 2ab - b^2,$
remainder $4b^3$.

Alternatively, the answer may be expressed as

$$2a^2 - 2ab - b^2 + \frac{4b^3}{2a - b}$$

Further problems on basic operations may be found in Section 3.7, Problems 1 to 14, page 37.

3.2 Laws of indices

The laws of indices are:

(i) $a^m \times a^n = a^{m+n}$

(ii) $\dfrac{a^m}{a^n} = a^{m-n}$

(iii) $(a^m)^n = a^{mn}$

(iv) $a^{m/n} = \sqrt[n]{a^m}$

(v) $a^{-n} = \dfrac{1}{a^n}$

(vi) $a^0 = 1$

Problem 13.   Simplify $a^3b^2c \times ab^3c^5$

Grouping like terms gives:

$$a^3 \times a \times b^2 \times b^3 \times c \times c^5$$

Using the first law of indices gives:

$$a^{3+1} \times b^{2+3} \times c^{1+5}$$

i.e. $a^4 \times b^5 \times c^6$, i.e. $\boldsymbol{a^4b^5c^6}$

Problem 14.   Simplify $a^{1/2}b^2c^{-2} \times a^{1/6}b^{1/2}c$

Using the first law of indices,

$$a^{1/2}b^2c^{-2} \times a^{(1/6)}b^{(1/2)}c$$
$$= a^{(1/2)+(1/6)} \times b^{2+(1/2)} \times c^{-2+1}$$
$$= \boldsymbol{a^{2/3}b^{5/2}c^{-1}}$$

Problem 15.   Simplify $\dfrac{a^3b^2c^4}{abc^{-2}}$ and evaluate when $a = 3$, $b = \frac{1}{8}$ and $c = 2$

Using the second law of indices,

$$\frac{a^3}{a} = a^{3-1} = a^2, \quad \frac{b^2}{b} = b^{2-1} = b$$

and $\quad \dfrac{c^4}{c^{-2}} = c^{4--2} = c^6$. Thus $\dfrac{a^3 b^2 c^4}{abc^{-2}} = a^2 bc^6$

When $a = 3$, $b = \dfrac{1}{8}$ and $c = 2$,

$$a^2 bc^6 = (3)^2 \left(\frac{1}{8}\right)(2)^6 = (9)\left(\frac{1}{8}\right)(64) = 72$$

Problem 16. Simplify $\dfrac{p^{1/2} q^2 r^{2/3}}{p^{1/4} q^{1/2} r^{1/6}}$ and evaluate when $p = 16$, $q = 9$ and $r = 4$, taking positive roots only

---

Using the second law of indices gives:

$$p^{(1/2)-(1/4)} q^{2-(1/2)} r^{(2/3)-(1/6)} = p^{1/4} q^{3/2} r^{1/2}$$

When $p = 16$, $q = 9$ and $r = 4$,

$$p^{1/4} q^{3/2} r^{1/2} = (16)^{1/4}(9)^{3/2}(4)^{1/2}$$

$$= (\sqrt[4]{16})(\sqrt{9^3})(\sqrt{4})$$

$$= (2)(3^3)(2) = \mathbf{108}$$

Problem 17. Simplify $\dfrac{x^2 y^3 + xy^2}{xy}$

Algebraic expressions of the form $\dfrac{a+b}{c}$ can be split into $\dfrac{a}{c} + \dfrac{b}{c}$. Thus

$$\frac{x^2 y^3 + xy^2}{xy} = \frac{x^2 y^3}{xy} + \frac{xy^2}{xy}$$

$$= x^{2-1} y^{3-1} + x^{1-1} y^{2-1}$$

$$= xy^2 + y$$

(since $x^0 = 1$, from the sixth law of indices).

Problem 18. Simplify $\dfrac{x^2 y}{xy^2 - xy}$

The highest common factor (HCF) of each of the three terms comprising the numerator and denominator is xy. Dividing each term by xy gives:

$$\frac{x^2 y}{xy^2 - xy} = \frac{\dfrac{x^2 y}{xy}}{\dfrac{xy^2}{xy} - \dfrac{xy}{xy}} = \frac{x}{y-1}$$

Problem 19. Simplify $\dfrac{a^2 b}{ab^2 - a^{1/2} b^3}$

The HCF of each of the three terms is $a^{1/2} b$. Dividing each term by $a^{1/2} b$ gives:

$$\frac{a^2 b}{ab^2 - a^{1/2} b^3} = \frac{\dfrac{a^2 b}{a^{1/2} b}}{\dfrac{ab^2}{a^{1/2} b} - \dfrac{a^{1/2} b^3}{a^{1/2} b}} = \frac{a^{3/2}}{a^{1/2} b - b^2}$$

Problem 20. Simplify $(p^3)^{1/2}(q^2)^4$

Using the third law of indices gives:

$$p^{3 \times (1/2)} q^{2 \times 4}$$

i.e. $\quad p^{(3/2)} q^8$

Problem 21. Simplify $\dfrac{(mn^2)^3}{(m^{(1/2)} n^{(1/4)})^4}$

The brackets indicate that each letter in the bracket must be raised to the power outside.
Using the third law of indices gives:

$$\frac{(mn^2)^3}{(m^{1/2} n^{1/4})^4} = \frac{m^{1 \times 3} n^{2 \times 3}}{m^{(1/2) \times 4} n^{(1/4) \times 4}} = \frac{m^3 n^6}{m^2 n^1}$$

Using the second law of indices gives:

$$\frac{m^3 n^6}{m^2 n^1} = m^{3-2} n^{6-1} = \mathbf{mn^5}$$

Problem 22. Simplify $(a^3 \sqrt{b} \sqrt{c^5})(\sqrt{a} \sqrt[3]{b^2} c^3)$ and evaluate when $a = \frac{1}{4}$, $b = 64$ and $c = 1$

Using the fourth law of indices, the expression can be written as:

$$(a^3 b^{1/2} c^{5/2})(a^{1/2} b^{2/3} c^3)$$

Using the first law of indices gives:

$$a^{3+(1/2)} b^{(1/2)+(2/3)} c^{(5/2)+3} = a^{7/2} b^{7/6} c^{11/2}$$

It is usual to express the answer in the same form as the question. Hence

$$a^{7/2} b^{7/6} c^{11/2} = \sqrt{a^7} \sqrt[6]{b^7} \sqrt{c^{11}}$$

When $a = \dfrac{1}{4}$, $b = 64$ and $c = 1$,

$$\sqrt{a^7} \sqrt[6]{b^7} \sqrt{c^{11}} = \sqrt{\left(\frac{1}{4}\right)^7} (\sqrt[6]{64^7}) \sqrt{1^{11}}$$

$$= \left(\frac{1}{2}\right)^7 (2)^7 (1) = \mathbf{1}$$

Problem 23. Simplify $(a^3 b)(a^{-4} b^{-2})$, expressing the answer with positive indices only

---

Using the first law of indices gives:

$$a^{3+-4} b^{1+-2} = a^{-1} b^{-1}$$

Using the fifth law of indices gives:

$$a^{-1} b^{-1} = \frac{1}{a^{+1} b^{+1}} = \frac{1}{ab}$$

Problem 24. Simplify $\dfrac{d^2 e^2 f^{(1/2)}}{(d^{(3/2)} e f^{(5/2)})^2}$ expressing the answer with positive indices only

---

Using the third law of indices gives:

$$\frac{d^2 e^2 f^{1/2}}{d^{(3/2) \times 2} e^{1 \times 2} f^{(5/2) \times 2}} = \frac{d^2 e^2 f^{1/2}}{d^3 e^2 f^5}$$

Using the second law of indices gives:

$$d^{2-3} e^{2-2} f^{(1/2)-5}$$

$$= d^{-1} e^0 f^{-9/2}$$

$$= d^{-1} f^{(-9/2)} \text{ since } e^0 = 1$$

from the sixth law of indices

$$= \frac{1}{df^{(9/2)}} \text{ from the fifth law of indices}$$

Problem 25. Simplify $\dfrac{(x^2 y^{1/2})(\sqrt{x}\sqrt[3]{y^2})}{(x^5 y^3)^{1/2}}$

Using the third and fourth laws of indices gives:

$$\frac{(x^2 y^{1/2})(x^{1/2} y^{2/3})}{x^{5/2} y^{3/2}}$$

Using the first and second laws of indices gives:

$$x^{2+(1/2)-(5/2)} y^{(1/2)+(2/3)-(3/2)} = x^0 y^{-1/3}$$

$$= y^{-1/3} \text{ or } \frac{1}{y^{1/3}}$$

from the fifth and sixth laws of indices.

Further problems on laws of indices may be found in Section 3.7, Problems 15 to 25, page 37.

3.3 Brackets and factorization

When two or more terms in an algebraic expression contain a common factor, then this factor can be shown outside of a bracket. For example

$$ab + ac = a(b + c)$$

which is simply the reverse of law (v) of indices, and

$$6px + 2py - 4pz = 2p(3x + y - 2z)$$

This process is called **factorization**.

Problem 26. Remove the brackets and simplify the expression
$(3a + b) + 2(b + c) - 4(c + d)$

Both b and c in the second bracket have to be multiplied by 2, and c and d in the third bracket by -4 when the brackets are removed. Thus:

$$(3a + b) + 2(b + c) - 4(c + d)$$

$$= 3a + b + 2b + 2c - 4c - 4d$$

Collecting similar terms together gives:

$$3a + 3b - 2c - 4d$$

Problem 27.   Simplify
$a^2 - (2a - ab) - a(3b + a)$

When the brackets are removed, both $2a$ and $-ab$ in the first bracket must be multiplied by -1 and both $3b$ and a in the second bracket by $-a$. Thus

$$a^2 - (2a - ab) - a(3b + a)$$

$$= a^2 - 2a + ab - 3ab - a^2$$

Collecting similar terms together gives: $-2a - 2ab$. Since $-2a$ is a common factor the answer can be expressed as $-2a(1 + b)$.

Problem 28.   Simplify $(a + b)(a - b)$

Each term in the second bracket has to be multiplied by each term in the first bracket. Thus:

$$(a + b)(a - b) = a(a - b) + b(a - b)$$

$$= a^2 - ab + ab - b^2$$

$$= a^2 - b^2$$

Alternatively

$$\begin{array}{r} a + b \\ a - b \\ \hline \end{array}$$

Multiplying by $a \rightarrow \quad a^2 + ab$

Multiplying by $-b \rightarrow \quad - ab - b^2$

Adding gives: $\quad \dfrac{a^2 \qquad - b^2}{}$

Problem 29.   Remove the brackets from the expression $(x - 2y)(3x + y^2)$

$$(x - 2y)(3x + y^2) = x(3x + y^2) - 2y(3x + y^2)$$

$$= 3x^2 + xy^2 - 6xy - 2y^3$$

Problem 30.   Simplify $(2x - 3y)^2$

$$(2x - 3y)^2 = (2x - 3y)(2x - 3y)$$

$$= 2x(2x - 3y) - 3y(2x - 3y)$$

$$= 4x^2 - 6xy - 6xy + 9y^2$$

$$= 4x^2 - 12xy + 9y^2$$

Alternatively,

$$\begin{array}{r} 2x - 3y \\ 2x - 3y \\ \hline \end{array}$$

Multiplying by $2x \rightarrow \quad 4x^2 - 6xy$

Multiplying by $-3y \rightarrow \quad - 6xy + 9y^2$

Adding gives: $\quad \dfrac{4x^2 - 12xy + 9y^2}{}$

Problem 31.   Remove the brackets from the expression $2[p^2 - 3(q + r) + q^2]$

In this problem there are two brackets and the 'inner' one is removed first.
Hence

$$2[p^2 - 3(q + r) + q^2] = 2[p^2 - 3q - 3r + q^2]$$

$$= 2p^2 - 6q - 6r + 2q^2$$

Problem 32.   Remove the brackets and simplify the expression
$2a - [3\{2(4a - b) - 5(a + 2b)\} + 4a]$

Removing the innermost brackets gives:

$$2a - [3\{8a - 2b - 5a - 10b\} + 4a]$$

Collecting together similar terms gives:

$$2a - [3\{3a - 12b\} + 4a]$$

Removing the 'curly' brackets gives:

$$2a - [9a - 36b + 4a]$$

Collecting together similar terms gives:

$$2a - [13a - 36b]$$

Removing the outer brackets gives:

$$2a - 13a + 36b$$

i.e. $-11a + 36b$ or $36b - 11a$

(see law (iii), page 23).

Problem 33.   Simplify
$x(2x - 4y) - 2x(4x + y)$

Removing brackets gives:

$$2x^2 - 4xy - 8x^2 - 2xy$$

Collecting together similar terms gives:

$$-6x^2 - 6xy$$

Factorizing gives:

$$-6x(x + y)$$

since $-6x$ is common to both terms.

Problem 34. Factorize (a) $xy - 3xz$
(b) $4a^2 + 16ab^3$ (c) $3a^2b - 6ab^2 + 15ab$

For each part of this problem, the HCF of the terms will become one of the factors. Thus:

(a) $xy - 3xz = x(y - 3z)$
(b) $4a^2 + 16ab^3 = 4a(a + 4b^3)$
(c) $3a^2b - 6ab^2 + 15ab = 3ab(a - 2b + 5)$

Problem 35. Factorize $ax - ay + bx - by$

The first two terms have a common factor of a and the last two terms a common factor of b. Thus:

$$ax - ay + bx - by = a(x - y) + b(x - y)$$

The two newly formed terms have a common factor of $(x - y)$. Thus:

$$a(x - y) + b(x - y) = (x - y)(a + b)$$

Problem 36. Factorize
$2ax - 3ay + 2bx - 3by$

---

a is a common factor of the first two terms and b a common factor of the last two terms. Thus:

$$2ax - 3ay + 2bx - 3by$$
$$= a(2x - 3y) + b(2x - 3y)$$

$(2x - 3y)$ is now a common factor thus:

$$a(2x - 3y) + b(2x - 3y) = (2x - 3y)(a + b)$$

Alternatively, $2x$ is a common factor of the original first and third terms and $-3y$ is a common factor of the second and fourth terms. Thus:

$$2ax - 3ay + 2bx - 3by = 2x(a + b) - 3y(a + b)$$

$(a + b)$ is now a common factor thus:

$$2x(a + b) - 3y(a + b) = (a + b)(2x - 3y)$$

as before.

Problem 37. Factorize $x^3 + 3x^2 - x - 3$

---

x^2 is a common factor of the first two terms, thus:

$$x^3 + 3x^2 - x - 3 = x^2(x + 3) - x - 3$$

-1 is a common factor of the last two terms, thus:

$$x^2(x + 3) - x - 3 = x^2(x + 3) - 1(x + 3)$$

$(x + 3)$ is now a common factor, thus:

$$x^2(x + 3) - 1(x + 3) = (x + 3)(x^2 - 1)$$

Further problems on brackets and factorization may be found in Section 3.7, Problems 26 to 42, page 37.

3.4 Fundamental laws and precedence

The **laws of precedence** which apply to arithmetic also apply to algebraic expressions. The order is <u>B</u>rackets, <u>O</u>f, <u>D</u>ivision, <u>M</u>ultiplication, <u>A</u>ddition and <u>S</u>ubtraction (i.e. **BODMAS**).

Problem 38. Simplify $2a + 5a \times 3a - a$

Multiplication is performed before addition and subtraction thus:

$$2a + 5a \times 3a - a = 2a + 15a^2 - a$$
$$= a + 15a^2 = a(1 + 15a)$$

Problem 39. Simplify $(a + 5a) \times 2a - 3a$

The order of precedence is brackets, multiplication, then subtraction. Hence

$$(a + 5a) \times 2a - 3a = 6a \times 2a - 3a$$
$$= 12a^2 - 3a$$
$$= 3a(4a - 1)$$

Problem 40. Simplify $a + 5a \times (2a - 3a)$

The order of precedence is brackets, multiplication, then subtraction. Hence

$a + 5a \times (2a - 3a) = a + 5a \times -a = a + -5a^2$

$$= a - 5a^2 = \mathbf{a(1 - 5a)}$$

Problem 41. Simplify $a \div 5a + 2a - 3a$

The order of precedence is division, then addition and subtraction. Hence

$$a \div 5a + 2a - 3a = \frac{a}{5a} + 2a - 3a$$

$$= \frac{1}{5} + 2a - 3a$$

$$= \frac{1}{5} - a$$

Problem 42. Simplify $a \div (5a + 2a) - 3a$

The order of precedence is brackets, division and subtraction. Hence

$$a \div (5a + 2a) - 3a = a \div 7a - 3a = \frac{a}{7a} - 3a$$

$$= \frac{1}{7} - 3a$$

Problem 43. Simplify $a \div (5a + 2a - 3a)$

The order of precedence is brackets, then division. Hence:

$$a \div (5a + 2a - 3a) = a \div 4a = \frac{a}{4a} = \frac{1}{4}$$

Problem 44. Simplify
$3c + 2c \times 4c + c \div 5c - 8c$

The order of precedence is division, multiplication, addition and subtraction.
  Hence:

$3c + 2c \times 4c + c \div 5c - 8c$

$$= 3c + 2c \times 4c \times \left(\frac{c}{5c}\right) - 8c$$

$$= 3c + 8c^2 + \frac{1}{5} - 8c$$

$$= \mathbf{8c^2 - 5c + \frac{1}{5}} \text{ or } \mathbf{c(8c - 5) + \frac{1}{5}}$$

Problem 45. Simplify
$(3c + 2c)4c + c \div 5c - 8c$

The order of precedence is brackets, division, multiplication, addition and subtraction. Hence

$(3c + 2c)4c + c \div 5c - 8c$

$$= 5c \times 4c + c \div 5c - 8c$$

$$= 5c \times 4c + \frac{c}{5c} - 8c$$

$$= \mathbf{20c^2 + \frac{1}{5} - 8c} \text{ or } \mathbf{4c(5c - 2) + \frac{1}{5}}$$

Problem 46. Simplify
$3c + 2c \times 4c + c \div (5c - 8c)$

The order of precedence is brackets, division, multiplication and addition.
  Hence:

$3c + 2c \times 4c + c \div (5c - 8c)$

$$= 3c + 2c \times 4c + c \div -3c$$

$$= 3c + 2c \times 4c + \frac{c}{-3c}$$

Now $\dfrac{c}{-3c} = \dfrac{1}{-3}$

Multiplying numerator and denominator by -1 gives $\dfrac{1 \times -1}{-3 \times -1}$, i.e. $-\dfrac{1}{3}$

Hence: $3c + 2c \times 4c + \dfrac{c}{-3c} = 3c + 2c \times 4c - \dfrac{1}{3}$

$$= 3c + 8c^2 - \frac{1}{3} \text{ or}$$

$$\mathbf{c(3 + 8c) - \frac{1}{3}}$$

Problem 47. Simplify
$(3c + 2c)(4c + c) \div (5c - 8c)$

The order of precedence is brackets, division and multiplication. Hence

$$(3c + 2c)(4c + c) \div (5c - 8c) = 5c \times 5c \div -3c$$

$$= 5c \times \frac{5c}{-3c}$$

$$= 5c \times -\frac{5}{3}$$

$$= -\frac{25}{3}c$$

Problem 48.   Simplify
$(2a - 3) \div 4a + 5 \times 6 - 3a$

The bracket around the $(2a - 3)$ shows that both $2a$ and -3 have to be divided by $4a$, and to remove the bracket the expression is written in fraction form.
   Hence:

$$(2a - 3) \div 4a + 5 \times 6 - 3a = \frac{2a - 3}{4a} + 5 \times 6 - 3a$$

$$= \frac{2a - 3}{4a} + 30 - 3a = \frac{2a}{4a} - \frac{3}{4a} + 30 - 3a$$

$$= \frac{1}{2} - \frac{3}{4a} + 30 - 3a = 30\frac{1}{2} - \frac{3}{4a} - 3a \cdot$$

Problem 49.   Simplify
$\frac{1}{3}$ of $3p + 4p(3p - p)$

Applying BODMAS, the expression becomes $\frac{1}{3}$ of $3p + 4p \times 2p$, and changing 'of' to 'x' gives:

$\frac{1}{3} \times 3p + 4p \times 2p$

i.e. $p + 8p^2$ or $p(1 + 8p)$

Further problems on fundamental laws and precedence may be found in Section 3.7, Problems 43 to 54, page 38.

3.5   Partial fractions

By algebraic addition,

$$\frac{1}{x - 2} + \frac{3}{x + 1} = \frac{(x + 1) + 3(x - 2)}{(x - 2)(x + 1)}$$

$$= \frac{4x - 5}{x^2 - x - 2}$$

The reverse process of moving from $\dfrac{4x - 5}{x^2 - x - 2}$ to $\dfrac{1}{x - 2} + \dfrac{3}{x + 1}$ is called resolving into **partial fractions**.
   In order to resolve an algebraic expression into partial fractions:

(i) the denominator must factorize (in the above example, $x^2 - x - 2$ factorizes as $(x - 2)(x + 1)$), and

(ii) the numerator must be at least one degree less than the denominator (in the above example $(4x - 5)$ is of degree 1 since the highest powered x term is x^1 and $(x^2 - x - 2)$ is of degree 2)

When the degree of the numerator is equal to or higher than the degree of the denominator, the numerator must be divided by the denominator until the remainder is of less degree than the denominator (see Problems 52 and 53).
   There are basically three types of partial fraction and the form of partial fraction used is summarized below, where $f(x)$ is assumed to be of less degree than the relevant denominator and A, B and C are constants to be determined.

Type	Denominator containing	Expression	Form of partial fraction
1	Linear factors (see Problems 50 to 53)	$\dfrac{f(x)}{(x + a)(x - b)(x + c)}$	$\dfrac{A}{(x + a)} + \dfrac{B}{(x - b)} + \dfrac{C}{(x + c)}$
2	Repeated linear factors (see Problems 54 to 56)	$\dfrac{f(x)}{(x + a)^3}$	$\dfrac{A}{(x + a)} + \dfrac{B}{(x + a)^2} + \dfrac{C}{(x + a)^3}$
3	Quadratic factors (see Problems 57 and 58)	$\dfrac{f(x)}{(ax^2 + bx + c)(x + d)}$	$\dfrac{Ax + B}{(ax^2 + bx + c)} + \dfrac{C}{(x + d)}$

(In the latter type, $ax^2 + bx + c$ is a quadratic expression which does not factorize without containing surds or imaginary terms.)

Resolving an algebraic expression into partial fractions is used as a preliminary to integrating certain functions.

Problem 50. Resolve $\dfrac{11 - 3x}{x^2 + 2x - 3}$ into partial fractions

---

The denominator factorizes as $(x - 1)(x + 3)$ and the numerator is of less degree than the denominator. Thus $\dfrac{11 - 3x}{x^2 + 2x - 3}$ may be resolved into partial fractions.

Let $\quad \dfrac{11 - 3x}{(x - 1)(x + 3)} \equiv \dfrac{A}{(x - 1)} + \dfrac{B}{(x + 3)}$,

where A and B are constants to be determined,

i.e. $\dfrac{11 - 3x}{(x - 1)(x + 3)} \equiv \dfrac{A(x + 3) + B(x - 1)}{(x - 1)(x + 3)}$,

by algebraic addition.

Since the denominators are the same on each side of the identity then the numerators are equal to each other.

Thus, $11 - 3x \equiv A(x + 3) + B(x - 1)$.

To determine constants A and B, values of x are chosen to make the term in A or B equal to zero.

When $x = 1$, then $11 - 3(1) \equiv A(1 + 3) + B(0)$

i.e. $\quad 8 = 4A$

i.e. $\quad A = 2$

When $x = -3$, then $11 - 3(-3) \equiv A(0) + B(-3 - 1)$

i.e. $\quad 20 = -4B$

i.e. $\quad B = -5$

Thus $\dfrac{11 - 3x}{x^2 + 2x - 3} \equiv \dfrac{2}{(x - 1)} + \dfrac{-5}{(x + 3)}$

$\equiv \dfrac{2}{(x - 1)} - \dfrac{5}{(x + 3)}$

(Check: $\dfrac{2}{x - 1} - \dfrac{5}{x - 3} = \dfrac{2(x + 3) - 5(x - 1)}{(x - 1)(x + 3)}$

$= \dfrac{11 - 3x}{x^2 + 2x - 3}$)

Problem 51. Convert $\dfrac{2x^2 - 9x - 35}{(x + 1)(x - 2)(x + 3)}$ into the sum of three partial fractions

---

Let $\dfrac{2x^2 - 9x - 35}{(x + 1)(x - 2)(x + 3)}$

$\equiv \dfrac{A}{(x + 1)} + \dfrac{B}{(x - 2)} + \dfrac{C}{(x + 3)}$

$\equiv \dfrac{A(x - 2)(x + 3) + B(x + 1)(x + 3) + C(x + 1)(x - 2)}{(x + 1)(x - 2)(x + 3)}$

by algebraic addition.

Equating the numerators gives:

$2x^2 - 9x - 35 \equiv A(x - 2)(x + 3)$
$\qquad\qquad\qquad + B(x + 1)(x + 3)$
$\qquad\qquad\qquad + C(x + 1)(x - 2)$

Let $x = -1$. Then

$2(-1)^2 - 9(-1) - 35 \equiv A(-3)(2) + B(0)(2)$
$\qquad\qquad\qquad\qquad + C(0)(-3)$

i.e. $\quad -24 = -6A$

i.e. $\quad A = \dfrac{-24}{-6} = 4$

Let $x = 2$. Then

$2(2)^2 - 9(2) - 35 \equiv A(0)(5) + B(3)(5) + C(3)(0)$

i.e. $\quad -45 = 15B$

i.e. $\quad B = \dfrac{-45}{15} = -3$

Let $x = -3$. Then

$2(-3)^2 - 9(-3) - 35 \equiv A(-5)(0) + B(-2)(0)$
$\qquad\qquad\qquad\qquad + C(-2)(-5)$

i.e. $\quad 10 = 10C$

i.e. $\quad C = 1$

Thus $\dfrac{2x^2 - 9x - 35}{(x + 1)(x - 2)(x + 3)}$

$\equiv \dfrac{4}{(x + 1)} - \dfrac{3}{(x - 2)} + \dfrac{1}{(x + 3)}$

Problem 52. Resolve $\dfrac{x^2+1}{x^2-3x+2}$ into partial fractions

The denominator is of the same degree as the numerator. Thus dividing out gives:

$$
\begin{array}{r}
1 \\
x^2-3x+2 \overline{) x^2 +1} \\
x^2-3x+2 \\
\hline
3x-1
\end{array}
$$

Hence $\dfrac{x^2+1}{x^2-3x+2} \equiv 1+\dfrac{3x-1}{x^2-3x+2}$

$$\equiv 1+\dfrac{3x-1}{(x-1)(x-2)}$$

Let $\dfrac{3x-1}{(x-1)(x-2)} \equiv \dfrac{A}{(x-1)}+\dfrac{B}{(x-2)}$

$$\equiv \dfrac{A(x-2)+B(x-1)}{(x-1)(x-2)}$$

Equating numerators gives:

$$3x-1 \equiv A(x-2)+B(x-1)$$

Let $x=1$. Then $2=-A$

i.e. $A=-2$

Let $x=2$. Then $5=B$

Hence $\dfrac{3x-1}{(x-1)(x-2)} \equiv \dfrac{-2}{(x-1)}+\dfrac{5}{(x-2)}$

Thus $\dfrac{x^2+1}{x^2-3x+2} \equiv 1-\dfrac{2}{(x-1)}+\dfrac{5}{(x-2)}$

Problem 53. Express $\dfrac{x^3-2x^2-4x-4}{x^2+x-2}$ in partial fractions

The numerator is of higher degree than the denominator. Thus dividing out gives:

$$
\begin{array}{r}
x- 3 \\
x^2+x-2 \overline{) x^3-2x^2-4x- 4} \\
x^3+ x^2-2x \\
\hline
-3x^2-2x- 4 \\
-3x^2-3x+ 6 \\
\hline
x- 10
\end{array}
$$

Thus $\dfrac{x^3-2x^2-4x-4}{x^2+x-2} \equiv x-3+\dfrac{x-10}{x^2+x-2}$

$$\equiv x-3+\dfrac{x-10}{(x+2)(x-1)}$$

Let $\dfrac{x-10}{(x+2)(x-1)} \equiv \dfrac{A}{(x+2)}+\dfrac{B}{(x-1)}$

$$\equiv \dfrac{A(x-1)+B(x+2)}{(x+2)(x-1)}$$

Equating the numerators gives:

$$x-10 \equiv A(x-1)+B(x+2)$$

Let $x=-2$. Then $-12=-3A$

i.e. $A=4$

Let $x=1$. Then $-9=3B$

i.e. $B=-3$

Hence $\dfrac{x-10}{(x+2)(x-1)} \equiv \dfrac{4}{(x+2)}-\dfrac{3}{(x-1)}$

Thus $\dfrac{x^3-2x^2-4x-4}{x^2+x-2}$

$$\equiv x-3+\dfrac{4}{(x+2)}-\dfrac{3}{(x-1)}$$

Problem 54. Resolve $\dfrac{2x+3}{(x-2)^2}$ into partial fractions

The denominator contains a repeated linear factor, $(x-2)^2$.

Let $\dfrac{2x+3}{(x-2)^2} \equiv \dfrac{A}{(x-2)}+\dfrac{B}{(x-2)^2} \equiv \dfrac{A(x-2)+B}{(x-2)^2}$

Equating the numerators gives:

$$2x+3 \equiv A(x-2)+B$$

Let $x=2$. Then $7=A(0)+B$, i.e. $B=7$

$$2x+3 \equiv A(x-2)+B \equiv Ax-2A+B$$

Since an identity is true for all values of the unknown, the coefficients of similar terms may be equated.

Hence, equating the coefficients of x gives: $2=A$.

(Also, as a check, equating the constant terms gives:

$$3=-2A+B$$

When $A=2$ and $B=7$,

$$\text{RHS}=-2(2)+7=3=\text{LHS})$$

Hence $\dfrac{2x+3}{(x-2)^2} \equiv \dfrac{2}{(x-2)} + \dfrac{7}{(x-2)^2}$

Problem 55. Express $\dfrac{5x^2 - 2x - 19}{(x+3)(x-1)^2}$ as the sum of three partial fractions

---

The denominator is a combination of a linear factor and a repeated linear factor.

Let $\dfrac{5x^2 - 2x - 19}{(x+3)(x-1)^2}$

$\equiv \dfrac{A}{(x+3)} + \dfrac{B}{(x-1)} + \dfrac{C}{(x-1)^2}$

$\equiv \dfrac{A(x-1)^2 + B(x+3)(x-1) + C(x+3)}{(x+3)(x-1)^2}$,

by algebraic addition
Equating the numerators gives:

$5x^2 - 2x - 19 \equiv A(x-1)^2 + B(x+3)(x-1)$
$\qquad\qquad\qquad + C(x+3) \qquad (1)$

Let $x = -3$. Then

$5(-3)^2 - 2(-3) - 19 \equiv A(-4)^2 + B(0)(-4) + C(0)$

i.e. $32 = 16A$; i.e. $A = 2$

Let $x = 1$. Then

$5(1)^2 - 2(1) - 19 \equiv A(0)^2 + B(4)(0) + C(4)$

i.e. $-16 = 4C$; i.e. $C = -4$

Without expanding the RHS of equation (1) it can be seen that equating the coefficients of x^2 gives: $5 = A + B$. Since $A = 2, B = 3$.
(Check: Identity (1) may be expressed as:

$5x^2 - 2x - 19 \equiv A(x^2 - 2x + 1)$
$\qquad\qquad + B(x^2 + 2x - 3) + C(x+3)$

i.e. $5x^2 - 2x - 19 \equiv Ax^2 - 2Ax + A + Bx^2 + 2Bx$
$\qquad\qquad\qquad - 3B + Cx + 3C$

Equating the x term coefficients gives:

$-2 \equiv -2A + 2B + C$

When $A = 2, B = 3$ and $C = -4$ then

$-2A + 2B + C = -2(2) + 2(3) - 4 = -2 = \text{LHS}$

Equating the constant term gives:

$-19 \equiv A - 3B + 3C$

RHS $= 2 - 3(3) + 3(-4) = 2 - 9 - 12 = -19$
$\quad = \text{LHS})$

Hence $\dfrac{5x^2 - 2x - 19}{(x+3)(x-1)^2}$

$\equiv \dfrac{2}{(x+3)} + \dfrac{3}{(x-1)} - \dfrac{4}{(x-1)^2}$

Problem 56. Resolve $\dfrac{3x^2 + 16x + 15}{(x+3)^3}$ into partial fractions

---

Let

$\dfrac{3x^2 + 16x + 15}{(x+3)^3} \equiv \dfrac{A}{(x+3)} + \dfrac{B}{(x+3)^2} + \dfrac{C}{(x+3)^3}$

$\equiv \dfrac{A(x+3)^2 + B(x+3) + C}{(x+3)^3}$

Equating the numerators gives:

$3x^2 + 16x + 15 \equiv A(x+3)^2 + B(x+3) + C \qquad (1)$

Let $x = -3$. Then

$3(-3)^2 + 16(-3) + 15 \equiv A(0)^2 + B(0) + C$

i.e. $-6 = C$

Identity (1) may be expanded as:

$3x^2 + 16x + 15 \equiv A(x^2 + 6x + 9) + B(x+3) + C$

i.e.

$3x^2 + 16x + 15 \equiv Ax^2 + 6Ax + 9A + Bx + 3B + C$

Equating the coefficients of x^2 terms gives: $3 = A$.
Equating the coefficients of x terms gives:

$16 = 6A + B$

Since $A = 3, B = -2$.
(Check: equating the constant terms gives:

$15 = 9A + 3B + C$

When $A = 3, B = -2$ and $C = -6$,

$9A + 3B + C = 9(3) + 3(-2) + (-6)$
$\qquad\qquad = 27 - 6 - 6 = 15 = \text{LHS})$

Thus $\dfrac{3x^2 + 16x + 15}{(x+3)^3}$

$$\equiv \frac{3}{(x+3)} - \frac{2}{(x+3)^2} - \frac{6}{(x+3)^3}$$

Problem 57. Express $\dfrac{7x^2 + 5x + 13}{(x^2 + 2)(x+1)}$ in partial fractions

---

The denominator is a combination of a quadratic factor, $(x^2 + 2)$ (which does not factorize without introducing imaginary surd terms), and a linear factor, $(x+1)$.

Let $\dfrac{7x^2 + 5x + 13}{(x^2 + 2)(x+1)} \equiv \dfrac{Ax + B}{(x^2 + 2)} + \dfrac{C}{(x+1)}$

$$= \frac{(Ax + B)(x+1) + C(x^2 + 2)}{(x^2 + 2)(x+1)}$$

Equating numerators gives:

$$7x^2 + 5x + 13 \equiv (Ax + B)(x+1) + C(x^2 + 2)$$

$$(1)$$

Let $x = -1$. Then

$$7(-1)^2 + 5(-1) + 13 \equiv (Ax + B)(0) + C(1+2)$$

i.e. $15 = 3C$; i.e. $C = 5$

Identity (1) may be expanded as:

$$7x^2 + 5x + 13 \equiv Ax^2 + Ax + Bx + B + Cx^2 + 2C$$

Equating the coefficients of x^2 terms gives:

$$7 = A + C; \text{ since } C = 5, A = 2$$

Equating the coefficients of x terms gives:

$$5 = A + B; \text{ since } A = 2, B = 3$$

(Check: equating the constant terms gives:

$$13 = B + 2C$$

When $B = 3$ and $C = 5$,

$$B + 2C = 3 + 10 = 13 = \text{LHS})$$

Hence $\dfrac{7x^2 + 5x + 13}{(x^2 + 2)(x+1)} \equiv \dfrac{2x + 3}{(x^2 + 2)} + \dfrac{5}{(x+1)}$

Problem 58. Resolve $\dfrac{3 + 6x + 4x^2 - 2x^3}{x^2(x^2 + 3)}$ into partial fractions

---

Terms such as x^2 may be treated as $(x + 0)^2$, i.e. they are repeated linear factors.

Let $\dfrac{3 + 6x + 4x^2 - 2x^3}{x^2(x^2 + 3)} \equiv \dfrac{A}{x} + \dfrac{B}{x^2} + \dfrac{Cx + D}{(x^2 + 3)}$

$$\equiv \frac{Ax(x^2 + 3) + B(x^2 + 3) + (Cx + D)x^2}{x^2(x^2 + 3)}$$

Equating the numerators gives:

$$3 + 6x + 4x^2 - 2x^3 \equiv Ax(x^2 + 3) + B(x^2 + 3)$$
$$+ (Cx + D)x^2$$
$$\equiv Ax^3 + 3Ax + Bx^2 + 3B$$
$$+ Cx^3 + Dx^2$$

Let $x = 0$. Then $3 = 3B$

i.e. $B = 1$

Equating the coefficients of x^3 terms gives:

$$-2 = A + C \qquad (1)$$

Equating the coefficients of x^2 terms gives:

$$4 = B + D$$

Since $B = 1$, $D = 3$

Equating the coefficients of x terms gives: $6 = 3A$

i.e. $A = 2$

From equation (1), since $A = 2$, $C = -4$

Hence $\dfrac{3 + 6x + 4x^2 - 2x^3}{x^2(x^2 + 3)}$

$$\equiv \frac{2}{x} + \frac{1}{x^2} + \frac{-4x + 3}{x^2 + 3}$$

$$\equiv \frac{2}{x} + \frac{1}{x^2} + \frac{3 - 4x}{x^2 + 3}$$

Further problems on partial fractions may be found in Section 3.7, Problems 55 to 70, pages 38 and 39.

3.6 Direct and inverse proportionality

An expression such as $y = 3x$ contains two variables. For every value of x there is a corresponding value of y. The variable x is called the **independent variable** and y is called the **dependent variable**.

When an increase or decrease in an independent variable leads to an increase or decrease of the same proportion in the dependent variable this is termed **direct proportion**. If $y = 3x$ then y is directly proportional to x, which may be written as $y \propto x$ or $y = kx$, where k is called the **coefficient of proportionality** (in this case, k being equal to 3).

When an increase in an independent variable leads to a decrease of the same proportion in the dependent variable (or vice versa) this is termed **inverse proportion**. If y is inversely proportional to x then $y \propto (1/x)$ or $y = k/x$. Alternatively, $k = xy$, that is, for inverse proportionality the product of the variables is constant.

Examples of laws involving direct and inverse proportionality in science include:

(i) **Hooke's law**, which states that within the elastic limit of a material, the strain ε produced is directly proportional to the stress, σ, producing it, i.e. $\varepsilon \propto \sigma$ or $\varepsilon = k\sigma$

(ii) **Charles's law**, which states that for a given mass of gas at constant pressure the volume V is directly proportional to its thermodynamic temperature T, i.e. $V \propto T$ or $V = kT$

(iii) **Ohm's law**, which states that the current I flowing through a fixed resistor is directly proportional to the applied voltage V, i.e. $I \propto V$ or $I = kV$

(iv) **Boyle's law**, which states that for a gas at constant temperature, the volume V of a fixed mass of gas is inversely proportional to its absolute pressure p, i.e. $p \propto (1/V)$ or $p = k/V$, i.e. $pV = k$

Problem 59. If y is directly proportional to x and $y = 2.48$ when $x = 0.4$, determine (a) the coefficient of proportionality and (b) the value of y when $x = 0.65$

(a) $y \propto x$, i.e. $y = kx$. When $y = 2.48$ and $x = 0.4$, $2.48 = k(0.4)$. Hence the coefficient of proportionality,

$$k = \frac{2.48}{0.4} = \mathbf{6.2}$$

(b) $y = kx$. Hence, when $x = 0.65$, $y = (6.2)(0.65) = \mathbf{4.03}$

Problem 60. Hooke's law states that stress σ is directly proportional to strain ε within the elastic limit of a material. When, for mild steel, the stress is 25×10^6 pascals, the strain is 0.000125 Determine (a) the coefficient of proportionality and (b) the value of strain when the stress is 18×10^6 pascals

---

(a) $\sigma \propto \varepsilon$, i.e. $\sigma = k\varepsilon$, from which $k = \sigma/\varepsilon$. Hence the coefficient of proportionality,

$$k = \frac{25 \times 10^6}{0.000125} = \mathbf{200 \times 10^9 \ pascals}$$

(The coefficient of proportionality k in this case is called Young's Modulus of Elasticity.)

(b) Since $\sigma = k\varepsilon$, $\varepsilon = \sigma/k$.
Hence when $\sigma = 18 \times 10^6$,

$$\text{strain } \varepsilon = \frac{18 \times 10^6}{200 \times 10^9} = \mathbf{0.00009}$$

Problem 61. The electrical resistance R of a piece of wire is inversely proportional to the cross-sectional area A. When $A = 5$ mm^2, $R = 7.02$ ohms. Determine (a) the coefficient of proportionality and (b) the cross-sectional area when the resistance is 4 ohms

---

(a) $R \propto 1/A$, i.e. $R = k/A$ or $k = RA$. Hence, when $R = 7.2$ and $A = 5$, the coefficient of proportionality, $k = (7.2)(5) = \mathbf{36}$

(b) Since $k = RA$ then $A = k/R$.
When $R = 4$, the cross-sectional area,

$$A = \frac{36}{4} = \mathbf{9 \ mm^2}$$

Problem 62. Boyle's law states that at constant temperature, the volume V of a fixed mass of gas is inversely proportional to its absolute pressure p. If a gas occupies a volume of 0.08 m^3 at a pressure of 1.5×10^6 pascals determine (a) the coefficient of proportionality and (b) the volume if the pressure is changed to 4×10^6 pascals

---

(a) $V \propto 1/p$, i.e. $V = k/p$ or $k = pV$.

Hence the coefficient of proportionality,

$$k = (1.5 \times 10^6)(0.08)$$

$$= \mathbf{0.12 \times 10^6}$$

(b)   Volume $V = \dfrac{k}{p} = \dfrac{0.12 \times 10^6}{4 \times 10^6} = \mathbf{0.03 \ m^3}$

Further problems on direct and inverse proportionality may be found in the following Section 3.7, Problems 71 to 75, page 39.

3.7   Further problems on algebra

Basic operations

1   Find the value of $2xy + 3yz - xyz$, when $x = 2$, $y = -2$ and $z = 4$. $[-16]$

2   Evaluate $3pq^2r^3$ when $p = \dfrac{2}{3}$, $q = -2$ and $r = -1$. $[-8]$

3   Find the sum of $3a$, $-2a$, $-6a$, $5a$ and $4a$. $[4a]$

4   Simplify $\dfrac{4}{3}c + 2c - \dfrac{1}{6}c - c$. $\left[2\dfrac{1}{6}c\right]$

5   Find the sum of $3x$, $2y$, $-5x$, $2z$, $-\dfrac{1}{2}y$, $-\dfrac{1}{4}x$.
$$\left[-2\dfrac{1}{4}x + 1\dfrac{1}{2}y + 2z\right]$$

6   Add together $2a + 3b + 4c$, $-5a - 2b + c$, $4a - 5b - 6c$. $[a - 4b - c]$

7   Add together $3d + 4e$, $-2e + f$, $2d - 3f$, $4d - e + 2f - 3e$. $[9d - 2e]$

8   From $4x - 3y + 2z$ subtract $x + 2y - 3z$.
$$[3x - 5y + 5z]$$

9   Subtract $\dfrac{3}{2}a - \dfrac{b}{3} + c$ from $\dfrac{b}{2} - 4a - 3c$.
$$\left[-5\dfrac{1}{2}a + \dfrac{5}{6}b - 4c\right]$$

10   Multiply $3x + 2y$ by $x - y$.
$$[3x^2 - xy - 2y^2]$$

11   Multiply $2a - 5b + c$ by $3a + b$.
$$[6a^2 - 13ab + 3ac - 5b^2 + bc]$$

12   Simplify (i) $3a \div 9ab$ (ii) $4a^2b \div 2a$.
$$\left[(i) \ \dfrac{1}{3b} \ (ii) \ 2ab\right]$$

13   Divide $2x^2 + xy - y^2$ by $x + y$. $[2x - y]$

14   Divide $p^3 + q^3$ by $p + q$. $[p^2 - pq + q^2]$

Laws of indices

15   Simplify $(x^2y^3z)(x^3yz^2)$ and evaluate when $x = \dfrac{1}{2}$, $y = 2$ and $z = 3$. $\left[x^5y^4z^3, \ 13\dfrac{1}{2}\right]$

16   Simplify $(a^{3/2}bc^{-3})(a^{1/2}b^{-1/2}c)$ and evaluate when $a = 3$, $b = 4$ and $c = 2$.
$$\left[a^2b^{1/2}c^{-2}, \ 4\dfrac{1}{2}\right]$$

17   Simplify $\dfrac{a^5bc^3}{a^2b^3c^2}$ and evaluate when $a = \dfrac{3}{2}$, $b = \dfrac{1}{2}$ and $c = \dfrac{2}{3}$. $[a^3b^{-2}c, \ 9]$

In Problems 18 to 25, simplify the given expressions:

18   $\dfrac{x^{1/5}y^{1/2}z^{1/3}}{x^{-1/2}y^{1/3}z^{-1/6}}$ $[x^{7/10}y^{1/6}z^{1/2}]$

19   $\dfrac{a^2b + a^3b}{a^2b^2}$ $\left[\dfrac{1+a}{b}\right]$

20   $\dfrac{p^3q^2}{pq^2 - p^2q}$ $\left[\dfrac{p^2q}{q - p}\right]$

21   $(a^2)^{1/2}(b^2)^3(c^{1/2})^3$ $[ab^6c^{3/2}]$

22   $\dfrac{(abc)^2}{(a^2b^{-1}c^{-3})^3}$ $[a^{-4}b^5c^{11}]$

23   $(\sqrt{x}\sqrt{y^3}\sqrt[3]{z^2})(\sqrt{x}\sqrt{y^3}\sqrt{z^3})$ $[xy^3\sqrt[6]{z^{13}}]$

24   $(e^2f^3)(e^{-3}f^{-5})$, expressing the answer with positive indices only. $\left[\dfrac{1}{ef^2}\right]$

25   $\dfrac{(a^3b^{1/2}c^{-1/2})(ab)^{1/3}}{(\sqrt{a^3}\sqrt{b}\ c)}$
$$\left[a^{11/6}b^{1/3}c^{-3/2} \ \text{or} \ \dfrac{\sqrt[6]{a^{11}}\sqrt[3]{b}}{\sqrt{c^3}}\right]$$

Brackets and factorization

In Problems 26 to 38, remove the brackets and simplify where possible:

26   $(x + 2y) + (2x - y)$ $[3x + y]$

27   $(4a + 3y) - (a - 2y)$ $[3a + 5y]$

28   $2(x - y) - 3(y - x)$ $[5(x - y)]$

29   $2x^2 - 3(x - xy) - x(2y - x)$ $[x(3x - 3 + y)]$

30   $2(p + 3q - r) - 4(r - q + 2p) + p$
$$[-5p + 10q - 6r]$$

31   $(a+b)(a+2b)$ $[a^2+3ab+2b^2]$

32   $(p+q)(3p-2q)$ $[3p^2+pq-2q^2]$

33   (i) $(x-2y)^2$ (ii) $(3a-b)^2$
$[$(i) $x^2-4xy+4y^2$ (ii) $9a^2-6ab+b^2]$

34   $3a(b+c)+4c(a-b)$ $[3ab+7ac-4bc]$

35   $2x+(y-(2x+y))$ $[0]$

36   $3a+2(a-(3a-2))$ $[4-a]$

37   $2-5(a(a-2b)-(a-b)^2)$ $[2+5b^2]$

38   $24p-[2(3(5p-q)-2(p+2q))+3q]$
$[11q-2p]$

In Problems 39 to 42, factorize:

39   (i) $pb+2pc$ (ii) $2l^2+8ln$
$[$(i) $p(b+2c)$ (ii) $2l(l+4n)]$

40   (i) $21a^2b^2-28ab$ (ii) $2xy^2+6x^2y+8x^3y$
$[$(i) $7ab(3ab-4)$ (ii) $2xy(y+3x+4x^2)]$

41   (i) $ay+by+a+b$ (ii) $px+qx+py+qy$
$[$(i) $(a+b)(y+1)$ (ii) $(p+q)(x+y)]$

42   (i) $ax-ay+bx-by$
(ii) $2ax+3ay-4bx-6by$
$[$(i) $(x-y)(a+b)$ (ii) $(a-2b)(2x+3y)]$

Fundamental laws and precedence

In Problems 43 to 54, simplify:

43   $2x\div 4x+6x$ $\left[\dfrac{1}{2}+6x\right]$

44   $2x\div(4x+6x)$ $\left[\dfrac{1}{5}\right]$

45   $3a-2a\times 4a+a$ $[4a(1-2a)]$

46   $(3a-2a)4a+a$ $[a(4a+1)]$

47   $3a-2a(4a+a)$ $[a(3-10a)]$

48   $2y+4\div 6y+3\times 4-5y$ $\left[\dfrac{2}{3y}-3y+12\right]$

49   $(2y+4)\div 6y+3\times 4-5y$
$\left[\dfrac{2}{3y}+12\dfrac{1}{3}-5y\right]$

50   $2y+4\div 6y+3(4-5y)$ $\left[\dfrac{2}{3y}+12-13y\right]$

51   $3\div y+2\div y+1$ $\left[\dfrac{5}{y}+1\right]$

52   $p^2-3pq\times 2p\div 6q+pq$ $[pq]$

53   $(x+1)(x-4)\div(2x+2)$ $\left[\dfrac{1}{2}(x-4)\right]$

54   $\dfrac{1}{4}$ of $2y+3y(2y-y)$ $\left[y\left(\dfrac{1}{2}+3y\right)\right]$

Partial fractions

Resolve the following into partial fractions:

55   $\dfrac{12}{x^2-9}$ $\left[\dfrac{2}{(x-3)}-\dfrac{2}{(x+3)}\right]$

56   $\dfrac{4(x-4)}{x^2-2x-3}$ $\left[\dfrac{5}{(x+1)}-\dfrac{1}{(x-3)}\right]$

57   $\dfrac{x^2-3x+6}{x(x-2)(x-1)}$ $\left[\dfrac{3}{x}+\dfrac{2}{(x-2)}-\dfrac{4}{(x-1)}\right]$

58   $\dfrac{3(2x^2-8x-1)}{(x+4)(x+1)(2x-1)}$
$\left[\dfrac{7}{(x+4)}-\dfrac{3}{(x+1)}-\dfrac{2}{(2x-1)}\right]$

59   $\dfrac{x^2+9x+8}{x^2+x-6}$ $\left[1+\dfrac{2}{(x+3)}+\dfrac{6}{(x-2)}\right]$

60   $\dfrac{x^2-x-14}{x^2-2x-3}$ $\left[1-\dfrac{2}{(x-3)}+\dfrac{3}{(x+1)}\right]$

61   $\dfrac{3x^3-2x^2-16x+20}{(x-2)(x+2)}$
$\left[3x-2+\dfrac{1}{(x-2)}-\dfrac{5}{(x+2)}\right]$

62   $\dfrac{4x-3}{(x+1)^2}$ $\left[\dfrac{4}{(x+1)}-\dfrac{7}{(x+1)^2}\right]$

63   $\dfrac{x^2+7x+3}{x^2(x+3)}$ $\left[\dfrac{1}{x^2}+\dfrac{2}{x}-\dfrac{1}{(x+3)}\right]$

64   $\dfrac{5x^2-30x+44}{(x-2)^3}$
$\left[\dfrac{5}{(x-2)}-\dfrac{10}{(x-2)^2}+\dfrac{4}{(x-2)^3}\right]$

65   $\dfrac{18+21x-x^2}{(x-5)(x+2)^2}$
$\left[\dfrac{2}{(x-5)}-\dfrac{3}{(x+2)}+\dfrac{4}{(x+2)^2}\right]$

66   $\dfrac{x^2-x-13}{(x^2+7)(x-2)}$ $\left[\dfrac{2x+3}{(x^2+7)}-\dfrac{1}{(x-2)}\right]$

67 $\dfrac{6x-5}{(x-4)(x^2+3)}$ $\left[\dfrac{1}{(x-4)}+\dfrac{2-x}{(x^2+3)}\right]$

68 $\dfrac{15+5x+5x^2-4x^3}{x^2(x^2+5)}$ $\left[\dfrac{1}{x}+\dfrac{3}{x^2}+\dfrac{2-5x}{(x^2+5)}\right]$

69 $\dfrac{x^3+4x^2+20x-7}{(x-1)^2(x^2+8)}$

$\left[\dfrac{3}{(x-1)}+\dfrac{2}{(x-1)^2}+\dfrac{1-2x}{(x^2+8)}\right]$

70 When solving the differential equation $\dfrac{d^2\theta}{dt^2} -$ $6\dfrac{d\theta}{dt}-10\theta = 20-e^{2t}$ by Laplace transforms, given that when $t=0$, $\theta=4$ and $\dfrac{d\theta}{dt}=\dfrac{25}{2}$, the following expression for $\mathcal{L}\{\theta\}$ results:

$$\mathcal{L}\{\theta\} = \dfrac{4s^3 - \dfrac{39}{2}s^2 + 42s - 40}{s(s-2)(s^2-6s+10)}$$

Show that the expression can be resolved into partial fractions to give:

$$\mathcal{L}\{\theta\} = \dfrac{2}{s} - \dfrac{1}{2(s-2)} + \dfrac{5s-3}{2(s^2-6s+10)}$$

Direct and inverse proportionality

71 If p is directly proportional to q and $p=37.5$ when $q=2.5$, determine (a) the constant of proportionality and (b) the value of p when q is 5.2. [(a) 15 (b) 78]

72 Charles's law states that for a given mass of gas at constant pressure the volume is directly proportional to its thermodynamic temperature. A gas occupies a volume of 2.25 litres at 300 K. Determine (a) the constant of proportionality, (b) the volume at 420 K and (c) the temperature when the volume is 2.625 litres.
[(a) 0.0075 (b) 3.15 l (c) 350 K]

73 Ohm's law states that the current flowing in a fixed resistor is directly proportional to the applied voltage. When 30 volts is applied across a resistor the current flowing through the resistor is 2.4×10^{-3} amperes. Determine (a) the constant of proportionality, (b) the current when the voltage is 52 volts and (c) the voltage when the current is 3.6×10^{-3} amperes.
[(a) 0.00008 (b) 4.16×10^{-3} A (c) 45 V]

74 If y is inversely proportional to x and $y=15.3$ when $x=0.6$, determine (a) the coefficient of proportionality, (b) the value of y when x is 1.5, and (c) the value of x when y is 27.2.
[(a) 9.18 (b) 6.12 (c) 0.3375]

75 Boyle's law states that for a gas at constant temperature, the volume of a fixed mass of gas is inversely proportional to its absolute pressure. If a gas occupies a volume of 1.5 m^3 at a pressure of 200×10^3 pascals, determine (a) the constant of proportionality, (b) the volume when the pressure is 800×10^3 pascals and (c) the pressure when the volume is 1.25 m^3.
[(a) 300×10^3 (b) 0.375 m^3 (c) 240×10^3 Pa]

4

Simple equations

4.1 Expressions, equations and identities

$(3x - 5)$ is an example of an **algebraic expression**, whereas $3x - 5 = 1$ is an example of an **equation** (i.e. it contains an 'equals' sign).

An equation is simply a statement that two quantities are equal. For example, 1 m $=$ 1000 mm or $y = mx + c$.

An **identity** is a relationship which is true for all values of the unknown, whereas an equation is only true for particular values of the unknown. For example, $3x - 5 = 1$ is an equation, since it is only true when $x = 2$, whereas $3x \equiv 8x - 5x$ is an identity since it is true for all values of x. (Note '$\equiv$' means 'is identical to'.)

Simple linear equations (or equations of the first degree) are those in which an unknown quantity is raised only to the power 1.

To **'solve an equation'** means 'to find the value of the unknown'.

Any arithmetic operation may be applied to an equation **as long as the equality of the equation is maintained**.

4.2 Worked problems on simple equations

Problem 1.   Solve the equation $4x = 20$

Dividing each side of the equation by 4 gives:

$$\frac{4x}{4} = \frac{20}{4}$$

(Note that the same operation has been applied to both the left-hand side (LHS) and the right-hand side (RHS) of the equation so the equality has been maintained.)

Cancelling gives $x = 5$, which is the solution to the equation. Solutions to simple equations should always be checked and this is accomplished by substituting the solution into the original equation. In this case, LHS $= 4(5) = 20 =$ RHS.

Problem 2.   Solve $\dfrac{2x}{5} = 6$

The LHS is a fraction and this can be removed by multiplying both sides of the equation by 5.

Hence $5 \left(\dfrac{2x}{5} \right) = 5(6)$

Cancelling gives: $2x = 30$

Dividing both sides of the equation by 2 gives:

$$\frac{2x}{2} = \frac{30}{2}, \text{ i.e. } x = 15$$

Problem 3.   Solve $a - 5 = 8$

Adding 5 to both sides of the equation gives:

$$a - 5 + 5 = 8 + 5$$

i.e. $a = 13$

The result of the above procedure is to move the '-5' from the LHS of the original equation, across the equals sign, to the RHS, but the sign is changed to $+$.

Problem 4.   Solve $x + 3 = 7$

Subtracting 3 from both sides of the equation gives:

$$x + 3 - 3 = 7 - 3$$

i.e. $x = 4$

The result of the above procedure is to move the '$+3$' from the LHS of the original equation, across the equals sign, to the RHS, but the sign is changed to $-$. Thus a term can be moved from one side of an equation to the other as long as a change in sign is made.

Problem 5.   Solve $6x + 1 = 2x + 9$

In such equations the terms containing x are grouped on one side of the equation and the remaining terms grouped on the other side of the equation. As in Problems 3 and 4, changing from one side of an equation to the other must be accompanied by a change of sign.

Thus since $6x + 1 = 2x + 9$

then $6x - 2x = 9 - 1$

$$4x = 8$$
$$\frac{4x}{4} = \frac{8}{4}$$

i.e. $x = 2$

Check: LHS of original equation $= 6(2) + 1 = 13$
RHS of original equation $= 2(2) + 9 = 13$
Hence the solution $x = 2$ is correct.

Problem 6.   Solve $4 - 3p = 2p - 11$

In order to keep the p term positive the terms in p are moved to the RHS and the constant terms to the LHS.

Hence $4 + 11 = 2p + 3p$

$$15 = 5p$$
$$\frac{15}{5} = \frac{5p}{5}$$

Hence $p = 3$

Check: LHS $= 4 - 3(3) = 4 - 9 = -5$
RHS $= 2(3) - 11 = 6 - 11 = -5$
Hence the solution $p = 3$ is correct.

If, in this example, the unknown quantities had been grouped initially on the LHS instead of the RHS then:

$$-3p - 2p = -11 - 4$$

i.e. $-5p = -15$

$$\frac{-5p}{-5} = \frac{-15}{-5}$$

and $p = 3$, as before

It is often easier, however, to work with positive values where possible.

Problem 7.   Solve $3(x - 2) = 9$

Removing the bracket gives: $3x - 6 = 9$

Rearranging gives: $3x = 9 + 6$

$$3x = 15$$
$$\frac{3x}{3} = \frac{15}{3}$$

i.e. $x = 5$

Check: LHS $= 3(5 - 2) = 3(3) = 9 = $ RHS
Hence the solution $x = 5$ is correct.

Problem 8.   Solve
$4(2r - 3) - 2(r - 4) = 3(r - 3) - 1$

Removing brackets gives:

$$8r - 12 - 2r + 8 = 3r - 9 - 1$$

Rearranging gives:

$$8r - 2r - 3r = -9 - 1 + 12 - 8$$

i.e. $3r = -6$

$$r = \frac{-6}{3} = -2$$

Check: LHS $= 4(-4 - 3) - 2(-2 - 4)$
$$= -28 + 12 = -16$$
RHS $= 3(-2 - 3) - 1 = -15 - 1 = -16$

Hence the solution $r = -2$ is correct.

Problem 9.   Solve $\dfrac{3}{x} = \dfrac{4}{5}$

The lowest common multiple (LCM) of the denominators, i.e. the lowest algebraic expression that both x and 5 will divide into, is $5x$.
Multiplying both sides by $5x$ gives:

$$5x\left(\frac{3}{x}\right) = 5x\left(\frac{4}{5}\right)$$

Cancelling gives:

$$15 = 4x \qquad (1)$$
$$\frac{15}{4} = \frac{4x}{4}$$

i.e. $x = \dfrac{15}{4}$ or $3\dfrac{3}{4}$

Check: LHS $= \dfrac{3}{3\frac{3}{4}} = \dfrac{3}{\frac{15}{4}} = 3\left(\dfrac{4}{15}\right)$

$$= \frac{12}{15} = \frac{4}{5} = \text{RHS}$$

(Note that when there is only one fraction on each side of an equation, 'cross-multiplication' can be applied. In this example, if $3/x = \frac{4}{5}$ then $(3)(5) = 4x$, which is a quicker way of arriving at equation (1) above.)

Problem 10.   Solve $\dfrac{2y}{5} + \dfrac{3}{4} + 5 = \dfrac{1}{20} - \dfrac{3y}{2}$

The LCM of the denominators is 20. Multiplying each term by 20 gives:

$$20\left(\frac{2y}{5}\right) + 20\left(\frac{3}{4}\right) + 20(5)$$

$$= 20\left(\frac{1}{20}\right) - 20\left(\frac{3y}{2}\right)$$

Cancelling gives: $4(2y) + 5(3) + 100 = 1 - 10(3y)$

i.e. $8y + 15 + 100 = 1 - 30y$

Rearranging gives:

$$8y + 30y = 1 - 15 - 100$$

$$38y = -114$$

$$y = \frac{-114}{38} = -3$$

Check: LHS $= \dfrac{2(-3)}{5} + \dfrac{3}{4} + 5 = \dfrac{-6}{5} + \dfrac{3}{4} + 5$

$$= \frac{-9}{20} + 5 = 4\frac{11}{20}$$

RHS $= \dfrac{1}{20} - \dfrac{3(-3)}{2} = \dfrac{1}{20} + \dfrac{9}{2} = 4\dfrac{11}{20}$

Hence the solution $y = -3$ is correct.

Problem 11.   Solve $\dfrac{3}{t-2} = \dfrac{4}{3t+4}$

By 'cross-multiplication': $3(3t + 4) = 4(t - 2)$

Removing brackets gives: $9t + 12 = 4t - 8$

Rearranging gives: $9t - 4t = -8 - 12$

i.e. $5t = -20$

$$t = \frac{-20}{5} = -4$$

Check: LHS $= \dfrac{3}{-4-2} = \dfrac{3}{-6} = -\dfrac{1}{2}$

RHS $= \dfrac{4}{3(-4)+4} = \dfrac{4}{-12+4}$

$$= \frac{4}{-8} = -\frac{1}{2}$$

Hence the solution $t = -4$ is correct.

Problem 12.   Solve $\sqrt{x} = 2$

Wherever square root signs are involved with the unknown quantity, both sides of the equation must be squared. Hence

$$(\sqrt{x})^2 = (2)^2$$

i.e. $x = 4$

Problem 13.   Solve $2\sqrt{d} = 8$

To avoid possible errors it is usually best to arrange the term containing the square root on its own. Thus

$$\frac{2\sqrt{d}}{2} = \frac{8}{2}$$

i.e. $\sqrt{d} = 4$

Squaring both sides gives: $d = 16$, which may be checked in the original equation.

Problem 14.   Solve $\left(\dfrac{\sqrt{b}+3}{\sqrt{b}}\right) = 2$

To remove the fraction each term is multiplied by $\sqrt{b}$. Hence

$$\sqrt{b}\left(\frac{\sqrt{b}+3}{\sqrt{b}}\right) = \sqrt{b}(2)$$

Cancelling gives: $\sqrt{b} + 3 = 2\sqrt{b}$

Rearranging gives: $3 = 2\sqrt{b} - \sqrt{b} = \sqrt{b}$

Squaring both sides gives: $9 = b$

Check: LHS $= \dfrac{\sqrt{9}+3}{\sqrt{9}} = \dfrac{3+3}{3} = \dfrac{6}{3} = 2 =$ RHS

Problem 15.   Solve $x^2 = 25$

This problem involves a square term and thus is not a simple equation (it is, in fact, a quadratic equation). However, the solution of such an equation is often required and is therefore included for completeness.

Whenever a square of the unknown is involved, the square root of both sides of the equation is taken. Hence

$$\sqrt{x^2} = \sqrt{25}$$

i.e. $x = 5$

However, $x = -5$ is also a solution of the equation because $(-5) \times (-5) = +25$. Therefore, whenever the square root of a number is required there are always two answers, one positive, the other negative.

The solution of $x^2 = 25$ is thus written as $x = \pm 5$.

Problem 16.   Solve $\dfrac{15}{4t^2} = \dfrac{2}{3}$

'Cross-multiplying' gives: $15(3) = 2(4t^2)$

$$45 = 8t^2$$

$$\frac{45}{8} = t^2$$

i.e. $t^2 = 5\dfrac{5}{8} = 5.625$

Hence $t = \sqrt{(5.625)} = \pm 2.372$, correct to 4 significant figures.

Further problems involving simple equations may be found in Section 4.4, Problems 1 to 44, pages 45 and 46.

4.3   Practical problems involving simple equations

Problem 17.   A copper wire has a length l of 1.5 km, a resistance R of 5 Ω and a resistivity ρ of 17.2×10^{-6} Ω mm. Find the cross-sectional area, a, of the wire, given that $R = \rho l / a$

Since $R = \rho l / a$ then

$$5\ \Omega = \frac{(17.2 \times 10^{-6}\ \Omega\ \text{mm})(1500 \times 10^3\ \text{mm})}{a}$$

From the units given, a is measured in mm^2. Thus

$$5a = 17.2 \times 10^{-6} \times 1500 \times 10^3$$

$$a = \frac{17.2 \times 10^{-6} \times 1500 \times 10^3}{5}$$

$$= \frac{17.2 \times 1500 \times 10^3}{10^6 \times 5}$$

$$= \frac{17.2 \times \cancel{15}^3}{10 \times \cancel{5}_1} = 5.16$$

Hence the cross-sectional area of the wire is 5.16 mm^2.

Problem 18.   A rectangular box with square ends has its length 15 cm greater than its breadth and the total length of its edges is 2.04 m. Find the width of the box and its volume

---

Let x cm = width = height of box. Then the length of the box is $(x + 15)$ cm. The length of the edges of the box is $2(4x) + 4(x + 15)$ cm.

Hence $204 = 2(4x) + 4(x + 15)$

$$204 = 8x + 4x + 60$$

$$204 - 60 = 12x$$

i.e. $144 = 12x$

and $x = 12$ cm

Hence the width of the box is 12 cm.

Volume of box = length × width × height

$$= (x + 15)(x)(x) = (27)(12)(12)$$

$$= \mathbf{3888\ cm^3}$$

Problem 19.   The temperature coefficient of resistance α may be calculated from the formula $R_t = R_0(1 + \alpha t)$. Find α given $R_t = 0.928$, $R_0 = 0.8$ and $t = 40$

Since $R_t = R_0(1 + \alpha t)$

then $0.928 = 0.8[1 + \alpha(40)]$

$$0.928 = 0.8 + (0.8)(\alpha)(40)$$

$$0.928 - 0.8 = 32\alpha$$

$$0.128 = 32\alpha$$

Hence
$$\alpha = \frac{0.128}{32} = 0.004$$

Problem 20. The distance s metres travelled in time t seconds is given by the formula $s = ut + \frac{1}{2}at^2$, where u is the initial velocity in m/s and a is the acceleration in m/s^2. Find the acceleration of the body if it travels 168 m in 6 s, with an initial velocity of 10 m/s

---

$s = ut + \dfrac{1}{2}at^2$, and $s = 168$, $u = 10$ and $t = 6$

Hence
$$168 = (10)(6) + \frac{1}{2}a(6)^2$$
$$168 = 60 + 18a$$
$$168 - 60 = 18a$$
$$108 = 18a$$
$$a = \frac{108}{18} = 6$$

Hence the acceleration of the body is 6 m/s^2.

Problem 21. When three resistors in an electrical circuit are connected in parallel the total resistance R_T is given by:

$$\frac{1}{R_T} = \frac{1}{R_1} + \frac{1}{R_2} + \frac{1}{R_3}$$

Find the total resistance when $R_1 = 5\ \Omega$, $R_2 = 10\ \Omega$ and $R_3 = 30\ \Omega$

$$\frac{1}{R_T} = \frac{1}{5} + \frac{1}{10} + \frac{1}{30} = \frac{6+3+1}{30} = \frac{10}{30} = \frac{1}{3}$$

Taking the reciprocal of both sides gives: $R_T = 3\ \Omega$.

Alternatively, if $\dfrac{1}{R_T} = \dfrac{1}{5} + \dfrac{1}{10} + \dfrac{1}{30}$ the LCM of the denominators is $30\,R_T$.

Hence $30R_T\left(\dfrac{1}{R_T}\right) = 30R_T\left(\dfrac{1}{5}\right) + 30R_T\left(\dfrac{1}{10}\right) +$
$30R_T\left(\dfrac{1}{30}\right)$.

Cancelling gives:
$$30 = 6R_T + 3R_T + R_T$$
$$30 = 10R_T$$
$$R_T = \frac{30}{10} = 3\ \Omega, \text{ as above}$$

Problem 22. The extension x m of an aluminium tie bar of length l m and cross-sectional area A m^2 when carrying a load of F newtons is given by the modulus of elasticity $E = Fl/Ax$. Find the extension of the tie bar (in mm) if $E = 70 \times 10^9$ N/m^2, $F = 20 \times 10^6$ N, $A = 0.1$ m^2 and $l = 1.4$ m

---

$E = Fl/Ax$. Hence

$$70 \times 10^9\,\frac{\text{N}}{\text{m}^2} = \frac{(20 \times 10^6\ \text{N})(1.4\ \text{m})}{(0.1\ \text{m}^2)(x)}$$

(the unit of x is thus metres)

$$70 \times 10^9 \times 0.1 \times x = 20 \times 10^6 \times 1.4$$
$$x = \frac{20 \times 10^6 \times 1.4}{70 \times 10^9 \times 0.1}$$

Cancelling gives:
$$x = \frac{2 \times 1.4}{7 \times 100}\ \text{m}$$
$$= \frac{2 \times 1.4}{7 \times 100} \times 1000\ \text{mm}$$

Hence the extension of the tie bar, $x = 4$ mm.

Problem 23. Power in a d.c. circuit is given by $P = V^2/R$ where V is the supply voltage and R is the circuit resistance. Find the supply voltage if the circuit resistance is 1.25 Ω and the power measured is 320 W

Since $\quad P = \dfrac{V^2}{R}\quad$ then $320 = \dfrac{V^2}{1.25}$

$$(320)(1.25) = V^2$$

i.e. $\qquad\qquad\qquad V^2 = 400$

Supply voltage, $\qquad V = \sqrt{400} = \pm 20$ **V**

Problem 24. A painter is paid £4.20 per hour for a basic 36 hour week, and overtime is paid at one and a third times this rate. Determine how many hours the painter has to work in a week to earn £212.80

Basic rate per hour = £4.20;

overtime rate per hour = $1\dfrac{1}{3} \times$ £4.20 = £5.60

Let the number of overtime hours worked = x. Then

$$(36)(4.20) + (x)(5.60) = 212.80$$

$$151.2 + 5.6x = 212.80$$

$$5.6x = 212.80 - 151.2 = 61.6$$

$$x = \frac{61.6}{5.6} = 11$$

Thus 11 hours overtime would have to be worked to earn £212.80 per week. Hence the total number of hours worked is $36 + 11$, i.e. **47 hours**.

Problem 25.   A formula relating initial and final states of pressures, P_1 and P_2, volumes V_1 and V_2, and absolute temperatures, T_1 and T_2, of an ideal gas is

$$\frac{P_1 V_1}{T_1} = \frac{P_2 V_2}{T_2}$$

Find the value of P_2 given $P_1 = 100 \times 10^3$, $V_1 = 1.0$, $V_2 = 0.266$, $T_1 = 423$ and $T_2 = 293$

Since $\dfrac{P_1 V_1}{T_1} = \dfrac{P_2 V_2}{T_2}$

then $\dfrac{(100 \times 10^3)(1.0)}{423} = \dfrac{P_2(0.266)}{293}$

'Cross-multiplying' gives:

$$(100 \times 10^3)(1.0)(293) = P_2(0.266)(423)$$

$$P_2 = \frac{(100 \times 10^3)(1.0)(293)}{(0.266)(423)}$$

Hence $\qquad P_2 = \mathbf{260 \times 10^3}$ **or** $\mathbf{2.6 \times 10^5}$

Problem 26.   The stress f in a material of a thick cylinder can be obtained from

$\dfrac{D}{d} = \sqrt{\left(\dfrac{f + p}{f - p}\right)}$. Calculate the stress given that $D = 21.5$, $d = 10.75$ and $p = 1800$

Since $\dfrac{D}{d} \quad = \sqrt{\left(\dfrac{f + p}{f - p}\right)}$

then $\dfrac{21.5}{10.75} = \sqrt{\left(\dfrac{f + 1800}{f - 1800}\right)}$

i.e. $\quad 2 \quad = \sqrt{\left(\dfrac{f + 1800}{f - 1800}\right)}$

Squaring both sides gives:

$$4 = \frac{f + 1800}{f - 1800}$$

$$4(f - 1800) = f + 1800$$

$$4f - 7200 = f + 1800$$

$$4f - f = 1800 + 7200$$

$$3f = 9000$$

$$f = \frac{9000}{3} = 3000$$

Hence the stress, f, is 3000.

Problem 27.   12 workmen employed on a building site earn between them a total of £2015 per week. Labourers are paid £158 per week and craftsmen are paid £175 per week. How many craftsmen and how many labourers are employed?

Let the number of craftsmen be c. The number of labourers is therefore $(12 - c)$. The wage bill equation is:

$$175c + 158(12 - c) = 2015$$

$$175c + 1896 - 158c = 2015$$

$$175c - 158c = 2015 - 1896$$

$$17c = 119$$

$$c = \frac{119}{17} = 7$$

Hence there are 7 craftsmen and $(12 - 7)$, i.e. 5 labourers on the site.

Further examples on practical problems involving simple equations may be found in the following Section 4.4, Problems 45 to 58, pages 46 and 47.

4.4   Further problems on simple equations

Simple equations

Solve the following equations:

1   $2x + 5 = 7$ [1]

2   $8 - 3t = 2$ [2]

3    $\dfrac{2}{3}c - 1 = 3$ [6]

4    $2x - 1 = 5x + 11$ [−4]

5    $7 - 4p = 2p - 3$ $\left[1\dfrac{2}{3}\right]$

6    $2.6x - 1.3 = 0.9x + 0.4$ [1]

7    $2a + 6 - 5a = 0$ [2]

8    $3x - 2 - 5x = 2x - 4$ $\left[\dfrac{1}{2}\right]$

9    $20d - 3 + 3d = 11d + 5 - 8$ [0]

10   $2(x - 1) = 4$ [3]

11   $16 = 4(t + 2)$ [2]

12   $5(f - 2) - 3(2f + 5) + 15 = 0$ [−10]

13   $2x = 4(x - 3)$ [6]

14   $6(2 - 3y) - 42 = -2(y - 1)$ [−2]

15   $2(3g - 5) - 5 = 0$ $\left[2\dfrac{1}{2}\right]$

16   $4(3x + 1) = 7(x + 4) - 2(x + 5)$ [2]

17   $10 + 3(r - 7) = 16 - (r + 2)$ $\left[6\dfrac{1}{4}\right]$

18   $8 + 4(x - 1) - 5(x - 3) = 2(5 - 2x)$ [−3]

19   $\dfrac{1}{5}d + 3 = 4$ [5]

20   $2 + \dfrac{3}{4}y = 1 + \dfrac{2}{3}y + \dfrac{5}{6}$ [−2]

21   $\dfrac{1}{4}(2x - 1) + 3 = \dfrac{1}{2}$ $\left[-4\dfrac{1}{2}\right]$

22   $\dfrac{1}{5}(2f - 3) + \dfrac{1}{6}(f - 4) + \dfrac{2}{15} = 0$ [2]

23   $\dfrac{1}{3}(3m - 6) - \dfrac{1}{4}(5m + 4) + \dfrac{1}{5}(2m - 9) = -3$ [12]

24   $\dfrac{x}{3} - \dfrac{x}{5} = 2$ [15]

25   $1 - \dfrac{y}{3} = 3 + \dfrac{y}{3} - \dfrac{y}{6}$ [−4]

26   $\dfrac{2}{a} = \dfrac{3}{8}$ $\left[5\dfrac{1}{3}\right]$

27   $\dfrac{1}{3n} + \dfrac{1}{4n} = \dfrac{7}{24}$ [2]

28   $\dfrac{x + 3}{4} = \dfrac{x - 3}{5} + 2$ [13]

29   $\dfrac{3t}{20} = \dfrac{6 - t}{12} + \dfrac{2t}{15} - \dfrac{3}{2}$ [−10]

30   $\dfrac{y}{5} + \dfrac{7}{20} = \dfrac{5 - y}{4}$ [2]

31   $\dfrac{v - 2}{2v - 3} = \dfrac{1}{3}$ [3]

32   $\dfrac{2}{a - 3} = \dfrac{3}{2a + 1}$ [−11]

33   $\dfrac{1}{3m - 2} + \dfrac{1}{5m + 3} = 0$ $\left[-\dfrac{1}{8}\right]$

34   $\dfrac{x}{4} - \dfrac{x + 6}{5} = \dfrac{x + 3}{2}$ [−6]

35   $\dfrac{2c - 3}{4} - \dfrac{1 - c}{5} - 1 = \dfrac{2c + 3}{3} + \dfrac{43}{60}$ [110]

36   $3\sqrt{t} = 9$ [9]

37   $2\sqrt{y} = 5$ $\left[6\dfrac{1}{4}\right]$

38   $4 = \sqrt{\left(\dfrac{3}{a}\right)} + 3$ [3]

39   $\dfrac{3\sqrt{x}}{1 - \sqrt{x}} = -6$ [4]

40   $10 = 5\sqrt{\left(\dfrac{x}{2} - 1\right)}$ [10]

41   $16 = \dfrac{t^2}{9}$ [±12]

42   $\sqrt{\left(\dfrac{y + 2}{y - 2}\right)} = \dfrac{1}{2}$ $\left[-3\dfrac{1}{3}\right]$

43   $\dfrac{6}{a} = \dfrac{2a}{3}$ [±3]

44   $\dfrac{11}{2} = 5 + \dfrac{8}{x^2}$ [±4]

Practical problems involving simple equations

45   A formula used for calculating resistance of a cable is $R = (\rho l)/a$. Given $R = 1.25$, $l = 2500$ and $a = 2 \times 10^{-4}$ find the value of ρ.   [10^{-7}]

46   Force F newtons is given by $F = ma$, where m is the mass in kilograms and a is the acceleration in metres per second squared. Find the acceleration when a force of 4 kN is applied to a mass of 500 kg.   [8 m/s²]

47   $PV = mRT$ is the characteristic gas equation. Find the value of m when $P = 100 \times 10^3$, $V = 3.00$, $R = 288$ and $T = 300$.   [3.472]

48 When three resistors R_1, R_2 and R_3 are connected in parallel the total resistance R_T is determined from

$$\frac{1}{R_T} = \frac{1}{R_1} + \frac{1}{R_2} + \frac{1}{R_3}$$

(a) Find the total resistance when $R_1 = 3\ \Omega$, $R_2 = 6\ \Omega$ and $R_3 = 18\ \Omega$

(b) Find the value of R_3 given that $R_T = 3\ \Omega$, $R_1 = 5\ \Omega$ and $R_2 = 10\ \Omega$

[(a) 1.8 Ω (b) 30 Ω]

49 Five pens and two rulers cost 94p. If a ruler costs 5p more than a pen, find the cost of each.

[12p, 17p]

50 Ohm's law may be represented by $I = V/R$, where I is the current in amperes, V is the voltage in volts and R is the resistance in ohms. A soldering iron takes a current of 0.30 A from a 240 V supply. Find the resistance of the element.

[800 Ω]

51 A rectangle has a length of 20 cm and a width b cm. When its width is reduced by 4 cm its area becomes 160 cm^2. Find the original width and area of the rectangle.

[12 cm, 240 cm^2]

52 Given $R_2 = R_1(1 + \alpha t)$, find α given $R_1 = 5.0$, $R_2 = 6.03$ and $t = 51.5$.

[0.004]

53 If $v^2 = u^2 + 2as$, find u given $v = 24$, $a = -40$ and $s = 4.05$.

[30]

54 The relationship between the temperature on a Fahrenheit scale and that on a Celsius scale is given by $F = \frac{9}{5}C + 32$. Express 113 °F in degrees Celsius.

[45 °C]

55 If $t = 2\pi\sqrt{(w/Sg)}$, find the value of S given $w = 1.219$, $g = 9.81$ and $t = 0.3132$.

[50]

56 Two joiners and five mates earn £1134 between them for a particular job. If a joiner earns £28 more than a mate, calculate the earnings for a joiner and for a mate.

[£182, £154]

57 An alloy contains 60% by weight of copper, the remainder being zinc. How much copper must be mixed with 50 kg of this alloy to give an alloy containing 75% copper?

[30 kg]

58 A rectangular laboratory has a length equal to one and a half times its width and a perimeter of 40 m. Find its length and width.

[12 m, 8 m]

5

Simultaneous equations

5.1 Introduction to simultaneous equations

Only one equation is necessary when finding the value of a **single unknown quantity** (as with simple equations in Chapter 4).

When an equation contains **two unknown quantities** it has an infinite number of solutions. When two equations are available connecting the same two unknown values then a unique solution is possible. Similarly, for three unknown quantities it is necessary to have three equations in order to solve for a particular value of each of the unknown quantities, and so on.

Equations which have to be solved together to find the unique values of the unknown quantities, which are true for each of the equations, are called **simultaneous equations**.

There are two methods of solving simultaneous equations analytically: (a) by **substitution**, and (b) by **elimination**.

A graphical solution of simultaneous equations is shown in Chapter 16.

5.2 Worked problems on simultaneous equations in two unknowns

Problem 1. Solve the following equations for x and y, (a) by substitution, and (b) by elimination:

$$x + 2y = -1 \tag{1}$$
$$4x - 3y = 18 \tag{2}$$

(a) **By substitution**

From equation (1): $x = -1 - 2y$
Substituting this expression for x into equation (2) gives:

$$4(-1 - 2y) - 3y = 18$$

This is now a simple equation in y.
Removing the bracket gives:

$$-4 - 8y - 3y = 18$$
$$-11y = 18 + 4 = 22$$
$$y = \frac{22}{-11} = -2$$

Substituting $y = -2$ into equation (1) gives:

$$x + 2(-2) = -1$$
$$x - 4 = -1$$
$$x = -1 + 4 = 3$$

Thus $x = 3$ and $y = -2$ is the solution to the simultaneous equations.
(Check: In equation (2), since $x = 3$ and $y = -2$,
LHS $= 4(3) - 3(-2)$
$= 12 + 6 = 18 =$ RHS)

(b) **By elimination**

$$x + 2y = -1 \tag{1}$$
$$4x - 3y = 18 \tag{2}$$

If equation (1) is multiplied throughout by 4 the coefficient of x will be the same as in equation (2), giving:

$$4x + 8y = -4 \tag{3}$$

Subtracting equation (3) from equation (2) gives:

$$4x - 3y = 18 \tag{2}$$
$$4x + 8y = -4 \tag{3}$$
$$\overline{0 - 11y = 22}$$

Hence $y = \dfrac{22}{-11} = -2$

(Note, in the above subtraction, $18 - -4 = 18 + 4 = 22$.)

Substituting $y = -2$ into either equation (1) or equation (2) will give $x = 3$ as in method (a). The solution $x = 3$, $y = -2$ is the only pair of values that satisfies both of the original equations.

Problem 2.   Solve, by a substitution method, the simultaneous equations

$$3x - 2y = 12 \tag{1}$$
$$x + 3y = -7 \tag{2}$$

From equation (2), $x = -7 - 3y$
Substituting for x in equation (1) gives:

$$3(-7 - 3y) - 2y = 12$$

i.e. $-21 - 9y - 2y = 12$

$$-11y = 12 + 21 = 33$$

Hence $y = \dfrac{33}{-11} = -3$

Substituting $y = -3$ in equation (2) gives:

$$x + 3(-3) = -7$$

i.e. $x - 9 = -7$

Hence $x = -7 + 9 = 2$

Thus $x = 2$, $y = -3$ is the solution of the simultaneous equations.

(Such solutions should always be checked by substituting values into each of the original two equations.)

Problem 3.   Use an elimination method to solve the simultaneous equations

$$3x + 4y = 5 \tag{1}$$
$$2x - 5y = -12 \tag{2}$$

If equation (1) is multiplied throughout by 2 and equation (2) by 3, then the coefficient of x will be the same in the newly formed equations. Thus

2 × equation (1) gives: $6x + 8y = 10$ (3)

3 × equation (2) gives: $6x - 15y = -36$ (4)

Equation (3) − equation (4) gives:

$$0 + 23y = 46$$

i.e. $y = \dfrac{46}{23} = 2$

(Note $+8y - -15y = 8y + 15y = 23y$ and $10 - -36 = 10 + 36 = 46$. Alternatively, 'change the signs of the bottom line and add'.)
Substituting $y = 2$ in equation (1) gives:

$$3x + 4(2) = 5$$

from which $3x = 5 - 8 = -3$

and $x = -1$

Checking in equation (2),

left-hand side $= 2(-1) - 5(2)$

$$= -2 - 10 = -12$$

$$= \text{right-hand side}$$

Hence $x = -1$ and $y = 2$ is the solution of the simultaneous equations.

The elimination method is the most common method of solving simultaneous equations.

Problem 4.   Solve $7x - 2y = 26$ (1)
$6x + 5y = 29$ (2)

When equation (1) is multiplied by 5 and equation (2) by 2 the coefficients of y in each equation are numerically the same, i.e. 10, but are of opposite sign.

5 × equation (1) gives: $35x - 10y = 130$ (3)

2 × equation (2) gives: $12x + 10y = 58$ (4)

Adding equation (3) and (4) gives: $\left.\right\}$ $47x + 0 = 188$

Hence $x = \dfrac{188}{47} = 4$

(Note that when the signs of common coefficients are **different** the two equations are **added**, and when the signs of common coefficients are the **same** the two equations are **subtracted** (as in Problems 1 and 3).)
Substituting $x = 4$ in equation (1) gives:

$$7(4) - 2y = 26$$
$$28 - 2y = 26$$
$$28 - 26 = 2y$$
$$2 = 2y$$

Hence $y = 1$

Checking, by substituting $x = 4$, $y = 1$ in equation (2), gives:

$$\text{LHS} = 6(4) + 5(1) = 24 + 5 = 29 = \text{RHS}$$

Thus the solution is $x = 4$, $y = 1$, since these values maintain the equality when substituted in both equations.

Problem 5. Solve $3p = 2q$ (1)
$\qquad\qquad\quad 4p + q + 11 = 0$ (2)

Rearranging gives:

$$3p - 2q = 0 \tag{3}$$

$$4p + q = -11 \tag{4}$$

Multiplying equation (4) by 2 gives:

$$8p + 2q = -22 \tag{5}$$

Adding equations (3) and (5) gives:

$$11p + 0 = -22$$

$$p = \frac{-22}{11} = -2$$

Substituting $p = -2$ into equation (1) gives:

$$3(-2) = 2q$$

$$-6 = 2q$$

$$q = \frac{-6}{2} = -3$$

Checking, by substituting $p = -2$ and $q = -3$ into equation (2) gives:

$$\text{LHS} = 4(-2) + (-3) + 11$$

$$= -8 - 3 + 11 = 0 = \text{RHS}$$

Hence the solution is $p = -2$, $q = -3$.

Problem 6. Solve $\dfrac{x}{8} + \dfrac{5}{2} = y$ (1)

$\qquad\qquad\qquad 13 - \dfrac{y}{3} = 3x$ (2)

Whenever fractions are involved in simultaneous equations it is usual first to remove them. Thus,

multiplying equation (1) by 8 gives:

$$8\left(\frac{x}{8}\right) + 8\left(\frac{5}{2}\right) = 8y$$

i.e. $\qquad\qquad x + 20 = 8y \tag{3}$

Multiplying equation (2) by 3 gives:

$$39 - y = 9x \tag{4}$$

Rearranging equations (3) and (4) gives:

$$x - 8y = -20 \tag{5}$$

$$9x + y = 39 \tag{6}$$

Multiplying equation (6) by 8 gives:

$$72x + 8y = 312 \tag{7}$$

Adding equations (5) and (7) gives:

$$73x + 0 = 292$$

$$x = \frac{292}{73} = 4$$

Substituting $x = 4$ into equation (5) gives:

$$4 - 8y = -20$$

$$4 + 20 = 8y$$

$$24 = 8y$$

$$y = \frac{24}{8} = 3$$

Checking: substituting $x = 4$, $y = 3$ in the original equations gives:

Equation (1):

$$\text{LHS} = \frac{4}{8} + \frac{5}{2} = \frac{1}{2} + 2\frac{1}{2} = 3 = y = \text{RHS}$$

Equation (2):

$$\text{LHS} = 13 - \frac{3}{3} = 13 - 1 = 12$$

$$\text{RHS} = 3x = 3(4) = 12$$

Hence the solution is $x = 4$, $y = 3$.

Problem 7. Solve $2.5x + 0.75 - 3y = 0$ (1)
$\qquad\qquad\qquad 1.6x = 1.08 - 1.2y$ (2)

It is often easier to remove decimal fractions. Thus multiplying equations (1) and (2) by 100 gives:

$$250x + 75 - 300y = 0 \qquad (1')$$

$$160x = 108 - 120y \qquad (2')$$

Rearranging gives:

$$250x - 300y = -75 \qquad (3)$$

$$160x + 120y = 108 \qquad (4)$$

Multiplying equation (3) by 2 gives:

$$500x - 600y = -150 \qquad (5)$$

Multiplying equation (4) by 5 gives:

$$800x + 600y = 540 \qquad (6)$$

Adding equations (5) and (6) gives:

$$1300x + 0 = 390$$

$$x = \frac{390}{1300} = \frac{39}{130} = \frac{3}{10} = 0.3$$

Substituting $x = 0.3$ into equation (1') gives:

$$250(0.3) + 75 - 300y = 0$$

$$75 + 75 = 300y$$

$$150 = 300y$$

$$y = \frac{150}{300} = 0.5$$

Checking: $x = 0.3$, $y = 0.5$ in equation (2') gives:

$$LHS = 160(0.3) = 48$$

$$RHS = 108 - 120(0.5)$$

$$= 108 - 60 = 48$$

Hence the solution is $x = 0.3$, $y = 0.5$.

Problem 8.   Solve $\dfrac{2}{x} + \dfrac{3}{y} = 7$ $\qquad (1)$

$\dfrac{1}{x} - \dfrac{4}{y} = -2$ $\qquad (2)$

In this type of equation a substitution can initially be made.

Let $\dfrac{1}{x} = a$ and $\dfrac{1}{y} = b$

Thus equation (1) becomes: $2a + 3b = 7$ $\qquad (3)$

and equation (2) becomes: $a - 4b = -2$ $\qquad (4)$

Multiplying equation (4) } by 2 gives: $\qquad 2a - 8b = -4$ $\qquad (5)$

Subtracting equation (5) from equation (3) gives:

$$0 + 11b = 11$$

i.e. $\qquad b = 1$

Substituting $b = 1$ in equation (3) gives:

$$2a + 3 = 7$$

$$2a = 7 - 3 = 4$$

i.e. $\qquad a = 2$

Checking: substituting $a = 2$, $b = 1$ in equation (4) gives:

$$LHS = 2 - 4(1) = 2 - 4 = -2 = RHS$$

Hence $a = 2$, $b = 1$

However, since $\qquad \dfrac{1}{x} = a$ then $x = \dfrac{1}{a} = \dfrac{1}{2}$

and since $\qquad \dfrac{1}{y} = b$ then $y = \dfrac{1}{b} = \dfrac{1}{1} = 1$

Hence the solution is $x = \frac{1}{2}$, $y = 1$, which may be checked in the original equations.

Problem 9.   Solve $\dfrac{1}{2a} + \dfrac{3}{5b} = 4$ $\qquad (1)$

$\dfrac{4}{a} + \dfrac{1}{2b} = 10.5$ $\qquad (2)$

Let $\dfrac{1}{a} = x$ and $\dfrac{1}{b} = y$

Then $\dfrac{x}{2} + \dfrac{3}{5}y = 4$ $\qquad (3)$

$4x + \dfrac{1}{2}y = 10.5$ $\qquad (4)$

To remove fractions, equation (3) is multiplied by 10 giving:

$$10\left(\frac{x}{2}\right) + 10\left(\frac{3}{5}y\right) = 10(4)$$

i.e. $\qquad 5x + 6y = 40$ $\qquad (5)$

Multiplying equation (4) by 2 gives:

$$8x + y = 21 \qquad (6)$$

Multiplying equation (6) by 6 gives:

$$48x + 6y = 126 \qquad (7)$$

Subtracting equation (5) from equation (7) gives:

$$43x + 0 = 86$$

$$x = \frac{86}{43} = 2$$

Substituting $x = 2$ into equation (3) gives:

$$\frac{2}{2} + \frac{3}{5}y = 4$$

$$\frac{3}{5}y = 4 - 1 = 3$$

$$y = \frac{5}{3}(3) = 5$$

Since $\dfrac{1}{a} = x$ then $a = \dfrac{1}{x} = \dfrac{1}{2}$

and since $\dfrac{1}{b} = y$ then $b = \dfrac{1}{y} = \dfrac{1}{5}$

Hence the solution is $a = \frac{1}{2}$, $b = \frac{1}{5}$, which may be checked in the original equations.

Problem 10.   Solve $\dfrac{1}{x + y} = \dfrac{4}{27}$ $\qquad (1)$

$\dfrac{1}{2x - y} = \dfrac{4}{33}$ $\qquad (2)$

To eliminate fractions, both sides of equation (1) are multiplied by $27(x + y)$ giving

$$27(x + y)\left(\frac{1}{x + y}\right) = 27(x + y)\left(\frac{4}{27}\right)$$

i.e. $\qquad\qquad 27(1) = 4(x + y)$

$$27 = 4x + 4y \qquad (3)$$

Similarly, in equation (2) $33 = 4(2x - y)$

i.e. $\qquad\qquad 33 = 8x - 4y \qquad (4)$

Equation (3) + equation (4) gives:

$$60 = 12x, \quad \text{i.e. } x = \frac{60}{12} = 5$$

Substituting $x = 5$ in equation (3) gives:

$$27 = 4(5) + 4y$$

from which $\quad 4y = 27 - 20 = 7$

and $\qquad\qquad y = \frac{7}{4} = 1\frac{3}{4}$

Hence $x = 5$, $y = 1\frac{3}{4}$ is the required solution, which may be checked in the original equations.

Problem 11.   Solve

$\dfrac{x - 1}{3} + \dfrac{y + 2}{5} = \dfrac{2}{15}$ $\qquad (1)$

$\dfrac{1 - x}{6} + \dfrac{5 + y}{2} = \dfrac{5}{6}$ $\qquad (2)$

Before equations (1) and (2) can be simultaneously solved, the fractions need to be removed and the equations rearranged.

Multiplying equation (1) by 15 gives:

$$15\left(\frac{x - 1}{3}\right) + 15\left(\frac{y + 2}{5}\right) = 15\left(\frac{2}{15}\right)$$

i.e. $\qquad\qquad 5(x - 1) + 3(y + 2) = 2$

$$5x - 5 + 3y + 6 = 2$$

$$5x + 3y = 2 + 5 - 6$$

Hence $\qquad\qquad 5x + 3y = 1 \qquad (3)$

Multiplying equation (2) by 6 gives:

$$6\left(\frac{1 - x}{6}\right) + 6\left(\frac{5 + y}{2}\right) = 6\left(\frac{5}{6}\right)$$

i.e. $\qquad\qquad (1 - x) + 3(5 + y) = 5$

$$1 - x + 15 + 3y = 5$$

$$-x + 3y = 5 - 1 - 15$$

Hence $\qquad\qquad -x + 3y = -11 \qquad (4)$

Thus the initial problem containing fractions can be expressed as:

$$5x + 3y = 1 \qquad (3)$$

$$-x + 3y = -11 \qquad (4)$$

Subtracting equation (4) from equation (3) gives:

$$6x + 0 = 12$$

$$x = \frac{12}{6} = 2$$

Substituting $x = 2$ into equation (3) gives:

$$5(2) + 3y = 1$$

$$10 + 3y = 1$$

$$3y = 1 - 10 = -9$$

$$y = \frac{-9}{3} = -3$$

Checking, substituting $x = 2$, $y = -3$ in equation (4) gives:

$$\text{LHS} = -2 + 3(-3) = -2 - 9$$
$$= -11 = \text{RHS}$$

Hence the solution is $x = 2$, $y = -3$, which may be checked in the original equations.

Further problems on simultaneous equations may be found in Section 5.4, Problems 1 to 27, pages 56 and 57.

5.3   Practical problems involving simultaneous equations in two unknowns

---

Problem 12.   The law connecting friction F and load L for an experiment is of the form $F = aL + b$, where a and b are constants. When $F = 5.6$, $L = 8.0$ and when $F = 4.4$, $L = 2.0$. Find the values of a and b and the value of F when $L = 6.5$

Substituting $F = 5.6$, $L = 8.0$ into $F = aL + b$ gives:

$$5.6 = 8.0a + b \tag{1}$$

Substituting $F = 4.4$, $L = 2.0$ into $F = aL + b$ gives:

$$4.4 = 2.0a + b \tag{2}$$

Subtracting equation (2) from equation (1) gives:

$$1.2 = 6.0a$$

$$a = \frac{1.2}{6.0} = \frac{1}{5}$$

Substituting $a = \frac{1}{5}$ into equation (1) gives:

$$5.6 = 8.0\left(\frac{1}{5}\right) + b$$

$$5.6 = 1.6 + b$$

$$5.6 - 1.6 = b$$

i.e. $$b = 4$$

Checking, substituting $a = \frac{1}{5}$, $b = 4$ in equation (2), gives:

$$\text{RHS} = 2.0\left(\frac{1}{5}\right) + 4 = 0.4 + 4 = 4.4 = \text{LHS}$$

Hence $a = \frac{1}{5}$ and $b = 4$.
When $L = 6.5$, $F = aL + b = \frac{1}{5}(6.5) + 4 = 1.3 + 4$, i.e. **$F = 5.30$**.

Problem 13.   The equation of a straight line, of slope m and intercept on the y-axis c, is $y = mx + c$. If a straight line passes through the point where $x = 1$ and $y = -2$, and also through the point where $x = 3\frac{1}{2}$ and $y = 10\frac{1}{2}$, find the values of the slope and the y-axis intercept

---

Substituting $x = 1$ and $y = -2$ into $y = mx + c$ gives:

$$-2 = m + c \tag{1}$$

Substituting $x = 3\frac{1}{2}$ and $y = 10\frac{1}{2}$ into $y = mx + c$ gives:

$$10\frac{1}{2} = 3\frac{1}{2}m + c \tag{2}$$

Subtracting equation (1) from equation (2) gives:

$$12\frac{1}{2} = 2\frac{1}{2}m$$

$$m = \frac{12\frac{1}{2}}{2\frac{1}{2}} = 5$$

Substituting $m = 5$ into equation (1) gives:

$$-2 = 5 + c$$

$$c = -2 - 5 = -7$$

Checking: substituting $m = 5$, $c = -7$ in equation (2) gives:

$$\text{RHS} = \left(3\frac{1}{2}\right)(5) + (-7)$$

$$= 17\frac{1}{2} - 7 = 10\frac{1}{2} = \text{LHS}$$

Hence the slope, $m = 5$ and the y-axis intercept, $c = -7$.

Problem 14.   When Kirchhoff's laws are applied to a particular electrical circuit the currents I_1 and I_2 are connected by the equations:

$$27 = 1.5I_1 + 8(I_1 - I_2) \qquad (1)$$
$$-26 = 2I_2 - 8(I_1 - I_2) \qquad (2)$$

Solve the equations to find the values of currents I_1 and I_2

Removing the brackets from equation (1) gives:

$$27 = 1.5I_1 + 8I_1 - 8I_2$$

Rearranging gives: $9.5I_1 - 8I_2 = 27 \qquad (3)$

Removing the brackets from equation (2) gives:

$$-26 = 2I_2 - 8I_1 + 8I_2$$

Rearranging gives: $-8I_1 + 10I_2 = -26 \qquad (4)$

Multiplying equation (3) by 5 gives:

$$47.5I_1 - 40I_2 = 135 \qquad (5)$$

Multiplying equation (4) by 4 gives:

$$-32I_1 + 40I_2 = -104 \qquad (6)$$

Adding equations (5) and (6) gives:

$$15.5I_1 + 0 = 31$$

$$I_1 = \frac{31}{15.5} = 2$$

Substituting $I_1 = 2$ into equation (3) gives:

$$9.5(2) - 8I_2 = 27$$
$$19 - 8I_2 = 27$$
$$19 - 27 = 8I_2$$
$$-8 = 8I_2$$
$$I_2 = -1$$

Hence the solution is $I_1 = 2$ **and** $I_2 = -1$ (which may be checked in the original equations).

Problem 15.   The distance s metres from a fixed point of a vehicle travelling in a straight line with constant acceleration, a m/s², is given by $s = ut + \frac{1}{2}at^2$, where u is the initial velocity in m/s and t the time in seconds. Determine the initial velocity and the acceleration given that $s = 42$ m when $t = 2$ s and $s = 144$ m when $t = 4$ s. Find also the distance travelled after 3 s

---

Substituting $s = 42$, $t = 2$ into $s = ut + \frac{1}{2}at^2$ gives:

$$42 = 2u + \frac{1}{2}a(2)^2$$

i.e. $42 = 2u + 2a \qquad (1)$

Substituting $s = 144$, $t = 4$ into $s = ut + \frac{1}{2}at^2$ gives:

$$144 = 4u + \frac{1}{2}a(4)^2$$

i.e. $144 = 4u + 8a \qquad (2)$

Multiplying equation (1) by 2 gives:

$$84 = 4u + 4a \qquad (3)$$

Subtracting equation (3) from equation (2) gives:

$$60 = 0 + 4a$$

$$a = \frac{60}{4} = 15$$

Substituting $a = 15$ into equation (1) gives:

$$42 = 2u + 2(15)$$
$$42 - 30 = 2u$$
$$u = \frac{12}{2} = 6$$

Substituting $a = 15$, $u = 6$ in equation (2) gives:

$$\text{RHS} = 4(6) + 8(15)$$
$$= 24 + 120 = 144 = \text{LHS}$$

Hence the initial velocity, $u = 6$ m/s and the acceleration, $a = 15$ m/s².
Distance travelled after 3 s is given by $s = ut + \frac{1}{2}at^2$ where $t = 3$, $u = 6$ and $a = 15$.

Hence $s = 6(3) + \dfrac{1}{2}(15)(3)^2$

$\qquad = 18 + 67\dfrac{1}{2}$

i.e. distance travelled after 3 $s = 85\dfrac{1}{2}$ m

Problem 16.   A craftsman and 4 labourers together earn £737 per week, whilst 4 craftsmen and 9 labourers earn £1982 basic per week. Determine the basic weekly wage of a craftsman and a labourer

---

Let C represent the wage of a craftsman and L that of a labourer. Thus

$\qquad C + 4L = 737 \qquad\qquad\qquad$ (1)

$\qquad 4C + 9L = 1982 \qquad\qquad\quad$ (2)

Multiplying equation (1) by 4 gives:

$\qquad 4C + 16L = 2948 \qquad\qquad\;$ (3)

Subtracting equation (2) from equation (3) gives:

$\qquad 7L = 966$

$\qquad L = \dfrac{966}{7} = 138$

Substituting $L = 138$ into equation (1) gives:

$\qquad C + 4(138) = 737$

$\qquad\quad C + 552 = 737$

$\qquad\qquad C = 737 - 552 = 185$

Checking: substituting $C = 185$, $L = 138$ into equation (2) gives:

$\qquad$ LHS $= 4(185) + 9(138)$

$\qquad\qquad = 740 + 1242 = 1982 =$ RHS

Thus the solution is that the basic weekly wage of a craftsman is £185 and that of a labourer is £138.

Problem 17.   The resistance $R\ \Omega$ of a length of wire at $t\,^\circ$C is given by $R = R_0(1 + \alpha t)$, where R_0 is the resistance at $0\,^\circ$C and α is the temperature coefficient of resistance in $/^\circ$C. Find the values of α and R_0 if $R = 30$ ohms at $50\,^\circ$C and $R = 35\ \Omega$ at $100\,^\circ$C

Substituting $R = 30$, $t = 50$ into $R = R_0(1 + \alpha t)$ gives:

$\qquad 30 = R_0(1 + 50\alpha) \qquad\qquad$ (1)

Substituting $R = 35$, $t = 100$ into $R = R_0(1 + \alpha t)$ gives:

$\qquad 35 = R_0(1 + 100\alpha) \qquad\qquad$ (2)

Although these equations may be solved by the conventional substitution method, an easier way is to eliminate R_0 by division. Thus, dividing equation (1) by equation (2) gives:

$$\frac{30}{35} = \frac{R_0(1 + 50\alpha)}{R_0(1 + 100\alpha)} = \frac{1 + 50\alpha}{1 + 100\alpha}$$

'Cross-multiplying' gives:

$\qquad 30(1 + 100\alpha) = 35(1 + 50\alpha)$

$\qquad 30 + 3000\alpha = 35 + 1750\alpha$

$\qquad 3000\alpha - 1750\alpha = 35 - 30$

$\qquad\qquad 1250\alpha = 5$

i.e.$\quad \alpha = \dfrac{5}{1250} = \dfrac{1}{250}$ or 0.004

Substituting $\alpha = \dfrac{1}{250}$ into equation (1) gives:

$$30 = R_0\left[1 + \frac{1}{250}(50)\right]$$

$\qquad 30 = R_0(1.2)$

$\qquad R_0 = \dfrac{30}{1.2} = 25$

Checking: substituting $\alpha = \dfrac{1}{250}$, $R_0 = 25$ in equation (2) gives:

$$\text{RHS} = 25\left[1 + \frac{1}{250}(100)\right]$$

$\qquad\qquad = 25(1.4) = 35 =$ LHS

Thus the solution is $\alpha = 0.004/^\circ$C and $R_0 = 25\ \Omega$.

Problem 18.   The molar heat capacity of a solid compound is given by the equation $c = a + bT$, where a and b are constants. When $c = 52$, $T = 100$ and when $c = 172$, $T = 400$. Determine the values of a and b

When $c = 52$, $T = 100$. Hence

$$52 = a + 100b \tag{1}$$

When $c = 172$, $T = 400$. Hence

$$172 = a + 400b \tag{2}$$

Equation (2) – equation (1) gives:

$$120 = 300b$$

from which

$$b = \frac{120}{300} = 0.4$$

Substituting $b = 0.4$ in equation (1) gives:

$$52 = a + 100(0.4)$$

$$a = 52 - 40 = 12$$

Hence $a = 12$ and $b = 0.4$.

Further examples on practical problems involving simultaneous equations in two unknowns may be found in the following Section 5.4, Problems 28 to 35, page 57.

5.4   Further problems involving simultaneous equations in two unknowns

In Problems 1 to 27, solve the simultaneous equations and verify the results:

1   $a + b = 7$
$a - b = 3$ $[a = 5,\ b = 2]$

2   $2x + 5y = 7$
$x + 3y = 4$ $[x = 1,\ y = 1]$

3   $3s + 2t = 12$
$4s - t = 5$ $[s = 2,\ t = 3]$

4   $3x - 2y = 13$
$2x + 5y = -4$ $[x = 3,\ y = -2]$

5   $5m - 3n = 11$
$3m + n = 8$ $\left[m = 2\frac{1}{2},\ n = \frac{1}{2}\right]$

6   $8a - 3b = 51$
$3a + 4b = 14$ $[a = 6,\ b = -1]$

7   $5x = 2y$
$3x + 7y = 41$ $[x = 2,\ y = 5]$

8   $5c = 1 - 3d$
$2d + c + 4 = 0$ $[c = 2,\ d = -3]$

9   $7p + 11 + 2q = 0$
$-1 = 3q - 5p$ $[p = -1,\ q = -2]$

10   $\dfrac{x}{2} + \dfrac{y}{3} = 4$

$\dfrac{x}{6} - \dfrac{y}{9} = 0$ $[x = 4,\ y = 6]$

11   $\dfrac{a}{2} - 7 = -2b$

$12 = 5a + \dfrac{2}{3}b$ $[a = 2,\ b = 3]$

12   $\dfrac{3}{2}s - 2t = 8$

$\dfrac{s}{4} + 3t = -2$ $[s = 4,\ t = -1]$

13   $\dfrac{x}{5} + \dfrac{2y}{3} = \dfrac{49}{15}$

$\dfrac{3x}{7} - \dfrac{y}{2} + \dfrac{5}{7} = 0$ $[x = 3,\ y = 4]$

14   $v - 1 = \dfrac{u}{12}$

$u + \dfrac{v}{4} - \dfrac{25}{2} = 0$ $[u = 12,\ v = 2]$

15   $1.5x - 2.2y = -18$
$2.4x + 0.6y = 33$ $[x = 10,\ y = 15]$

16   $3b - 2.5a = 0.45$
$1.6a + 0.8b = 0.8$ $[a = 0.30,\ b = 0.40]$

17   $10.1 + 1.7y = 0.8x$
$2.5x + 1.5 + 1.3y = 0$ $[x = 2,\ y = -5]$

18   $0.4b - 0.7 = 0.5a$
$1.2a - 3.6 = 0.3b$ $[a = 5,\ b = 8]$

19   $2.30c - 1.70d = 9.11$
$3.68 + 8.80c + 4.20d = 0$
                              $[c = 1.3,\ d = -3.6]$

20   $\dfrac{3}{x} + \dfrac{2}{y} = 14$

$\dfrac{5}{x} - \dfrac{3}{y} = -2$ $\left[x = \dfrac{1}{2},\ y = \dfrac{1}{4}\right]$

21   $\dfrac{4}{a} - \dfrac{3}{b} = 18$

$\dfrac{2}{a} + \dfrac{5}{b} = -4$ $\left[a = \dfrac{1}{3},\ b = -\dfrac{1}{2}\right]$

22   $\dfrac{1}{2p} + \dfrac{3}{5q} = 5$

$\dfrac{5}{p} - \dfrac{1}{2q} = \dfrac{35}{2}$ $\left[p = \dfrac{1}{4}, q = \dfrac{1}{5}\right]$

23   $\dfrac{5}{x} + \dfrac{3}{y} = 1.1$

$\dfrac{3}{x} - \dfrac{7}{y} = -1.1$ $[x = 10, y = 5]$

24   $\dfrac{c+1}{4} - \dfrac{d+2}{3} + 1 = 0$

$\dfrac{1-c}{5} + \dfrac{3-d}{4} + \dfrac{13}{20} = 0$ $[c = 3, d = 4]$

25   $\dfrac{3r+2}{5} - \dfrac{2s-1}{4} = \dfrac{11}{5}$

$\dfrac{3+2r}{4} + \dfrac{5-s}{3} = \dfrac{15}{4}$ $\left[r = 3, s = \dfrac{1}{2}\right]$

26   $\dfrac{5}{x+y} = \dfrac{20}{27}$

$\dfrac{4}{2x-y} = \dfrac{16}{33}$ $\left[x = 5, y = 1\dfrac{3}{4}\right]$

27   If $5x - \dfrac{3}{y} = 1$ and $x + \dfrac{4}{y} = \dfrac{5}{2}$ find the value

of $\dfrac{xy+1}{y}$.   [1]

Practical problems involving simultaneous equations in two unknowns

28   In a system of pulleys, the effort P required to raise a load W is given by $P = aW + b$, where a and b are constants. If $W = 40$ when $P = 12$ and $W = 90$ when $P = 22$, find the values of a and b.

$\left[a = \dfrac{1}{5}, b = 4\right]$

29   Applying Kirchhoff's laws to an electrical circuit produces the following equations:

$5 = 0.2I_1 + 2(I_1 - I_2)$

$12 = 3I_2 + 0.4I_2 - 2(I_1 - I_2)$

Determine the values of currents I_1 and I_2.
$[I_1 = 6.47, I_2 = 4.62]$

30   Velocity v is given by the formula $v = u + at$. If $v = 20$ when $t = 2$ and $v = 40$ when $t = 7$ find the values of u and a. Hence find the velocity when $t = 3.5$.
$[u = 12; a = 4, v = 26]$

31   Three new cars and 4 new vans supplied to a dealer together cost £83 700 and 5 new cars and 2 new vans of the same models cost £89 100. Find the cost of a car and a van.
[£13 500, £10 800]

32   $y = mx + c$ is the equation of a straight line of slope m and y-axis intercept c. If the line passes through the point where $x = 2$ and $y = 2$, and also through the point where $x = 5$ and $y = \frac{1}{2}$, find the slope and y-axis intercept of the straight line.

$\left[m = -\dfrac{1}{2}, c = 3\right]$

33   The resistance R ohms of copper wire at $t\,°C$ is given by $R = R_0(1 + \alpha t)$, where R_0 is the resistance at $0\,°C$ and α is the temperature coefficient of resistance. If $R = 25.44\ \Omega$ at $30\,°C$ and $R = 32.17\ \Omega$ at $100\,°C$, find α and R_0.
$[\alpha = 0.00426, R_0 = 22.56\ \Omega]$

34   The molar heat capacity of a solid compound is given by the equation $c = a + bT$. When $c = 52$, $T = 100$ and when $c = 172$, $T = 400$. Find the values of a and b.
$[a = 12, b = 0.40]$

35   In an engineering process two variables p and q are related by: $q = ap + b/p$, where a and b are constants. Evaluate a and b if $q = 13$ when $p = 2$ and $q = 22$ when $p = 5$.
$[a = 4, b = 10]$

6

Transposition of formulae

6.1 Introduction to transposition of formulae

When a symbol other than the subject is required to be calculated it is usual to rearrange the formula to make a new subject. This rearranging process is called **transposing the formula** or **transposition**.

The rules used for transposition of formulae are the same as those used for the solution of simple equations (see Chapter 4)–basically, **that the equality of an equation must be maintained.**

6.2 Worked problems on transposition of formulae

Problem 1. Transpose $p = q + r + s$ to make r the subject

The aim is to obtain r on its own on the left-hand side (LHS) of the equation. Changing the equation around so that r is on the LHS gives:

$$q + r + s = p \qquad (1)$$

Subtracting $(q + s)$ from both sides of the equation gives:

$$q + r + s - (q + s) = p - (q + s)$$

Thus $\quad q + r + s - q - s = p - q - s$

i.e. $\quad r = p - q - s \qquad (2)$

It is shown with simple equations that a quantity can be moved from one side of an equation to the other with an appropriate change of sign. Thus equation (2) follows immediately from equation (1) above.

Problem 2. If $a + b = w - x + y$, express x as the subject

Rearranging gives:

$$w - x + y = a + b \text{ and}$$

$$-x = a + b - w - y$$

Multiplying both sides by -1 gives:

$$(-1)(-x) = (-1)(a + b - w - y)$$

i.e. $\qquad x = -a - b + w + y$

The result of multiplying each side of the equation by -1 is to change all the signs in the equation.

It is conventional to express answers with positive quantities first. Hence rather than $x = -a - b + w + y$, $x = w + y - a - b$, since the order of terms connected by $+$ and $-$ signs is immaterial.

Problem 3. Transpose $v = f\lambda$ to make λ the subject

Rearranging gives: $f\lambda = v$

Dividing both sides by f gives: $\dfrac{f\lambda}{f} = \dfrac{v}{f}$,

i.e. $\lambda = \dfrac{v}{f}$

Problem 4. When a body falls freely through a height h, the velocity v is given by $v^2 = 2gh$. Express this formula with h as the subject

---

Rearranging gives: $2gh = v^2$

Dividing both sides by $2g$ gives: $\dfrac{2gh}{2g} = \dfrac{v^2}{2g}$,

i.e. $h = \dfrac{v^2}{2g}$

Problem 5. If $I = V/R$, rearrange to make V the subject

Rearranging gives: $\dfrac{V}{R} = I$

Multiplying both sides by R gives:

$$R\left(\dfrac{V}{R}\right) = R(I)$$

Hence $\qquad V = IR$

Problem 6. Transpose $a = \dfrac{F}{m}$ for m

Rearranging gives:

$$\dfrac{F}{m} = a$$

Multiplying both sides by m gives:

$$m\left(\dfrac{F}{m}\right) = m(a), \text{ i.e. } F = ma$$

Rearranging gives:

$$ma = F$$

Dividing both sides by a gives:

$$\dfrac{ma}{a} = \dfrac{F}{a}$$

i.e. $\boldsymbol{m} = \dfrac{\boldsymbol{F}}{\boldsymbol{a}}$

Problem 7. Rearrange the formula $R = (\rho l)/a$ to make (i) a the subject and (ii) l the subject

---

(i) Rearranging gives:

$$\dfrac{\rho l}{a} = R$$

Multiplying both sides by a gives:

$$a\left(\dfrac{\rho l}{a}\right) = a(R), \text{ i.e. } \rho l = aR$$

Rearranging gives:

$$aR = \rho l$$

Dividing both sides by R gives:

$$\dfrac{aR}{R} = \dfrac{\rho l}{R}$$

i.e. $\qquad \boldsymbol{a} = \dfrac{\boldsymbol{\rho l}}{\boldsymbol{R}}$

(ii) $\rho l/a = R$

Multiplying both sides by a gives:

$$\rho l = aR$$

Dividing both sides by ρ gives:

$$\dfrac{\rho l}{\rho} = \dfrac{aR}{\rho}$$

i.e. $\qquad \boldsymbol{l} = \dfrac{\boldsymbol{aR}}{\boldsymbol{\rho}}$

Problem 8. Transpose the formula $v = u + (ft)/m$ to make f the subject

---

Rearranging gives: $u + \dfrac{ft}{m} = v$, and $\dfrac{ft}{m} = v - u$

Multiplying each side by m gives:

$$m\left(\dfrac{ft}{m}\right) = m(v - u), \text{ i.e. } ft = m(v - u)$$

Dividing both sides by t gives:

$$\dfrac{ft}{t} = \dfrac{m}{t}(v - u), \text{ i.e. } \boldsymbol{f} = \dfrac{\boldsymbol{m}}{\boldsymbol{t}}(\boldsymbol{v} - \boldsymbol{u})$$

Problem 9. The final length l_2 of a piece of wire heated through $\theta\,°C$ is given by the formula $l_2 = l_1(1 + \alpha\theta)$. Make the coefficient of expansion, α, the subject

---

Rearranging gives:

$$l_1(1 + \alpha\theta) = l_2$$

Removing the bracket gives:

$$l_1 + l_1\alpha\theta = l_2$$

Rearranging gives:

$$l_1\alpha\theta = l_2 - l_1$$

Dividing both sides by $l_1\theta$ gives:

$$\dfrac{l_1\alpha\theta}{l_1\theta} = \dfrac{l_2 - l_1}{l_1\theta}$$

i.e. $\qquad \alpha = \dfrac{l_2 - l_1}{l_1\theta}$

Problem 10. A formula for the distance moved by a body is given by $s = \frac{1}{2}(v + u)t$. Rearrange the formula to make u the subject

Rearranging gives: $\dfrac{1}{2}(v + u)t = s$

Multiplying both sides by 2 gives: $(v + u)t = 2s$

Dividing both sides by t gives: $\dfrac{(v + u)t}{t} = \dfrac{2s}{t}$

i.e. $v + u = \dfrac{2s}{t}$

Hence $u = \dfrac{2s}{t} - v$ or $\dfrac{2s - vt}{t}$

Problem 11. A formula for kinetic energy is $k = \frac{1}{2}mv^2$. Transpose the formula to make v the subject

Rearranging gives: $\dfrac{1}{2}mv^2 = k$

Whenever the prospective new subject is a squared term, that term is isolated on the LHS, and then the square root of both sides of the equation is taken.

Multiplying both sides by 2 gives: $mv^2 = 2k$

Dividing both sides by m gives: $\dfrac{mv^2}{m} = \dfrac{2k}{m}$

i.e. $v^2 = \dfrac{2k}{m}$

Taking the square root of both sides gives:

$$\sqrt{v^2} = \sqrt{\left(\dfrac{2k}{m}\right)}$$

i.e. $v = \sqrt{\left(\dfrac{2k}{m}\right)}$

Problem 12. In a right angled triangle having sides x, y and hypotenuse z, Pythagoras' theorem states $z^2 = x^2 + y^2$. Transpose the formula to find x

Rearranging gives: $x^2 + y^2 = z^2$

and $x^2 = z^2 - y^2$

Taking the square root of both sides gives:

$$x = \sqrt{(z^2 - y^2)}$$

Problem 13. Given $t = 2\pi\sqrt{(1/g)}$, find g in terms of t, l and π

Whenever the prospective new subject is within a square root sign, it is best to isolate that term on the LHS and then to square both sides of the equation.

Rearranging gives: $2\pi\sqrt{\left(\dfrac{l}{g}\right)} = t$

Dividing both sides by 2π gives: $\sqrt{\left(\dfrac{l}{g}\right)} = \dfrac{t}{2\pi}$

Squaring both sides gives: $\dfrac{l}{g} = \left(\dfrac{t}{2\pi}\right)^2 = \dfrac{t^2}{4\pi^2}$

Cross-multiplying, i.e. multiplying each term by $4\pi^2 g$, gives:

$4\pi^2 l = gt^2$

or $gt^2 = 4\pi^2 l$

Dividing both sides by t^2 gives:

$\dfrac{gt^2}{t^2} = \dfrac{4\pi^2 l}{t^2}$

i.e. $g = \dfrac{4\pi^2 l}{t^2}$

Problem 14. The impedance of an a.c. circuit is given by $Z = \sqrt{(R^2 + X^2)}$. Make the reactance, X, the subject

Rearranging gives: $\sqrt{(R^2 + X^2)} = Z$

Squaring both sides gives: $R^2 + X^2 = Z^2$

Rearranging gives: $X^2 = Z^2 - R^2$

Taking the square root of both sides gives:

$$X = \sqrt{(Z^2 - R^2)}$$

Problem 15.   The volume V of a hemisphere is given by $V = \frac{2}{3}\pi r^3$. Find r in terms of V

Rearranging gives: $\dfrac{2}{3}\pi r^3 = V$

Multiplying both sides by 3 gives: $2\pi r^3 = 3V$

Dividing both sides by 2π gives: $\dfrac{2\pi r^3}{2\pi} = \dfrac{3V}{2\pi}$,

i.e. $r^3 = \dfrac{3V}{2\pi}$

Taking the cube root of both sides gives:

$$\sqrt[3]{r^3} = \sqrt[3]{\left(\frac{3V}{2\pi}\right)}, \text{ i.e. } \boldsymbol{r = \sqrt[3]{\left(\frac{3V}{2\pi}\right)}}$$

Problem 16.   Transpose the formula $p = \dfrac{a^2x + a^2y}{r}$ to make a the subject

Rearranging gives: $\dfrac{a^2x + a^2y}{r} = p$

Multiplying both sides by r gives: $a^2x + a^2y = rp$

Factorizing the LHS gives: $a^2(x + y) = rp$

Dividing both sides by $(x + y)$ gives:

$$\frac{a^2(x + y)}{(x + y)} = \frac{rp}{(x + y)}$$

i.e. $a^2 = \dfrac{rp}{(x + y)}$

Taking the square root of both sides gives:

$$\boldsymbol{a = \sqrt{\left(\frac{rp}{x + y}\right)}}$$

Problem 17.   Make b the subject of the formula $a = \dfrac{x - y}{\sqrt{(bd + be)}}$

Rearranging gives: $\dfrac{x - y}{\sqrt{(bd + be)}} = a$

Multiplying both sides by $\sqrt{(bd + be)}$ gives:

$$x - y = a\sqrt{(bd + be)}$$

or $a\sqrt{(bd + be)} = x - y$

Dividing both sides by a gives:

$$\sqrt{(bd + be)} = \frac{x - y}{a}$$

Squaring both sides gives:

$$bd + be = \left(\frac{x - y}{a}\right)^2$$

Factorizing the LHS gives:

$$b(d + e) = \left(\frac{x - y}{a}\right)^2$$

Dividing both sides by $(d + e)$ gives:

$$b = \frac{\left(\dfrac{x - y}{a}\right)^2}{(d + e)}$$

i.e. $\boldsymbol{b = \dfrac{(x - y)^2}{a^2(d + e)}}$

Problem 18.   If $cd = 3d + e - ad$, express d in terms of a, c and e

Rearranging to obtain the terms in d on the LHS gives:

$$cd - 3d + ad = e$$

Factorizing the LHS gives:

$$d(c - 3 + a) = e$$

Dividing both sides by $(c - 3 + a)$ gives:

$$\boldsymbol{d = \dfrac{e}{c - 3 + a}}$$

Problem 19.   If $a = b/(1 + b)$, make b the subject of the formula

Rearranging gives: $\dfrac{b}{1 + b} = a$

Multiplying both sides by $(1 + b)$ gives:

$$b = a(1 + b)$$

Removing the bracket gives:

$$b = a + ab$$

Rearranging to obtain terms in b on the LHS gives:

$$b - ab = a$$

Factorizing the LHS gives:

$$b(1 - a) = a$$

Dividing both sides by $(1 - a)$ gives:

$$b = \frac{a}{1 - a}$$

Problem 20.   Transpose the formula
$V = \dfrac{Er}{R + r}$ to make r the subject

Rearranging gives: $\dfrac{Er}{R + r} = V$

Multiplying both sides by $(R + r)$ gives:

$$Er = V(R + r)$$

Removing the bracket gives:

$$Er = VR + Vr$$

Rearranging to obtain terms in r on the LHS gives:

$$Er - Vr = VR$$

Factorizing gives:

$$r(E - V) = VR$$

Dividing both sides by $(E - V)$ gives:

$$r = \frac{VR}{E - V}$$

Problem 21.   Transpose the formula
$y = (pq^2/(r + q^2)) - t$ to make q the subject

Rearranging gives: $\dfrac{pq^2}{r + q^2} - t = y$

and $\dfrac{pq^2}{r + q^2} = y + t$

Multiplying both sides by $(r + q^2)$ gives:

$$pq^2 = (r + q^2)(y + t)$$

Removing brackets gives:

$$pq^2 = ry + rt + q^2 y + q^2 t$$

Rearranging to obtain terms in q on the LHS gives:

$$pq^2 - q^2 y - q^2 t = ry + rt$$

Factorizing gives:

$$q^2(p - y - t) = r(y + t)$$

Dividing both sides by $(p - y - t)$ gives:

$$q^2 = \frac{r(y + t)}{(p - y - t)}$$

Taking the square root of both sides gives:

$$q = \sqrt{\left(\frac{r(y + t)}{(p - y - t)} \right)}$$

Problem 22.   Given that $\dfrac{D}{d} = \sqrt{\left(\dfrac{f + p}{f - p} \right)}$,
express p in terms of D, d and f

Rearranging gives: $\sqrt{\left(\dfrac{f + p}{f - p} \right)} = \dfrac{D}{d}$

Squaring both sides gives: $\dfrac{f + p}{f - p} = \dfrac{D^2}{d^2}$

Cross-multiplying, i.e. multiplying each term by $d^2(f - p)$, gives:

$$d^2(f + p) = D^2(f - p)$$

Removing brackets gives:

$$d^2 f + d^2 p = D^2 f - D^2 p$$

Rearranging, to obtain terms in p on the LHS, gives:

$$d^2 p + D^2 p = D^2 f - d^2 f$$

Factorizing gives:

$$p(d^2 + D^2) = f(D^2 - d^2)$$

Dividing both sides by $(d^2 + D^2)$ gives:

$$p = \frac{f(D^2 - d^2)}{(d^2 + D^2)}$$

Further problems on transposing formulae may be found in the following Section 6.3, Problems 1 to 33.

6.3 Further problems on transposition of formulae

Make the symbol indicated the subject of each of the formulae shown in Problems 1 to 28, and express each in its simplest form:

1   $a + b = c - d - e$ (d)

$$[d = c - e - a - b]$$

2   $x + 3y = t$ (y)

$$\left[y = \frac{1}{3}(t - x) \right]$$

3   $c = 2\pi r$ (r)

$$\left[r = \frac{c}{2\pi} \right]$$

4   $y = mx + c$ (x)

$$\left[x = \frac{y - c}{m} \right]$$

5   $I = PRT$ (T)

$$\left[T = \frac{I}{PR} \right]$$

6   $I = \dfrac{E}{R}$ (R)

$$\left[R = \frac{E}{I} \right]$$

7   $S = \dfrac{a}{1 - r}$ (r)

$$\left[r = \frac{S - a}{S} \right]$$

8   $F = \dfrac{9}{5}C + 32$ (C)

$$\left[C = \frac{5}{9}(F - 32) \right]$$

9   $y = \dfrac{\lambda(x - d)}{d}$ (x)

$$\left[x = \frac{d}{\lambda}(y + \lambda) \right]$$

10   $A = \dfrac{3(F - f)}{L}$ (f)

$$\left[f = \frac{3F - AL}{3} \right]$$

11   $y = \dfrac{Ml^2}{8EI}$ (E)

$$\left[E = \frac{Ml^2}{8yI} \right]$$

12   $R = R_0(1 + \alpha t)$ (t)

$$\left[t = \frac{R - R_0}{R_0\alpha} \right]$$

13   $\dfrac{1}{R} = \dfrac{1}{R_1} + \dfrac{1}{R_2}$ (R_2)

$$\left[R_2 = \frac{RR_1}{R_1 - R} \right]$$

14   $I = \dfrac{E - e}{R + r}$ (R)

$$\left[R = \frac{E - e - Ir}{I} \right]$$

15   $y = 4ab^2c^2$ (b)

$$\left[b = \sqrt{\left(\frac{y}{4ac^2} \right)} \right]$$

16   $\dfrac{a^2}{x^2} + \dfrac{b^2}{y^2} = 1$ (x)

$$\left[x = \frac{ay}{\sqrt{(y^2 - b^2)}} \right]$$

17   $t = 2\pi\sqrt{\left(\dfrac{1}{g} \right)}$ (l)

$$\left[l = \frac{t^2 g}{4\pi^2} \right]$$

18   $v^2 = u^2 + 2as$ (u)

$$[u = \sqrt{(v^2 - 2as)}]$$

19   $A = \dfrac{\pi R^2 \theta}{360}$ (R)

$$\left[R = \sqrt{\left(\frac{360A}{\pi\theta} \right)} \right]$$

20   $N = \sqrt{\left(\dfrac{a + x}{y} \right)}$ (a)

$$[a = N^2 y - x]$$

21    $Z = \sqrt{[R^2 + (2\pi f L)^2]}$ (L)

$$\left[L = \frac{\sqrt{(Z^2 - R^2)}}{2\pi f} \right]$$

22    $y = \dfrac{a^2 m - a^2 n}{x}$ (a)

$$\left[a = \sqrt{\left(\frac{xy}{m-n} \right)} \right]$$

23    $M = \pi(R^4 - r^4)$ (R)

$$\left[R = \sqrt[4]{\left(\frac{M}{\pi} + r^4 \right)} \right]$$

24    $x + y = \dfrac{r}{3+r}$ (r)

$$\left[r = \frac{3(x+y)}{(1 - x - y)} \right]$$

25    $m = \dfrac{\mu L}{L + rCR}$ (L)

$$\left[L = \frac{mrCR}{\mu - m} \right]$$

26    $a^2 = \dfrac{b^2 - c^2}{b^2}$ (b)

$$\left[b = \frac{c}{\sqrt{(1 - a^2)}} \right]$$

27    $\dfrac{x}{y} = \dfrac{1 + r^2}{1 - r^2}$ (r)

$$\left[r = \sqrt{\left(\frac{x-y}{x+y} \right)} \right]$$

28    $\dfrac{p}{q} = \sqrt{\left(\dfrac{a + 2b}{a - 2b} \right)}$ (b)

$$\left[b = \frac{a(p^2 - q^2)}{2(p^2 + q^2)} \right]$$

29   A formula for the focal length, f, of a convex lens is $\dfrac{1}{f} = \dfrac{1}{u} + \dfrac{1}{v}$. Transpose the formula to make v the subject and evaluate v when $f = 5$ and $u = 6$.

$$\left[v = \frac{uf}{u - f}, 30 \right]$$

30   The quantity of heat, Q, is given by the formula $Q = mc(t_2 - t_1)$. Make t_2 the subject of the formula and evaluate t_2 when $m = 10$, $t_1 = 15$, $c = 4$ and $Q = 1600$.

$$\left[t_2 = t_1 + \frac{Q}{mc}, 55 \right]$$

31   The velocity, v, of water in a pipe appears in the formula $h = \dfrac{0.03Lv^2}{2dg}$. Express v as the subject of the formula and evaluate v when $h = 0.712$, $L = 150$, $d = 0.30$ and $g = 9.81$.

$$\left[v = \sqrt{\left(\frac{2dgh}{0.03L} \right)}, 0.965 \right]$$

32   The sag S at the centre of a wire is given by the formula $S = \sqrt{\left(\dfrac{3d(l - d)}{8} \right)}$. Make l the subject of the formula and evaluate l when $d = 1.75$ and $S = 0.80$.

$$\left[l = \frac{8S^2}{3d} + d, 2.725 \right]$$

33   In an electrical alternating current circuit the impedance Z is given by

$$Z = \sqrt{\left\{ R^2 + \left(\omega L - \frac{1}{\omega C} \right)^2 \right\}}$$

Transpose the formula to make C the subject and hence evaluate C when $Z = 130$, $R = 120$, $\omega = 314$ and $L = 0.32$.

$$\left[C = \frac{1}{\omega \{ \omega L - \sqrt{(Z^2 - R^2)} \}}, 63.1 \times 10^{-6} \right]$$

7

Quadratic equations

7.1 Introduction to quadratic equations

An **equation** is a statement that two quantities are equal. To **'solve an equation'** means 'to find the value of the unknown'. The value of the unknown is called the **root** of the equation.

A **quadratic equation** is one in which the highest power of the unknown quantity is 2. For example, $x^2 - 3x + 1 = 0$ is a quadratic equation.

There are four methods of **solving quadratic equations**.

These are: (i) by factorization (where possible), (ii) by 'completing the square', (iii) by using the 'quadratic formula', or (iv) graphically (see Chapter 16).

7.2 Solution of quadratic equations by factorization

Multiplying out $(2x+1)(x-3)$ gives $2x^2-6x+x-3$, i.e. $2x^2-5x-3$. The reverse process of moving from $2x^2 - 5x - 3$ to $(2x+1)(x-3)$ is called **factorizing**.

If the quadratic expression can be factorized this provides the simplest method of solving a quadratic equation. For example, if $2x^2 - 5x - 3 = 0$, then, by factorizing:

$$(2x + 1)(x - 3) = 0$$

Hence either $(2x + 1) = 0$, i.e. $x = -\dfrac{1}{2}$

or $(x - 3) = 0$, i.e. $x = 3$

The technique of factorizing is often one of 'trial and error'.

> **Problem 1.** Solve the equations
> (a) $x^2 + 2x - 8 = 0$ and
> (b) $3x^2 - 11x - 4 = 0$ by factorization

(a) $x^2 + 2x - 8 = 0$. The factors of x^2 are x and x. These are placed in brackets thus: $(x\ \ )(x\ \ )$ The factors of -8 are $+8$ and -1, or -8 and $+1$, or $+4$ and -2, or -4 and $+2$. The only combination to give a middle term of $+2x$ is $+4$ and -2, i.e.

$$x^2 + 2x - 8 = (x - 2)(x + 4)$$

(Note that the product of the two inner terms added to the product of the two outer terms must equal the middle term, $+2x$ in this case.) The quadratic equation $x^2 + 2x - 8 = 0$ thus becomes $(x + 4)(x - 2) = 0$. Since the only way that this can be true is for either the first or the second or both factors to be zero, then

either $(x + 4) = 0$ i.e. $x = -4$

or $(x - 2) = 0$ i.e. $x = 2$

Hence the roots of $x^2 + 2x - 8 = 0$ are $x = -4$ and 2.

(b) $3x^2 - 11x - 4 = 0$

The factors of $3x^2$ are $3x$ and x. These are placed in brackets thus: $(3x\ \ )(x\ \ )$ The factors of -4 are -4 and $+1$, or $+4$ and -1, or -2 and 2.

Remembering that the product of the two inner terms added to the product of the two outer terms must equal $-11x$, the only combination to give this is -4 and $+1$, i.e.

$$3x^2 - 11x - 4 = (3x + 1)(x - 4)$$

The quadratic equation $3x^2 - 11x - 4 = 0$ thus becomes

$$(3x + 1)(x - 4) = 0.\quad \text{Hence,}$$

either $(3x + 1) = 0,$ i.e. $x = -\dfrac{1}{3}$

or $(x - 4) = 0,$ i.e. $x = 4$

and both solutions may be checked in the original equation.

Problem 2.   Determine the roots of (a) $x^2 - 6x + 9 = 0$, and (b) $4x^2 - 25 = 0$, by factorization

(a) $x^2 - 6x + 9 = 0$. Hence $(x - 3)(x - 3) = 0$, i.e. $(x - 3)^2 = 0$ (the left-hand side is known as a **perfect square**). Hence $x = 3$ is the only root of the equation $x^2 - 6x + 9 = 0$

(b) $4x^2 - 25 = 0$ (the left-hand side is **the difference of two squares**, $(2x)^2$ and $(5)^2$). Thus $(2x + 5)(2x - 5) = 0$

Hence either $(2x + 5) = 0$,   i.e. $x = -\dfrac{5}{2}$

or $(2x - 5) = 0$,   i.e. $x = \dfrac{5}{2}$

Problem 3.   Solve the following quadratic equations by factorizing:
(a) $4x^2 + 8x + 3 = 0$
(b) $15x^2 + 2x - 8 = 0$

(a) $4x^2 + 8x + 3 = 0$. The factors of $4x^2$ are $4x$ and x or $2x$ and $2x$. The factors of 3 are 3 and 1, or -3 and -1. Remembering that the product of the inner terms added to the product of the two outer terms must equal $+8x$, the only combination that is true (by trial and error) is $(4x^2 + 8x + 3) = (2x + 3)(2x + 1)$.
Hence $(2x + 3)(2x + 1) = 0$ from which either $(2x + 3) = 0$ or $(2x + 1) = 0$

Thus $2x = -3$, from which, $x = -\dfrac{3}{2}$

or $2x = -1$, from which, $x = -\dfrac{1}{2}$

which may be checked in the original equation

(b) $15x^2 + 2x - 8 = 0$. The factors of $15x^2$ are $15x$ and x or $5x$ and $3x$. The factors of -8 are -4 and $+2$, or 4 and -2, or -8 and $+1$, or 8 and -1. By trial and error the only combination that works is $15x^2 + 2x - 8 = (5x + 4)(3x - 2)$.
Hence $(5x + 4)(3x - 2) = 0$ from which either

$5x + 4 = 0$

or $3x - 2 = 0$

Hence $x = -\dfrac{4}{5}$ or $x = \dfrac{2}{3}$

which may be checked in the original equation

Problem 4.   The roots of a quadratic equation are $\frac{1}{3}$ and -2. Determine the equation

If the roots of a quadratic equation are α and β then $(x - \alpha)(x - \beta) = 0$. Hence if $\alpha = \frac{1}{3}$ and $\beta = -2$, then

$$\left(x - \frac{1}{3}\right)(x - (-2)) = 0$$

$$\left(x - \frac{1}{3}\right)(x + 2) = 0$$

$$x^2 - \frac{1}{3}x + 2x - \frac{2}{3} = 0$$

$$x^2 + \frac{5}{3}x - \frac{2}{3} = 0$$

Hence $\mathbf{3x^2 + 5x - 2 = 0}$

Problem 5.   Find the equations in x whose roots are (a) 5 and -5 (b) 1.2 and -0.4

(a)   If 5 and -5 are the roots of a quadratic equation then

$$(x - 5)(x + 5) = 0$$

i.e. $x^2 - 5x + 5x - 25 = 0$

i.e. $\mathbf{x^2 - 25 = 0}$

(b)   If 1.2 and -0.4 are the roots of a quadratic equation then

$$(x - 1.2)(x + 0.4) = 0$$

i.e. $x^2 - 1.2x + 0.4x - 0.48 = 0$

i.e. $\mathbf{x^2 - 0.8x - 0.48 = 0}$

Further problems on solving quadratic equations by factorization may be found in Section 7.7, Problems 1 to 6, page 72.

7.3   Solution of quadratic equations by 'completing the square'

An expression such as x^2 or $(x + 2)^2$ or $(x - 3)^2$ is called a perfect square.

If $x^2 = 3$ then $x = \pm\sqrt{3}$
If $(x + 2)^2 = 5$ then $x + 2 = \pm\sqrt{5}$
and $x = -2 \pm \sqrt{5}$
If $(x - 3)^2 = 8$ then $x - 3 = \pm\sqrt{8}$
and $x = 3 \pm \sqrt{8}$

Hence if a quadratic equation can be rearranged so that one side of the equation is a perfect square and the other side of the equation is a number, then the solution of the equation is readily obtained by taking the square roots of each side as in the above examples. The process of rearranging one side of a quadratic equation into a perfect square before solving is called '**completing the square**'.

$$(x + a)^2 = x^2 + 2ax + a^2$$

Thus in order to make the quadratic expression $x^2 + 2ax$ into a perfect square it is necessary to add (half the coefficient of x)^2, i.e.

$$\left(\frac{2a}{2}\right)^2 \text{ or } a^2$$

For example, $x^2 + 3x$ becomes a perfect square by adding $(3/2)^2$, i.e.

$$x^2 + 3x + \left(\frac{3}{2}\right)^2 = \left(x + \frac{3}{2}\right)^2$$

The method is demonstrated in the following worked problems.

Problem 6.   Solve $2x^2 + 5x = 3$ by 'completing the square'

The procedure is as follows:

1   Rearrange the equation so that all terms are on the same side of the equals sign (and the coefficient of the x^2 term is positive). Hence $2x^2 + 5x - 3 = 0$

2   Make the coefficient of the x^2 term unity. In this case this is achieved by dividing throughout by 2. Hence

$$\frac{2x^2}{2} + \frac{5x}{2} - \frac{3}{2} = 0$$

i.e. $x^2 + \frac{5}{2}x - \frac{3}{2} = 0$

3   Rearrange the equations so that the x^2 and x terms are on one side of the equals sign and the

constant is on the other side. Hence

$$x^2 + \frac{5}{2}x = \frac{3}{2}$$

4   Add to both sides of the equation (half the coefficient of x)^2. In this case the coefficient of x is 5/2. Half the coefficient squared is therefore $(5/4)^2$. Thus

$$x^2 + \frac{5}{2}x + \left(\frac{5}{4}\right)^2 = \frac{3}{2} + \left(\frac{5}{4}\right)^2$$

The LHS is now a perfect square, i.e.

$$\left(x + \frac{5}{4}\right)^2 = \frac{3}{2} + \left(\frac{5}{4}\right)^2$$

5   Evaluate the RHS. Thus

$$\left(x + \frac{5}{4}\right)^2 = \frac{3}{2} + \frac{25}{16} = \frac{24 + 25}{16} = \frac{49}{16}$$

6   Taking the square root of both sides of the equation (remembering that the square root of a number gives a $\pm$ answer). Thus

$$\sqrt{\left(x + \frac{5}{4}\right)^2} = \sqrt{\left(\frac{49}{16}\right)}$$

i.e. $x + \frac{5}{4} = \pm\frac{7}{4}$

7   Solve the simple equation. Thus

$$x = -\frac{5}{4} \pm \frac{7}{4}$$

i.e. $x = -\frac{5}{4} + \frac{7}{4} = \frac{2}{4} = \frac{1}{2}$

and $x = -\frac{5}{4} - \frac{7}{4} = -\frac{12}{4} = -3$

Hence $x = \frac{1}{2}$ or -3 are the roots of the equation $2x^2 + 5x = 3$.

Problem 7.   Solve $2x^2 + 9x + 8 = 0$, correct to 3 significant figures, by 'completing the square'

Making the coefficient of x^2 unity gives:

$$x^2 + \frac{9}{2}x + 4 = 0$$

and rearranging gives:

$$x^2 + \frac{9}{2}x = -4$$

Adding to both sides (half the coefficient of $x)^2$ gives:

$$x^2 + \frac{9}{2}x + \left(\frac{9}{4}\right)^2 = \left(\frac{9}{4}\right)^2 - 4$$

The LHS is now a perfect square, thus

$$\left(x + \frac{9}{4}\right)^2 = \frac{81}{16} - 4 = \frac{17}{16}$$

Taking the square root of both sides gives:

$$x + \frac{9}{4} = \sqrt{\left(\frac{17}{16}\right)} = \pm 1.031$$

Hence $x = -\frac{9}{4} \pm 1.031$

i.e. $x = -3.28$ or -1.22, correct to 3 significant figures.

Problem 8.   By 'completing the square', solve the quadratic equation
$4.6y^2 + 3.5y - 1.75 = 0$, correct to 3 decimal places

$$4.6y^2 + 3.5y - 1.75 = 0$$

Making the coefficient of y^2 unity gives:

$$y^2 + \frac{3.5}{4.6}y - \frac{1.75}{4.6} = 0$$

and rearranging gives:

$$y^2 + \frac{3.5}{4.6}y = \frac{1.75}{4.6}$$

Adding to both sides (half the coefficient of $y)^2$ gives:

$$y^2 + \frac{3.5}{4.6}y + \left(\frac{3.5}{9.2}\right)^2 = \frac{1.75}{4.6} + \left(\frac{3.5}{9.2}\right)^2$$

The LHS is now a perfect square, thus

$$\left(y + \frac{3.5}{9.2}\right)^2 = 0.5251654$$

Taking the square root of both sides gives:

$$y + \frac{3.5}{9.2} = \sqrt{0.5251654}$$

$$= \pm 0.7246830$$

Hence $y = -\dfrac{3.5}{9.2} \pm 0.7246830$

i.e. $y = 0.344$ or -1.105

Further problems on solving quadratic equations by 'completing the square' may be found in Section 7.7, Problems 7 and 8, page 72.

7.4   Solution of quadratic equations by formula

Let the general form of a quadratic equation be given by:

$$ax^2 + bx + c = 0$$

where a, b and c are constants.
Dividing $ax^2 + bx + c = 0$ by a gives:

$$x^2 + \frac{b}{a}x + \frac{c}{a} = 0$$

Rearranging gives:

$$x^2 + \frac{b}{a}x = -\frac{c}{a}$$

Adding to each side of the equation the square of half the coefficient of the term in x to make the LHS a perfect square gives:

$$x^2 + \frac{b}{a}x + \left(\frac{b}{2a}\right)^2 = \left(\frac{b}{2a}\right)^2 - \frac{c}{a}$$

Rearranging gives:

$$\left(x + \frac{b}{2a}\right)^2 = \frac{b^2}{4a^2} - \frac{c}{a} = \frac{b^2 - 4ac}{4a^2}$$

Taking the square root of both sides gives:

$$x + \frac{b}{2a} = \sqrt{\left(\frac{b^2 - 4ac}{4a^2}\right)} = \frac{\pm\sqrt{(b^2 - 4ac)}}{2a}$$

Hence $x = -\dfrac{b}{2a} \pm \dfrac{\sqrt{(b^2 - 4ac)}}{2a}$

i.e. the quadratic formula is $x = \dfrac{-b \pm \sqrt{(b^2 - 4ac)}}{2a}$

(This method of solution is 'completing the square' — as shown in Section 7.3.)

Summarizing:

if $ax^2 + bx + c = 0$

then $x = \dfrac{-b \pm \sqrt{(b^2 - 4ac)}}{2a}$

This is known as the **quadratic formula**.

Problem 9. Solve (a) $x^2 + 2x - 8 = 0$ and (b) $3x^2 - 11x - 4 = 0$ by using the quadratic formula

---

(a)   Comparing $x^2 + 2x - 8 = 0$ with $ax^2 + bx + c = 0$ gives $a = 1$, $b = 2$ and $c = -8$.
Substituting these values into the quadratic formula

$$x = \frac{-b \pm \sqrt{(b^2 - 4ac)}}{2a}$$

gives

$$x = \frac{-2 \pm \sqrt{[(2)^2 - 4(1)(-8)]}}{2(1)}$$

$$= \frac{-2 \pm \sqrt{(4 + 32)}}{2} = \frac{-2 \pm \sqrt{36}}{2}$$

$$= \frac{-2 \pm 6}{2} = \frac{-2 + 6}{2} \text{ or } \frac{-2 - 6}{2}$$

Hence $x = \dfrac{4}{2} = 2$ or $\dfrac{-8}{2} = -4$ (as in Problem 1(a))

(b)   Comparing $3x^2 - 11x - 4 = 0$ with $ax^2 + bx + c = 0$ gives $a = 3$, $b = -11$ and $c = -4$. Hence

$$x = \frac{-(-11) \pm \sqrt{[(-11)^2 - 4(3)(-4)]}}{2(3)}$$

$$= \frac{+11 \pm \sqrt{(121 + 48)}}{6} = \frac{11 \pm \sqrt{169}}{6}$$

$$= \frac{11 \pm 13}{6} = \frac{11 + 13}{6} \text{ or } \frac{11 - 13}{6}$$

Hence $x = \dfrac{24}{6} = 4$ or $\dfrac{-2}{6} = -\dfrac{1}{3}$ (as in Problem 1(b))

Problem 10. Solve $4x^2 + 7x + 2 = 0$ giving the roots correct to 2 decimal places

---

Comparing $4x^2 + 7x + 2 = 0$ with $ax^2 + bx + c$ gives $a = 4$, $b = 7$ and $c = 2$. Hence

$$x = \frac{-7 \pm \sqrt{[(7)^2 - 4(4)(2)]}}{2(4)} = \frac{-7 \pm \sqrt{17}}{8}$$

$$= \frac{-7 \pm 4.123}{8} = \frac{-7 + 4.123}{8} \text{ or } \frac{-7 - 4.123}{8}$$

Hence $x = -0.36$ or -1.39, **correct to 2 decimal places.**

Problem 11. Use the quadratic formula to solve $\dfrac{x + 2}{4} + \dfrac{3}{x - 1} = 7$ correct to 4 significant figures

---

Multiplying throughout by $4(x - 1)$ gives:

$$4(x - 1)\frac{(x + 2)}{4} + 4(x - 1)\frac{3}{(x - 1)}$$

$$= 4(x - 1)(7)$$

i.e. $(x - 1)(x + 2) + (4)(3) = 28(x - 1)$

$x^2 + x - 2 + 12 = 28x - 28$

Hence $x^2 - 27x + 38 = 0$

Using the quadratic formula:

$$x = \frac{-(-27) \pm \sqrt{[(-27)^2 - 4(1)(38)]}}{2}$$

$$= \frac{27 \pm \sqrt{577}}{2} = \frac{27 \pm 24.0208}{2}$$

Hence $x = \dfrac{27 + 24.0208}{2} = 25.5104$

or $x = \dfrac{27 - 24.0208}{2} = 1.4896$

Hence $x = 25.51$ or 1.490, **correct to 4 significant figures.**

Further problems on solving quadratic equations by formula may be found in Section 7.7, Problems 9 and 10, page 72.

7.5   Practical problems involving quadratic equations

There are many **practical problems** where a quadratic equation has first to be obtained, from given information, before it is solved.

Problem 12.   The area of a rectangle is 23.6 cm^2 and its width is 3.10 cm shorter than its length. Determine the dimensions of the rectangle, correct to 3 significant figures

---

Let the length of the rectangle be x cm. Then the width is $(x - 3.10)$ cm. Area = length × width = $x(x - 3.10) = 23.6$, i.e.

$$x^2 - 3.10x - 23.6 = 0$$

Using the quadratic formula,

$$x = \frac{-(-3.10) \pm \sqrt{[(-3.10)^2 - 4(1)(-23.6)]}}{2(1)}$$

$$= \frac{3.10 \pm \sqrt{(9.61 + 94.4)}}{2} = \frac{3.10 \pm 10.20}{2}$$

$$= \frac{13.30}{2} \text{ or } \frac{-7.10}{2}$$

Hence $x = 6.65$ cm or -3.55 cm. The latter solution is neglected since length cannot be negative.
Thus length $x = 6.65$ cm and width $= x - 3.10 = 6.65 - 3.10 = 3.55$ cm.
Hence the dimensions of the rectangle are 6.65 cm by 3.55 cm.
(Check: Area $= 6.65 \times 3.55 = 23.6$ cm^2, correct to 3 significant figures.)

Problem 13.   Calculate the diameter of a solid cylinder which has a height of 82.0 cm and a total surface area of 2.0 m^2

Total surface area of a cylinder = curved surface area $+2$ circular ends $= 2\pi rh + 2\pi r^2$ (where r = radius and h = height).
  Since the total surface area $= 2.0$ m^2 and the height $h = 82$ cm or 0.82 m, then

$$2.0 = 2\pi r(0.82) + 2\pi r^2$$

i.e. $2\pi r^2 + 2\pi r(0.82) - 2.0 = 0$

Dividing throughout by 2π gives:

$$r^2 + 0.82r - \frac{1}{\pi} = 0$$

Using the quadratic formula:

$$r = \frac{-0.82 \pm \sqrt{\left[(0.82)^2 - 4(1)\left(-\frac{1}{\pi}\right)\right]}}{2(1)}$$

$$= \frac{-0.82 \pm \sqrt{1.9456}}{2} = \frac{-0.82 \pm 1.3948}{2}$$

$$= 0.2874 \text{ or } -1.1074$$

Thus the radius r of the cylinder is 0.2874 m (the negative solution being neglected).

Hence the diameter of the cylinder $= 2 \times 0.2874$

= 0.5748 m   or   57.5 cm
correct to 3 significant figures

Problem 14.   The height s metres of a mass projected vertically upwards at time t seconds is $s = ut - \frac{1}{2}gt^2$. Determine how long the mass will take after being projected to reach a height of 16 m (a) on the ascent and (b) on the descent, when $u = 30$ m/s and $g = 9.81$ m/s^2

When height $s = 16$ m, $16 = 30t - \frac{1}{2}(9.81)t^2$
i.e. $4.905t^2 - 30t + 16 = 0$
Using the quadratic formula:

$$t = \frac{-(-30) \pm \sqrt{[(-30)^2 - 4(4.905)(16)]}}{2(4.905)}$$

$$= \frac{30 \pm \sqrt{586.1}}{9.81} = \frac{30 \pm 24.21}{9.81} = 5.53 \text{ or } 0.59$$

Hence the mass will reach a height of 16 m after 0.59 s on the ascent and after 5.53 s on the descent.

Problem 15.   A shed is 4.0 m long and 2.0 m wide. A concrete path of constant width is laid all the way around the shed. If the area of the path is 9.50 m^2 calculate its width to the nearest centimetre

---

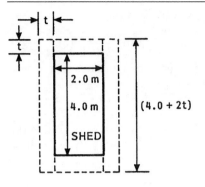

Figure 7.1

Figure 7.1 shows a plan view of the shed with its surrounding path of width t metres.

Area of path $= 2(2.0 \times t) + 2t(4.0 + 2t)$

i.e. $\quad 9.50 = 4.0t + 8.0t + 4t^2$

or $\quad 4t^2 + 12.0t - 9.50 = 0$

Hence $\quad t = \dfrac{-(12.0) \pm \sqrt{[(12.0)^2 - 4(4)(-9.50)]}}{2(4)}$

$= \dfrac{-12.0 \pm \sqrt{296.0}}{8}$

$= \dfrac{-12.0 \pm 17.20465}{8}$

Hence $\quad t = 0.6506$ m or -3.65058 m

Neglecting the negative result which is meaningless, the width of the path, $t = $ **0.651 m or 65 cm**, correct to the nearest centimetre.

Problem 16. If the total surface area of a solid cone is 482.2 cm^2 and its slant height is 15.3 cm, determine its base diameter

---

From Chapter 9, page 91, the total surface area A of a solid cone is given by:

$$A = \pi r l + \pi r^2$$

where l is the slant height and r the base radius.
If $A = 482.2$ and $l = 15.3$, then $482.2 = \pi r(15.3) + \pi r^2$

i.e. $\quad \pi r^2 + 15.3\pi r - 482.2 = 0$

or $\quad r^2 + 15.3r - \dfrac{482.2}{\pi} = 0$

Using the quadratic formula,

$$r = \dfrac{-15.3 \pm \sqrt{\left[(15.3)^2 - 4\left(\dfrac{-482.2}{\pi}\right)\right]}}{2}$$

$$= \dfrac{-15.3 \pm \sqrt{848.0461}}{2} = \dfrac{-15.3 \pm 29.12123}{2}$$

Hence radius $r = 6.9106$ cm (or -22.21 cm, which is meaningless, and is thus ignored.)
Thus **the diameter of the base** $= 2r = 2(6.9106)$ = **13.82 cm**.

Further problems on practical problems involving quadratic equations may be found in Section 7.7, Problems 11 to 20, page 73.

7.6 The solution of linear and quadratic equations simultaneously

Sometimes a linear equation and a quadratic equation need to be solved simultaneously. An algebraic method of solution is shown in Problem 17; a graphical solution is shown in Chapter 16, page 162.

Problem 17. Determine the values of x and y which simultaneously satisfy the equations:

$y = 5x - 4 - 2x^2$ and $y = 6x - 7$

For a simultaneous solution the values of y must be equal, hence the RHS of each equation is equated. Thus

$$5x - 4 - 2x^2 = 6x - 7$$

Rearranging gives:

$$5x - 4 - 2x^2 - 6x + 7 = 0$$

i.e. $\quad -x + 3 - 2x^2 = 0$

or $\quad 2x^2 + x - 3 = 0$

Factorizing gives:

$$(2x + 3)(x - 1) = 0$$

i.e. $\quad x = -\dfrac{3}{2}$ or $x = 1$

In the equation $y = 6x - 7$,

when $x = -\dfrac{3}{2}$, $y = 6\left(-\dfrac{3}{2}\right) - 7 = -16$

and when $x = 1$, $y = 6 - 7 = -1$.
(Checking the result in $y = 5x - 4 - 2x^2$: when $x = -\dfrac{3}{2}$,

$y = 5\left(-\dfrac{3}{2}\right) - 4 - 2\left(-\dfrac{3}{2}\right)^2$

$= -\dfrac{15}{2} - 4 - \dfrac{9}{2} = -16$

as above; and when $x = 1$, $y = 5 - 4 - 2 = -1$ as above.)
Hence the simultaneous solutions occur when $x = -\dfrac{3}{2}$, $y = -16$ and when $x = 1$, $y = -1$.

Further problems on solving linear and quadratic equations simultaneously may be found in the following Section 7.7, Problems 21 to 23, page 73.

7.7   Further problems on quadratic equations

Solving quadratic equations by factorization

In Problems 1 to 4, solve the given equations by factorization:

1   (a) $x^2 + 4x - 32 = 0$
    (b) $x^2 - 16 = 0$
    (c) $(x + 2)^2 = 16$
$$[(a)\ 4, -8\ (b)\ 4, -4\ (c)\ 2, -6]$$

2   (a) $2x^2 - x - 3 = 0$
    (b) $6x^2 - 5x + 1 = 0$
    (c) $10x^2 + 3x - 4 = 0$
$$\left[(a)\ -1, 1\dfrac{1}{2}\ (b)\ \dfrac{1}{2}, \dfrac{1}{3}\ (c)\ \dfrac{1}{2}, -\dfrac{4}{5}\right]$$

3   (a) $x^2 - 4x + 4 = 0$
    (b) $21x^2 - 25x = 4$
    (c) $8x^2 + 13x - 6 = 0$
$$\left[(a)\ 2\ (b)\ 1\dfrac{1}{3}, -\dfrac{1}{7}\ (c)\ \dfrac{3}{8}, -2\right]$$

4   (a) $5x^2 + 13x - 6 = 0$
    (b) $6x^2 - 5x - 4 = 0$

(c) $8x^2 + 2x - 15 = 0$
$$\left[(a)\ \dfrac{2}{5}, -3\ (b)\ \dfrac{4}{3}, -\dfrac{1}{2}\ (c)\ \dfrac{5}{4}, -\dfrac{3}{2}\right]$$

5   Determine the quadratic equations in x whose roots are
    (a) 3 and 1, (b) 2 and -5, (c) -1 and -4.
$$\left[\begin{array}{l}(a)\ x^2 - 4x + 3 = 0 \\ (b)\ x^2 + 3x - 10 = 0 \\ (c)\ x^2 + 5x + 4 = 0\end{array}\right]$$

6   Determine the quadratic equations in x whose roots are
    (a) $2\dfrac{1}{2}$ and $-\dfrac{1}{2}$, (b) 6 and -6, (c) 2.4 and -0.7.
$$\left[\begin{array}{l}(a)\ 4x^2 - 8x - 5 = 0\ (b)\ x^2 - 36 = 0 \\ (c)\ x^2 - 1.7x - 1.68 = 0\end{array}\right]$$

Solving quadratic equations by 'completing the square'

In Problems 7 and 8, solve the given equations by completing the square, correct to 3 decimal places:

7   (a) $x^2 + 4x + 1 = 0$
    (b) $2x^2 + 5x - 4 = 0$
    (c) $3x^2 - x - 5 = 0$
$$\left[\begin{array}{l}(a)\ -3.732, -0.268\ (b)\ -3.137, 0.637 \\ (c)\ 1.468, -1.135\end{array}\right]$$

8   (a) $5x^2 - 8x + 2 = 0$
    (b) $4x^2 - 11x + 3 = 0$
    (c) $2x^2 + 5x = 2$
$$\left[\begin{array}{l}(a)\ 1.290, 0.310\ (b)\ 2.443, 0.307 \\ (c)\ -2.851, 0.351\end{array}\right]$$

Solving quadratic equations by formula

In Problems 9 and 10 solve the given equations by using the quadratic formula, correct to 3 decimal places:

9   (a) $2x^2 + 5x - 4 = 0$
    (b) $5.76x^2 + 2.86x - 1.35 = 0$
    (c) $2x^2 - 7x + 4 = 0$
$$\left[\begin{array}{l}(a)\ 0.637, -3.137\ (b)\ 0.296, -0.792 \\ (c)\ 2.781, 0.719\end{array}\right]$$

10   (a) $4x + 5 = \dfrac{3}{x}$

    (b) $(2x + 1) = \dfrac{5}{x - 3}$

    (c) $\dfrac{x + 1}{x - 1} = x - 3$
$$\left[\begin{array}{l}(a)\ 0.443, -1.693\ (b)\ 3.608, -1.108 \\ (c)\ 4.562, 0.438\end{array}\right]$$

Practical problems involving quadratic equations

11 The angle a rotating shaft turns through in t seconds is given by $\theta = \omega t + \frac{1}{2}\alpha t^2$. Determine the time taken to complete 4 radians if ω is 3.0 rad/s and α is 0.60 rad/s^2.

[1.191 s]

12 The power P developed in an electrical circuit is given by $P = 10I - 8I^2$, where I is the current in amperes. Determine the current necessary to produce a power of 2.5 watts in the circuit.

[0.345 A or 0.905 A]

13 The area of a triangle is 47.6 cm^2 and its perpendicular height is 4.3 cm more than its base length. Determine the length of the base correct to 3 significant figures.

[7.84 cm]

14 The sag l metres in a cable stretched between two supports, distance x m apart, is given by: $l = \dfrac{12}{x} + x$. Determine the distance between supports when the sag is 20 m.

[0.619 m or 19.38 m]

15 The acid dissociation constant K_a of ethanoic acid is 1.8×10^{-5} mol dm^{-3} for a particular solution. Using the Ostwald dilution law

$$K_a = \frac{x^2}{v(1-x)}$$

determine x, the degree of ionization, given that $v = 10$ dm^3.

[0.0133]

16 A rectangular building is 15 m long by 11 m wide. A concrete path of constant width is laid all the way around the building. If the area of the path is 60.0 m^2, calculate its width correct to the nearest millimetre.

[1.066 m]

17 The total surface area of a closed cylindrical container is 20.0 m^3. Calculate the radius of the cylinder if its height is 2.80 m^2.

[86.78 cm]

18 The bending moment M at a point in a beam is given by

$$M = \frac{3x(20 - x)}{2}$$

where x metres is the distance from the point of support. Determine the value of x when the bending moment is 50 Nm.

[1.835 m or 18.165 m]

19 A tennis court measures 24 m by 11 m. In the layout of a number of courts an area of ground must be allowed for at the ends and at the sides of each court. If a border of constant width is allowed around each court and the total area of the court and its border is 950 m^2, find the width of the borders.

[7 m]

20 Two resistors, when connected in series, have a total resistance of 40 ohms. When connected in parallel their total resistance is 8.4 ohms. If one of the resistors has a resistance R_x ohms:
(a) show that $R_x^2 - 40R_x + 336 = 0$ and
(b) calculate the resistance of each

[12 ohms, 28 ohms]

Solving linear and quadratic equations simultaneously

In Problems 21 to 23 determine the solutions of the simultaneous equations:

21 $y = x^2 + x + 1$
    $y = 4 - x$

[$x = 1$, $y = 3$ and $x = -3$, $y = 7$]

22 $y = 15x^2 + 21x - 11$
    $y = 2x - 1$

$\left[x = \dfrac{2}{5}, y = -\dfrac{1}{5} \text{ and } x = -1\dfrac{2}{3}, y = -4\dfrac{1}{3} \right]$

23 $2x^2 + y = 4 + 5x$
    $x + y = 4$

[$x = 0$, $y = 4$ and $x = 3$, $y = 1$]

8

Logarithms and exponential functions

8.1 Logarithms

With the use of calculators firmly established, logarithmic tables are now rarely used for calculation. However, the theory of logarithms is important, for there are several scientific and engineering laws that involve the rules of logarithms.

If a number y can be written in the form a^x, then the index x is called the 'logarithm of y to the base of a',

i.e. $\boxed{\text{if } y = a^x \text{ then } x = \log_a y}$

Thus, since $1000 = 10^3$, then $3 = \log_{10} 1000$

Check this using the 'log' button on your calculator.

(a) Logarithms having a base of 10 are called **common logarithms** and $\log_{10}$ is usually abbreviated to lg. The following values may be checked by using a calculator:
$\lg 17.9 = 1.2528\ldots$, $\lg 462.7 = 2.6652\ldots$ and $\lg 0.0173 = -1.7619\ldots$

(b) Logarithms having a base of e (where 'e' is a mathematical constant approximately equal to 2.7183) are called **hyperbolic, Napierian** or **natural logarithms**, and $\log_e$ is usually abbreviated to ln. The following values may be checked by using a calculator:
$\ln 3.15 = 1.1474\ldots$, $\ln 362.7 = 5.8935\ldots$ and $\ln 0.156 = -0.18578\ldots$

8.2 Laws of logarithms

There are three laws of logarithms, which apply to any base:

(i) To multiply two numbers:

$$\boxed{\log(A \times B) = \log A + \log B}$$

The following may be checked by using a calculator:

$\lg 10 = 1$, also $\lg 5 + \lg 2$

$$= 0.69897\ldots + 0.301029\ldots = 1$$

Hence $\lg(5 \times 2) = \lg 10 = \lg 5 + \lg 2$

(ii) To divide two numbers:

$$\boxed{\log \frac{A}{B} = \log A - \log B}$$

The following may be checked using a calculator:

$$\ln \frac{5}{2} = \ln 2.5 = 0.91629\ldots$$

Also $\ln 5 - \ln 2 = 1.60943\ldots - 0.69314\ldots$

$$= 0.91629\ldots$$

Hence $\ln \dfrac{5}{2} = \ln 5 - \ln 2$

(iii) To raise a number to a power:

$$\boxed{\lg A^n = n \log A}$$

The following may be checked using a calculator:

$\lg 5^2 = \lg 25 = 1.39794\ldots$

Also $2 \lg 5 = 2 \times 0.69897\ldots = 1.39794\ldots$

Hence $\lg 5^2 = 2 \lg 5$

The laws of logarithms may be used to solve certain equations involving powers–called **indicial equations**. For example, to solve, say, $3^x = 27$, logarithms to a base of 10 are taken of both sides,

i.e. $\log_{10} 3^x = \log_{10} 27$

and $x \log_{10} 3 = \log_{10} 27$ by the third law of logarithms

Rearranging gives $x = \dfrac{\log_{10} 27}{\log_{10} 3} = \dfrac{1.43136\ldots}{0.4771\ldots} = 3$

which may be readily checked.
(Note: $(\log 8 / \log 2)$ is **not** equal to $\lg(8/2)$.)

> **Problem 1.** Evaluate (a) $\log_3 9$ (b) $\log_{10} 10$ (c) $\log_{16} 8$

(a) Let $x = \log_3 9$ then $3^x = 9$ from the definition of a logarithm, i.e. $3^x = 3^2$, from which $x = 2$. Hence $\log_3 9 = 2$

(b) Let $x = \log_{10} 10$ then $10^x = 10$ from the definition of a logarithm, i.e. $10^x = 10^1$, from which $x = 1$.
Hence $\log_{10} 10 = 1$ (which may be checked by a calculator)

(c) Let $x = \log_{16} 8$ then $16^x = 8$, from the definition of a logarithm, i.e. $(2^4)^x = 2^3$, i.e. $2^{4x} = 2^3$ from the laws of indices, from which $4x = 3$ and $x = \dfrac{3}{4}$.
Hence $\log_{16} 8 = \dfrac{3}{4}$

> **Problem 2.** Evaluate (a) $\lg 0.001$ (b) $\ln e$ (c) $\log_3 \dfrac{1}{81}$

(a) Let $x = \lg 0.001 = \log_{10} 0.001$ then $10^x = 0.001$, i.e. $10^x = 10^{-3}$, from which $x = -3$.
Hence $\lg 0.001 = -3$ (which may be checked by a calculator)

(b) Let $x = \ln e = \log_e e$ then $e^x = e$, i.e. $e^x = e^1$ from which $x = 1$. Hence $\ln e = 1$ (which may be checked by a calculator)

(c) Let $x = \log_3 \dfrac{1}{81}$ then $3^x = \dfrac{1}{81} = \dfrac{1}{3^4} = 3^{-4}$, from which $x = -4$.
Hence $\log_3 \dfrac{1}{81} = -4$

> **Problem 3.** Solve the following equations:
> (a) $\lg x = 3$ (b) $\log_2 x = 3$ (c) $\log_5 x = -2$

(a) If $\lg x = 3$ then $\log_{10} x = 3$ and $x = 10^3$, i.e. $x = 1000$

(b) If $\log_2 x = 3$ then $x = 2^3 = 8$

(c) If $\log_5 x = -2$ then $x = 5^{-2} = \dfrac{1}{5^2} = \dfrac{1}{25}$

> **Problem 4.** Write (a) $\log 30$ (b) $\log 450$ in terms of $\log 2$, $\log 3$ and $\log 5$ to any base

(a)
$$\log 30 = \log(2 \times 15) = \log(2 \times 3 \times 5)$$
$$= \log 2 + \log 3 + \log 5$$
by the first law of logarithms

(b)
$$\log 450 = \log(2 \times 225) = \log(2 \times 3 \times 75)$$
$$= \log(2 \times 3 \times 3 \times 25)$$
$$= \log(2 \times 3^2 \times 5^2)$$
$$= \log 2 + \log 3^2 + \log 5^2$$
by the first law of logarithms
i.e. $\log 450 = \log 2 + 2\log 3 + 2\log 5$
by the third law of logarithms

> **Problem 5.** Write $\log \left(\dfrac{8 \times \sqrt[4]{5}}{81} \right)$ in terms of $\log 2$, $\log 3$ and $\log 5$ to any base

$\log \left(\dfrac{8 \times \sqrt[4]{5}}{81} \right) = \log 8 + \log \sqrt[4]{5} - \log 81$ (by the first and second laws of logarithms)
$= \log 2^3 + \log 5^{(1/4)} - \log 3^4$ by the laws of indices,
i.e. $\log \left(\dfrac{8 \times \sqrt[4]{5}}{81} \right) = 3\log 2 + \dfrac{1}{4}\log 5 - 4\log 3$ by the third law of logarithms.

> **Problem 6.** Simplify
> $\log 64 - \log 128 + \log 32$

$$64 = 2^6, \quad 128 = 2^7 \text{ and } 32 = 2^5$$

Hence
$$\log 64 - \log 128 + \log 32$$
$$= \log 2^6 - \log 2^7 + \log 2^5$$
$$= 6\log 2 - 7\log 2 + 5\log 2$$
by the third law of logarithms
$$= 4\log 2$$

> **Problem 7.** Evaluate:
> $$\dfrac{\log 25 - \log 125 + \dfrac{1}{2}\log 625}{3\log 5}$$

$$\frac{\log 25 - \log 125 + \frac{1}{2}\log 625}{3\log 5}$$

$$= \frac{\log 5^2 - \log 5^3 + \frac{1}{2}\log 5^4}{3\log 5}$$

$$= \frac{2\log 5 - 3\log 5 + \frac{4}{2}\log 5}{3\log 5} = \frac{1\log 5}{3\log 5} = \frac{1}{3}$$

Problem 8. Solve the equation:
$\log(x-1) + \log(x+1) = 2\log(x+2)$

$\log(x-1) + \log(x+1)$

$= \log(x-1)(x+1)$ from the first law of
logarithms

$= \log(x^2 - 1)$

$2\log(x+2) = \log(x+2)^2 = \log(x^2 + 4x + 4)$

Hence if $\log(x^2 - 1) = \log(x^2 + 4x + 4)$

then $x^2 - 1 = x^2 + 4x + 4$

i.e. $-1 = 4x + 4$

i.e. $-5 = 4x$

i.e. $x = \frac{-5}{4}$ or $-1\frac{1}{4}$

Problem 9. Solve the equation $2^x = 3$,
correct to 4 significant figures

Taking logarithms to base 10 of both sides of $2^x = 3$
gives:

$$\log_{10} 2^x = \log_{10} 3$$

i.e. $x\log_{10} 2 = \log_{10} 3$

Rearranging gives:

$$x = \frac{\log_{10} 3}{\log_{10} 2} = \frac{0.47712125\ldots}{0.30102999\ldots} = \mathbf{1.5850}$$

correct to 4 significant figures.

Problem 10. Solve the equation
$2^{x+1} = 3^{2x-5}$ correct to 2 decimal places

Taking logarithms to base 10 of both sides gives:

$$\log_{10} 2^{x+1} = \log_{10} 3^{2x-5}$$

i.e. $(x+1)\log_{10} 2 = (2x-5)\log_{10} 3$

$x\log_{10} 2 + \log_{10} 2 = 2x\log_{10} 3 - 5\log_{10} 3$

$x(0.3010) + (0.3010) = 2x(0.4771) - 5(0.4771)$

i.e. $0.3010x + 0.3010 = 0.9542x - 2.3855$

Hence $2.3855 + 0.3010 = 0.9542x - 0.3010x$

$2.6865 = 0.6532x$

from which $x = \frac{2.6865}{0.6532} = \mathbf{4.11}$, correct to 2
decimal places.

Problem 11. Solve the equation
$x^{3.2} = 41.15$, correct to 4 significant figures

Taking logarithms to base 10 of both sides gives:

$$\log_{10} x^{3.2} = \log_{10} 41.15$$

$$3.2\log_{10} x = \log_{10} 41.15$$

Hence $\log_{10} x = \frac{\log_{10} 41.15}{3.2} = 0.50449$

Thus $x =$ antilog $0.50449 = 10^{0.50449} = \mathbf{3.195}$
correct to 4 significant figures.

Further problems on logarithms may be found in
Section 8.8, Problems 1 to 38, pages 82 and 83.

8.3 The exponential function

An exponential function is one which contains e^x, e
being a constant called the exponent and having an
approximate value of 2.7183. The exponent arises
from the natural laws of growth and decay and is
used as a base for natural or Napierian logarithms.

8.4 Evaluating exponential functions

The value of e^x may be determined by using:

(a) the power series for e^x, or
(b) a calculator, or
(c) tables of exponential functions

The most common method of evaluating an exponential function is by using a scientific notation calculator, this now having replaced the use of tables. However, let us first look at the power series for e^x.

The power series for e^x

The value of e^x can be calculated to any required degree of accuracy since it is defined in terms of the following **power series**:

$$e^x = 1 + x + \frac{x^2}{2!} + \frac{x^3}{3!} + \frac{x^4}{4!} + \dots \qquad (1)$$

(where $3! = 3 \times 2 \times 1$ and is called 'factorial 3'). The series is valid for all values of x.

The series is said to **converge**, i.e. if all the terms are added, an actual value for e^x (where x is a real number) is obtained. The more terms that are taken, the closer will be the value of e^x to its actual value. The value of the exponent e, correct to say 4 decimal places, may be determined by substituting $x = 1$ in the power series of equation (1). Thus

$$e^1 = 1 + 1 + \frac{(1)^2}{2!} + \frac{(1)^3}{3!} + \frac{(1)^4}{4!} + \frac{(1)^5}{5!} + \frac{(1)^6}{6!}$$

$$+ \frac{(1)^7}{7!} + \frac{(1)^8}{8!} + \dots$$

$$= 1 + 1 + 0.5 + 0.16667 + 0.04167 + 0.00833$$

$$+ 0.00139 + 0.00020 + 0.00002 + \dots$$

$$= 2.71828$$

i.e. $e = 2.7183$ correct to 4 decimal places.

The value of $e^{0.05}$, correct to say 8 significant figures, is found by substituting $x = 0.05$ in the power series for e^x. Thus

$$e^{0.05} = 1 + 0.05 + \frac{(0.05)^2}{2!} + \frac{(0.05)^3}{3!}$$

$$+ \frac{(0.05)^4}{4!} + \frac{(0.05)^5}{5!} + \dots$$

$$= 1 + 0.05 + 0.00125 + 0.000020833$$

$$+ 0.000000260 + 0.000000003$$

and by adding,

$e^{0.05} = 1.0512711$, correct to 8 significant figures.

In this example, successive terms in the series grow smaller very rapidly and it is relatively easy to determine the value of $e^{0.05}$ to a high degree of accuracy. However, when x is nearer to unity or larger than unity, a very large number of terms are required for an accurate result.

If in the series of equation (1), x is replaced by $-x$, then

$$e^{-x} = 1 + (-x) + \frac{(-x)^2}{2!} + \frac{(-x)^3}{3!} + \dots$$

$$e^{-x} = 1 - x + \frac{(x)^2}{2!} - \frac{(x)^3}{3!} + \dots$$

In a similar manner the power series for e^x may be used to evaluate any exponential function of the form ae^{kx}, where a and k are constants.

In the series of equation (1), let x be replaced by kx. Then

$$ae^{kx} = a\left\{1 + (kx) + \frac{(kx)^2}{2!} + \frac{(kx)^3}{3!} + \dots\right\}$$

Thus $5e^{2x} = 5\left\{1 + (2x) + \frac{(2x)^2}{2!} + \frac{(2x)^3}{3!} + \dots\right\}$

$$= 5\left\{1 + 2x + \frac{4x^2}{2} + \frac{8x^3}{6} + \dots\right\}$$

i.e. $5e^{2x} = 5\left\{1 + 2x + 2x^2 + \frac{4}{3}x^3 + \dots\right\}$

Problem 12. Determine the value of $5e^{0.5}$, correct to 5 significant figures, by using the power series for e^x

---

$$e^x = 1 + x + \frac{x^2}{2!} + \frac{x^3}{3!} + \dots$$

Hence $e^{0.5} = 1 + 0.5 + \frac{(0.5)^2}{(2)(1)} + \frac{(0.5)^3}{(3)(2)(1)}$

$$+ \frac{(0.5)^4}{(4)(3)(2)(1)} + \frac{(0.5)^5}{(5)(4)(3)(2)(1)}$$

$$+ \frac{(0.5)^6}{(6)(5)(4)(3)(2)(1)}$$

$$= 1 + 0.5 + 0.125 + 0.020833$$

$$+ 0.0026042 + 0.0002604$$

$$+ 0.0000217$$

i.e. $e^{0.5} = 1.64872$ correct to 6 significant figures.

Hence $5e^{0.5} = 5(1.64872) = \mathbf{8.2436}$, correct to 5 significant figures.

Problem 13. Determine the value of $3e^{-1}$, correct to 4 decimal places, using the power series for e^x

Substituting $x = -1$ in the power series

$$e^x = 1 + x + \frac{x^2}{2!} + \frac{x^3}{3!} + \frac{x^4}{4!} + \ldots$$

gives $e^{-1} = 1 + (-1) + \frac{(-1)^2}{2!}$

$$+ \frac{(-1)^3}{3!} + \frac{(-1)^4}{4!} + \ldots$$

$$= 1 - 1 + 0.5 - 0.166667$$

$$+ 0.041667 - 0.008333$$

$$+ 0.001389 - 0.000198 + \ldots$$

$$= 0.367858 \text{ correct to 6 decimal places}$$

Hence $3e^{-1} = (3)(0.367858) = \mathbf{1.1036}$ correct to 4 decimal places.

Problem 14. Expand $e^x(x^2 - 1)$ as far as the term in x^5

The power series for e^x is $1 + x + \frac{x^2}{2!} + \frac{x^3}{3!} + \frac{x^4}{4!} + \ldots$
Hence

$$e^x(x^2 - 1) = \left(1 + x + \frac{x^2}{2!} + \frac{x^3}{3!} + \frac{x^4}{4!} \right.$$

$$\left. + \frac{x^5}{5!} + \ldots \right) (x^2 - 1)$$

$$= \left(x^2 + x^3 + \frac{x^4}{2!} + \frac{x^5}{3!} + \ldots \right)$$

$$- \left(1 + x + \frac{x^2}{2!} + \frac{x^3}{3!} \right.$$

$$\left. + \frac{x^4}{4!} + \frac{x^5}{5!} + \ldots \right)$$

Grouping like terms gives:

$$e^x(x^2 - 1) = -1 - x + \left(x^2 - \frac{x^2}{2!} \right) + \left(x^3 - \frac{x^3}{3!} \right)$$

$$+ \left(\frac{x^4}{2!} - \frac{x^4}{4!} \right) + \left(\frac{x^5}{3!} - \frac{x^5}{5!} \right) + \ldots$$

$$= -1 - x + \frac{x^2}{2} + \frac{5}{6}x^3 + \frac{11}{24}x^4 + \frac{19}{120}x^5$$

when expanded as far as the term in x^5.

Use of a calculator

Most scientific notation calculators contain an 'e^x' function which enables all practical values of e^x and e^{-x} to be determined, correct to 8 or 9 significant figures. For example

$$e^1 = 2.7182818$$

$$e^{2.4} = 11.023176$$

$$e^{-1.618} = 0.19829489$$

correct to 8 significant figures.

In practical situations the degree of accuracy given by a calculator is often far greater than is appropriate. The accepted convention is that the final result is stated to one significant figure greater than the least significant measured value. Use your calculator to check the following values:

$$e^{0.12} = 1.1275, \text{ correct to 5 significant figures}$$

$$e^{-1.47} = 0.22993, \text{ correct to 5 decimal places}$$

$$e^{-0.431} = 0.6499, \text{ correct to 4 decimal places}$$

$$e^{9.32} = 11159, \text{ correct to 5 significant figures}$$

$$e^{-2.785} = 0.0617291, \text{ correct to 7 decimal places}$$

Problem 15. Using a calculator, evaluate, correct to 5 significant figures:

(a) $e^{2.731}$ (b) $e^{-3.162}$ (c) $\frac{5}{3}e^{5.253}$

(a) $e^{2.731} = 15.348227 \ldots = \mathbf{15.348}$, correct to 5 significant figures

(b) $e^{-3.162} = 0.04234097 \ldots = \mathbf{0.042341}$, correct to 5 significant figures

(c) $\frac{5}{3}e^{5.253} = \frac{5}{3}(191.138825 \ldots) = \mathbf{318.56}$, correct to 5 significant figures

Problem 16. Use a calculator to determine the following, each correct to 4 significant figures:

(a) $3.72e^{0.18}$ (b) $53.2e^{-1.4}$ (c) $\frac{5}{122}e^7$

(a) $3.72e^{0.18} = (3.72)(1.197217\ldots) = \mathbf{4.454}$,
     correct to 4 significant figures

(b) $53.2e^{-1.4} = (53.2)(0.246596\ldots) = \mathbf{13.12}$,
     correct to 4 significant figures

(c) $\dfrac{5}{122}e^7 = \left(\dfrac{5}{122}\right)(1096.6331\ldots) = \mathbf{44.94}$,
     correct to 4 significant figures

Problem 17. Evaluate the following, correct to 4 decimal places, using a calculator:

(a) $0.0256(e^{5.21} - e^{2.49})$

(b) $5\left(\dfrac{e^{0.25} - e^{-0.25}}{e^{0.25} + e^{-0.25}}\right)$

(a) $0.0256(e^{5.21} - e^{2.49})$

$= 0.0256(183.094058\ldots - 12.0612761\ldots)$

$= \mathbf{4.3784}$, correct to 4 decimal places

(b) $5\left(\dfrac{e^{0.25} - e^{-0.25}}{e^{0.25} + e^{-0.25}}\right)$

$= 5\left(\dfrac{1.28402541\ldots - 0.77880078\ldots}{1.28402541\ldots + 0.77880078\ldots}\right)$

$= 5\left(\dfrac{0.5052246\ldots}{2.0628261\ldots}\right)$

$= \mathbf{1.2246}$, correct to 4 decimal places

Problem 18. The instantaneous voltage v in a capacitive circuit is related to time t by the equation $v = Ve^{-t/CR}$ where V, C and R are constants. Determine v, correct to 4 significant figures, when $t = 30 \times 10^{-3}$ seconds, $C = 10 \times 10^{-6}$ farads, $R = 47 \times 10^3$ ohms and $V = 200$ volts

---

$v = Ve^{-t/CR}$

$= 200e^{(-30 \times 10^{-3})/(10 \times 10^{-6} \times 47 \times 10^3)}$

Using a calculator, $v = 200e^{-0.0638297\ldots}$

$= 200(0.9381646\ldots)$

$= \mathbf{187.6 \ volts}$

Further problems on evaluating exponential functions may be found in Section 8.8, Problems 39 to 49, pages 83 and 84.

8.5 Napierian logarithms

Logarithms having a base of e are called **hyperbolic**, **Napierian** or **natural logarithms** and the Napierian logarithm of x is written as $\log_e x$, or more commonly, $\ln x$.

8.6 Evaluating Napierian logarithms

The value of a Napierian logarithm may be determined by using:

(a) a calculator, or
(b) a relationship between common and Napierian logarithms, or
(c) Napierian logarithm tables

The most common method of evaluating a Napierian logarithm is by a scientific notation calculator, this now having replaced the use of four-figure tables, and also the relationship between common and Napierian logarithms,

$$\log_e y = 2.3026 \log_{10} y$$

Most scientific notation calculators contain a '$\ln x$' function which displays the value of the Napierian logarithm of a number when the appropriate key is pressed.

Using a calculator,

$\ln 4.692 = 1.5458589\ldots$

$= 1.5459$, correct to 4 decimal places

and $\ln 35.78 = 3.57738907\ldots$

$= 3.5774$, correct to 4 decimal places

Use your calculator to check the following values:

$\ln 1.732 = 0.54928$, correct to 5 significant figures

$\ln 1 = 0$

$\ln 0.52 = -0.6539$, correct to 4 decimal places

$\ln 593 = 6.3852$, correct to 5 significant figures

$\ln 1750 = 7.4674$, correct to 4 decimal places

$\ln 0.17 = -1.772$, correct to 4 significant figures

$\ln 0.00032 = -8.04719$, correct to 6 significant figures

$\ln e^3 = 3$

$\ln e^1 = 1$

From the last two examples we can conclude that

$$\log e^x = x$$

This is useful when solving equations involving exponential functions. For example, to solve $e^{3x} = 8$, take Napierian logarithms of both sides, which gives

$$\ln e^{3x} = \ln 8$$

i.e. $3x = \ln 8$

from which $x = \dfrac{1}{3} \ln 8 = \mathbf{0.6931}$, correct to 4 decimal places.

Problem 23.   Using a calculator evaluate correct to 5 significant figures:
(a) $\ln 47.291$ (b) $\ln 0.06213$ (c) $3.2 \ln 762.923$

(a) $\ln 47.291 = 3.8563200\ldots = \mathbf{3.8563}$, correct to 5 significant figures
(b) $\ln 0.06213 = -2.7785263\ldots = \mathbf{-2.7785}$, correct to 5 significant figures
(c) $3.2 \ln 762.923 = 3.2(6.6371571\ldots) = \mathbf{21.239}$, correct to 5 significant figures

---

Problem 24.   Use a calculator to evaluate the following, each correct to 5 significant figures:

(a) $\dfrac{1}{4} \ln 4.7291$ (b) $\dfrac{\ln 7.8693}{7.8693}$

(c) $\dfrac{5.29 \ln 24.07}{e^{-0.1762}}$

(a) $\dfrac{1}{4} \ln 4.7291 = \dfrac{1}{4}(1.5537349\ldots) = \mathbf{0.38843}$, correct to 5 significant figures

(b) $\dfrac{\ln 7.8693}{7.8693} = \dfrac{2.06296911\ldots}{7.8693} = \mathbf{0.26215}$, correct to 5 significant figures

(c) $\dfrac{5.29 \ln 24.07}{e^{-0.1762}} = \dfrac{5.29(3.18096625\ldots)}{(0.83845027\ldots)}$
$= \mathbf{20.070}$, correct to 5 significant figures

Problem 25.   Evaluate the following:

(a) $\dfrac{\ln e^{2.5}}{\lg 10^{0.5}}$ (b) $\dfrac{4e^{2.23} \lg 2.23}{\ln 2.23}$
(correct to 3 decimal places)

(a) $\dfrac{\ln e^{2.5}}{\lg 10^{0.5}} = \dfrac{2.5}{0.5} = 5$

(b) $\dfrac{4e^{2.23} \lg 2.23}{\ln 2.23}$
$= \dfrac{4(9.29986607\ldots)(0.34830486\ldots)}{(0.80200158\ldots)}$
$= \mathbf{16.156}$, correct to 3 decimal places

Problem 26.   Solve the equation $7 = 4e^{-3x}$ to find x, correct to 4 significant figures

---

Rearranging $7 = 4e^{-3x}$ gives:

$$\dfrac{7}{4} = e^{-3x}$$

Taking the reciprocal of both sides gives:

$$\dfrac{4}{7} = \dfrac{1}{e^{-3x}} = e^{3x}$$

Taking Napierian logarithms of both sides gives:

$$\ln\left(\dfrac{4}{7}\right) = \ln(e^{3x})$$

Since $\log_e e^{\alpha} = \alpha$, then $\ln\dfrac{4}{7} = 3x$

Hence $x = \dfrac{1}{3}\ln\left(\dfrac{4}{7}\right) = \dfrac{1}{3}(-0.55962) = \mathbf{-0.1865}$, correct to 4 significant figures.

Problem 27.   Given $20 = 60(1 - e^{-t/2})$ determine the value of t, correct to 3 significant figures

---

Rearranging $20 = 60(1 - e^{-t/2})$ gives:

$$\dfrac{20}{60} = 1 - e^{-1/2}$$

and

$$e^{-t/2} = 1 - \dfrac{20}{60} = \dfrac{2}{3}$$

Taking the reciprocal of both sides gives:

$$e^{t/2} = \dfrac{3}{2}$$

Taking Napierian logarithms of both sides gives:

$$\ln e^{t/2} = \ln \frac{3}{2}$$

i.e. $$\frac{t}{2} = \ln \frac{3}{2}$$

from which, $t = 2 \ln \dfrac{3}{2} = \textbf{0.811}$, correct to 3 significant figures.

Problem 28.   Solve the equation
$3.72 = \ln \left(\dfrac{5.14}{x} \right)$ to find x

From the definition of a logarithm, since
$3.72 = \ln \left(\dfrac{5.14}{x} \right)$ then $e^{3.72} = \dfrac{5.14}{x}$

Rearranging gives: $x = \dfrac{5.14}{e^{3.72}} = 5.14 e^{-3.72}$

i.e. $x = \textbf{0.1246}$, correct to 4 significant figures.

Further problems on evaluating Napierian logarithms may be found in Section 8.8, Problems 50 to 59, page 84.

8.7 Laws of growth and decay

The laws of exponential growth and decay are of the form $y = Ae^{kx}$ and $y = A(1 - e^{kx})$, where A and k are constants. The laws occur frequently in engineering and science and examples of quantities related by a natural law include:

(i)   Linear expansion $l = l_0 e^{\alpha\theta}$
(ii) Change in electrical resistance
      with temperature $R_\theta = R_0 e^{\alpha\theta}$
(iii) Tension in belts $T_1 = T_0 e^{\mu\theta}$
(iv) Newton's law of cooling $\theta = \theta_0 e^{-kt}$
(v)   Biological growth $y = y_0 e^{kt}$
(vi) Discharge of a capacitor $q = Qe^{-t/CR}$
(vii) Atmospheric pressure $p = p_0 e^{-h/c}$
(viii) Radioactive decay $N = N_0 e^{-\lambda t}$
(ix) Decay of current in an inductive
      circuit $i = Ie^{-Rt/L}$
(x)   Growth of current in a
      capacitive circuit $i = I(1 - e^{-t/CR})$

Problem 29.   The resistance R of an electrical conductor at temperature $\theta\,°C$ is given by $R = R_0 e^{\alpha\theta}$, where α is a constant and $R_0 = 5 \times 10^3$ ohms. Determine the value of α when $R = 6 \times 10^3$ ohms and $\theta = 1500\,°C$. Also, find the temperature when the resistance R is 5.4×10^3 ohms

---

Transposing $R = R_0 e^{\alpha\theta}$ gives $\dfrac{R}{R_0} = e^{\alpha\theta}$

Taking Napierian logarithms of both sides gives:

$$\ln \left(\frac{R}{R_0} \right) = \ln e^{\alpha\theta} = \alpha\theta$$

Hence

$$\alpha = \frac{1}{\theta} \ln \left(\frac{R}{R_0} \right) = \frac{1}{1500} \ln \left(\frac{6 \times 10^3}{5 \times 10^3} \right)$$

$$= \frac{1}{1500}(0.1823)$$

Hence $\alpha = \textbf{1.215} \times \textbf{10}^{-4}$.

From above, $\ln \left(\dfrac{R}{R_0} \right) = \alpha\theta$ hence $\theta = \dfrac{1}{\alpha} \ln \left(\dfrac{R}{R_0} \right)$

When $R = 5.4 \times 10^3$, $\alpha = 1.215 \times 10^{-4}$ and $R_0 = 5 \times 10^3$

$$\theta = \frac{1}{1.215 \times 10^{-4}} \ln \left(\frac{5.4 \times 10^3}{5 \times 10^3} \right)$$

$$= \frac{10^4}{1.215}(7.696 \times 10^{-2}) = \textbf{633.4}\,°\textbf{C}$$

Problem 30.   In an experiment involving Newton's law of cooling, the temperature $\theta(°C)$ is given by $\theta = \theta_0 e^{-kt}$. Find the value of constant k when $\theta_0 = 56.6\,°C$, $\theta = 16.5\,°C$ and $t = 83.0$ seconds

---

Transposing $\theta = \theta_0 e^{-kt}$ gives $\dfrac{\theta}{\theta_0} = e^{-kt}$ from which

$$\frac{\theta_0}{\theta} = \frac{1}{e^{-kt}} = e^{kt}$$

Taking Napierian logarithms of both sides gives:

$$\ln \left(\frac{\theta_0}{\theta} \right) = kt$$

from which,

$$k = \frac{1}{t} \ln \left(\frac{\theta_0}{\theta}\right) = \frac{1}{83.0} \ln \left(\frac{56.6}{16.5}\right)$$

$$= \frac{1}{83.0}(1.2326)$$

Hence $k = 1.485 \times 10^{-2}$.

Problem 31.   The current i amperes flowing in a capacitor at time t seconds is given by $i = 8.0(1 - e^{-t/CR})$, where the circuit resistance R is 25×10^3 ohms and capacitance C is 16×10^{-6} farads. Determine (a) the current i after 0.5 seconds and (b) the time for the current to reach 6.0 A

(a)   Current $i = 8.0(1 - e^{-t/CR})$

$$= 8.0[1 - e^{-0.5/(16 \times 10^{-6})(25 \times 10^3)}]$$

$$= 8.0(1 - e^{-1.25})$$

$$= 8.0(1 - 0.2865) = 8.0(0.7135)$$

$$= \mathbf{5.708 \ amperes}$$

(b)   Transposing $i = 8.0(1 - e^{-t/CR})$

gives $\dfrac{i}{8.0} = 1 - e^{-t/CR}$

from which, $e^{-t/CR} = 1 - \dfrac{i}{8.0} = \dfrac{8.0 - i}{8.0}$

Taking the reciprocal of both sides gives:

$$e^{t/CR} = \frac{8.0}{8.0 - i}$$

Taking Napierian logarithms of both sides gives:

$$\frac{t}{CR} = \ln \left(\frac{8.0}{8.0 - i}\right)$$

Hence $t = CR \ln \left(\dfrac{8.0}{8.0 - i}\right)$

$$= (16 \times 10^{-6})(25 \times 10^3) \ln \left(\frac{8.0}{8.0 - 6.0}\right)$$

when $i = 6.0$ amperes, i.e.

$$t = \frac{400}{10^3} \ln \left(\frac{8.0}{2.0}\right) = 0.4 \ln 4.0 = 0.4(1.3863)$$

$$= \mathbf{0.5545 \ s}$$

Problem 32.   The temperature θ_2 of a winding which is being heated electrically at time t is given by: $\theta_2 = \theta_1(1 - e^{-t/\tau})$ where θ_1 is the temperature (in degrees Celsius) at time $t = 0$ and τ is a constant. Calculate:

(a) θ_1, correct to the nearest degree, when θ_2 is 50°C, t is 30 s and τ is 60 s
(b)   the time t, correct to 1 decimal place, for θ_2 to be half the value of θ_1

(a)   Transposing the formula to make θ_1 the subject gives:

$$\theta_1 = \frac{\theta_2}{(1 - e^{-t/\tau})} = \frac{50}{1 - e^{-30/60}}$$

$$= \frac{50}{1 - e^{-1/2}} = \frac{50}{0.3935}$$

i.e. $\theta_1 = \mathbf{127 °C}$, correct to the nearest degree

(b)   Transposing to make t the subject of the formula gives:

$$\frac{\theta_2}{\theta_1} = 1 - e^{-t/\tau}$$

from which, $e^{-t/\tau} = 1 - \dfrac{\theta_2}{\theta_1}$

Hence $-\dfrac{t}{\tau} = \ln \left(1 - \dfrac{\theta_2}{\theta_1}\right)$

i.e. $t = -\tau \ln \left(1 - \dfrac{\theta_2}{\theta_1}\right)$

Since $\theta_2 = \dfrac{1}{2}\theta_1$

$$t = -60 \ln \left(1 - \frac{1}{2}\right) = -60 \ln 0.5 = 41.59 \ s$$

Hence the time for the temperature to fall to one-half of its original value is 41.6 s, correct to 1 decimal place.

Further problems on the laws of growth and decay may be found in the following Section 8.8, Problems 60 to 70, pages 84 and 85.

8.8   Further problems on logarithms and exponential functions

Logarithms

In Problems 1 to 11, evaluate the given expression:

1   $\log_{10} 10\,000$ [4]

2   $\log_2 16$ [4]

3   $\log_5 125$ [3]

4   $\log_2 \dfrac{1}{8}$ [−3]

5   $\log_8 2$ $\left[\dfrac{1}{3}\right]$

6   $\log_7 343$ [3]

7   $\lg 100$ [2]

8   $\lg 0.01$ [−2]

9   $\log_4 8$ $\left[1\dfrac{1}{2}\right]$

10   $\log_{27} 3$ $\left[\dfrac{1}{3}\right]$

11   $\ln e^2$ [2]

In Problems 12 to 18 solve the equations:

12   $\log_{10} x = 4$ [10 000]

13   $\lg x = 5$ [100 000]

14   $\log_3 x = 2$ [9]

15   $\log_4 x = -2\dfrac{1}{2}$ $\left[\pm\dfrac{1}{32}\right]$

16   $\lg x = -2$ [0.01]

17   $\log_8 x = -\dfrac{4}{3}$ $\left[\dfrac{1}{16}\right]$

18   $\ln x = 3$ $[e^3]$

In Problems 19 to 22 write the give expressions in terms of $\log 2$, $\log 3$ and $\log 5$ to any base:

19   $\log 60$ $[2\log 2 + \log 3 + \log 5]$

20   $\log 300$ $[2\log 2 + \log 3 + 2\log 5]$

21   $\log\left(\dfrac{16 \times \sqrt[4]{5}}{27}\right)$ $\left[4\log 2 + \dfrac{1}{4}\log 5 - 3\log 3\right]$

22   $\log\left(\dfrac{125 \times \sqrt[4]{16}}{\sqrt[4]{81^3}}\right)$ $[\log 2 - 3\log 3 + 3\log 5]$

Simplify the expressions given in Problems 23 to 25:

23   $\log 27 - \log 9 + \log 81$ $[5\log 3]$

24   $\log 64 + \log 32 - \log 128$ $[4\log 2]$

25   $\log 8 - \log 4 + \log 32$ $[6\log 2]$

Evaluate the expressions given in Problems 26 and 27:

26   $\dfrac{\dfrac{1}{2}\log 16 - \dfrac{1}{3}\log 8}{\log 4}$ $\left[\dfrac{1}{2}\right]$

27   $\dfrac{\log 9 - \log 3 + \dfrac{1}{2}\log 81}{2\log 3}$ $\left[\dfrac{3}{2}\right]$

Solve the equations given in Problems 28 to 30:

28   $\log x^4 - \log x^3 = \log 5x - \log 2x$ $\left[x = 2\dfrac{1}{2}\right]$

29   $\log 2t^3 - \log t = \log 16 + \log t$ $[t = 8]$

30   $2\log b^2 - 3\log b = \log 8b - \log 4b$ $[b = 2]$

In Problems 31 to 38 solve the indicial equations for x, each correct to 4 significant figures:

31   $3^x = 6.4$ [1.691]

32   $2^x = 9$ [3.170]

33   $2^{x-1} = 3^{2x-1}$ [0.2696]

34   $x^{1.5} = 14.91$ [6.058]

35   $25.28 = 4.2^x$ [2.251]

36   $4^{2x-1} = 5^{x+2}$ [3.959]

37   $x^{-0.25} = 0.792$ [2.542]

38   $0.027^x = 3.26$ [−0.3272]

Evaluating exponential functions

In Problems 39 and 40 use a calculator to evaluate the given functions correct to 4 significant figures:

39   (a) $e^{4.4}$ (b) $e^{-0.25}$ (c) $e^{0.92}$

[(a) 81.45 (b) 0.7788 (c) 2.509]

40   (a) $e^{-1.8}$ (b) $e^{-0.78}$ (c) e^{10}

[(a) 0.1653 (b) 0.4584 (c) 22 030]

41   Evaluate, correct to 5 significant figures:

(a) $3.5e^{2.8}$ (b) $-\dfrac{6}{5}e^{-1.5}$ (c) $2.16e^{5.7}$

[(a) 57.556 (b) −0.26776 (c) 645.55]

42   Use a calculator to evaluate the following, correct to 5 significant figures:

(a) $e^{1.629}$ (b) $e^{-2.7483}$ (c) $0.62e^{4.178}$

[(a) 5.0988 (b) 0.064037 (c) 40.446]

In Problems 43 and 44, evaluate correct to 5 decimal places:

43   (a) $\dfrac{1}{7}e^{3.4629}$ (b) $8.52e^{-1.2651}$ (c) $\dfrac{5e^{2.6921}}{3e^{1.1171}}$

[(a) 4.55848 (b) 2.40444 (c) 8.05124]

44 (a) $\dfrac{5.6823}{e^{-2.1347}}$ (b) $\dfrac{e^{2.1127} - e^{-2.1127}}{2}$

(c) $\dfrac{4(e^{-1.7295} - 1)}{e^{3.6817}}$

[(a) 48.04106 (b) 4.07482 (c) −0.08286]

45 The length of a bar, l, at a temperature θ is given by $l = l_0 e^{\alpha\theta}$, where l_0 and α are constants. Evaluate l, correct to 4 significant figures, where $l_0 = 2.587$, $\theta = 321.7$ and $\alpha = 1.771 \times 10^{-4}$. [2.739]

46 Evaluate $5.6e^{-1}$, correct to 4 decimal places, using the power series for e^x. [2.0601]

47 Use the power series for e^x to determine, correct to 4 significant figures, (a) e^2 (b) $e^{-0.3}$ and check your result by using a calculator.

[(a) 7.389 (b) 0.7408]

48 Expand $(1 - 2x)e^{2x}$ as far as the term in x^4.

$$\left[1 - 2x^2 - \frac{8x^3}{3} - 2x^4 \right]$$

49 Expand $(2e^{x^2})(x^{1/2})$ to six terms.

$$\left[\begin{array}{l} 2x^{1/2} + 2x^{5/2} + x^{9/2} + \dfrac{1}{3}x^{13/2} \\ + \dfrac{1}{12}x^{17/2} + \dfrac{1}{60}x^{21/2} \end{array} \right]$$

Evaluating Napierian logarithms

In Problems 50 to 52 use a calculator to evaluate the given functions, correct to 4 decimal places:

50 (a) $\ln 1.73$ (b) $\ln 5.413$ (c) $\ln 9.412$

[(a) 0.5481 (b) 1.6888 (c) 2.2420]

51 (a) $\ln 17.3$ (b) $\ln 541.3$ (c) $\ln 9412$

[(a) 2.8507 (b) 6.2940 (c) 9.1497]

52 (a) $\ln 0.173$ (b) $\ln 0.005413$ (c) $\ln 0.09412$

[(a) −1.7545 (b) −5.2190 (c) −2.3632]

In Problems 53 and 54, evaluate correct to 5 significant figures:

53 (a) $\dfrac{1}{6} \ln 5.2932$ (b) $\dfrac{\ln 82.473}{4.829}$

(c) $\dfrac{5.62 \ln 321.62}{e^{1.2942}}$

[(a) 0.27774 (b) 0.91374 (c) 8.8941]

54 (a) $\dfrac{2.946 \ln e^{1.76}}{\lg 10^{1.41}}$ (b) $\dfrac{5e^{-0.1629}}{2 \ln 0.00165}$

(c) $\dfrac{\ln 4.8629 - \ln 2.4711}{5.173}$

[(a) 3.6773 (b) −0.33154 (c) 0.13087]

In Problems 55 to 59 solve the given equations, each correct to 4 significant figures:

55 $1.5 = 4e^{2t}$ [−0.4904]

56 $7.83 = 2.91e^{-1.7x}$ [−0.5822]

57 $16 = 24(1 - e^{-t/2})$ [2.197]

58 $5.17 = \ln\left(\dfrac{x}{4.64}\right)$ [816.2]

59 $3.72 \ln\left(\dfrac{1.59}{x}\right) = 2.43$ [0.8274]

Laws of growth and decay

60 Two quantities x and y are related by the equation $y = ae^{-kx}$, where a and k are constants. Determine the value of (a) y when $a = 2.114$, $k = -3.20$ and $x = 1.429$, (b) x when $y = 115.4$, $a = 17.8$ and $k = 4.65$.

[(a) 204.7 (b) −0.4020]

61 The pressure p pascals at height h metres above ground level is given by $p = p_0 e^{-h/C}$, where p_0 is the pressure at ground level and C is a constant. When p_0 is 1.012×10^5 Pa and the pressure at a height of 1420 m is 9.921×10^4 Pa, determine the value of C. [71 500]

62 The length l metres of a metal bar at temperature $t\,°C$ is given by $l = l_0 e^{\alpha t}$, where l_0 and α are constants. Determine (a) the value of α when $l = 1.993$ m, $l_0 = 1.894$ m and $t = 250\,°C$, and (b) the value of l_0 when $l = 2.416$ m, $t = 310\,°C$ and $\alpha = 1.682 \times 10^{-4}$.

[(a) 2.038×10^{-4} (b) 2.293 m]

63 The temperature $\theta_2\,°C$ of an electrical conductor at time t seconds is given by

$$\theta_2 = \theta_1(1 - e^{-t/T})$$

where θ_1 is the initial temperature and T seconds is a constant. Determine (a) θ_1 when $\theta_2 = 50\,°C$, $t = 30$ s and $T = 80$ s, and (b) the time t for θ_2 to fall to half the value of θ_1 if T remains at 80 s.

[(a) 159.9°C (b) 55.45 s]

64 Quantities x and y are related by

$$y = 8.317(1 - e^{cx/t})$$

where c and t are constants. Determine (a) the value of y when $c = 2.9 \times 10^{-3}$, $x = 841.2$ and $t = 4.379$, and (b) the value of t when $y = -83.68$, $x = 841.2$ and $c = 2.9 \times 10^{-2}$.
$$[\text{(a)} -6.201 \text{ (b) } 10.15]$$

65 The voltage drop, v volts, across an inductor L henrys at time t seconds is given by

$$v = 200e^{-Rt/L}$$

where $R = 150 \ \Omega$ and $L = 12.5 \times 10^{-3}$ H. Determine (a) the voltage when $t = 160 \times 10^{-6}$ s, and (b) the time for the voltage to reach 85 V.   [(a) 29.32 volts (b) 71.31×10^{-6} s]

66 A belt is in contact with a pulley for a sector of $\theta = 1.12$ radians and the coefficient of friction between these two surfaces is $\mu = 0.26$.

(a) Determine the tension on the taut side of the belt, T newtons, when the tension on the slack side is given by $T_0 = 22.7$ newtons, given that these quantities are related by the law $T = T_0 e^{\mu\theta}$

(b) It is required that the transmitted force $(T - T_0)$ be increased to 24.0 newtons. Assuming that T_0 remains at 22.7 newtons and θ at 1.12 radians, find the coefficient of friction.   [(a) 30.4 N (b) 0.644]

67 The instantaneous current i at time t is given by:

$$i = 10e^{-t/CR}$$

when a capacitor is being charged. The capacitance C is 7×10^{-6} farads and the resistance R

is 0.3×10^6 ohms. Determine:

(a) the instantaneous current when t is 2.5 seconds, and

(b) the time for the instantaneous current to fall to 5 amperes

Sketch a curve of current against time from $t = 0$ to $t = 6$ seconds.
$$[\text{(a) } 3.04 \text{ A (b) } 1.46 \text{ s}]$$

68 The amount of product x (in mol/cm^3) found in a chemical reaction starting with 2.5 mol/cm^3 of reactant is given by $x = 2.5(1 - e^{-4t})$ where t is the time, in minutes, to form product x. Plot a graph at 30 second intervals up to 2.5 minutes and determine x after 1 minute.
$$[2.45 \text{ mol/cm}^3]$$

69 The current i flowing in a capacitor at time t is given by:

$$i = 12.5(1 - e^{-t/CR})$$

where resistance R is 30 kilohms and the capacitance C is 20 micro-farads. Determine

(a) the current flowing after 0.5 seconds, and
(b) the time for the current to reach 10 amperes
$$[\text{(a) } 7.07 \text{ A (b) } 0.966 \text{ s}]$$

70 The amount A after n years of a sum invested P is given by the compound interest law: $A = Pe^{rn/100}$ when the per unit interest rate r is added continuously. Determine, correct to the nearest pound, the amount after 8 years for a sum of £1500 invested if the interest rate is 6% per annum.   [£2424]

9

Areas and volumes

9.1 Mensuration

Mensuration is a branch of mathematics concerned with the determination of lengths, areas and volumes.

9.2 Properties of quadrilaterals

Polygon

A **polygon** is a closed plane figure bounded by straight lines. A polygon which has:

(i)   3 sides is called a **triangle**
(ii) 4 sides is called a **quadrilateral**
(iii) 5 sides is called a **pentagon**
(iv) 6 sides is called a **hexagon**
(v)   7 sides is called a **heptagon**
(vi) 8 sides is called an **octagon**

There are five types of **quadrilateral**, these being

(i)   rectangle
(ii) square
(iii) parallelogram
(iv) rhombus
(v)   trapezium

(The properties of these are given below.)
   If the opposite corners of any quadrilateral are joined by a straight line, two triangles are produced. Since the sum of the angles of a triangle is 180°, the sum of the angles of a quadrilateral is 360°.
In a **rectangle**, shown in Fig. 9.1:

(i)   all four angles are right angles
(ii) opposite sides are parallel and equal in length
(iii) diagonals *AC* and *BD* are equal in length and bisect one another

In a **square**, shown in Fig. 9.2:

(i)   all four angles are right angles
(ii) opposite sides are parallel
(iii) all four sides are equal in length

(iv) diagonals *PR* and *QS* are equal in length and bisect one another at right angles

Figure 9.1

Figure 9.2

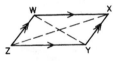

Figure 9.3

In a **parallelogram**, shown in Fig. 9.3:

(i)   opposite angles are equal
(ii) opposite sides are parallel and equal in length
(iii) diagonals *WY* and *XZ* bisect one another

In a **rhombus**, shown in Fig. 9.4:

(i)   opposite angles are equal
(ii) opposite angles are bisected by a diagonal
(iii) opposite sides are parallel
(iv) all four sides are equal in length
(v)   diagonals *AC* and *BD* bisect one another at right angles

In a **trapezium**, shown in Fig. 9.5:

(i)   only one pair of sides is parallel

Figure 9.4   **Figure 9.5**

9.3   Areas of plane figures

Table 9.1

(i) Square	$\text{Area} = x^2$
(ii) Rectangle	$\text{Area} = l \times b$
(ii) Parallelogram	$\text{Area} = b \times h$
(iv) Triangle	$\text{Area} = \frac{1}{2} \times b \times h$
(v) Trapezium	$\text{Area} = \frac{1}{2}(a+b)h$
(vi) Circle	$\text{Area} = \pi r^2$ or $\frac{\pi d^2}{4}$
(vii) Semicircle	$\text{Area} = \frac{1}{2}\pi r^2$ or $\frac{\pi d^2}{8}$
(viii) Sector of a circle	$\text{Area} = \frac{\theta°}{360°}(\pi r^2)$ or $\frac{1}{2}r^2\theta$ (θ in rads)

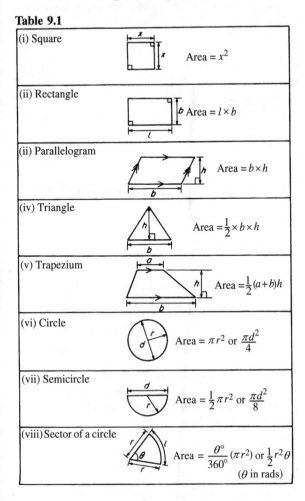

Problem 1.   State the types of quadrilateral shown in Fig. 9.6 and determine the angles marked a to l

---

(i)   **ABCD is a square**
The diagonals of a square bisect each of the right angles, hence

$$a = \frac{90°}{2} = \mathbf{45°}$$

(ii)   **EFGH is a rectangle**
In triangle FGH, $40° + 90° + b = 180°$ (angles in a triangle add up to $180°$) from which, $b = \mathbf{50°}$. Also $c = \mathbf{40°}$ (alternate angles between parallel lines EF and HG).
(Alternatively, b and c are complementary, i.e. add up to $90°$) $d = 90° + c$ (external angle

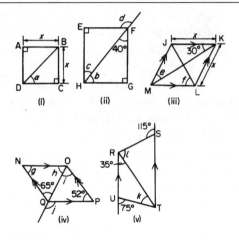

Figure 9.6

of a triangle equals the sum of the interior opposite angles), hence

$$d = 90° + 40° = \mathbf{130°}$$

(iii)   **JKLM is a rhombus**
The diagonals of a rhombus bisect the interior angles and opposite internal angles are equal. Thus $\angle JKM = \angle MKL = \angle JMK = \angle LMK = 30°$, hence

$$e = \mathbf{30°}$$

In triangle KLM, $30° + \angle KLM + 30° = 180°$ (angles in a triangle add up to $180°$), hence $\angle KLM = 120°$.
The diagonal JL bisects $\angle KLM$, hence

$$f = \frac{120°}{2} = \mathbf{60°}$$

(iv)   **NOPQ is a parallelogram**
$g = \mathbf{52°}$ (since opposite interior angles of a parallelogram are equal). In triangle NOQ, $g + h + 65° = 180°$ (angles in a triangle add up to $180°$), from which

$$h = 180° - 65° - 52° = \mathbf{63°}$$

$i = \mathbf{65°}$ (alternate angles between parallel lines NQ and OP)
$j = 52° + i = 52° + 65° = \mathbf{117°}$ (external angle of a triangle equals the sum of the interior opposite angles)

(v)   **RSTU is a trapezium**
$35° + k = 75°$ (external angle of a triangle equals the sum of the interior opposite angles), hence

$$k = \mathbf{40°}$$

$\angle STR = 35°$ (alternate angles between parallel lines RU and ST).

$l + 35° = 115°$ (external angle of a triangle equals the sum of the interior opposite angles), hence

$$l = 115° - 35° = \mathbf{80°}$$

Problem 2. A rectangular tray is 820 mm long and 400 mm wide. Find its area in (a) mm², (b) cm², (c) m²

(a)   Area = length × width = 820 × 400
    = **328 000 mm²**

(b)   1 cm² = 100 mm². Hence

$$328\,000 \text{ mm}^2 = \frac{328\,000}{100} \text{ cm}^2 = \mathbf{3280 \text{ cm}^2}$$

(c)   1 m² = 10 000 cm². Hence

$$3280 \text{ cm}^2 = \frac{3280}{10\,000} \text{ m}^2 = \mathbf{0.3280 \text{ m}^2}$$

Problem 3. Find (a) the cross-sectional area of the girder shown in Fig. 9.7(a) and (b) the area of the path shown in Fig. 9.7(b)

(a)   The girder may be divided into three separate rectangles as shown.
    Area of rectangle $A = 50 \times 5 = 250$ mm²
    Area of rectangle $B = (75 - 8 - 5) \times 6$
    $$= 62 \times 6 = 372 \text{ mm}^2$$
    Area of rectangle $C = 70 \times 8 = 560$ mm²
    Total area of girder = 250 + 372 + 560
    $$= \mathbf{1182 \text{ mm}^2} \text{ or}$$
    $$\mathbf{11.82 \text{ cm}^2}$$

(b)   Area of path = area of large rectangle − area of small rectangle = $(25 \times 20) - (21 \times 16)$
    $= 500 - 336 = \mathbf{164 \text{ m}^2}$

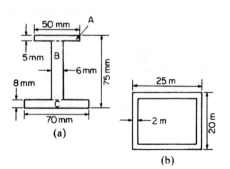

Figure 9.7

Problem 4. Find the area of the parallelogram shown in Fig. 9.8 (dimensions are in mm)

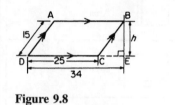

Figure 9.8

Area of parallelogram = base × perpendicular height. The perpendicular height h is found using Pythagoras' theorem.

$$BC^2 = CE^2 + h^2$$

i.e. $15^2 = (34 - 25)^2 + h^2$

$$h^2 = 15^2 - 9^2 = 225 - 81 = 144$$

Hence, $h = \sqrt{144} = 12$ mm

$(-12$ can be neglected)

Hence, area of $ABCD = 25 \times 12 = \mathbf{300 \text{ mm}^2}$.

Problem 5. Figure 9.9 shows the gable end of a building. Determine the area of brickwork in the gable end

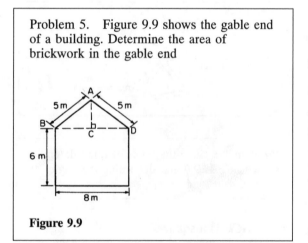

Figure 9.9

The shape is that of a rectangle and a triangle.
Area of rectangle = $6 \times 8 = 48$ m²

$$\text{Area of triangle} = \frac{1}{2} \times \text{base} \times \text{height}$$

$CD = 4$ m, $AD = 5$ m, hence $AC = 3$ m (since it is a 3, 4, 5 triangle).

Hence, area of triangle $ABD = \frac{1}{2} \times 8 \times 3 = 12$ m².

Total area of brickwork = $48 + 12 = \mathbf{60 \text{ m}^2}$.

Problem 6.   Determine the area of the shape shown in Fig. 9.10

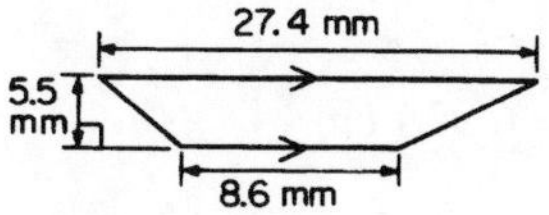

27.4 mm

5.5 mm

8.6 mm

Figure 9.10

The shape shown is a trapezium.

Area of trapezium $= \dfrac{1}{2}$ (sum of parallel sides)

$\times$ (perpendicular distance between them)

$$= \dfrac{1}{2}(27.4 + 8.6)(5.5)$$

$$= \dfrac{1}{2} \times 36 \times 5.5 = \mathbf{99 \ mm^2}$$

Problem 7.   Find the areas of the circles having (a) a radius of 5 cm, (b) a diameter of 15 mm, (c) a circumference of 70 mm

Area of a circle $= \pi r^2$ or $\pi d^2/4$

(a)   Area $= \pi r^2 = \pi(5)^2 = 25\pi = \mathbf{78.54 \ cm^2}$

(b)   Area $= \pi d^2/4 = \pi(15)^2/4 = 225\pi/4$
$= \mathbf{176.7 \ mm^2}$

(c)   Circumference, $c = 2\pi r$, hence

$$r = \dfrac{c}{2\pi} = \dfrac{70}{2\pi} = \dfrac{35}{\pi} \ mm$$

Area of circle $= \pi r^2 = \pi \left(\dfrac{35}{\pi}\right)^2 = \dfrac{35^2}{\pi}$

$$= \mathbf{389.9 \ mm^2} \text{ or } \mathbf{3.899 \ cm^2}$$

Problem 8.   Calculate the areas of the following sectors of circles:

(a)   having radius 6 cm with angle subtended at centre 50°
(b)   having diameter 80 mm with angle subtended at centre 107°42′
(c)   having radius 8 cm with angle subtended at centre 1.15 radians

Area of sector of a circle $= \dfrac{\theta^2}{360}(\pi r^2)$ or

$\dfrac{1}{2}r^2\theta$ (θ in radians)

(a)   Area of sector $= \dfrac{50}{360}(\pi 6^2) = \dfrac{50 \times \pi \times 36}{360}$

$$= 5\pi = \mathbf{15.71 \ cm^2}$$

(b)   If diameter $= 80$ mm, then radius, $r = 40$ mm, and area of sector

$$= \dfrac{107°42'}{360}(\pi 40^2) = \dfrac{107\frac{42}{60}}{360}(\pi 40^2)$$

$$= \dfrac{107.7}{360}(\pi 40^2)$$

$$= \mathbf{1504 \ mm^2} \text{ or } \mathbf{15.04 \ cm^2}$$

(c)   Area of sector $= \dfrac{1}{2}r^2\theta = \dfrac{1}{2} \times 8^2 \times 1.15$

$$= \mathbf{36.8 \ cm^2}$$

Problem 9.   A hollow shaft has an outside diameter of 5.45 cm and an inside diameter of 2.25 cm. Calculate the cross-sectional area of the shaft

The cross-sectional area of the shaft is shown by the shaded part in Fig. 9.11 (often called an **annulus**).

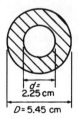

$d = 2.25$ cm

$D = 5.45$ cm

Figure 9.11

Area of shaded part $=$ area of large circle $-$ area of small circle

$$= \dfrac{\pi D^2}{4} - \dfrac{\pi d^2}{4} = \dfrac{\pi}{4}(D^2 - d^2)$$

$$= \dfrac{\pi}{4}(5.45^2 - 2.25^2)$$

$$= \mathbf{19.35 \ cm^2}$$

Problem 10.   Calculate the area of a regular octagon, if each side is 5 cm and the width across the flats is 12 cm

An octagon is an 8-sided polygon. If radii are drawn from the centre of the polygon to the vertices then 8 equal triangles are produced (see Fig. 9.12).

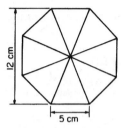

Figure 9.12

$$\text{Area of one triangle} = \frac{1}{2} \times \text{base} \times \text{height}$$

$$= \frac{1}{2} \times 5 \times \frac{12}{2} = 15 \text{ cm}^2$$

$$\text{Area of octagon} = 8 \times 15 = \textbf{120 cm}^2$$

Problem 11.   Determine the area of a regular hexagon which has sides 8 cm long

A hexagon is a 6-sided polygon which may be divided into 6 equal triangles as shown in Fig. 9.13. The angle subtended at the centre of each triangle is $360°/6 = 60°$. The other two angles in the triangle add up to $120°$ and are equal to each other. Hence each of the triangles is equilateral with each angle $60°$ and each side 8 cm.

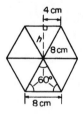

Figure 9.13

$$\text{Area of one triangle} = \frac{1}{2} \times \text{base} \times \text{height} = \frac{1}{2} \times 8 \times h$$

h is calculated using Pythagoras' theorem:

$$8^2 = h^2 + 4^2$$

from which

$$h = \sqrt{(8^2 - 4^2)} = 6.928 \text{ cm}$$

$$\text{Hence area of one triangle} = \frac{1}{2} \times 8 \times 6.928$$

$$= 27.71 \text{ cm}^2$$

$$\text{Area of hexagon} = 6 \times 27.71 = \textbf{166.3 cm}^2$$

Problem 12.   Figure 9.14 shows a plan of a floor of a building which is to be carpeted. Calculate the area of the floor in square metres. Calculate the cost, correct to the nearest pound, of carpeting the floor with carpet costing £16.80 per m², assuming 30% extra carpet is required due to wastage in fitting

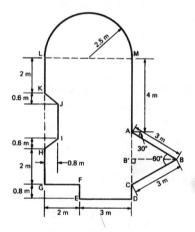

Figure 9.14

$$\text{Area of floor plan} = \text{area of triangle } ABC$$
$$+ \text{ area of semicircle} + \text{area of rectangle } CGLM$$
$$+ \text{ area of rectangle } CDEF$$
$$- \text{ area of trapezium } HIJK$$

Triangle ABC is equilateral since $AB = BC = 3$ m and hence angle $B'CB = 60°$. $\sin B'CB = BB'/3$, i.e.

$$BB' = 3\sin 60° = 2.598 \text{ m}$$

Area of triangle $ABC = \frac{1}{2}(AC)(BB')$

$$= \frac{1}{2}(3)(2.598)$$

$$= 3.897 \text{ m}^2$$

Area of semicircle $= \frac{1}{2}\pi r^2 = \frac{1}{2}\pi(2.5)^2$

$$= 9.817 \text{ m}^2$$

Area of $CGLM = 5 \times 7 = 35 \text{ m}^2$

Area of $CDEF = 0.8 \times 3 = 2.4 \text{ m}^2$

Area of $HIJK = \frac{1}{2}(KH + IJ)(0.8)$

Since $MC = 7$ m then $LG = 7$ m, hence
$JI = 7 - 5.2 = 1.8$ m

Hence area of $HIJK = \frac{1}{2}(3 + 1.8)(0.8)$

$$= 1.92 \text{ m}^2$$

Total floor area $= 3.897 + 9.817 + 35 + 2.4 - 1.92$

$$= 49.194 \text{ m}^2$$

To allow for 30% wastage, amount of carpet required
$= 1.3 \times 49.194 = 63.95 \text{ m}^2$.
Cost of carpet at £16.80 per m^2 = $63.95 \times 16.80 =$
£1074, correct to the nearest pound.

Further problems on areas of plane figures may be
found in Section 9.9, Problems 1 to 15, pages 104
and 105.

9.4   Volumes and surface areas of regular solids

For a summary of volumes and surface areas of
regular solids, see Table 9.2.

Problem 13.   A water tank is the shape of a
rectangular prism having length 2 m, breadth
75 cm and height 50 cm. Determine the
capacity of the tank in (a) m^3 (b) cm^3
(c) litres

Volume of rectangular prism $= l \times b \times h$ (see
Table 9.2).

(a)   Volume of tank $= 2 \times 0.75 \times 0.5 = \textbf{0.75 m}^3$

(b)   1 m^3 = 10^6 cm^3.

Table 9.2

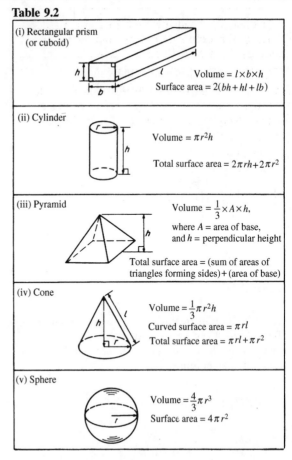

(i) Rectangular prism (or cuboid)

Volume $= l \times b \times h$
Surface area $= 2(bh + hl + lb)$

(ii) Cylinder

Volume $= \pi r^2 h$
Total surface area $= 2\pi rh + 2\pi r^2$

(iii) Pyramid

Volume $= \frac{1}{3} \times A \times h$,
where A = area of base,
and h = perpendicular height
Total surface area = (sum of areas of triangles forming sides) + (area of base)

(iv) Cone

Volume $= \frac{1}{3}\pi r^2 h$
Curved surface area $= \pi rl$
Total surface area $= \pi rl + \pi r^2$

(v) Sphere

Volume $= \frac{4}{3}\pi r^3$
Surface area $= 4\pi r^2$

Hence 0.75 m^3 = 0.75×10^6 cm^3

$$= \textbf{750 000 cm}^3$$

(c)   1 litre = 1000 cm^3

Hence 750 000 cm^3 = $\dfrac{750\,000}{1000}$ litres

$$= \textbf{750 litres}$$

Problem 14.   Find the volume and total
surface area of a cylinder of length 15 cm
and diameter 8 cm

Volume of cylinder $= \pi r^2 h$ (see Table 9.2).
Since diameter = 8 cm, then radius $r = 4$ cm.
Hence

$$\text{volume} = \pi \times 4^2 \times 15 = \textbf{754 cm}^3$$

Total surface area (i.e. including the two ends)

$$= 2\pi rh + 2\pi r^2$$

$$= (2 \times \pi \times 4 \times 15) + (2 \times \pi \times 4^2)$$

$$= \textbf{477.5 cm}^2$$

Problem 15. Determine the volume (in cm^3) of the shape shown in Fig. 9.15

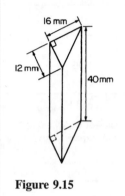
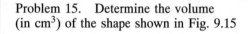

Figure 9.15

The solid shown in Fig. 9.15 is a triangular prism. The volume V of any prism is given by: $V = Ah$, where A is the cross-sectional area and h is the perpendicular height. Hence

$$\text{volume} = \left(\frac{1}{2} \times 16 \times 12 \right) \times 40$$

$$= 3840 \text{ mm}^3$$

$$= \textbf{3.840 cm}^3$$

(since 1 cm^3 = 1000 mm^3)

Problem 16. Calculate the volume and total surface area of the solid prism shown in Fig. 9.16

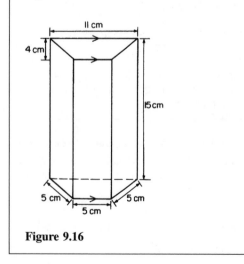

Figure 9.16

The solid shown in Fig. 9.16 is a trapezoidal prism.

Volume = cross-sectional area × height

$$= \left[\frac{1}{2}(11 + 5)4 \right] \times 15 = 32 \times 15$$

$$= \textbf{480 cm}^3$$

Surface area = sum of two trapeziums

$$+ 4 \text{ rectangles}$$

$$= (2 \times 32) + (5 \times 15) + (11 \times 15) + 2(5 \times 15)$$

$$= 64 + 75 + 165 + 150 = \textbf{454 cm}^2$$

Problem 17. Determine the volume and the total surface area of the square pyramid shown in Fig. 9.17 if its perpendicular height is 12 cm

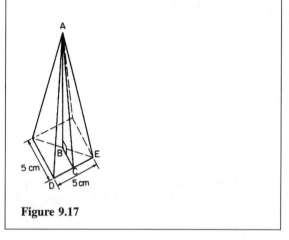

Figure 9.17

Volume of pyramid

$$= \frac{1}{3} \text{ (area of base)} \times \text{perpendicular height}$$

$$= \frac{1}{3}(5 \times 5) \times 12 = \textbf{100 cm}^3$$

The total surface area consists of a square base and 4 equal triangles.

Area of triangle ADE

$$= \frac{1}{2} \times \text{base} \times \text{perpendicular height}$$

$$= \frac{1}{2} \times 5 \times AC$$

The length AC may be calculated using Pythagoras' theorem on triangle ABC, where $AB = 12$ cm, $BC = \frac{1}{2} \times 5 = 2.5$ cm

$$AC = \sqrt{(AB^2 + BC^2)} = \sqrt{(12^2 + 2.5^2)}$$
$$= 12.26 \text{ cm}$$

Hence area of triangle $ADE = \frac{1}{2} \times 5 \times 12.26$

$$= 30.65 \text{ cm}^2$$

Total surface area of pyramid $= (5 \times 5) + 4(30.65)$

$$= \textbf{147.6 cm}^2$$

Problem 18. Determine the volume and total surface area of a cone of radius 5 cm and perpendicular height 12 cm

---

The cone is shown in Fig. 9.18.

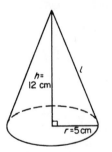

Figure 9.18

Volume of cone $= \frac{1}{3}\pi r^2 h = \frac{1}{3} \times \pi \times 5^2 \times 12$

$$= \textbf{314.2 cm}^3$$

Total surface area

$$= \text{curved surface area} + \text{area of base}$$
$$= \pi r l + \pi r^2$$

From Fig. 9.18, slant height l may be calculated using Pythagoras' theorem

$$l = \sqrt{(12^2 + 5^2)} = 13 \text{ cm}$$

Hence total surface area

$$= (\pi \times 5 \times 13) + (\pi \times 5^2)$$
$$= \textbf{282.7 cm}^2$$

Problem 19. Find the volume and surface area of a sphere of diameter 8 cm

---

Since diameter $= 8$ cm, then radius, $r = 4$ cm.

Volume of sphere $= \frac{4}{3}\pi r^3 = \frac{4}{3} \times \pi \times 4^3$

$$= \textbf{268.1 cm}^3$$

Surface area of sphere $= 4\pi r^2 = 4 \times \pi \times 4^2$

$$= \textbf{201.1 cm}^2$$

Problem 20. A wooden section is shown in Fig. 9.19. Find (a) its volume (in m³), and (b) its total surface area

---

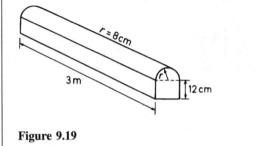

Figure 9.19

The section of wood is a prism whose end comprises a rectangle and a semicircle. Since the radius of the semicircle is 8 cm, the diameter is 16 cm. Hence the rectangle has dimensions 12 cm by 16 cm.

Area of end $= (12 \times 16) + \frac{1}{2}\pi 8^2 = 292.5 \text{ cm}^2$

Volume of wooden section

$$= \text{area of end} \times \text{perpendicular height}$$
$$= 292.5 \times 300 = 87\,750 \text{ cm}^3$$
$$= \frac{87\,750 \text{ m}^3}{10^6}$$
$$= \textbf{0.08775 m}^3$$

The total surface area comprises the two ends (each of area 292.5 cm²), three rectangles and a curved surface (which is half a cylinder), hence total surface area

$$= (2 \times 292.5) + 2(12 \times 300) + (16 \times 300)$$
$$+ \frac{1}{2}(2\pi \times 8 \times 300)$$

$$= 585 + 7200 + 4800 + 2400\pi$$

$$= \mathbf{20\,125\ cm^2}\ \text{or}\ \mathbf{2.0125\ m^2}$$

Problem 21. A pyramid has a rectangular base 3.60 cm by 5.40 cm. Determine the volume and total surface area of the pyramid if each of its sloping edges is 15.0 cm

---

The pyramid is shown in Fig. 9.20. To calculate the volume of the pyramid the perpendicular height EF is required. Diagonal BD is calculated using Pythagoras' theorem:
i.e. $BD = \sqrt{\{(3.60)^2 + (5.40)^2\}} = 6.490$ cm

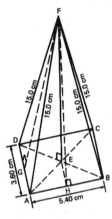

Figure 9.20

Hence $EB = \dfrac{1}{2}BD = \dfrac{6.490}{2} = 3.245$ cm

Using Pythagoras' theorem on triangle BEF gives $BF^2 = EB^2 + EF^2$, from which, $EF = \sqrt{(BF^2 - EB^2)}$
$= \sqrt{\{(15.0)^2 - (3.245)^2\}} = 14.64$ cm.

Volume of pyramid

$$= \frac{1}{3}(\text{area of base})(\text{perpendicular height})$$

$$= \frac{1}{3}(3.60 \times 5.40)(14.64) = \mathbf{94.87\ cm^3}$$

Area of triangle ADF (which equals triangle BCF)
$= \frac{1}{2}(AD)(FG)$, where G is the midpoint of AD. Using Pythagoras' theorem on triangle FGA gives

$$FG = \sqrt{\{(15.0)^2 - (1.80)^2\}} = 14.89\ \text{cm}$$

Hence area of triangle $ADF = \frac{1}{2}(3.60)(14.89) = 26.80$ cm^2.
Similarly, if H is the midpoint of AB, then $FH = \sqrt{\{(15.0)^2 - (2.70)^2\}} = 14.75$ cm, hence area of triangle ABF (which equals triangle CDF)

$$= \frac{1}{2}(5.40)(14.75) = 39.83\ \text{cm}^2$$

Total surface area of pyramid

$$= 2(26.80) + 2(39.83) + (3.60)(5.40)$$

$$= 53.60 + 79.66 + 19.44$$

$$= \mathbf{152.7\ cm^2}$$

Problem 22. Calculate the volume and total surface area of a hemisphere of diameter 5.0 cm

---

Volume of hemisphere $= \dfrac{1}{2}(\text{volume of sphere})$

$$= \frac{2}{3}\pi r^3 = \frac{2}{3}\pi\left(\frac{5.0}{2}\right)^3$$

$$= \mathbf{32.7\ cm^3}$$

Total surface area

$$= \text{curved surface area} + \text{area of circle}$$

$$= \frac{1}{2}(\text{surface area of sphere}) + \pi r^2$$

$$= 2\pi r^2 + \pi r^2 = 3\pi r^2$$

$$= 3\pi\left(\frac{5.0}{2}\right)^2 = \mathbf{58.9\ cm^2}$$

Problem 23. A rectangular piece of metal having dimensions 4 cm by 3 cm by 12 cm is melted down and recast into a pyramid having a rectangular base measuring 2.5 cm by 5 cm. Calculate the perpendicular height of the pyramid

---

Volume of rectangular prism of metal

$$= 4 \times 3 \times 12 = 144\ \text{cm}^3$$

Volume of pyramid

$$= \frac{1}{3}(\text{area of base})(\text{perpendicular height})$$

Assuming no waste of metal,

$$144 = \frac{1}{3}(2.5 \times 5)(\text{height})$$

i.e. perpendicular height $= \dfrac{144 \times 3}{2.5 \times 5} = \mathbf{34.56\ cm}$

Problem 24. A rivet consists of a cylindrical head, of diameter 1 cm and depth 2 mm, and a shaft of diameter 2 mm and length 1.5 cm. Determine the volume of metal in 2000 such rivets

---

Radius of cylindrical head $= \frac{1}{2}$ cm $= 0.5$ cm and height of cylindrical head $= 2$ mm $= 0.2$ cm.
Hence, volume of cylindrical head $= \pi r^2 h = \pi(0.5)^2(0.2) = 0.1571\ \text{cm}^3$
Volume of cylindrical shaft $= \pi r^2 h$

$$= \pi \left(\frac{1}{10}\right)^2 (1.5)$$

$$= 0.0471\ \text{cm}^3$$

Total volume of 1 rivet $= 0.1571 + 0.0471 = 0.2042\ \text{cm}^3$.
Volume of metal in 2000 such rivets $= 2000 \times 0.2042 = 408.4\ \text{cm}^3$.

Problem 25. A solid metal cylinder of radius 6 cm and height 15 cm is melted down and recast into a shape comprising a hemisphere surmounted by a cone. Assuming that 8% of the metal is wasted in the process, determine the height of the conical portion, if its diameter is to be 12 cm

---

Volume of cylinder $= \pi r^2 h = \pi \times 6^2 \times 15$

$$= 540\pi\ \text{cm}^3$$

If 8% of metal is lost then 92% of 540π gives the volume of the new shape (shown in Fig. 9.21). Hence the volume of (hemisphere + cone) $= 0.92 \times 540\pi\ \text{cm}^3$, i.e.

$$\frac{1}{2}\left(\frac{4}{3}\pi r^3\right) + \frac{1}{3}\pi r^2 h = 0.92 \times 540\pi$$

Dividing throughout by π gives:

$$\frac{2}{3}r^3 + \frac{1}{3}r^2 h = 0.92 \times 540$$

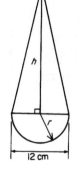

Figure 9.21

Since the diameter of the new shape is to be 12 cm, then radius $r = 6$ cm, hence $\frac{2}{3}(6)^3 + \frac{1}{3}(6)^2 h = 0.92 \times 540$

$$144 + 12h = 496.8$$

i.e. height of conical portion,

$$h = \frac{496.8 - 144}{12} = \mathbf{29.4\ cm}$$

Problem 26. A block of copper having a mass of 50 kg is drawn out to make 500 m of wire of uniform cross-section. Given that the density of copper is 8.91 g/cm³, calculate (a) the volume of copper, (b) the cross-sectional area of the wire, and (c) the diameter of the cross-section of the wire

---

(a) A density of 8.91 g/cm³ means that 8.91 g of copper has a volume of 1 cm³, or 1 g of copper has a volume of $(1/8.91)$ cm³.
Hence 50 kg, i.e. 50 000 g, has a volume
$$\frac{50\,000}{8.91}\ \text{cm}^3 = \mathbf{5612\ cm^3}$$

(b) Volume of wire = area of circular cross-section × length of wire.
Hence 5612 cm³ = area × (500 × 100 cm),
from which, area $= \dfrac{5612}{500 \times 100}\ \text{cm}^2$
$$= \mathbf{0.1122\ cm^2}$$

(c) Area of circle $= \pi r^2$ or $\pi d^2/4$, hence $0.1122 = \dfrac{\pi d^2}{4}$, from which $d = \sqrt{\left(\dfrac{4 \times 0.1122}{\pi}\right)} = 0.3780\ \text{cm}$

i.e. **diameter of cross-section is 3.780 mm**

Problem 27. A boiler consists of a cylindrical section of length 8 m and diameter 6 m, on one end of which is surmounted a hemispherical section of diameter 6 m, and on the other end a conical section of height 4 m. Calculate the volume of the boiler and the total surface area

The boiler is shown in Fig. 9.22.

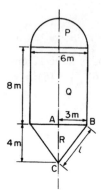

Figure 9.22

Volume of hemisphere,

$$P = \frac{2}{3}\pi r^3 = \frac{2}{3} \times \pi \times 3^3 = 18\pi \ \text{m}^3$$

Volume of cylinder,

$$Q = \pi r^2 h = \pi \times 3^2 \times 8 = 72\pi \ \text{m}^3$$

Volume of cone,

$$R = \frac{1}{3}\pi r^2 h = \frac{1}{3} \times \pi \times 3^2 \times 4 = 12\pi \ \text{m}^3$$

Total volume of boiler $= 18\pi + 72\pi + 12\pi$

$$= 102\pi = \mathbf{320.4 \ m^3}$$

Surface area of hemisphere,

$$P = \frac{1}{2}(4\pi r^2) = 2 \times \pi \times 3^2 = 18\pi \ \text{m}^2$$

Curved surface area of cylinder,

$$Q = 2\pi r h = 2 \times \pi \times 3 \times 8 = 48\pi \ \text{m}^2$$

The slant height of the cone, l, is obtained by Pythagoras' theorem on triangle ABC

$$l = \sqrt{(4^2 + 3^2)} = 5$$

Curved surface area of cone,

$$R = \pi r l = \pi \times 3 \times 5 = 15\pi \ \text{m}^2$$

Total surface area of boiler $= 18\pi + 48\pi + 15\pi = 81\pi = \mathbf{254.5 \ m^2}$.

Further problems on volumes and surface areas of regular solids may be found in Section 9.9, Problems 16 to 37, pages 105 to 107.

9.5 Volumes and surface areas of frusta of pyramids and cones

The **frustum** of a pyramid or cone is the portion remaining when a part containing the vertex is cut off by a plane parallel to the base. The **volume of a frustum of a pyramid or cone** is given by the volume of the whole pyramid or cone minus the volume of the small pyramid or cone cut off.

The **surface area of the sides of a frustum of a pyramid or cone** is given by the surface area of the whole pyramid or cone minus the surface area of the small pyramid or cone cut off. This gives the lateral surface area of the frustum. If the total surface area of the frustum is required then the surface area of the two parallel ends are added to the lateral surface area.

Figure 9.23

There is an alternative method for finding the volume and surface area of a **frustum of a cone**. With reference to Fig. 9.23:

$$\mathbf{Volume} = \frac{1}{3}\pi h (R^2 + Rr + r^2)$$

Curved surface area $= \pi l (R + r)$
Total surface area $= \pi l (R + r) + \pi r^2 + \pi R^2$

Problem 28. Determine the volume of a frustum of a cone if the diameter of the ends are 6.0 cm and 4.0 cm and its perpendicular height is 3.6 cm

Method 1
A section through the vertex of a complete cone is shown in Fig. 9.24.

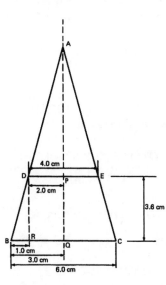

Figure 9.24

Using similar triangles $\dfrac{AP}{DP} = \dfrac{DR}{BR}$

Hence $\dfrac{AP}{2.0} = \dfrac{3.6}{1.0}$, from which $AP = \dfrac{(2.0)(3.6)}{1.0} = 7.2$ cm.

The height of the large cone $= 3.6 + 7.2 = 10.8$ cm.

Volume of frustum of cone

$\qquad$ = volume of large cone

$\qquad\qquad$ − volume of small cone cut off

$$= \frac{1}{3}\pi(3.0)^2(10.8) - \frac{1}{3}\pi(2.0)^2(7.2)$$

$$= 101.79 - 30.16 = \textbf{71.6 cm}^3$$

Method 2
From above, volume of the frustum of a cone

$$= \frac{1}{3}\pi h(R^2 + Rr + r^2)$$

where $R = 3.0$ cm, $r = 2.0$ cm and $h = 3.6$ cm.

Hence volume of frustum

$$= \frac{1}{3}\pi(3.6)[(3.0)^2 + (3.0)(2.0) + (2.0)^2]$$

$$= \frac{1}{3}\pi(3.6)(19.0) = \textbf{71.6 cm}^3$$

Problem 29. Find the total surface area of the frustum of the cone in Problem 28

Method 1
Curved surface area of frustum

$\qquad$ = curved surface area of large cone

$\qquad\qquad$ − curved surface area of small cone cut off

From Fig. 9.24, using Pythagoras' theorem:

$$AB^2 = AQ^2 + BQ^2, \text{ from which}$$

$$AB = \sqrt{\{(10.8)^2 + (3.0)^2\}} = 11.21 \text{ cm}$$

and

$$AD^2 = AP^2 + DP^2, \text{ from which}$$

$$AD = \sqrt{\{(7.2)^2 + (2.0)^2\}} = 7.47 \text{ cm}$$

Curved surface area of large cone

$$= \pi r l = \pi(BQ)(AB) = \pi(3.0)(11.21)$$

$$= 105.7 \text{ cm}^2$$

and curved surface area of small cone

$$= \pi(DP)(AD) = \pi(2.0)(7.47) = 46.94 \text{ cm}^2$$

Hence, curved surface area of frustum

$$= 105.7 - 46.94 = 58.76 \text{ cm}^2$$

Total surface area of frustum

$\qquad$ = curved surface area + area of two

$\qquad\quad$ circular ends

$$= 58.76 + \pi(2.0)^2 + \pi(3.0)^2$$

$$= 58.76 + 12.57 + 28.27 = \textbf{99.6 cm}^2$$

Method 2
From page 96, total surface area of frustum $= \pi l(R + r) + \pi r^2 + \pi R^2$, where $l = BD = 11.21 - 7.47 = 3.74$ cm, $R = 3.0$ cm and $r = 2.0$ cm.

Hence total surface area of frustum

$$= \pi(3.74)(3.0 + 2.0) + \pi(2.0)^2 + \pi(3.0)^2$$

$$= \textbf{99.6 cm}^2$$

Problem 30. A storage hopper is in the shape of a frustum of a pyramid. Determine its volume if the ends of the frustum are squares of sides 8.0 m and 4.6 m, respectively, and the perpendicular height between its ends is 3.6 m

The frustum is shown shaded in Fig. 9.25(a) as part of a complete pyramid.

A section perpendicular to the base through the vertex is shown in Fig. 9.25(b).

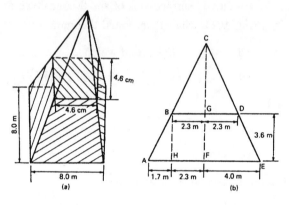

Figure 9.25

By similar triangles: $\dfrac{CG}{BG} = \dfrac{BH}{AH}$

Height $CG = BG\left(\dfrac{BH}{AH}\right) = \dfrac{(2.3)(3.6)}{(1.7)} = 4.87$ m

Height of complete pyramid $= 3.6 + 4.87 = 8.47$ m

Volume of large pyramid $= \dfrac{1}{3}(8.0)^2(8.47)$

$$= 180.7 \text{ m}^3$$

Volume of small pyramid cut off $= \dfrac{1}{3}(4.6)^2(4.87)$

$$= 34.35 \text{ m}^3$$

Hence volume of storage hopper $= 180.7 - 34.35$

$$= \mathbf{146.4 \text{ m}^3}$$

Problem 31. Determine the lateral surface area of the storage hopper in Problem 30

The lateral surface area of the storage hopper consists of four equal trapeziums.

From Fig. 9.26, area of trapezium $PRSU$

$$= \frac{1}{2}(PR + SU)(QT)$$

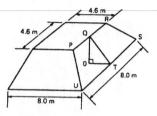

Figure 9.26

$OT = 1.7$ m (same as AH in Fig. 9.25(a)) and $OQ = 3.6$ m. By Pythagoras' theorem,

$$QT = \sqrt{(OQ^2 + OT^2)} = \sqrt{\{(3.6^2 + 1.7)^2\}}$$

$$= 3.98 \text{ m}$$

Area of trapezium $PRSU = \dfrac{1}{2}(4.6 + 8.0)(3.98)$

$$= 25.07 \text{ m}^2$$

Lateral surface area of hopper $= 4(25.07)$

$$= \mathbf{100.3 \text{ m}^2}$$

Problem 32. A lampshade is in the shape of a frustum of a cone. The vertical height of the shade is 25.0 cm and the diameters of the ends are 20.0 cm and 10.0 cm, respectively. Determine the area of the material needed to form the lampshade, correct to 3 significant figures

The curved surface area of a frustum of a cone $= \pi l(R + r)$ from page 96.

Since the diameters of the ends of the frustum are 20.0 cm and 10.0 cm, then from Fig. 9.27,

$$r = 5.0 \text{ cm}, R = 10.0 \text{ cm and}$$

$$l = \sqrt{\{(25.0)^2 + (5.0)^2\}} = 25.50 \text{ cm},$$

from Pythagoras' theorem.

Hence curved surface area $= \pi(25.50)(10.0 + 5.0) = 1201.7$ cm², i.e. the area of material needed to form the lampshade is **1200 cm²**, correct to 3 significant figures.

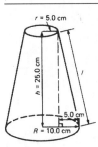

Figure 9.27

Problem 33. A cooling tower is in the form of a cylinder surmounted by a frustum of a cone as shown in Fig. 9.28. Determine the volume of air space in the tower if 40% of the space is used for pipes and other structures

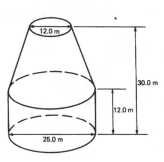

Figure 9.28

Volume of cylindrical portion

$$= \pi r^2 h = \pi \left(\frac{25.0}{2}\right)^2 (12.0) = 5890 \text{ m}^3$$

Volume of frustum of cone $= \frac{1}{3}\pi h(R^2 + Rr + r^2)$, where $h = 30.0 - 12.0 = 18.0$ m, $R = 25.0/2 = 12.5$ m and $r = 12.0/2 = 6.0$ m.
Hence volume of frustum of cone

$$= \frac{1}{3}\pi(18.0)[(12.5)^2 + (12.5)(6.0) + (6.0)^2]$$

$$= 5038 \text{ m}^3$$

Total volume of cooling tower $= 5890 + 5038 = 10\,928$ m^3.
If 40% of space is occupied then volume of air space $= 0.6 \times 10\,928 = \textbf{6557 m}^3$.

Further problems on volumes and surface areas of frustra of pyramids and cones may be found in Section 9.9, Problems 38 to 44, page 107.

9.6 The frustum and zone of a sphere

Volume of sphere $= \frac{4}{3}\pi r^3$
Surface area of sphere $= 4\pi r^2$

A **frustum of a sphere** is the portion contained between two parallel planes. In Fig. 9.29, *PQRS* is a frustum of the sphere. A **zone of a sphere** is the curved surface of a frustum. With reference to Fig. 9.29:

Surface area of a zone of a sphere $= 2\pi r h$

Volume of frustum of sphere

$$= \frac{\pi h}{6}(h^2 + 3r_1^2 + 3r_2^2)$$

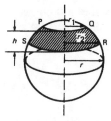

Figure 9.29

Problem 34. Determine the volume of a frustum of a sphere of diameter 49.74 cm if the diameter of the ends of the frustum are 24.0 cm and 40.0 cm, and the height of the frustum is 7.00 cm

---

From above, volume of frustum of a sphere $= (\pi h/6)(h^2 + 3r_1^2 + 3r_2^2)$ where $h = 7.00$ cm, $r_1 = 24.0/2 = 12.0$ cm and $r_2 = 40.0/2 = 20.0$ cm.
Hence volume of frustum

$$= \pi\frac{(7.00)}{6}[(7.00)^2 + 3(12.0)^2 + 3(20.0)^2]$$

$$= \textbf{6161 cm}^3$$

> Problem 35.   Determine for the frustum of
> Problem 34 the curved surface area of the
> frustum

The curved surface area of the frustum = surface
area of zone = $2\pi rh$ (from above), where r = radius
of sphere = $49.74/2 = 24.87$ cm and $h = 7.00$ cm.
Hence, surface area of zone = $2\pi(24.87)(7.00) =$
1094 cm^2.

> Problem 36.   The diameters of the ends of
> the frustum of a sphere are 14.0 cm and
> 26.0 cm, respectively, and the thickness of
> the frustum is 5.0 cm. Determine, correct to
> 3 significant figures. (a) the volume of the
> frustum of the sphere, (b) the radius of the
> sphere and (c) the area of the zone formed

The frustum is shown shaded in the cross-section of
Fig. 9.30.

(a)   Volume of frustum of sphere = $(\pi h/6)(h^2 +
3r_1^2 + 3r_2^2)$ from above, where $h = 5.0$ cm, $r_1 =
14.0/2 = 7.0$ cm and $r_2 = 26.0/2 = 13.0$ cm.

Hence volume of frustum of sphere

$$= \frac{\pi(5.0)}{6}[(5.0)^2 + 3(7.0)^2 + 3(13.0)^2]$$

$$= \frac{\pi(5.0)}{6}[25.0 + 147.0 + 507.0]$$

$$= \textbf{1780 cm}^3, \text{ correct to 3 significant}$$
$$\text{figures}$$

(b)   The radius, r, of the sphere may be calculated
using Fig. 9.30.

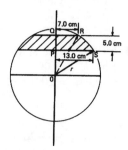

Figure 9.30

Using Pythagoras' theorem:

$$OS^2 = PS^2 + OP^2$$

i.e. $r^2 = (13.0)^2 + OP^2$ (1)

$$OR^2 = QR^2 + OQ^2$$

i.e. $r^2 = (7.0)^2 + OQ^2$

However, $OQ = QP + OP = 5.0 + OP$,
therefore

$$r^2 = (7.0)^2 + (5.0 + OP)^2$$ (2)

Equating equations (1) and (2) gives:

$$(13.0)^2 + OP^2 = (7.0)^2 + (5.0 + OP)^2$$

$$169.0 + OP^2 = 49.0 + 25.0 + 10.0(OP) + OP^2$$

$$169.0 = 74.0 + 10.0(OP)$$

Hence $OP = \dfrac{169.0 - 74.0}{10.0} = 9.50$ cm

Substituting $OP = 9.50$ cm into equation (1)
gives:

$$r^2 = (13.0)^2 + (9.50)^2$$

from which $r = \sqrt{[(13.0)^2 + (9.50)^2]}$
i.e. **radius of sphere, r = 16.1 cm**

(c)   Area of zone of sphere = $2\pi rh$

$$= 2\pi(16.1)(5.0)$$

$$= \textbf{506 cm}^2,$$

correct to 3 significant figures

> Problem 37.   A frustum of a sphere of
> diameter 12.0 cm is formed by two parallel
> planes, one through the diameter and the
> other distance h from the diameter. The
> curved surface area of the frustum is
> required to be $\frac{1}{4}$ of the total surface area of
> the sphere. Determine (a) the volume and
> surface area of the sphere, (b) the thickness h
> of the frustum, (c) the volume of the frustum
> and (d) the volume of the frustum expressed
> as a percentage of the sphere

(a)   Volume of sphere

$$V = \frac{4}{3}\pi r^3 = \frac{4}{3}\pi\left(\frac{12.0}{2}\right)^3 = \textbf{904.8 cm}^3$$

Surface area of sphere $= 4\pi r^2$

$$= 4\pi \left(\frac{12.0}{2}\right)^2$$

$$= \textbf{452.4 cm}^2$$

(b) Curved surface area of frustum

$$= \frac{1}{4} \times \text{surface area of sphere}$$

$$= \frac{1}{4} \times 452.4 = 113.1 \text{ cm}^2$$

From above, $113.1 = 2\pi rh = 2\pi \left(\frac{12.0}{2}\right) h$

Hence thickness of frustum $h = \dfrac{113.1}{2\pi(6.0)}$

$= \textbf{3.0 cm}$

(c) Volume of frustum, $V = \dfrac{\pi h}{6}(h^2 + 3r_1^2 + 3r_2^2)$

where $h = 3.0$ cm, $r_2 = 6.0$ cm and $r_1 = \sqrt{(OQ^2 - OP^2)}$, from Fig. 9.31,

i.e. $r_1 = \sqrt{[(6.0)^2 - (3.0)^2]} = 5.196$ cm

Hence volume of frustum

$$= \frac{\pi(3.0)}{6}[(3.0)^2 + 3(5.196)^2 + 3(6.0)^2]$$

$$= \frac{\pi}{2}[9.0 + 81 + 108.0] = \textbf{311.0 cm}^3$$

(d) $\dfrac{\text{Volume of frustum}}{\text{Volume of sphere}}$

$$= \frac{311.0}{904.8} \times 100\%$$

$$= \textbf{34.37\%}$$

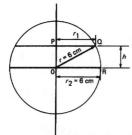

Figure 9.31

Problem 38.   A spherical storage tank is filled with liquid to a depth of 20 cm. If the internal diameter of the vessel is 30 cm, determine the number of litres of liquid in the container (1 litre = 1000 cm^3)

The liquid is represented by the shaded area in the section shown in Fig. 9.32. The volume of liquid comprises a hemisphere and a frustum of thickness 5 cm.

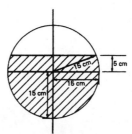

Figure 9.32

Hence volume of liquid $= \dfrac{2}{3}\pi r^3 + \dfrac{\pi h}{6}[h^2 + 3r_1^2 + 3r_2^2]$,

where $r_2 = 30/2 = 15$ cm and $r_1 = \sqrt{[(15)^2 - (5)^2]}$ $= 14.14$ cm.

Volume of liquid

$$= \frac{2}{3}\pi(15)^3 + \frac{\pi(5)}{6}[5^2 + 3(14.14)^2 + 3(15)^2]$$

$$= 7069 + 3403 = 10\,470 \text{ cm}^3$$

Since 1 litre $= 1000$ cm^3, the number of litres of liquid $= 10\,470/1000 = \textbf{10.47 litres}$.

Further problems on frustums and zones of spheres may be found in Section 9.9, Problems 45 to 49, page 107.

9.7   Prismoidal rule for finding volumes

The prismoidal rule applies to a solid of length x divided by only three equidistant plane areas, A_1, A_2 and A_3 as shown in Fig. 9.33 and is merely an extension of Simpson's rule–but for volumes (see Chapter 18).

With reference to Fig. 9.33

$$\text{Volume, } V = \frac{x}{6}[A_1 + 4A_2 + A_3]$$

The prismoidal rule gives precise values of volume for regular solids such as pyramids, cones, spheres and prismoids.

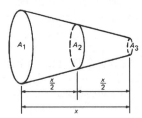

Figure 9.33

Problem 39.   A container is in the shape of a frustum of a cone. Its diameter at the bottom is 18 cm and at the top 30 cm. If the depth is 24 cm determine the capacity of the container, correct to the nearest litre, by the prismoidal rule (1 litre = 1000 cm³)

The container is shown in Fig. 9.34. At the mid-point, i.e. at a distance of 12 cm from one end, the radius r_2 is $(9 + 15)/2 = 12$ cm, since the sloping sides change uniformly.

Volume of container by the prismoidal rule = $(x/6)[A_1+4A_2+A_3]$, from above, where $x = 24$ cm, $A_1 = \pi(15)^2$ cm², $A_2 = \pi(12)^2$ cm² and $A_3 = \pi(9)^2$ cm².

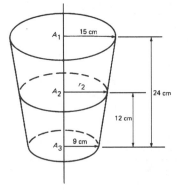

Figure 9.34

Hence volume of container

$$= \frac{24}{6}[\pi(15)^2 + 4\pi(12)^2 + \pi(9)^2]$$

$$= 4[706.86 + 1809.56 + 254.47]$$

$$= 11\,080 \text{ cm}^3 = \frac{11\,080}{1000} \text{ litres}$$

$$= \textbf{11 litres, correct to the nearest litre}$$

(Check: Volume of frustum of cone

$$= \frac{1}{3}\pi h[R^2 + Rr + r^2] \text{ from Section 9.5}$$

$$= \frac{1}{3}\pi(24)[(15)^2 + (15)(9) + (9)^2]$$

$$= 11\,080 \text{ cm}^3 \text{ as shown above})$$

Problem 40.   A frustum of a sphere of radius 13 cm is formed by two parallel planes on opposite sides of the centre, each at a distance of 5 cm from the centre. Determine the volume of the frustum (a) by using the prismoidal rule, and (b) by using the formula for the volume of a frustum of a sphere

---

The frustum of the sphere is shown by the section in Fig. 9.35. Radius $r_1 = r_2 = PQ = \sqrt{(13^2 - 5^2)} = 12$ cm, by Pythagoras' theorem.

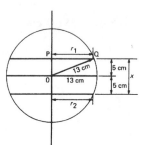

Figure 9.35

(a)   Using the prismodial rule,

Volume of frustum, V

$$= \frac{x}{6}[A_1 + 4A_2 + A_3]$$

$$= \frac{10}{6}[\pi(12)^2 + 4\pi(13)^2 + \pi(12)^2]$$

$$= \frac{10\pi}{6}[144 + 676 + 144]$$

$$= \textbf{5047 cm}^3$$

(b) Using the formula for the volume of a frustum of a sphere:

$$\text{Volume } V = \frac{\pi h}{6}(h^2 + 3r_1^2 + 3r_2^2)$$

$$= \frac{\pi(10)}{6}[10^2 + 3(12)^2 + 3(12)^2]$$

$$= \frac{10\pi}{6}(100 + 432 + 432)$$

$$= \textbf{5047 cm}^3$$

Problem 41. A hole is to be excavated in the form of a prismoid. The bottom is to be a rectangle 16 m long by 12 m wide; the top is also a rectangle, 26 m long by 20 m wide. Find the volume of earth to be removed, correct to 3 significant figures, if the depth of the hole is 6.0 m

---

The hole is shown in Fig. 9.36. Let A_1 represent the area of the top of the hole, i.e. $A_1 = 20 \times 26 = 520$ m^2. Let A_3 represent the area of the bottom of the hole, i.e. $A_3 = 12 \times 16 = 192$ m^2. Let A_2 represent the rectangular area through the middle of the hole parallel to areas A_1 and A_2. The length of this rectangle is $(26 + 16)/2 = 21$ m and the width is $(20 + 12)/2 = 16$ m, assuming the sloping edges are uniform. Thus area $A_2 = 21 \times 16 = 336$ m^2.

Using the prismoidal rule,

$$\text{volume of hole} = \frac{x}{6}[A_1 + 4A_2 + A_3]$$

$$= \frac{6}{6}[520 + 4(336) + 192]$$

$$= 2056 \text{ m}^3 = \textbf{2060 m}^3,$$

correct to 3 significant figures.

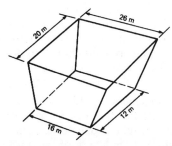

Figure 9.36

Problem 42. The roof of a building is in the form of a frustum of a pyramid with a square base of side 5.0 m. The flat top is a square of side 1.0 m and all the sloping sides are pitched at the same angle. The vertical height of the flat top above the level of the eaves is 4.0 m. Calculate, using the prismoidal rule, the volume enclosed by the roof

Let area of top of frustum be $A_1 = (1.0)^2 = 1.0$ m^2.
Let area of bottom of frustum be $A_3 = (5.0)^2 = 25.0$ m^2.
Let area of section through the middle of the frustum parallel to A_1 and A_3 be A_2. The length of the side of the square forming A_2 is the average of the sides forming A_1 and A_3, i.e. $(1.0 + 5.0)/2 = 3.0$ m.
Hence $A_2 = (3.0)^2 = 9.0$ m^2.
Using the prismoidal rule, volume of frustum

$$= \frac{x}{6}[A_1 + 4A_2 + A_3]$$

$$= \frac{4.0}{6}[1.0 + 4(9.0) + 25.0]$$

Hence volume enclosed by roof = 41.3 m^3.

Further problems on the prismoidal rule may be found in Section 9.9, Problems 50 to 53, pages 107 and 108.

9.8 Areas and volumes of similar shapes

The areas of similar shapes are proportional to the squares of corresponding linear dimensions. For example, Fig. 9.37 shows two squares, one of which has sides three times as long as the other.

$$\text{Area of Fig. 9.37(a)} = (x)(x) = x^2$$

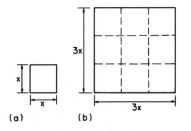

(a)   (b)

Figure 9.37

Area of Fig. 9.37(b) $= (3x)(3x) = 9x^2$

Hence Fig. 9.37(b) has an area $(3)^2$, i.e. 9 times the area of Fig. 9.37(a).

The volumes of similar bodies are proportional to the cubes of corresponding linear dimensions. For example, Fig. 9.38 shows two cubes, one of which has sides three times as long as those of the other.

Volume of Fig. 9.38(a) $= (x)(x)(x) = x^3$

Volume of Fig. 9.38(b) $= (3x)(3x)(3x) = 27x^3$

Hence Fig. 9.38(b) has a volume $(3)^3$, i.e. 27 times the volume of Fig. 9.38(a).

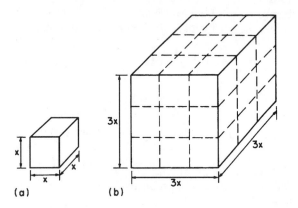

(a)   (b)

Figure 9.38

Problem 43.   A rectangular garage is shown on a building plan having dimensions 10 mm by 20 mm. If the plan is drawn to a scale of 1 to 250, determine the true area of the garage in square metres

---

Area of garage on the plan = 10 mm × 20 mm = 200 mm^2.
Since the areas of similar shapes are proportional to the squares of corresponding dimensions then:

$$\text{True area of garage} = 200 \times (250)^2$$

$$= 12.5 \times 10^6 \text{ mm}^2$$

$$= \frac{12.5 \times 10^6}{10^6} \text{ m}^2$$

$$= \textbf{12.5 m}^2$$

Problem 44.   A car has a mass of 1000 kg. A model of the car is made to a scale of 1 to 50. Determine the mass of the model if the car and its model are made of the same material

---

$$\frac{\text{Volume of model}}{\text{Volume of car}} = \left(\frac{1}{50}\right)^3$$ since the volume of similar bodies are proportional to the cube of corresponding dimensions.

Mass = density × volume, and since both car and model are made of the same material then:

$$\frac{\text{Mass of model}}{\text{Mass of car}} = \left(\frac{1}{50}\right)^3$$

Hence mass of model = (mass of car) $\left(\dfrac{1}{50}\right)^3 =$

$$\frac{1000}{50^3} = \textbf{0.008 kg or 8 g}.$$

Further problems on areas and volumes of similar shapes may be found in the following Section 9.9, Problems 54 to 56, page 108.

9.9   Further problems on areas and volumes

Areas of plane figures

1   A rectangular plate is 85 mm long and 42 mm wide. Find its area in square centimetres.
[35.7 cm^2]

2   A rectangular field has an area of 1.2 hectares and a length of 150 m. Find (a) its width and (b) the length of a diagonal (1 hectare = 10 000 m^2).   [(a) 80 m (b) 170 m]

3   Determine the area of each of the angle iron sections shown in Fig. 9.39.
[(a) 29 cm^2 (b) 650 mm^2]

4   A rectangular garden measures 40 m by 15 m. A 1 m flower border is made round the two shorter sides and one long side. A circular swimming pool of diameter 8 m is constructed in the middle of the garden. Find, correct to the nearest square metre, the area remaining.
[482 m^2]

5   The area of a trapezium is 13.5 cm^2 and the perpendicular distance between its parallel sides is 3 cm. If the length of one of the parallel sides is 5.6 cm, find the length of the other parallel side.   [3.4 cm]

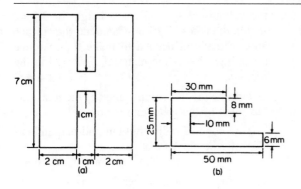

Figure 9.39

6 Find the angles p, q, r, s and t in Fig. 9.40(a) to (c).

$$\left[\begin{array}{l} p = 105°, q = 35°, r = 142°, \\ s = 95°, t = 146° \end{array}\right]$$

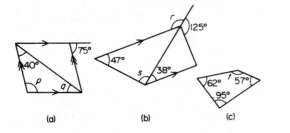

Figure 9.40

7 Name the types of quadrilateral shown in Fig. 9.41(i) to (iv), and determine (a) the area, and (b) the perimeter of each.

$$\left[\begin{array}{l} \text{(i) rhombus (a) 14 cm}^2 \text{ (b) 16 cm} \\ \text{(ii) parallelogram (a) 180 mm}^2 \text{ (b) 80 mm} \\ \text{(iii) rectangle (a) 3600 mm}^2 \text{ (b) 300 mm} \\ \text{(iv) trapezium (a) 190 cm}^2 \text{ (b) 62.91 cm} \end{array}\right]$$

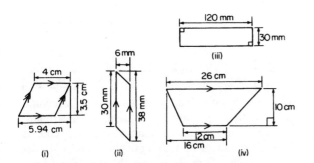

Figure 9.41

8 Determine the area of circles having (a) a radius of 4 cm, (b) a diameter of 30 mm and (c) a circumference of 200 mm.
   [(a) 50.27 cm^2 (b) 706.9 mm^2 (c) 3183 mm^2]

9 An annulus has an outside diameter of 60 mm and an inside diameter of 20 mm. Determine its area.   [2513 mm^2]

10 If the area of a circle is 320 mm^2, find (a) its diameter, and (b) its circumference.
   [(a) 20.19 mm (b) 63.41 mm]

11 Calculate the areas of the following sectors of circles: (a) radius 9 cm, angle subtended at centre 75°, (b) diameter 35 mm, angle subtended at centre 48°37′ and (c) diameter 5 cm, angle subtended at centre 2.19 radians.
   [(a) 53.01 cm^2 (b) 129.9 mm^2 (c) 6.84 cm^2]

12 Calculate the area of a regular octagon if each side is 20 mm and the width across the flats is 48.3 mm.   [1932 mm^2]

13 Determine the area of a regular hexagon which has sides 25 mm.   [1624 mm^2]

14 Find the area of triangle ABC if $\angle B = 90°$, $a = 4.87$ cm and $b = 7.54$ cm.   [14.02 cm^2]

15 A plot of land is in the shape shown in Fig. 9.42. Determine (a) its area in hectares (1 ha $= 10^4$ m^2), and (b) the length of fencing required, to the nearest metre, to completely enclose the plot of land.
   [(a) 0.918 ha (b) 456 m]

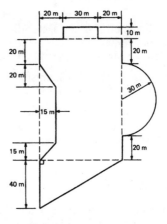

Figure 9.42

Volumes and surface areas of regular solids

16 A rectangular block of metal has dimensions of 40 mm by 25 mm by 15 mm. Determine its

volume. Find also its mass if the metal has a density of 9 g/cm^3.   [15 cm^3, 135 g]

17 Determine the maximum capacity, in litres, of a fish tank measuring 50 cm by 40 cm by 2.5 m (1 litre = 1000 cm^3).   [500 l]

18 Determine how many cubic metres of concrete are required for a 120 m long path, 150 mm wide and 80 mm deep.   [1.44 m^3]

19 Calculate the volume of a metal tube whose outside diameter is 8 cm and whose inside diameter is 6 cm, if the length of the tube is 4 m.   [8796 cm^3]

20 The volume of a cylinder is 400 cm^3. If its radius is 5.20 cm, find its height. Determine also its curved surface area.
   [(a) 4.709 cm, 153.9 cm^2]

21 If a cone has a diameter of 80 mm and a perpendicular height of 120 mm calculate its volume in cm^3 and its curved surface area.
   [201.1 cm^3, 159.0 cm^3]

22 A cylinder is cast from a rectangular piece of alloy 5 cm by 7 cm by 12 cm. If the length of the cylinder is to be 60 cm, find its diameter.
   [2.99 cm]

23 Find the volume and the total surface area of a regular hexagonal bar of metal of length 3 m if each side of the hexagon is 6 cm.
   [28 060 cm^3, 1.099 m^2]

24 A square pyramid has a perpendicular height of 4 cm. If a side of the base is 2.4 cm long find the volume and total surface area of the pyramid.   [7.68 cm^3, 25.81 cm^2]

25 A sphere has a diameter of 6 cm. Determine its volume and surface area.
   [113.1 cm^3, 113.1 cm^2]

26 Find the total surface area of a hemisphere of diameter 50 mm.   [5890 mm^2 or 58.90 cm^2]

27 Determine the mass of a hemispherical copper container whose external and internal radii are 12 cm and 10 cm. Assume that 1 cm^3 of copper weighs 8.9 g.   [13.57 kg]

28 If the volume of a sphere is 566 cm^3, find its radius.   [5.131 cm]

29 A metal plumb bob comprises a hemisphere surmounted by a cone. If the diameter of the hemisphere and cone are each 4 cm and the total length is 5 cm, find its total volume.
   [29.32 cm^3]

30 A marquee is in the form of a cylinder surmounted by a cone. The total height is 6 m and

the cylindrical portion has a height of 3.5 m, with a diameter of 15 m. Calculate the surface area of material needed to make the marquee assuming 12% of the material is wasted in the process.   [393.4 m^2]

31 Determine (a) the volume and (b) the total surface area of the following solids:

   (i)   a cone of radius 8.0 cm and perpendicular height 10 cm
         [(a) 670 cm^3 (b) 523 cm^2]

   (ii) a sphere of diameter 7.0 cm^2
         [(a) 180 cm^3 (b) 154 cm^2]

   (iii) a hemisphere of radius 3.0 cm
         [(a) 56.5 cm^3 (b) 84.8 cm^2]

   (iv) a 2.5 cm by 2.5 cm square pyramid of perpendicular height 5.0 cm
         [(a) 10.4 cm^3 (b) 32.0 cm^2]

   (v)   a 4.0 cm by 6.0 cm rectangular pyramid of perpendicular height 12.0 cm
         [(a) 96.0 cm^3 (b) 146 cm^2]

   (vi) a 4.2 cm by 4.2 cm square pyramid whose sloping edges are each 15.0 cm
         [(a) 86.5 cm^3 (b) 142 cm^2]

   (vii) a pyramid having an octagonal base of side 5.0 cm and perpendicular height 20 cm   [(a) 805 cm^3 (b) 539 cm^2]

32 The volume of a sphere is 325 cm^3. Determine its diameter.   [8.53 cm]

33 A metal sphere weighing 24 kg is melted down and recast into a solid cone of base radius 8.0 cm. If the density of the metal is 8000 kg/m^3 determine (a) the diameter of the metal sphere and (b) the perpendicular height of the cone, assuming that 15% of the metal is lost in the process.   [(a) 17.9 cm (b) 38.0 cm]

34 Find the volume of a regular hexagonal pyramid if the perpendicular height is 16.0 cm and the side of the base is 3.0 cm.   [125 cm^3]

35 A buoy consists of a hemisphere surmounted by a cone. The diameter of the cone and hemisphere is 2.5 m and the slant height of the cone is 4.0 m. Determine the volume and surface area of the buoy.   [(a) 10.3 m^3, 25.5 m^2]

36 A petrol container is in the form of a central cylindrical portion 5.0 m long with a hemispherical section surmounted on each end. If the diameters of the hemisphere and cylinder are both 1.2 m determine the capacity of the tank in litres (1 litre = 1000 cm^3).   [6560 l]

37 Figure 9.43 shows a metal rod section. Determine its volume and total surface area.
$$[657.1 \text{ cm}^3, 1027 \text{ cm}^2]$$

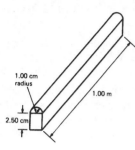

Figure 9.43

Volumes and surface areas of frustra of pyramids and cones

38 The radii of the faces of a frustum of a cone are 2.0 cm and 4.0 cm and the thickness of the frustum is 5.0 cm. Determine its volume and total surface area. $[147 \text{ cm}^3, 164 \text{ cm}^2]$

39 A frustum of a pyramid has square ends, the squares having sides 9.0 cm and 5.0 cm, respectively. Calculate the volume and total surface area of the frustum if the perpendicular distance between its ends is 8.0 cm.
$$[403 \text{ cm}^3, 337 \text{ cm}^2]$$

40 A cooling tower is in the form of a frustum of a cone. The base has a diameter of 32.0 m, the top has a diameter of 14.0 m and the vertical height is 24.0 m. Calculate the volume of the tower and the curved surface area.
$$[10\,480 \text{ m}^3, 1852 \text{ m}^2]$$

41 A loudspeaker diaphragm is in the form of a frustum of a cone. If the end diameters are 28.0 cm and 6.00 cm and the vertical distance between the ends is 30.0 cm, find the area of material needed to cover the curved surface of the speaker. $[1707 \text{ cm}^2]$

42 A rectangular prism of metal having dimensions 4.3 cm by 7.2 cm by 12.4 cm is melted down and recast into a frustum of a square pyramid, 10% of the metal being lost in the process. If the ends of the frustum are squares of side 3 cm and 8 cm, respectively, find the thickness of the frustum. $[10.69 \text{ cm}]$

43 Determine the volume and total surface area of a bucket consisting of an inverted frustum of a cone, of slant height 36.0 cm and end diameters 55.0 cm and 35.0 cm.
$$[55\,910 \text{ cm}^3, 8427 \text{ cm}^2]$$

44 A cylindrical tank of diameter 2.0 m and perpendicular height 3.0 m is to be replaced by a tank of the same capacity but in the form of a frustum of a cone. If the diameters of the ends of the frustum are 1.0 m and 2.0 m, respectively, determine the vertical height required.
$$[5.14 \text{ m}]$$

Frustums and zones of spheres

45 Determine the volume and surface area of a frustum of a sphere of diameter 47.85 cm, if the radii of the ends of the frustum are 14.0 cm and 22.0 cm and the height of the frustum is 10.0 cm. $[11\,210 \text{ cm}^3, 1503 \text{ cm}^2]$

46 Determine the volume (in cm^3) and the surface area (in cm^2) of a frustum of a sphere if the diameter of the ends are 80.0 mm and 120.0 mm and the thickness is 30.0 mm.
$$[259.2 \text{ cm}^3, 118.3 \text{ cm}^2]$$

47 A sphere has a radius of 6.50 cm. Determine its volume and surface area. A frustum of the sphere is formed by two parallel planes, one through the diameter and the other at a distance h from the diameter.
  If the curved surface area of the frustum is to be $\frac{1}{5}$ of the surface area of the sphere, find the height h and the volume of the frustum.
$$[1150 \text{ cm}^3, 531 \text{ cm}^2, 2.60 \text{ cm}, 326.7 \text{ cm}^3]$$

48 A sphere has a diameter of 32.0 mm. Calculate the volume (in cm^3) of the frustum of the sphere contained between two parallel planes distances 12.0 mm and 10.00 mm from the centre and on opposite sides of it.
$$[14.84 \text{ cm}^3]$$

49 A spherical storage tank is filled with liquid to a depth of 30.0 cm. If the inner diameter of the vessel is 45.0 cm determine the number of litres of liquid in the container (1 litre $=$ 1000 cm^3). $[35.34 \text{ l}]$

The prismoidal rule

50 Use the prismoidal rule to find the volume of a frustum of a sphere contained between two parallel planes on opposite sides of the centre each of radius 7.0 cm and each 4.0 cm from the centre. $[1500 \text{ cm}^3]$

51 Determine the volume of a cone of perpendicular height 16.0 cm and base diameter 10.0 cm by using the prismoidal rule. $[418.9 \text{ cm}^3]$

52 A bucket is in the form of a frustum of a cone. The diameter of the base is 28.0 cm and the diameter of the top is 42.0 cm. If the length is 32.0 cm, determine the capacity of the bucket (in litres) using the prismoidal rule (1 litre $= 1000$ cm^3).

[31.20 litres]

53 Determine the capacity of a water reservoir, in litres, the top being a 30.0 m by 12.0 m rectangle, the bottom being a 20.0 m by 8.0 m rectangle and the depth being 5.0 m (1 litre $= 1000$ cm^3).   [1.267×10^6 litres]

Areas and volumes of similar shapes

54 The area of a park on a map is 500 mm^2. If the scale of the map is 1 to 40 000 determine the true area of the park in hectares (1 hectare $= 10^4$ m^2).   [80 ha]

55 The diameter of two spherical bearings are in the ratio 2 : 5. What is the ratio of their volumes?   [8 : 125]

56 An engineering component has a mass of 400 g. If each of its dimensions are reduced by 30% determine its new mass.   [137.2 g]

10

An introduction to trigonometry

10.1 Trigonometry

Trigonometry is the branch of mathematics which deals with the measurement of sides and angles of triangles, and their relationships with each other.

10.2 The theorem of Pythagoras

With reference to Fig. 10.1, the side opposite the right angle (side b) is called the **hypotenuse**. The **theorem** of **Pythagoras** states: 'In any right-angled triangle, the square on the hypotenuse is equal to the sum of the squares on the other two sides.' Hence $b^2 = a^2 + c^2$.

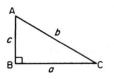

Figure 10.1

Problem 1.   In Fig. 10.2, find the length of BC

By Pythagoras' theorem: $a^2 = b^2 + c^2$

i.e. $\qquad a^2 = 4^2 + 3^2$

$\qquad\qquad = 16 + 9 = 25$

Hence $\qquad a = \sqrt{25} = \pm 5$

Figure 10.2

(-5 has no meaning in this context and is thus ignored)
Thus **BC = 5 cm**.

Problem 2.   In Fig. 10.3, find the length of EF

By Pythagoras' theorem: $e^2 = d^2 + f^2$

Hence $\qquad\qquad 13^2 = d^2 + 5^2$

$\qquad\qquad 169 = d^2 + 25$

$\qquad\qquad d^2 = 169 - 25 = 144$

Thus $\qquad\qquad d = \sqrt{144} = 12$ cm

Thus **EF = 12 cm**.

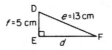

Figure 10.3

Problem 3.   Two aircraft leave an airfield at the same time. One travels due north at an average speed of 300 km/h and the other due west at an average speed of 220 km/h. Calculate their distance apart after 4 hours

---

After 4 hours, the first aircraft has travelled $4 \times 300 = 1200$ km due north, and the second aircraft has travelled $4 \times 220 = 880$ km due west, as shown in Fig. 10.4. Distance apart after 4 hours $= BC$.

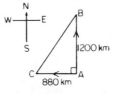

Figure 10.4

From Pythagoras' theorem:

$$BC^2 = 1200^2 + 880^2$$

$$= 1\,440\,000 + 774\,400 \text{ and}$$

$$BC = \sqrt{(2\,214\,400)}$$

Hence distance apart after 4 hours = 1488 km.

Further problems on the theorem of Pythagoras may be found in Section 10.8, Problems 1 to 7, page 116.

10.3   Trigonometric ratios of acute angles

(a)   With reference to the right-angled triangle shown in Fig. 10.5:

(i)   sine $\theta = \dfrac{\text{opposite side}}{\text{hypotenuse}}$, i.e. $\boldsymbol{\sin\theta = \dfrac{b}{c}}$

(ii)   cosine $\theta = \dfrac{\text{adjacent side}}{\text{hypotenuse}}$, i.e. $\boldsymbol{\cos\theta = \dfrac{a}{c}}$

(iii) tangent $\theta = \dfrac{\text{opposite side}}{\text{adjacent side}}$, i.e. $\boldsymbol{\tan\theta = \dfrac{b}{a}}$

Figure 10.5

Problem 4.   From Fig. 10.6, find $\sin D$, $\cos D$ and $\tan F$

By Pythagoras' theorem, $17^2 = 8^2 + EF^2$, from which, $EF = \sqrt{(17^2 - 8^2)} = 15$

$$\sin D = \frac{EF}{DF} = \frac{15}{17} \text{ or } \mathbf{0.8824}$$

Figure 10.6

$$\cos D = \frac{DE}{DF} = \frac{8}{17} \text{ or } \mathbf{0.4706}$$

$$\tan F = \frac{DE}{EF} = \frac{8}{15} \text{ or } \mathbf{0.5333}$$

Problem 5.   Determine the values of $\sin\theta$, $\cos\theta$ and $\tan\theta$ for the right-angled triangle ABC shown in Fig. 10.7

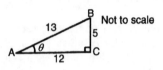

Figure 10.7

By definition:

$$\sin\theta = \frac{\text{opposite side}}{\text{hypotenuse}} = \frac{5}{13} = \mathbf{0.3846}$$

$$\cos\theta = \frac{\text{adjacent side}}{\text{hypotenuse}} = \frac{12}{13} = \mathbf{0.9231}$$

$$\tan\theta = \frac{\text{opposite side}}{\text{adjacent side}} = \frac{5}{12} = \mathbf{0.4167}$$

Problem 6.   If $\cos X = \dfrac{9}{41}$ determine the values of $\sin X$ and $\tan X$

Figure 10.8 shows a right-angled triangle.

Since $\cos X = \dfrac{9}{41}$, then $XY = 9$ units and $XZ = 41$ units.

Using Pythagoras' theorem: $41^2 = 9^2 + YZ^2$ from which $YZ = \sqrt{(41^2 - 9^2)} = 40$ units.

Thus

$$\sin X = \frac{40}{41} \text{ and } \tan X = \frac{40}{9} = 4\frac{4}{9}$$

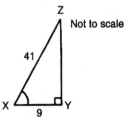

Figure 10.8

Problem 7.   Point A lies at coordinate $(2, 3)$ and point B at $(8, 7)$. Determine (a) the distance AB, (b) the gradient of the straight line AB, and (c) the angle AB makes with the horizontal

(a)   Points A and B are shown in Fig. 10.9(a). In Fig. 10.9(b), the horizontal and vertical lines AC and BC are constructed.

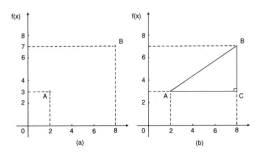

Figure 10.9

Since ABC is a right-angled triangle, and $AC = (8 - 2) = 6$ and $BC = (7 - 3) = 4$, then by Pythagoras' theorem

$$AB^2 = AC^2 + BC^2$$
$$= 6^2 + 4^2$$

and $AB = \sqrt{(6^2 + 4^2)} = \sqrt{52} = \mathbf{7.211}$, correct to 3 decimal places

(b)   The gradient of AB is given by $\tan\theta$,
i.e. **gradient** $= \tan\theta = \dfrac{BC}{AC} = \dfrac{4}{6} = \dfrac{2}{3}$

(c)   **The angle AB makes with the horizontal** is given by $\arctan\frac{2}{3} = \mathbf{33.69°}$

Further problems on the trigonometric ratios of acute angles may be found in Section 10.8, Problems 8 to 14, pages 116 and 117.

10.4   Evaluating trigonometric ratios

Four-figure tables are available which gives sines, cosines, and tangents, for angles between $0°$ and $90°$. However, the easiest method of evaluating trigonometric functions of any angle is by using a **calculator**.

The following values, correct to 4 decimal places, may be checked:

$\sin 18° = 0.3090$, $\text{cosine } 56° = 0.5592$
$\text{tangent } 29° = 0.5543$, $\text{sine } 172° = 0.1392$
$\text{cosine } 115° = -0.4226$, $\text{tangent } 178° = -0.0349$
$\text{sine } 241.63° = -0.8799$, $\text{cosine } 331.78° = 0.8811$
$\text{tangent } 296.42° = -2.0127$

To evaluate, say, sine $42°23'$ using a calculator means finding sine $42\dfrac{23°}{60}$ since there are 60 minutes in 1 degree

$$\frac{23}{60} = 0.383\dot{3}, \text{ thus } 42°23' = 42.383\dot{3}°$$

Thus sine $42°23' = $ sine $42.383\dot{3} = 0.6741$, correct to 4 decimal places.

Similarly, cosine $72°38' = $ cosine $72\dfrac{38°}{60} = 0.2985$, correct to 4 decimal places.

Problem 8.   Evaluate, correct to 4 decimal places:
(a) sine $11°$ (b) sine $121.68°$ (c) sine $259°10'$

(a)   sine $11° = \mathbf{0.1908}$
(b)   sine $121.68° = \mathbf{0.8510}$
(c)   sine $259°10' = $ sine $259\dfrac{10°}{60} = \mathbf{-0.9822}$

Problem 9.   Evaluate, correct to 4 decimal places:
(a) cosine $23°$ (b) cosine $159.32°$
(c) cosine $321°41'$

(a)   cosine $23° = \mathbf{0.9205}$
(b)   cosine $159.32° = \mathbf{-0.9356}$
(c)   cosine $321°41' = $ cosine $321\dfrac{41°}{60} = \mathbf{0.7846}$

Problem 10.   Evaluate, correct to 4 significant figures:
(a) tangent $276°$ (b) tangent $131.29°$
(c) tangent $76°58'$

(a)   tan $276° = \mathbf{-9.514}$
(b)   tan $131.29° = \mathbf{-1.139}$
(c)   tan $76°58' = $ tan $76\dfrac{58°}{60} = \mathbf{4.320}$

Problem 11. Evaluate, correct to 4 significant figures:

(a) sin 2.162 (b) cos(3π/8) (c) tan 1.16

(a) sin 2.162 means the sine of 2.162 radians. Hence a calculator needs to be on the radian function.
Hence sin 2.162 = **0.8303**
(b) cos(3π/8) = cos 1.178097... = **0.3827**
(c) tan 1.16 = **2.296**

Problem 12. Determine the acute angle:

(a) arcsin 0.7321 (b) arccos 0.4174 (c) arctan 1.4695

(a) Note that 'arcsin θ' is an abbreviation for 'the angle whose sine is equal to θ'. 0.7321 is entered into a calculator and then the inverse sine (or sin⁻¹) key is pressed. Hence arcsin 0.7321 = 47.06273...° Subtracting 47 leaves 0.06273...° Multiplying by 60 gives 4′ to the nearest minute.
Hence arcsin 0.7321 = **47.06°** or **47°4′**
Alternatively, in radians,
arcsin 0.7321 = **0.821 radians**
(b) arccos 0.4174 = **65.33°** or **65°20′** or **1.140 radians**
(c) arctan 1.4695 = **55.76°** or **55°46′** or **0.973 radians**

Problem 13. Evaluate

$$\frac{4.2 \tan 49°26' - 3.7 \sin 66°1'}{7.1 \cos 29°34'}$$

correct to 3 significant figures

By calculator, $\tan 49°26' = \tan\left(49\frac{26}{60}\right)^{\circ} = 1.1681$,
$\sin 66°1' = 0.9137$ and $\cos 29°34' = 0.8698$. Hence

$$\frac{4.2 \tan 49°26' - 3.7 \sin 66°1'}{7.1 \cos 29°34'}$$

$$= \frac{(4.2 \times 1.1681) - (3.7 \times 0.9137)}{(7.1 \times 0.8698)}$$

$$= \frac{4.9060 - 3.3807}{6.1756} = \frac{1.5253}{6.1756}$$

$$= 0.2470 = \mathbf{0.247},$$

correct to 3 significant figures.

Problem 14. Evaluate correct to 4 decimal places:

(a) sin(−112°) (b) tangent(−217.29°)
(c) cosine(−93°16′)

(a) Positive angles are shown anticlockwise and negative angles are shown clockwise. From Fig. 10.10, −112° is actually the same as +248° (i.e. 360° − 112°).

Hence, by calculator, sine(−112°) = sine 248°

$$= -0.9272$$

(b) tangent(−217.29°) = **−0.7615** (which is the same as tan(360° − 217.29°), i.e. tan 142.71°)

(c) cosine(−93°16′) = cosine $\left(-93\frac{16}{60}\right)^{\circ}$

$$= -0.0570$$

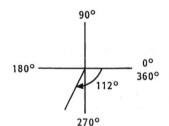

Figure 10.10

Further problems on evaluating trigonometric ratios may be found in Section 10.8, Problems 15 to 23, page 117.

10.5 Fractional and surd forms of trigonometric ratios

In Fig. 10.11, *ABC* is an equilateral triangle of side 2 units. *AD* bisects angle *A* and bisects the side *BC*. Using Pythagoras' theorem on triangle *ABD* gives:

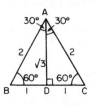

Figure 10.11

$AD = \sqrt{(2^2 - 1^2)} = \sqrt{3}$. Hence

$$\sin 30° = \frac{BD}{AB} = \frac{1}{2} \qquad \cos 30° = \frac{AD}{AB} = \frac{\sqrt{3}}{2}$$

$$\tan 30° = \frac{BD}{AD} = \frac{1}{\sqrt{3}}$$

$$\sin 60° = \frac{AD}{AB} = \frac{\sqrt{3}}{2} \qquad \cos 60° = \frac{BD}{AB} = \frac{1}{2}$$

$$\tan 60° = \frac{AD}{BD} = \sqrt{3}$$

In Fig. 10.12, PQR is an isosceles triangle with $PQ = QR = 1$ unit. By Pythagoras' theorem, $PR = \sqrt{(1^2 + 1^2)} = \sqrt{2}$

Figure 10.12

Hence

$$\sin 45° = \frac{1}{\sqrt{2}} \qquad \cos 45° = \frac{1}{\sqrt{2}} \qquad \tan 45° = 1$$

A quantity which is not exactly expressible as a rational number is called a **surd**. For example, $\sqrt{2}$ and $\sqrt{3}$ are called surds because they cannot be expressed as a fraction and the decimal part may be continued indefinitely.
For example, $\sqrt{2} = 1.4142135\ldots$
From above, $\sin 30° = \cos 60°$, $\sin 45° = \cos 45°$ and $\sin 60° = \cos 30°$.

In general, $\sin \theta = \cos (90° - \theta)$

and $\cos \theta = \sin (90° - \theta)$

For example, it may be checked by calculator that $\sin 25° = \cos 65°$, $\sin 42° = \cos 48°$, $\cos 84°10' = \sin 5°50'$, and so on.

Problem 15.   Using surd forms, evaluate
$$\frac{3 \tan 60° - 2 \cos 30°}{\tan 30°}$$

From above, $\tan 60° = \sqrt{3}$, $\cos 30° = \sqrt{3}/2$ and $\tan 30° = 1/\sqrt{3}$. Hence

$$\frac{3 \tan 60° - 2 \cos 30°}{\tan 30°} = \frac{3(\sqrt{3}) - 2\left(\dfrac{\sqrt{3}}{2}\right)}{\dfrac{1}{\sqrt{3}}}$$

$$= \frac{3\sqrt{3} - \sqrt{3}}{\dfrac{1}{\sqrt{3}}}$$

$$= \frac{2\sqrt{3}}{\dfrac{1}{\sqrt{3}}} = 2\sqrt{3}\left(\frac{\sqrt{3}}{1}\right)$$

$$= 2(3) = \mathbf{6}$$

A further problem on the fractional and surd forms of trigonometric ratios may be found in Section 10.8, Problem 24, pages 117 and 118.

10.6   Solution of right-angled triangles

To 'solve a right-angled triangle' means 'to find the unknown sides and angles'. This is achieved by using (i) the theorem of Pythagoras, and/or (ii) trigonometric ratios.

Problem 16.   Sketch a right-angled triangle ABC such that $B = 90°$, $AB = 5$ cm and $BC = 12$ cm. Determine the length of AC and hence evaluate $\sin A$, $\cos C$ and $\tan A$

Triangle ABC is shown in Fig. 10.13. By Pythagoras' theorem, $AC = \sqrt{(5^2 + 12^2)} = 13$. By definition:

$$\sin A = \frac{\text{opposite side}}{\text{hypotenuse}} = \frac{12}{13} \text{ or } \mathbf{0.9231}$$

$$\cos C = \frac{\text{adjacent side}}{\text{hypotenuse}} = \frac{12}{13} \text{ or } \mathbf{0.9231}$$

Figure 10.13

$$\tan A = \frac{\text{opposite side}}{\text{adjacent side}} = \frac{12}{5} \text{ or } \textbf{2.400}$$

> **Problem 17.**   In triangle PQR shown in Fig. 10.14, find the lengths of PQ and PR

$$\tan 38° = \frac{PQ}{QR} = \frac{PQ}{7.5},$$

Figure 10.14

hence $PQ = 7.5 \tan 38° = 7.5(0.7813) = \textbf{5.860 cm}$

$$\cos 38° = \frac{QR}{PR} = \frac{7.5}{PR}.$$

Hence $PR = \dfrac{7.5}{\cos 38°} = \dfrac{7.5}{0.7880} = \textbf{9.518 cm}.$

Check: Using Pythagoras' theorem $(7.5)^2 + (5.860)^2 = 90.59 = (9.518)^2$.

> **Problem 18.**   Solve the triangle ABC shown in Fig. 10.15

To 'solve triangle ABC' means 'to find the length AC and angles B and C'.

$$\sin C = \frac{35}{37} = 0.9459$$

Figure 10.15

Hence

$$C = \arcsin 0.9459 = \textbf{71°4}'$$

$B = 180° - 90° - 71°4' = \textbf{18°56}'$ (since angles in a triangle add up to $180°$)

$$\sin B = \frac{AC}{37}$$

hence

$$AC = 37 \sin 18°56' = 37(0.3245) = \textbf{12.0 mm}$$

Check: Using Pythagoras' theorem $37^2 = 35^2 + 12^2$.

> **Problem 19.**   Solve triangle XYZ given $\angle X = 90°$, $\angle Y = 23°17'$ and $YZ = 20.0$ mm. Determine also its area

It is always advisable to make a reasonably accurate sketch so as to visualize the expected magnitudes of unknown sides and angles. Such a sketch is shown in Fig. 10.16.

$$\angle Z = 180° - 90° - 23°17' = \textbf{66°43}'$$

$$\sin 23°17' = \frac{XZ}{20.0}$$

Figure 10.16

Hence

$$XZ = 20.0 \sin 23°17'$$
$$= 20.0(0.3953) = \textbf{7.906 mm}$$
$$\cos 23°17' = \frac{XY}{20.0}$$

Hence

$$XY = 20.0 \cos 23°17'$$
$$= 20.0(0.9186) = \textbf{18.37 mm}$$

Check: Using Pythagoras' theorem $(18.37)^2 + (7.906)^2 = 400.0 = (20.0)^2$.

Area of triangle $XYZ = \frac{1}{2}(\text{base})(\text{perpendicular height})$

$$= \frac{1}{2}(XY)(XZ) = \frac{1}{2}(18.37)(7.906) = \textbf{72.62 mm}^2$$

Further problems on the solution of right-angled triangles may be found in Section 10.8, Problems 25 to 27, page 118.

10.7 Angles of elevation and depression

(a) If, in Fig. 10.17, BC represents horizontal ground and AB a vertical flagpole, then the **angle of elevation** of the top of the flagpole, A, from the point C is the angle that the imaginary straight line AC must be raised (or elevated) from the horizontal CB, i.e. angle θ

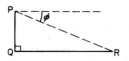

Figure 10.17

(b) If, in Fig. 10.18, PQ represents a vertical cliff and R a ship at sea, then the **angle of depression** of the ship from point P is the angle through which the imaginary straight line PR must be lowered (or depressed) from the horizontal to the ship, i.e. angle ϕ.
(Note, $\angle PRQ$ is also ϕ — alternate angles between parallel lines)

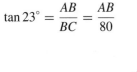

Figure 10.18

Problem 20.   An electricity pylon stands on horizontal ground. At a point 80 m from the base of the pylon, the angle of elevation of the top of the pylon is 23°. Calculate the height of the pylon to the nearest metre

Figure 10.19 shows the pylon AB and the angle of elevation of A from point C is 23°.

$$\tan 23^\circ = \frac{AB}{BC} = \frac{AB}{80}$$

Figure 10.19

Hence height of pylon

$$AB = 80 \tan 23^\circ$$
$$= 80(0.4245) = 33.96 \text{ m}$$
$$= \textbf{34 m to the nearest metre}$$

Problem 21.   A surveyor measures the angle of elevation of the top of a perpendicular building as 19°. He moves 120 m nearer the building and finds the angle of elevation is now 47°. Determine the height of the building

The building PQ and the angles of elevation are shown in Fig. 10.20.

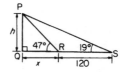

Figure 10.20

In triangle PQS

$$\tan 19^\circ = \frac{h}{x + 120}$$

hence $h = \tan 19^\circ (x + 120)$, i.e.

$$h = 0.3443(x + 120) \qquad (1)$$

In triangle PQR

$$\tan 47^\circ = \frac{h}{x}$$

hence $h = \tan 47^\circ (x)$, i.e.

$$h = 1.0724x \qquad (2)$$

Equating equations (1) and (2) gives:

$$0.3443(x + 120) = 1.0724x$$
$$0.3443x + (0.3443)(120) = 1.0724x$$
$$(0.3443)(120) = (1.0724 - 0.3443)x$$
$$41.316 = 0.7281x$$
$$x = \frac{41.316}{0.7281}$$
$$= 56.74 \text{ m}$$

From equation (2), height of building $h = 1.0724x = 1.0724(56.74) = \textbf{60.85 m}$.

Problem 22. The angle of depression of a ship viewed at a particular instant from the top of a 75 m vertical cliff is 30°. Find the distance of the ship from the base of the cliff at this instant. The ship is sailing away from the cliff at constant speed and 1 minute later its angle of depression from the top of the cliff is 20°. Determine the speed of the ship in km/h

Figure 10.21 shows the cliff AB, the initial position of the ship at C and the final position at D. Since the angle of depression is initially 30° then $\angle ACB = 30°$ (alternate angles between parallel lines).

$$\tan 30° = \frac{AB}{BC} = \frac{75}{BC}$$

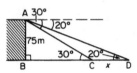

Figure 10.21

Hence

$$BC = \frac{75}{\tan 30°} = \frac{75}{0.5774}$$

$$= \textbf{129.9 m} = \textbf{initial position of ship}$$

In triangle ABD,

$$\tan 20° = \frac{AB}{BD} = \frac{75}{BC + CD} = \frac{75}{129.9 + x}$$

Hence

$$129.9 + x = \frac{75}{\tan 20°} = \frac{75}{0.3640} = 206.0 \text{ m}$$

from which $x = 206.0 - 129.9 = 76.1$ m.
Thus the ship sails 76.1 m in 1 minute, i.e. 60 s, hence speed of ship

$$= \frac{\text{distance}}{\text{time}} = \frac{76.1}{60} \text{ m/s}$$

$$= \frac{76.1 \times 60 \times 60}{60 \times 1000} \text{ km/h} = \textbf{4.566 km/h}$$

Further problems on angles of elevation and depression may be found in the following Section 10.8, Problems 28 to 36, page 118.

10.8 Further problems on an introduction to trigonometry

Theorem of Pythagoras

1   In a triangle ABC, $\angle B$ is a right angle, $AB = 6.92$ cm and $BC = 8.78$ cm. Find the length of the hypotenuse.   [11.18 cm]

2   In a triangle CDE, $\angle D = 90°$, $CD = 14.83$ mm and $CE = 28.31$ mm. Determine the length of DE.   [24.11 mm]

3   Show that if a triangle has sides of 8, 15 and 17 cm it is right angled.

4   Triangle PQR is isosceles, $\angle Q$ being a right angle. If the hypotenuse is 38.47 cm find (a) the lengths of sides PQ and QR, and (b) the value of $\angle QPR$.
                    [(a) 27.20 cm each (b) 45°]

5   A man cycles 24 km due south and then 20 km due east. Another man, starting at the same time as the first man, cycles 32 km due east and then 7 km due south. Find the distance between the two men.   [20.81 km]

6   A ladder 3.5 m long is placed against a perpendicular wall with its foot 1.0 m from the wall. How far up the wall (to the nearest centimetre) does the ladder reach? If the foot of the ladder is now moved 30 cm further away from the wall, how far does the top of the ladder fall?   [3.35 m, 10 cm]

7   Two ships leave a port at the same time. One travels due west at 18.4 km/h and the other due south at 27.6 km/h. Calculate how far apart the two ships are after 4 hours.   [132.7 km]

Trigonometric ratios of acute angles

8   Sketch a triangle XYZ such that $\angle Y = 90°$, $XY = 9$ cm and $YZ = 40$ cm. Determine $\sin Z$, $\cos Z$, $\tan X$ and $\cos X$.

$$\left[\begin{array}{c} \sin Z = \dfrac{9}{41}, \quad \cos Z = \dfrac{40}{41}, \\[2mm] \tan X = \dfrac{40}{9}, \quad \cos X = \dfrac{9}{41} \end{array} \right]$$

9   In triangle ABC shown in Fig. 10.22, find $\sin A$, $\cos A$, $\tan A$, $\sin B$, $\cos B$ and $\tan B$.

$$\left[\begin{array}{l} \sin A = \dfrac{3}{5}, \quad \cos A = \dfrac{4}{5}, \quad \tan A = \dfrac{3}{4} \\[3mm] \sin B = \dfrac{4}{5}, \quad \cos B = \dfrac{3}{5}, \quad \tan B = \dfrac{4}{3} \end{array}\right]$$

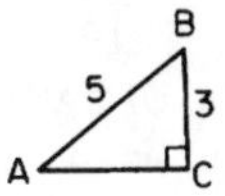

Figure 10.22

10   If $\cos A = \dfrac{15}{17}$ find $\sin A$ and $\tan A$, in fraction form.

$$\left[\sin A = \dfrac{8}{17}, \quad \tan A = \dfrac{8}{15}\right]$$

11   If $\tan X = \dfrac{15}{112}$, find $\sin X$ and $\cos X$, in fraction form.

$$\left[\sin X = \dfrac{15}{113}, \quad \cos X = \dfrac{112}{113}\right]$$

12   For the right-angled triangle shown in Fig. 10.23, find (a) $\sin \alpha$ (b) $\cos \theta$ (c) $\tan \theta$.

$$\left[(a) \ \dfrac{15}{17} \ (b) \ \dfrac{15}{17} \ (c) \ \dfrac{8}{15}\right]$$

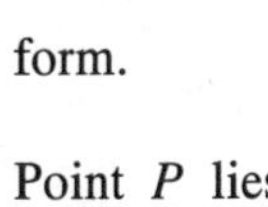

Figure 10.23

13   If $\tan \theta = \dfrac{7}{24}$, find $\sin \theta$ and $\cos \theta$ in fraction form.

$$\left[(a) \ \sin \theta = \dfrac{7}{25}, \quad \cos \theta = \dfrac{24}{25}\right]$$

14   Point P lies at coordinate $(-3, 1)$ and point Q at $(5, -4)$. Determine (a) the distance PQ, (b) the gradient of the straight line PQ and (c) the angle PQ makes with the horizontal.

[(a) 9.434 (b) -0.625 (c) $-32°$]

Evaluating trigonometric ratios

In Problems 15 to 18, evaluate correct to 4 decimal places:

15   (a) sine $27°$ (b) sine $172.41°$ (c) sine $302°52'$

[(a) 0.4540 (b) 0.1321 (c) -0.8399]

16   (a) cosine $124°$ (b) cosine $21.46°$ (c) cosine $284°10'$.

[(a) -0.5592 (b) 0.9307 (c) 0.2447]

17   (a) tangent $145°$ (b) tangent $310.59°$ (c) tangent $49°16'$.

[(a) -0.7002 (b) -1.1671 (c) 1.1612]

18   (a) sine $\dfrac{2\pi}{3}$ (b) cos 1.681 (c) tan 3.672.

[(a) 0.8660 (b) -0.1010 (c) 0.5865]

In Problem 19, determine the acute angle in degrees (correct to 2 decimal places), degrees and minutes, and in radians (correct to 3 decimal places):

19   (a) arcsin 0.2341 (b) arccos 0.8271 (c) arctan 0.8106.

$$\left[\begin{array}{l} (a) \ 13.54°, \ 13°32', \ 0.236 \text{ rad} \\ (b) \ 34.20°, \ 34°12', \ 0.597 \text{ rad} \\ (c) \ 39.03°, \ 39°2', \ 0.681 \text{ rad} \end{array}\right]$$

20   Evaluate the following, each correct to 4 significant figures:
(a) $4\cos 56°19' - 3\sin 21°57'$

(b) $\dfrac{11.5\tan 49°11' - \sin 90°}{3\cos 45°}$

(c) $\dfrac{5\sin 86°3'}{3\tan 14°29' - 2\cos 31°9'}$

[(a) 1.097 (b) 5.805 (c) -5.325]

21   Determine the acute angle, in degrees and minutes, correct to the nearest minute, given by

$$\arcsin \left(\dfrac{4.32\sin 42°16'}{7.86}\right)$$

[21°42']

22   Evaluate correct to 4 decimal places

$$\dfrac{(\sin 34°27')(\cos 69°2')}{(2\tan 53°39')}$$

[0.0745]

23   Evaluate (a) sine$(-125°)$ (b) tan$(-241°)$ (c) cos$(-49°15')$.

[(a) -0.8192 (b) -1.8040 (c) 0.6528]

Fractional and surd forms of trigonometric ratios

24   Evaluate the following without using tables or calculators, leaving, where necessary, in surd form:
(a) $3\sin 30° - 2\cos 60°$
(b) $5\tan 60° - 3\sin 60°$

(c) $\dfrac{\tan 60°}{3 \tan 30°}$

(d) $(\tan 45°)(4 \cos 60° - 2 \sin 60°)$

(e) $\dfrac{\tan 60° - \tan 30°}{1 + \tan 30° \tan 60°}$

$$\left[\text{(a) } \frac{1}{2} \text{ (b) } \frac{7}{2}\sqrt{3} \text{ (c) } 1 \text{ (d) } 2 - \sqrt{3} \text{ (e) } \frac{1}{\sqrt{3}} \right]$$

Solution of right-angled triangles

25 Solve the triangles shown in Fig. 10.24.

$$\left[\begin{array}{l} \text{(i)} \quad BC = 3.50 \text{ cm, } AB = 6.10 \text{ cm,} \\ \qquad \angle B = 55° \\ \text{(ii)} \quad FE = 5 \text{ cm, } \angle E = 53°8', \\ \qquad \angle F = 36°52' \\ \text{(iii)} \quad GH = 9.841 \text{ mm, } GI = 11.32 \text{ mm,} \\ \qquad H = 49° \end{array} \right]$$

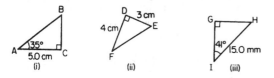

Figure 10.24

26 Solve the triangles shown in Fig. 10.25 and find their areas.

$$\left[\begin{array}{l} \text{(i)} \quad KL = 5.43 \text{ cm, } JL = 8.62 \text{ cm,} \\ \qquad \angle J = 39°, \text{ area} = 18.19 \text{ cm}^2 \\ \text{(ii)} \quad MN = 28.86 \text{ mm, } NO = 13.82 \text{ mm,} \\ \qquad \angle O = 64°25', \text{ area} = 199.4 \text{ mm}^2 \\ \text{(iii)} \quad PR = 7.934 \text{ m, } \angle Q = 65°3', \\ \qquad \angle R = 24°57', \text{ area} = 14.64 \text{ m}^2 \end{array} \right]$$

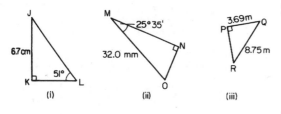

Figure 10.25

27 A ladder rests against the top of the perpendicular wall of a building and makes an angle of 67° with the ground. If the foot of the ladder is 5 m from the wall, calculate the height of the building.   [11.78 m]

Angles of elevation and depression

28 A vertical tower stands on level ground. At a point 105 m from the foot of the tower the angle of elevation of the top is 19°. Find the height of the tower.   [36.15 m]

29 If the angle of elevation of the top of a vertical 30 m high aerial is 32°, how far is it to the aerial?   [48 m]

30 From the top of a vertical cliff 90 m high the angle of depression of a boat is 19°50′. Determine the distance of the boat from the cliff.   [249.5 m]

31 From the top of a vertical cliff 80.0 m high the angles of depression of two buoys lying due west of the cliff are 23° and 15°, respectively. How far are the buoys apart?   [110.1 m]

32 From a point on horizontal ground a surveyor measures the angle of elevation of the top of a flagpole as 18°40′. He moves 50 m nearer to the flagpole and measures the angle of elevation as 26°22′. Determine the height of the flagpole.   [53.0 m]

33 A flagpole stands on the edge of the top of a building. At a point 200 m from the building the angles of elevation of the top and bottom of the pole are 32° and 30°, respectively. Calculate the height of the flagpole.   [9.50 m]

34 From a ship at sea, the angles of elevation of the top and bottom of a vertical lighthouse standing on the edge of a vertical cliff are 31° and 26°, respectively. If the lighthouse is 25.0 m high, calculate the height of the cliff.   [107.8 m]

35 From a window 4.2 m above horizontal ground the angle of depression of the foot of a building across the road is 24° and the angle of elevation of the top of the same building is 34°. Determine, correct to the nearest centimetre, the width of the road and the height of the building.   [9.43 m, 10.56 m]

36 The elevation of a tower from two points, one due east of the tower and the other due west of it are 20° and 24°, respectively, and the two points of observation are 300 m apart. Find the height of the tower to the nearest metre.   [60 m]

11

The solution of triangles and their areas

11.1 Sine and cosine rules

To **'solve a triangle'** means 'to find the values of unknown sides and angles'. If a triangle is **right angled**, trigonometric ratios and the theorem of Pythagoras may be used for its solution. However, for a **non-right-angled triangle**, trigonometric ratios and Pythagoras' theorem **cannot** be used. Instead, two rules, called the sine rule and the cosine rule, are used.

Sine rule

With reference to triangle ABC of Fig. 11.1, the **sine rule** states:

$$\frac{a}{\sin A} = \frac{b}{\sin B} = \frac{c}{\sin C}$$

The rule may be used only when:

(i) 1 side and any 2 angles are initially given, or
(ii) 2 sides and an angle (not the included angle) are initially given

Figure 11.1

Cosine rule

With reference to triangle ABC of Fig. 11.1, the **cosine rule** states:

$$a^2 = b^2 + c^2 - 2bc \cos A$$
$$\text{or } b^2 = a^2 + c^2 - 2ac \cos B$$
$$\text{or } c^2 = a^2 + b^2 - 2ab \cos C$$

The rule may be used only when:

(i) 2 sides and the included angle are initially given, or
(ii) 3 sides are initially given

11.2 Area of any triangle

The **area of any triangle** such as ABC of Fig. 11.1 is given by:

(i) $\frac{1}{2} \times \text{base} \times \text{perpendicular height, or}$

(ii) $\frac{1}{2}ab \sin C$, or $\frac{1}{2}ac \sin B$ or $\frac{1}{2}bc \sin A$, or

(iii) $\sqrt{[s(s-a)(s-b)(s-c)]}$, where

$$s = \frac{a+b+c}{2}$$

11.3 Worked problems on the solution of triangles and their areas

> **Problem 1.** In the triangle XYZ, $X = 51°$, $Y = 67°$ and $YZ = 15.2$ cm. Solve the triangle and find its area

The triangle XYZ is shown in Fig. 11.2. Since the angles in a triangle add up to 180°, then $Z =$

Figure 11.2

$180° - 51° - 67° = \mathbf{62°}$. Applying the sine rule:

$$\frac{15.2}{\sin 51°} = \frac{y}{\sin 67°} = \frac{z}{\sin 62°}$$

Using $\dfrac{15.2}{\sin 51°} = \dfrac{y}{\sin 67°}$ and transposing gives:

$$y = \frac{15.2 \sin 67°}{\sin 51°} = \mathbf{18.00 \ cm} = XZ$$

Using $\dfrac{15.2}{\sin 51°} = \dfrac{z}{\sin 62°}$ and transposing gives:

$$z = \frac{15.2 \sin 62°}{\sin 51°} = \mathbf{17.27 \ cm} = XY$$

Area of triangle $XYZ = \dfrac{1}{2} xy \sin z$

$$= \frac{1}{2}(15.2)(18.00) \sin 62°$$

$$= \mathbf{120.8 \ cm^2}$$

(or area $= \dfrac{1}{2} xz \sin Y = \dfrac{1}{2}(15.2)(17.27) \sin 67°$

$$= \mathbf{120.8 \ cm^2})$$

It is always worth checking with triangle problems that the longest side is opposite the largest angle, and vice versa. In this problem, Y is the largest angle and thus XZ should be the longest of the three sides.

Problem 2.   Solve the triangle ABC given $B = 78°51'$, $AC = 22.31$ mm and $AB = 17.92$ mm. Find also its area

---

Triangle ABC is shown in Fig. 11.3.

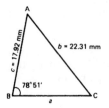

Figure 11.3

Applying the sine rule:

$$\frac{22.31}{\sin 78°51'} = \frac{17.92}{\sin C}$$

from which,

$$\sin C = \frac{17.92 \sin 78°51'}{22.31} = 0.7881$$

Hence $C = \arcsin 0.7881 = 52°0'$ or $128°0'$ (see Chapter 37). Since $B = 78°51'$, C cannot be $128°0'$, since $128°0' + 78°51'$ is greater than $180°$. Thus only $C = 52°0'$ is valid. Angle $A = 180° - 78°51' - 52°0' = 49°9'$.
Applying the sine rule:

$$\frac{a}{\sin 49°9'} = \frac{22.31}{\sin 78°51'}$$

from which,

$$a = \frac{22.31 \sin 49°9'}{\sin 78°51'} = 17.20 \ mm$$

Hence $A = 49°9'$, $C = 52°0'$ and $BC = 17.20$ mm.

Area of triangle $ABC = \dfrac{1}{2} ac \sin B$

$$= \frac{1}{2}(17.20)(17.92) \sin 78°51'$$

$$= \mathbf{151.2 \ mm^2}$$

Problem 3.   Solve the triangle PQR and find its area given that $QR = 36.5$ mm, $PR = 29.6$ mm and $Q = 36°$

Triangle PQR is shown in Fig. 11.4. Applying the sine rule:

$$\frac{29.6}{\sin 36°} = \frac{36.5}{\sin P}$$

from which,

$$\sin P = \frac{36.5 \sin 36°}{29.6} = 0.7248$$

Hence $P = \arcsin 0.7248 = 46°27'$ or $133°33'$. When $P = 46°27'$ and $Q = 36°$ then $R = 180° - 46°27' - 36° = 97°33'$.

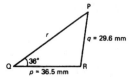

Figure 11.4

When $P = 133°33'$ and $Q = 36°$ then $R = 180° - 133°33' - 36° = 10°27'$.

Thus, in this problem, there are **two** separate sets of results and both are feasible solutions. Such a situation is called the **ambiguous case**.

Case 1. $P = 46°27'$, $Q = 36°$, $R = 97°33'$, $p = 36.5$ mm and $q = 29.6$ mm.
From the sine rule:

$$\frac{r}{\sin 97°33'} = \frac{29.6}{\sin 36°}$$

from which,

$$r = \frac{29.6 \sin 97°33'}{\sin 36°} = \textbf{49.92 mm}$$

Area $= \frac{1}{2}pq \sin R = \frac{1}{2}(36.5)(29.6) \sin 97°33' = $ **535.5 mm^2**.

Case 2. $P = 133°33'$, $Q = 36°$, $R = 10°27'$, $p = 36.5$ mm and $q = 29.6$ mm.
From the sine rule:

$$\frac{r}{\sin 10°27'} = \frac{29.6}{\sin 36°}$$

from which,

$$r = \frac{29.6 \sin 10°27'}{\sin 36°} = \textbf{9.134 mm}$$

Area $= \frac{1}{2}pq \sin R$

$= \frac{1}{2}(36.5)(29.6) \sin 10°27'$

$= \textbf{97.98 mm}^2$

Triangle PQR for case 2 is shown in Fig. 11.5.

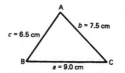

Figure 11.5

Problem 4.   Solve triangle DEF and find its area given that $EF = 35.0$ mm, $DE = 25.0$ mm and $E = 64°$

Triangle DEF is shown in Fig. 11.6. Applying the cosine rule:

$$e^2 = d^2 + f^2 - 2df \cos E$$

i.e. $e^2 = (35.0)^2 + (25.0)^2$

$- [2(35.0)(25.0) \cos 64°]$

$= 1225 + 625 - 767.1 = 1083$

$e = \sqrt{1083} = \textbf{32.91 mm}$

Figure 11.6

Applying the sine rule:

$$\frac{32.91}{\sin 64°} = \frac{25.0}{\sin F}$$

from which,

$$\sin F = \frac{25.0 \sin 64°}{32.91} = 0.6828$$

Thus $F = \arcsin 0.6828 = 43°4'$ or $136°56'$

$F = 136°56'$ is not possible in this case since $136°56' + 64°$ is greater than $180°$. Thus only $F = \textbf{43°4'}$ is valid

$$D = 180° - 64° - 43°4' = \textbf{72°56'}$$

Area of triangle $DEF = \frac{1}{2}df \sin E$

$= \frac{1}{2}(35.0)(25.0) \sin 64°$

$= \textbf{393.2 mm}^2$

Problem 5.   A triangle ABC has sides $a = 9.0$ cm, $b = 7.5$ cm and $c = 6.5$ cm. Determine its three angles and its area

---

Triangle ABC is shown in Fig. 11.7. It is usual first to calculate the largest angle to determine whether the triangle is acute or obtuse. In this case the largest angle is A (i.e. opposite the longest side).

Figure 11.7

Applying the cosine rule:

$$a^2 = b^2 + c^2 - 2bc\cos A$$

from which,

$$2bc\cos A = b^2 + c^2 - a^2$$

and

$$\cos A = \frac{b^2 + c^2 - a^2}{2bc} = \frac{7.5^2 + 6.5^2 - 9.0^2}{2(7.5)(6.5)}$$

$$= 0.1795$$

Hence $A = \arccos 0.1795 = $ **79°40′** or 280°20′, which is obviously impossible.

The triangle is thus acute angled since $\cos A$ is positive. (If $\cos A$ had been negative, angle A would be obtuse, i.e. lie between 90° and 180°.)
Applying the sine rule:

$$\frac{9.0}{\sin 79°40'} = \frac{7.5}{\sin B}$$

from which,

$$\sin B = \frac{7.5 \sin 79°40'}{9.0} = 0.8198$$

Hence $B = \arcsin 0.8198 = $ **55°4′**

$$C = 180° - 79°40' - 55°4' = \mathbf{45°16'}$$

Area $= \sqrt{[s(s - a)(s - b)(s - c)]}$, where

$$s = \frac{a + b + c}{2} = \frac{9.0 + 7.5 + 6.5}{2} = 11.5 \text{ cm}$$

Hence area $= \sqrt{[11.5(11.5 - 9.0)(11.5 - 7.5)}$

$$(11.5 - 6.5)]$$

$$= \sqrt{[11.5(2.5)(4.0)(5.0)]}$$

$$= \mathbf{23.98 \text{ cm}^2}$$

Alternatively, area $= \dfrac{1}{2}ab \sin C$

$$= \frac{1}{2}(9.0)(7.5)\sin 45°16'$$

$$= \mathbf{23.98 \text{ cm}^2}$$

Problem 6.   Solve triangle XYZ (Fig. 11.8) and find its area given that $Y = 128°$, $XY = 7.2$ cm and $YZ = 4.5$ cm

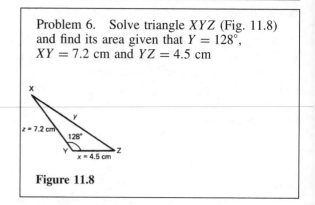

Figure 11.8

Applying the cosine rule:

$$y^2 = x^2 + z^2 - 2xz \cos Y$$

$$= (4.5)^2 + (7.2)^2 - \{2(4.5)(7.20)\cos 128°\}$$

$$= 20.25 + 51.84 - \{-39.89\}$$

$$= 20.25 + 51.84 + 39.89 = 112.0$$

$$y = \sqrt{(112.0)} = \mathbf{10.58 \text{ cm}}$$

Applying the sine rule:

$$\frac{10.58}{\sin 128°} = \frac{7.2}{\sin Z}$$

from which,

$$\sin Z = \frac{7.2 \sin 128°}{10.58} = 0.5363$$

Hence $Z = \arcsin 0.5363 = $ **32°26′** (or 147°34′ which, here, is impossible). $X = 180° - 128° - 32°26' = $ **19°34′**

$$\text{Area} = \frac{1}{2}xz \sin Y = \frac{1}{2}(4.5)(7.2)\sin 128°$$

$$= \mathbf{12.77 \text{ cm}^2}$$

Further problems on the solution of triangles and their areas may be found in Section 11.6, Problems 1 to 5, pages 126 and 127.

11.4   Lengths and areas on an inclined plane

In Fig. 11.9, rectangle $ADEF$ is a plane inclined at an angle of θ to the horizontal plane $ABCD$.

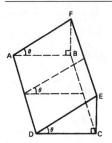

Figure 11.9

$$\cos\theta = \frac{DC}{DE}, \text{ from which } DE = \frac{DC}{\cos\theta}$$

Hence the line of greatest slope on an inclined plane is given by:

$\left(\dfrac{1}{\cos\theta}\right)$ **(its projection on to the horizontal plane)**

Area of $ADEF = (AD)(DE)$

$$= (AD)\left(\frac{DC}{\cos\theta}\right)$$

$$= \left(\frac{1}{\cos\theta}\right) \text{ (area of horizontal plane)}$$

Problem 7. A vertical, cylindrical ventilation shaft of diameter 36.0 cm has its end at an angle of 20° to the horizontal as shown in Fig. 11.10. Determine the area of the end cover plate

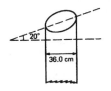

Figure 11.10

Cover plate area

$$= \left(\frac{1}{\cos\theta}\right) \text{ (horizontal plane area)}, \text{ from above}$$

$$= \left(\frac{1}{\cos 20°}\right)\left[\pi\left(\frac{36.0}{2}\right)^2\right] = \textbf{1083 cm}^2$$

Problem 8. A rectangular chimney stack having dimensions of 1.2 m by 0.70 m passes through a roof that has a pitch of 35°. Determine the area of the void in the roof through which the stack passes

---

Area of void in roof $= \left(\dfrac{1}{\cos\theta}\right)$ (area of horizontal plane)

$$= \left(\frac{1}{\cos 35°}\right)(1.2 \times 0.70)$$

$$= \textbf{1.025 m}^2$$

Further problems on the lengths and areas on an inclined plane may be found in Section 11.6, Problems 6 and 7, page 127.

11.5 Practical situations involving trigonometry

There are a number of **practical situations** where the use of trigonometry is needed to find unknown sides and angles of triangles. This is demonstrated in Problems 9 to 15.

Problem 9. A room 8.0 m wide has a span roof which slopes at 33° on one side and 40° on the other. Find the length of the roof slopes, correct to the nearest centimetre

---

A section of the roof is shown in Fig. 11.11. Angle at ridge, $B = 180° - 33° - 40° = 107°$. From the sine rule:

$$\frac{8.0}{\sin 107°} = \frac{a}{\sin 33°}$$

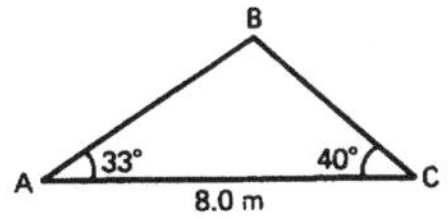

Figure 11.11

from which, $a = \dfrac{8.0\sin 33°}{\sin 107°} = 4.556$ m

Also from the sine rule:

$$\frac{8.0}{\sin 107°} = \frac{c}{\sin 40°}$$

from which,

$$c = \frac{8.0 \sin 40°}{\sin 107°} = 5.377 \text{ m}$$

Hence the roof slopes are 4.56 m and 5.38 m, correct to the nearest centimetre.

Problem 10.   A man leaves a point walking at 6.5 km/h in a direction E 20° N (i.e. a bearing of 70°). A cyclist leaves the same point at the same time in a direction E 40° S (i.e. a bearing of 130°) travelling at a constant speed. Find the average speed of the cyclist if the walker and cyclist are 80 km apart after 5 hours

---

After 5 hours the walker has travelled $5 \times 6.5 = 32.5$ km (shown as AB in Fig. 11.12). If AC is the distance the cyclist travels in 5 hours then $BC = 80$ km. Applying the sine rule:

$$\frac{80}{\sin 60°} = \frac{32.5}{\sin C}$$

from which,

$$\sin C = \frac{32.5 \sin 60°}{80} = 0.3518$$

Hence $C = \arcsin 0.3518 = 20°36'$ (or $159°24'$, which is impossible in this case). $B = 180° - 60° - 20°36' = 99°24'$.

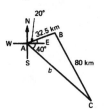

Figure 11.12

Applying the sine rule:

$$\frac{80}{\sin 60°} = \frac{b}{\sin 99°24'}$$

from which,

$$b = \frac{80 \sin 99°24'}{\sin 60°} = 91.14 \text{ km}$$

Since the cyclist travels 91.14 km in 5 hours then:

$$\text{average speed} = \frac{\text{distance}}{\text{time}} = \frac{91.14}{5}$$

$$= \textbf{18.23 km/h}$$

Problem 11.   Two voltage phasors are shown in Fig. 11.13. If $V_1 = 40$ V and $V_2 = 100$ V determine the value of their resultant (i.e. length OA) and the angle the resultant makes with V_1

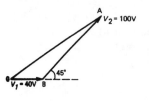

Figure 11.13

Angle $OBA = 180° - 45° = 135°$.
Applying the cosine rule:

$$OA^2 = V_1^2 + V_2^2 - 2V_1V_2 \cos OBA$$

$$= 40^2 + 100^2 - \{2(40)(100) \cos 135°\}$$

$$= 1600 + 10\,000 - \{-5657\}$$

$$= 1600 + 10\,000 + 5657$$

$$= 17\,257$$

The resultant $OA = \sqrt{(17\,257)} = 131.4$ V.
Applying the sine rule:

$$\frac{131.4}{\sin 135°} = \frac{100}{\sin AOB}$$

from which,

$$\sin AOB = \frac{100 \sin 135°}{131.4} = 0.5381$$

Hence angle $AOB = \arcsin 0.5381 = 32°33'$ (or $147°27'$, which is impossible in this case).
Hence the resultant voltage is 131.4 volts at 32°33' to V_1.

Problem 12.   In Fig. 11.14, *PR* represents the inclined jib of a crane and is 10.0 long. *PQ* is 4.0 m long. Determine the inclination of the jib to the vertical and the length of tie *QR*

Applying the sine rule:

$$\frac{PR}{\sin 120°} = \frac{PQ}{\sin R}$$

from which,

$$\sin R = \frac{PQ \sin 120°}{PR} = \frac{(4.0) \sin 120°}{10.0}$$

$$= 0.3464$$

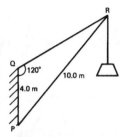

Figure 11.14

Hence $R = \arcsin 0.3464 = 20°16'$ (or $159°44'$, which is impossible in this case).
$P = 180° - 120° - 20°16' = \mathbf{39°44'}$, **which is the inclination of the jib to the vertical**.
Applying the sine rule:

$$\frac{10.0}{\sin 120°} = \frac{QR}{\sin 39°44'}$$

from which,

$$QR = \frac{10.0 \sin 39°44'}{\sin 120°} = \mathbf{7.38\ m}$$

$$= \textbf{length of tie}$$

Problem 13.   A vertical aerial stands on horizontal ground. A surveyor positioned due east of the aerial measures the elevation of the top as 48°. He moves due south 30.0 m and measures the elevation as 44°. Determine the height of the aerial

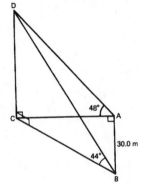

Figure 11.15

In Fig. 11.15, *DC* represents the aerial, *A* is the initial position of the surveyor and *B* his final position. From triangle *ACD*, $\tan 48° = DC/AC$, from which $AC = DC/\tan 48°$.
Similarly, from triangle *BCD*, $BC = DC/\tan 44°$.
For triangle *ABC*, using Pythagoras' theorem:

$$BC^2 = AB^2 + AC^2$$

$$\left(\frac{DC}{\tan 44°}\right)^2 = (30.0)^2 + \left(\frac{DC}{\tan 48°}\right)^2$$

$$DC^2 \left(\frac{1}{\tan^2 44°} - \frac{1}{\tan^2 48°}\right) = 30.0^2$$

$$DC^2 (1.072323 - 0.810727) = 30.0^2$$

$$DC^2 = \frac{30.0^2}{0.261596} = 3440$$

Hence, height of aerial, $DC = \sqrt{3440} = \mathbf{58.65\ m.}$

Problem 14.   A crank mechanism of a petrol engine is shown in Fig. 11.16. Arm *OA* is 10.0 cm long and rotates clockwise about 0. The connecting rod *AB* is 30.0 cm long and end *B* is constrained to move horizontally.

(a)   For the position shown in Fig. 11.16 determine the angle between the connecting rod *AB* and the horizontal and the length of *OB*

(b)   How far does *B* move when angle *AOB* changes from 50° to 120°?

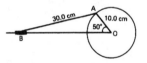

Figure 11.16

(a) Applying the sine rule:

$$\frac{AB}{\sin 50°} = \frac{AO}{\sin B}$$

from which,

$$\sin B = \frac{AO \sin 50°}{AB} = \frac{10.0 \sin 50°}{30.0}$$
$$= 0.2553$$

Hence $B = \arcsin 0.2553 = 14°47'$ (or $165°13'$, which is impossible in this case).
Hence the connecting rod AB makes an angle of $14°47'$ with the horizontal.
Angle $OAB = 180° - 50° - 14°47' = 115°13'$.
Applying the sine rule:

$$\frac{30.0}{\sin 50°} = \frac{OB}{\sin 115°13'}$$

from which,

$$\boldsymbol{OB = \frac{30.0 \sin 115°13'}{\sin 50°} = 35.43 \text{ cm}}$$

(b) Figure 11.17 shows the initial and final positions of the crank mechanism. In triangle $OA'B'$, applying the sine rule:

$$\frac{30.0}{\sin 120°} = \frac{10.0}{\sin A'B'O}$$

from which,

$$\sin A'B'O = \frac{10.0 \sin 120°}{30.0} = 0.2887$$

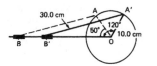

Figure 11.17

Hence $A'B'O = \arcsin 0.2887 = 16°47'$ (or $163°13'$ which is impossible in this case).

Angle $OA'B' = 180° - 120° - 16°47'$
$$= 43°13'$$

Applying the sine rule:

$$\frac{30.0}{\sin 120°} = \frac{OB'}{\sin 43°13'}$$

from which,

$$OB' = \frac{30.0 \sin 43°13'}{\sin 120°} = 23.72 \text{ cm}$$

Since $OB = 35.43$ cm and $OB' = 23.72$ cm
then $BB' = 35.43 - 23.72 = 11.71$ cm
Hence B moves 11.71 cm when angle AOB changes from $50°$ to $120°$

Problem 15.   The area of a field is in the form of a quadrilateral $ABCD$ as shown in Fig. 11.18. Determine its area

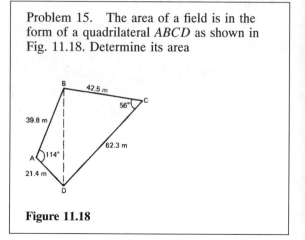

Figure 11.18

A diagonal drawn from B to D divides the quadrilateral into two triangles. Area of quadrilateral $ABCD$

$$= \text{area of triangle } ABD + \text{area of triangle } BCD$$
$$= \frac{1}{2}(39.8)(21.4)\sin 114° + \frac{1}{2}(42.5)(62.3)\sin 56°$$
$$= 389.04 + 1097.5$$
$$= \mathbf{1487 \ m^2}$$

Further problems on practical situations involving trigonometry may be found in the following Section 11.6, Problems 8 to 20, pages 127 and 128.

11.6   Further problems on the solution of triangles and their areas

Solution of triangles and their areas

1   Use the sine rule to solve the following triangles ABC and find their areas:
    (a) $A = 29°$, $B = 68°$, $b = 27$ mm
    (b) $B = 71°26'$, $C = 56°32'$, $b = 8.60$ cm
    (c) $A = 117°$, $C = 24°30'$, $a = 15.2$ mm

$$\begin{bmatrix} \text{(a) } C = 83°, a = 14.1 \text{ mm,} \\ \quad c = 28.9 \text{ mm, area} = 189 \text{ mm}^2 \\ \text{(b) } A = 52°2', c = 7.568 \text{ cm,} \\ \quad a = 7.152 \text{ cm, area} = 25.65 \text{ cm}^2 \\ \text{(c) } B = 38°30', b = 10.62 \text{ mm,} \\ \quad c = 7.074 \text{ mm, area} = 33.47 \text{ mm}^2 \end{bmatrix}$$

2   Use the sine rule to solve the following triangles *DEF* and find their areas:

(a) $d = 17$ cm, $f = 22$ cm, $F = 26°$
(b) $e = 4.20$ m, $f = 7.10$ m, $F = 81°$
(c) $d = 32.6$ mm, $e = 25.4$ mm, $D = 104°22'$

$$\begin{bmatrix} \text{(a) } D = 19°48', E = 134°12', \\ \quad e = 36.0 \text{ cm, area} = 134 \text{ cm}^2 \\ \text{(b) } E = 35°45', D = 63°15', \\ \quad d = 6.419 \text{ m, area} = 13.31 \text{ m}^2 \\ \text{(c) } E = 49°0', F = 26°38', \\ \quad f = 15.08 \text{ mm, area} = 185.6 \text{ mm}^2 \end{bmatrix}$$

3   Use the sine rule to solve the following triangles *JKL* and find their areas:

(a) $j = 3.85$ cm, $k = 3.23$ cm, $K = 36°$
(b) $k = 46$ mm, $l = 36$ mm, $L = 35°$
(c) $j = 2.92$ m, $l = 3.24$ m, $J = 27°30'$

$$\begin{bmatrix} \text{(a) } J = 44°29', L = 99°31', \\ \quad l = 5.420 \text{ cm, area} = 6.132 \text{ cm}^2 \\ \text{OR } J = 135°31', L = 8°29', \\ \quad l = 0.810 \text{ cm, area} = 0.916 \text{ cm}^2 \\ \text{(b) } K = 47°8', J = 97°52', \\ \quad j = 62.2 \text{ mm, area} = 820.2 \text{ mm}^2 \\ \text{OR } K = 132°52', J = 12°8', \\ \quad j = 13.19 \text{ mm, area} = 174.0 \text{ mm}^2 \\ \text{(c) } L = 30°49', K = 121°41', \\ \quad k = 5.382 \text{ m, area} = 4.026 \text{ m}^2 \\ \text{OR } L = 149°11', K = 3°19', \\ \quad k = 0.366 \text{ m, area} = 0.274 \text{ m}^2 \end{bmatrix}$$

4   Use the cosine and sine rules to solve the following triangles *PQR* and find their areas:

(a) $q = 12$ cm, $r = 16$ cm, $P = 54°$
(b) $p = 56$ mm, $q = 38$ mm, $R = 64°$
(c) $q = 3.25$ m, $r = 4.42$ m, $P = 105°$

$$\begin{bmatrix} \text{(a) } p = 13.2 \text{ cm}, Q = 47°21', \\ \quad R = 78°39', \text{area} = 77.7 \text{ cm}^2 \\ \text{(b) } r = 52.1 \text{ mm}, Q = 40°58', \\ \quad P = 75°2', \text{area} = 956 \text{ mm}^2 \\ \text{(c) } p = 6.127 \text{ m}, Q = 30°49', \\ \quad R = 44°11', \text{area} = 6.938 \text{ m}^2 \end{bmatrix}$$

5   Use the cosine and sine rules to solve the following triangles *XYZ* and find their areas:

(a) $x = 10.0$ cm, $y = 8.0$ cm, $z = 7.0$ cm
(b) $x = 2.4$ m, $y = 3.6$ m, $z = 1.5$ m

(c) $x = 21$ mm, $y = 34$ mm, $z = 42$ mm

$$\begin{bmatrix} \text{(a) } X = 83°20', Y = 52°37', \\ \quad Z = 44°3', \text{area} = 27.8 \text{ cm}^2 \\ \text{(b) } X = 28°57', Y = 133°26', \\ \quad Z = 17°37', \text{area} = 1.31 \text{ m}^2 \\ \text{(c) } Z = 29°46', Y = 53°31', \\ \quad Z = 96°43', \text{area} = 355 \text{ mm}^2 \end{bmatrix}$$

Lengths and areas on an inclined plane

6   A vertical 35.0 cm by 35.0 cm ventilation shaft has an end covered by a plate that makes an angle of 21°15′ with the horizontal. Determine the area of the plate.

[1314 cm²]

7   A chimney stack has a diameter of 1.5 m and passes through a roof that has a pitch of 36°30′. Determine the area of the resulting void in the roof covering.

[2.20 m²]

Practical situations involving trigonometry

8   A ship *P* sails at a steady speed of 45 km/h in a direction of W 32° N (i.e. a bearing of 302°) from a port. At the same time another ship *Q* leaves the port at a steady speed of 35 km/h in a direction N 15° E (i.e. a bearing of 015°). Determine their distance apart after 4 hours.

[193 km]

9   Two sides of a triangular plot of land are 52.0 m and 34.0 m, respectively. If the area of the plot is 620 m² find (a) the length of fencing required to enclose the plot, and (b) the angles of the triangular plot.

[(a) 122.6 m (b) 94°49′, 40°39′, 44°32′]

10   A jib crane is shown in Fig. 11.19. If the tie rod *PR* is 8.0 long and *PQ* is 4.5 m long determine (a) the length of jib *RQ*, and (b) the

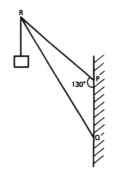

Figure 11.19

angle between the jib and the tie rod.

[(a) 11.4 m (b) 17°33′]

11   A building site is in the form of a quadrilateral as shown in Fig. 11.20, and its area is 1510 m². Determine the length of the perimeter of the site   [163.4 m]

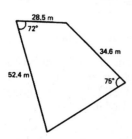

Figure 11.20

12   Determine the length of members *BF* and *EB* in the roof truss shown in Fig. 11.21.

[*BF* = 3.9 m, *EB* = 4.0 m]

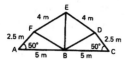

Figure 11.21

13   A laboratory 9.0 m wide has a span roof which slopes at 36° on one side and 44° on the other. Determine the lengths of the roof slopes.

[6.35 m, 5.37 m]

14   *PQ* and *QR* are the phasors representing the alternating currents in two branches of a circuit. Phasor *PQ* is 20.0 A and is horizontal. Phasor *QR* (which is joined to the end of *PQ* to form triangle *PQR*) is 14.0 A and is at an angle of 35° to the horizontal. Determine the resultant phasor *PR* and the angle it makes with phasor *PQ*.   [32.48 A, 14°19′]

15   A vertical aerial *AB*, 9.60 m high, stands on ground which is inclined 12° to the horizontal.

A stay connects the top of the aerial *A* to a point *C* on the ground 10.0 m downhill from *B*, the foot of the aerial. Determine (a) the length of the stay, and (b) the angle the stay makes with the ground.

[(a) 15.23 m (b) 38°4′]

16   Three forces acting on a fixed point are represented by the sides of a triangle of dimensions 7.2 cm, 9.6 cm and 11.0 cm. Determine the angles between the lines of action and the three forces.

[80°25′, 59°23′, 40°12′]

17   A reciprocating engine mechanism is shown in Fig. 11.22. The crank *AB* is 12.0 cm long and the connecting rod *BC* is 32.0 cm long. For the position shown determine the length of *AC* and the angle between the crank and the connecting rod.   [40.25 cm, 126°3′]

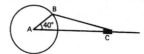

Figure 11.22

18   From Fig. 11.22, determine how far *C* moves, correct to the nearest millimetre when angle *CAB* changes from 40° to 160°, *B* moving in an anticlockwise direction.   [19.8 cm]

19   A surveyor, standing W 25° S of a tower measures the angle of elevation of the top of the tower as 46°30′. From a position E 23° S from the tower the elevation of the top is 37°15′. Determine the height of the tower if the distance between the two observations is 75 m.   [36.2 m]

20   An aeroplane is sighted due east from a radar station at an elevation of 40° and a height of 8000 m and later at an elevation of 35° and height 5500 m in a direction E 70° S. If it is descending uniformly, find the angle of descent. Determine also the speed of the aeroplane in km/h if the time between the two observations is 45 s.   [13°57′, 829.9 km/h]

12

Cartesian and polar coordinates

12.1 Introduction

There are two ways in which the position of a point in a plane can be represented. These are

(a) by **Cartesian coordinates**, i.e. (x, y), and
(b) by **polar coordinates**, i.e. (r, θ), where r is a 'radius' from a fixed point and θ is an angle from a fixed point

12.2 Changing from Cartesian into polar coordinates

In Fig. 12.1, if lengths x and y are known, then the length of r can be obtained from Pythagoras' theorem (see Chapter 10) since OPQ is a right-angled triangle. Hence

$$r^2 = (x^2 + y^2)$$

from which $\boxed{r = \sqrt{(x^2 + y^2)}}$

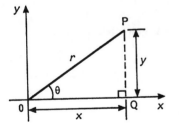

Figure 12.1

From trigonometric ratios (see Chapter 10),

$$\tan \theta = \frac{y}{x}$$

from which $\boxed{\theta = \arctan \dfrac{y}{x}}$

$r = \sqrt{(x^2 + y^2)}$ and $\theta = \arctan(y/x)$ are the two formulae we need to change from Cartesian to polar coordinates. The angle θ, which may be expressed in degrees or radians, must **always** be measured from the positive x-axis, i.e. measured from the line OQ in Fig. 12.1. It is suggested that when changing from Cartesian to polar coordinates a diagram should always be sketched.

> Problem 1. Change the Cartesian coordinates $(3, 4)$ into polar coordinates

A diagram representing the point $(3, 4)$ is shown in Fig. 12.2.

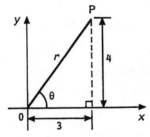

Figure 12.2

From Pythagoras' theorem, $r = \sqrt{(3^2 + 4^2)} = 5$ (note that -5 has no meaning in this context). By trigonometric ratios, $\theta = \arctan \frac{4}{3} = 53.13°$ or 0.927 rad (note that $53.13° = 53.13 \times (\pi/180)$ rad $= 0.927$ rad).
Hence $(3, 4)$ in Cartesian coordinates corresponds to $(5, 53.13°)$ or $(5, 0.927$ rad) in polar coordinates.

> Problem 2. Express in polar coordinates the position $(-4, 3)$

A diagram representing the point using the Cartesian coordinates $(-4, 3)$ is shown in Fig. 12.3.
From Pythagoras' theorem, $r = \sqrt{(4^2 + 3^2)} = 5$.

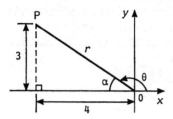

Figure 12.3

By trigonometric ratios, $\alpha = \arctan\frac{3}{4} = 36.87°$ or 0.644 rad.

Hence $\theta = 180° - 36.87° = 143.13°$

or $\theta = \pi - 0.644 = 2.498$ rad

Hence the position of point P in polar coordinates form is $(5, 143.13°)$ or $(5, 2.498\ \text{rad})$.

Problem 3.   Express $(-5, -12)$ in polar coordinates

A sketch showing the position $(-5, -12)$ is shown in Fig. 12.4.

$$r = \sqrt{(5^2 + 12^2)} = 13$$

$$\alpha = \arctan\frac{12}{5} = 67.38° \text{ or } 1.176 \text{ rad}$$

Hence $\theta = 180° + 67.38° = 247.38°$

or $\theta = \pi + 1.176 = 4.318$ rad

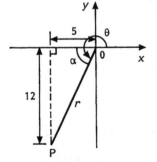

Figure 12.4

Thus $(-5, -12)$ in Cartesian coordinates corresponds to $(13, 247.38°)$ or $(13, 4.318\ \text{rad})$ in polar coordinates.

Problem 4.   Express $(2, -5)$ in polar coordinates

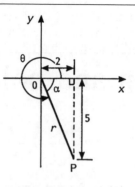

Figure 12.5

A sketch showing the position $(2, -5)$ is shown in Fig. 12.5.

$$r = \sqrt{(2^2 + 5^2)} = \sqrt{29} = 5.385 \text{ correct to}$$
3 decimal places

$$\alpha = \arctan\frac{5}{2} = 68.20° \text{ or } 1.190 \text{ rad}$$

Hence $\theta = 360° - 68.20° = 291.80°$

or $\theta = 2\pi - 1.190 = 5.093$ rad

Thus $(2, -5)$ in Cartesian coordinates corresponds to $(5.385, 291.80°)$ or $(5.385, 5.093\ \text{rad})$ in polar coordinates.

12.3   Changing from polar into Cartesian coordinates

From the right-angled triangle OPQ in Fig. 12.6

$$\cos\theta = \frac{x}{r} \text{ and}$$

$$\sin\theta = \frac{y}{r}, \text{ from trigonometric ratios}$$

Hence $\boxed{x = r\cos\theta}$ and $\boxed{y = r\sin\theta}$

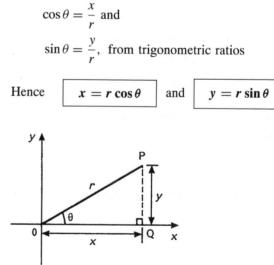

Figure 12.6

If length r and angle θ are known then $x = r\cos\theta$ and $y = r\sin\theta$ are the two formulae we need to change from polar to Cartesian coordinates.

> **Problem 5.** Change $(4, 32°)$ into Cartesian coordinates

A sketch showing the position $(4, 32°)$ is shown in Fig. 12.7.

Now $x = r\cos\theta = 4\cos 32° = 3.39$

and $y = r\sin\theta = 4\sin 32° = 2.12$

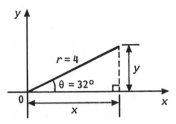

Figure 12.7

Hence $(4, 32°)$ **in polar coordinates corresponds to** $(3.39, 2.12)$ **in Cartesian coordinates.**

> **Problem 6.** Express $(6, 137°)$ in Cartesian coordinates

A sketch showing the position $(6, 137°)$ is shown in Fig. 12.8.

$$x = r\cos\theta = 6\cos 137° = -4.388$$

which corresponds to length OA in Fig. 12.8

$$y = r\sin\theta = 6\sin 137° = 4.092$$

which corresponds to length AB in Fig. 12.8.

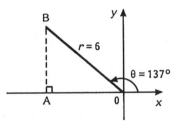

Figure 12.8

Thus $(6, 137°)$ in polar coordinates corresponds to $(-4.388, 4.092)$ in Cartesian coordinates.
(Note that when changing from polar to Cartesian coordinates it is not quite so essential to draw a sketch. Use of $x = r\cos\theta$ and $y = r\sin\theta$ automatically produces the correct signs.)

> **Problem 7.** Express $(4.5, 5.16 \text{ rad})$ in Cartesian coordinates

A sketch showing the position $(4.5, 5.16 \text{ rad})$ is shown in Fig. 12.9.

$$x = r\cos\theta = 4.5\cos 5.16 = 1.948$$

which corresponds to length OA in Fig. 12.9

$$y = r\sin\theta = 4.5\sin 5.16 = -4.057$$

which corresponds to length AB in Fig. 12.9.

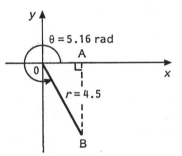

Figure 12.9

Thus $(1.948, -4.057)$ in Cartesian coordinates corresponds to $(4.5, 5.16 \text{ rad})$ in polar coordinates.

12.4 Use of R → P and P → R functions on calculators

Another name for Cartesian coordinates is **rectangular** coordinates. Many scientific notation calculators possess R → P and P → R functions. The R is the first letter of the word rectangular and the P is the first letter of the word polar. Check the operation manual for your particular calculator to determine how to use these two functions. They make changing from Cartesian to polar coordinates, and vice versa, so much quicker and easier.

12.5   Further problems on Cartesian and polar coordinates

In Problems 1 to 8, express the given Cartesian coordinates as polar coordinates, correct to 2 decimal places, in both degrees and in radians:

1   (3, 5)

[(5.83, 59.04°) or (5.83, 1.03 rad)]

2   (6.18, 2.35)

[(6.61, 20.82°) or (6.61, 0.36 rad)]

3   (−2, 4)

[(4.47, 116.57°) or (4.47, 2.03 rad)]

4   (−5.4, 3.7)

[(6.55, 145.58°) or (6.55, 2.54 rad)]

5   (−7, −3)

[(7.62, 203.20°) or (7.62, 3.55 rad)]

6   (−2.4, −3.6)

[(4.33, 236.31°) or (4.33, 4.12 rad)]

7   (5, −3)

[(5.83, 329.04°) or (5.83, 5.74 rad)]

8   (9.6, −12.4)

[(15.68, 307.75°) or (15.68, 5.37 rad)]

In Problems 9 to 16, express the given polar coordinates as Cartesian coordinates, correct to 3 decimal places:

9   (5, 75°)   [(1.294, 4.830)]

10   (4.4, 1.12 rad)   [(1.917, 3.960)]

11   (7, 140°)   [(−5.362, 4.500)]

12   (3.6, 2.5 rad)   [(−2.884, 2.154)]

13   (10.8, 210°)   [(−9.353, −5.400)]

14   (4, 4 rad)   [(−2.615, −3.027)]

15   (1.5, 300°)   [(0.750, −1.299)]

16   (6, 5.5 rad)   [(4.252, −4.233)]

13

Straight line graphs

13.1 Introduction to graphs

A **graph** is a pictorial representation of information showing how one quantity varies with another related quantity.

The most common method of showing a relationship between two sets of data is to use **Cartesian** or **rectangular axes** as shown in Fig. 13.1.

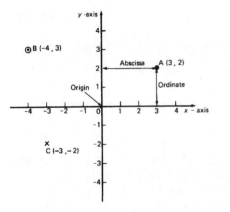

Figure 13.1

The points on a graph are called **coordinates**. Point A in Fig. 13.1 has the coordinates $(3, 2)$, i.e. 3 units in the x direction and 2 units in the y direction. Similarly, point B has coordinates $(-4, 3)$ and C has coordinates $(-3, -2)$. The origin has coordinates $(0, 0)$.

The horizontal distance of a point from the vertical axis is called the **abscissa** and the vertical distance from the horizontal axis is called the **ordinate**.

13.2 The straight line graph

Let a relationship between two variables x and y be $y = 3x + 2$.
When $x = 0$, $y = 3(0) + 2 = 2$. When $x = 1$, $y = 3(1) + 2 = 5$.
When $x = 2$, $y = 3(2) + 2 = 8$, and so on.

Thus coordinates $(0, 2)$, $(1, 5)$ and $(2, 8)$ have been produced from the equation by selecting arbitrary values of x, and are shown plotted in Fig. 13.2. When the points are joined together a **straight line graph** results.

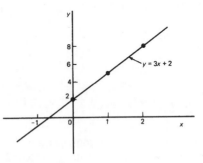

Figure 13.2

The **gradient** or **slope** of a straight line is the ratio of the change in the value of y to the change in the value of x between any two points on the line. If, as x increases ($\rightarrow$), y also increases ($\uparrow$), then the gradient is positive.
In Fig. 13.3(a) the gradient of AC

$$= \frac{\text{change in } y}{\text{change in } x} = \frac{CB}{BA} = \frac{7-3}{3-1} = \frac{4}{2} = 2$$

If as x increases ($\rightarrow$), y decreases ($\downarrow$), then the gradient is negative.
In Fig. 13.3(b), the gradient of DF

$$= \frac{\text{change in } y}{\text{change in } x} = \frac{FE}{ED} = \frac{11-2}{-3-0}$$

$$= \frac{9}{-3} = -3$$

Figure 13.3(c) shows a straight line graph $y = 3$. Since the straight line is horizontal the gradient is zero.

The value of y when $x = 0$ is called the **y-axis intercept**. In Fig. 13.3(a) the y-axis intercept is 1 and in Fig. 13.3(b) it is 2.

If the equation of a graph is of the form $y = mx + c$, where m and c are constants, the graph

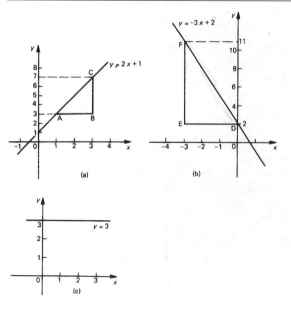

(a)

(b)

(c)

Figure 13.3

will always be a straight line, m representing the gradient and c the y-axis intercept. Thus $y = 5x + 2$ represents a straight line of gradient 5 and y-axis intercept 2. Similarly, $y = -3x - 4$ represents a straight line of gradient -3 and y-axis intercept -4.

Summary of general rules to be applied when drawing graphs

(i)   Give the graph a title clearly explaining what is being illustrated

(ii)   Choose scales such that the graph occupies as much space as possible on the graph paper being used

(iii) Choose scales so that interpolation is made as easy as possible. Usually scales such as 1 cm = 1 unit, or 1 cm = 2 units, or 1 cm = 10 units are used. Awkward scales such as 1 cm = 3 units or 1 cm = 7 units should not be used

(iv)   The scales need not start at zero, as this produces an accumulation of points within a small area of the graph paper

(v)   The coordinates, or points, should be clearly marked. This may be done either by a cross, or a dot and circle, or just by a dot (see Fig. 13.1)

(vi)   A statement should be made next to each axis explaining the numbers represented with their appropriate units

(vii) Sufficient numbers should be written next to each axis without cramping

---

Problem 1.   Plot the graph $y = 4x + 3$ in the range $x = -3$ to $x = +4$. From the graph, find (a) the value of y when $x = 2.2$, and (b) the value of x when $y = -3$

Whenever an equation is given and a graph is required, a table giving corresponding values of the variable is necessary. The table is achieved as follows:

When $x = -3$, $y = 4x + 3 = 4(-3) + 3 = -12 + 3 = -9$.

When $x = -2$, $y = 4(-2) + 3 = -8 + 3 = -5$, and so on.

Such a table is shown below:

x	-3	-2	-1	0	1	2	3	4
y	-9	-5	-1	3	7	11	15	19

The coordinates $(-3, -9)$, $(-2, -5)$, $(-1, -1)$, and so on, are plotted and joined together to produce the straight line shown in Fig. 13.4. (Note that the scales used on the x- and y-axes do not have to be the same.) From the graph:

(a)   when $x = 2.2$, $y = \mathbf{11.8}$, and

(b)   when $y = -3$, $x = \mathbf{-1.5}$

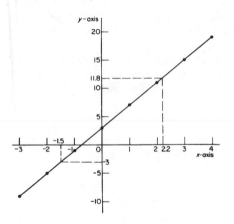

Figure 13.4

Problem 2.   Plot the following graphs on the same axes between the range $x = -4$ to $x = +4$, and determine the gradients of each:
(a) $y = x$ (b) $y = x + 2$ (c) $y = x + 5$
(d) $y = x - 3$

A table of coordinates is produced for each graph.

(a) $y = x$

x	−4	−3	−2	−1	0	1	2	3	4
y	−4	−3	−2	−1	0	1	2	3	4

(b) $y = x + 2$

x	−4	−3	−2	−1	0	1	2	3	4
y	−2	−1	0	1	2	3	4	5	6

(c) $y = x + 5$

x	−4	−3	−2	−1	0	1	2	3	4
y	1	2	3	4	5	6	7	8	9

(d) $y = x - 3$

x	−4	−3	−2	−1	0	1	2	3	4
y	−7	−6	−5	−4	−3	−2	−1	0	1

The coordinates are plotted and joined for each graph. The results are shown in Fig. 13.5. Each of the straight lines produced is parallel to each other, i.e. the slope or gradient is the same for each.

To find the gradient of any straight line, say, $y = x - 3$, a horizontal and vertical component needs to be constructed. In Fig. 13.5, AB is constructed vertically at $x = 4$ and BC constructed horizontally at $y = -3$. The gradient of AC

$$= \frac{AB}{BC} = \frac{1 - (-3)}{4 - 0} = \frac{4}{4} = 1$$

i.e. the gradient of the straight line $y = x - 3$ is 1. The actual positioning of AB and BC is unimportant for the gradient is also given by

$$\frac{DE}{EF} = \frac{-1 - (-2)}{2 - 1} = \frac{1}{1} = 1$$

The slope or gradient of each of the straight lines in Fig. 13.5 is thus 1 since they are all parallel to each other.

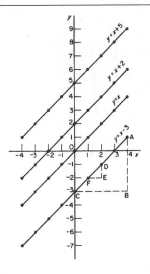

Figure 13.5

Problem 3.   Plot the following graphs on the same axes between the values $x = -3$ to $x = +3$ and determine the gradient and y-axis intercept of each:
(a) $y = 3x$ (b) $y = 3x + 7$ (c) $y = -4x + 4$
(d) $y = -4x - 5$

A table of coordinates is drawn up for each equation.

(a) $y = 3x$

x	−3	−2	−1	0	1	2	3
y	−9	−6	−3	0	3	6	9

(b) $y = 3x + 7$

x	−3	−2	−1	0	1	2	3
y	−2	1	4	7	10	13	16

(c) $y = -4x + 4$

x	−3	−2	−1	0	1	2	3
y	16	12	8	4	0	−4	−8

(d) $y = -4x - 5$

x	−3	−2	−1	0	1	2	3
y	7	3	−1	−5	−9	−13	−17

Each of the graphs is plotted as shown in Fig. 13.6, and each is a straight line. $y = 3x$ and $y = 3x + 7$

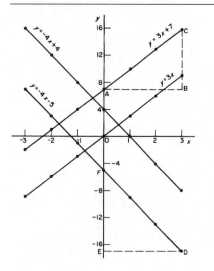

Figure 13.6

are parallel to each other and thus have the same gradient. The gradient of AC is given by

$$\frac{BC}{AC} = \frac{16 - 7}{3 - 0} = \frac{9}{3} = 3$$

Hence the gradient of both $y = 3x$ and $y = 3x + 7$ is 3.

$y = -4x + 4$ and $y = -4x - 5$ are parallel to each other and thus have the same gradient. The gradient of DF is given by

$$\frac{EF}{ED} = \frac{-5 - (-17)}{0 - 3} = \frac{12}{-3} = -4$$

Hence the gradient of both $y = -4x + 4$ and $y = -4x - 5$ is -4.

The y-axis intercept means the value of y where the straight line cuts the y-axis. From Fig. 13.6,

$$y = 3x \text{ cuts the } y\text{-axis at } y = 0$$

$$y = 3x + 7 \text{ cuts the } y\text{-axis at } y = +7$$

$$y = -4x + 4 \text{ cuts the } y\text{-axis at } y = +4$$

and $\quad y = -4x - 5 \text{ cuts the } y\text{-axis at } y = -5$

Some general conclusions can be drawn from the graphs shown in Figs 13.4, 13.5 and 13.6.

When an equation is of the form $y = mx + c$, where m and c are constants, then

(i) a graph of y against x produces a straight line,
(ii) m represents the slope or gradient of the line, and
(iii) c represents the y-axis intercept

Thus, given an equation such as $y = 3x + 7$, it may be deduced 'on sight' that its gradient is $+3$ and its y-axis intercept is $+7$, as shown in Fig. 13.6. Similarly, if $y = -4x - 5$, then the gradient is -4 and the y-axis intercept is -5, as shown in Fig. 13.6.

When plotting a graph of the form $y = mx + c$, only two coordinates need be determined. When the coordinates are plotted a straight line is drawn between the two points. Normally, three coordinates are determined, the third one acting as a check.

Problem 4. The following equations represent straight lines. Determine, without plotting graphs, the gradient and y-axis intercept for each: (a) $y = 3$ (b) $y = 2x$ (c) $y = 5x - 1$ (d) $2x + 3y = 3$

(a) $y = 3$ (which is of the form $y = 0x + 3$) represents a horizontal straight line intercepting the y-axis at 3. Since the line is horizontal its **gradient is zero**

(b) $y = 2x$ is of the form $y = mx + c$, where c is zero. Hence **gradient = 2** and **y-axis intercept = 0** (i.e. the origin)

(c) $y = 5x - 1$ is of the form $y = mx + c$. Hence **gradient = 5** and **y-axis intercept = -1**

(d) $2x + 3y = 3$ is not in the form $y = mx + c$ as it stands. Transposing to make y the subject gives $3y = 3 - 2x$, i.e.

$$y = \frac{3 - 2x}{3} = \frac{3}{3} - \frac{2x}{3}$$

i.e. $\quad y = -\frac{2x}{3} + 1$

which is of the form $y = mx + c$.

Hence **gradient** $= -\dfrac{2}{3}$ and **y-axis intercept = +1**

Problem 5. Without plotting graphs, determine the gradient and y-axis intercept values of the following equations:
(a) $y = 7x - 3$ (b) $3y = -6x + 2$
(c) $y - 2 = 4x + 9$ (d) $\dfrac{y}{3} = \dfrac{x}{3} - \dfrac{1}{5}$
(e) $2x + 9y + 1 = 0$

(a) $y = 7x - 3$ is of the form $y = mx + c$, hence **gradient, $m = 7$** and **y-axis intercept, $c = -3$**

(b) Rearranging $3y = -6x + 2$ gives

$$y = -\frac{6x}{3} + \frac{2}{3},$$

i.e. $y = -2x + \frac{2}{3}$

which is of the form $y = mx + c$. Hence **gradient $m = -2$ and y-axis intercept, $c = \frac{2}{3}$**

(c) Rearranging $y - 2 = 4x + 9$ gives $y = 4x + 11$, hence **gradient $= 4$** and **y-axis intercept $= 11$**

(d) Rearranging $\frac{y}{3} = \frac{x}{2} - \frac{1}{5}$ gives

$$y = 3\left(\frac{x}{2} - \frac{1}{5}\right) = \frac{3}{2}x - \frac{3}{5}$$

Hence **gradient $= \frac{3}{2}$** and **y-axis intercept $= -\frac{3}{5}$**

(e) Rearranging $2x + 9y + 1 = 0$ gives

$$9y = -2x - 1$$

i.e. $y = -\frac{2}{9}x - \frac{1}{9}$

Hence **gradient $= -\frac{2}{9}$** and **y-axis intercept $= -\frac{1}{9}$**

Problem 6. Determine the gradient of the straight line graph passing through the coordinates (a) $(-2, 5)$ and $(3, 4)$, and (b) $(-2, -3)$ and $(-1, 3)$

A straight line graph passing through coordinates (x_1, y_1) and (x_2, y_2) has a gradient given by

$$m = \frac{y_2 - y_1}{x_2 - x_1} \quad \text{(see Fig. 13.7)}$$

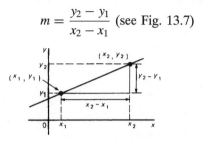

Figure 13.7

(a) A straight line passes through $(-2, 5)$ and $(3, 4)$, hence $x_1 = -2$, $y_1 = 5$, $x_2 = 3$ and $y_2 = 4$;

hence gradient $m = \dfrac{y_2 - y_1}{x_2 - x_1}$

$$= \frac{4 - 5}{3 - (-2)} = -\frac{1}{5}$$

(b) A straight line passes through $(-2, -3)$ and $(-1, 3)$, hence $x_1 = -2$, $y_1 = -3$, $x_2 = -1$ and $y_2 = 3$;

hence gradient $m = \dfrac{y_2 - y_1}{x_2 - x_1} = \dfrac{3 - (-3)}{-1 - (-2)}$

$$= \frac{3 + 3}{-1 + 2} = \frac{6}{1} = 6$$

Problem 7. Plot the graph $3x + y + 1 = 0$ and $2y - 5 = x$ on the same axes and find their point of intersection

---

Rearranging $3x + y + 1 = 0$ gives $y = -3x - 1$.
Rearranging $2y - 5 = x$ gives $2y = x + 5$ and $y = \frac{1}{2}x + 2\frac{1}{2}$.

Since both equations are of the form $y = mx + c$ both are straight lines. Knowing an equation is a straight line means that only two coordinates need to be plotted and a straight line drawn through them. A third coordinate is usually determined to act as a check. A table of values is produced for each equation as shown below.

x	1	0	-1
$-3x - 1$	-4	-1	2

x	2	0	-3
$\frac{1}{2}x + 2\frac{1}{2}$	$3\frac{1}{2}$	$2\frac{1}{2}$	1

The graphs are plotted as shown in Fig. 13.8. **The two straight lines are seen to intersect at $(-1, 2)$.**

Further problems on straight line graphs may be found in Section 13.4. Problems 1 to 11, pages 141 and 142.

13.3 Practical problems involving straight line graphs

When a set of coordinate values are given or are obtained experimentally and it is believed that they

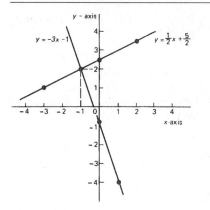

Figure 13.8

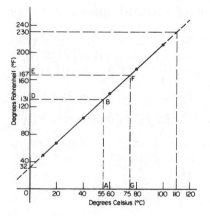

Figure 13.9

follow a law of the form $y = mx + c$, then if a straight line can be drawn reasonably close to most of the coordinate values when plotted, this verifies that a law of the form $y = mx + c$ exists. From the graph, constants m (i.e. gradient) and c (i.e. y-axis intercept) can be determined. This technique is called **determination of law** (see Chapter 14).

Problem 8.   The temperature in degrees Celsius and the corresponding values in degrees Fahrenheit are shown in the table below. Construct rectangular axes, choose a suitable scale and plot a graph of degrees Celsius (on the horizontal axis) against degrees Fahrenheit (on the vertical scale).

°C	10	20	40	60	80	100
°F	50	68	104	140	176	212

From the graph find (a) the temperature in degrees Fahrenheit at 55 °C, (b) the temperature in degrees Celsius at 167 °F, (c) the Fahrenheit temperature at 0 °C, and (d) the Celsius temperature at 230 °F

The coordinates (10, 50), (20, 68), (40, 104), and so on are plotted as shown in Fig. 13.9. When the coordinates are joined, a straight line is produced. Since a straight line results there is a linear relationship between degrees Celsius and degrees Fahrenheit.

(a)   To find the Fahrenheit temperature at 55 °C a vertical line AB is constructed from the horizontal axis to meet the straight line at B. The point where the horizontal line BD

meets the vertical axis indicates the equivalent Fahrenheit temperature.
Hence 55 °C is equivalent to 131 °F
This process of finding an equivalent value in between the given information in the above table is called **interpolation**

(b)   To find the Celsius temperature at 167 °F, a horizontal line EF is constructed as shown in Fig. 13.9. The point where the vertical line FG cuts the horizontal axis indicates the equivalent Celsius temperature.
Hence 167 °F is equivalent to 75 °C

(c)   If the graph is assumed to be linear even outside of the given data, then the graph may be extended at both ends (shown by broken lines in Fig. 13.9).
From Fig. 13.9, **0 °C corresponds to 32 °F**

(d)   **230 °F is seen to correspond to 110 °C.**
The process of finding equivalent values outside of the given range is called **extrapolation**

Problem 9.   In an experiment on Charles's law, the value of the volume of gas, V m³, was measured for various temperatures T °C. Results are shown below.

V m³	25.0	25.8	26.6	27.4	28.2	29.0
T °C	60	65	70	75	80	85

Plot a graph of volume (vertical) against temperature (horizontal) and from it find (a) the temperature when the volume is 28.6 m³, and (b) the volume when the temperature is 67 °C

If a graph is plotted with both the scales starting at zero then the result is as shown in Fig. 13.10. All of the points lie in the top right-hand corner of the graph, making interpolation difficult. A more accurate graph is obtained if the temperature axis starts at 55 °C and the volume axis starts at 24.5 m³. The axes corresponding to these values is shown by the broken lines in Fig. 13.10 and are called **false axes**, since the origin is not now at zero. A magnified version of this relevant part of the graph is shown in Fig. 13.11. From the graph:

(a)   when the volume is 28.6 m³, the equivalent temperature is **82.5 °C**, and

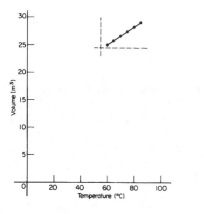

Figure 13.10

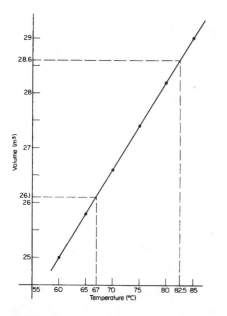

Figure 13.11

(b)   when the temperature is 67 °C, the equivalent volume is **26.1 m³**

Problem 10.   In an experiment demonstrating Hooke's law, the strain in an aluminium wire was measured for various stresses. The results were:

Stress N/mm²	4.9	8.7	15.0
Strain	0.00007	0.00013	0.00021

Stress N/mm²	18.4	24.2	27.3
Strain	0.00027	0.00034	0.00039

Plot a graph of stress (vertically) against strain (horizontally). Find:

(a)   Young's Modulus of Elasticity for aluminium which is given by the gradient of the graph,

(b)   the value of the strain at a stress of 20 N/mm², and

(c)   the value of the stress when the strain is 0.00020

The coordinates (0.00007, 4.9), (0.00013, 8.7), and so on, are plotted as shown in Fig. 13.12. The graph produced is the best straight line which can be drawn corresponding to these points. (With experimental results it is unlikely that all the points will lie exactly on a straight line.) The graph, and each of its axes, are labelled. Since the straight line passes through the origin, then stress is directly proportional to strain for the given range of values.

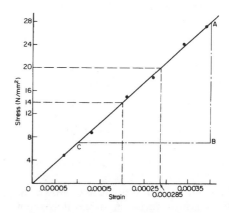

Figure 13.12

(a)   The gradient of the straight line AC is given by

$$\frac{AB}{BC} = \frac{28 - 7}{0.00040 - 0.00010} = \frac{21}{0.00030}$$

$$= \frac{21}{3 \times 10^{-4}} = \frac{7}{10^{-4}} = 7 \times 10^{4}$$

$$= 70\,000 \text{ N/mm}^2$$

Thus Young's Modulus of Elasticity for aluminium is 70 000 N/mm².

Since $1 \text{ m}^2 = 10^6 \text{ mm}^2$, $70\,000 \text{ N/mm}^2$ is equivalent to $70\,000 \times 10^6 \text{ N/m}^2$, i.e. **$70 \times 10^9 \text{ N/m}^2$ (or Pascals)**.

From Fig. 13.12:

(b)   the value of the strain at a stress of 20 N/mm^2 is **0.000285**, and

(c)   the value of the stress when the strain is 0.00020 is **14 N/mm²**

Problem 11.   The following values of resistance R ohms and corresponding voltage V volts are obtained from a test on a filament lamp.

R ohms	30	48.5	73	107	128
V volts	16	29	52	76	94

Choose suitable scales and plot a graph with R representing the vertical axis and V the horizontal axis. Determine (a) the slope of the graph, (b) the R axis intercept value, (c) the equation of the graph, (d) the value of resistance when the voltage is 60 V, and (e) the value of the voltage when the resistance is 40 ohms. (f) If the graph were to continue in the same manner, what value of resistance would be obtained at 110 V?

The coordinates (16, 30), (29, 48.5), and so on, are shown plotted in Fig. 13.13 where the best straight line is drawn through the points.

(a)   The slope or gradient of the straight line AC is given by

$$\frac{AB}{BC} = \frac{135 - 10}{100 - 0} = \frac{125}{100} = \textbf{1.25}$$

(Note that the vertical line AB and the horizontal line BC may be constructed anywhere along the length of the straight line. However, calculations are made easier if the horizontal line BC is carefully chosen, in this case, 100)

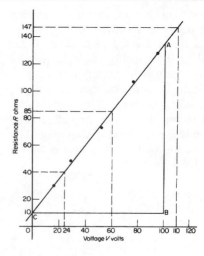

Figure 13.13

(b)   The R-axis intercept is at **$R = 10 \, \Omega$** (by extrapolation)

(c)   The equation of a straight line is $y = mx + c$, when y is plotted on the vertical axis and x on the horizontal axis. m represents the gradient and c the y-axis intercept. In this case, R corresponds to y, V corresponds to x, $m = 1.25$ and $c = 10$. Hence the equation of the graph is **$R = (1.25\,V + 10) \, \Omega$**

From Fig. 13.13,

(d)   when the voltage is 60 V, the resistance is **85 Ω**,

(e)   when the resistance is 40 ohms, the voltage is **24 V**, and

(f)   by extrapolation, when the voltage is 110 V, the resistance is **147 Ω**

Problem 12.   Experimental tests to determine the breaking stress σ of rolled copper at various temperatures t gave the following results.

Stress σ N/cm²	8.46	8.02	7.75	7.35	7.06	6.63
Temperature t °C		70	200	280	410	500	640

Show that the values obey the law $\sigma = at + b$, where a and b are constants and determine approximate values for a and b. Use the law to determine the stress at 250 °C and the temperature when the stress is 7.54 N/cm²

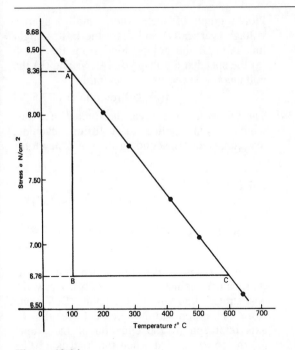

Figure 13.14

The coordinates $(70, 8.46)$, $(200, 8.04)$, and so on, are plotted as shown in Fig. 13.14. Since the graph is a straight line then the values obey the law $\sigma = at + b$, and the gradient of the straight line is

$$a = \frac{AB}{BC} = \frac{8.36 - 6.76}{100 - 600} = \frac{1.60}{-500} = -0.0032$$

Vertical axis intercept, $b = 8.68$.

Hence the law of the graph is $\sigma = -0.0032t + 8.68$.

When the temperature is $250\,°C$, stress σ is given by

$$\sigma = -0.0032(250) + 8.68 = 7.88 \text{ N/cm}^2$$

Rearranging $\sigma = -0.0032t + 8.68$ gives

$$0.0032t = 8.68 - \sigma, \text{ i.e. } t = \frac{8.68 - \sigma}{0.0032}$$

Hence when the stress $\sigma = 7.54 \text{ N/cm}^2$,

$$\text{temperature } t = \frac{8.68 - 7.54}{0.0032} = 356.3\,°C$$

Further practical problems involving straight line graphs may be found in the following Section 13.4, Problems 12 to 22, pages 142 and 143.

13.4 Further problems on straight line graphs

The straight line graph

1 Corresponding values obtained experimentally for two quantities are:

x	-2.0	-0.5	0	1.0	2.5	3.0	5.0
y	-13.0	-5.5	-3.0	2.0	9.5	12.0	22.0

Use a horizontal scale for x of 1 cm $= \frac{1}{2}$ unit and a vertical scale for y of 1 cm $= 2$ units and draw a graph of x against y. Label the graph and each of its axes. By interpolation, find from the graph the value of y when x is 3.5. [14.5]

2 The equation of a line is $4y = 2x+5$. A table of corresponding values is produced and is shown below. Complete the table and plot a graph of y against x. Find the gradient of the graph.

x	-4	-3	-2	-1	0	1	2	3	4
y		-0.25			1.25				3.25

$$\left[\frac{1}{2}\right]$$

3 Determine the gradient and intercept on the y-axis for each of the following equations:
(a) $y = 4x - 2$ (b) $y = -x$ (c) $y = -3x - 4$ (d) $y = 4$

$$[\text{(a) } 4, -2 \text{ (b) } -1, 0 \text{ (c) } -3, -4 \text{ (d) } 0, 4]$$

4 Find the gradient and intercept on the y-axis for each of the following equations:
(a) $2y - 1 = 4x$ (b) $6x - 2y = 5$
(c) $3(2y - 1) = \dfrac{x}{4}$

$$\left[\text{(a) } 2, \frac{1}{2} \text{ (b) } 3, -2\frac{1}{2} \text{ (c) } \frac{1}{24}, \frac{1}{2}\right]$$

Determine the gradient and y-axis intercept for each of the equations in Problems 5 and 6 and sketch the graphs.

5 (a) $y = 6x - 3$ (b) $y = -2x + 4$ (c) $y = 3x$ (d) $y = 7$

$$[\text{(a) } 6, -3 \text{ (b) } -2, 4 \text{ (c) } 3, 0 \text{ (d) } 0, 7]$$

6 (a) $2y + 1 = 4x$ (b) $2x + 3y + 5 = 0$
(c) $3(2y - 4) = \dfrac{x}{3}$ (d) $5x - \dfrac{y}{2} - \dfrac{7}{3} = 0$

$$\left[\text{(a) } 2, -\frac{1}{2} \text{ (b) } -\frac{2}{3}, -1\frac{2}{3} \text{ (c) } \frac{1}{18}, 2 \text{ (d) } 10, -4\frac{2}{3}\right]$$

7   Determine the gradient of the straight line graphs passing through the coordinates
(a) $(2, 7)$ and $(-3, 4)$ (b) $(-4, -1)$ and $(-5, 3)$
(c) $\left(\dfrac{1}{4}, -\dfrac{3}{4}\right)$ and $\left(-\dfrac{1}{2}, \dfrac{5}{8}\right)$

$$\left[\text{(a)}\dfrac{3}{5} \text{ (b) } -4 \text{ (c) } -1\dfrac{5}{6}\right]$$

8   State which of the following equations will produce graphs which are parallel to one another:
(a) $y - 4 = 2x$ (b) $4x = -(y + 1)$
(c) $x = \dfrac{1}{2}(y + 5)$ (d) $1 + \dfrac{1}{2}y = \dfrac{3}{2}x$
(e) $2x = \dfrac{1}{2}(7 - y)$

[(a) and (c), (b) and (e)]

9   Draw a graph of $y - 3x + 5 = 0$ over a range of $x = -3$ to $x = 4$. Hence determine (a) the value of y when $x = 1.3$, and (b) the value of x when $y = -9.2$.

[(a) -1.1 (b) -1.4]

10   Draw on the same axes the graphs of $y = 3x - 5$ and $3y + 2x = 7$. Find the coordinates of the point of intersection. Check the result obtained by solving the two simultaneous equations algebraically.   [(2, 1)]

11   Plot the graphs $y = 2x + 3$ and $2y = 15 - 2x$ on the same axes and determine their point of intersection.

$$\left[1\dfrac{1}{2}, 6\right]$$

Practical problems involving straight line graphs

12   The resistance R ohms of a copper winding is measured at various temperatures $t\,°C$ and the results are as follows:

R ohms	112	120	126	131	136
$t\,°C$	20	36	48	58	64

Plot a graph of R (vertically) against t (horizontally) and find from it (a) the temperature when the resistance is 122 Ω, and (b) the resistance when the temperature is $52\,°C$.

[(a) $40\,°C$ (b) 128 Ω]

13   The speed of a motor varies with armature voltage as shown by the following experimental results:

n (rev/min)	285	517	615	750	917	1050
V volts	60	95	110	130	155	175

Plot a graph of speed (horizontally) against voltage (vertically) and draw the best straight line through the points. Find from the graph (a) the speed at a voltage of 145 V, and (b) the voltage at a speed of 400 rev/min.

[(a) 850 rev/min (b) 77.5 V]

14   The following table gives the force F newtons which, when applied to a lifting machine, overcomes a corresponding load of L newtons.

Force F newtons	25	47	64	120	149	187
Load L newtons	50	140	210	430	550	700

Choose suitable scales and plot a graph of F (vertically) against L (horizontally). Draw the best straight line through the points. Determine from the graph (a) the gradient, (b) the F-axis intercept, (c) the equation of the graph, (d) the force applied when the load is 310 N, and (e) the load that a force of 160 N will overcome. (f) If the graph were to continue in the same manner, what value of force will be needed to overcome an 800 N load?

$$\left[\begin{array}{l}\text{(a) } 0.25 \text{ (b) } 12 \text{ (c) } F = 0.25\,L + 12 \\ \text{(d) } 89.5 \text{ N (e) } 592 \text{ N (f) } 212 \text{ N}\end{array}\right]$$

15   The following table gives the results of tests carried out to determine the breaking stress σ of rolled copper at various temperatures, t:

Stress σ (N/cm^2)	8.51	8.07	7.80	7.47	7.23	6.78
Temperature $t(°C)$	75	220	310	420	500	650

Plot a graph of stress (vertically) against temperature (horizontally). Draw the best straight line through the plotted coordinates. Determine the slope of the graph and the vertical axis intercept.   [-0.003, 8.73]

16   The velocity v of a body after varying time intervals t was measured as follows:

t (seconds)	2	5	8	11	15	18
v (m/s)	16.9	19.0	21.1	23.2	26.0	28.1

Plot v vertically and t horizontally and draw a graph of velocity against time. Determine from the graph (a) the velocity after 10 s, (b) the time at 20 m/s, and (c) the equation of the graph.

$$[\text{(a) } 22.5 \text{ m/s (b) } 6.43 \text{ s (c) } v = 0.7t + 15.5]$$

17 The mass m of a steel joist varies with length l as follows:

mass, m (kg)	80	100	120	140	160
length, l (m)	3.00	3.74	4.48	5.23	5.97

Plot a graph of mass (vertically) against length (horizontally). Determine the equation of the graph. $[m = 26.9l - 0.63]$

18 The crushing strength of mortar varies with the percentage of water used in its preparation, as shown below:

Crushing strength, F (tonnes)	1.64	1.36	1.07	0.78	0.50	0.22
% of water used, $w\%$	6	9	12	15	18	21

Plot a graph of F (vertically) against w (horizontally).
(a) Interpolate and determine the crushing strength when 10% of water is used
(b) Assuming the graph continues in the same manner extrapolate and determine the percentage of water used when the crushing strength is 0.15 tonnes
(c) What is the equation of the graph?

$$\begin{bmatrix} \text{(a) } 1.26t \text{ (b) } 21.68\% \\ \text{(c) } F = -0.095w + 2.21 \end{bmatrix}$$

19 The velocity v of a body after varying time intervals t was measured as follows:

t seconds	2	5	7	10	14	17
v m/s	15.5	17.3	18.5	20.3	22.7	24.5

Plot a graph with velocity vertical and time horizontal. Determine from the graph (a) the gradient, (b) the vertical axis intercept, (c) the equation of the graph, (d) the velocity after 12.5 s, and (e) the time when the velocity is 18 m/s.

$$\begin{bmatrix} \text{(a) } 0.6 \text{ (b) } 14.3 \text{ (c) } v = 0.6t + 14.3 \\ \text{(d) } 21.8 \text{ m/s (e) } 6.17 \text{ s} \end{bmatrix}$$

20 In an experiment demonstrating Hooke's law, the strain in a copper wire was measured for various stresses. The results were:

Stress (pascals)	10.6×10^6	18.2×10^6	24.0×10^6
Strain	0.00011	0.00019	0.00025

Stress (pascals)	30.7×10^6	39.4×10^6
Strain	0.00032	0.00041

Plot a graph of stress (vertically) against strain (horizontally). Determine (a) Young's Modulus of Elasticity for copper, which is given by the gradient of the graph, (b) the value of strain at a stress of 21×10^6 Pa, and (c) the value of stress when the strain is 0.00030.

$$\begin{bmatrix} \text{(a) } 96 \times 10^9 \text{ Pa (b) } 0.00022 \\ \text{(c) } 28.8 \times 10^6 \text{ Pa} \end{bmatrix}$$

21 An experiment with a set of pulley blocks gave the following results:

Effort, E (newtons)	9.0	11.0	13.6	17.4	20.8	23.6
Load, L (newtons)	15	25	38	57	74	88

Plot a graph of effort (vertically) against load (horizontally) and determine (a) the gradient, (b) the vertical axis intercept, (c) the law of the graph, (d) the effort when the load is 30 N, and (e) the load when the effort is 19 N.

$$\begin{bmatrix} \text{(a) } \dfrac{1}{5} \text{ (b) } 6 \text{ (c) } E = \dfrac{1}{5}L + 6 \\ \text{(d) } 12 \text{ N (e) } 65 \text{ N} \end{bmatrix}$$

22 The variation of pressure p in a vessel with temperature T is believed to follow a law of the form $p = aT + b$, where a and b are constants. Verify this law for the results given below and determine the approximate values of a and b. Hence determine the pressures at temperatures of 285 K and 310 K and the temperature at a pressure of 250 kPa.

pressure, p kPa	244	247	252	258	262	267
temperature, T K	273	277	282	289	294	300

$$\begin{bmatrix} a = 0.85, \ b = 12, \\ 254.3 \text{ kPa}, \ 275.5 \text{ kPa}, \ 280 \text{ K} \end{bmatrix}$$

14

Reduction of non-linear laws to linear form

14.1 Determination of law

Frequently, the relationship between two variables, say x and y, is not a linear one, i.e. when x is plotted against y a curve results. In such cases the non-linear equation may be modified to the linear form, $y = mx + c$, so that the constants, and thus the law relating the variables, can be determined.

This technique is called '**determination of law**'.

Some examples of the reduction of equations to linear form include:

(i) $y = ax^2 + b$ compares with $Y = mX + c$, where $m = a$, $c = b$ and $X = x^2$. Hence y is plotted vertically against x^2 horizontally to produce a straight line graph of gradient 'a' and y-axis intercept 'b'

(ii) $y = \dfrac{a}{x} + b$

y is plotted vertically against l/x horizontally to produce a straight line graph of gradient 'a' and y-axis intercept 'b'

(iii) $y = ax^2 + bx$

Dividing both sides by x gives $y/x = ax + b$. Comparing with $Y = mX + c$ shows that y/x is plotted vertically against x horizontally to produce a straight line graph of gradient 'a' and y/x axis intercept 'b'

(iv) $y = ax^n$

Taking logarithms to a base of 10 of both sides gives:

$$\lg y = \lg(ax^n) = \lg a + \lg x^n$$

i.e. $\lg y = n \lg x + \lg a$

which compares with $Y = mX + c$ and shows that $\lg y$ is plotted vertically against $\lg x$ horizontally to produce a straight line graph of gradient n and $\lg y$-axis intercept $\lg a$

(v) $y = ab^x$

Taking logarithms to a base of 10 of both sides gives:

$$\lg y = \lg(ab^x)$$

i.e. $\lg y = \lg a + \lg b^x$

i.e. $\lg y = x \lg b + \lg a$

or $\lg y = (\lg b)x + \lg a$

which compares with $Y = mX + c$ and shows that $\lg y$ is plotted vertically against x horizontally to produce a straight line graph of gradient $\lg b$ and $\lg y$-axis intercept $\lg a$

(vi) $y = ae^{bx}$

Taking logarithms to a base of e of both sides gives:

$$\ln y = \ln(ae^{bx})$$

i.e. $\ln y = \ln a + \ln e^{bx}$

i.e. $\ln y = \ln a + bx \ln e$

i.e. $\ln y = bx + \ln a$

which compares with $Y = mX + c$ and shows that $\ln y$ is plotted vertically against x horizontally to produce a straight line graph of gradient b and $\ln y$-axis intercept $\ln a$

14.2 Worked problems on reducing non-linear laws to linear form

Problem 1. Experimental values of x and y, shown below, are believed to be related by the law $y = ax^2 + b$. By plotting a suitable graph verify this law and determine approximate values of a and b

x	1	2	3	4	5
y	9.8	15.2	24.2	36.5	53.0

If y is plotted against x a curve results and it is not possible to determine the values of constants a and b from the curve. Comparing $y = ax^2 + b$ with $Y = mX + c$ shows that y is to be plotted vertically against x^2 horizontally. A table of values is drawn up as shown below.

x	1	2	3	4	5
x^2	1	4	9	16	25
y	9.8	15.2	24.2	36.5	53.0

A graph of y against x^2 is shown in Fig. 14.1, with the best straight line drawn through the points. Since a straight line graph results, the law is verified. From the graph, gradient

$$a = \frac{AB}{BC} = \frac{53 - 17}{25 - 5} = \frac{36}{20} = 1.8$$

and the y-axis intercept, $b = 8.0$.
Hence the law of the graph is $y = 1.8x^2 + 8.0$.

Problem 2. Values of load L newtons and distance d metres obtained experimentally are shown in the following table:

Load, L N	32.3	29.6	27.0	23.2
distance, d m	0.75	0.37	0.24	0.17

Load, L N	18.3	12.8	10.0	6.4
distance, d m	0.12	0.09	0.08	0.07

Verify that load and distance are related by a law of the form $L = \dfrac{a}{d} + b$ and determine approximate values of a and b. Hence calculate the load when the distance is 0.20 m and the distance when the load is 20 N

Comparing $L = \dfrac{a}{d} + b$, i.e. $L = a\dfrac{1}{d} + b$ with $Y = mX + c$ shows that L is to be plotted vertically against l/d horizontally. Another table of values is drawn up as shown below.

L	32.3	29.6	27.0	23.2	18.3	12.8	10.0	6.4
d	0.75	0.37	0.24	0.17	0.12	0.09	0.08	0.07
l/d	1.33	2.70	4.17	5.88	8.33	11.1	12.5	14.3

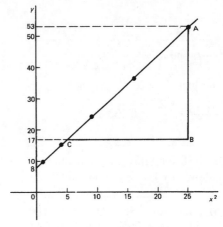

Figure 14.1

A graph of L against l/d is shown in Fig. 14.2. A straight line can be drawn through the points, which verifies that load and distance are related by a law of the form $L = \dfrac{a}{d} + b$.

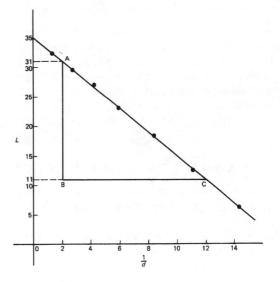

Figure 14.2

Gradient of straight line,

$$a = \frac{AB}{BC} = \frac{31 - 11}{2 - 12} = \frac{20}{-10} = -2$$

L-axis intercept, $b = 35$.

Hence the law of the graph is $L = -\dfrac{2}{d} + 35$.
When the distance $d = 0.20$ m,

$$\text{load } L = \frac{-2}{0.20} + 35 = \mathbf{25.0\ N}$$

Rearranging $L = -\dfrac{2}{d} + 35$ gives

$$\dfrac{2}{d} = 35 - L \text{ and } d = \dfrac{2}{35 - L}$$

Hence when the load $L = 20$ N, distance

$$d = \dfrac{2}{35 - 20} = \dfrac{2}{15} = \mathbf{0.133 \text{ m}}$$

Problem 3.   The solubility s of potassium chlorate is shown by the following table:

$t\,^{\circ}$C	10	20	30	40	50	60	80	100
s	4.9	7.6	11.1	15.4	20.4	26.4	40.6	58.0

The relationship between s and t is thought to be of the form $s = 3 + at + bt^2$. Plot a graph to test the supposition and use the graph to find approximate values of a and b. Hence calculate the solubility of potassium chlorate at $70\,^{\circ}$C

Rearranging $s = 3 + at + bt^2$ gives $s - 3 = at + bt^2$ and $(s-3)/t = a + bt$ or $(s-3)/t = bt + a$ which is of the form $Y = mX + c$, showing that $(s-3)/t$ is to be plotted vertically and t horizontally. Another table of values is drawn up as shown below.

t	10	20	30	40	50	60	80	100
s	4.9	7.6	11.1	15.4	20.4	26.4	40.6	58.0
$\left(\dfrac{s-3}{t}\right)$	0.19	0.23	0.27	0.31	0.35	0.39	0.47	0.55

A graph of $(s-3)/t$ against t is shown plotted in Fig. 14.3.
A straight line fits the points which shows that s and t are related by $s = 3 + at + bt^2$.
Gradient of straight line,

$$b = \dfrac{AB}{BC} = \dfrac{0.39 - 0.19}{60 - 10} = \dfrac{0.20}{50} = \mathbf{0.004}$$

Vertical axis intercept, $a = \mathbf{0.15}$.
Hence the law of the graph is

$$s = 3 + 0.15t + 0.004t^2$$

The solubility of potassium chlorate at $70\,^{\circ}$C is given by

$$s = 3 + 0.15(70) + 0.004(70)^2$$
$$= 3 + 10.5 + 19.6 = \mathbf{33.1}$$

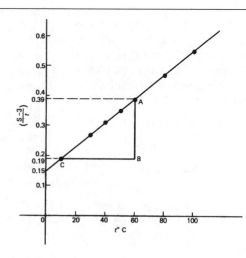

Figure 14.3

Problem 4.   The current flowing in, and the power dissipated by, a resistor are measured experimentally for various values and the results are as shown below.

Current, I amperes	2.2	3.6	4.1	5.6	6.8
Power, P watts	116	311	403	753	1110

Show that the law relating current and power is of the form $P = RI^n$, where R and n are constants, and determine the law

Taking logarithms to a base of 10 of both sides of $P = RI^n$ gives:

$$\lg P = \lg(RI^n) = \lg R + \lg I^n = \lg R + n \lg I$$

i.e. $\lg P = n \lg I + \lg R$,

which is of the form $Y = mX + c$, showing that $\lg P$ is to be plotted vertically against $\lg I$ horizontally. A table of values for $\lg I$ and $\lg P$ is drawn up as shown below.

I	2.2	3.6	4.1	5.6	6.8
$\lg I$	0.342	0.556	0.613	0.748	0.833
P	116	311	403	753	1110
$\lg P$	2.064	2.493	2.605	2.877	3.045

A graph of $\lg P$ against $\lg I$ is shown in Fig. 14.4 and since a straight line results the law $P = RI^n$ is verified.

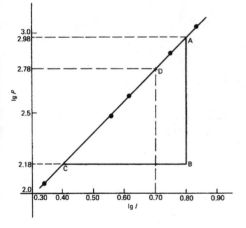

Figure 14.4

Gradient of straight line,

$$n = \frac{AB}{BC} = \frac{2.98 - 2.18}{0.8 - 0.4} = \frac{0.80}{0.4} = 2$$

It is not possible to determine the vertical axis intercept on sight since the horizontal axis scale does not start at zero. Selecting any point from the graph, say point D, where $\lg I = 0.70$ and $\lg P = 2.78$, and substituting values into $\lg P = n \lg I + \lg R$ gives

$$2.78 = (2)(0.70) + \lg R$$

from which $\quad \lg R = 2.78 - 1.40 = 1.38$

Hence $\quad R = \text{antilog } 1.38(= 10^{1.38}) = 24.0$

Hence the law of the graph is $P = 24.0\, I^2$.

Problem 5. The periodic time, T, of oscillation of a pendulum is believed to be related to its length, l, by a law of the form $T = kl^n$, where k and n are constants. Values of T were measured for various lengths of the pendulum and the results are as shown below.

Periodic time, T s	1.0	1.3	1.5	1.8	2.0	2.3
Length, l m	0.25	0.42	0.56	0.81	1.0	1.32

Show that the law is true and determine the approximate values of k and n. Hence find the periodic time when the length of the pendulum is 0.75 m

From para (iv) of Section 14.1, if $T = kl^n$ then

$$\lg T = n \lg l + \lg k$$

and comparing with

$$Y = mX + c$$

shows that $\lg T$ is plotted vertically against $\lg l$ horizontally.
A table of values for $\lg T$ and $\lg l$ is drawn up as shown below.

T	1.0	1.3	1.5	1.8	2.0	2.3
$\lg T$	0	0.114	0.176	0.255	0.301	0.362
l	0.25	0.42	0.56	0.81	1.0	1.32
$\lg l$	−0.602	−0.377	−0.252	−0.092	0	0.121

A graph of $\lg T$ against $\lg l$ is shown in Fig. 14.5 and the law $T = kl^n$ is true since a straight line results.

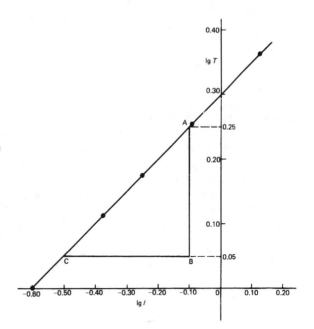

Figure 14.5

From the graph, gradient of straight line,

$$n = \frac{AB}{BC} = \frac{0.25 - 0.05}{-0.10 - (-0.50)} = \frac{0.20}{0.40} = \frac{1}{2}$$

Vertical axis intercept, $\lg k = 0.30$.
Hence $k = \text{antilog } 0.30(= 10^{0.30}) = 2.0$.

Hence the law of the graph is $T = 2.0\ l^{1/2}$

$$\text{or } T = 2.0\sqrt{l}$$

When length $l = 0.75$ m then $T = 2.0\sqrt{(0.75)}$

$$= \mathbf{1.73 \ s}$$

Problem 6.   Quantities x and y are believed to be related by a law of the form $y = ab^x$, where a and b are constants. Values of x and corresponding values of y are:

x	0	0.6	1.2	1.8	2.4	3.0
y	5.0	9.67	18.7	36.1	69.8	135.0

Verify the law and determine the approximate values of a and b. Hence determine (a) the value of y when x is 2.1, and (b) the value of x when y is 100

From para (v) of Section 14.1, if $y = ab^x$ then

$$\lg y = (\lg b)x + \lg a$$

and comparing with

$$Y = mX + c$$

shows that $\lg y$ is plotted vertically and x horizontally.
Another table is drawn up as shown below.

x	0	0.6	1.2	1.8	2.4	3.0
y	5.0	9.67	18.7	36.1	69.8	135.0
$\lg y$	0.70	0.99	1.27	1.56	1.84	2.13

A graph of $\lg y$ against x is shown in Fig. 14.6 and since a straight line results, the law $y = ab^x$ is verified.
Gradient of straight line,

$$\lg b = \frac{AB}{BC} = \frac{2.13 - 1.17}{3.0 - 1.0} = \frac{0.96}{2.0} = 0.48$$

Hence $b = $ antilog $0.48 (= 10^{0.48}) = \mathbf{3.0}$, correct to 2 significant figures.
Vertical axis intercept, $\lg a = 0.70$, from which

$$a = \text{antilog } 0.70 (= 10^{0.70})$$

$$= \mathbf{5.0}, \text{ correct to 2 significant figures}$$

Hence the law of the graph is $y = 5.0(3.0)^x$.

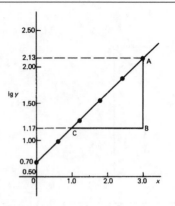

Figure 14.6

(a)   When $x = 2.1$, $y = 5.0(3.0)^{2.1} = \mathbf{50.2}$
(b)   When $y = 100$, $100 = 5.0(3.0)^x$, from which $100/5.0 = (3.0)^x$, i.e. $20 = (3.0)^x$

Taking logarithms of both sides gives

$$\lg 20 = \lg(3.0)^x = x \lg 3.0$$

Hence $x = \dfrac{\lg 20}{\lg 3.0} = \dfrac{1.3010}{0.4771} = \mathbf{2.73}$

Problem 7.   The current i mA flowing in a capacitor which is being discharged varies with time t ms as shown below.

i mA	203	61.14	22.49	6.13	2.49	0.615
t ms	100	160	210	275	320	390

Show that these results are related by a law of the form $i = Ie^{t/T}$, where I and T are constants. Determine the approximate values of I and T

Taking Napierian logarithms of both sides of $i = Ie^{t/T}$ gives

$$\ln i = \ln(Ie^{t/T}) = \ln I + \ln e^{t/T}$$

i.e. $\ln i = \ln I + \dfrac{t}{T}$ (since $\ln e = 1$)

or $\ln i = \left(\dfrac{1}{T}\right)t + \ln I$

which compares with $y = mx + c$, showing that $\ln i$ is plotted vertically against t horizontally. (For methods of evaluating Napierian logarithms see

Chapter 8.) Another table of values is drawn up as shown below.

t	100	160	210	275	320	390
i	203	61.14	22.49	6.13	2.49	0.615
$\ln i$	5.31	4.11	3.11	1.81	0.91	−0.49

A graph of $\ln i$ against t is shown in Fig. 14.7 and since a straight line results the law $i = Ie^{t/T}$ is verified.

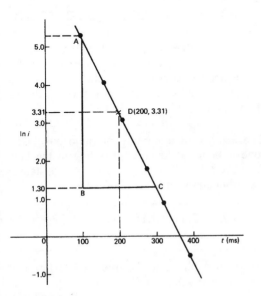

Figure 14.7

Gradient of straight line,

$$\frac{1}{T} = \frac{AB}{BC} = \frac{5.30 - 1.30}{100 - 300} = \frac{4.0}{-200} = -0.02$$

Hence $T = \dfrac{1}{-0.02} = \mathbf{-50}$

Selecting any point on the graph, say point D, where $t = 200$ and $\ln i = 3.31$, and substituting into

$$\ln i = \left(\frac{1}{T}\right) t + \ln I$$

gives

$$3.31 = -\left(\frac{1}{50}\right)(200) + \ln I$$

from which

$$\ln I = 3.31 + 4.0 = 7.31$$

and $I = $ antilog $7.31(= e^{7.31}) = 1495$ or **1500** correct to 3 significant figures.

Hence the law of the graph is $i = 1500e^{-t/50}$.

14.3 Further problems on reducing non-linear laws to linear form

In Problems 1 to 4, x and y are two related variables and all other letters denote constants. For the stated laws to be verified it is necessary to plot graphs of the variables in a modified form. State for each (a) what should be plotted on the vertical axis, (b) what should be plotted on the horizontal axis, (c) the gradient, and (d) the vertical axis intercept.

1 (i) $y = d + cx^2$
   (ii) $y - a = b\sqrt{x}$

$$\left[\begin{array}{l} \text{(i) (a) } y \text{ (b) } x^2 \text{ (c) } c \text{ (d) } d \\ \text{(ii) (a) } y \text{ (b) } \sqrt{x} \text{ (c) } b \text{ (d) } a \end{array}\right]$$

2 (i) $y - e = \dfrac{f}{x}$
   (ii) $y - cx = bx^2$

$$\left[\begin{array}{l} \text{(i) (a) } y \text{ (b) } \dfrac{1}{x} \text{ (c) } f \text{ (d) } e \\ \text{(ii) (a) } \dfrac{y}{x} \text{ (b) } x \text{ (c) } b \text{ (d) } c \end{array}\right]$$

3 (i) $y = \dfrac{a}{x} + bx$
   (ii) $y = ba^x$

$$\left[\begin{array}{l} \text{(i) (a) } \dfrac{y}{x} \text{ (b) } \dfrac{1}{x^2} \text{ (c) } a \text{ (d) } b \\ \text{(ii) (a) } \lg y \text{ (b) } x \text{ (c) } \lg a \text{ (d) } \lg b \end{array}\right]$$

4 (i) $y = kx^l$
   (ii) $\dfrac{y}{m} = e^{nx}$

$$\left[\begin{array}{l} \text{(i) (a) } \lg y \text{ (b) } \lg x \text{ (c) } l \text{ (d) } \lg k \\ \text{(ii) (a) } \ln y \text{ (b) } x \text{ (c) } n \text{ (d) } \ln m \end{array}\right]$$

5 In an experiment the resistance of wire is measured for wires of different diameters with the following results:

R ohms	1.64	1.14	0.89	0.76	0.63
d mm	1.10	1.42	1.75	2.04	2.56

It is thought that R is related to d by the law $R = (a/d^2) + b$, where a and b are constants. Verify this and find the approximate values for a and b. Determine the cross-sectional area needed for a resistance reading of 0.50 ohms.

$$[a = 1.5,\ b = 0.4,\ 11.78 \text{ mm}^2]$$

6   Corresponding experimental values of two quantities x and y are given below.

x	1.5	3.0	4.5	6.0	7.5	9.0
y	11.5	25.0	47.5	79.0	119.5	169.0

By plotting a suitable graph verify that y and x are connected by a law of the form $y = kx^2 + c$, where k and c are constants. Determine the law of the graph and hence find the value of x when y is 60.0.   [$y = 2x^2 + 7$, 5.15]

7   Experimental results of the safe load L kN, applied to girders of varying spans, d m, are shown below.

Span, d m	2.0	2.8	3.6	4.2	4.8
Load, L kN	475	339	264	226	198

It is believed that the relationship between load and span is $L = c/d$, where c is a constant. Determine (a) the value of constant c, and (b) the safe load for a span of 3.0 m.
                [(a) 950 (b) 317 kN]

8   The following results give corresponding values of two quantities x and y which are believed to be related by a law of the form $y = ax^2 + bx$ where a and b are constants.

y	33.86	55.54	72.80	84.10	111.4	168.1
x	3.4	5.2	6.5	7.3	9.1	12.4

Verify the law and determine approximate values of a and b. Hence determine (i) the value of y when x is 8.0, and (ii) the value of x when y is 146.5.
                [$a = 0.4$, $b = 8.6$ (i) 94.4 (ii) 11.2]

9   The luminosity I of a lamp varies with the applied voltage V and the relationship between I and V is thought to be $I = kV^n$. Experimental results obtained are:

I candelas	1.92	4.32	9.72	15.87	23.52	30.72
V volts	40	60	90	115	140	160

Verify that the law is true and determine the law of the graph. Determine also the luminosity when 75 V is applied across the lamp.
                [$I = 0.0012\,V^2$, 6.75 candelas]

10   The head of pressure h and the flow velocity v are measured and are believed to be connected by the law $v = ah^b$, where a and b are constants. The results are as shown below.

h	10.6	13.4	17.2	24.6	29.3
v	9.77	11.0	12.44	14.88	16.24

Verify that the law is true and determine values of a and b.   [$a = 3.0$, $b = 0.5$]

11   Experimental values of x and y are measured as follows:

x	0.4	0.9	1.2	2.3	3.8
y	8.35	13.47	17.94	51.32	215.20

The law relating x and y is believed to be of the form $y = ab^x$, where a and b are constants. Determine the approximate values of a and b. Hence find the value of y when x is 2.0 and the value of x when y is 100.
                [$a = 5.7$, $b = 2.6$, 38.53, 3.0]

12   The activity of a mixture of radioactive isotope is believed to vary according to the law $R = R_0 t^{-c}$, where R_0 and c are constants. Experimental results are shown below.

R	9.72	2.65	1.15	0.47	0.32	0.23
t	2	5	9	17	22	28

Verify that the law is true and determine approximate values of R_0 and c.
                [$R_0 = 26.0$, $c = 1.42$]

13   Determine the law of the form $y = ae^{kx}$ which relates the following values:

y	0.0306	0.285	0.841	5.21	173.2	1181
x	−4.0	5.3	9.8	17.4	32.0	40.0

                [$y = 0.08e^{0.24x}$]

14   The tension T in a belt passing round a pulley wheel and in contact with the pulley over an angle of θ radians is given by $T = T_0 e^{\mu\theta}$, where T_0 and μ are constants. Experimental results obtained are:

T newtons	47.9	52.8	60.3	70.1	80.9
θ radians	1.12	1.48	1.97	2.53	3.06

Determine approximate values of T_0 and μ. Hence find the tension when θ is 2.25 radians and the value of θ when the tension is 50.0 newtons.
[$T_0 = 35.4$ N, $\mu = 0.27$, 65.0 N, 1.28 radians]

15

Graphs with logarithmic scales

15.1 Logarithmic scales

Graph paper is available where the scale markings along the horizontal and vertical axes are proportional to the logarithms of the numbers. Such graph paper is called **log-log graph paper**.

A **logarithmic scale** is shown in Fig. 15.1 where the distance between, say 1 and 2, is proportional to $\lg 2 - \lg 1$, i.e. 0.3010 of the total distance from 1 to 10. Similarly, the distance between 7 and 8 is proportional to $\lg 8 - \lg 7$, i.e. 0.05799 of the total distance from 1 to 10. Thus the distance between markings progressively decreases as the numbers increase from 1 to 10.

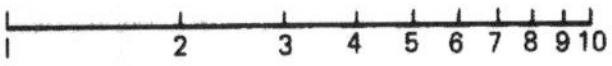

Figure 15.1

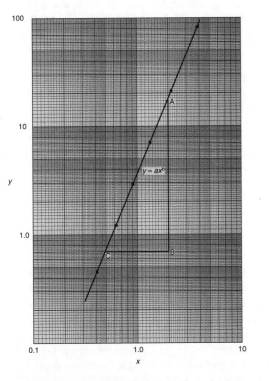

Figure 15.2 Graph to verify a law of the form $y = ax^b$

With log-log graph paper the scale markings are from 1 to 9, and this pattern can be repeated several times. The number of times the pattern of markings is repeated on an axis signifies the number of **cycles**. When the vertical axis has, say, 3 sets of values from 1 to 9, and the horizontal axis has, say, 2 sets of values from 1 to 9, then this log-log graph paper is called 'log 3 cycle × 2 cycle' (see Fig. 15.2). Many different arrangements are available ranging from 'log 1 cycle × 1 cycle' through to 'log 5 cycle × 5 cycle'.

To depict a set of values, say, from 0.4 to 161, on an axis of log-log graph paper, 4 cycles are required, from 0.1 to 1, 1 to 10, 10 to 100 and 100 to 1000.

15.2 Graphs of the form $y = ax^n$

Taking logarithms to a base of 10 of both sides of $y = ax^n$ gives:

$$\lg y = \lg(ax^n) = \lg a + \lg x^n$$

i.e.
$$\lg y = n \lg x + \lg a$$

which compares with $\quad Y = mX + c$

Thus, by plotting $\lg y$ vertically against $\lg x$ horizontally, a straight line results, i.e. the equation $y = ax^n$ is reduced to linear form. With log-log graph paper available x and y may be plotted directly, without having first to determine their logarithms, as shown in Chapter 14.

Problem 1. Experimental values of two related quantities x and y are shown below.

x	0.41	0.63	0.92	1.36	2.17	3.95
y	0.45	1.21	2.89	7.10	20.79	82.46

The law relating x and y is believed to be $y = ax^b$, where a and b are constants. Verify that this law is true and determine the approximate values of a and b

If $y = ax^b$ then $\lg y = b \lg x + \lg a$, from above, which is of the form $Y = mX + c$, showing that to produce a straight line graph $\lg y$ is plotted vertically against $\lg x$ horizontally. x and y may be plotted directly on to log-log graph paper as shown in Fig. 15.2. The values of y range from 0.45 to 82.46 and 3 cycles are needed (i.e. 0.1 to 1, 1 to 10 and 10 to 100). The values of x range from 0.41 to 3.95 and 2 cycles are needed (i.e. 0.1 to 1 and 1 to 10). Hence 'log 3 cycle × 2 cycle' is used as shown in Fig. 15.2 where the axes are marked and the points plotted. Since the points lie on a straight line the law $y = ax^b$ is verified.

To evaluate constants a and b

Method 1. Any two points on the straight line, say points A and C, are selected, and AB and BC are measured (say in centimetres).

Then, gradient, $b = \dfrac{AB}{BC} = \dfrac{11.5 \text{ units}}{5 \text{ units}} = 2.3$

Since $\lg y = b \lg x + \lg a$, when $x = 1$, $\lg x = 0$ and $\lg y = \lg a$.

The straight line crosses the ordinate $x = 1.0$ at $y = 3.5$.

Hence $\lg a = \lg 3.5$, i.e. $a = 3.5$.

Method 2. Any two points on the straight line, say points A and C, are selected. A has coordinates (2, 17.25) and C has coordinates (0.5, 0.7).

Since $y = ax^b$ then $17.25 = a(2)^b$ (1)

and $0.7 = a(0.5)^b$ (2)

i.e. two simultaneous equations are produced and may be solved for a and b.

Dividing equation (1) by equation (2) to eliminate a gives

$$\frac{17.25}{0.7} = \frac{(2)^b}{(0.5)^b} = \left(\frac{2}{0.5}\right)^b$$

i.e. $24.643 = (4)^b$

Taking logarithms of both sides gives $\lg 24.643 = b \lg 4$, i.e.

$$b = \frac{\lg 24.643}{\lg 4}$$

$= 2.3$,   correct to 2 significant figures

Substituting $b = 2.3$ in equation (1) gives: $17.25 = a(2)^{2.3}$, i.e.

$$a = \frac{17.25}{(2)^{2.3}} = \frac{17.25}{4.925}$$

$= 3.5$,   correct to 2 significant figures

Hence the law of the graph is $y = 3.5x^{2.3}$.

Problem 2.   The power dissipated by a resistor was measured for varying values of current flowing in the resistor and the results are as shown:

Current, I amperes	1.4	4.7	6.8	9.1	11.2	13.1
Power, P watts	49	552	1156	2070	3136	4290

Prove that the law relating current and power is of the form $P = RI^n$, where R and n are constants, and determine the law. Hence calculate the power when the current is 12 amperes and the current when the power is 1000 W

Since $P = RI^n$ then $\lg P = n \lg I + \lg R$, which is of the form $Y = mX + c$, showing that to produce a straight line graph $\lg P$ is plotted vertically against $\lg I$ horizontally. Power values range from 49 to 4290, hence 3 cycles of log-log graph paper are needed (10 to 100, 100 to 1000 and 1000 to 10 000). Current values range from 1.4 to 11.2, hence 2 cycles of log-log graph paper are needed (1 to 10 and 10 to 100).

Thus 'log 3 cycles × 2 cycles' is used as shown in Fig. 15.3 (or, if not available, graph paper having a larger number of cycles per axis can be used). The coordinates are plotted and a straight line results which proves that the law relating current and power is of the form $P = RI^n$.

Gradient of straight line

$$n = \frac{AB}{BC} = \frac{14 \text{ units}}{7 \text{ units}} = 2$$

At point C, $I = 2$ and $P = 100$. Substituting these values into $P = RI^n$ gives: $100 = R(2)^2$. Hence $R = 100/(2)^2 = 25$ which may have been found from the intercept on the $I = 1.0$ axis in Fig. 15.3.

Hence the law of the graph is $P = 25I^2$.

When current $I = 12$, power $P = 25(12)^2 = $ **3600 watts** (which may be read from the graph).

When power $P = 1000$, $1000 = 25I^2$.

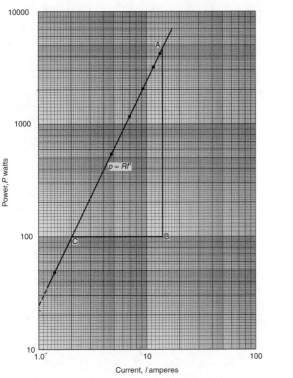

Figure 15.3 Variation of power with current

Hence
$$I^2 = \frac{1000}{25} = 40$$

from which, $I = \sqrt{40} = \mathbf{6.32\ A}$

Problem 3. The pressure p and volume v of a gas are believed to be related by a law of the form $p = cv^n$, where c and n are constants. Experimental values of p and corresponding values of v obtained in a laboratory are:

p pascals	2.28×10^5	8.04×10^5	2.03×10^6
v m³	3.2×10^{-2}	1.3×10^{-2}	6.7×10^{-3}

p pascals	5.05×10^6	1.82×10^7
v m³	3.5×10^{-3}	1.4×10^{-3}

Verify that the law is true and determine approximate values of c and n

---

Since $p = cv^n$, then $\lg p = n \lg v + \lg c$, which is of the form $Y = mX + c$, showing that to produce a straight line graph $\lg p$ is plotted vertically against $\lg v$ horizontally. The coordinates are plotted on 'log 3 cycle × 2 cycle' graph paper as shown in Fig. 15.4. With the data expressed in standard form, the axes are marked in standard form also. Since a straight line results the law $p = cv^n$ is verified.

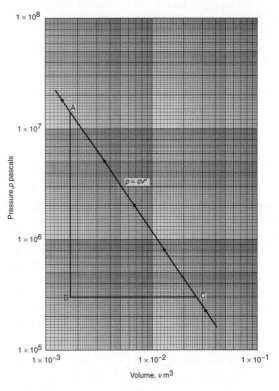

Figure 15.4 Variation of pressure with volume

The straight line has a negative gradient and the value of the gradient is given by:

$$\frac{AB}{BC} = \frac{14 \text{ units}}{10 \text{ units}} = 1.4$$

Hence $n = \mathbf{-1.4}$.

Selecting any point on the straight line, say point C, having coordinates $(2.63 \times 10^{-2}, 3 \times 10^5)$, and substituting these values in $p = cv^n$, gives:

$$3 \times 10^5 = c(2.63 \times 10^{-2})^{-1.4}$$

Hence
$$c = \frac{3 \times 10^5}{(2.63 \times 10^{-2})^{-1.4}} = \frac{3 \times 10^5}{(0.0263)^{-1.4}}$$

$$= \frac{3 \times 10^5}{1.63 \times 10^2}$$

$$= \mathbf{1840}, \text{ correct to 3 significant figures}$$

Hence the law of the graph is

$$p = 1840v^{-1.4} \text{ or } pv^{1.4} = 1840$$

Further problems on graphs of the form $y = ax^n$ may be found in Section 15.5, Problems 1 to 3, pages 156 and 157.

15.3 Graphs of the form $y = ab^x$

Taking logarithms to a base of 10 of both sides of $y = ab^x$ gives:

$$\lg y = \lg(ab^x) = \lg a + \lg b^x = \lg a + x \lg b$$

i.e.
$$\lg y = (\lg b)x + \lg a$$

which compares with $\quad Y = mX + c$

Thus, by plotting $\lg y$ vertically against x horizontally a straight line results, i.e. the graph $y = ab^x$ is reduced to linear form. In this case, graph paper having a linear horizontal scale and a logarithmic vertical scale may be used. This type of graph paper is called **log-linear graph paper**, and is specified by the number of cycles on the logarithmic scale. For example, graph paper having 3 cycles on the logarithmic scale is called 'log 3 cycle × linear' graph paper.

Problem 4. Experimental values of quantities x and y are believed to be related by a law of the form $y = ab^x$, where a and b are constants. The values of x and corresponding values of y are:

x	0.7	1.4	2.1	2.9	3.7	4.3
y	18.4	45.1	111	308	858	1850

Verify the law and determine the approximate values of a and b.
Hence evaluate (i) the value of y when x is 2.5, and (ii) the value of x when y is 1200

Since $y = ab^x$ then $\lg y = (\lg b)x + \lg a$ (from above), which is of the form $Y = mX + c$, showing that to produce a straight line graph $\lg y$ is plotted vertically against x horizontally. Using log-linear graph paper, values of x are marked on the horizontal scale to cover the range 0.7 to 4.3. Values of y range from 18.4 to 1850 and 3 cycles are

needed (i.e. 10 to 100, 100 to 1000 and 1000 to 10 000). Thus 'log 3 cycles × linear' graph paper is used as shown in Fig. 15.5. A straight line is drawn through the coordinates, hence the law $y = ab^x$ is verified.

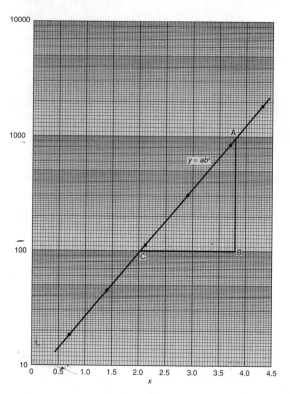

Figure 15.5 Graph to verify a law of the form $y = ab^x$

Gradient of straight line, $\lg b = AB/BC$. Direct measurement (say in centimetres) is not made with log-linear graph paper since the vertical scale is logarithmic and the horizontal scale is linear. Hence

$$\frac{AB}{BC} = \frac{\lg 1000 - \lg 100}{3.82 - 2.02} = \frac{3 - 2}{1.80}$$

$$= \frac{1}{1.80} = 0.5556$$

Hence $b = $ antilog $0.5556(= 10^{0.5556}) = \mathbf{3.6}$, correct to 2 significant figures.

Point A has coordinates (3.82, 1000).
Substituting these values into $y = ab^x$ gives: $1000 = a(3.6)^{3.82}$, i.e.

$$a = \frac{1000}{(3.6)^{3.82}}$$

$$= \mathbf{7.5}, \text{ correct to 2 significant figures}$$

Hence the law of the graph is $y = 7.5(3.6)^x$.

(i) When $x = 2.5$, $y = 7.5(3.6)^{2.5} = \mathbf{184}$
(ii) When $y = 1200$, $1200 = 7.5(3.6)^x$, hence

$$(3.6)^x = \frac{1200}{7.5} = 160$$

Taking logarithms gives $x \lg 3.6 = \lg 160$

i.e. $x = \dfrac{\lg 160}{\lg 3.6} = \dfrac{2.2041}{0.5563}$

$= \mathbf{3.96}$

A further problem on graphs of the form $y = ab^x$ may be found in Section 15.5, Problem 4, page 157.

15.4   Graphs of the form $y = ae^{kx}$

Taking logarithms to a base of e of both sides of $y = ae^{kx}$ gives:

$$\ln y = \ln(ae^{kx}) = \ln a + \ln e^{kx} = \ln a + kx \ln e$$

i.e. $\ln y = kx + \ln a$ (since $\ln e = 1$)

which compares with $Y = mX + c$

Thus, by plotting $\ln y$ vertically against x horizontally, a straight line results, i.e. the equation $y = ae^{kx}$ is reduced to linear form. Since $\ln y = 2.3026 \lg y$, i.e. $\ln y = $ (a constant)$(\lg y)$, the same log-linear graph paper can be used for Napierian logarithms as for logarithms to a base of 10.

Problem 5.   The data given below is believed to be related by a law of the form $y = ae^{kx}$, where a and b are constants. Verify that the law is true and determine approximate values of a and b. Also determine the value of y when x is 3.8 and the value of x when y is 85.

| x | -1.2 | 0.38 | 1.2 | 2.5 | 3.4 | 4.2 | 5.3 |
| y | 9.3 | 22.2 | 34.8 | 71.2 | 117 | 181 | 332 |

Since $y = ae^{kx}$ then $\ln y = kx + \ln a$ (from above), which is of the form $Y = mX + c$, showing that to produce a straight line graph $\ln y$ is plotted vertically against x horizontally. The value of y ranges from 9.3 to 332 hence 'log 3 cycle $\times$ linear' graph paper is

used. The plotted coordinates are shown in Fig. 15.6 and since a straight line passes through the points the law $y = ae^{kx}$ is verified.

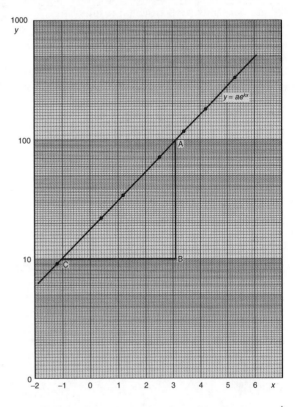
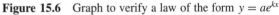

Figure 15.6   Graph to verify a law of the form $y = ae^{kx}$

Gradient of straight line,

$$k = \frac{AB}{BC} = \frac{\ln 100 - \ln 10}{3.12 - (-1.08)} = \frac{2.3026}{4.20}$$

$= 0.55$, correct to 2 significant figures

Since $\ln y = kx + \ln a$, when $x = 0$, $\ln y = \ln a$, i.e. $y = a$.
The vertical axis intercept value at $x = 0$ is 18, hence $a = 18$.
The law of the graph is thus $y = 18e^{0.55x}$.
When x is 3.8,
$y = 18e^{0.55(3.8)} = 18e^{2.09} = 18(8.0849) = \mathbf{146}$
When y is 85, $85 = 18e^{0.55x}$.

Hence, $e^{0.55x} = \dfrac{85}{18} = 4.7222$

and $0.55x = \ln 4.7222 = 1.5523$.

Hence $x = \dfrac{1.5523}{0.55} = \mathbf{2.82}$

Problem 6. The voltage, v volts, across an inductor is believed to be related to time, t ms, by the law $v = Ve^{t/T}$, where V and T are constants. Experimental results obtained are:

v volts	883	347	90	55.5	18.6	5.2
t ms	10.4	21.6	37.8	43.6	56.7	72.0

Show that the law relating voltage and time is as stated and determine the approximate values of V and T. Find also the value of voltage after 25 ms and the time when the voltage is 30.0 V

Since $v = Ve^{t/T}$ then $\ln v = \dfrac{1}{T}t + \ln V$

which is of the form $Y = mX + c$.

Using 'log 3 cycle × linear' graph paper, the points are plotted as shown in Fig. 15.7.

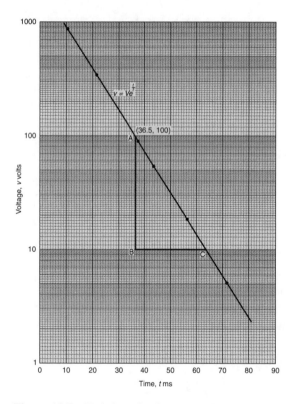

1000

100

Voltage, v volts

$v = Ve^{\frac{t}{T}}$

(36.5, 100)

A

B

C

10

1

0   10   20   30   40   50   60   70   80   90

Time, t ms

Figure 15.7   Variation of voltage with time

Since the points are joined by a straight line the law $v = Ve^{t/T}$ is verified.

Gradient of straight line,

$$\frac{1}{T} = \frac{AB}{BC} = \frac{\ln 100 - \ln 10}{36.5 - 64.2} = \frac{2.3026}{-27.7}$$

Hence $T = \dfrac{-27.7}{2.3026} = -\mathbf{12.0}$, correct to 3 significant figures.

Since the straight line does not cross the vertical axis at $t = 0$ in Fig. 15.7, the value of V is determined by selecting any point, say A, having coordinates (36.5, 100) and substituting these values into $v = Ve^{t/T}$. Thus

$$100 = Ve^{36.5/-12.0}$$

i.e. $V = \dfrac{100}{e^{-36.5/12.0}} = \mathbf{2090}$ **volts**,

correct to 3 significant figures

Hence the law of the graph is $v = 2090e^{-t/12.0}$.

When time $t = 25$ ms, voltage $v = 2090e^{-25/12.0} = \mathbf{260\ V}$.

When the voltage is 30.0 volts, $30.0 = 2090e^{-t/12.0}$, hence

$$e^{-t/12.0} = \frac{30.0}{2090} \text{ and } e^{t/12.0} = \frac{2090}{30.0} = 69.67$$

Taking Napierian logarithms gives:

$$\frac{t}{12.0} = \ln 69.67 = 4.2438$$

from which, time $t = (12.0)(4.2438) = \mathbf{50.9\ ms}$.

Further problems on graphs of the form $y = ae^{kx}$ may be found in the following Section 15.5, Problems 5 and 6, page 157.

15.5   Further problems on graphs having logarithmic scales

Graphs of the form $y = ax^n$

1   Quantities x and y are believed to be related by a law of the form $y = ax^n$, where a and n are constants. Experimental values of x and corresponding values of y are:

x	0.8	2.3	5.4	11.5	21.6	42.9
y	8	54	250	974	3028	10 410

Show that the law is true and determine the values of a and n. Hence determine the value of y when x is 7.5 and the value of x when y is 5000. $\qquad [a = 12, n = 1.8, 451, 28.5]$

2 Show from the following results of voltage V and admittance Y of an electrical circuit that the law connecting the quantities is of the form $V = kY^n$, and determine the values of k and n.

Voltage, V volts	2.88	2.05	1.60	1.22	0.96
Admittance, Y siemens	0.52	0.73	0.94	1.23	1.57

$$[k = 1.5, n = -1]$$

3 Quantities x and y are believed to be related by a law of the form $y = mn^x$. The values of x and corresponding values of y are:

x	0	0.5	1.0	1.5	2.0	2.5	3.0
y	1.0	3.2	10	31.6	100	316	1000

Verify the law and find the values of m and n.
$$[m = 1, n = 10]$$

Graphs of the form $y = ab^x$

4 Experimental values of p and corresponding values of q are shown below.

p	−13.2	−27.9	−62.2	−383.2	−1581	−2931
q	0.30	0.75	1.23	2.32	3.17	3.54

Show that the law relating p and q is $p = ab^q$, where a and b are constants. Determine

(i) values of a and b, and state the law, (ii) the value of p when q is 2.0, and (iii) the value of q when p is −2000.

$$\left[\begin{array}{l} \text{(i) } a = -8, b = 5.3, p = -8(5.3)^q \\ \text{(ii) } -224.7 \text{ (iii) } 3.31 \end{array} \right]$$

Graphs of the form $y = ae^{kx}$

5 Atmospheric pressure p is measured at varying altitudes h and the results are as shown below.

Altitude, h m	500	1500	3000	5000	8000
pressure, p cm	73.39	68.42	61.60	53.56	43.41

Show that the quantities are related by the law $p = ae^{kh}$, where a and k are constants. Determine the values of a and k and state the law. Find also the atmospheric pressure at 10 000 m.

$$\left[\begin{array}{l} a = 76, k = -7 \times 10^{-5}, \\ p = 76e^{-7 \times 10^{-5}h}, 37.74 \text{ cm} \end{array} \right]$$

6 At particular times, t minutes, measurements are made of the temperature, $\theta°$, of a cooling liquid and the following results are obtained:

Temperature $\theta°$C	92.2	55.9	33.9	20.6	12.5
Time t minutes	10	20	30	40	50

Prove that the quantities follow a law of the form $\theta = \theta_0 e^{kt}$, where θ_0 and k are constants, and determine the approximate value of θ_0 and k.
$$[\theta_0 = 152, k = -0.05]$$

16

Graphical solution of equations

16.1 Graphical solution of simultaneous equations

Linear simultaneous equations in two unknowns may be solved graphically by:

(i) plotting the two straight lines on the same axes, and
(ii) noting their point of intersection

The coordinates of the point of intersection give the required solution.

Problem 1. Solve graphically the simultaneous equations

$$2x - y = 4$$

$$x + y = 5$$

Rearranging each equation into $y = mx + c$ form gives:

$$y = 2x - 4 \qquad (1)$$

$$y = -x + 5 \qquad (2)$$

Only three coordinates need be calculated for each graph since both are straight lines.

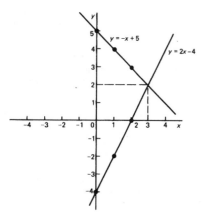

Figure 16.1

x	0	1	2
$y = 2x - 4$	−4	−2	0

x	0	1	2
$y = -x + 5$	5	4	3

Each of the graphs is plotted as shown in Fig. 16.1. The point of intersection is at $(3, 2)$ and since this is the only point which lies simultaneously on both lines then $x = 3$, $y = 2$ is the solution of the simultaneous equations.

Problem 2. Solve graphically the equations

$$1.20x + y = 1.80$$

$$x - 5.0y = 8.50$$

Rearranging each equation into $y = mx + c$ form gives:

$$y = -1.20x + 1.80 \qquad (1)$$

$$y = \frac{x}{5.0} - \frac{8.5}{5.0}$$

i.e. $\quad y = 0.20x - 1.70 \qquad (2)$

Three coordinates are calculated for each equation as shown below.

x	0	1	2
$y = -1.20x + 1.80$	1.80	0.60	−0.60

x	0	1	2
$y = 0.20x - 1.70$	−1.70	−1.50	−1.30

The two lines are plotted as shown in Fig. 16.2. The point of intersection is $(2.50, -1.20)$. Hence the solution of the simultaneous equations is $x = 2.50$, $y = -1.20$. (It is sometimes useful initially to sketch the two straight lines to determine the region where the point of intersection is. Then, for greater accuracy, a graph having a smaller range of values can be drawn to 'magnify' the point of intersection.)

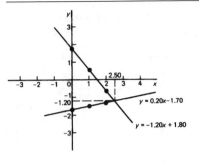

Figure 16.2

Further problems on the graphical solution of simultaneous equations may be found in Section 16.5, Problems 1 to 6, page 164.

16.2 Graphical solutions of quadratic equations

A general **quadratic equation** is of the form $y = ax^2 + bx + c$, where a, b and c are constants and a is not equal to zero.

A graph of a quadratic equation always produces a shape called a **parabola**.

The gradient of the curve between O and A and between B and C in Fig. 16.3 is positive, whilst the gradient between A and B is negative. Points such as A and B are called **turning points**. At A the gradient is zero and, as x increases, the gradient of the curve changes from positive just before A to negative just after. Such a point is called a **maximum value**. At B the gradient is also zero, and, as x increases, the gradient of the curve changes from negative just before B to positive just after. Such a point is called a **minimum value**.

(More on maximum and minimum values may be found in Chapters 19 and 22.)

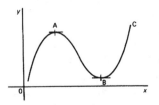

Figure 16.3

Quadratic graphs

(i) $y = ax^2$

Graphs of $y = x^2$, $y = 3x^2$ and $y = \frac{1}{2}x^2$ are shown in Fig. 16.4.
All have minimum values at the origin $(0, 0)$.

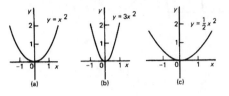

Figure 16.4

Graphs of $y = -x^2$, $y = -3x^2$ and $y = -\frac{1}{2}x^2$ are shown in Fig. 16.5.
All have maximum values at the origin $(0, 0)$.

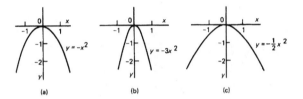

Figure 16.5

When $y = ax^2$,
  (a) curves are symmetrical about the y-axis,
  (b) the magnitude of 'a' affects the gradient of the curve,
and (c) the sign of 'a' determines whether it has a maximum or minimum value

(ii) $y = ax^2 + c$
Graphs of $y = x^2 + 3$, $y = x^2 - 2$, $y = -x^2 + 2$ and $y = -2x^2 - 1$ are shown in Fig. 16.6.
When $y = ax^2 + c$:
  (a) curves are symmetrical about the y-axis,
  (b) the magnitude of 'a' affects the gradient of the curve,
and (c) the constant 'c' is the y-axis intercept

(iii) $y = ax^2 + bx + c$
Whenever 'b' has a value other than zero the curve is displaced to the right or left of the y-axis. When b/a is positive, the curve is displaced $b/2a$ to the left of the y-axis, as shown in Fig. 16.7(a). When b/a is negative the curve is displaced $b/2a$ to the right of the y-axis, as shown in Fig. 16.7(b)

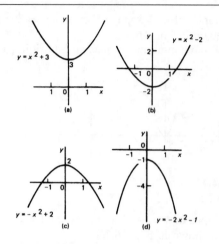

(a)   (b)

(c)   (d)

Figure 16.6

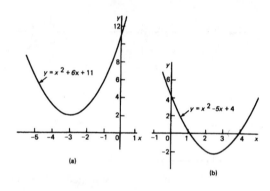

(a)   (b)

Figure 16.7

Quadratic equations of the form $ax^2 + bx + c = 0$ may be solved graphically by:

(i)   plotting the graph $y = ax^2 + bx + c$, and

(ii) noting the points of intersection on the x-axis (i.e. where $y = 0$)

The x values of the points of intersection give the required solutions since at these points both $y = 0$ and $ax^2 + bx + c = 0$. The number of solutions, or roots of a quadratic equation, depends on how many times the curve cuts the x-axis and there can be no real roots (as in Fig. 16.7(a)) or one root (as in Figs 16.4 and 16.5) or two roots (as in Fig. 16.7(b)).

Let $y = 4x^2 + 4x - 15$. A table of values is drawn up as shown below.

x				-3	-2	-1	0	1	2
$4x^2$				36	16	4	0	4	16
$4x$				-12	-8	-4	0	4	8
-15				-15	-15	-15	-15	-15	-15
$y = 4x^2 + 4x - 15$				9	-7	-15	-15	-7	9

A graph of $y = 4x^2 + 4x - 15$ is shown in Fig. 16.8. The only points where $y = 4x^2 + 4x - 15$ and $y = 0$ are the points marked A and B. This occurs at $x = -2.5$ and $x = 1.5$ and these are the solutions of the quadratic equation $4x^2 + 4x - 15 = 0$. (By substituting $x = -2.5$ and $x = 1.5$ into the original equation the solutions may be checked.) The curve has a turning point at $(-0.5, -16)$ and the nature of the point is a **minimum**.

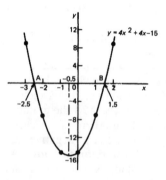

Figure 16.8

An alternative graphical method of solving $4x^2 + 4x - 15 = 0$ is to rearrange the equation as $4x^2 = -4x + 15$ and then plot two separate graphs–in this case $y = 4x^2$ and $y = -4x + 15$. Their points of intersection give the roots of equation $4x^2 = -4x + 15$, i.e. $4x^2 + 4x - 15 = 0$. This is shown in Fig. 16.9, where the roots are $x = -2.5$ and $x = 1.5$ as before.

Problem 3.   Solve the quadratic equation $4x^2 + 4x - 15 = 0$ graphically given that the solutions lie in the range $x = -3$ to $x = 2$. Determine also the coordinates and nature of the turning point of the curve

Problem 4.   Solve graphically the quadratic equation $-5x^2 + 9x + 7.2 = 0$ given that the solutions lie between $x = -1$ and $x = 3$. Determine also the coordinates of the turning point and state its nature

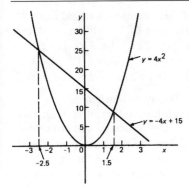

Figure 16.9

Let $y = -5x^2 + 9x + 7.2$. A table of values is drawn up as shown below. A graph of $y = -5x^2 + 9x + 7.2$ is shown plotted in Fig. 16.10. The graph crosses the x-axis (i.e. where $y = 0$) at $x = -0.6$ and $x = 2.4$ and these are the solutions of the quadratic equation $-5x^2 + 9x + 7.2 = 0$. The turning point is a **maximum** having coordinates $(0.9, 11.25)$.

x	-1	-0.5	0	1
$-5x^2$	-5	-1.25	0	-5
$+9x$	-9	-4.5	0	9
$+7.2$	7.2	7.2	7.2	7.2
$y = -5x^2 + 9x + 7.2$	-6.8	1.45	7.2	11.2

x	2	2.5	3
$-5x^2$	-20	-31.25	-45
$+9x$	18	22.5	27
$+7.2$	7.2	7.2	7.2
$y = -5x^2 + 9x + 7.2$	5.2	-1.55	-10.8

Problem 5. Plot a graph of $y = 2x^2$ and hence solve the equations:
(a) $2x^2 - 8 = 0$, and (b) $2x^2 - x - 3 = 0$

A graph of $y = 2x^2$ is shown in Fig. 16.11.

(a) Rearranging $2x^2 - 8 = 0$ gives $2x^2 = 8$ and the solution of this equation is obtained from the points of intersection of $y = 2x^2$ and $y = 8$, i.e. at coordinates $(-2, 8)$ and

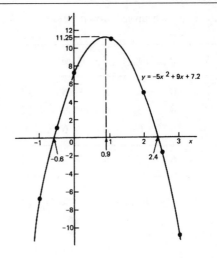

Figure 16.10

$(2, 8)$, shown as A and B, respectively, in Fig. 16.11. Hence the solutions of $2x^2 - 8 = 0$ are $x = -2$ and $x = +2$

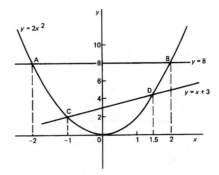

Figure 16.11

(b) Rearranging $2x^2 - x - 3 = 0$ gives $2x^2 = x + 3$ and the solution of this equation is obtained from the points of intersection of $y = 2x^2$ and $y = x + 3$, i.e. at C and D in Fig. 16.11. Hence the solutions of $2x^2 - x - 3 = 0$ are $x = -1$ and $x = 1.5$

Problem 6. Plot the graph of $y = -2x^2 + 3x + 6$ for values of x from $x = -2$ to $x = 4$. Use the graph to find the roots of the following equations:
(a) $-2x^2 + 3x + 6 = 0$ (b) $-2x^2 + 3x + 2 = 0$
(c) $-2x^2 + 3x + 9 = 0$ (d) $-2x^2 + x + 5 = 0$

A table of values is drawn up as shown below.

x	-2	-1	0	1	2	3	4
$-2x^2$	-8	-2	0	-2	-8	-18	-32
$+3x$	-6	-3	0	3	6	9	12
$+6$	6	6	6	6	6	6	6
y	-8	1	6	7	4	-3	-14

A graph of $y = -2x^2 + 3x + 6$ is shown in Fig. 16.12.

(a) The parabola $y = -2x^2 + 3x + 6$ and the straight line $y = 0$ intersect at A and B, where $x = -1.13$ and $x = 2.63$ and these are the roots of the equation $-2x^2 + 3x + 6 = 0$

(b) Comparing $y = -2x^2 + 3x + 6$ (1)

with $0 = -2x^2 + 3x + 2$ (2)

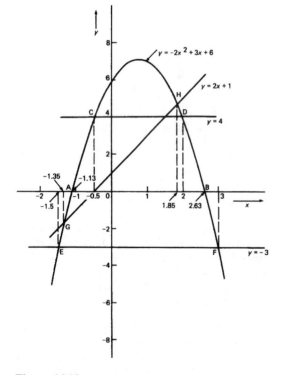

Figure 16.12

shows that if 4 is added to both sides of equation (2), the right-hand side of both equations will be the same. Hence $4 = -2x^2 + 3x + 6$.

The solution of this equation is found from the points of intersection of the line $y = 4$ and the parabola $y = -2x^2 + 3x + 6$, i.e. points C and D in Fig. 16.12. Hence the roots of $-2x^2 + 3x + 2 = 0$ are $x = -0.5$ and $x = 2$

(c) $-2x^2 + 3x + 9 = 0$ may be rearranged as $-2x^2 + 3x + 6 = -3$, and the solution of this equation is obtained from the points of intersection of the line $y = -3$ and the parabola $y = -2x^2 + 3x + 6$, i.e. at points E and F in Fig. 16.12. Hence the roots of $-2x^2 + 3x + 9 = 0$ are $x = -1.5$ and $x = 3$

(d) Comparing $y = -2x^2 + 3x + 6$ (3)

with $0 = -2x^2 + x + 5$ (4)

shows that if $2x + 1$ is added to both sides of equation (4) the right-hand side of both equations will be the same. Hence equation (4) may be written as $2x + 1 = -2x^2 + 3x + 6$. The solution of this equation is found from the points of intersection of the line $y = 2x + 1$ and the parabola $y = -2x^2 + 3x + 6$, i.e. points G and H in Fig. 16.12. Hence the roots of $-2x^2 + x + 5 = 0$ are $x = -1.35$ and $x = 1.85$

Further problems on the graphical solution of quadratic equations may be found in Section 16.5, Problems 7 to 16, pages 164 and 165.

16.3 Graphical solution of linear and quadratic equations simultaneously

The solution of **linear and quadratic equations simultaneously** may be achieved graphically by: (i) plotting the straight line and parabola on the same axes, and (ii) noting the points of intersection. The coordinates of the points of intersection give the required solutions.

> Problem 7. Determine graphically the values of x and y which simultaneously satisfy the equations $y = 2x^2 - 3x - 4$ and $y = 2 - 4x$

$y = 2x^2 - 3x - 4$ is a parabola and a table of values is drawn up as shown below:

x	-2	-1	0	1	2	3
$2x^2$	8	2	0	2	8	18
$-3x$	6	3	0	-3	-6	-9
-4	-4	-4	-4	-4	-4	-4
y	10	1	-4	-5	-2	5

$y = 2 - 4x$ is a straight line and only three coordinates need be calculated:

x	0	1	2
y	2	-2	-6

The two graphs are plotted in Fig. 16.13 and the points of intersection, shown as A and B, are at co-ordinates $(-2, 10)$ and $(1\frac{1}{2}, -4)$. Hence the simultaneous solutions occur when $x = -2$, $y = 10$ and when $x = 1\frac{1}{2}$, $y = -4$. (These solutions may be checked by substituting into each of the original equations.)

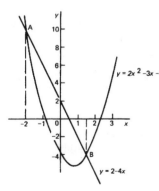

Figure 16.13

Further problems on the graphical solution of linear and quadratic equations simultaneously may be found in Section 16.5, Problems 17 and 18, page 165.

16.4 Graphical solution of cubic equations

A **cubic equation** of the form $ax^3 + bx^2 + cx + d = 0$ may be solved graphically by: (i) plotting the graph $y = ax^3 + bx^2 + cx + d$, and (ii) noting the points of intersection on the x-axis (i.e. where $y = 0$). The x-values of the points of intersection give the required solution since at these points both $y = 0$ and $ax^3 + bx^2 + cx + d = 0$.

The number of solutions, or roots of a cubic equation depends on how many times the curve cuts the x-axis and there can be one, two or three possible roots, as shown in Fig. 16.14.

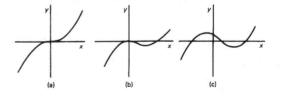

Figure 16.14

Problem 8. Solve graphically the cubic equation $4x^3 - 8x^2 - 15x + 9 = 0$ given that the roots lie between $x = -2$ and $x = 3$. Determine also the coordinates of the turning points and distinguish between them

Let $y = 4x^3 - 8x^2 - 15x + 9$. A table of values is drawn up as shown below.

x	-2	-1	0	1	2	3
$4x^3$	-32	-4	0	4	32	108
$-8x^2$	-32	-8	0	-8	-32	-72
$-15x$	30	15	0	-15	-30	-45
$+9$	9	9	9	9	9	9
y	-25	12	9	-10	-21	0

A graph of $y = 4x^3 - 8x^2 - 15x + 9$ is shown in Fig. 16.15.

The graph crosses the x-axis (where $y = 0$) at $x = -1\frac{1}{2}$, $x = \frac{1}{2}$ and $x = 3$ and these are the solutions to the cubic equation $4x^3 - 8x^2 - 15x + 9 = 0$. The turning points occur at $(-0.6, 14.2)$, which is a **maximum**, and $(2, -21)$, which is a **minimum**.

Problem 9. Plot the graph of $y = 2x^3 - 7x^2 + 4x + 4$ for values of x between $x = -1$ and $x = 3$. Hence determine the roots of the equation $2x^3 - 7x^2 + 4x + 4 = 0$

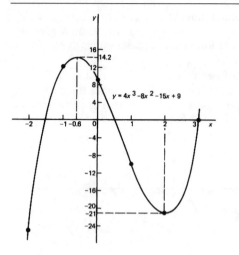

Figure 16.15

A table of values is drawn up as shown below.

x	-1	0	1	2	3
$2x^3$	-2	0	2	16	54
$-7x^2$	-7	0	-7	-28	-63
$+4x$	-4	0	4	8	12
$+4$	4	4	4	4	4
y	-9	4	3	0	7

A graph of $y = 2x^3 - 7x^2 + 4x + 4$ is shown in Fig. 16.16. The graph crosses the x-axis at $x = -0.5$ and touches the x-axis at $x = 2$.

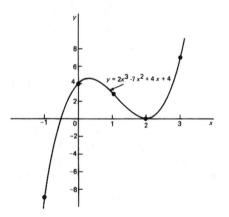

Figure 16.16

Hence the solutions of the equation $2x^3 - 7x^2 + 4x + 4 = 0$ are $x = -0.5$ and $x = 2$.

Further problems on the graphical solutions of cubic equations may be found in the following Section 16.5, Problems 19 to 25, page 165.

16.5   Further problems on the graphical solution of equations

Simultaneous equations

In Problems 1 to 5, solve the simultaneous equations graphically:

1   $x + y = 2$
    $3y - 2x = 1$ $[x = 1, \ y = 1]$

2   $y = 5 - x$
    $x - y = 2$ $\left[x = 3\frac{1}{2}, \ y = 1\frac{1}{2}\right]$

3   $3x + 4y = 5$
    $2x - 5y + 12 = 0$ $[x = -1, \ y = 2]$

4   $1.4x - 7.06 = 3.2y$
    $2.1x - 6.7y = 12.87$ $[x = 2.3, \ y = -1.2]$

5   $3x - 2y = 0$
    $4x + y + 11 = 0$ $[x = -2, \ y = -3]$

6   The friction force F newtons and load L newtons are connected by a law of the form $F = aL + b$, where a and b are constants. When $F = 4$ newtons, $L = 6$ newtons and when $F = 2.4$ newtons, $L = 2$ newtons. Determine graphically the values of a and b.
                                     $[a = 0.4, \ b = 1.6]$

Quadratic equations

7   Sketch the following graphs and state the nature and coordinates of their turning points.
    (a) $y = 4x^2$ (b) $y = 2x^2 - 1$
    (c) $y = -x^2 + 3$ (d) $y = -\frac{1}{2}x^2 - 1$
    $\left[\begin{array}{l}\text{(a) Minimum } (0, 0) \quad \text{(b) Minimum } (0, -1) \\ \text{(c) Maximum } (0, 3) \quad \text{(d) Maximum } (0, -1)\end{array}\right]$

Solve graphically the quadratic equations in Problems 8 to 11 by plotting the curves between the given limits. Give answers correct to 1 decimal place:

8   $4x^2 - x - 1 = 0$; $x = -1$ to $x = 1$
                                     $[-0.4 \text{ or } 0.6]$

9   $x^2 - 3x = 27$; $x = -5$ to $x = 8$
                                     $[-3.9 \text{ or } 6.9]$

10   $2x^2 - 6x - 9 = 0$; $x = -2$ to $x = 5$
                                     $[-1.1 \text{ or } 4.1]$

11   $2x(5x - 2) = 39.6$; $x = -2$ to $x = 3$
                                     $[-1.8 \text{ to } 2.2]$

12 Solve the quadratic equation $2x^2 + 7x + 6 = 0$ graphically, given that the solutions lie in the range $x = -3$ to $x = 1$. Determine also the nature and coordinates of its turning point.

$$\left[x = -1\frac{1}{2} \text{ or } -2, \text{ Minimum at } \left(-1\frac{3}{4}, -\frac{1}{8} \right) \right]$$

13 Solve graphically the quadratic equation $10x^2 - 9x - 11.2 = 0$, given that the roots lie between $x = -1$ and $x = 2$. $[x = -0.7 \text{ or } 1.6]$

14 Plot a graph of $y = 3x^2$ and hence solve the equations
(a) $3x^2 - 8 = 0$, and (b) $3x^2 - 2x - 1 = 0$.

$$\left[\text{(a) } \pm 1.63 \text{ (b) } 1 \text{ or } -\frac{1}{3} \right]$$

15 Plot the graphs $y = 2x^2$ and $y = 3 - 4x$ on the same axes and find the coordinates of the points of intersection. Hence determine the roots of the equation $2x^2 + 4x - 3 = 0$.

$$[(-2.58, 13.31), (0.58, 0.67);$$
$$x = -2.58 \text{ or } 0.58]$$

16 Plot a graph of $y = 10x^2 - 13x - 30$ for values of x between $x = -2$ and $x = 3$.
Solve the equation $10x^2 - 13x - 30 = 0$ and from the graph determine
(a) the value of y when x is 1.3,
(b) the value of x when y is 10, and
(c) the roots of the equation $10x^2 - 15x - 18 = 0$

$$\begin{bmatrix} x = -1.2 \text{ or } 2.5 \\ \text{(a) } -30 \text{ (b) } 2.75 \text{ and } -1.45 \\ \text{(c) } 2.29 \text{ or } -0.79 \end{bmatrix}$$

Linear and quadratic equations simultaneously

17 Determine graphically the values of x and y which simultaneously satisfy the equations $y = 2(x^2 - 2x - 4)$ and $y + 4 = 3x$.

$$\left[x = 4, \ y = 8 \text{ and } x = -\frac{1}{2}, \ y = -5\frac{1}{2} \right]$$

18 Plot the graph of $y = 4x^2 - 8x - 21$ for values of x from -2 to $+4$. Use the graph to find the roots of the following equations:
(a) $4x^2 - 8x - 21 = 0$ (b) $4x^2 - 8x - 16 = 0$
(c) $4x^2 - 6x - 18 = 0$

$$\begin{bmatrix} \text{(a) } x = -1.5 \text{ or } 3.5 \\ \text{(b) } x = -1.24 \text{ or } 3.24 \\ \text{(c) } x = -1.5 \text{ or } 3.0 \end{bmatrix}$$

Cubic equations

19 Plot the graph $y = 4x^3 + 4x^2 - 11x - 6$ between $x = -3$ and $x = 2$ and use the graph to solve the cubic equation $4x^3 + 4x^2 - 11x - 6 = 0$.

$$[x = -2.0, -0.5 \text{ or } 1.5]$$

20 By plotting a graph of $y = x^3 - 2x^2 - 5x + 6$ between $x = -3$ and $x = 4$ solve the equation $x^3 - 2x^2 - 5x + 6 = 0$. Determine also the coordinates of the turning points and distinguish between them.

$$\begin{bmatrix} x = -2, 1 \text{ or } 3, \\ \text{Minimum at } (2.12, -4.10) \\ \text{Maximum at } (-0.79, 8.21) \end{bmatrix}$$

In Problems 21 to 24, solve graphically the cubic equations given, each correct to 2 significant figures.

21 $x^3 - 1 = 0$ $[x = 1]$

22 $x^3 - x^2 - 5x + 2 = 0$
$$[x = -2.0, 0.38 \text{ or } 2.6]$$

23 $x^3 - 2x^2 = 2x - 2$ $[x = 0.69 \text{ or } 2.5]$

24 $2x^3 - x^2 - 9.08x + 8.28 = 0$
$$[x = -2.3, 1.0 \text{ or } 1.8]$$

25 Show that the cubic equation $8x^3 + 36x^2 + 54x + 27 = 0$ has only one real root and determine its value. $[x = -1.5]$

17

Graphs of exponential, logarithmic and trigonometric functions and polar curves

17.1 Graphs of exponential functions

Values of e^x and e^{-x}, obtained from a calculator, correct to 2 decimal places, over a range $x = -3$ to $x = 3$, are shown in the table below.

x	−3.0	−2.5	−2.0	−1.5	−1.0	−0.5	0
e^x	0.05	0.08	0.14	0.22	0.37	0.61	1.00
e^{-x}	20.09	12.18	7.39	4.48	2.72	1.65	1.00

x	0.5	1.0	1.5	2.0	2.5	3.0
e^x	1.65	2.72	4.48	7.39	12.18	20.09
e^{-x}	0.61	0.37	0.22	0.14	0.08	0.05

Figure 17.1 shows graphs of $y = e^x$ and $y = e^{-x}$. A similar table may be drawn up for $y = 5e^{(1/2)x}$, a graph of which is shown in Fig. 17.2. The gradient of the curve at any point, dy/dx, is obtained by drawing a tangent to the curve at that point and measuring the gradient of the tangent. For example:

when $x = 0$, $y = 5$ and

$$\frac{dy}{dx} = \frac{BC}{AB} = \frac{(6.2 - 3.7)}{1} = 2.5$$

and when $x = 2$, $y = 13.6$ and

$$\frac{dy}{dx} = \frac{EF}{DE} = \frac{(16.8 - 10)}{1} = 6.8$$

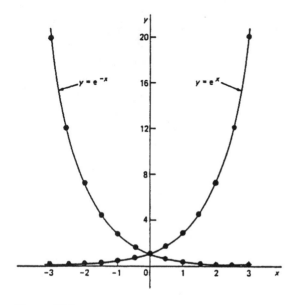

Figure 17.1

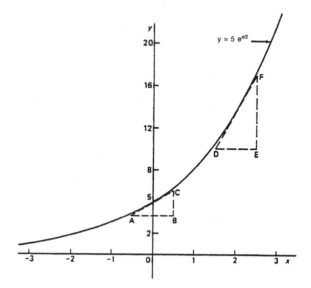

Figure 17.2

These two results each show that $dy/dx = \frac{1}{2}y$ and further determinations of the gradients of $y = 5e^{(1/2)x}$ would give the same result for each. In general, for all natural growth and decay laws of the form $y = Ae^{kx}$, where k is a positive constant for growth laws (as in Fig. 17.2) and a negative constant for decay curves, $dy/dx = ky$, i.e. **the rate of change of the variable, y, is proportional to the variable itself**.

> **Problem 1.** Plot a graph of $y = 2e^{0.3x}$ over a range of $x = -2$ to $x = 3$. Hence determine the value of y when $x = 2.2$ and the value of x when $y = 1.6$

A table of values is drawn up as shown below.

x	-3	-2	-1	0	1	2	3
$0.3x$	-0.9	-0.6	-0.3	0	0.3	0.6	0.9
$e^{0.3x}$	0.407	0.549	0.741	1.00	1.350	1.822	2.460
$2e^{0.3x}$	0.81	1.10	1.48	2.00	2.70	3.64	4.92

A graph of $y = 2e^{0.3x}$ is shown plotted in Fig. 17.3. When $x = 2.2$, $y = $ **3.87** and when $y = 1.6$, $x = $ **-0.74**.

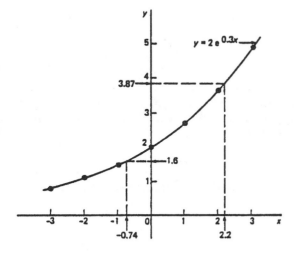

Figure 17.3

> **Problem 2.** Plot the graph of $y = \frac{1}{3}e^{-2x}$ over the range $x = -1.5$ to $x = 1.5$. Determine, from the graph, the value of y when $x = -1.2$ and the value of x when $y = 1.4$

A table of values is drawn up as shown below.

x	-1.5	-1.0	-0.5	0	0.5	1.0	1.5
$-2x$	3	2	1	0	-1	-2	-3
e^{-2x}	20.086	7.389	2.718	1.00	0.368	0.135	0.050
$\frac{1}{3}e^{-2x}$	6.70	2.46	0.91	0.33	0.12	0.05	0.02

A graph of $y = \frac{1}{3}e^{-2x}$ is shown in Fig. 17.4.

When $x = -1.2$, $y = $ **3.67** and when $y = 1.4$, $x = $ **-0.72**.

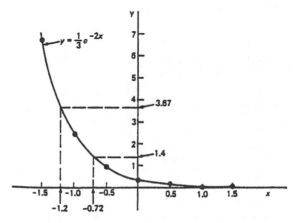

Figure 17.4

> **Problem 3.** A natural law of growth is of the form $y = 4e^{0.2x}$. Plot a graph depicting this law for values of x from $x = -3$ to $x = 3$. From the graph determine (a) the value of y when x is 2.2, (b) the value of x when y is 3.4, and (c) the rate of change of y with respect to x (i.e. dy/dx) at $x = -2$

A table of values is drawn up as shown below.

x	-3	-2	-1	0	1	2	3
$0.2x$	-0.6	-0.4	-0.2	0	0.2	0.4	0.6
$e^{0.2x}$	0.549	0.670	0.819	1.00	1.221	1.492	1.822
$4e^{0.2x}$	2.20	2.68	3.28	4.00	4.88	5.97	7.29

A graph of $4e^{0.2x}$ is shown in Fig. 17.5. From the graph:

(a)   when $x = 2.2$, $y = $ **6.2**
(b)   when $y = 3.4$, $x = $ **-0.8**

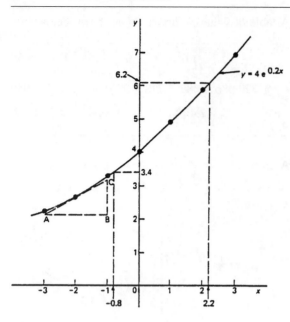

Figure 17.5

(c)   at $x = 2$, gradient, i.e.

$$\frac{dy}{dx} = \frac{BC}{AB} = \frac{1.08}{2} = \mathbf{0.54}$$

Problem 4.   The decay of voltage, v volts, across a capacitor at time t seconds is given by $v = 250e^{-t/3}$. Draw a graph showing the natural decay curve over the first 6 s. From the graph find (a) the voltage after 3.4 s, and (b) the time when the voltage is 150 V. (c) Determine the rate of change of voltage after 2 s and after 4 s

---

A table of values is drawn up as shown below.

t	0	1	2	3
$e^{-t/3}$	1.00	0.7165	0.5134	0.3679
$v = 250e^{-t/3}$	250.0	179.1	128.4	91.97

t	4	5	6
$e^{-t/3}$	0.2636	0.1889	0.1353
$v = 250e^{-t/3}$	65.90	47.22	33.83

The natural decay curve of $v = 250e^{-t/3}$ is shown in Fig. 17.6. From the graph:

(a)   when time $t = 3.4$ s, **voltage $v = 80$ volts**, and

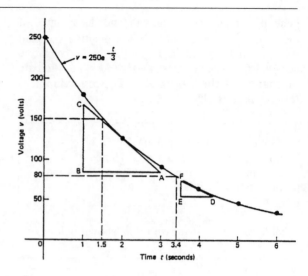

Figure 17.6

(b)   when voltage $v = 150$ volts, **time $t = 1.5$ s**
(c)   Rate of change of voltage (i.e. dv/dt) after 2 s is given by BC/AB

$$\frac{BC}{AC} = \frac{(170 - 85)}{(1 - 3)} = \frac{-85}{2} = \mathbf{-42.5\ V/s}$$

After 4 s, $\dfrac{dv}{dt} = \dfrac{EF}{DE} = \dfrac{(77 - 55)}{(3.5 - 4.5)}$

$$= \frac{22}{-1} = \mathbf{-22\ V/s}$$

Further problems on graphs of exponential functions may be found in Section 17.6, Problems 1 to 8, page 175.

17.2   Graphs of logarithmic functions

A graph of $y = \log_{10} x$ is shown in Fig. 17.7 and a graph of $y = \log_e x$ is shown in Fig. 17.8. Both are seen to be of similar shapes; in fact, the same general shape occurs for a logarithm to any base.

In general, with a logarithm to any base a, it is noted that:

(i)   **$\log_a 1 = 0$**

Let $\log_a 1 = x$, then $a^x = 1$ from the definition of the logarithm (see Chapter 8). If $a^x = 1$ then $x = 0$ from the laws of indices. Hence $\log_a 1 = 0$. In the above graphs it is seen that $\log_{10} 1 = 0$ and $\log_e 1 = 0$

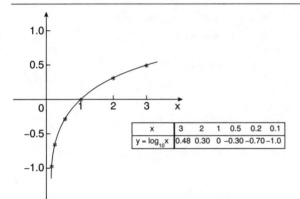

x	3	2	1	0.5	0.2	0.1
$y = \log_{10} x$	0.48	0.30	0	−0.30	−0.70	−1.0

Figure 17.7

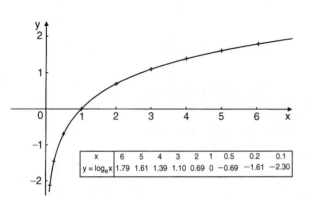

x	6	5	4	3	2	1	0.5	0.2	0.1
$y = \log_e x$	1.79	1.61	1.39	1.10	0.69	0	−0.69	−1.61	−2.30

Figure 17.8

(ii) **$\log_a a = 1$**

Let $\log_a a = x$ then $a^x = a$ from the definition of a logarithm. If $a^x = a$ then $x = 1$.
Hence $\log_a a = 1$. (Check with a calculator that $\log_{10} 10 = 1$ and $\log_e e = 1$)

(iii) **$\log_a 0 \to -\infty$**

Let $\log_a 0 = x$ then $a^x = 0$ from the definition of a logarithm. If $a^x = 0$, and a is a positive real number, then x must approach minus infinity. (For example, check with a calculator, $2^{-2} = 0.25$, $2^{-20} = 9.54 \times 10^{-7}$, $2^{-200} = 6.22 \times 10^{-61}$, and so on.)
Hence $\log_a 0 \to -\infty$

17.3 The production of a sine and cosine wave

In Fig. 17.9, let OR be a vector 1 unit long and free to rotate anticlockwise about O. In one revolution a circle is produced and is shown with 15° sectors.

Each radius arm has a vertical and a horizontal component. For example, at 30°, the vertical component is TS and the horizontal component is OS.
From trigonometric ratios,

$$\sin 30° = \frac{TS}{TO} = \frac{TS}{1}, \text{ i.e. } TS = \sin 30° \text{ and}$$

$$\cos 30° = \frac{OS}{TO} = \frac{OS}{1}, \text{ i.e. } OS = \cos 30°$$

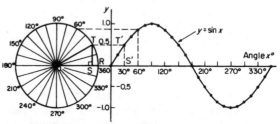

Figure 17.9

The vertical component TS may be projected across to $T'S'$, which is the corresponding value of 30° on the graph of y against angle $x°$. If all such vertical components as TS are projected on to the graph, then a **sine wave** is produced as shown in Fig. 17.9. If all horizontal components such as OS are projected on to a graph of y against angle $x°$, then a **cosine wave** is produced. It is easier to visualize these projections by redrawing the circle with the radius arm OR initially in a vertical position as shown in Fig. 17.10. From Figs 17.9 and 17.10 it is seen that a cosine curve is of the same form as the sine curve but is displaced by 90° (or $\pi/2$ radians).

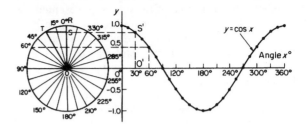

Figure 17.10

17.4 Sinusoidal form $A \sin(\omega t \pm \alpha)$

In Fig. 17.11, let OR represent a vector that is free to rotate anticlockwise about O at a velocity of ω rad/s. A rotating vector is called a **phasor**. After a time

t seconds OR will have turned through an angle ωt radians (shown as angle TOR in Fig. 17.11). If ST is constructed perpendicular to OR, then $\sin \omega t = ST/TO$, i.e. $ST = TO \sin \omega t$.

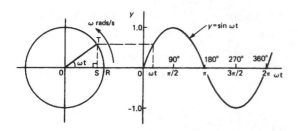

Figure 17.11

If all such vertical components are projected on to a graph of y against ωt, a sine wave results of amplitude OR (as shown in the previous section). If phasor OR makes one revolution (i.e. 2π radians) in T seconds, then the angular velocity, $\omega = 2\pi/T$ rad/s, from which, $T = 2\pi/\omega$ **seconds**. T is known as the **periodic time**.

The number of complete cycles occurring per second is called the **frequency**, f:

$$\text{Frequency} = \frac{\text{number of cycles}}{\text{second}}$$

$$= \frac{1}{T} = \frac{\omega}{2\pi} \text{ Hz, i.e. } f = \frac{\omega}{2\pi} \text{ Hz}$$

Hence angular velocity, $\omega = 2\pi f$ **rad/s**.

Amplitude is the name given to the maximum or peak value of a sine wave. The amplitude of the sine wave shown in Fig. 17.11 has an amplitude of 1.

A sine or cosine wave may not always start at $0°$. To show this a periodic function is represented by $y = \sin(\omega t \pm \alpha)$ or $y = \cos(\omega t \pm \alpha)$, where α is a phase displacement compared with $y = \sin \omega t$ or $y = \cos \omega t$. A graph of $y = \sin(\omega t - \alpha)$ **lags** $y = \sin \omega t$ by angle α, and a graph of $y = \sin(\omega t + \alpha)$ **leads** $y = \sin \omega t$ by angle α.

The angle ωt is measured in **radians** (i.e. $(\omega \text{ rad/s})(t \text{ s}) = \omega t$ radians) hence angle α should also be in radians.

The relationship between degrees and radians is:

$$360° = 2\pi \text{ radians or } \boxed{180° = \pi \text{ radians}}$$

Hence 1 rad $= \dfrac{180}{\pi} = 57.30°$ and, for example,

$$71° = 71 \times \frac{\pi}{180} \text{ rad} = 1.239 \text{ rad}$$

Given a general sinusoidal function
$$y = A \sin(\omega t \pm \alpha), \text{ then}$$

(i) $A = $ amplitude

(ii) $\omega = $ angular velocity $= 2\pi f$ rad/s

(iii) $\dfrac{2\pi}{\omega} = $ periodic time T seconds

(iv) $\dfrac{\omega}{2\pi} = $ frequency, f hertz

(v) $\alpha = $ angle of lead or lag (compared with $y = A \sin \omega t$)

Problem 5.   An alternating current is given by $i = 30 \sin(100\pi t + 0.27)$ amperes. Find the amplitude, periodic time, frequency and phase angle (in degrees and minutes)

$i = 30 \sin(100\pi t + 0.27)$ A

Amplitude $= \textbf{30 A}$

Angular velocity $\omega = 100\pi$

Hence periodic time, $T = \dfrac{2\pi}{\omega} = \dfrac{2\pi}{100\pi} = \dfrac{1}{50}$

$$= \textbf{0.02 s or 20 ms}$$

Frequency, $f = \dfrac{1}{T} = \dfrac{1}{0.02} = \textbf{50 Hz}$

Phase angle,
$$\alpha = 0.27 \text{ rad} = \left(0.27 \times \frac{180}{\pi}\right)^{\circ}$$

$$= \textbf{15°28}' \textbf{ leading } i = 30 \sin(100\pi t)$$

Problem 6.   An oscillating mechanism has a maximum displacement of 2.5 m and a frequency of 60 Hz. At time $t = 0$ the displacement is 90 cm. Express the displacement in the general form $A \sin(\omega t \pm \alpha)$

Amplitude $=$ maximum displacement $= 2.5$ m

Angular velocity, $\omega = 2\pi f = 2\pi(60) = 120\pi$ rad/s

Hence displacement $= 2.5 \sin(120\pi t + \alpha)$ m

When $t = 0$, displacement $= 90$ cm $= 0.90$ m

Hence $0.90 = 2.5 \sin(0 + \alpha)$

i.e. $\sin \alpha = \dfrac{0.90}{2.5} = 0.36$

Hence $\alpha = \arcsin 0.36 = 21°6' = 0.368$ rad

Thus **displacement** $= 2.5 \sin(120\pi t + 0.368)$ **m.**

Further problems on the sinusoidal form $A \sin(\omega t \pm \alpha)$ may be found in Section 17.6, Problems 9 to 15, pages 175 and 176.

17.5 Polar curves

With Cartesian coordinates the equation of a curve is expressed as a general relationship between x and y, i.e. $y = f(x)$.

Similarly, with polar coordinates the equation of a curve is expressed in the form $r = f(\theta)$. When a graph of $r = f(\theta)$ is required a table of values needs to be drawn up and the coordinates (r, θ) plotted.

Problem 7.   Plot the polar graph of $r = 5 \sin \theta$ between $\theta = 0°$ and $\theta = 360°$ using increments of $30°$

A table of values at $30°$ intervals is produced as shown below.

θ	0	30°	60°	90°	120°
$r = 5 \sin \theta$	0	2.50	4.33	5.00	4.33

θ	150°	180°	210°	240°	270°
$r = 5 \sin \theta$	2.50	0	−2.50	−4.33	−5.00

θ	300°	330°	360°
$r = 5 \sin \theta$	−4.33	−2.50	0

The graph is plotted as shown in Fig. 17.12.

Initially the zero line OA is constructed and then the broken lines in Fig. 17.12 at $30°$ intervals are produced. The maximum value of r is 5.00, hence OA is scaled and circles drawn as shown with the largest at a radius of 5 units. The polar coordinates $(0, 0°)$, $(2.50, 30°)$, $(4.33, 60°)$, $(5.00, 90°)$ ... are plotted and shown as points O, B, C, D, ... in Fig. 17.12. When polar coordinate $(0, 180°)$ is plotted and the points joined with a smooth curve a complete circle is seen to have been produced.

When plotting the next point, $(-2.50, 210°)$, since r is negative it is plotted in the opposite direction to $210°$, i.e. 2.50 units long on the $30°$ axis. Hence the point $(-2.50, 210°)$ is equivalent to the point

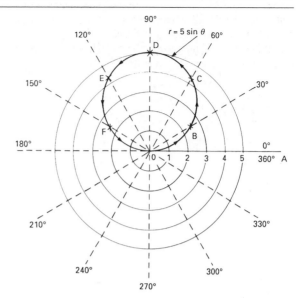

Figure 17.12

$(2.50, 30°)$. Similarly, $(-4.33, 240°)$ is the same point as $(4.33, 60°)$.

When all the coordinates are plotted the graph $r = 5 \sin \theta$ appears as a single circle; it is, in fact, two circles, one on top of the other.

In general, a polar curve $r = a \sin \theta$ is as shown in Fig. 17.13.

In a similar manner to that explained in Problem 7, it may be shown that the polar curve $r = a \cos \theta$ is as sketched in Fig. 17.14.

Problem 8.   Plot the polar graph of $r = 4 \sin^2 \theta$ between $\theta = 0$ and $\theta = 2\pi$ radians using intervals of $\dfrac{\pi}{6}$

A table of values is produced as shown below.

θ	0	$\dfrac{\pi}{6}$	$\dfrac{\pi}{3}$	$\dfrac{\pi}{2}$	$\dfrac{2\pi}{3}$
$\sin \theta$	0	0.50	0.866	1.00	0.866
$r = 4 \sin^2 \theta$	0	1	3	4	3

θ	$\dfrac{5\pi}{6}$	π	$\dfrac{7\pi}{6}$	$\dfrac{4\pi}{3}$	$\dfrac{3\pi}{2}$
$\sin \theta$	0.50	0	−0.50	−0.866	−1.00
$r = 4 \sin^2 \theta$	1	0	1	3	4

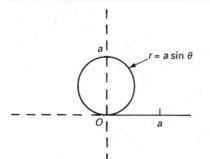

Figure 17.13

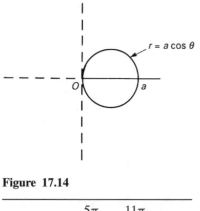

Figure 17.14

θ	$\dfrac{5\pi}{3}$	$\dfrac{11\pi}{6}$	2π
$\sin\theta$	-0.866	-0.50	0
$r = 4\sin^2\theta$	3	1	0

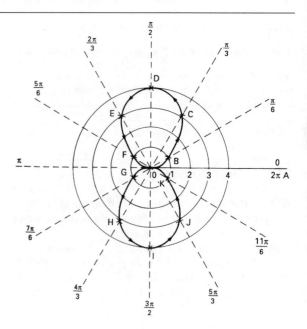

Figure 17.15

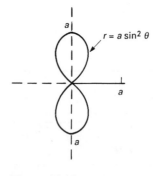

Figure 17.16

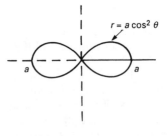

Figure 17.17

The zero line OA is first constructed and then the broken lines at intervals of $\pi/6$ rad (or $30°$) are produced. The maximum value of r is 4 hence OA is scaled and circles produced as shown with the largest at a radius of 4 units. The polar coordinates $(0, 0)$, $(1, \pi/6)$, $(3, \pi/3)$, $\ldots$ $(0, \pi)$ are plotted and shown as points O, B, C, D, E, F, O, respectively. Then $(1, 7\pi/6)$, $(3, 4\pi/3)$, $\ldots$ $(0, 0)$ are plotted as shown by points G, H, I, J, K, O, respectively. Thus two distinct loops are produced as shown in Fig. 17.15.

In general, a polar curve $r = a\sin^2\theta$ is as shown in Fig. 17.16. In a similar manner it may be shown that the polar curve $r = a\cos^2\theta$ is as sketched in Fig. 17.17.

> Problem 9.   Plot the polar graph of $r = 3\sin 2\theta$ between $\theta = 0°$ and $\theta = 360°$, using $15°$ intervals

A table of values is produced as shown below.

θ	0	15°	30°	45°	60°	75°	90°
$r = 3\sin 2\theta$	0	1.5	2.6	3.0	2.6	1.5	0

θ	105°	120°	135°	150°	165°	180°
$r = 3\sin 2\theta$	−1.5	−2.6	−3.0	−2.6	−1.5	0

θ	195°	210°	225°	240°	255°	270°
$r = 3\sin 2\theta$	1.5	2.6	3.0	2.6	1.5	0

θ	285°	300°	315°	330°	345°	360°
$r = 3\sin 2\theta$	−1.5	−2.6	−3.0	−2.6	−1.5	0

The polar graph $r = 3\sin 2\theta$ is plotted as shown in Fig. 17.18 and is seen to contain four similar shaped loops displaced at 90° from each other.

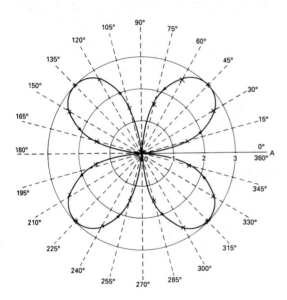

Figure 17.18

In general, a polar curve $r = a\sin 2\theta$ is as shown in Fig. 17.19. In a similar manner it may be shown that polar curves of $r = a\cos 2\theta$, $r = a\sin 3\theta$ and $r = a\cos 3\theta$ are as sketched in Fig. 17.20.

> **Problem 10.** Sketch the polar curve $r = 2\theta$ between $\theta = 0$ and $\theta = 5\pi/2$ rad at intervals of $\pi/6$

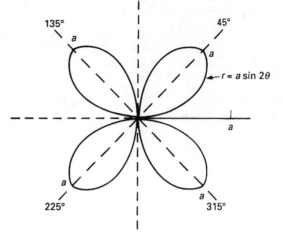

Figure 17.19

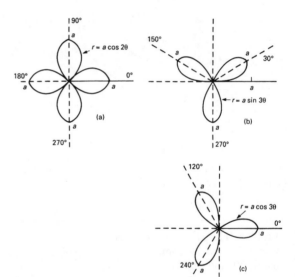

Figure 17.20

A table of values is produced as shown below.

θ	0	$\dfrac{\pi}{6}$	$\dfrac{\pi}{3}$	$\dfrac{\pi}{2}$	$\dfrac{2\pi}{3}$	$\dfrac{5\pi}{6}$
$r = 2\theta$	0	1.05	2.09	3.14	4.19	5.24

θ	π	$\dfrac{7\pi}{6}$	$\dfrac{4\pi}{3}$	$\dfrac{3\pi}{2}$	$\dfrac{5\pi}{3}$	$\dfrac{11\pi}{6}$
$r = 2\theta$	6.28	7.33	8.38	9.42	10.47	11.52

θ	2π	$\dfrac{13\pi}{6}$	$\dfrac{7\pi}{3}$	$\dfrac{5\pi}{2}$
$r = 2\theta$	12.57	13.61	14.66	15.71

The polar graph of $r = 2\theta$ is shown in Fig. 17.21 and is seen to be an ever-increasing spiral.

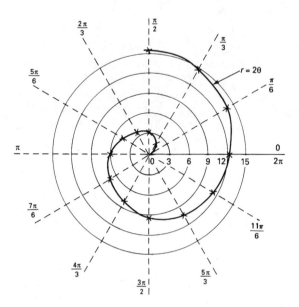

Figure 17.21

Problem 11.   Plot the polar curve $r = 5(1 + \cos\theta)$ from $\theta = 0°$ to $\theta = 360°$, using 30° intervals

---

A table of values is shown below.

θ	0	30°	60°	90°	120°
$r = 5(1 + \cos\theta)$	10.00	9.33	7.50	5.00	2.50

θ	150°	180°	210°	240°	270°
$r = 5(1 + \cos\theta)$	0.67	0	0.67	2.50	5.00

θ	300°	330°	360°
$r = 5(1 + \cos\theta)$	7.50	9.33	10.00

The polar curve $r = 5(1 + \cos\theta)$ is shown in Fig. 17.22.

In general, a polar curve $r = a(1 + \cos\theta)$ is as shown in Fig. 17.23 and the shape is called a **cardioid**.

In a similar manner it may be shown that the polar curve $r = a + b\cos\theta$ varies in shape according to the relative values of a and b. When $a = b$ the polar curve shown in Fig. 17.23 results.

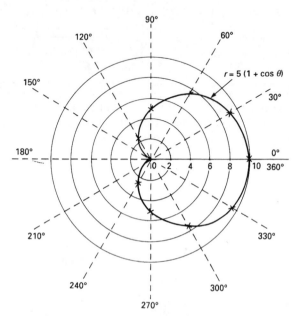

Figure 17.22

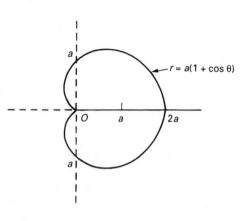

Figure 17.23

When $a < b$ the general shape shown in Fig. 17.24(a) results and when $a > b$ the general shape shown in Fig. 17.24(b) results.

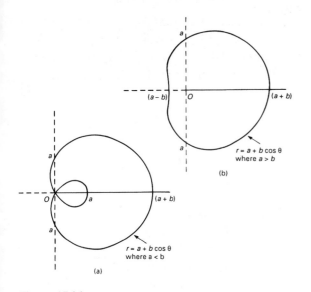

Figure 17.24

Further problems on polar curves may be found in Section 17.6, Problems 16 to 21, page 176.

17.6 Further problems on graphs of exponential, logarithmic and trigonometric functions and polar curves

Graphs of exponential functions

1 Plot a graph of $y = 3e^{0.2x}$ over the range $x = -3$ to $x = 3$. Hence determine the value of y when $x = 1.4$ and the value of x when $y = 4.5$.   [3.97, 2.03]

2 Plot a graph of $y = \frac{1}{2}e^{-1.5x}$ over a range $x = -1.5$ to $x = 1.5$ and hence determine the value of y when $x = -0.8$ and the value of x when $y = 3.5$.   [1.66, -1.30]

3 Plot a graph of $y = 2.5e^{-0.15x}$ over a range $x = -8$ to $x = 8$. Determine from the graph the value of y when $x = -6.2$ and the value of x when $y = 5.4$.   [6.34, -5.13]

4 Draw a graph of $y = 2(2e^{-x} - 3e^{2x})$ over a range of $x = -3$ to $x = 3$. Determine the value

of y when $x = -2.2$ and the value of x when $y = 17.4$.   [36.0, -1.49]

5 In a chemical reaction the amount of starting material C cm^3 left after t minutes is given by $C = 40e^{-0.006t}$. Plot a graph of C against t and determine (a) the concentration C after 1 hour, (b) the time taken for the concentration to decrease by half. (c) Determine the rate of change of C with t after 40 min.

$$\begin{bmatrix} \text{(a) } 27.9 \text{ cm}^3 \\ \text{(b) } 115.5 \text{ min} \\ \text{(c) } -0.189 \text{ cm}^3/\text{min} \end{bmatrix}$$

6 The rate at which a body cools is given by $\theta = 250e^{-0.05t}$ where the excess of temperature of a body above its surroundings at time t minutes is $\theta\,^\circ$C. Plot a graph showing this natural decay curve for the first hour of cooling and hence determine the rate of cooling after (a) 15 minutes, (b) 45 minutes.

[(a) $-5.90\,^\circ$C/min (b) $-1.32\,^\circ$C/min]

7 The tensions in two sides of a belt, T and T_0 newtons, passing round a pulley wheel and in contact with the pulley for an angle θ radians is given by $T = T_0 e^{0.3\theta}$. Plot a graph depicting this relationship over a range $\theta = 0$ to $\theta = 2.0$ radians, given $T_0 = 50$ N. From the graph determine the value of $dT/d\theta$ when $\theta = 1.2$ radians.   [21.5 N/rad]

8 The voltage drop, v volts, across an inductor is related to time t ms, by $v = 30 \times 10^3 e^{-t/10}$. Plot a graph of v against t from $t = 0$ to $t = 10$ ms. Use the graph to determine the rate of change of voltage with time (i.e. dv/dt) when $t = 5.5$ ms.

[-1731 V/ms]

Sinusoidal form $A \sin(\omega t \pm \alpha)$

In Problems 9 to 12 find the amplitude, periodic time, frequency and phase angle (stating whether it is leading or lagging $\sin \omega t$) of the alternating quantities given:

9 $i = 40 \sin(50\pi t + 0.29)$ mA

$$\begin{bmatrix} 40 \text{ mA, } 0.04 \text{ s, } 25 \text{ Hz, } 0.29 \text{ rad} \\ \text{(or } 16^\circ 37') \text{ leading } 40 \sin 50\pi t \end{bmatrix}$$

10 $y = 75 \sin(40t - 0.54)$ cm

$$\begin{bmatrix} 75 \text{ cm, } 0.157 \text{ s, } 6.37 \text{ Hz, } 0.54 \text{ rad} \\ \text{(or } 30^\circ 56') \text{ lagging } 75 \sin 40t \end{bmatrix}$$

11 $v = 300 \sin(200\pi t - 0.412)$ V

$$\begin{bmatrix} 300 \text{ V, } 0.01 \text{ s, } 100 \text{ Hz, } 0.412 \text{ rad} \\ \text{(or } 23^\circ 36') \text{ lagging } 300 \sin 200\pi t \end{bmatrix}$$

12   $x = 15 \sin(314.2t + 0.732)$ m

$$\begin{bmatrix} 15 \text{ m, } 0.02 \text{ s, } 50 \text{ Hz, } 0.732 \text{ rad} \\ (\text{or } 41°56') \text{ leading } 15 \sin 314.2t \end{bmatrix}$$

13   A sinusoidal voltage has a maximum value of 120 V and a frequency of 50 Hz. At time $t = 0$, the voltage is (a) zero, and (b) 50 V. Express the instantaneous voltage v in the form $v = A \sin(\omega t \pm \alpha)$.

$$\begin{bmatrix} (a) \ v = 120 \sin 100\pi t \text{ volts} \\ (b) \ v = 120 \sin(100\pi t + 0.43) \text{ volts} \end{bmatrix}$$

14   An alternating current has a periodic time of 25 ms and a maximum value of 20 A. When time $= 0$, current $i = -10$ amperes. Express the current i in the form $i = A \sin(\omega t \pm \alpha)$.

$$\left[i = 20 \sin\left(80\pi t - \frac{\pi}{6} \right) \text{ amperes} \right]$$

15   An oscillating mechanism has a maximum displacement of 3.2 m and a frequency of 50 Hz. At time $t = 0$ the displacement is 150 cm. Express the displacement in the general form $A \sin(\omega t \pm \alpha)$.

$$[3.2 \sin(100\pi t + 0.488) \text{ m}]$$

Polar curves

In Problems 16 to 21, sketch the given polar curves:

16   (a) $r = 3 \sin \theta$ (b) $r = 4 \cos \theta$

$$\begin{bmatrix} (a) \text{ Similar to Fig. 17.13, where } a = 3 \\ (b) \text{ Similar to Fig. 17.14, where } a = 4 \end{bmatrix}$$

17   (a) $r = 2 \sin^2 2\theta$ (b) $r = 7 \cos^2 2\theta$

$$\begin{bmatrix} (a) \text{ Similar to Fig. 17.16, where } a = 2 \\ (b) \text{ Similar to Fig. 17.17, where } a = 7 \end{bmatrix}$$

18   (a) $r = 5 \sin 2\theta$ (b) $r = 4 \cos 2\theta$

$$\begin{bmatrix} (a) \text{ Similar to Fig. 17.19, where } a = 5 \\ (b) \text{ Similar to Fig. 17.20(a), where } a = 4 \end{bmatrix}$$

19   (a) $r = 3 \cos 3\theta$ (b) $r = 9 \sin 3\theta$

$$\begin{bmatrix} (a) \text{ Similar to Fig. 17.20(c), where } a = 3 \\ (b) \text{ Similar to Fig. 17.20(b), where } a = 9 \end{bmatrix}$$

20   (a) $r = 5\theta$ (b) $r = 2(1 + \cos \theta)$

$$\begin{bmatrix} (a) \text{ Similar to Fig. 17.21} \\ (b) \text{ Similar to Fig. 17.23, where } a = 2 \end{bmatrix}$$

21   (a) $r = 1 + 2 \cos \theta$ (b) $r = 3 + \cos \theta$

$$\begin{bmatrix} (a) \text{ Similar to Fig. 17.24(a),} \\ \text{where } a = 1 \text{ and } b = 2 \\ (b) \text{ Similar to Fig. 17.24(b),} \\ \text{where } a = 3 \text{ and } b = 1 \end{bmatrix}$$

18

Irregular areas and mean values of waveforms

18.1 Areas of irregular figures

Areas of irregular plane surfaces may be approximately determined by using (a) a planimeter, (b) the trapezoidal rule, (c) the mid-ordinate rule, and (d) Simpson's rule. Such methods may be used, for example, by engineers estimating areas of indicator diagrams of steam engines, surveyors estimating areas of plots of land or naval architects estimating areas of water planes or transverse sections of ships.

(a) **A planimeter** is an instrument for directly measuring small areas bounded by an irregular curve

(b) **Trapezoidal rule**
To determine the area $PQRS$ in Fig. 18.1:

   (i) Divide base PS into any number of equal intervals, each of width d (the greater the number of intervals, the greater the accuracy)

  (ii) Accurately measure ordinates y_1, y_2, y_3, etc.

 (iii) Area $PQRS$

$$= d \left[\frac{y_1 + y_7}{2} + y_2 + y_3 \right.$$
$$\left. + y_4 + y_5 + y_6 \right]$$

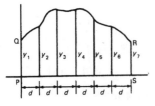

Figure 18.1

In general, the trapezoidal rule states:

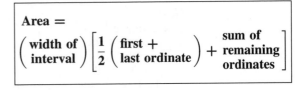

$$\text{Area} = \left(\begin{array}{c} \text{width of} \\ \text{interval} \end{array} \right) \left[\frac{1}{2} \left(\begin{array}{c} \text{first} + \\ \text{last ordinate} \end{array} \right) + \begin{array}{c} \text{sum of} \\ \text{remaining} \\ \text{ordinates} \end{array} \right]$$

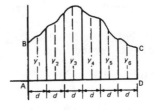

Figure 18.2

(c) **Mid-ordinate rule**
To determine the area $ABCD$ of Fig. 18.2:

   (i) Divide base AD into any number of equal intervals, each of width d (the greater the number of intervals, the greater the accuracy)

  (ii) Erect ordinates in the middle of each interval (shown by broken lines in Fig. 18.2)

 (iii) Accurately measure ordinates y_1, y_2, y_3, etc.

 (iv) Area $ABCD = d(y_1 + y_2 + y_3 + y_4 + y_5 + y_6 + y_7)$

In general, the mid-ordinate rule states:

Area = (width of interval) (sum of mid-ordinates)

(d) **Simpson's rule**
To determine the area $PQRS$ of Fig. 18.1:

   (i) Divide base PS into an **even** number of intervals, each of width d (the greater the number of intervals, the greater the accuracy)

(ii)   Accurately measure ordinates y_1, y_2, y_3, etc.

(iii)   Area $PQRS = \dfrac{d}{3}[(y_1 + y_7) + 4(y_2 + y_4 + y_6) + 2(y_3 + y_5)]$

In general, Simpson's rule states:

$$\text{Area} = \frac{1}{3}\begin{pmatrix}\text{width of}\\\text{interval}\end{pmatrix}\left[\begin{pmatrix}\text{first } +\\\text{last ordinate}\end{pmatrix} + 4\begin{pmatrix}\text{sum of even}\\\text{ordinates}\end{pmatrix} + 2\begin{pmatrix}\text{sum of}\\\text{remaining}\\\text{odd ordinates}\end{pmatrix}\right]$$

Problem 1.   A car starts from rest and its speed is measured every second for 6 s.

Time t (s)	0	1	2	3
Speed v (m/s)	0	2.5	5.5	8.75

Time t (s)	4	5	6
Speed v (m/s)	12.5	17.5	24.0

Determine the distance travelled in 6 seconds (i.e. the area under the v/t graph) by (a) the trapezoidal rule, (b) the mid-ordinate rule, and (c) Simpson's rule

A graph of speed/time is shown in Fig. 18.3.

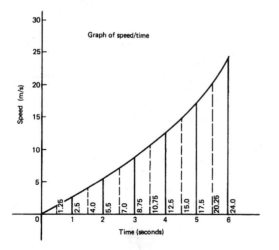

Figure 18.3

(a)   **Trapezoidal rule** (see para. (b) above)
The time base is divided into 6 strips each of width 1 s, and the length of the ordinates measured. Thus

$$\text{area} = (1)\left[\left(\frac{0 + 24.0}{2}\right) + 2.5 + 5.5 + 8.75 + 12.5 + 17.5\right]$$
$$= \textbf{58.75 m}$$

(b)   **Mid-ordinate rule** (see para. (c) above)
The time base is divided into 6 strips each of width 1 second. Mid-ordinates are erected as shown in Fig. 18.3 by the broken lines. The length of each mid-ordinate is measured. Thus

$$\text{area} = (1)[1.25 + 4.0 + 7.0 + 10.75 + 15.0 + 20.25)$$
$$= \textbf{58.25 m}$$

(c)   **Simpson's rule** (see para. (d) above)
The time base is divided into 6 strips each of width 1 s, and the length of the ordinates measured. Thus

$$\text{area} = \frac{1}{3}(1)[(0 + 24.0) + 4(2.5 + 8.75 + 17.5) + 2(5.5 + 12.5)]$$
$$= \textbf{58.33 m}$$

Problem 2.   A river is 15 m wide. Soundings of the depth are made at equal intervals of 3 m across the river and are as shown below.

Depth (m)	0	2.2	3.3	4.5	4.2	2.4	0

Calculate the cross-sectional area of the flow of water at this point using Simpson's rule

From para. (d) above,

$$\text{Area} = \frac{1}{3}(3)[(0 + 0) + 4(2.2 + 4.5 + 2.4) + 2(3.3 + 4.2)]$$
$$= (1)[0 + 36.4 + 15]$$
$$= \textbf{51.4 m}^2$$

Further problems on areas of irregular figures may be found in Section 18.3, Problems 1 to 5, page 182.

18.2 The mean or average value of a waveform

The mean or average value, y, of the waveform shown in Fig. 18.4 is given by:

$$y = \frac{\text{area under curve}}{\text{length of base, } b}$$

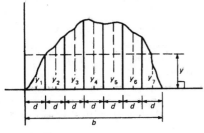

Figure 18.4

If the mid-ordinate rule is used to find the area under the curve, then:

$$y = \frac{\text{sum of mid-ordinates}}{\text{number of mid-ordinates}}$$

$$\left(= \frac{y_1 + y_2 + y_3 + y_4 + y_5 + y_6 + y_7}{7} \right.$$

$$\text{for Fig. 18.4})$$

For a **sine wave**, the mean or average value:

(i) over one complete cycle is zero (see Fig. 18.5(a)),

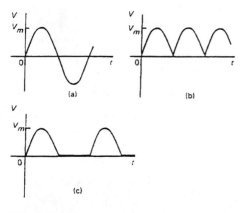

Figure 18.5

(ii) over half a cycle is **0.637 × maximum value**, or **2/π × maximum value**,

(iii) of a full-wave rectified waveform (see Fig. 18.5(b)) is **0.637 × maximum value**,

(iv) of a half-wave rectified waveform (see Fig. 18.5(c)) is **0.318 × maximum value**, or **1/π maximum value**

Problem 3. Determine the average values over half a cycle of the periodic waveforms shown in Fig. 18.6.

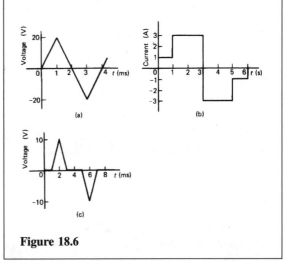

Figure 18.6

(a) Area under triangular waveform (a) for a half cycle is given by:

$$\text{Area} = \frac{1}{2}\,(\text{base})(\text{perpendicular height})$$

$$= \frac{1}{2}(2 \times 10^{-3})(20) = 20 \times 10^{-3}\ \text{Vs}$$

Average value of waveform

$$= \frac{\text{area under curve}}{\text{length of base}}$$

$$= \frac{20 \times 10^{-3}\ \text{Vs}}{2 \times 10^{-3}\ \text{s}} = \textbf{10 V}$$

(b) Area under waveform (b) for a half cycle $= (1 \times 1) + (3 \times 2) = 7$ As

Average value of waveform

$$= \frac{\text{area under curve}}{\text{length of base}}$$

$$= \frac{7\ \text{As}}{3\ \text{s}} = \textbf{2.33 A}$$

(c)   A half cycle of the voltage waveform (c) is completed in 4 ms.

$$\text{Area under curve} = \frac{1}{2}\{(3-1)10^{-3}\}(10)$$

$$= 10 \times 10^{-3} \text{ Vs}$$

Average value of waveform

$$= \frac{\text{area under curve}}{\text{length of base}} = \frac{10 \times 10^{-3} \text{ Vs}}{4 \times 10^{-3} \text{ s}}$$

$$= \textbf{2.5 V}$$

Problem 4.   Determine the mean value of current over one complete cycle of the periodic waveforms shown in Fig. 18.7

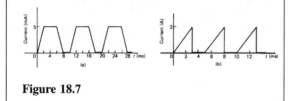

Figure 18.7

(a)   One cycle of the trapezoidal waveform (a) is completed in 10 ms (i.e. the periodic time is 10 ms).
Area under curve = area of trapezium

$$= \frac{1}{2} \text{ (sum of parallel sides)}$$

$$\times \text{ (perpendicular distance between parallel sides)}$$

$$= \frac{1}{2}\{(4+8) \times 10^{-3}\}(5 \times 10^{-3})$$

$$= 30 \times 10^{-6} \text{ As}$$

Mean value over one cycle

$$= \frac{\text{area under curve}}{\text{length of base}} = \frac{30 \times 10^{-6} \text{ As}}{10 \times 10^{-3} \text{ s}}$$

$$= \textbf{3 mA}$$

(b)   One cycle of the sawtooth waveform (b) is completed in 5 ms.

$$\text{Area under curve} = \frac{1}{2}(3 \times 10^{-3})(2)$$

$$= 3 \times 10^{-3} \text{ As}$$

Mean value over one cycle

$$= \frac{\text{area under curve}}{\text{length of base}} = \frac{3 \times 10^{-3} \text{ As}}{5 \times 10^{-3} \text{ s}} = \textbf{0.6 A}$$

Problem 5.   The power used in a manufacturing process during a 6 hour period is recorded at intervals of 1 hour as shown below.

Time (h)	0	1	2	3	4	5	6
Power (kW)	0	14	29	51	45	23	0

Plot a graph of power against time and, by using the mid-ordinate rule, determine (a) the area under the curve, and (b) the average value of the power

---

The graph of power/time is shown in Fig. 18.8.

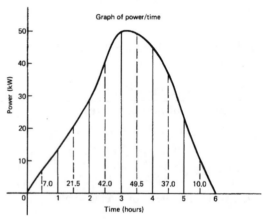

Figure 18.8

(a)   The time base is divided into 6 equal intervals, each of width 1 hour. Mid-ordinates are erected (shown by broken lines in Fig. 18.8) and measured. The values are shown in Fig. 18.8.

$$\text{Area under curve} = \text{(width of interval)}$$

$$\times \text{(sum of mid-ordinates)}$$

$$= (1)(7.0 + 21.5 + 42.0$$

$$+ 49.5 + 37.0 + 10.0)$$

$$= \textbf{167 kWh}$$

(i.e. a measure of electrical energy)

(b)   Average value of waveform

$$= \frac{\text{area under curve}}{\text{length of base}} = \frac{167 \text{ kWh}}{6 \text{ h}}$$

$$= \textbf{27.83 kW}$$

Alternatively, average value

$$= \frac{\text{sum of mid-ordinates}}{\text{number of mid-ordinates}}$$

Problem 6.   Figure 18.9 shows a sinusoidal output voltage of a full-wave rectifier. Determine, using the mid-ordinate rule with 6 intervals, the mean output voltage

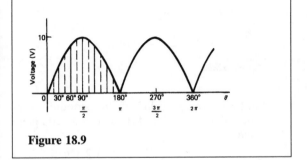

Figure 18.9

One cycle of the output voltage is completed in π radians or 180°. The base is divided into 6 intervals, each of width 30°. The mid-ordinate of each interval will lie at 15°, 45°, 75°, etc.
At 15° the height of the mid-ordinate is $10\sin 15° = 2.588$ V.
At 45° the height of the mid-ordinate is $10\sin 45° = 7.071$ V, and so on.
The results are tabulated below:

Mid-ordinate	Height of mid-ordinate		
15°	$10\sin 15°$	$= 2.588$	V
45°	$10\sin 45°$	$= 7.071$	V
75°	$10\sin 75°$	$= 9.659$	V
105°	$10\sin 105°$	$= 9.659$	V
135°	$10\sin 135°$	$= 7.071$	V
165°	$10\sin 165°$	$= 2.588$	V
	Sum of mid-ordinates	$= 38.636$	V

Mean or average value of output voltage

$$= \frac{\text{sum of mid-ordinates}}{\text{number of mid-ordinates}} = \frac{38.636}{6}$$

$$= \textbf{6.439 V}$$

(With a larger number of intervals a more accurate answer may be obtained.) For a sine wave the actual mean value is $0.637 \times$ maximum value, which in this problem gives 6.37 V.

Problem 7.   An indicator diagram for a steam engine is shown in Fig. 18.10. The base line has been divided into 6 equally spaced intervals and the lengths of the 7 ordinates measured with the results shown in centimetres. Determine (a) the area of the indicator diagram using Simpson's rule, and (b) the mean pressure in the cylinder given that 1 cm represents 100 kPa

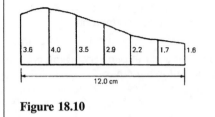

Figure 18.10

(a)   The width of each interval is 12.0/6 cm. Using Simpson's rule,

$$\text{area} = \frac{1}{3}(2.0)[(3.6 + 1.6) + 4(4.0$$

$$+ 2.9 + 1.7) + 2(3.5 + 2.2)]$$

$$= \frac{2}{3}[5.2 + 34.4 + 11.4]$$

$$= \textbf{34 cm}^2$$

(b)   Mean height of ordinates

$$= \frac{\text{area of diagram}}{\text{length of base}} = \frac{34}{12} = 2.83 \text{ cm}$$

Since 1 cm represents 100 kPa, the mean pressure in the cylinder $= 2.83 \times 100$ kPa/cm $= \textbf{283 kPa}$

Further problems on mean or average values of waveforms may be found in the following Section 18.3, Problems 6 to 10, pages 182 and 183.

18.3   Further problems on irregular areas and mean values of waveforms

Areas of irregular figures

1   Plot a graph of $y = 3x - x^2$ by completing a table of values of y from $x = 0$ to $x = 3$. Determine the area enclosed by the curve, the x-axis and ordinates $x = 0$ and $x = 3$ by (a) the trapezoidal rule, (b) the mid-ordinate rule, and (c) by Simpson's rule.

$$\left[4\frac{1}{2} \text{ square units} \right]$$

2   Plot the graph of $y = 2x^2 + 3$ between $x = 0$ and $x = 4$. Estimate the area enclosed by the curve, the ordinates $x = 0$ and $x = 4$, and the x-axis by an approximate method.

[54.7 square units]

3   The velocity of a car at one second intervals is given in the following table:

time t (s)	0	1	2	3	4	5	6
velocity v (m/s)	0	2.0	4.5	8.0	14.0	21.0	29.0

Determine the distance travelled in 6 seconds (i.e. the area under the v/t graph) using an approximate method.   [63 m]

4   The shape of a piece of land is shown in Fig. 18.11. To estimate the area of the land, a surveyor takes measurements at intervals of 50 m, perpendicular to the straight portion with the results shown (the dimensions being in metres). Estimate the area of the land in hectares (1 ha = 10^4 m^2).   [4.70 ha]

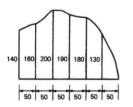

140 160 200 190 180 130

50 50 50 50 50 50

Figure 18.11

5   The deck of a ship is 35 m long. At equal intervals of 5 m the width is given by the following table:

Width 0 2.8 5.2 6.5 5.8 4.1 3.0 2.3
(m)

Estimate the area of the deck.

[143 m^2]

Mean or average values of waveforms

6   Determine the mean value of the periodic waveforms shown in Fig. 18.12 over a half cycle.

[(a) 2 A (b) 50 V (c) 2.5 A]

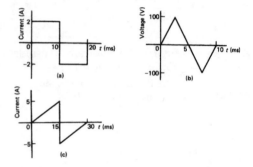

Figure 18.12

7   Find the average value of the periodic waveforms shown in Fig. 18.13 over one complete cycle.

[(a) 2.5 V (b) 3 A]

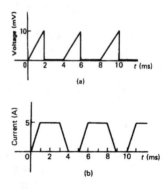

Figure 18.13

8   An alternating current has the following values at equal intervals of 5 ms:

Time (ms)	0	5	10	15	20	25	30
Current (A)	0	0.9	2.6	4.9	5.8	3.5	0

Plot a graph of current against time and estimate the area under the curve over the 30 ms period using the mid-ordinate rule and determine its mean value. [0.093 As, 3.1 A]

9 Determine, using an approximate method, the average value of a sine wave of maximum value 50 V for (a) a half cycle, and (b) a complete cycle. [(a) 31.83 V (b) 0]

10 An indicator diagram of a steam engine is 12 cm long. Seven evenly spaced ordinates, including the end ordinates, are measured as follows:

5.90, 5.52, 4.22, 3.63, 3.32, 3.24, 3.16 cm

Determine the area of the diagram and the mean pressure in the cylinder if 1 cm represents 90 kPa.

[49.13 cm^2, 368.5 kPa]

19

Introduction to differentiation

19.1 Introduction to calculus

Calculus is a branch of mathematics involving or leading to calculations dealing with continuously varying functions.

Calculus is a subject which falls into two parts:

(i) **differential calculus** (or **differentiation**), and
(ii) **integral calculus** (or **integration**)

Differentiation is used in calculations involving velocity and acceleration, rates of change and maximum and minimum values of curves.

19.2 Functional notation

In an equation such as $y = 3x^2 + 2x - 5$, y is said to be a function of x and may be written as $y = f(x)$.

An equation written in the form $f(x) = 3x^2 + 2x - 5$ is termed **functional notation**.

The value of $f(x)$ when $x = 0$ is denoted by $f(0)$, and the value of $f(x)$ when $x = 2$ is denoted by $f(2)$, and so on. Thus when $f(x) = 3x^2 + 2x - 5$, then

$$f(0) = 3(0)^2 + 2(0) - 5 = -5$$

and $f(2) = 3(2)^2 + 2(2) - 5 = 11$

and so on.

Problem 1. If $f(x) = 4x^2 - 3x + 2$ find $f(0)$, $f(3)$, $f(-1)$ and $f(3) - f(-1)$

$f(x) = 4x^2 - 3x + 2$

$f(0) = 4(0)^2 - 3(0) + 2 = \mathbf{2}$

$f(3) = 4(3)^2 - 3(3) + 2 = 36 - 9 + 2 = \mathbf{29}$

$f(-1) = 4(-1)^2 - 3(-1) + 2 = 4 + 3 + 2 = \mathbf{9}$

$f(3) - f(-1) = 29 - 9 = \mathbf{20}$

Problem 2. Given that $f(x) = 5x^2 + x - 7$ determine

(i) $f(2) \div f(1)$ (ii) $f(3 + a)$

(iii) $f(3 + a) - f(3)$ (iv) $\dfrac{f(3 + a) - f(3)}{a}$

$f(x) = 5x^2 + x - 7$

(i) $f(2) = 5(2)^2 + 2 - 7 = 15$

$f(1) = 5(1)^2 + 1 - 7 = -1$

$f(2) + f(1) = \dfrac{15}{-1} = \mathbf{-15}$

(ii) $f(3 + a) = 5(3 + a)^2 + (3 + a) - 7$

$= 5(9 + 6a + a^2) + (3 + a) - 7$

$= 45 + 30a + 5a^2 + 3 + a - 7$

$= \mathbf{41 + 31a + 5a^2}$

(iii) $f(3) = 5(3)^2 + 3 - 7 = 41$

$f(3 + a) - f(3) = (41 + 31a + 5a^2) - (41)$

$= \mathbf{31a + 5a^2}$

(iv) $\dfrac{f(3 + a) - f(3)}{a} = \dfrac{31a + 5a^2}{a} = \mathbf{31 + 5a}$

Further problems on functional notation may be found in Section 19.13, Problems 1 to 4, page 196.

19.3 The gradient of a curve

(a) If a tangent is drawn at a point P on a curve, then the gradient of this tangent is said to be the **gradient of the curve** at P. In Fig. 19.1, the gradient of the curve at P is equal to the gradient of the tangent PQ

(b) For the curve shown in Fig. 19.2, let the points A and B have coordinates (x_1, y_1) and (x_2, y_2), respectively. In functional notation, $y_1 = f(x_1)$ and $y_2 = f(x_2)$ as shown.

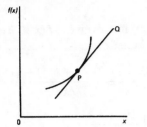

Figure 19.1

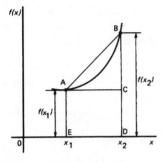

Figure 19.2

The gradient of the chord AB

$$= \frac{BC}{AC} = \frac{BD - CD}{ED}$$

$$= \frac{f(x_2) - f(x_1)}{(x_2 - x_1)}$$

(c)   For the curve $f(x) = x^2$ shown in Fig. 19.3:

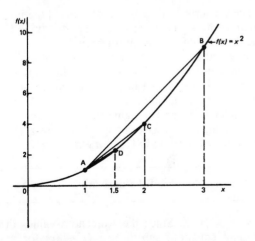

Figure 19.3

(i)   the gradient of chord AB

$$= \frac{f(3) - f(1)}{3 - 1} = \frac{9 - 1}{2} = 4$$

(ii)   the gradient of chord AC

$$= \frac{f(2) - f(1)}{2 - 1} = \frac{4 - 1}{1} = 3$$

(iii)   the gradient of chord AD

$$= \frac{f(1.5) - f(1)}{1.5 - 1} = \frac{2.25 - 1}{0.5} = 2.5$$

(iv)   if E is the point on the curve $(1.1, f(1.1))$ then the gradient of chord AE

$$= \frac{f(1.1 - f(1)}{1.1 - 1} = \frac{1.21 - 1}{0.1} = 2.1$$

(v)   if F is the point on the curve $(1.01, f(1.01))$ then the gradient of chord AF

$$= \frac{f(1.01) - f(1)}{1.01 - 1} = \frac{1.0201 - 1}{0.01}$$

$$= 2.01$$

Thus as point B moves closer and closer to point A the gradient of the chord approaches nearer and nearer to the value 2. This is called the **limiting value** of the gradient of the chord AB and when B coincides with A the chord becomes the tangent to the curve.

A further problem on the gradient of a curve may be found in Section 19.13, Problem 5, page 197.

19.4   Differentiation from first principles

(i)   In Fig. 19.4, A and B are two points very close together on a curve, δx (delta x) and δy (delta y) representing small increments in the x and y directions, respectively.
Gradient of chord $AB = \delta y / \delta x$,
however $\delta y = f(x + \delta x) - f(x)$
Hence $\dfrac{\delta y}{\delta x} = \dfrac{f(x + \delta x) - f(x)}{\delta x}$
As δx approaches zero, $\delta y / \delta x$ approaches a limiting value and the gradient of the chord approaches the gradient of the tangent at A

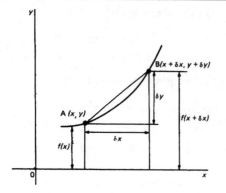

Figure 19.4

(ii) When determining the gradient of a tangent to a curve there are two notations used. The gradient of the curve at A in Fig. 19.4 can either be written as

$$\lim_{\delta x \to 0} \frac{\delta y}{\delta x} \text{ or } \lim_{\delta x \to 0} \left\{ \frac{f(x + \delta x) - f(x)}{\delta x} \right\}$$

In **Leibniz notation,** $\dfrac{dy}{dx} = \lim_{\delta x \to 0} \dfrac{\delta y}{\delta x}$

In **functional notation,**

$$f'(x) = \lim_{\delta x \to 0} \left\{ \frac{f(x + \delta x) - f(x)}{\delta x} \right\}$$

(iii) dy/dx is the same as $f'(x)$ and is called the **differential coefficient** or the **derivative.** The process of finding the differential coefficient is called **differentiation.**
Summarizing, the differential coefficient,

$$\frac{dy}{dx} = f'(x)$$

$$= \lim_{\delta x \to 0} \frac{\delta y}{\delta x}$$

$$= \lim_{\delta x \to 0} \left\{ \frac{f(x + \delta x) - f(x)}{\delta x} \right\}$$

Problem 3.   Differentiate from first principle $f(x) = x^2$ and determine the value of the gradient of the curve at $x = 2$

To 'differentiate from first principles' means 'to find $f'(x)$' by using the expression

$$f'(x) = \lim_{\delta x \to 0} \left\{ \frac{f(x + \delta x) - f(x)}{\delta x} \right\}$$

$f(x) = x^2$.
Substituting $(x + \delta x)$ for x gives $f(x + \delta x) = (x + \delta x)^2 = x^2 + 2x\ \delta x + \delta x^2$,

hence $f'(x) = \lim_{\delta x \to 0} \left\{ \dfrac{(x^2 + 2x\ \delta x + \delta x^2) - (x^2)}{\delta x} \right\}$

$$= \lim_{\delta x \to 0} \left\{ \frac{2x\ \delta x + \delta x^2}{\delta x} \right\}$$

$$= \lim_{\delta x \to 0} [2x + \delta x]$$

As $\delta x \to 0$, $[2x + \delta x] \to [2x + 0]$. Thus $f'(x) = 2x$, i.e. the differential coefficient of x^2 is $2x$. At $x = 2$, the gradient of the curve, $f'(x) = 2(2) = 4$.

Problem 4.   Find the differential coefficient of $y = 5x$

By definition,

$$\frac{dy}{dx} = f'(x) = \lim_{\delta x \to 0} \left\{ \frac{f(x + \delta x) - f(x)}{\delta x} \right\}$$

The function being differentiated is $y = f(x) = 5x$. Substituting $(x + \delta x)$ for x gives $f(x + \delta x) = 5(x + \delta x) = 5x + 5\ \delta x$. Hence

$$\frac{dy}{dx} = f'(x)$$

$$= \lim_{\delta x \to 0} \left\{ \frac{(5x + 5\delta x) - (5x)}{\delta x} \right\}$$

$$= \lim_{\delta x \to 0} \left\{ \frac{5\ \delta x}{\delta x} \right\} = \lim_{\delta x \to 0} [5]$$

Since the term δx does not appear in [5] the limiting value as $\delta x \to 0$ of [5] is 5. Thus $dy/dx = 5$, i.e. the differential coefficient of $5x$ is 5.
The equation $y = 5x$ represents a straight line of gradient 5 (see Chapter 13). The 'differential coefficient' (i.e. dy/dx or $f'(x)$) means 'the gradient of the curve', and since the slope of the line $y = 5x$ is 5 this result can be obtained by inspection. Hence, in general, if $y = kx$ (where k is a constant), then the slope of the line is k and dy/dx or $f'(x) = k$.

Problem 5.   Find the derivative of $y = 8$

$y = f(x) = 8$. Since there are no x-values in the original equation, substituting $(x + \delta x)$ for x still gives $f(x + \delta x) = 8$. Hence

$$\frac{dy}{dx} = f'(x) = \underset{\delta x \to 0}{\text{limit}} \left\{ \frac{f(x + \delta x) - f(x)}{\delta x} \right\}$$

$$= \underset{\delta x \to 0}{\text{limit}} \left\{ \frac{8 - 8}{\delta x} \right\} = 0$$

Thus, when $y = 8$, $dy/dx = 0$.
The equation $y = 8$ represents a straight horizontal line and the gradient of a horizontal line is zero, hence the result could have been determined by inspection. 'Finding the derivative' means 'finding the gradient', hence, in general, for any horizontal line if $y = k$ (where k is a constant) then $dy/dx = 0$.

Problem 6.   Differentiate from first principles $f(x) = 2x^3$

Substituting $(x + \delta x)$ for x gives
$f(x + \delta x) = 2(x + \delta x)^3$

$$= 2(x + \delta x)(x^2 + 2x \, \delta x + \delta x^2)$$

$$= 2(x^3 + 3x^2 \, \delta x + 3x \, \delta x^2 + \delta x^3)$$

$$= 2x^3 + 6x^2 \, \delta x + 6x \, \delta x^2 + 2 \, \delta x^3$$

$$f'(x) = \underset{\delta x \to 0}{\text{limit}} \left\{ \frac{f(x + \delta x) - f(x)}{\delta x} \right\}$$

$$= \underset{\delta x \to 0}{\text{limit}} \left\{ \frac{(2x^3 + 6x^2 \, \delta x + 6x \, \delta x^2 + 2 \, \delta x^3)}{\delta x} \frac{-(2x^3)}{} \right\}$$

$$= \underset{\delta x \to 0}{\text{limit}} \left\{ \frac{(6x^2 \, \delta x + 6x \, \delta x^2 - 2 \, \delta x^3)}{\delta x} \right\}$$

$$= \underset{\delta x \to 0}{\text{limit}} \{6x^2 + 6x \, \delta x + 2 \, \delta x^2\}$$

Hence $f'(x) = 6x^2$, i.e. the differential coefficient of $2x^3$ is $6x^2$.

Problem 7.   Find the differential coefficient of $y = 4x^2 + 5x - 3$ and determine the gradient of the curve at $x = -3$

$y = f(x) = 4x^2 + 5x - 3$

$f(x + \delta x) = 4(x + \delta x)^2 + 5(x + \delta x) - 3$

$$= 4(x^2 + 2x \, \delta x + \delta x^2) + 5x + 5 \, \delta x - 3$$

$$= 4x^2 + 8x \, \delta x + 4 \, \delta x^2 + 5x + 5 \, \delta x - 3$$

$$\frac{dy}{dx} = f'(x)$$

$$= \underset{\delta x \to 0}{\text{limit}} \left\{ \frac{f(x + \delta x) - f(x)}{\delta x} \right\}$$

$$= \underset{\delta x \to 0}{\text{limit}} \left\{ \frac{(4x^2 + 8x \, \delta x + 4 \, \delta x^2 + 5x + 5 \, \delta x - 3)}{\delta x} \frac{-(4x^2 + 5x - 3)}{} \right\}$$

$$= \underset{\delta x \to 0}{\text{limit}} \left\{ \frac{8x \, \delta x + 4 \, \delta x^2 + 5 \, \delta x}{\delta x} \right\}$$

$$= \underset{\delta x \to 0}{\text{limit}} \{8x + 4 \, \delta x + 5\}$$

i.e. $dy/dx = f'(x) = 8x + 5$
At $x = -3$, the gradient of the curve $= dy/dx = f'(x) = 8(-3) + 5 = -19$.

Further problems on differentiation from first principles may be found in Section 19.13, Problems 6 to 19, page 197.

19.5   Differentiation of $y = ax^n$ by the general rule

From differentiation by first principles, a general rule for differentiating ax^n emerges where a and n are any constants. This rule is:

$$\text{if } y = ax^n \text{ then } \frac{dy}{dx} = anx^{n-1}$$

or,   if $f(x) = ax^n$ then $f'(x) = anx^{n-1}$

(Each of the results obtained in Problems 3 and 7 may be deduced by using this general rule.)
When differentiating, results can be expressed in a number of ways. For example:

(i)   if $y = 3x^2$ then $dy/dx = 6x$,
(ii)   if $f(x) = 3x^2$ then $f'(x) = 6x$,
(iii) the differential coefficient of $3x^2$ is $6x$,
(iv)   the derivative of $3x^2$ is $6x$, and
(v) $\dfrac{d}{dx}(3x^2) = 6x$

Problem 8.   Using the general rule, differentiate the following with respect to x:
(a) $y = 5x^7$ (b) $y = 3\sqrt{x}$ (c) $y = 4/x^2$

(a)   Comparing $y = 5x^7$ with $y = ax^n$ shows that $a = 5$ and $n = 7$. Using the general rule,

$$\frac{dy}{dx} = anx^{n-1} = (5)(7)x^{7-1} = 35x^6$$

(b) $y = 3\sqrt{x} = 3x^{1/2}$. Hence $a = 3$ and $n = \dfrac{1}{2}$

$$\frac{dy}{dx} = anx^{n-1} = (3)\left(\frac{1}{2}\right)x^{(1/2)-1}$$

$$= \frac{3}{2}x^{-1/2} = \frac{3}{2x^{1/2}} = \frac{3}{2\sqrt{x}}$$

(c) $y = 4/x^2 = 4x^{-2}$. Hence $a = 4$ and $n = -2$

$$\frac{dy}{dx} = anx^{n-1} = (4)(-2)x^{-2-1} = -8x^{-3}$$

$$= -\frac{8}{x^3}$$

Problem 9.   Find the differential coefficient
of $y = \dfrac{2}{5}x^3 - \dfrac{4}{x^3} + 4\sqrt{x^5} + 7$

$$y = \frac{2}{5}x^3 - \frac{4}{x^3} + 4\sqrt{x^5} + 7$$

i.e.

$$y = \frac{2}{5}x^3 - 4x^{-3} + 4x^{5/2} + 7$$

$$\frac{dy}{dx} = \left(\frac{2}{5}\right)(3)x^{3-1} - (4)(-3)x^{-3-1}$$

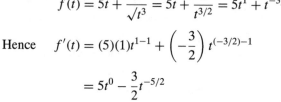

$$\left(+ (4)\left(\frac{5}{2}\right)x^{(5/2)-1}\right) + 0$$

$$= \frac{6}{5}x^2 + 12x^{-4} + 10x^{3/2}$$

i.e.

$$\frac{dy}{dx} = \frac{6}{5}x^2 + \frac{12}{x^4} + 10\sqrt{x^3}$$

Problem 10.   If $f(t) = 5t + \dfrac{1}{\sqrt{t^3}}$ find $f'(t)$

$$f(t) = 5t + \frac{1}{\sqrt{t^3}} = 5t + \frac{1}{t^{3/2}} = 5t^1 + t^{-3/2}$$

Hence $f'(t) = (5)(1)t^{1-1} + \left(-\dfrac{3}{2}\right)t^{(-3/2)-1}$

$$= 5t^0 - \frac{3}{2}t^{-5/2}$$

i.e. $$f'(t) = 5 - \frac{3}{2t^{5/2}} = 5 - \frac{3}{2\sqrt{t^5}}$$

Problem 11.   Differentiate $y = \dfrac{(x+2)^2}{x}$
with respect to x

$$y = \frac{(x+2)^2}{x} = \frac{x^2 + 4x + 4}{x} = \frac{x^2}{x} + \frac{4x}{x} + \frac{4}{x}$$

i.e. $$y = x + 4 + 4x^{-1}$$

Hence $$\frac{dy}{dx} = 1 + 0 + (4)(-1)x^{-1-1}$$

$$= 1 - 4x^{-2} = 1 - \frac{4}{x^2}$$

Further problems on differentiation of $y = ax^n$ by
the general rule may be found in Section 19.13,
Problems 20 to 31, page 197.

19.6   Differentiation of sine and cosine functions

Figure 19.5(a) shows a graph of $y = \sin\theta$. The
gradient is continually changing as the curve moves
from O to A to B to C to D. The gradient, given by

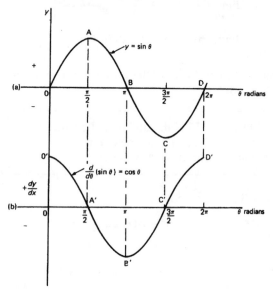

Figure 19.5

$dy/d\theta$, may be plotted in a corresponding position below $y = \sin\theta$, as shown in Fig. 19.5(b).

(i) At 0, the gradient is positive and is at its steepest. Hence $0'$ is a maximum positive value

(ii) Between O and A the gradient is positive but is decreasing in value until at A the gradient is zero, shown as A'

(iii) Between A and B the gradient is negative but is increasing in value until at B the gradient is at its steepest. Hence B' is a maximum negative value

(iv) If the gradient of $y = \sin\theta$ is further investigated between B and C and C and D then the resulting graph of $dy/d\theta$ is seen to be a cosine wave.

Hence the rate of change of $\sin\theta$ is $\cos\theta$, i.e.

if $y = \sin\theta$ then $\dfrac{dy}{d\theta} = \cos\theta$

It may also be shown that:

if $y = \sin a\theta$, $\dfrac{dy}{d\theta} = a\cos a\theta$ (where a is a constant)

and if $y = \sin(a\theta + \alpha)$, $\dfrac{dy}{d\theta} = a\cos(a\theta + \alpha)$

(where a and α are constants)

If a similar exercise is followed for $y = \cos\theta$ then the graphs of Fig. 19.6 result, showing $dy/d\theta$ to be a graph of $\sin\theta$, but displaced by π radians. If each point on the curve $y = \sin\theta$

(as shown in Fig. 19.5(a)) were to be made negative (i.e. $+\pi/2$ is made $-\pi/2$, $-3\pi/2$ is made $-(-3\pi/2)$, or $3\pi/2$, and so on) then the graph shown in Fig. 19.6(b) would result. This latter graph therefore represents the curve of $-\sin\theta$.

Thus, if $y = \cos\theta$, $\dfrac{dy}{d\theta} = -\sin\theta$

It may also be shown that:

if $y = \cos a\theta$, $\dfrac{dy}{d\theta} = -a\sin a\theta$

(where a is a constant)

and   if $y = \cos(a\theta + \alpha)$,

$\dfrac{dy}{d\theta} = -a\sin(a\theta + \alpha)$

(where a and α are constants)

Problem 12. Differentiate the following with respect to the variable:
(a) $y = 2\sin 5\theta$ (b) $f(t) = 3\cos 2t$

(a) $y = 2\sin 5\theta$

mistake ???

$\dfrac{dy}{d\theta} = (2)(5)\cos 5\theta = \mathbf{10\cos\theta}$

(b) $f(t) = 3\cos 2t$

$f'(t) = (3)(-2)\sin 2t = \mathbf{-6\sin 2t}$

Problem 13. Find the differential coefficient of $y = 7\sin 2x - 3\cos 4x$

$y = 7\sin 2x - 3\cos 4x$

$\dfrac{dy}{dx} = (7)(2)\cos 2x - (3)(-4)\sin 4x$

$= \mathbf{14\cos 2x + 12\sin 4x}$

Problem 14. Differentiate the following with respect to the variable:
(a) $f(\theta) = 5\sin(100\pi\theta - 0.40)$
(b) $f(t) = 2\cos(5t + 0.20)$

(a) If $f(\theta) = 5\sin(100\pi\theta - 0.40)$

$f'(\theta) = 5[100\pi\cos(100\pi\theta - 0.40)]$

$= \mathbf{500\pi\cos(100\pi\theta - 0.40)}$

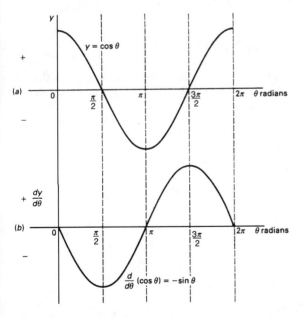

(a)

$y = \cos\theta$

2π θ radians

(b)

$\dfrac{dy}{d\theta}$

2π θ radians

$\dfrac{d}{d\theta}(\cos\theta) = -\sin\theta$

Figure 19.6

(b)   If $f(t) = 2\cos(5t + 0.20)$

$$f'(t) = 2[-5\sin(5t + 0.20)]$$

$$= -10\sin(5t + 0.20)$$

Problem 15.   An alternating voltage is given by $v = 100\sin 200t$ volts, where t is the time in seconds. Calculate the rate of change of voltage when (a) $t = 0.005$ s and (b) $t = 0.01$ s

---

$v = 100\sin 200t$ volts. The rate of change of v is given by dv/dt

$$\frac{dv}{dt} = (100)(200)\cos 200t = 20\,000\cos 200t$$

(a)   When $t = 0.005$ s,

$$\frac{dv}{dt} = 20\,000\cos(200)(0.005) = 20\,000\cos 1$$

cos 1 means 'the cosine of 1 radian' (make sure your calculator is on radians–not degrees). Hence

$$\frac{dv}{dt} = \mathbf{10\,806\ volts\ per\ second}$$

(b)   When $t = 0.01$ s,

$$\frac{dv}{dt} = 20\,000\cos(200)(0.01)$$

$$= 20\,000\cos 2$$

$$= \mathbf{-8323\ volts\ per\ second}$$

Further problems on the differentiation of sine and cosine functions may be found in Section 19.13, Problems 32 to 38, page 197.

19.7   Differentiation of e^{ax} and $\ln ax$

A graph of $y = e^x$ is shown in Fig. 19.7(a). The gradient of the curve at any point is given by dy/dx and is continually changing. By drawing tangents to the curve at many points on the curve and measuring the gradient of the tangents, values of dy/dx for corresponding values of x may be obtained. These values are shown graphically in Fig. 19.7(b). The graph of dy/dx against x is identical to the original graph of $y = e^x$. It follows that:

$$\text{if } \boldsymbol{y = e^x}, \text{ then } \frac{dy}{dx} = e^x$$

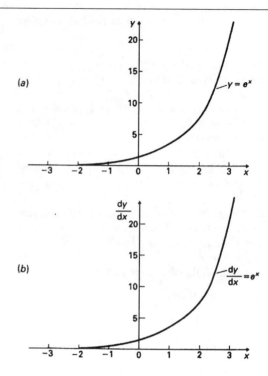

(a)

(b)

Figure 19.7

It may also be shown that

$$\text{if } y = e^{ax}, \text{ then } \frac{dy}{dx} = ae^{ax}$$

Therefore if $y = 2e^{6x}$, then

$$\frac{dy}{dx} = (2)(6e^{6x}) = \mathbf{12e^{6x}}$$

A graph of $y = \ln x$ is shown in Fig. 19.8(a). The gradient of the curve at any point is given by dy/dx and is continually changing. By drawing tangents to the curve at many points on the curve and measuring the gradient of the tangents, values of dy/dx for corresponding values of x may be obtained. These values are shown graphically in Fig. 19.8(b).

The graph of $\dfrac{dy}{dx}$ against x is the graph of $\dfrac{dy}{dx} = \dfrac{1}{x}$.

It follows that:

$$\text{if } \boldsymbol{y = \ln x}, \text{ then } \frac{dy}{dx} = \frac{1}{x}$$

It may also be shown that

$$\text{if } \boldsymbol{y = \ln ax}, \text{ then } \frac{dy}{dx} = \frac{1}{x}$$

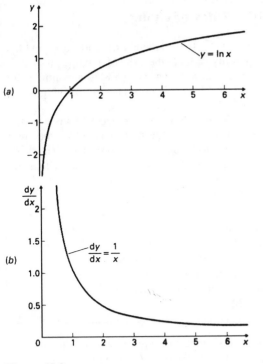

Figure 19.8

(Note that in the latter expression 'a' does not appear in the dy/dx term.) Thus if $y = \ln 4x$, then $dy/dx = 1/x$.

Problem 16.   Differentiate the following with respect to the variable:

(a) $y = 3e^{2x}$ (b) $f(t) = \dfrac{4}{3e^{5t}}$

(a)   If $y = 3e^{2x}$ then $dy/dx = (3)(2e^{3x}) = 6e^{2x}$

(b)   If $f(t) = \dfrac{4}{3e^{5t}} = \dfrac{4}{3}e^{-5t}$, then

$$f'(t) = \left(\frac{4}{3}\right)(-5e^{-5t}) = \frac{-20}{3}e^{-5t} = \frac{-20}{3e^{5t}}$$

Problem 17.   Differentiate $y = 5\ln 3x$

If $y = 5\ln 3x$, $\dfrac{dy}{dx} = (5)\left(\dfrac{1}{x}\right) = \dfrac{5}{x}$

Further problems on the differentiation of e^{ax} and $\ln ax$ may be found in Section 19.13, Problems 39 to 42, pages 197 and 198.

19.8   Summary of standard derivatives

The differential coefficients used are summarized in Table 19.1.

Table 19.1

y or $f(x)$	$\dfrac{dy}{dx}$ or $f'(x)$
ax^n	$an\,x^{n-1}$
$\sin ax$	$a\cos ax$
$\cos ax$	$-a\sin ax$
$\sin(ax + \alpha)$	$a\cos(ax + \alpha)$
$\cos(ax + \alpha)$	$-a\sin(ax + \alpha)$
e^{ax}	ae^{ax}
$\ln ax$	$1/x$

Problem 18.   Find the gradient of the curve $y = 3x^2 - 7x + 2$ at the point $(1, -2)$

If $y = 3x^2 - 7x + 2$, then gradient $= dy/dx = 6x - 7$. At the point $(1, -2)$, $x = 1$, hence **gradient =** $6(1) - 7 = -1$.

Problem 19.   If

$$y = \frac{3}{x^2} - 2\sin 4x + \frac{2}{e^x} + \ln 5x \text{ find } \frac{dy}{dx}$$

$$y = \frac{3}{x^2} - 2\sin 4x + \frac{2}{e^x} + \ln 5x$$

$$= 3x^{-2} - 2\sin 4x + 2e^{-x} + \ln 5x$$

$$\frac{dy}{dx} = (3)(-2x^{-3}) - 2(4\cos 4x) + (2)(-e^{-x}) + \frac{1}{x}$$

$$= \frac{-6}{x^3} - 8\cos 4x - \frac{2}{e^x} + \frac{1}{x}$$

Further problems on standard derivatives may be found in Section 19.13, Problems 43 and 44, page 198.

19.9   Successive differentiation

When a function $y = f(x)$ is differentiated with respect to x the differential coefficient is written as dy/dx or $f'(x)$. If the expression is differentiated again, the second differential coefficient is obtained and is written as dy^2/dx^2 (pronounced dee two y by dee x squared) or $f''(x)$ (pronounced f double-dash x). By successive diffentiation further higher derivatives such as d^3y/dx^3 and d^4y/dx^4 may be obtained.

Thus if $\qquad y = 5x^4$

$$\frac{dy}{dx} = 20x^3, \frac{d^2y}{dx^2} = 60x^2, \frac{d^3y}{dx^3} = 120x$$

$$\frac{d^4y}{dx^4} = 120 \text{ and } \frac{d^5y}{dx^5} = 0$$

Problem 20. If $f(x) = 4x^5 - 2x^3 + x - 3$, find $f''(x)$

$$f(x) = 4x^5 - 2x^3 + x - 3$$

$$f'(x) = 20x^4 - 6x^2 + 1$$

$$f''(x) = 80x^3 - 12x = 4x(20x^2 - 3)$$

Problem 21. Given
$$y = \frac{2}{3}x^3 - \frac{4}{x^2} + \frac{1}{2x} - \sqrt{x} \text{ determine } \frac{d^2y}{dx^2}$$

$$y = \frac{2}{3}x^3 - \frac{4}{x^2} + \frac{1}{2x} - \sqrt{x}$$

$$= \frac{2}{3}x^3 - 4x^{-2} + \frac{1}{2}x^{-1} - x^{1/2}$$

$$\frac{dy}{dx} = \left(\frac{2}{3}\right)(3)x^2 - 4(-2)x^{-3} + \left(\frac{1}{2}\right)(-1)x^{-2}$$
$$- \left(\frac{1}{2}\right)x^{-1/2}$$

i.e.

$$\frac{dy}{dx} = 2x^2 + 8x^{-3} - \frac{1}{2}x^{-2} - \frac{1}{2}x^{-1/2}$$

$$\frac{d^2y}{dx^2} = 4x - 24x^{-4} + x^{-3} + \frac{1}{4}x^{-3/2}$$

$$= 4x - \frac{24}{x^4} + \frac{1}{x^3} + \frac{1}{4\sqrt{x^3}}$$

Further problems on successive differentiation may be found in Section 19.13, Problems 45 to 47, page 198.

19.10 Rates of change

(i) If a quantity y depends on and varies with a quantity x then the rate of change of y with respect to x is dy/dx. Thus, for example, the rate of change of pressure p with height h is dp/dh

(ii) A rate of change with respect to time is usually just called 'the rate of change', the 'with respect to time' being assumed. Thus, for example, a rate of change of voltage v is dv/dt and a rate of change of temperature θ is $d\theta/dt$, and so on

---

Problem 22. The length l metres of a certain metal rod at temperature $t\,°C$ is given by $l = 1 + 0.00003t + 0.0000003t^2$. Determine the rate of change of length, in mm/°C, when the temperature is (a) 100°C, and (b) 250°C

The rate of change of length means dl/dt.
Since $l = 1 + 0.00003t + 0.0000003t^2$, then

$$\frac{dl}{dt} = 0.00003 + 0.0000006t$$

(a) When $t = 100\,°C$,

$$\frac{dl}{dt} = 0.00003 + (0.0000006)(100)$$

$$= 0.00009 \text{ m/°C} = \mathbf{0.09 \text{ mm/°C}}$$

(b) When $t = 250\,°C$,

$$\frac{dl}{dt} = 0.00003 + (0.0000006)(250)$$

$$= 0.00018 \text{ m/°C} = \mathbf{0.18 \text{ mm/°C}}$$

Problem 23. The luminous intensity I candelas of a lamp at varying voltage V is given by $I = 5 \times 10^{-4}V^2$. Determine the voltage at which the light is increasing at a rate of 0.4 candelas per volt

---

The rate of change of light with respect to voltage is given by dI/dV.
Since $I = 5 \times 10^{-4}V^2$,

$$\frac{dI}{dV} = (5 \times 10^{-4})(2)V = 10 \times 10^{-4}V = 10^{-3}V$$

When the light is increasing at 0.4 candelas per volt then $+0.4 = 10^{-3}V$, from which

$$\text{voltage } V = \frac{0.4}{10^{-3}} = 0.4 \times 10^{+3} = \textbf{400 volts}$$

Problem 24. Newton's law of cooling is given by $\theta = \theta_0 e^{-kt}$, where the excess of temperature at zero time is $\theta_0\,°C$ and at time t seconds is $\theta\,°C$. Determine the rate of change of temperature after 50 s, given that $\theta° = 15\,°C$ and $k = -0.02$

The rate of change of temperature is $d\theta/dt$. Since $\theta = \theta_0 e^{-kt}$ then

$$\frac{d\theta}{dt} = (\theta_0)(-k)e^{-kt}$$

$$= -k\theta_0 e^{-kt}$$

Where $\theta_0 = 15$, $k = -0.02$ and $t = 50$, then

$$\frac{d\theta}{dt} = -(-0.02)(15)e^{-(-0.02)(50)}$$

$$= 0.3e^1 = \textbf{0.815}\,°\textbf{C/s}$$

Problem 25. The pressure p of the atmosphere at height h above ground level is given by $p = p_0 e^{-h/c}$, where p_0 is the pressure at ground level and c is a constant. Determine the rate of change of pressure with height when $p_0 = 10^5$ pascals and $c = 6 \times 10^4$ at 1500 m

---

The rate of change of pressure with height is dp/dh. Since $p = p_0 e^{-h/c}$, then

$$\frac{dp}{dh} = (p_0)\left(-\frac{1}{c}\right)e^{-h/c} = -\frac{p_0}{c}e^{-h/c}$$

Where $p_0 = 10^5$, $c = 6 \times 10^4$ and $h = 1500$, then

$$\frac{dp}{dh} = -\frac{10^5}{6 \times 10^4}e^{(-1500/6 \times 10^4)}$$

$$= -\frac{5}{3}e^{-0.025} = \textbf{1.63 Pa/m}$$

Further problems on rates of change may be found in Section 19.13, Problems 48 to 52, page 198.

19.11 Velocity and acceleration

When a car moves a distance x metres in a time t seconds along a straight road, if the velocity v is constant then $v = x/t$ m/s, i.e. the gradient of the distance/time graph shown in Fig. 19.9 is constant.

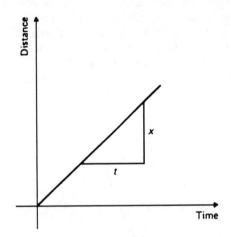

Figure 19.9

If, however, the velocity of the car is not constant then the distance/time graph will not be a straight line. It may be as shown in Fig. 19.10.

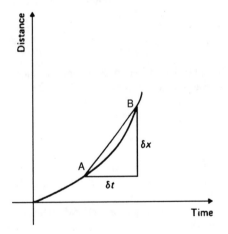

Figure 19.10

The average velocity over a small time δt and distance δx is given by the gradient of the chord AB, i.e. the average velocity over time δt is $\delta x/\delta t$. As $\delta t \to 0$, the chord AB becomes a tangent, such that at point A the velocity is given by:

$$v = \frac{dx}{dt}$$

Hence the velocity of the car at any instant is given by the gradient of the distance/time graph. If an expression for the distance x is known in terms of time t then the velocity is obtained by differentiating the expression. The acceleration a of the car is defined as the rate of change of velocity. A velocity/time graph is shown in Fig. 19.11. If δv is the change in v and δt the corresponding change in time, then $a = \frac{\delta v}{\delta t}$.

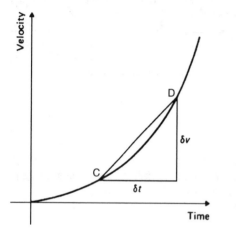

Figure 19.11

As $\delta t \rightarrow 0$, the chord CD becomes a tangent such that at point C the acceleration is given by:

$$a = \frac{dv}{dt}$$

Hence the acceleration of the car at any instant is given by the gradient of the velocity/time graph. If an expression for velocity v is known in terms of time t then the acceleration is obtained by differentiating the expression.

Acceleration $a = \frac{dv}{dt}$. However, $v = \frac{dx}{dt}$

Hence $a = \frac{d}{dt}\left(\frac{dx}{dt}\right) = \frac{d^2x}{dt^2}$

Thus acceleration is given by the second differential coefficient of distance x with respect to time t.

Problem 26.   The distance x metres moved by a body in a time t seconds is given by: $x = 4t^3 - 2t^2 + 3t + 5$. Express the velocity and acceleration in terms of t and determine their values when $t = 2$ s

Distance, $x = 4t^3 - 2t^2 + 3t + 5$ m

Velocity, $v = \frac{dx}{dt} = 12t^2 - 4t + 3$ m/s

Acceleration, $a = \frac{d^2x}{dt^2} = 24t - 4$ m/s^2

After 2 seconds,

$$\text{velocity} = 12(2)^2 - 4(2) + 3 = 43 \text{ m/s}$$

and $\text{acceleration} = 24(2) - 4 = 44 \text{ m/s}^2$

Problem 27.   Supplies are dropped from a helicopter and the distance fallen in a time t seconds is given by $x = \frac{1}{2}gt^2$, where $g = 9.8$ m/s^2. Determine the velocity and acceleration of the supplies after it has fallen for 2 seconds

Distance $x = \frac{1}{2}gt^2 = \frac{1}{2}(9.8)t^2 = 4.9t^2$ m

$$\text{velocity, } v = \frac{dv}{dt} = 9.8t \text{ m/s}$$

and   acceleration, $a = \frac{d^2x}{dt^2} = 9.8$ m/s^2

When time $t = 2$ s,

$$\text{velocity, } v = (9.8)(2) = 19.6 \text{ m/s}$$

and   acceleration, $a = 9.8$ m/s^2

(which is the acceleration due to gravity)

Further problems on velocity and acceleration may be found in Section 19.13, Problems 53 and 54, page 198.

19.12   Maximum and minimum points

In Fig. 19.12, the gradient (or rate of change) of the curve changes from positive between O and

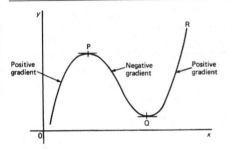

Figure 19.12

P to negative between P and Q and then positive again between Q and R. At point P the gradient is zero and, as x increases, the gradient of the curve changes from positive just before P to negative just after. Such a point is called a **maximum point** and appears as the 'crest of a wave'. At point Q, the gradient is also zero and, as x increases, the gradient of the curve changes from negative just before Q to positive just after. Such a point is called a **minimum value**, and appears as the 'bottom of a valley'. Points such as P and Q are given the general name of **turning points**, or **stationary points**.

Procedure for finding and distinguishing between stationary points

(i)   Given $y = f(x)$, determine dy/dx (i.e. $f'(x)$)
(ii)   Let $dy/dx = 0$ and solve for the values of x
(iii)   Substitute the values of x into the original equation, $y = f(x)$, to find the corresponding y-ordinate values. This establishes the coordinates of the stationary points

To determine the nature of the stationary points:

Either
(iv)   Determine the sign of the gradient of the curve just before and just after the stationary points. If the sign change for the gradient of the curve is:
(a) **positive to negative**–the point is a **maximum** one;
(b) **negative to positive**–the point is a **minimum** one

or
(v)   Find d^2y/dx^2 and substitute into it the values of x found in (ii).
If the result is:
(a) **positive**–the point is a **minimum** one;
(b) **negative**–the point is a **maximum** one

Consider the equation $y = x^2 - 2x + 3$. Gradient of the curve,

$$\frac{dy}{dx} = 2x - 2$$

At the turning point, the gradient is zero, hence $2x - 2 = 0$, from which $2x = 2$ and $x = 1$. When $x = 1$, $y = (1)^2 - 2(1) + 3 = 2$, hence at the coordinates $(1, 2)$ a turning point occurs. To determine the nature of the turning point:

Method 1

Consider the gradient of the curve at a value of x just less than 1, say 0.9. At $x = 0.9$, gradient $= 2x - 2 = 2(0.9) - 2 = -0.2$.

Now consider the gradient of the curve at a value of x just greater than 1, say 1.1.

At $x = 1.1$, gradient $= 2x - 2 = 2(1.1) - 2 = 0.2$.

Hence the gradient has changed from negative just before the turning point at $x = 1$, to positive just after. This indicates a **minimum value**.

Method 2

If the gradient of the curve, $dy/dx = 2x - 2$, then $d^2y/dx^2 = 2$, which is positive, hence the turning point is a **minimum**.

A graph of $y = x^2 - 2x + 3$ with the minimum point at $(1, 2)$ is shown in Fig. 19.13.

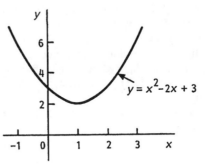

Figure 19.13

> **Problem 28.**   Locate the turning point on the curve $y = 2x^2 - 4x$ and determine its nature by examining the sign of the gradient on either side

Following the above procedure:

(i)   Since $y = 2x^2 - 4x$, $dy/dx = 4x - 4$

(ii) At a turning point, $dy/dx = 0$. Hence $4x - 4 = 0$ from which, $x = 1$

(iii) When $x = 1$, $y = 2(1)^2 - 4(1) = -2$
Hence the coordinates of the turning point are (1, −2)

(iv) If x is slightly less than 1, say 0.9, then $dy/dx = 4(0.9) - 4 = -0.4$, i.e. negative. If x is slightly greater than 1, say 1.1, then $dy/dx = 4(1.1) - 4 = 0.4$, i.e. positive. Since the gradient of the curve is negative just before the turning point and positive just after **(1, −2) is a minimum point.**

Problem 29.   Find the maximum and minimum values of the curve $y = x^3 - 3x + 5$ by

(a)   examining the gradient on either side of the turning points, and

(b)   determining the sign of the second derivative

Since $y = x^3 - 3x + 5$ then $dy/dx = 3x^2 - 3$.
For a maximum or minimum value $dy/dx = 0$.
Hence $3x^2 - 3 = 0$, from which, $3x^2 = 3$ and $x = \pm 1$.
When $x = 1$, $y = (1)^3 - 3(1) + 5 = 3$.
When $x = -1$, $y = (-1)^3 - 3(-1) + 5 = 7$.
Hence $(1, 3)$ and $(-1, 7)$ are the coordinates of the turning points.

(a)   Considering the point $(1, 3)$:
If x is slightly less than 1, say 0.9, then $dy/dx = 3(0.9)^2 - 3$, which is negative.
If x is slightly more than 1, say 1.1, then $dy/dx = 3(1.1)^2 - 3$, which is positive.
Since the gradient changes from negative to positive, **the point (1, 3) is a minimum point**.
Considering the point $(-1, 7)$:
If x is slightly less than -1, say -1.1, then $dy/dx = 3(-1.1)^2 - 3$, which is positive.
If x is slightly more than -1, say -0.9, then $dy/dx = 3(-0.9)^2 - 3$, which is negative.
Since the gradient changes from positive to negative, **the point (−1, 7) is a maximum point**

(b)   Since $dy/dx = 3x^2 - 3$, then $dy^2/dx^2 = 6x$.
When $x = 1$, dy^2/dx^2 is positive, hence $(1, 3)$ is a **minimum value**.
When $x = -1$, d^2y/dx^2 is negative, hence $(-1, 7)$ is a **maximum value**.
Thus the maximum value is 7 and the minimum value is 3.
It can be seen that the second differential method of determining the nature of the turning points is, in this case, quicker than investigating the gradient

---

Problem 30.   Locate the turning point on the following curve and determine whether it is a maximum or minimum point: $y = 4\theta + e^{-\theta}$

Since $y = 4\theta + e^{-\theta}$ then $dy/d\theta = 4 - e^{-\theta} = 0$ for a maximum or minimum value.
Hence $4 = e^{-\theta}$, $\frac{1}{4} = e^{\theta}$, giving $\theta = \ln(\frac{1}{4}) = -1.3863$ (see Chapter 9).
When $\theta = 1.3863$,

$$y = 4(-1.3863) + e^{-(-1.3863)}$$

$$= 5.5452 + 4.0000 = -1.5452$$

Thus $(-1.3863, -1.5452)$ are the coordinates of the turning point.

$$\frac{d^2y}{d\theta^2} = e^{-\theta}$$

When

$$\theta = -1.3863, \quad \frac{d^2y}{d\theta^2} = e^{+1.3863} = 4.0$$

which is positive, hence **(−1.3863, −1.5452) is a minimum point**.

Further problems on maximum and minimum points may be found in the following Section 19.13, Problems 55 to 62, pages 198 and 199.

19.13   Further problems on the introduction to differentiation

Functional notation

1   If $f(x) = 6x^2 - 2x + 1$ find $f(0)$, $f(1)$, $f(2)$, $f(-1)$ and $f(-3)$.   [1, 5, 21, 9, 61]

2   If $f(x) = 2x^2 + 5x - 7$ find $f(1)$, $f(2)$, $f(-1)$, $f(2) - f(-1)$.   [0, 11, −10, 21]

3   Given $f(x) = 3x^3 + 2x^2 - 3x + 2$ prove that $f(1) = \frac{1}{7}f(2)$.

4   If $f(x) = -x^2 + 3x + 6$ find $f(2)$, $f(2 + a)$, $f(2 + a) - f(2)$ and $\{f(2 + a) - f(2)\}/a$.
                   [8; $-a^2 - a + 8$; $-a^2 - a$; $-a - 1$]

The gradient of a curve

5 Plot the curve $f(x) = 4x^2 - 1$ for values of x from $x = -1$ to $x = +4$. Label the coordinates $(3, f(3))$ and $(1, f(1))$ as J and K, respectively. Join points J and K to form the chord JK. Determine the gradient of chord JK. By moving J nearer and nearer to K determine the gradient of the tangent of the curve at K.

[16; 8]

Differentiation from first principles

In Problems 6 to 17, differentiate from first principles:

6 $y = x$ [1]

7 $y = 7x$ [7]

8 $y = 4x^2$ [8x]

9 $y = 5x^3$ $[15x^2]$

10 $y = -2x^2 + 3x - 12$ $[-4x + 3]$

11 $y = 23$ [0]

12 $f(x) = 9x$ [9]

13 $f(x) = \dfrac{2x}{3}$ $\left[\dfrac{2}{3}\right]$

14 $f(x) = 9x^2$ [18x]

15 $f(x) = -7x^3$ $[-21x^2]$

16 $f(x) = x^2 + 15x - 4$ $[2x + 15]$

17 $f(x) = 4$ [0]

18 Determine $\dfrac{d}{dx}(4x^3)$ from first principles.

$[12x^2]$

19 Find $\dfrac{d}{dx}(3x^2 + 5)$ from first principles.

[6x]

Differentiation of $y = ax^n$ by the general rule

20 Using the general rule for ax^n check the results of Problems 6 to 19.

In Problems 21 to 28, determine the differential coefficients with respect to the variable:

21 $y = 7x^4$ $[28x^3]$

22 $y = \sqrt{x}$ $\left[\dfrac{1}{2\sqrt{x}}\right]$

23 $y = \sqrt{t^3}$ $\left[\dfrac{3}{2}\sqrt{t}\right]$

24 $y = 6 + \dfrac{1}{x^3}$ $\left[\dfrac{-3}{x^4}\right]$

25 $y = 3x - \dfrac{1}{\sqrt{x}} + \dfrac{1}{x}$ $\left[3 + \dfrac{1}{2\sqrt{x^3}} - \dfrac{1}{x^2}\right]$

26 $y = \dfrac{5}{x^2} - \dfrac{1}{\sqrt{x^7}} + 2$ $\left[-\dfrac{10}{x^3} + \dfrac{7}{2\sqrt{x^9}}\right]$

27 $y = 3(t - 2)^2$ $[6t - 12]$

28 $y = (x + 1)^3$ $[3x^2 + 6x + 3]$

29 Differentiate $f(x) = 6x^2 - 3x + 5$ and find the gradient of the curve at (a) $x = -1$, and (b) $x = 2$. $[12x - 3 \text{ (a) } -15 \text{ (b) } 21]$

30 Find the differential coefficient of $y = 2x^3 + 3x^2 - 4x - 1$ and determine the gradient of the curve at $x = 2$. $[6x^2 + 6x - 4; 32]$

31 Determine the derivative of $y = -2x^3 + 4x + 7$ and determine the gradient of the curve at $x = -1.5$. $[-6x^2 + 4; -9.5]$

Differentiation of sine and cosine functions

32 Show graphically that the rate of change of $\sin \theta$ is $\cos \theta$.

33 Show graphically that the rate of change of $\cos \theta$ is $-\sin \theta$.

34 Differentiate with respect to x:
(a) $y = 4 \sin 3x$ (b) $y = 2 \cos 6x$
$[(a) \ 12 \cos 3x \ (b) \ -12 \sin 6x]$

35 Given $f(\theta) = 2 \sin 3\theta - 5 \cos 2\theta$, find $f'(\theta)$.
$[6 \cos 3\theta + 10 \sin 2\theta]$

36 An alternating current is given by $i = 5 \sin 100t$ amperes, where t is the time in seconds. Determine the rate of change of current when $t = 0.01$ seconds.   [270.2 A/s]

37 $v = 50 \sin 40t$ volts represents an alternating voltage where t is the time in seconds. At a time of 20×10^{-3} seconds, find the rate of change of voltage.   [1393.4 V/s]

38 If $f(t) = 3 \sin(4t + 0.12) - 2 \cos(3t - 0.72)$ determine $f'(t)$.
$[12 \cos(4t + 0.12) + 6 \sin(3t - 0.72)]$

Differentiation of e^{ax} and $\ln ax$

39 Differentiate with respect to x (a) $y = 5e^{3x}$ (b) $y = \dfrac{2}{7e^{2x}}$. $\left[(a) \ 15e^{3x} \ (b) \ \dfrac{-4}{7e^{2x}}\right]$

40. Given $f(\theta) = 5\ln 2\theta - 4\ln 3\theta$ determine $f'(\theta)$.

$$\left[\frac{5}{\theta} - \frac{4}{\theta} = \frac{1}{\theta}\right]$$

41 If $f(t) = 4\ln t + 2$ evaluate $f'(t)$ when $t = 0.25$.   [16]

42 Evaluate dy/dx when $x = 1$, given $y = 3e^{4x} - (5/2e^{3x}) + 8\ln 5x$. Give the answer correct to 3 significant figures.   [664]

Differentiation of standard derivatives

43 Find the gradient of the curve $y = 2x^4 + 3x^3 - x + 4$ at the points $(0, 4)$ and $(1, 8)$.

$$[-1, 16]$$

44 Differentiate

$$y = \frac{2}{x^2} + 2\ln 2x - 2(\cos 5x + 3\sin 2x) - \frac{2}{e^{3x}}.$$

$$\left[\frac{-4}{x^3} + \frac{2}{x} + 10\sin 5x - 12\cos 2x + \frac{6}{e^{3x}}\right]$$

Successive differentiation

45 If $y = 3x^4 + 2x^3 - 3x + 2$ find

(a) $\dfrac{d^2y}{dx^2}$ (b) $\dfrac{d^3y}{dx^3}$

$$[\text{(a) } 36x^2 + 12x \text{ (b) } 72x + 12]$$

46 (a) Given $f(t) = \dfrac{2}{5}t^2 - \dfrac{1}{t^3} + \dfrac{3}{t} - \sqrt{t} + 1$
determine $f''(t)$

(b) Evaluate $f''(t)$ when $t = 1$

$$\left[\text{(a) } \frac{4}{5} - \frac{12}{t^5} + \frac{6}{t^3} + \frac{1}{4\sqrt{t^3}} \text{ (b) } -4.95\right]$$

47 Find the second differential coefficient with respect to the variable of the following:
(a) $3\sin 2t + \cos t$ (b) $2\ln 4\theta$

$$[\text{(a) } -(12\sin 2t + \cos t) \text{ (b) } -2/\theta^2]$$

Rates of change

48 The length, l metres, of a rod of metal at temperature $\theta\,°C$ is given by $l = 1 + 2\times10^{-4}\theta + 4\times10^{-6}\theta^2$. Determine the rate of change of l, in cm/°C, when the temperature is (a) $100\,°C$ (b) $400\,°C$.   [(a) 0.1 cm/°C (b) 0.34 cm/°C]

49 An alternating current, i amperes, is given by $i = 100\sin 2\pi ft$, where f is the frequency in hertz and t the time in seconds. Determine the rate of change of current when $t = 12$ ms, given that $f = 50$ Hz.   [−25 420 A/s]

50 The luminous intensity, I candelas, of a lamp is given by $I = 8\times10^{-4}V^2$, where V is the voltage. Find (a) the rate of change of luminous intensity with voltage when $V = 100$ volts, and (b) the voltage at which the light is increasing at a rate of 0.5 candelas per volt.

$$[\text{(a) } 0.16 \text{ cd/V (b) } 312.5 \text{ V}]$$

51 The voltage across the plates of a capacitor at any time t seconds is given by $v = Ve^{-t/CR}$, where V, C and R are constants.
Given $V = 200$ volts, $C = 0.1\times10^{-6}$ farads and $R = 2\times10^6$ ohms find
(a) the initial rate of change of voltage, and
(b) the rate of change of voltage after 0.2 s

$$[\text{(a) } -1000 \text{ V/s (b) } 367.9 \text{ V/s}]$$

52 Newton's law of cooling is given by $\theta = \theta_0 e^{-kt}$, where the excess of temperature at zero time is $\theta_0\,°C$ and at time t seconds is $\theta\,°C$.
Given $\theta_0 = 15\,°C$ and $k = -0.02$, find the time when the rate of change of temperature is $1\,°C/s$.   [60.2 s]

Velocity and acceleration

53 A body obeys the equation $x = 5t - 25t^2$ where x is in metres and t is in seconds. Determine expressions for velocity and acceleration. Find also its velocity and acceleration when $t = 1$ s.

$$\left[\begin{array}{ll} v = (5 - 50t) \text{ m/s}, & a = -50 \text{ m/s}^2, \\ v_1 = -45 \text{ m/s}, & a_1 = -50 \text{ m/s}^2 \end{array}\right]$$

54 An object moves in a straight line so that after t seconds its distance x metres from a fixed point on the line is given by:

$$x = \frac{2}{3}t^3 - t^2 + 4t - 5$$

Obtain an expression for the velocity and acceleration of the object after t seconds and their values when $t = 2$ s.

$$\left[\begin{array}{l} v = (2t^2 - 2t + 4) \text{ m/s}, \\ a = (4t - 2) \text{ m/s}^2, \\ v_2 = 8 \text{ m/s}, \\ a_2 = 6 \text{ m/s}^2 \end{array}\right]$$

Maximum and minimum points

55 Define (a) a maximum point, and (b) a minimum point.

56 Sketch the following curves and state for each:
(a) the value, and the coordinates, of the first maximum point, and

(b) the value, and the coordinates, of the first minimum point

(i) $y = 4\sin 2\theta$ (ii) $y = 3\cos 3\theta$

$$\left[\begin{array}{l} \text{(i) (a) 4 at } \left(\dfrac{\pi}{4}, 4\right) \;\; \text{(b)} -4 \text{ at } \left(\dfrac{3\pi}{4}, -4\right) \\[3mm] \text{(ii) (a) 3 at } (0, 3) \quad \text{(b)} -3 \text{ at } \left(\dfrac{\pi}{3}, -3\right) \end{array}\right]$$

In Problems 57 to 61, find the turning points and distinguish between them:

57   $y = 3x^2 - 4x + 2$ $\left[\text{Minimum at } \left(\dfrac{2}{3}, \dfrac{2}{3}\right)\right]$

58   $x = \theta(6 - \theta)$ [Maximum at $(3, 9)$]

59   $y = 5x - 2\ln x$

[Minimum at $(0, 4000, 3.8326)$]

60   $y = 2x - e^x$

[Maximum at $(0.6931, -0.6137)$]

61   $x = 8t + \dfrac{1}{2t^2}$ $\left[\text{Minimum at } \left(\dfrac{1}{2}, 6\right)\right]$

62   The speed, v, of a car (in m/s) is related to time t s by the equation $v = 3 + 12t - 3t^2$. Determine the maximum speed of the car in km/h.

[54 km/h]

Multiple choice questions on Part 1 Mathematics for Engineering

All questions have only one correct answer

Suggested time allowed: 1 hour

1 The resistance R_2 of a material is given by $R_2 = R_1(1 + \alpha t)$. If $R_2 = 180$, $\alpha = 2.7 \times 10^{-4}$ and $t = 50$, the value of R_1 (correct to 3 significant figures) is

  A 182   B 178
  C 180   D 159

2 Correct to 3 decimal places, sin (−1.4 rad) is

  A −0.024   B 0.985
  C 0.024   D −0.985

3 $(-3, 4)$ in polar coordinates is

  A (5, 2.214 rad)   B (7, 53.13°)
  C (5, 53.13°)   D (5, 306.87°)

4 The quadratic equation in x whose roots are −3 and +2 is

  A $x^2 - x - 6$ B $x^2 - 5x + 6$
  C $x^2 + x - 6$ D $x^2 + 5x + 6$

5 A pylon stands on horizontal ground. At a point 50 m from the base of the pylon, the angle of elevation of the top of the pylon is 20°. The height of the pylon is

  A 18.2 m   B 137.4 m
  C 17.1 m   D 18.8 m

6 The gradient of the curve $y = -2x^3 + x^2 - 3x + 4$ at $x = 2$ is

  A −6   B 25
  C −31   D −23

7 The relationship between two related engineering variables θ and t is $t - a\theta = b\theta^2$ where a and b are constants. To produce a straight line graph it is necessary to plot

  A t vertically against θ horizontally
  B θ vertically against t horizontally
  C θ horizontally against t/θ vertically
  D θ^2 horizontally against t vertically

8 The current i amperes flowing in a capacitor at time t seconds is given by $i = 5(1 - e^{-t/CR})$, where resistance R is 20×10^3 ohms and capacitance C is 10×10^{-6} farads. When current i reaches 2 amperes, the time is

  A 0.10 s   B 0.18 s
  C 0.22 s   D 1.02 s

9 The area of triangle ABC in Fig. 1 is

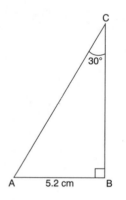

Figure 1

  A 27.04 cm^2 B 46.83 cm^2
  C 15.61 cm^2 D 23.42 cm^2

10 In a system of pulleys the effort P required to raise a load W is given by $P = aW + b$, where a and b are constants. If $W = 52$ when $P = 18$,

and $W = 80$ when P is 25, the values of a and b are

A $a = 1, b = -34$ B $a = \dfrac{1}{4}, b = 5$

C $a = \dfrac{1}{2}, b = -15$ D $a = 4, b = \dfrac{1}{5}$

11 An indicator diagram for a steam engine is as shown in Fig. 2. The base has been divided into 6 equally spaced intervals and the lengths of the 7 ordinates measured with the results shown in millimetres. Using Simpson's rule the area of the indicator diagram is

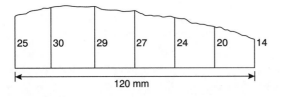

Figure 2

A   4060 mm²   B   9060 mm²

C   3540 mm²   D   3020 mm²

12 A triangular plot of land ABC has sides $a = 8.0$ m, $b = 7.0$ m, and $c = 6.0$ m. Angle A is equal to

A   46.57°   B   53.13°

C   75.52°   D   86.59°

13 $a^3 b^2 c \times ab^{-5}c^2 \div a^2 bc^{-3}$ is equivalent to

A $a^6 b^{-2}$ B $\left(\dfrac{a}{b}\right)^2$

C $a^2 b^{-4} c^6$ D $\dfrac{a^2}{bc}$

14 The equation of a straight line graph is $2y = 7 - 6x$. The gradient of the straight line is

A   7   B -3

C -6 D $\dfrac{7}{2}$

15 An alternating voltage is given by $v = 120 \sin(100\pi t - 0.32)$ volts. When time $t = 4$ ms, the voltage v has a value of

A   96.67 V   B   2.63 V
C   1.96 V   D   120 V

16 Transposing $v = u + \dfrac{ft}{m}$ to make t the subject gives

A $\dfrac{mv}{ut}$ B $\dfrac{m(u - v)}{f}$

C $\dfrac{m}{f}(v - u)$ D $\dfrac{f(v - u)}{m}$

17 A graph of resistance against voltage for an electrical circuit is shown in Fig. 3. The equation relating resistance R and voltage V is

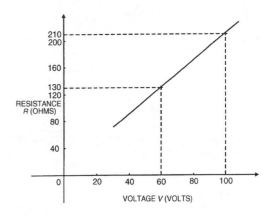

Figure 3

A $R = 2\,V + 20$ B $R = 0.5\,V + 10$
C $R = 0.5\,V + 20$ D $R = 2\,V + 10$

18 Which is the correct formula for calculating b in triangle ABC shown in Fig. 4?

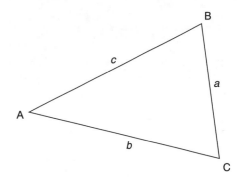

Figure 4

A $b = a + c - 2ac \cos B$

B $b^2 = a^2 - c^2 + 2ac \cos B$

C $b^2 = a^2 + c^2 - 2ac \cos B$

D $b = a - c + 2ac \cos B$

19 Some currency exchange rates are shown in the table.

France £1 = 7.50 fr
Spain £1 = 190 pes
Germany £1 = 2.50 Dm
U.S.A. £1 = $1.50

Which of the following is correct?

A £15 is equivalent to 2 French francs
B $3000 is equivalent to £2000
C 400 German marks is equivalent to £1000
D 19 Spanish pesetas is equivalent to £10

20 Differentiating $y = 3x^4$ gives

A $\dfrac{dy}{dx} = 12x^3$

B $\dfrac{dy}{dx} = \dfrac{3}{5}x^5$

C $\dfrac{dy}{dx} = 3x^5$

D $\dfrac{dy}{dx} = 4x^3$

Part 2

Further Mathematics for Engineering

20

Methods of differentiation

20.1 Differentiation of common functions

The standard derivatives summarized below were derived in Chapter 19 and are true for all real values of x.

y or $f(x)$	$\dfrac{dy}{dx}$ or $f'x$
ax^n	anx^{n-1}
$\sin ax$	$a\cos ax$
$\cos ax$	$-a\sin ax$
e^{ax}	ae^{ax}
$\ln ax$	$\dfrac{1}{x}$

The **differential coefficient of a sum or difference** is the sum or difference of the differential coefficients of the separate terms.
Thus, if $f(x) = p(x) + q(x) - r(x)$ (where f, p, q and r are functions), then
$f'(x) = p'(x) + q'(x) - r'(x)$
Differentiation of common functions is demonstrated in the following worked problems.

Problem 1. Find the differential coefficients of (a) $y = 12x^3$ (b) $y = \dfrac{12}{x^3}$

If $y = ax^n$ then $\dfrac{dy}{dx} = anx^{n-1}$

(a) Since $y = 12x^3$, $a = 12$ and $n = 3$ thus
$\dfrac{dy}{dx} = (12)(3)x^{3-1} = 36x^2$

(b) $y = 12/x^3$ is rewritten in the standard ax^n form as $y = 12x^{-3}$ and in the general rule $a = 12$ and $n = -3$.

Thus $\dfrac{dy}{dx} = (12)(-3)x^{-3-1}$

$= -36x^{-4} = \dfrac{-36}{x^4}$

Problem 2. Differentiate
(a) $y = 6$ (b) $y = 6x$

(a) $y = 6$ may be written as $y = 6x^0$, i.e. in the general rule $a = 6$ and $n = 0$. Hence
$\dfrac{dy}{dx} = (6)(0)x^{0-1} = 0$
In general, the differential coefficient of a constant is always zero

(b) Since $y = 6x$, in the general rule $a = 6$ and $n = 1$.
Hence $\dfrac{dy}{dx} = (6)(1)x^{1-1} = 6x^0 = 6$
In general, the differential coefficient of kx, where k is a constant, is always k

---

Problem 3. Find the derivatives of
(a) $y = 3\sqrt{x}$ (b) $y = \dfrac{5}{\sqrt[3]{x^4}}$

(a) $y = 3\sqrt{x}$ is rewritten in the standard differential form as $y = 3x^{1/2}$. In the general rule, $a = 3$ and $n = \frac{1}{2}$

Thus $\dfrac{dy}{dx} = (3)\left(\dfrac{1}{2}\right)x^{(1/2)-1} = \dfrac{3}{2}x^{-1/2}$

$= \dfrac{3}{2x^{1/2}} = \dfrac{3}{2\sqrt{x}}$

(b) $y = \dfrac{5}{\sqrt[3]{x^4}} = \dfrac{5}{x^{4/3}} = 5x^{-4/3}$ in the standard differential form.
In the general rule, $a = 5$ and $n = -\frac{4}{3}$

Thus $\dfrac{dy}{dx} = (5)\left(-\dfrac{4}{3}\right)x^{(-4/3)-1}$

$= \dfrac{-20}{3}x^{-7/3} = \dfrac{-20}{3x^{7/3}} = \dfrac{-20}{3\sqrt[3]{x^7}}$

Problem 4. Differentiate
$y = 5x^4 + 4x - \dfrac{1}{2x^2} + \dfrac{1}{\sqrt{x}} - 3$ with respect
to x

$y = 5x^4 + 4x - \dfrac{1}{2x^2} + \dfrac{1}{\sqrt{x}} - 3$ is rewritten as

$y = 5x^4 + 4x - \dfrac{1}{2}x^{-2} + x^{-1/2} - 3$

When differentiating a sum, each term is differentiated in turn.

Thus $\dfrac{dy}{dx} = (5)(4)x^{4-1} + (4)(1)x^{1-1}$

$- \dfrac{1}{2}(-2)x^{-2-1} + (1)\left(-\dfrac{1}{2}\right)x^{(-1/2)-1} - 0$

$= 20x^3 + 4 + x^{-3} - \dfrac{1}{2}x^{-3/2}$

i.e. $\dfrac{dy}{dx} = 20x^3 + 4 - \dfrac{1}{x^3} - \dfrac{1}{2\sqrt{x^3}}$

Problem 5. Differentiate (a) $s = (t+1)^2$
(b) $y = 3.5\theta^{1.2}$ with respect to the variable

(a) $s = (t+1)^2 = t^2 + 2t + 1$. Hence $\dfrac{ds}{dt} = 2t + 2$
$= 2(t+1)$

(b) $y = 3.5\theta^{1.2}$. Hence $\dfrac{dy}{d\theta} = (3.5)(1.2)\theta^{1.2-1}$
$= 4.2\theta^{0.2}$

Problem 6. Find the differential coefficients of (a) $y = 3\sin 4x$ (b) $f(t) = 2\cos 3t$ with respect to the variable

(a) When $y = 3\sin 4x$
then $\dfrac{dy}{dx} = (3)(4\cos 4x) = \mathbf{12\cos 4x}$

(b) When $f(t) = 2\cos 3t$
then $f'(t) = (2)(-3\sin 3t) = \mathbf{-6\sin 3t}$

Problem 7. Determine the derivatives of
(a) $y = 3e^{5x}$ (b) $f(\theta) = \dfrac{2}{e^{3\theta}}$ (c) $y = 6\ln 2x$

(a) When $y = 3e^{5x}$ then $\dfrac{dy}{dx} = (3)(5)e^{5x} = \mathbf{15e^{5x}}$

(b) $f(\theta) = \dfrac{2}{e^{3\theta}} = 2e^{-3\theta}$. Thus
$f'(\theta) = (2)(-3)e^{-3\theta} = -6e^{-3\theta} = \dfrac{-6}{e^{3\theta}}$

(c) When $y = 6\ln 2x$ then $\dfrac{dy}{dx} = 6\left(\dfrac{1}{x}\right) = \dfrac{6}{x}$

Problem 8. Find the gradient of the curve
$y = 3x^4 - 2x^2 + 5x - 2$ at the points $(0, -2)$
and $(1, 4)$

The gradient of a curve at a given point is given by the corresponding value of the derivative. Thus, since $y = 3x^4 - 2x^2 + 5x - 2$
then the gradient $= \dfrac{dy}{dx} = 12x^3 - 4x + 5$
At the point $(0, -2)$, $x = 0$. Thus the gradient $= 12(0)^3 - 4(0) + 5 = \mathbf{5}$.
At the point $(1, 4)$, $x = 1$. Thus the gradient $= 12(1)^3 - 4(1) + 5 = \mathbf{13}$.

Problem 9. Determine the coordinates of the point on the graph $y = 3x^2 - 7x + 2$ where the gradient is -1

The gradient of the curve is given by the derivative.
When $y = 3x^2 - 7x + 2$ then $\dfrac{dy}{dx} = 6x - 7$.
Since the gradient is -1 then $6x - 7 = -1$, from which, $x = 1$.
When $x = 1$, $y = 3(1)^2 - 7(1) + 2 = -2$.
Hence the gradient is -1 at the point $(1, -2)$.

Problem 10. (a) Given
$f(x) = \dfrac{2x^3 - 3\sqrt{x} + 4\sqrt[4]{x^3}}{x^2}$, determine $f'(x)$
(b) Evaluate $f'(x)$ when $x = 1$

(a) $f(x) = \dfrac{2x^3 - 3\sqrt{x} + 4\sqrt[4]{x^3}}{x^2}$

$= \dfrac{2x^3 - 3x^{1/2} + 4x^{3/4}}{x^2}$

$= \dfrac{2x^3}{x^2} - \dfrac{3x^{1/2}}{x^2} + \dfrac{4x^{3/4}}{x^2}$

$= 2x^{3-2} - 3x^{(1/2)-2} + 4x^{(3/4)-2}$,
by the laws of indices,

i.e. $f(x) = 2x - 3x^{-3/2} + 4x^{-5/4}$

Hence $f'(x)$

$$= 2 - (3)\left(-\frac{3}{2}\right)x^{-5/2} + (4)\left(-\frac{5}{4}\right)x^{-9/4}$$

$$= 2 + \frac{9}{2}x^{-5/2} - 5x^{-9/4} = 2 + \frac{9}{2x^{5/2}} - \frac{5}{x^{9/4}}$$

$$= 2 + \frac{9}{2\sqrt{x^5}} - \frac{5}{\sqrt[4]{x^9}}$$

(b)   When $x = 1$, $f'(x) = 2 + \frac{9}{2\sqrt{1^5}} - \frac{5}{\sqrt[4]{1^9}}$

$$= 2 + \frac{9}{2} - 5 = 1\frac{1}{2}$$

Problem 11.

(a)   Find $\dfrac{dx}{d\theta}$ when $x =$

$\dfrac{3}{\theta} - 4(\sin 5\theta - 2\cos 2\theta) + 2\ln 3\theta - \dfrac{15}{2e^{4\theta}}$

(b)   Evaluate $\dfrac{dx}{d\theta}$ when $\theta = \dfrac{\pi}{2}$, correct to 4 significant figures

---

(a) $x = \dfrac{3}{\theta} - 4(\sin 5\theta - 2\cos 2\theta) + 2\ln 3\theta - \dfrac{15}{2e^{4\theta}}$

i.e. $x = 3\theta^{-1} - 4\sin 5\theta + 8\cos 2\theta$

$$+ 2\ln 3\theta - \frac{15}{2}e^{-4\theta}$$

$$\frac{dx}{d\theta} = (3)(-1)\theta^{-2} - (4)(5\cos 5\theta)$$

$$+ (8)(-2\sin 2\theta) + \frac{2}{\theta} - \frac{15}{2}(-4)e^{-4\theta}$$

$$= -3\theta^{-2} - 20\cos 5\theta - 16\sin 2\theta$$

$$+ \frac{2}{\theta} + 30e^{-4\theta}$$

i.e. $\dfrac{dx}{d\theta} = \dfrac{-3}{\theta^2} - 20\cos 5\theta - 16\sin 2\theta + \dfrac{2}{\theta} + \dfrac{30}{e^{4\theta}}$

(b)   When $\theta = \dfrac{\pi}{2}$, $\dfrac{dx}{d\theta} = \dfrac{-3}{(\pi/2)^2} - 20\cos 5\left(\dfrac{\pi}{2}\right)$

$$- 16\sin 2\left(\frac{\pi}{2}\right) + \frac{2}{(\pi/2)} + \frac{30}{e^{4(\pi/2)}}$$

$\cos\dfrac{5\pi}{2} = 0$ and $\sin\pi = 0$.

Hence $\dfrac{dx}{d\theta} = \dfrac{-3}{(\pi/2)^2} + \dfrac{2}{(\pi/2)} + \dfrac{30}{e^{2\pi}}$

$$= -1.21585 + 1.27324 + 0.05602$$

Hence when $\theta = \dfrac{\pi}{2}$, $\dfrac{dx}{d\theta} = \mathbf{0.1134}$, correct to 4 significant figures

---

Further problems on differentiating common functions may be found in Section 20.6, Problems 1 to 12, pages 212 and 213.

20.2   Differentiation of a product

When $y = uv$, and u and v are both functions of x,

then $$\boxed{\frac{dy}{dx} = u\frac{dv}{dx} + v\frac{du}{dx}}$$

This is known as the **product rule**.

Problem 12.   Find the differential coefficient of $y = 3x^2\sin 2x$

$3x^2\sin 2x$ is a product of two terms $3x^2$ and $\sin 2x$. Let $u = 3x^2$ and $v = \sin 2x$.

Using the product rule:

$$\frac{dy}{dx} = u\frac{dv}{dx} + v\frac{du}{dx}$$

$$\downarrow \qquad \downarrow \qquad\qquad \downarrow \qquad \downarrow$$

gives: $$\frac{dy}{dx} = (3x^2)\ (2\cos 2x)\ +\ (\sin 2x)\ (6x)$$

i.e. $$\frac{dy}{dx} = 6x^2\cos 2x + 6x\sin 2x$$

$$= 6x(x\cos 2x + \sin 2x)$$

Note that the differential coefficient of a product is **not** obtained by merely differentiating each term and multiplying the two answers together. The product rule formula **must** be used when differentiating products.

Problem 13.   Find the rate of change of y with respect to x given $y = 3\sqrt{x}\ln 2x$

The rate of change of y with respect to x is given by $\dfrac{dy}{dx}$

$y = 3\sqrt{x}\ln 2x = 3x^{1/2}\ln 2x$, which is a product.

Let $u = 3x^{1/2}$ and $v = \ln x$.

Then

$$\frac{dy}{dx} = u \quad \frac{dv}{dx} + v \quad \frac{du}{dx}$$

$$\downarrow \qquad \downarrow \qquad \downarrow \qquad \downarrow$$

$$= (3x^{1/2}) \left(\frac{1}{x}\right) + (\ln 2x) \left[3\left(\frac{1}{2}\right)x^{(1/2)-1}\right]$$

$$= 3x^{(1/2)-1} + (\ln 2x) \left(\frac{3}{2}x^{-1/2}\right)$$

$$= 3x^{-1/2} \left(1 + \frac{1}{2}\ln 2x\right)$$

i.e. $\dfrac{dy}{dx} = \dfrac{3}{\sqrt{x}} \left(1 + \dfrac{1}{2}\ln 2x\right)$

Problem 14. If $x = 3e^{2\theta}\sin 5\theta$, find $\dfrac{dx}{d\theta}$

Let $u = 3e^{2\theta}$ and $v = \sin 5\theta$.

Then $\dfrac{dx}{d\theta} = u\dfrac{dv}{d\theta} + v\dfrac{du}{d\theta}$

$$= (3e^{2\theta})(5\cos 5\theta) + (\sin 5\theta)(6e^{2\theta})$$

$$= 15e^{2\theta}\cos 5\theta + 6e^{2\theta}\sin 5\theta$$

Hence $\dfrac{dx}{d\theta} = 3e^{2\theta}(5\cos 5\theta + 2\sin 5\theta).$

Problem 15. Differentiate $y = x^3 \cos 3x \ln x$

Let $u = x^3 \cos 3x$ (i.e. a product) and $v = \ln x$.

Then $\dfrac{dy}{dx} = u\dfrac{dv}{dx} + v\dfrac{du}{dx}$

$$\frac{du}{dx} = (x^3)(-3\sin 3x) + (\cos 3x)(3x^2) \text{ and}$$

$$\frac{dv}{dx} = \frac{1}{x}$$

Hence $\dfrac{dy}{dx} = (x^3 \cos 3x)\left(\dfrac{1}{x}\right)$

$$+ (\ln x)[-3x^3 \sin 3x + 3x^2 \cos 3x]$$

$$= x^2 \cos 3x + 3x^2 \ln x(\cos 3x - x\sin 3x)$$

i.e. $\dfrac{dy}{dx} = x^2\{\cos 3x + 3\ln x(\cos 3x - x\sin 3x)\}$

Problem 16. Determine the rate of change of voltage, given $v = 5t\sin 2t$ volts when $t = 0.2$ s

---

Rate of change of voltage $= \dfrac{dv}{dt}$

$$= (5t)(2\cos 2t) + (\sin 2t)(5)$$

$$= 10t\cos 2t + 5\sin 2t$$

When $t = 0.2$,

$$\frac{dv}{dt} = 10(0.2)\cos 2(0.2) + 5\sin 2(0.2)$$

$$= 2\cos 0.4 + 5\sin 0.4 \text{ (where } \cos 0.4$$

means the cosine of 0.4 radians,

i.e. $\left(0.4 \times \dfrac{180}{\pi}\right)^{\circ}$ or $22.92°$)

Hence $\dfrac{dv}{dt} = 2\cos 22.92° + 5\sin 22.92°$

$$= 1.8421 + 1.9472 = 3.7893$$

i.e. the rate of change of voltage when $t = 0.2$ s is 3.79 volts/s, correct to 3 significant figures.

Further problems on differentiating products may be found in Section 20.6, Problems 13 to 20, page 213.

20.3   Differentiation of a quotient

When $y = \dfrac{u}{v}$, and u and v are both functions of x

then $\boxed{\dfrac{dy}{dx} = \dfrac{v\dfrac{du}{dx} - u\dfrac{dv}{dx}}{v^2}}$

This is known as the **quotient rule**.

Problem 17. Find the differential coefficient of $y = \dfrac{4\sin 5x}{5x^4}$

$\dfrac{4\sin 5x}{5x^4}$ is a quotient. Let $u = 4\sin 5x$ and $v = 5x^4$

$$\frac{dy}{dx} = \frac{v\dfrac{du}{dx} - u\dfrac{dv}{dx}}{v^2}$$

where $\dfrac{du}{dx} = (4)(5)\cos 5x = 20\cos 5x$

and $\dfrac{dv}{dx} = (5)(4)x^3 = 20x^3$

Hence $\dfrac{dy}{dx} = \dfrac{(5x^4)(20\cos 5x) - (4\sin 5x)(20x^3)}{(5x^4)^2}$

$= \dfrac{100x^4 \cos 5x - 80x^3 \sin 5x}{25x^8}$

$= \dfrac{20x^3[5x\cos 5x - 4\sin 5x]}{25x^8}$

i.e. $\dfrac{dy}{dx} = \dfrac{4}{5x^5}(5x\cos 5x - 4\sin 5x)$

The differential coefficient is **not** obtained by merely differentiating each term in turn and then dividing the numerator by the denominator. The quotient formula **must** be used when differentiating quotients.

Problem 18.   Determine the differential coefficient of $y = \tan ax$

$y = \tan ax = \sin ax/\cos ax$. Differentiation of $\tan ax$ is thus treated as a quotient with $u = \sin ax$ and $v = \cos ax$

$\dfrac{dy}{dx} = \dfrac{v\dfrac{du}{dx} - u\dfrac{dv}{dx}}{v^2}$

$= \dfrac{(\cos ax)(a\cos ax) - (\sin ax)(-a\sin ax)}{(\cos ax)^2}$

$= \dfrac{a\cos^2 ax + a\sin^2 ax}{(\cos ax)^2}$

$= \dfrac{a(\cos^2 ax + \sin^2 ax)}{\cos^2 ax} = \dfrac{a}{\cos^2 ax},$

since $\cos^2 ax + \sin^2 ax = 1$ (see Chapter 37).

Hence $\dfrac{dy}{dx} = a\sec^2 ax$ (since $\sec^2 ax = \dfrac{1}{\cos^2 ax}$.

For further reciprocal trigonometric ratios, see Chapter 37.)

Problem 19.   Differentiate $y = \cot 2x$

$y = \cot 2x = \dfrac{1}{\tan 2x} = \dfrac{\cos 2x}{\sin 2x}$.

Let $u = \cos 2x$ and $v = \sin 2x$.

$\dfrac{dy}{dx} = \dfrac{v\dfrac{du}{dx} - u\dfrac{dv}{dx}}{v^2}$

$= \dfrac{(\sin 2x)(-2\sin 2x) - (\cos 2x)(2\cos 2x)}{(\sin 2x)^2}$

$= \dfrac{-2\sin^2 2x - 2\cos^2 2x}{\sin^2 2x}$

$= \dfrac{-2(\sin^2 2x + \cos^2 2x)}{\sin^2 2x}$

Hence $\dfrac{dy}{dx} = \dfrac{-2}{\sin^2 2x} = -2\csc^2 2x$.

Problem 20.   Find the derivative of $y = \sec ax$

$y = \sec ax = \dfrac{1}{\cos ax}$ (i.e. a quotient). Let $u = 1$ and $v = \cos ax$.

$\dfrac{dy}{dx} = \dfrac{v\dfrac{du}{dx} - u\dfrac{dv}{dx}}{v^2}$

$= \dfrac{(\cos ax)(0) - (1)(-a\sin ax)}{(\cos ax)^2}$

$= \dfrac{a\sin ax}{\cos^2 ax} = a\left(\dfrac{1}{\cos ax}\right)\left(\dfrac{\sin ax}{\cos ax}\right)$

i.e. $\dfrac{dy}{dx} = a\sec ax \tan ax$

Problem 21.   Show that
$\dfrac{d}{d\theta}(\csc a\theta) = -a\csc a\theta \cot a\theta$

Let $y = \csc a\theta = \dfrac{1}{\sin a\theta}$ (i.e. a quotient).

Let $u = 1$ and $v = \sin a\theta$.

$\dfrac{dy}{d\theta} = \dfrac{v\dfrac{du}{d\theta} - u\dfrac{dv}{d\theta}}{v^2}$

$= \dfrac{(\sin a\theta)(0) - (1)(a\cos a\theta)}{(\sin a\theta)^2}$

$= \dfrac{-a\cos a\theta}{\sin^2 a\theta}$

$= -a\left(\dfrac{1}{\sin a\theta}\right)\left(\dfrac{\cos a\theta}{\sin a\theta}\right)$

i.e. $\dfrac{d}{d\theta}(\csc a\theta) = -a\csc a\theta \cot a\theta$

Problem 22.   Differentiate $y = \dfrac{te^{2t}}{2\cos t}$

The function $te^{2t}/2\cos t$ is a quotient, whose numerator is a product.

Let $u = te^{2t}$ and $v = 2\cos t$.

then $\dfrac{du}{dt} = (t)(2e^{2t}) + (e^{2t})(1)$ and $\dfrac{dv}{dt} = -2\sin t$

Hence $\dfrac{dy}{dt} = \dfrac{v\dfrac{du}{dt} - u\dfrac{dv}{dt}}{v^2}$

$= \dfrac{(2\cos t)[2te^{2t} + e^{2t}] - (te^{2t})(-2\sin t)}{(2\cos t)^2}$

$= \dfrac{4te^{2t}\cos t + 2e^{2t}\cos t + 2te^{2t}\sin t}{4\cos^2 t}$

$= \dfrac{2e^{2t}[2t\cos t + \cos t + t\sin t]}{4\cos^2 t}$

i.e. $\dfrac{dy}{dt} = \dfrac{e^{2t}}{2\cos^2 t}(2t\cos t + \cos t + t\sin t)$

Problem 23.   Determine the gradient of the curve $y = \dfrac{5x}{2x^2 + 4}$ at the point $\left(\sqrt{3}, \dfrac{\sqrt{3}}{2}\right)$

Let $y = 5x$ and $v = 2x^2 + 4$.

$\dfrac{dy}{dx} = \dfrac{v\dfrac{du}{dx} - u\dfrac{dv}{dx}}{v^2}$

$= \dfrac{(2x^2 + 4)(5) - (5x)(4x)}{(2x^2 + 4)^2}$

$= \dfrac{10x^2 + 20 - 20x^2}{(2x^2 + 4)^2}$

$= \dfrac{20 - 10x^2}{(2x^2 + 4)^2}$

At the point $\left(\sqrt{3}, \dfrac{\sqrt{3}}{2}\right)$, $x = \sqrt{3}$.

Hence the gradient $= \dfrac{dy}{dx} = \dfrac{20 - 10(\sqrt{3})^2}{[2(\sqrt{3})^2 + 4]^2}$

$= \dfrac{20 - 30}{100} = -\dfrac{1}{10}$

Further problems on differentiating quotients may be found in Section 20.6, Problems 21 to 30, page 213.

20.4   Function of a function

It is often easier to make a substitution before differentiating.

If y is a function of x then $\boxed{\dfrac{dy}{dx} = \dfrac{dy}{du} \times \dfrac{du}{dx}}$

This is known as the **'function of a function'** rule (or sometimes the **chain rule**).

For example, if $y = (3x - 1)^9$ then, by making the substitution $u = (3x - 1)$, $y = u^9$, which is of the 'standard' form.

Hence $\dfrac{dy}{du} = 9u^8$ and $\dfrac{du}{dx} = 3$

Then $\dfrac{dy}{dx} = \dfrac{dy}{du} \times \dfrac{du}{dx}$

$= (9u^8)(3) = 27u^8 = \mathbf{27(3x - 1)^8}$

Since y is a function of u, and u is a function of x, then y is a function of a function of x.

Problem 24.   Differentiate $y = 3\cos(5x^2 + 2)$

Let $u = 5x^2 + 2$ then $y = 3\cos u$.

Hence $\dfrac{du}{dx} = 10x$ and $\dfrac{dy}{du} = -3\sin u$

Using the function of a function rule,

$\dfrac{dy}{dx} = \dfrac{dy}{du} \times \dfrac{du}{dx} = (-3\sin u)(10x) = -30x\sin u$

Rewriting u as $5x^2 + 2$ gives:

$\dfrac{dy}{dx} = \mathbf{-30x\sin(5x^2 + 2)}$

Problem 25.   Find the derivative of $y = (4t^3 - 3t)^6$

Let $u = 4t^3 - 3t$, then $y = u^6$.

Hence $\dfrac{du}{dt} = 12t^2 - 3$ and $\dfrac{dy}{du} = 6u^5$

Using the function of a function rule,

$$\dfrac{dy}{dt} = \dfrac{dy}{du} \times \dfrac{du}{dt} = (6u^5)(12t^2 - 3)$$

Rewriting u as $4t^3 - 3t$ gives

$$\frac{dy}{dt} = 6(4t^3 - 3t)^5(12t^2 - 3)$$

$$= 18(4t^2 - 1)(4t^3 - 3t)^5$$

Problem 26. Determine the differential coefficient of $y = \sqrt{(3x^2 + 4x - 1)}$

$y = \sqrt{(3x^2 + 4x - 1)} = (3x^2 + 4x - 1)^{1/2}$.
Let $u = 3x^2 + 4x - 1$ then $y = u^{1/2}$.

Hence $\dfrac{du}{dx} = 6x + 4$ and $\dfrac{dy}{du} = \dfrac{1}{2}u^{-1/2} = \dfrac{1}{2\sqrt{u}}$

Using the function of a function rule,

$$\frac{dy}{dx} = \frac{dy}{du} \times \frac{du}{dx} = \left(\frac{1}{2\sqrt{u}}\right)(6x+4) = \frac{3x + 2}{\sqrt{u}}$$

i.e. $\dfrac{dy}{dx} = \dfrac{3x + 2}{\sqrt{(3x^2 + 4x - 1)}}$

Problem 27. Differentiate $y = 3\tan^4 3x$

Let $u = \tan 3x$ then $y = 3u^4$

Hence $\dfrac{du}{dx} = 3\sec^2 3x$, (from Problem 18),

and $\dfrac{dy}{du} = 12u^3$

Then $\dfrac{dy}{dx} = \dfrac{dy}{du} \times \dfrac{du}{dx}$

$\qquad = (12u^3)(3\sec^2 3x)$

$\qquad = 12(\tan 3x)^3(3\sec^2 3x)$

i.e. $\dfrac{dy}{dx} = 36\tan^3 3x \sec^2 3x$

Problem 28. Find the differential coefficient of $y = \dfrac{2}{(2t^3 - 5)^4}$

$y = \dfrac{2}{(2t^3 - 5)^4} = 2(2t^3 - 5)^{-4}$.

Let $u = (2t^3 - 5)$, then $y = 2u^{-4}$

Hence $\dfrac{du}{dt} = 6t^2$ and $\dfrac{dy}{du} = -8u^{-5} = \dfrac{-8}{u^5}$

Then $\dfrac{dy}{dt} = \dfrac{dy}{du} \times \dfrac{du}{dt} = \left(\dfrac{-8}{u^5}\right)(6t^2)$

$\qquad = \dfrac{-48t^2}{(2t^3 - 5)^5}$

Problem 29. Differentiate $y = 4e^{(3\theta^2 - 2)}$

Let $u = 3\theta^2 - 2$ then $y = 4e^u$

Hence $\dfrac{du}{d\theta} = 6\theta$ and $\dfrac{dy}{du} = 4e^u$

Then $\dfrac{dy}{d\theta} = \dfrac{dy}{du} \times \dfrac{du}{d\theta}$

$\qquad = (4e^u)(6\theta) = 24\theta e^u = 24\theta e^{(3\theta^2 - 2)}$

Further problems on function of a function may be found in Section 20.6, Problems 31 to 41, pages 213 and 214.

20.5 Successive differentiation

When a function $y = f(x)$ is differentiated with respect to x the differential coefficient is written as $\dfrac{dy}{dx}$ or $f'(x)$. If the expression is differentiated again, the second differential coefficient is obtained and is written as $\dfrac{d^2y}{dx^2}$ (pronounced dee two y by dee x squared) or $f''(x)$ (pronounced f double-dash x). By successive differentiation further higher derivatives such as $\dfrac{d^3y}{dx^3}$ and $\dfrac{d^4y}{dx^4}$ may be obtained.

Thus if $y = 3x^4$, $\dfrac{dy}{dx} = 12x^3$, $\dfrac{d^2y}{dx^2} = 36x^2$,

$\dfrac{d^3y}{dx^3} = 72x$, $\dfrac{d^4y}{dx^4} = 72$ and $\dfrac{d^5y}{dx^5} = 0$

Problem 30. If $f(x) = 2x^5 - 4x^3 + 3x - 5$, find $f''(x)$

$$f(x) = 2x^5 - 4x^3 + 3x - 5$$
$$f'(x) = 10x^4 - 12x^2 + 3$$
$$f''(x) = 40x^3 - 24x = 4x(10x^2 - 6)$$

Problem 31. If $y = \cos x - \sin x$, evaluate x, in the range $0 \le x \le \dfrac{\pi}{2}$, when $\dfrac{d^2y}{dx^2}$ is zero

---

Since $y = \cos x - \sin x$, $\dfrac{dy}{dx} = -\sin x - \cos x$ and

$$\frac{d^2 y}{dx^2} = -\cos x + \sin x$$

When $\dfrac{d^2 y}{dx^2}$ is zero, $-\cos x + \sin x = 0$, i.e.

$\sin x = \cos x$ or $\dfrac{\sin x}{\cos x} = 1$.

Hence $\tan x = 1$ and $x = \arctan 1 = \mathbf{45°}$ or $\dfrac{\pi}{4}$ **rads**

in the range $0 \le x \le \dfrac{\pi}{2}$.

Problem 32. Given $y = 2xe^{-3x}$ show that

$$\frac{d^2 y}{dx^2} + 6\frac{dy}{dx} + 9y = 0$$

$y = 2xe^{-3x}$ (i.e. a product)

Hence $\dfrac{dy}{dx} = (2x)(-3e^{-3x}) + (e^{-3x})(2)$

$\qquad = -6xe^{-3x} + 2e^{-3x}$

$\dfrac{d^2 y}{dx^2} = [(-6x)(-3e^{-3x}) + (e^{-3x})(-6)]$

$\qquad + (-6e^{-3x})$

$\qquad = 18xe^{-3x} - 6e^{-3x} - 6e^{-3x}$

i.e. $\dfrac{d y^2}{dx^2} = 18xe^{-3x} - 12e^{-3x}$

Substituting values into $\dfrac{d^2 y}{dx^2} + 6\dfrac{dy}{dx} + 9y$ gives:

$(18xe^{-3x} - 12e^{-3x}) + 6(-6xe^{-3x} + 2e^{-3x}) + 9(2xe^{-3x})$

$= 18xe^{-3x} - 12e^{-3x} - 36xe^{-3x} + 12e^{-3x} + 18xe^{-3x} = 0$

Thus $\dfrac{d^2 y}{dx^2} + 6\dfrac{dy}{dx} + 9y = 0$ when $y = 2xe^{-3x}$.

Problem 33. Evaluate $\dfrac{d^2 y}{d\theta^2}$ when $\theta = 0$
given $y = 4 \sec 2\theta$

Since $y = 4 \sec 2\theta$, then

$\dfrac{dy}{d\theta} = (4)(2) \sec 2\theta \tan 2\theta$ (from Problem 20)

$\qquad = 8 \sec 2\theta \tan 2\theta$ (i.e. a product)

$$\frac{d^2 y}{d\theta^2} = (8 \sec 2\theta)(2 \sec^2 2\theta)$$

$\qquad + (\tan 2\theta)[(8)(2) \sec 2\theta \tan 2\theta]$

$\qquad = 16 \sec^3 2\theta + 16 \sec 2\theta \tan^2 2\theta$

When $\theta = 0$, $\dfrac{d^2 y}{d\theta^2} = 16 \sec^3 0 + 16 \sec 0 \tan^2 0$

$\qquad = 16(1) + 16(1)(0) = \mathbf{16}$

Further problems on successive differentiation may be found in the following Section 20.6, Problems 42 to 48, page 214.

20.6   Further problems on methods of differentiation

Differentiation of common functions

In Problems 1 to 6 find the differential coefficients of the given functions with respect to the variable:

1   (a) $5x^5$ (b) $2.4x^{3.5}$ (c) $\dfrac{1}{x}$

$\qquad \left[\text{(a) } 25x^4 \text{ (b) } 8.4x^{2.5} \text{ (c) } -\dfrac{1}{x^2} \right]$

2   (a) $\dfrac{-4}{x^2}$ (b) 6 (c) $2x$ $\left[\text{(a) } \dfrac{8}{x^3} \text{ (b) } 0 \text{ (c) } 2 \right]$

3   (a) $2\sqrt{x}$ (b) $3\sqrt[3]{x^5}$ (c) $\dfrac{4}{\sqrt{x}}$

$\qquad \left[\text{(a) } \dfrac{1}{\sqrt{x}} \text{ (b) } 5\sqrt[3]{x^2} \text{ (c) } -\dfrac{2}{\sqrt{x^3}} \right]$

4   (a) $\dfrac{-3}{\sqrt[3]{x}}$ (b) $(x - 1)^2$ (c) $2 \sin 3x$

$\qquad \left[\text{(a) } \dfrac{1}{\sqrt[3]{x^4}} \text{ (b) } 2(x - 1) \text{ (c) } 6 \cos 3x \right]$

5   (a) $-4 \cos 2x$ (b) $2e^{6x}$ (c) $\dfrac{3}{e^{5x}}$

$\qquad \left[\text{(a) } 8 \sin 2x \text{ (b) } 12e^{6x} \text{ (c) } \dfrac{-15}{e^{5x}} \right]$

6   (a) $4 \ln 9x$ (b) $\dfrac{e^x - e^{-x}}{2}$ (c) $\dfrac{1 - \sqrt{x}}{x}$

$\qquad \left[\text{(a) } \dfrac{4}{x} \text{ (b) } \dfrac{e^x + e^{-x}}{2} \text{ (c) } \dfrac{-1}{x^2} + \dfrac{1}{2\sqrt{x^3}} \right]$

7   Find the gradient of the curve $y = 2t^4 + 3t^3 - t + 4$ at the points $(0, 4)$ and $(1, 8)$.

$\qquad\qquad\qquad\qquad\qquad\qquad [-1, 16]$

8 If $f(t) = 4 \ln t + 2$ evaluate $f'(t)$ when $t = 0.25$. [16]

9 Find the coordinates of the point on the graph $y = 5x^2 - 3x + 1$ where the gradient is 2.

$$\left[\left(\frac{1}{2}, \frac{3}{4}\right)\right]$$

10 Given $f(x) = \dfrac{3x^4 - 4\sqrt{x} + 3\sqrt[4]{x^5}}{x^3}$ evaluate $f'(x)$ when $x = 1$.

$$\left[7\frac{3}{4}\right]$$

11 (a) Differentiate $y = \dfrac{2}{\theta^2} + 2 \ln 2\theta - 2(\cos 5\theta + 3 \sin 2\theta) - \dfrac{2}{e^{3\theta}}$

(b) Evaluate $\dfrac{dy}{d\theta}$ when $\theta = \dfrac{\pi}{2}$, correct to 4 significant figures

$$\left[\begin{array}{l}\text{(a) } \dfrac{-4}{\theta^3} + \dfrac{2}{\theta} + 10\sin 5\theta - 12\cos 2\theta + \dfrac{6}{e^{3\theta}} \\ \text{(b) } 22.30\end{array}\right]$$

12 Evaluate $\dfrac{ds}{dt}$, correct to 3 significant figures, when $t = \dfrac{\pi}{6}$ given $s = 3 \sin t - 3 + \sqrt{t}$.

[3.29]

Differentiation of products

In Problems 13 to 18 differentiate the given products with respect to the variable:

13 $2x^3 \cos 3x$ $\qquad [6x^2(\cos 3x - x \sin 3x)]$

14 $\sqrt{x^3} \ln 3x$ $\qquad \left[\sqrt{x}\left(1 + \dfrac{3}{2} \ln 3x\right)\right]$

15 $e^{3t} \sin 4t$ $\qquad [e^{3t}(4 \cos 4t + 3 \sin 4t)]$

16 $\sqrt{x} \sin 2x$ $\qquad \left[2\sqrt{x} \cos 2x + \dfrac{\sin 2x}{2\sqrt{x}}\right]$

17 $e^{4\theta} \ln 3\theta$ $\qquad \left[e^{4\theta}\left(\dfrac{1}{\theta} + 4 \ln 3\theta\right)\right]$

18 $e^t \ln t \cos t$ $\qquad \left[e^t\left\{\left(\dfrac{1}{t} + \ln t\right) \cos t - \ln t \sin t\right\}\right]$

19 Evaluate $\dfrac{di}{dt}$, correct to 4 significant figures, when $t = 0.1$, and $i = 15t \sin 3t$. [8.732]

20 Evaluate $\dfrac{dz}{dt}$, correct to 4 significant figures, when $t = 0.5$, given that $z = 2e^{3t} \sin 2t$.

[32.31]

Differentiation of quotients

In Problems 21 to 26, differentiate the quotients with respect to the variable:

21 $\dfrac{2 \cos 3x}{x^3}$ $\qquad \left[\dfrac{-6}{x^4}(x \sin 3x + \cos 3x)\right]$

22 $\dfrac{2x}{x^2 + 1}$ $\qquad \left[\dfrac{2(1 - x^2)}{(x^2 + 1)^2}\right]$

23 $\dfrac{3\sqrt{\theta^3}}{2 \sin 2\theta}$ $\qquad \left[\dfrac{3\sqrt{\theta}(3 \sin 2\theta - 4\theta \cos 2\theta)}{4 \sin^2 2\theta}\right]$

24 $\dfrac{\ln 2t}{\sqrt{t}}$ $\qquad \left[\dfrac{1 - \frac{1}{2} \ln 2t}{\sqrt{t^3}}\right]$

25 $\dfrac{3 \tan p}{e^{3p}}$ $\qquad \left[\dfrac{3(\sec^2 p - 3 \tan p)}{e^{3p}}\right]$

26 $\dfrac{2xe^{4x}}{\sin x}$ $\qquad \left[\dfrac{2e^{4x}}{\sin^2 x}\{(1 + 4x) \sin x - x \cos x\}\right]$

27 Find the gradient of the curve $y = \dfrac{2x}{x^2 - 5}$ at the point $(2, -4)$ $\qquad [-18]$

28 Evaluate $\dfrac{dy}{dx}$ at $x = 2.5$, correct to 3 significant figures, given $y = \dfrac{2x^2 + 3}{\ln 2x}$. $\qquad [3.82]$

29 Evaluate $f'\left(\dfrac{\pi}{3}\right)$ when $f(t) = 2 \tan 2t - \cot 4t$.

$$\left[21\frac{1}{3}\right]$$

30 Show that $\dfrac{\dfrac{d}{dz}(2 \sec 3z)}{\dfrac{d}{dz}(-\operatorname{cosec} 3z)} = 2 \tan^3 3z$.

Functions of a function

In Problems 31 to 40, find the differential coefficients with respect to the variable:

31 $(2x^3 - 5x)^5$ $\qquad [5(6x^2 - 5)(2x^3 - 5x)^4]$

32 $2 \sin(3\theta - 2)$ $\qquad [6 \cos(3\theta - 2)]$

33 $\sqrt{(2t^3 - 4)}$ $\qquad \left[\dfrac{3t^2}{\sqrt{(2t^3 - 4)}}\right]$

34 $2 \cos^5 \alpha$ $\qquad [-10\cos^4 \alpha \sin \alpha]$

35 $\dfrac{1}{(x^3 - 2x + 1)^5}$ $\qquad \left[\dfrac{5(2 - 3x^2)}{(x^3 - 2x + 1)^6}\right]$

36 $5e^{2t+1}$ $[10e^{2t+1}]$

37 $4\sec^3 x$ $[12\sec^3 x \tan x]$

38 $2\cot(5t^2 + 3)$ $[-20t\operatorname{cosec}^2(5t^2 + 3)]$

39 $6\tan(3y + 1)$ $[18\sec^2(3y + 1)]$

40 $2e^{\tan\theta}$ $[2\sec^2\theta e^{\tan\theta}]$

41 Differentiate $\theta\sin\left(\theta - \dfrac{\pi}{3}\right)$ with respect to θ, and evaluate, correct to 3 significant figures, when $\theta = \dfrac{\pi}{2}$. $[1.86]$

Successive differentiation

42 If $y = 3x^4 + 2x^3 - 3x + 2$ find (a) $\dfrac{d^2 y}{dx^2}$ (b) $\dfrac{d^3 y}{dx^3}$.

$[$(a) $36x^2 + 12x$ (b) $72x + 12]$

43 (a) Given $f(t) = \dfrac{2}{5}t^2 - \dfrac{1}{t^3} + \dfrac{3}{t} - \sqrt{t} + 1$ determine $f''(t)$

(b) Evaluate $f''(t)$ when $t = 1$

$\left[\text{(a)}\ \dfrac{4}{5} - \dfrac{12}{t^5} + \dfrac{6}{t^3} + \dfrac{1}{4\sqrt{t^3}}\ \text{(b)}\ -4.95\right]$

In Problems 44 and 45, find the second differential coefficient with respect to the variable:

44 (a) $3\sin 2t + \cos t$ (b) $2\ln 4\theta$

$\left[\text{(a)}\ -(12\sin 2t + \cos t)\ \text{(b)}\ \dfrac{-2}{\theta^2}\right]$

45 (a) $2\cos^2 x$ (b) $(2x - 3)^4$

$[$(a) $4(\sin^2 x - \cos^2 x)$ (b) $48(2x - 3)^2]$

46 If $y = Ae^{3t} + Be^{-2t}$ prove that
$$\frac{d^2 y}{dt^2} - \frac{dy}{dt} - 6y = 0.$$

47 Evaluate $f''(\theta)$ when $\theta = 0$ given $f(\theta) = 2\sec 3\theta$. $[18]$

48 Show that the differential equation
$$\frac{d^2 y}{dx^2} - 4\frac{dy}{dx} + 4y = 0 \text{ is satisfied when } y = xe^{2x}.$$

21

Differentiation of parametric equations

21.1 Introduction

Certain mathematical functions can be expressed more simply by expressing, say, x and y separately in terms of a third variable. For example, $y = r \sin \theta$, $x = r \cos \theta$. Then, any value given to θ will produce a pair of values for x and y, which may be plotted to provide a curve of $y = f(x)$.

The third variable, θ, is called a **parameter** and the two expressions for y and x are called **parametric equations**.

The above example of $y = r \sin \theta$ and $x = r \cos \theta$ are the parametric equations for a circle. The equation of any point on a circle, centre at the origin and of radius r is given by: $x^2 + y^2 = r^2$.

To show that $y = r \sin \theta$ and $x = r \cos \theta$ are suitable parametric equations for such a circle:

left-hand side of equation

$$= x^2 + y^2$$

$$= (r \cos \theta)^2 + (r \sin \theta)^2$$

$$= r^2 \cos^2 \theta + r^2 \sin^2 \theta$$

$$= r^2 (\cos^2 \theta + \sin^2 \theta)$$

$$= r^2 = \text{right hand side}$$

$$(\text{since } \cos^2 \theta + \sin^2 \theta = 1)$$

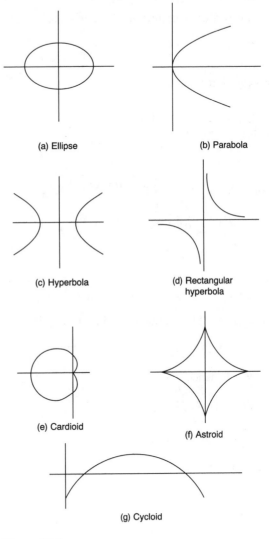

(a) Ellipse   (b) Parabola

(c) Hyperbola   (d) Rectangular hyperbola

(e) Cardioid   (f) Astroid

(g) Cycloid

Figure 21.1

21.2 Some common parametric equations

The following are some of the more common parametric equations, and Fig. 21.1 shows typical shapes of these curves.

(a) Ellipse $x = a \cos \theta, \ y = b \sin \theta$

(b) Parabola $x = at^2, \ y = 2at$

(c) Hyperbola $x = a \sec \theta, \ y = b \tan \theta$

(d) Rectangular hyperbola $x = ct, \ y = \dfrac{c}{t}$

(e) Cardioid $x = a(2 \cos \theta - \cos 2\theta),$ $y = a(2 \sin \theta - \sin 2\theta)$

(f) Astroid $x = a \cos^3 \theta, \ y = a \sin^3 \theta$

(g) Cycloid $x = a(\theta - \sin \theta),$ $y = a(1 - \cos \theta)$

21.3   Differentiation in parameters

When x and y are given in terms of a parameter, say θ, then by the function of a function rule of differentiation:

$$\frac{dy}{dx} = \frac{dy}{d\theta} \times \frac{d\theta}{dx}$$

It may be shown that this can be written as:

$$\frac{dy}{dx} = \frac{\dfrac{dy}{d\theta}}{\dfrac{dx}{d\theta}} \qquad (1)$$

For the second differential,

$$\frac{d^2y}{dx^2} = \frac{d}{dx}\left(\frac{dy}{dx}\right) = \frac{d}{d\theta}\left(\frac{dy}{dx}\right) \cdot \frac{d\theta}{dx}$$

$$\text{or} \quad \frac{d^2y}{dx^2} = \frac{\dfrac{d}{d\theta}\left(\dfrac{dy}{dx}\right)}{\dfrac{dx}{d\theta}} \qquad (2)$$

Problem 1. Given $x = 5\theta - 1$ and $y = 2\theta(\theta - 1)$, determine $\dfrac{dy}{dx}$ in terms of θ

$x = 5\theta - 1$, hence $\dfrac{dx}{d\theta} = 5$

$y = 2\theta(\theta - 1) = 2\theta^2 - 2\theta$,

hence $\dfrac{dy}{d\theta} = 4\theta - 2 = 2(2\theta - 1)$

From equation (1),

$$\frac{dy}{dx} = \frac{\dfrac{dy}{d\theta}}{\dfrac{dx}{d\theta}} = \frac{2(2\theta - 1)}{5} \quad \text{or} \quad \frac{2}{5}(2\theta - 1)$$

Problem 2. The parametric equations of a function are given by $y = 3\cos 2t$, $x = 2\sin t$. Determine expressions for

(a) $\dfrac{dy}{dx}$ (b) $\dfrac{d^2y}{dx^2}$

(a) $y = 3\cos 2t$, hence $\dfrac{dy}{dt} = -6\sin 2t$

$x = 2\sin t$, hence $\dfrac{dx}{dt} = 2\cos t$

From equation (1),

$$\frac{dy}{dx} = \frac{\dfrac{dy}{dt}}{\dfrac{dx}{dt}} = \frac{-6\sin 2t}{2\cos t}$$

$$= \frac{-6(2\sin t \cos t)}{2\cos t}$$

from double angles (see Chapter 38)

i.e. $\dfrac{dy}{dx} = -6\sin t$

(b)   From equation (2),

$$\frac{d^2y}{dx^2} = \frac{\dfrac{d}{dt}\left(\dfrac{dy}{dx}\right)}{\dfrac{dx}{dt}}$$

$$= \frac{\dfrac{d}{dt}(-6\sin t)}{2\cos t} = \frac{-6\cos t}{2\cos t}$$

i.e. $\dfrac{d^2y}{dx^2} = -3$

Problem 3. The equation of a tangent drawn to a curve at point (x_1, y_1) is given by:

$$y - y_1 = \frac{dy_1}{dx_1}(x - x_1)$$

Determine the equation of the tangent drawn to the parabola $x = 2t^2$, $y = 4t$ at the point t

---

At point t, $x_1 = 2t^2$, hence $\dfrac{dx_1}{dt} = 4t$

and $y_1 = 4t$, hence $\dfrac{dy_1}{dt} = 4$

From equation (1), $\dfrac{dy_1}{dx_1} = \dfrac{\dfrac{dy_1}{dt}}{\dfrac{dx_1}{dt}} = \dfrac{4}{4t} = \dfrac{1}{t}$

Hence the equation of the tangent is:

$$y - 4t = \frac{1}{t}(x - 2t^2)$$

Problem 4. The parametric equations of a cycloid are $x = 4(\theta - \sin\theta)$, $y = 4(1 - \cos\theta)$.

Determine (a) $\dfrac{dy}{dx}$ (b) $\dfrac{d^2y}{dx^2}$

(a) $x = 4(\theta - \sin\theta)$, hence

$$\frac{dx}{d\theta} = 4 - 4\cos\theta = 4(1 - \cos\theta)$$

$y = 4(1 - \cos\theta)$, hence $\dfrac{dy}{d\theta} = 4\sin\theta$

From equation (1),

$$\frac{dy}{dx} = \frac{\dfrac{dy}{d\theta}}{\dfrac{dx}{d\theta}} = \frac{4\sin\theta}{4(1 - \cos\theta)}$$

$$= \frac{\sin\theta}{(1 - \cos\theta)}$$

(b) From equation (2),

$$\frac{d^2y}{dx^2} = \frac{\dfrac{d}{d\theta}\left(\dfrac{dy}{dx}\right)}{\dfrac{dx}{d\theta}}$$

$$= \frac{\dfrac{d}{d\theta}\left(\dfrac{\sin\theta}{1 - \cos\theta}\right)}{4(1 - \cos\theta)}$$

$$= \frac{\dfrac{(1 - \cos\theta)(\cos\theta) - (\sin\theta)(\sin\theta)}{(1 - \cos\theta)^2}}{4(1 - \cos\theta)}$$

$$= \frac{\cos\theta - \cos^2\theta - \sin^2\theta}{4(1 - \cos\theta)^3}$$

$$= \frac{\cos\theta - (\cos^2\theta + \sin^2\theta)}{4(1 - \cos\theta)^3}$$

$$= \frac{\cos\theta - 1}{4(1 - \cos\theta)^3} = \frac{-(1 - \cos\theta)}{4(1 - \cos\theta)^3}$$

$$= \frac{-1}{4(1 - \cos\theta)^2}$$

Problem 5. The equation of the normal drawn to a curve at point (x_1, y_1) is given by:

$$y - y_1 = \frac{-1}{dy_1/dx_1}(x - x_1)$$

Determine the equation of the normal drawn to the astroid $x = 2\cos^3\theta$, $y = 2\sin^3\theta$ at the point $\theta = \dfrac{\pi}{4}$

$x = 2\cos^3\theta$, hence $\dfrac{dx}{d\theta} = -6\cos^2\theta\sin\theta$

$y = 2\sin^3\theta$, hence $\dfrac{dy}{d\theta} = 6\sin^2\theta\cos\theta$

From equation (1),

$$\frac{dy}{dx} = \frac{dy/d\theta}{dx/d\theta} = \frac{6\sin^2\theta\cos\theta}{-6\cos^2\theta\sin\theta}$$

$$= \frac{-\sin\theta}{\cos\theta} = -\tan\theta$$

When $\theta = \dfrac{\pi}{4}$, $\dfrac{dy}{dx} = -\tan\dfrac{\pi}{4} = -1$,

$$x_1 = 2\cos^3\frac{\pi}{4} = 0.7071 \text{ and}$$

$$y_1 = 2\sin^3\frac{\pi}{4} = 0.7071$$

Hence the equation of the normal is:

$$y - 0.7071 = \frac{-1}{-1}(x - 0.7071)$$

i.e. $y - 0.7071 = x - 0.7071$

i.e. $y = x$

Problem 6. The parametric equations for a hyperbola are $x = 2\sec\theta$, $y = 4\tan\theta$.

Evaluate $\dfrac{dy}{dx}$ and $\dfrac{d^2y}{dx^2}$, correct to 4 significant figures, when $\theta = 1$ radian

---

$x = 2\sec\theta$, hence $\dfrac{dx}{d\theta} = 2\sec\theta\tan\theta$

$y = 4\tan\theta$, hence $\dfrac{dy}{d\theta} = 4\sec^2\theta$

From equation (1),

$$\frac{dy}{dx} = \frac{dy/d\theta}{dx/d\theta} = \frac{4\sec^2\theta}{2\sec\theta\tan\theta} = \frac{2\sec\theta}{\tan\theta}$$

$$= \frac{2\left(\dfrac{1}{\cos\theta}\right)}{\left(\dfrac{\sin\theta}{\cos\theta}\right)} = \frac{2}{\sin\theta} \quad (\text{or } 2\cosec\theta)$$

When $\theta = 1$ rad, $\dfrac{dy}{dx} = \dfrac{2}{\sin 1} = \mathbf{2.377}$, correct to 4 significant figures.

From equation (2),

$$\frac{d^2y}{dx^2} = \frac{\dfrac{d}{d\theta}\left(\dfrac{dy}{dx}\right)}{\dfrac{dx}{d\theta}} = \frac{\dfrac{d}{d\theta}[2\cosec\theta]}{2\sec\theta\tan\theta}$$

$$= \frac{-2\cosec\theta\cot\theta}{2\sec\theta\tan\theta}$$

$$= \frac{-\left(\dfrac{1}{\sin\theta}\right)\left(\dfrac{\cos\theta}{\sin\theta}\right)}{\left(\dfrac{1}{\cos\theta}\right)\left(\dfrac{\sin\theta}{\cos\theta}\right)}$$

$$= -\left(\dfrac{\cos\theta}{\sin^2\theta}\right)\left(\dfrac{\cos^2\theta}{\sin\theta}\right)$$

$$= \frac{-\cos^3\theta}{\sin^3\theta} = -\cot^3\theta$$

When $\theta = 1$ rad, $\dfrac{d^2y}{dx^2} = -\cot^3 1 = \dfrac{-1}{(\tan 1)^3} = $ **−0.2647**, correct to 4 significant figures.

Problem 7.   When determining the surface tension of a liquid, the radius of curvature, ρ, of part of the surface is given by:

$$\rho = \frac{\sqrt{\left[1 + \left(\dfrac{dy}{dx}\right)^2\right]^3}}{d^2y/dx^2}$$

Find the radius of curvature of the part of the surface having the parametric equations $x = 3t^2$, $y = 6t$ at the point $t = 2$

$x = 3t^2$, hence $\dfrac{dx}{dt} = 6t$

$y = 6t$, hence $\dfrac{dy}{dt} = 6$

From equation (1),

$$\frac{dy}{dx} = \frac{\dfrac{dy}{dt}}{\dfrac{dx}{dt}} = \frac{6}{6t} = \frac{1}{t}$$

From equation (2),

$$\frac{d^2y}{dx^2} = \frac{\dfrac{d}{dt}\left(\dfrac{dy}{dx}\right)}{\dfrac{dx}{dt}} = \frac{\dfrac{d}{dt}\left(\dfrac{1}{t}\right)}{6t} = \frac{-\dfrac{1}{t^2}}{6t} = \frac{-1}{6t^3}$$

Hence radius of curvature, $\rho = \dfrac{\sqrt{\left[1 + \left(\dfrac{dy}{dx}\right)^2\right]^3}}{\dfrac{d^2y}{dx^2}}$

$$= \frac{\sqrt{\left[1 + \left(\dfrac{1}{t}\right)^2\right]^3}}{\dfrac{-1}{6t^3}}$$

When $t = 2$, $\rho = \dfrac{\sqrt{\left[1 + \left(\dfrac{1}{2}\right)^2\right]^3}}{\dfrac{-1}{6(2)^3}} = \dfrac{\sqrt{(1.25)^3}}{\dfrac{-1}{48}}$

$$= -48\sqrt{(1.25)^3} = \mathbf{-67.08}$$

Further problems on differentiation in parameters may be found in the following Section 21.4.

21.4   Further problems on differentiation of parametric equations

1   Given $x = 3t - 1$ and $y = t(t - 1)$, determine dy/dx in terms of t. $\left[\frac{1}{3}(2t - 1)\right]$

2   A parabola has parametric equations $x = t^2$, $y = 2t$. Evaluate dy/dx when $t = 0.5$.   [2]

3   The parametric equations for an ellipse are $x = 4\cos\theta$, $y = \sin\theta$. Determine

(a) $\dfrac{dy}{dx}$ (b) $\dfrac{d^2y}{dx^2}$.

$$\left[(a) \ -\frac{1}{4}\cot\theta \ (b) \ -\frac{1}{16}\csc^3\theta\right]$$

4   Evaluate dy/dx at $\theta = \pi/6$ radians for the hyperbola whose parametric equations are $x = 3\sec\theta$, $y = 6\tan\theta$.   [4]

5   The parametric equations for a rectangular hyperbola are $x = 2t$, $y = 2/t$. Evaluate dy/dx when $t = 0.40$.   [−6.25]

6   A cycloid has parametric equations $x = 2(\theta - \sin\theta)$, $y = 2(1 - \cos\theta)$. Evaluate at $\theta = 0.62$ radians, correct to 4 significant figures

(a) $\dfrac{dy}{dx}$ (b) $\dfrac{d^2y}{dx^2}$.   [(a) 3.122 (b) −14.43]

The equation of the tangent drawn to a curve at point (x_1, y_1) is: $y - y_1 = dy_1/dx_1(x - x_1)$. Use this in Problems 7 and 8:

7   Determine the equation of the tangent drawn to an ellipse $x = 3\cos\theta$, $y = 2\sin\theta$ at $\theta = \pi/6$.

$$[y = -1.155x + 4]$$

8   Determine the equation of the tangent drawn to the rectangular hyperbola $x = 5t$, $y = 5/t$ at $t = 2$.

$$\left[y = -\tfrac{1}{4}x + 5\right]$$

The equation of the normal drawn to a curve at point (x_1, y_1) is: $y - y_1 = \dfrac{-1}{dy_1/dx_1}(x - x_1)$. Use this in Problems 9 and 10:

9   Determine the equation of the normal drawn to the parabola $x = \tfrac{1}{4}t^2$, $y = \tfrac{1}{2}t$ at $t = 2$.

$$[y = -2x + 3]$$

10   Find the equation of the normal drawn to the cycloid at $x = 2(\theta - \sin\theta)$, $y = 2(1 - \cos\theta)$ at $\theta = \pi/2$ radians. $[y = -x + \pi]$

11   Determine the value of d^2y/dx^2, correct to 4 significant figures, at $\theta = \pi/6$ rad for the cardioid $x = 5(2\theta - \cos 2\theta)$, $y = 5(2\sin\theta - \sin 2\theta)$.   [0.8196]

12   The radius of curvature ρ of part of a surface when determining the surface tension of a liquid is given by:

$$\rho = \frac{[1 + (dy/dx)^2]^{3/2}}{d^2y/dx^2}$$

Find the radius of curvature (correct to 4 significant figures) of the part of the surface having parametric equations

(a) $x = 3t$, $y = 3/t$ at the point $t = \tfrac{1}{2}$

(b) $x = 4\cos^3 t$, $y = 4\sin^3 t$ at $t = \pi/6$ rad

$$[(a) \ 13.14 \ (b) \ 5.196]$$

22

Applications of differentiation

22.1 Rates of change

As stated in Chapter 19, if a quantity y depends on and varies with a quantity x then the rate of change of y with respect to x is dy/dx. Thus, for example, the rate of change of pressure p with height h is dp/dh. A rate of change with respect to time is usually just called 'the rate of change', the 'with respect to time' being assumed. Thus, for example, a rate of change of current, i, is di/dt and a rate of change of temperature, θ, is $d\theta/dt$, and so on.

Problem 1. The length l metres of a certain metal rod at temperature $\theta\,°C$ is given by $l = 1 + 0.00005\theta + 0.0000004\theta^2$. Determine the rate of change of length, in mm/°C, when the temperature is
(a) $100\,°C$, and (b) $400\,°C$

The rate of change of length means $\dfrac{dl}{d\theta}$

Since length $l = 1 + 0.00005\theta + 0.0000004\theta^2$,

then $\dfrac{dl}{d\theta} = 0.00005 + 0.0000008\theta$

(a)   When $\theta = 100\,°C$,

$\dfrac{dl}{d\theta} = 0.00005 + (0.0000008)(100)$

$= 0.00013\ \text{m/}°C$

$= \mathbf{0.13\ mm/°C}$

(b)   When $\theta = 400\,°C$,

$\dfrac{dl}{d\theta} = 0.00005 + (0.0000008)(400)$

$= 0.00037\ \text{m/}°C$

$= \mathbf{0.37\ mm/°C}$

Problem 2. The luminous intensity I candelas of a lamp at varying voltage V is given by $I = 4 \times 10^{-4}V^2$. Determine the voltage at which the light is increasing at a rate of 0.6 candelas per volt

---

The rate of change of light with respect to voltage is given by $\dfrac{dI}{dV}$

Since $I = 4 \times 10^{-4}V^2$,

$\dfrac{dI}{dV} = (4 \times 10^{-4})(2)V = 8 \times 10^{-4}V$

When the light is increasing at 0.6 candelas per volt then $+0.6 = 8 \times 10^{-4}V$, from which voltage

$V = \dfrac{0.6}{8 \times 10^{-4}} = 0.075 \times 10^{+4}$

$= \mathbf{750\ volts}$

Problem 3. Newton's law of cooling is given by $\theta = \theta_0 e^{-kt}$, where the excess of temperature at zero time is $\theta_0\,°C$ and at time t seconds is $\theta\,°C$. Determine the rate of change of temperature after 40 s, given that $\theta_0 = 16\,°C$ and $k = -0.03$

The rate of change of temperature is $\dfrac{d\theta}{dt}$

Since $\theta = \theta_0 e^{-kt}$

then $\dfrac{d\theta}{dt} = (\theta_0)(-k)e^{-kt}$

$= -k\theta_0 e^{-kt}$

When $\theta_0 = 16,\ k = -0.03$ and $t = 40$

then $\dfrac{d\theta}{dt} = -(-0.03)(16)e^{-(-0.03)(40)}$

$= 0.48e^{1.2}$

$= \mathbf{1.594\,°C/s}$

Problem 4. The displacement s cm of the end of a stiff spring at time t seconds is given by $s = ae^{-kt} \sin 2\pi f t$. Determine the velocity of the end of the spring after 1 s, if $a = 2$, $k = 0.9$ and $f = 5$

Velocity $v = \dfrac{ds}{dt}$ where $s = ae^{-kt} \sin 2\pi f t$ (i.e. a product).

Using the product rule,

$$\frac{ds}{dt} = (ae^{-kt})(2\pi f \cos 2\pi f t) + (\sin 2\pi f t)(-ake^{-kt})$$

When $a = 2$, $k = 0.9$, $f = 5$ and $t = 1$,

velocity, v $= (2e^{-0.9})(2\pi 5 \cos 2\pi 5)$

$\qquad + (\sin 2\pi 5)(-2)(0.9)e^{-0.9}$

$\qquad = 25.5455 \cos 10\pi - 0.7318 \sin 10\pi$

$\qquad = 25.5455(1) - 0.7318(0)$

$\qquad = \mathbf{25.55\ cm/s}$

(Note that $\cos 10\pi$ means 'the cosine of 10π radians, *not* degrees, and $\cos 10\pi \equiv \cos 2\pi = 1$.)

Further problems on rates of change may be found in Section 22.7, Problems 1 to 6, page 231.

22.2   Velocity and acceleration

As deduced in Chapter 19, if a body moves a distance x metres in a time t seconds then:

(i)   distance $x = f(t)$

(ii)   velocity $v = f'(t)$ or $\dfrac{dx}{dt}$, which is the gradient of the distance/time graph

(iii)   acceleration $a = \dfrac{dv}{dt} = f''(t)$ or $\dfrac{d^2x}{dt^2}$, which is the gradient of the velocity/time graph

Problem 5. The distance x metres moved by a car in a time t seconds is given by $x = 3t^3 - 2t^2 + 4t - 1$. Determine the velocity and acceleration when (a) $t = 0$ and (b) $t = 1.5$ s

Distance $x = 3t^3 - 2t^2 + 4t - 1$ m

Velocity $v = \dfrac{dx}{dt} = 9t^2 - 4t + 4$ m/s

Acceleration $a = \dfrac{d^2x}{dt^2} = 18t - 4$ m/s^2

(a)   When time $t = 0$,
velocity $v = 9(0)^2 - 4(0) + 4 = \mathbf{4\ m/s}$
and acceleration $a = 18(0) - 4 = \mathbf{-4\ m/s^2}$
(i.e. a deceleration)

(b)   When time $t = 1.5$ s, velocity $v = 9(1.5)^2 - 4(1.5) + 4 = \mathbf{18.25\ m/s}$
and acceleration $a = 18(1.5) - 4 = \mathbf{23\ m/s^2}$

Problem 6. The distance x metres travelled by a vehicle in time t seconds after the brakes are applied is given by $x = 20t - \dfrac{5}{3}t^2$. Determine (a) the speed of the vehicle (in km/h) at the instant the brakes are applied, and (b) the distance the car travels before it stops

(a)   Distance, $x = 20t - \dfrac{5}{3}t^2$.

Hence velocity $v = \dfrac{dx}{dt} = 20 - \dfrac{10}{3}t$

At the instant the brakes are applied, time $t = 0$

Hence velocity $v = 20$ m/s

$\qquad = \dfrac{20 \times 60 \times 60}{1000}$ km/h

$\qquad = \mathbf{72\ km/h}$

(Note: changing from m/s to km/h merely involves multiplying by 3.6.)

(b)   When the car finally stops, the velocity is zero,

i.e. $v = 20 - \dfrac{10}{3}t = 0$, from which, $20 = \dfrac{10}{3}t$,

giving $t = 6$ s.

Hence the distance travelled before the car stops is given by:

$$x = 20t - \frac{5}{3}t^2 = 20(6) - \frac{5}{3}(6)^2$$

$$= 120 - 60 = \mathbf{60\ m}$$

Problem 7. The angular displacement θ radians of a flywheel varies with time t seconds and follows the equation $\theta = 9t^2 - 2t^3$. Determine (a) the angular velocity and acceleration of the flywheel when time, $t = 1$ s, and (b) the time when the angular acceleration is zero

(a)   Angular displacement $\theta = 9t^2 - 2t^3$ rad

Angular velocity $\omega = \dfrac{d\theta}{dt} = 18t - 6t^2$ rad/s

When time $t = 1$ s,

$\omega = 18(1) - 6(1)^2 = \mathbf{12}$ **rad/s**

Angular acceleration $\alpha = \dfrac{d^2\theta}{dt^2} = 18 - 12t$ rad/s

When time $t = 1$ s,

$\alpha = 18 - 12(1) = \mathbf{6}$ **rad/s**^{2}

(b)   When the angular acceleration is zero, $18 - 12t = 0$, from which, $18 = 12t$, giving time, $t = \mathbf{1.5}$ **s**

Problem 8. The distance x metres moved by a body in t seconds is given by $x = 5t^3 - \dfrac{21}{2}t^2 + 6t - 4$. Determine (a) the initial velocity and the velocity after 3 s, (b) the values of t when the body came to rest, (c) its acceleration after 2 s, (d) the value of t when the acceleration is 24 m/s^2, and (e) the average velocity over the third second

---

(a)   Distance $x = 5t^3 - \dfrac{21}{2}t^2 + 6t - 4$

Velocity $v = \dfrac{dx}{dt} = 15t^2 - 21t + 6$

The initial velocity, i.e.

when $t = 0$, is $v_0 = \mathbf{6}$ **m/s**

Velocity after 3 s,

$v_3 = 15(3)^2 - 21(3) + 6 = 135 - 63 + 6$

$= \mathbf{78}$ **m/s**

(b)   When the body comes to rest, velocity $v = 0$, i.e. $15t^2 - 21t + 6 = 0$.

Rearranging and factorizing gives:

$3(5t^2 - 7t + 2) = 0$ and

$3(5t - 2)(t - 1) = 0$

Hence $(5t - 2) = 0$, from which, $t = \dfrac{2}{5}$ s

or $(t - 1) = 0$, from which, $t = \mathbf{1}$ **s**

(c)   Acceleration $a = \dfrac{d^2x}{dt^2} = 30t - 21$

Acceleration, after 2 s,

$a_2 = 30(2) - 21 = \mathbf{39}$ **m/s**^{2}

(d)   When the acceleration is 24 m/s^2 then $30t - 21 = 24$, from which $t = \dfrac{45}{30} = \mathbf{1.5}$ **s**

(e)   Distance travelled in the third second

$= $ (distance travelled after 3 s)

$-$ (distance travelled after 2 s)

$= \left[5(3)^3 - \dfrac{21}{2}(3)^2 + 6(3) - 4\right]$

$\quad - \left[5(2)^3 - \dfrac{21}{2}(2)^2 + 6(2) - 4\right]$

$= 54.5 - 6 = 48.5$ m

Average velocity over the third second

$= \dfrac{\text{distance travelled}}{\text{time taken}} = \dfrac{48.5 \text{ m}}{1 \text{ s}}$

$= \mathbf{48.5}$ **m/s**

Problem 9. The displacement x cm of the slide valve of an engine is given by $x = 2.2 \cos 5\pi t + 3.6 \sin 5\pi t$. Evaluate the velocity (in m/s) when time $t = 30$ ms

---

Displacement $x = 2.2 \cos 5\pi t + 3.6 \sin 5\pi t$

Velocity $v = \dfrac{dx}{dt} = (2.2)(-5\pi) \sin 5\pi t$

$\quad + (3.6)(5\pi) \cos 5\pi t$

$= -11\pi \sin 5\pi t + 18\pi \cos 5\pi t$ cm/s

When time $t = 30$ ms, velocity

$= -11\pi \sin\left(5\pi \dfrac{30}{10^3}\right) + 18\pi \cos\left(5\pi \dfrac{30}{10^3}\right)$

$= -11\pi \sin 0.4712 + 18\pi \cos 0.4712$

$= -11\pi \sin 27° + 18\pi \cos 27°$

$= -15.69 + 50.39 = 34.7$ cm/s $= \mathbf{0.347}$ **m/s**

Further problems on velocity and acceleration may be found in Section 22.7, Problems 7 to 14, pages 231 and 232.

22.3   Turning points

As stated in Chapter 19, in Fig. 22.1, the gradient (or rate of change) of the curve changes from

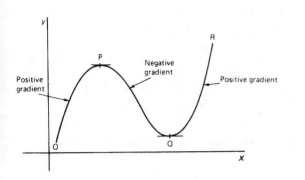

Figure 22.1

positive between O and P to negative between P and Q, and then positive again between Q and R. At point P, the gradient is zero and, as x increases, the gradient of the curve changes from positive just before P to negative just after. Such a point is called a **maximum point** and appears as the 'crest of a wave'. At point Q, the gradient is also zero and, as x increases, the gradient of the curve changes from negative just before Q to positive just after. Such a point is called a **minimum point**, and appears as the 'bottom of a valley'. Points such as P and Q are given the general name of **turning points**.

It is possible to have a turning point, the gradient on either side of which is the same. Such a point is given the special name of a **point of inflexion**, and examples are shown in Fig. 22.2.

Maximum and minimum points and points of inflexion are given the general term of **stationary points**.

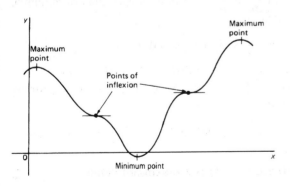

Figure 22.2

Procedure for finding and distinguishing between stationary points

(i)   Given $y = f(x)$, determine $\dfrac{dy}{dx}$ (i.e. $f'(x)$)

(ii)   Let $\dfrac{dy}{dx} = 0$ and solve for the values of x

(iii)   Substitute the values of x into the original equation, $y = f(x)$, to find the corresponding y-ordinate values. This establishes the coordinates of the stationary points

To determine the nature of the stationary points:

Either

(iv)   Find $\dfrac{d^2y}{dx^2}$ and substitute into it the values of x found in (ii).

If the result is:
   (a)   positive–the point is a minimum one,
   (b)   negative–the point is a maximum one,
   (c)   zero–the point is a point of inflexion

or

(v)   Determine the sign of the gradient of the curve just before and just after the stationary points. If the sign change for the gradient of the curve is:
   (a)   positive to negative–the point is a maximum one,
   (b)   negative to positive–the point is a minimum one,
   (c)   positive to positive or negative to negative–the point is a point of inflexion

(Refer to page 195 of Chapter 19 for some initial examples on turning points. Below are some further worked problems.)

> **Problem 10.** Locate the turning point on the curve $y = 3x^2 - 6x$ and determine its nature by examining the sign of the gradient on either side

Following the above procedure:

(i)   Since $y = 3x^2 - 6x$, $\dfrac{dy}{dx} = 6x - 6$

(ii)   At a turning point, $\dfrac{dy}{dx} = 0$. Hence $6x - 6 = 0$, from which, $x = 1$

(iii)   When $x = 1$, $y = 3(1)^2 - 6(1) = -3$
Hence the coordinates of the turning point are $(1, -3)$

(iv)   If x is slightly less than 1, say 0.9, then $\dfrac{dy}{dx} = 6(0.9) - 6 = -0.6$, i.e. negative

If x is slightly greater than 1, say, 1.1, then

$\dfrac{dy}{dx} = 6(1.1) - 6 = 0.6$, i.e. positive

Since the gradient of the curve is negative just before the turning point and positive just after (i.e. $-\vee+$), $(1, -3)$ **is a minimum point**

Problem 11. Determine the coordinates of the maximum and minimum values of the graph $y = \dfrac{x^3}{3} - \dfrac{x^2}{2} - 6x + \dfrac{5}{3}$ and distinguish between them. Sketch the graph

---

Following the given procedure:

(i) Since $y = \dfrac{x^3}{3} - \dfrac{x^2}{2} - 6x + \dfrac{5}{3}$

then $\dfrac{dy}{dx} = x^2 - x - 6$

(ii) At a turning point, $\dfrac{dy}{dx} = 0$.

Hence $x^2 - x - 6 = 0$,

i.e. $(x + 2)(x - 3) = 0$,

from which $x = -2$ or $x = 3$

(iii) When $x = -2$,

$y = \dfrac{(-2)^3}{3} - \dfrac{(-2)^2}{2} - 6(-2) + \dfrac{5}{3} = 9$

When $x = 3$,

$y = \dfrac{(3)^3}{3} - \dfrac{(3)^2}{2} - 6(3) + \dfrac{5}{3} = -11\dfrac{5}{6}$

Thus the coordinates of the turning points are $(-2, 9)$ and $\left(3, -11\dfrac{5}{6}\right)$

(iv) Since $\dfrac{dy}{dx} = x^2 - x - 6$ then $\dfrac{d^2y}{dx^2} = 2x - 1$

When $x = -2$, $\dfrac{d^2y}{dx^2} = 2(-2) - 1 = -5$, which is negative

Hence $(-2, 9)$ is a maximum point

When $x = 3$, $\dfrac{d^2y}{dx^2} = 2(3) - 1 = 5$, which is positive

Hence $\left(3, -11\dfrac{5}{6}\right)$ is a minimum point

Knowing $(-2, 9)$ is a maximum point (i.e. crest of a wave), and $\left(3, -11\dfrac{5}{6}\right)$ is a

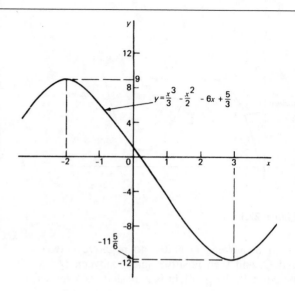

Figure 22.3

minimum point (i.e. bottom of a valley) and that when $x = 0$, $y = \dfrac{5}{3}$, a sketch may be drawn as shown in Fig. 22.3

Problem 12. Locate the turning point on the following curve and determine whether it is a maximum or minimum point: $y = 3(\ln \theta - \theta)$

Since $y = 3(\ln \theta - \theta) = 3 \ln \theta - 3\theta$

then $\dfrac{dy}{d\theta} = \dfrac{3}{\theta} - 3 = 0$, for a maximum or minimum value.

Hence $\dfrac{3}{\theta} = 3$, $3 = 3\theta$ and $\theta = 1$

When $\theta = 1$, $y = 3(\ln 1 - 1) = 3(0 - 1) = -3$

Hence $(1, -3)$ are the coordinates of the turning point.

$\dfrac{d^2y}{d\theta^2} = -\dfrac{3}{\theta^2}$. When $\theta = 1$, $\dfrac{d^2y}{d\theta^2} = -3$, which is negative

Hence $(1, -3)$ is a maximum value.

Problem 13. Determine the turning points on the curve $y = 4\sin x - 3\cos x$ in the range $x = 0$ to $x = 2\pi$ radians, and distinguish between them. Sketch the curve over one cycle

---

Since $y = 4\sin x - 3\cos x$ then $\dfrac{dy}{dx} = 4\cos x + 3\sin x = 0$, for a turning point from which, $4\cos x = -3\sin x$ and $\dfrac{-4}{3} = \dfrac{\sin x}{\cos x} = \tan x$

Hence $x = \arctan\left(\dfrac{-4}{3}\right) = 126°52'$ or $306°52'$, since tangent is negative in the second and fourth quadrants.

When $x = 126°52'$,
$$y = 4\sin 126°52' - 3\cos 126°52' = 5$$
When $x = 306°52'$,
$$y = 4\sin 306°52' - 3\cos 306°52' = -5$$

$$126°52' = \left(125°52' \times \dfrac{\pi}{180}\right) \text{ radians} = 2.214 \text{ rad}$$

$$306°52' = \left(306°52' \times \dfrac{\pi}{180}\right) \text{ radians} = 5.356 \text{ rad}$$

Hence $(2.214, 5)$ and $(5.356, -5)$ are the coordinates of the turning points

$$\dfrac{d^2y}{dx^2} = -4\sin x + 3\cos x$$

When $x = 2.214$ rad, $\dfrac{d^2y}{dx^2} = -4\sin 2.214 + 3\cos 2.214$, which is negative.

Hence $(2.214, 5)$ is a maximum point.

When $x = 5.356$ rad, $\dfrac{d^2y}{dx^2} = -4\sin 5.356 + 3\cos 5.356$, which is positive.

Hence $(5.356, -5)$ is a minimum point.

A sketch of $y = 4\sin x - 3\cos x$ is shown in Fig. 22.4.

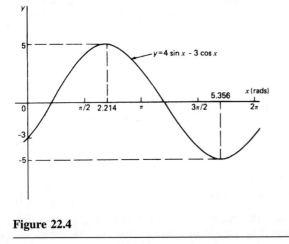

Figure 22.4

Further problems on turning points may be found in Section 22.7, Problems 15 to 23, page 232.

22.4 Practical problems involving maximum and minimum values

There are many **practical problems** involving maximum and minimum values which occur in science and engineering. Usually, an equation has to be determined from given data, and rearranged where necessary, so that it contains only one variable. Some examples are demonstrated in Problems 14 to 20.

Problem 14. A rectangular area is formed having a perimeter of 40 cm. Determine the length and breadth of the rectangle if it is to enclose the maximum possible area

---

Let the dimensions of the rectangle be x and y. Then the perimeter of the rectangle is $(2x + 2y)$. Hence

$$2x + 2y = 40, \text{ or } x + y = 20 \qquad (1)$$

Since the rectangle is to enclose the maximum possible area, a formula for area A must be obtained in terms of one variable only.

Area $A = xy$. From equation (1), $x = 20 - y$

Hence, area $A = (20 - y)y = 20y - y^2$

$\dfrac{dA}{dy} = 20 - 2y = 0$ for a turning point, from which,

$y = 10$ cm

$\dfrac{d^2A}{dy^2} = -2$, which is negative, giving a maximum point

When $y = 10$ cm, $x = 10$ cm, from equation (1)

Hence the length and breadth of the rectangle are each 10 cm, i.e. a square gives the maximum possible area. When the perimeter of a rectangle is 40 cm, the maximum possible area is $10 \times 10 = $ **100 cm^2**.

Problem 15. A rectangular sheet of metal having dimensions 20 cm by 12 cm has squares removed from each of the four corners and the sides bent upwards to form an open box. Determine the maximum possible volume of the box

---

The squares to be removed from each corner are shown in Fig. 22.5, having sides x cm. When the sides are bent upwards the dimensions of the box will be: length $(20 - 2x)$ cm, breadth $(12 - 2x)$ cm and height x cm.

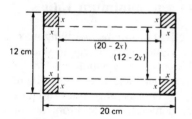

Figure 22.5

Volume of box, $V = (20 - 2x)(12 - 2x)(x) = 240x - 64x^2 + 4x^3$

$\dfrac{dV}{dx} = 240 - 128x + 12x^2 = 0$ for a turning point.

Hence $4(60 - 32x + 3x^2) = 0$, i.e. $3x^2 - 32x + 60 = 0$.

Using the quadratic formula,

$$x = \frac{32 \pm \sqrt{[(-32)^2 - 4(3)(60)]}}{(2)(3)} = 8.239 \text{ cm}$$

or 2.427 cm. Since the breadth is $(12 - 2x)$ cm then $x = 8.239$ cm is not possible and is neglected. Hence $x = 2.427$ cm.

$\dfrac{d^2V}{dx^2} = -128 + 24x$. When $x = 2.427$, $\dfrac{d^2V}{dx^2}$ is negative, giving a maximum value.

The dimensions of the box are:

length $= 20 - 2(2.427) = 15.146$ cm,

breadth $= 12 - 2(2.427) = 7.146$ cm,

height $= 2.427$ cm

Maximum volume $= (15.146)(7.146)(2.427) =$ **262.7 cm^3**.

Problem 16.   Determine the height and radius of a cylinder of volume 200 cm^3 which has the least surface area

Let the cylinder have radius r and perpendicular height h.

Volume of cylinder, $V = \pi r^2 h = 200$ (1)

Surface area of cylinder, $A = 2\pi rh + 2\pi r^2$

Least surface area means minimum surface area and a formula for the surface area in terms of one variable only is required.

From equation (1), $h = \dfrac{200}{\pi r^2}$ (2)

Hence surface area,

$$A = 2\pi r \left(\frac{200}{\pi r^2}\right) + 2\pi r^2 = \frac{400}{r} + 2\pi r^2$$

$$= 400r^{-1} + 2\pi r^2$$

$\dfrac{dA}{dr} = \dfrac{-400}{r^2} + 4\pi r = 0$, for a turning point.

Hence $4\pi r = \dfrac{400}{r^2}$ and $r^3 = \dfrac{400}{4\pi}$, from which,

$$r = \sqrt[3]{\left(\frac{100}{\pi}\right)} = 3.169 \text{ cm}.$$

$\dfrac{d^2A}{dr^2} = \dfrac{800}{r^3} + 4\pi$. When $r = 3.169$ cm, $\dfrac{d^2A}{dr^2}$ is positive, giving a minimum value.

From equation (2), when $r = 3.169$ cm,

$$h = \frac{200}{\pi(3.169)^2} = 6.339 \text{ cm}$$

Hence for the least surface area, a cylinder of volume 200 cm^3 has a radius of 3.169 cm and height of 6.339 cm.

Problem 17.   Determine the area of the largest piece of rectangular ground that can be enclosed by 100 m of fencing, if part of an existing straight wall is used as one side

Let the dimensions of the rectangle be x and y as shown in Fig. 22.6, where PQ represents the straight wall.

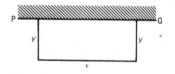

Figure 22.6

From Fig. 22.6, $x + 2y = 100$ (1)

Area of rectangle, $A = xy$ (2)

Since the maximum area is required, a formula for area A is needed in terms of one variable only.

From equation (1), $x = 100 - 2y$

Hence area $A = xy = (100 - 2y)y = 100y - 2y^2$

$\dfrac{dA}{dy} = 100 - 4y = 0$, for a turning point, from which,

$y = 25$ m

$\dfrac{d^2A}{dy^2} = -4$, which is negative, giving a maximum value

When $y = 25$ m, $x = 50$ m from equation (1)

Hence the maximum possible area

$$= xy = (50)(25) = \textbf{1250 m}^2$$

Problem 18.   Determine the height and radius of a cylinder of maximum volume which can be cut from a cone of height of 30 cm and base radius 7.5 cm

---

A cylinder of base radius r and height h is shown enclosed in a cone of height 30 cm and base radius 7.5 cm in Fig. 22.7.

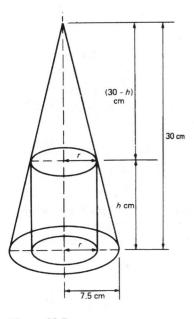

Figure 22.7

Volume of cylinder, $V = \pi r^2 h$ (1)

By similar triangles, $\dfrac{30 - h}{r} = \dfrac{30}{7.5}$ (2)

Since the maximum volume is required a formula for the volume V must be obtained in terms of one variable only.

From equation (2), $\dfrac{30 - h}{r} = 4$, from which,

$$30 - h = 4r \text{ and } h = 30 - 4r \tag{3}$$

Substituting for h in equation (1) gives:

$$V = \pi r^2 (30 - 4r) = 30\pi r^2 - 4\pi r^3$$

$\dfrac{dV}{dr} = 60\pi r - 12\pi r^2 = 0$, for a maximum or minimum value

Hence $12\pi r(5 - r) = 0$, from which $r = 0$ or $r = 5$ cm

$$\frac{d^2V}{dr^2} = 60\pi - 24\pi r$$

When $r = 0$, $\dfrac{d^2V}{dr^2}$ is positive, giving a minimum value (obviously)

When $r = 5$, $\dfrac{d^2V}{dr^2}$ is negative, giving a maximum value

From equation (3), height $h = 30 - 4r = 30 - 4(5) = 10$ cm

Hence the cylinder of maximum volume which can be cut from a cone of height 30 cm and radius 7.5 cm, has a height of 10 cm and a radius of 5 cm.

Problem 19.   An open rectangular box with square ends is fitted with an overlapping lid which covers the top and the front face. Determine the maximum volume of the box if 6 m^2 of metal are used in its construction

---

A rectangular box having square ends of side x and length y is shown in Fig. 22.8.

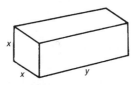

Figure 22.8

Surface area of box, A, consists of two ends and five faces (since the lid also covers the front face).

Hence $A = 2x^2 + 5xy = 6$ (1)

Since it is the maximum volume required, a formula for the volume in terms of one variable only is needed. Volume of box, $V = x^2 y$.

From equation (1), $y = \dfrac{6 - 2x^2}{5x} = \dfrac{6}{5x} - \dfrac{2x}{5}$ (2)

Hence volume $V = x^2 y = x^2 \left(\dfrac{6}{5x} - \dfrac{2x}{5} \right)$

$$= \dfrac{6x}{5} - \dfrac{2x^3}{5}$$

$\dfrac{dV}{dx} = \dfrac{6}{5} - \dfrac{6x^2}{5} = 0$ for a maximum or minimum value

Hence $6 = 6x^2$, giving $x = 1$ m ($x = -1$ is not possible, and is thus neglected)

$\dfrac{d^2V}{dx^2} = \dfrac{-12x}{5}$. When $x = 1$, $\dfrac{d^2V}{dx^2}$ is negative, giving a maximum value

From equation (2), when $x = 1$,

$$y = \dfrac{6}{5(1)} - \dfrac{2(1)}{5} = \dfrac{4}{5}$$

Hence the maximum volume of the box is given by

$$V = x^2 y = (1)^2 \left(\dfrac{4}{5} \right) = \dfrac{4}{5} \text{ m}^3$$

Problem 20.   Find the diameter and height of a cylinder of maximum volume which can be cut from a sphere of radius 12 cm

---

A cylinder of radius r and height h is shown enclosed in a sphere of radius $R = 12$ cm in Fig. 22.9.

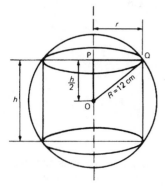

Figure 22.9

Volume of cylinder, $V = \pi r^2 h$ (1)

Using the right-angled triangle OPQ shown in Fig. 22.9,

$$r^2 + \left(\dfrac{h}{2} \right)^2 = R^2 \text{ by Pythagoras' theorem}$$

i.e. $r^2 + \dfrac{h^2}{4} = 144$ (2)

Since the maximum volume is required, a formula for the volume V is required in terms of one variable only.

From equation (2), $r^2 = 144 - \dfrac{h^2}{4}$

Substituting into equation (1) gives:

$$V = \pi \left(144 - \dfrac{h^2}{4} \right) h = 144\pi h - \dfrac{\pi h^3}{4}$$

$\dfrac{dV}{dh} = 144\pi - \dfrac{3\pi h^2}{4} = 0$, for a maximum or minimum value

Hence $144\pi = \dfrac{3\pi h^2}{4}$, from which,

$$h = \sqrt{\dfrac{(144)(4)}{3}} = 13.86 \text{ cm}$$

$\dfrac{d^2V}{dh^2} = \dfrac{-6\pi h}{4}$. When $h = 13.86$, $\dfrac{d^2V}{dh^2}$ is negative, giving a maximum value.

From equation (2), $r^2 = 144 - \dfrac{h^2}{4} = 144 - \dfrac{(13.86)^2}{4}$, from which, radius $r = 9.80$ cm

Diameter of cylinder $= 2r = 2(9.80) = 19.60$ cm

Hence the cylinder having the maximum volume that can be cut from a sphere of radius 12 cm is one in which the diameter is 19.60 cm and the height is 13.86 cm.

Further problems on practical maximum and minimum problems may be found in Section 22.7, Problems 24 to 35, pages 232 and 233.

22.5   Tangents and normals

Tangents

The equation of the tangent to a curve $y = f(x)$ at the points (x_1, y_1) is given by:

$$y - y_1 = m(x - x_1)$$

where $m = \dfrac{dy}{dx} =$ gradient of the curve at (x_1, y_1).

Problem 21. Find the equation of the tangent to the curve $y = x^2 - x - 2$ at the point $(1, -2)$

Gradient, $m = \dfrac{dy}{dx} = 2x - 1$

At the point $(1, -2)$,

$$x = 1 \text{ and } m = 2(1) - 1 = 1$$

Hence the equation of the tangent is:

$$y - y_1 = m(x - x_1)$$

i.e. $\quad y - -2 = 1(x - 1)$

i.e. $\quad y + 2 = x - 1$

or $\qquad y = x - 3$

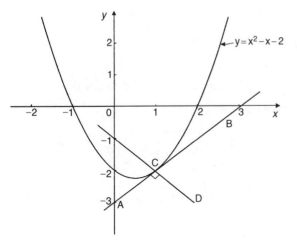

Figure 22.10

The graph of $y = x^2 - x - 2$ is shown in Fig. 22.10. The line AB is the tangent to the curve at the point C, i.e. $(1, -2)$, and the equation of this line is $y = x - 3$.

Normals

The normal at any point on a curve is the line which passes through the point and is at right angles to the tangent. Hence, in Fig. 22.10, the line CD is the normal.

It may be shown that if two lines are at right angles then the product of their gradients is -1. Thus if m is the gradient of the tangent, then the gradient of the normal is $-\dfrac{1}{m}$.

Hence the equation of the normal at the point (x_1, y_1) is given by:

$$y - y_1 = -\frac{1}{m}(x - x_1)$$

Problem 22. Find the equation of the normal to the curve $y = x^2 - x - 2$ at the point $(1, -2)$

$m = 1$ from Problem 21, hence the equation of the normal is

$$y - y_1 = -\frac{1}{m}(x - x_1)$$

i.e. $\quad y - -2 = -\dfrac{1}{1}(x - 1)$

i.e. $\quad y + 2 = -x + 1$

or $\qquad y = -x - 1$

Thus the line CD in Fig. 22.10 has the equation $y = -x - 1$.

Problem 23. Determine the equations of the tangent and normal to the curve $y = \dfrac{x^3}{5}$ at the point $\left(-1, -\dfrac{1}{5}\right)$

Gradient m of curve $y = \dfrac{x^3}{5}$ is given by

$$m = \frac{dy}{dx} = \frac{3x^2}{5}$$

At the point $\left(-1, -\dfrac{1}{5}\right)$,

$$x = -1 \text{ and } m = \frac{3(-1)^2}{5} = \frac{3}{5}$$

Equation of the tangent is:

$$y - y_1 = m(x - x_1)$$

i.e. $\quad y - -\dfrac{1}{5} = \dfrac{3}{5}(x - -1)$

i.e. $y + \dfrac{1}{5} = \dfrac{3}{5}(x+1)$

or $5y + 1 = 3x + 3$

or $\mathbf{5y - 3x = 2}$

Equation of the normal is:

$$y - y_1 = -\dfrac{1}{m}(x - x_1)$$

i.e. $y - -\dfrac{1}{5} = \dfrac{-1}{\left(\dfrac{3}{5}\right)}(x - -1)$

i.e. $y + \dfrac{1}{5} = -\dfrac{5}{3}(x+1)$

i.e. $y + \dfrac{1}{5} = -\dfrac{5}{3}x - \dfrac{5}{3}$

Multiplying each term by 15 gives:

$$15y + 3 = -25x - 25$$

Hence **equation of the normal** is:

$$\mathbf{15y + 25x + 28 = 0}$$

Further problems on tangents and normals may be found in Section 22.7, Problems 36 to 40, page 233.

22.6   Small changes

If y is a function of x, i.e. $y = f(x)$, and the approximate change in y corresponding to a small change δx in x is required, then:

$$\dfrac{\delta y}{\delta x} \approx \dfrac{dy}{dx}$$

and $\delta y \approx \dfrac{dy}{dx} \cdot \delta x$ or $\delta y \approx f'(x) \cdot \delta x$

> **Problem 24.** Given $y = 4x^2 - x$, determine the approximate change in y if x changes from 1 to 1.02

Since $y = 4x^2 - x$, then $\dfrac{dy}{dx} = 8x - 1$

Approximate change in y, $\delta y \approx \dfrac{dy}{dx} \cdot \delta x \approx (8x-1)\, \delta x$

When $x = 1$ and $\delta x = 0.02$, $\delta y \approx (8(1)-1)(0.02) \approx$ **0.14**

(Obviously, in this case, the exact value of δy may be obtained by evaluating y when $x = 1.02$, i.e. $y = 4(1.02)^2 - 1.02 = 3.1416$ and then subtracting from it the value of y when $x = 1$, i.e. $y = 4(1)^2 - 1 = 3$, giving $\delta y = 3.1416 - 3 = \mathbf{0.1416}$. Using $\delta y = \dfrac{dy}{dx} \cdot \delta x$ above gave 0.14, which shows that the formula gives the approximate change in y for a small change in x.)

> **Problem 25.** The time of swing T of a pendulum is given by $T = k\sqrt{1}$, where k is a constant. Determine the percentage change in the time of swing if the length of the pendulum l changes from 32.1 cm to 32.0 cm

If $T = k\sqrt{l} = kl^{1/2}$, then

$$\dfrac{dT}{dl} = k\left(\dfrac{1}{2}l^{-1/2}\right) = \dfrac{k}{2\sqrt{l}}$$

Approximate change in T,

$$\delta t \approx \dfrac{dT}{dl}\delta l \approx \left(\dfrac{k}{2\sqrt{l}}\right)\delta l$$

$$\approx \left(\dfrac{k}{2\sqrt{l}}\right)(-0.1)$$

(negative since l decreases)

Percentage error

$$= \left(\dfrac{\text{approximate change in } T}{\text{original value of } T}\right)100\%$$

$$= \dfrac{\left(\dfrac{k}{2\sqrt{l}}\right)(-0.1)}{k\sqrt{l}} \times 100\%$$

$$= \left(\dfrac{-0.1}{2l}\right)100\%$$

$$= \left(\dfrac{-0.1}{2(32.1)}\right)100\% = \mathbf{-0.156\%}$$

Hence the percentage change in the time of swing is 0.156% less.

> **Problem 26.** A circular template has a radius of 10 cm (± 0.02). Determine the possible error in calculating the area of the template. Find also the percentage error

Area of circular template, $A = \pi r^2$,

hence $\dfrac{dA}{dr} = 2\pi r$

Approximate change in area,

$$\delta A \approx \dfrac{dA}{dr} \cdot \delta r \approx (2\pi r)\,\delta r$$

When $r = 10$ cm and $\delta r = 0.02$,

$$\delta A = (2\pi 10)(0.02) \approx 0.4\pi \text{ cm}^2$$

i.e.

the possible error in calculating the template area is approximately 1.257 cm^2.

Percentage error $\approx \left(\dfrac{0.4\pi}{\pi(10)^2}\right) 100\% = \mathbf{0.40\%}$

Further problems on small changes may be found in the following Section 22.7, Problems 41 to 45, page 233.

22.7 Further problems on applications of differentiation

Rates of change

1 The length, l metres, of a rod of metal at temperature $\theta\,°$C is given by $l = 1 + 2 \times 10^{-4}\theta + 3 \times 10^{-6}\theta^2$. Determine the rate of change of l, in cm/°C, when the temperature is (a) $100\,°$C (b) $200\,°$C.

$$[(a)\ 0.08\ \text{cm/°C}\ (b)\ 0.14\ \text{cm/°C}]$$

2 An alternating current, i amperes, is given by $i = 10\sin 2\pi f t$, where f is the frequency in hertz and t the time in seconds. Determine the rate of change of current when $t = 20$ ms, given that $f = 150$ Hz. $\qquad$ $[3000\pi\ \text{A/s}]$

3 The luminous intensity, I candelas, of a lamp is given by $I = 6 \times 10^{-4}V^2$, where V is the voltage. Find (a) the rate of change of luminous intensity with voltage when $V = 200$ volts, and (b) the voltage at which the light is increasing at a rate of 0.3 candelas per volt.

$$[(a)\ 0.24\ \text{cd/V}\ (b)\ 250\ \text{V}]$$

4 The voltage across the plates of a capacitor at any time t seconds is given by $v = Ve^{-t/CR}$, where V, C and R are constants. Given $V = 300$ volts, $C = 0.12 \times 10^{-6}$ farads and $R = 4 \times 10^6$ ohms find (a) the initial rate of change of voltage, and (b) the rate of change of voltage after 0.5 s.

$$[(a)\ -625\ \text{V/s}\ (b)\ -220.5\ \text{V/s}]$$

5 The pressure p of the atmosphere at height h above ground level is given by $p = p_0 e^{-h/c}$, where p_0 is the pressure at ground level and c is a constant. Determine the rate of change of pressure with height when $p_0 = 1.013 \times 10^5$ pascals and $c = 6.05 \times 10^4$ at 1450 metres.

$$[-1.635\ \text{Pa/m}]$$

6 The displacement s cm of the end of a stiff spring at time t seconds is given by $s = ae^{-kt}\sin 2\pi f t$. Find the velocity of the end of the spring after 2 s, if $a = 3$, $k = 0.8$ and $f = 10$. $\qquad$ $[38.06\ \text{cm/s}]$

Velocity and acceleration

7 The distance x metres moved by a body in a time t seconds is given by $x = 3t^3 - 4t^2 + 6t - 5$. Determine the velocity and acceleration when (a) $t = 0$ and (b) $t = 2.5$ s.

$$\left[\begin{array}{l} (a)\ v = 6\ \text{m/s},\ a = -8\ \text{m/s}^2 \\ (b)\ v = 42\tfrac{1}{4}\ \text{m/s},\ a = 37\ \text{m/s}^2 \end{array}\right]$$

8 The distance s metres travelled by a car in t seconds after the brakes are applied is given by $s = 25t - 2.5t^2$. Find (a) the speed of the car (in km/h) when the brakes are applied, and (b) the distance the car travels before it stops.

$$[(a)\ 90\ \text{km/h}\ (b)\ 62.5\ \text{m}]$$

9 The equation $\theta = 10\pi + 24t - 3t^2$ gives the angle θ, in radians, through which a wheel turns in t seconds. Determine (a) the time the wheel takes to come to rest, and (b) the angle turned through in the last second of movement.

$$[(a)\ 4\ \text{s}\ (b)\ 3\ \text{rads}]$$

10 At any time t seconds the distance x metres of a particle moving in a straight line from a fixed point is given by $x = 4t + \ln(1 - t)$. Determine (a) the initial velocity and acceleration, (b) the velocity and acceleration after 1.5 s, and (c) the time when the velocity is zero.

$$\left[\begin{array}{l} (a)\ 3\ \text{m/s};\ -1\ \text{m/s}^2 \\ (b)\ 6\ \text{m/s};\ -4\ \text{m/s}^2 \\ (c)\ \dfrac{3}{4}\ \text{s} \end{array}\right]$$

11 A missile fired from ground level rises x metres vertically upwards in t seconds and $x = 100t - \dfrac{25}{2}t^2$. Find (a) the initial velocity of the missile, (b) the time when the height of the missile is a maximum, (c) the maximum height reached,

and (d) the velocity with which the missile strikes the ground.

$$\begin{bmatrix} \text{(a) 100 m/s (b) 4 s} \\ \text{(c) 200 m (d) } -100 \text{ m/s} \end{bmatrix}$$

12 The distance, in metres, of a body moving in a straight line from a fixed point on the line is given by $x = t^4 - \dfrac{8t^3}{3} + \dfrac{3t^2}{2} - 6$, where t is the time in seconds. Determine the velocity and acceleration after 2 seconds. Find also the time when the body is momentarily at rest.

$$\left[6 \text{ m/s, } 19 \text{ m/s}^2, t = 0, \frac{1}{2} \text{ or } 1\frac{1}{2} \text{ s} \right]$$

13 The angular displacement θ of a rotating disc is given by $\theta = 6 \sin \dfrac{t}{4}$, where t is the time in seconds. Determine (a) the angular velocity of the disc when t is 1.5 s, (b) the angular acceleration when t is 5.5 s, and (c) the first time when the angular velocity is zero.

$$\begin{bmatrix} \text{(a) } \omega = 1.40 \text{ rad/s} \\ \text{(b) } \alpha = -0.37 \text{ rad/s}^2 \\ \text{(c) } t = 6.28 \text{ s} \end{bmatrix}$$

14 $x = \dfrac{20t^3}{3} - \dfrac{23t^2}{2} + 6t + 5$ represents the distance, x metres, moved by a body in t seconds. Determine (a) the velocity and acceleration at the start, (b) the velocity and acceleration when $t = 3$ s, (c) the values of t when the body is at rest, (d) the value of t when the acceleration is 37 m/s^2, and (e) the distance travelled in the third second.

$$\begin{bmatrix} \text{(a) 6 m/s; } -23 \text{ m/s}^2 \\ \text{(b) 117 m/s; 97 m/s}^2 \\ \text{(c) } \frac{3}{4} \text{ s or } \frac{2}{5} \text{ s (d) } 1\frac{1}{2} \text{ s} \\ \text{(e) } 75\frac{1}{6} \text{ m} \end{bmatrix}$$

Turning points

15 (a) Define a turning point of a graph
   (b) Find the coordinates of the turning points on the curve $y = 4x^3 + 3x^2 - 60x - 12$. Distinguish between the turning points by examining the sign of the gradient on either side of the points

$$\begin{bmatrix} \text{(b) Minimum (2, } -88); \\ \text{Maximum } \left(-2\frac{1}{2}, 94\frac{1}{4} \right) \end{bmatrix}$$

In Problems 16 to 21, find the turning points and distinguish between them:

16 $y = 3x^2 - 4x + 2$ $\left[\text{Minimum at } \left(\dfrac{2}{3}, \dfrac{2}{3} \right) \right]$

17 $x = \theta(6 - \theta)$ [Maximum at (3, 9)]

18 $y = 5x - 2 \ln x$

[Minimum at (0.4000, 3.8326)]

19 $y = 2x - e^x$ [Maximum at (0.6931, -0.6137)]

20 $y = t^3 - \dfrac{t^2}{2} - 2t + 4$

$$\begin{bmatrix} \text{Minimum at } \left(1, 2\frac{1}{2} \right); \\ \text{Maximum at } \left(-\dfrac{2}{3}, 4\dfrac{22}{27} \right) \end{bmatrix}$$

21 $x = 8t + \dfrac{1}{2t^2}$ $\left[\text{Minimum at } \left(\dfrac{1}{2}, 6 \right) \right]$

22 Determine the maximum and minimum values on the graph $y = 12 \cos \theta - 5 \sin \theta$ in the range $x = 0$ to $x = 360°$. Sketch the graph over one cycle showing relevant points.

$$\begin{bmatrix} \text{Maximum of 13 at } 337°23', \\ \text{Minimum of } -13 \text{ at } 157°23' \end{bmatrix}$$

23 Show that the curve $y = \dfrac{2}{3}(t - 1)^3 + 2t(t - 2)$ has a maximum value of $\dfrac{2}{3}$ and a minimum value of -2. What is a point of inflexion?

Practical maximum and minimum problems

24 The speed, v, of a car (in m/s) is related to time t s by the equation $v = 3 + 12t - 3t^2$. Determine the maximum speed of the car in km/h.

[54 km/h]

25 Determine the maximum area of a rectangular piece of land that can be enclosed by 1200 m of fencing.   [90 000 m^2]

26 A shell is fired vertically upwards and its vertical height, x metres, is given by $x = 24t - 3t^2$, where t is the time in seconds. Determine the maximum height reached.   [48 m]

27 A lidless box with square ends is to be made from a thin sheet of metal. Determine the least area of the metal for which the volume of the box is 3.5 m^3.   [11.42 m^2]

28   A closed cylindrical container has a surface area of 400 cm². Determine the dimensions for maximum volume.

[radius = 4.607 cm; height = 9.212 cm]

29   Calculate the height of a cylinder of maximum volume which can be cut from a cone of height 20 cm and base radius 80 cm.   [6.67 cm]

30   The power developed in a resistor R by a battery of emf E and internal resistance r is given by $P = \dfrac{E^2R}{(R+r)^2}$. Differentiate P with respect to R and show that the power is a maximum when $R = r$.

31   Find the height and radius of a closed cylinder of volume 125 cm³ which has the least surface area.   [height = 5.42 cm; radius = 2.71 cm]

32   Resistance to motion, F, of a moving vehicle, is given by $F = \dfrac{5}{x} + 100x$. Determine the minimum value of resistance.   [44.72]

33   A right cylinder of maximum volume is to be cut from a sphere of radius 15 cm. Determine the base diameter and height of the cylinder.

[diameter = 24.49 cm; height = 17.32 cm]

34   An electrical voltage E is given by $E = (15 \sin 50\pi t + 40 \cos 50\pi t)$ volts, where t is the time in seconds. Determine the maximum value of voltage.   [42.72 volts]

35   A rectangular box with a lid which covers the top and front has a volume of 120 cm³ and the length of the box is to be one and a half times the height. Determine the dimensions of the box so that the surface area shall be a minimum.

[6.376 cm by 5.313 cm by 3.542 cm]

Tangents and normals

For the curves in Problems 36 to 40, at the points given, find (a) the equation of the tangent, and (b) the equation of the normal:

36   $y = 2x^2$ at the point (1, 2).

[(a) $y = 4x - 2$ (b) $4y + x = 9$]

37   $y = 3x^2 - 2x$ at the point (2, 8).

[(a) $y = 10x - 12$ (b) $10y + x = 82$]

38   $y = \dfrac{x^3}{2}$ at the point $\left(-1, -\dfrac{1}{2}\right)$.

$\left[\text{(a) } y = \dfrac{3}{2}x + 1 \text{ (b) } 6y + 4x + 7 = 0\right]$

39   $y = 1 + x - x^2$ at the point $(-2, -5)$.

[(a) $y = 5x + 5$ (b) $5y + x + 27 = 0$]

40   $\theta = \dfrac{1}{t}$ at the point $\left(3, \dfrac{1}{3}\right)$.

$\left[\begin{array}{l}\text{(a) } 9\theta + t = 6 \\ \text{(b) } \theta = 9t - 26\dfrac{2}{3} \text{ or } 3\theta = 27t - 80\end{array}\right]$

Small changes

41   If $y = 3x^2$ find the approximate change in y if x changes from 5.0 to 5.02.   [0.6]

42   Determine the change in y if x changes from 2.50 to 2.51 when (a) $y = 2x - x^2$ (b) $y = \dfrac{5}{x}$.

[(a) −0.03 (b) −0.008]

43   The pressure p and volume v of a mass of gas are related by the equation $pv = 50$. If the pressure increases from 25.0 to 25.4, determine the approximate change in the volume of the gas. Find also the percentage change in the volume of the gas.   [−0.032, −1.6%]

44   Determine the approximate increase in (a) the volume, and (b) the surface area of a cube of side x cm if x increases from 20.0 cm to 20.05 cm.   [(a) 60 cm³ (b) 12 cm²]

45   The radius of a sphere decreases from 6.0 cm to 5.96 cm. Determine the approximate change in (a) the surface area, and (b) the volume.

[(a) −6.03 cm² (b) −18.10 cm³]

23

Introduction to integration

23.1 The process of integration

The process of integration reverses the process of differentiation. In differentiation, if $f(x) = 2x^2$ then $f'(x) = 4x$. Thus the integral of $4x$ is $2x^2$, i.e. integration is the process of moving from $f'(x)$ to $f(x)$. By similar reasoning, the integral of $2t$ is t^2.

Integration is a process of summation or adding parts together and an elongated S, shown as $\int$, is used to replace the words 'the integral of'. Hence, from above, $\int 4x = 2x^2$ and $\int 2t$ is t^2.

In differentiation, the differential coefficient dy/dx indicates that a function of x is being differentiated with respect to x, the dx indicating that it is 'with respect to x'. In integration the variable of integration is shown by adding d (the variable) after the function to be integrated.

> Thus $\int 4x\,dx$ means 'the integral of $4x$ with respect to x',
> and $\int 2t\,dt$ means 'the integral of $2t$ with respect to t'

As stated above, the differential coefficient of $2x^2$ is $4x$, hence $\int 4x\,dx = 2x^2$. However, the differential coefficient of $2x^2 + 7$ is also $4x$. Hence $\int 4x\,dx$ could also be equal to $2x^2 + 7$. To allow for the possible presence of a constant, whenever the process of integration is performed, a constant 'c' is added to the result.

Thus $\int 4x\,dx = 2x^2 + c$ and $\int 2t\,dt = t^2 + c$

'c' is called the **arbitrary constant of integration**.

23.2 The general solution of integrals of the form ax^n

The general solution of integrals of the form $\int ax^n\,dx$, where a and n are constants, is given by:

$$\int ax^n\,dx = \frac{ax^{n+1}}{n+1} + c$$

This rule is true when n is fractional, zero, or a positive or negative integer, with the exception of $n = -1$.

Using this rule gives:

(i) $\displaystyle\int 3x^4\,dx = \frac{3x^{4+1}}{4+1} + c = \frac{3}{5}x^5 + c$

(ii) $\displaystyle\int \frac{2}{x^2}\,dx = \int 2x^{-2}\,dx = \frac{2x^{-2+1}}{-2+1} + c$

$\displaystyle = \frac{2x^{-1}}{-1} + c = \frac{-2}{x} + c$, and

(iii) $\displaystyle\int \sqrt{x}\,dx = \int x^{1/2}\,dx = \frac{x^{(1/2)+1}}{\frac{1}{2}+1} + c$

$\displaystyle = \frac{x^{3/2}}{\frac{3}{2}} + c = \frac{2}{3}\sqrt{x^3} + c$

Each of these three results may be checked by differentiation.

(a) The integral of a constant k is $kx + c$. For example, $\displaystyle\int 8\,dx = 8x + c$

(b) When a sum of several terms is integrated the result is the sum of the integrals of the separate terms. For example,

$$\int (3x + 2x^2 - 5)\,dx = \int 3x\,dx + \int 2x^2\,dx$$

$$- \int 5\,dx$$

$$= \frac{3x^2}{2} + \frac{2x^3}{3} - 5x + c$$

23.3 Standard integrals

Since integration is the reverse process of differentiation the **standard integrals** listed in Table 23.1 may be deduced and readily checked by differentiation.

Table 23.1 Standard integrals

(i) $\displaystyle\int ax^n\,dx \qquad = \dfrac{ax^{n+1}}{n+1} + c$

(except when $n = -1$)

(ii) $\displaystyle\int \cos ax\,dx \qquad = \dfrac{1}{a}\sin ax + c$

(iii) $\displaystyle\int \sin ax\,dx \qquad = -\dfrac{1}{a}\cos ax + c$

(iv) $\displaystyle\int \sec^2 ax\,dx \qquad = \dfrac{1}{a}\tan ax + c$

(v) $\displaystyle\int \operatorname{cosec}^2 ax\,dx \qquad = -\dfrac{1}{a}\cot ax + c$

(vi) $\displaystyle\int \operatorname{cosec} ax \cot ax\,dx \ = -\dfrac{1}{a}\operatorname{cosec} ax + c$

(vii) $\displaystyle\int \sec ax \tan ax\,dx \ = \dfrac{1}{a}\sec ax + c$

(viii) $\displaystyle\int e^{ax}\,dx \qquad = \dfrac{1}{a}e^{ax} + c$

(ix) $\displaystyle\int \dfrac{1}{x}\,dx \qquad = \ln x + c$

Problem 1.   Determine (a) $\displaystyle\int 5x^2\,dx$

(b) $\displaystyle\int 2t^3\,dt$

The standard integral of $\displaystyle\int ax^n\,dx = \dfrac{ax^{n+1}}{n+1} + c$

(a)   When $a = 5$ and $n = 2$ then

$$\int 5x^2\,dx = \dfrac{5x^{2+1}}{2+1} + c = \dfrac{5x^3}{3} + c$$

(b)   When $a = 2$ and $n = 3$ then

$$\int 2t^3\,dt = \dfrac{2t^{3+1}}{3+1} + c = \dfrac{2t^4}{4} + c = \dfrac{1}{2}t^4 + c$$

Each of these results may be checked by differentiating them.

Problem 2.   Find (a) $\displaystyle\int 3\,dx$ (b) $\displaystyle\int \dfrac{2}{5}x\,dx$

(a) $\int 3\,dx$ is the same as $\int 3x^0\,dx$, and using the standard integral $\int ax^n\,dx$ when $a = 3$ and $n = 0$ gives $\int 3x^0\,dx = \dfrac{3x^{0+1}}{0+1} + c = 3x + c$
In general, if k is a constant then $\int k\,dx = kx + c$

(b)   When $a = \dfrac{2}{5}$ and $n = 1$ then

$$\int \dfrac{2}{5}x\,dx = \left(\dfrac{2}{5}\right)\dfrac{x^{1+1}}{1+1} + c$$

$$= \left(\dfrac{2}{5}\right)\dfrac{x^2}{2} + c = \dfrac{1}{5}x^2 + c$$

Problem 3.   Determine

$\displaystyle\int \left(4 + \dfrac{3}{7}x - 6x^2\right) dx$

$\displaystyle\int \left(4 + \dfrac{3}{7}x - 6x^2\right) dx$ may be written as

$$\int 4\,dx + \int \dfrac{3}{7}x\,dx - \int 6x^2\,dx$$

i.e. each term is integrated separately. (This splitting up of terms only applies, however, for addition and subtraction.)

Hence $\displaystyle\int \left(4 + \dfrac{3}{7}x - 6x^2\right) dx$

$$= 4x + \left(\dfrac{3}{7}\right)\dfrac{x^{1+1}}{1+1} - (6)\dfrac{x^{2+1}}{2+1} + c$$

$$= 4x + \left(\dfrac{3}{7}\right)\dfrac{x^2}{2} - (6)\dfrac{x^3}{3} + c$$

$$= 4x + \dfrac{3}{14}x^2 - 2x^3 + c$$

Note that when an integral contains more than one term there is no need to have an arbitrary constant for each; just a single constant at the end is sufficient.

Problem 4.   Determine (a) $\displaystyle\int \dfrac{2x^3 - 3x}{4x}\,dx$

(b) $\displaystyle\int (1 - t)^2\,dt$

(a)   Rearranging into standard integral form gives:

$$\int \frac{2x^3 - 3x}{4x}\, dx = \int \left(\frac{2x^3}{4x} - \frac{3x}{4x}\right) dx$$

$$= \int \left(\frac{x^2}{2} - \frac{3}{4}\right) dx$$

$$= \left(\frac{1}{2}\right)\frac{x^{2+1}}{2+1} - \frac{3}{4}x + c$$

$$= \frac{1}{2}\left(\frac{x^3}{3}\right) - \frac{3}{4}x + c$$

$$= \frac{1}{6}x^3 - \frac{3}{4}x + c$$

(b)   Rearranging $\int (1-t)^2\, dt$ gives:

$$\int (1 - 2t + t^2)\, dt = t - \frac{2t^{1+1}}{1+1} + \frac{t^{2+1}}{2+1} + c$$

$$= t - \frac{2t^2}{2} + \frac{t^3}{3} + c$$

$$= t - t^2 + \frac{t^3}{3} + c$$

This problem shows that functions often have to be rearranged into the standard form of $\int ax^n\, dx$ before it is possible to integrate them.

Problem 5.   Determine (a) $\int \frac{3}{x^2}\, dx$

(b) $-\int \frac{2}{5x^3}\, dx$

(a) $\int \frac{3}{x^2}\, dx = \int 3x^{-2}.$

Using the standard integral $\int ax^n\, dx$ when $a = 3$ and $n = -2$ gives:

$$\int 3x^{-2}\, dx = \frac{3x^{-2+1}}{-2+1} + c = \frac{3x^{-1}}{-1} + c$$

$$= -3x^{-1} + c = \frac{-3}{x} + c$$

(b) $-\int \frac{2}{5x^3}\, dx = \int -\frac{2}{5}x^{-3}\, dx.$

Using the standard integral $\int ax^n\, dx$ when $a = -\frac{2}{5}$ and $n = -3$ gives:

$$\int -\frac{2}{5}x^{-3}\, dx = \left(-\frac{2}{5}\right)\frac{x^{-3+1}}{-3+1} + c$$

$$= \left(-\frac{2}{5}\right)\frac{x^{-2}}{-2} + c$$

$$= \frac{1}{5}x^{-2} + c = \frac{1}{5x^2} + c$$

Problem 6.   Determine (a) $\int 3\sqrt{x}\, dx$

(b) $\int \frac{2}{7}\sqrt[3]{x^5}\, dx$

For fractional powers it is necessary to appreciate $\sqrt[n]{a^m} = a^{m/n}$.

(a) $\int 3\sqrt{x}\, dx = \int 3x^{1/2}\, dx = \frac{(3)x^{(1/2)+1}}{\frac{1}{2}+1} + c$

$$= \frac{(3)x^{3/2}}{\frac{3}{2}} + c = 2x^{3/2} + c$$

$$= 2\sqrt{x^3} + c$$

(b) $\int \left(\frac{2}{7}\right)\sqrt[3]{x^5}\, dx = \int \left(\frac{2}{7}\right)x^{5/3}\, dx$

$$= \left(\frac{2}{7}\right)\frac{x^{(5/3)+1}}{\left(\frac{5}{3}+1\right)} + c$$

$$= \left(\frac{2}{7}\right)\frac{x^{8/3}}{\left(\frac{8}{3}\right)} + c$$

$$= \left(\frac{2}{7}\right)\left(\frac{3}{8}\right)x^{8/3} + c$$

$$= \frac{3}{28}\sqrt[3]{x^8} + c$$

Problem 7.   Determine

(a) $\int \frac{2}{\sqrt{x}}\, dx$ (b) $\int \frac{-5}{9\sqrt[4]{t^3}}\, dt$

(a) $\displaystyle\int \frac{2}{\sqrt{x}}\,dx = \int \frac{2}{x^{1/2}}\,dx = \int 2x^{-1/2}\,dx$

$\displaystyle = (2)\frac{x^{(-1/2)+1}}{\left(-\dfrac{1}{2}+1\right)} + c = \frac{2x^{1/2}}{\left(\dfrac{1}{2}\right)} + c$

$\displaystyle = 4\sqrt{x} + c$

(b) $\displaystyle\int \frac{-5}{9\sqrt[4]{t^3}}\,dt = \int \frac{-5}{9t^{3/4}}\,dt = \int \frac{-5}{9}t^{-3/4}\,dt$

$\displaystyle = \left(\frac{-5}{9}\right)\frac{t^{(-3/4)+1}}{\left(\dfrac{-3}{4}+1\right)} + c$

$\displaystyle = \left(\frac{-5}{9}\right)\frac{t^{1/4}}{\left(\dfrac{1}{4}\right)} + c$

$\displaystyle = \left(\frac{-5}{9}\right)\left(\frac{4}{1}\right)t^{1/4} + c$

$\displaystyle = \frac{-20}{9}\sqrt[4]{t} + c$

$\displaystyle\int \frac{(1+\theta)^2}{\sqrt{\theta}}\,d\theta = \int \frac{(1+2\theta+\theta^2)}{\sqrt{\theta}}\,d\theta$

$\displaystyle = \int \left(\frac{1}{\theta^{1/2}} + \frac{2\theta}{\theta^{1/2}} + \frac{\theta^2}{\theta^{1/2}}\right)d\theta$

$\displaystyle = \int (\theta^{-1/2} + 2\theta^{1-(1/2)} + \theta^{2-(1/2)})\,d\theta$

$\displaystyle = \int (\theta^{-1/2} + 2\theta^{1/2} + \theta^{3/2})\,d\theta$

$\displaystyle = \frac{\theta^{(-1/2)+1}}{\left(-\dfrac{1}{2}+1\right)} + \frac{(2)\theta^{(1/2)+1}}{\left(\dfrac{1}{2}+1\right)} + \frac{\theta^{(3/2)+1}}{\left(\dfrac{3}{2}+1\right)} + c$

$\displaystyle = \frac{\theta^{1/2}}{\dfrac{1}{2}} + \frac{2\theta^{3/2}}{\dfrac{3}{2}} + \frac{\theta^{5/2}}{\dfrac{5}{2}} + c$

$\displaystyle = 2\theta^{1/2} + \frac{4}{3}\theta^{3/2} + \frac{2}{5}\theta^{5/2} + c$

$\displaystyle = 2\sqrt{\theta} + \frac{4}{3}\sqrt{\theta^3} + \frac{2}{5}\sqrt{\theta^5} + c$

Problem 8.   Find (a) $\displaystyle\int 2\theta^{2.6}\,d\theta$

(b) $\displaystyle\int \frac{3}{x^{1.2}}\,dx$

(a) $\displaystyle\int 2\theta^{2.6}\,d\theta = \frac{(2)\theta^{2.6+1}}{2.6+1} + c = \frac{2\theta^{3.6}}{3.6} + c$

$\displaystyle = \frac{1}{1.8}\theta^{3.6} + c$

(b) $\displaystyle\int \frac{3}{x^{1.2}}\,dx = \int 3x^{-1.2}\,dx = \frac{(3)x^{-1.2+1}}{(-1.2+1)} + c$

$\displaystyle = \frac{3x^{-0.2}}{-0.2} + c$

$\displaystyle = -15x^{-0.2} + c = \frac{-15}{x^{0.2}} + c$

$\displaystyle \left(= \frac{-15}{x^{1/5}} + c = \frac{-15}{\sqrt[5]{x}} + c\right)$

Problem 9.   Determine $\displaystyle\int \frac{(1+\theta)^2}{\sqrt{\theta}}\,d\theta$

Problem 10.   Determine (a) $\displaystyle\int 4\cos 3x\,dx$

(b) $\displaystyle\int 5\sin 2\theta\,d\theta$

(a)   From Table 23.1(ii),

$\displaystyle\int 4\cos 3x\,dx = (4)\left(\frac{1}{3}\right)\sin 3x + c$

$\displaystyle = \frac{4}{3}\sin 3x + c$

(b)   From Table 23.1(iii),

$\displaystyle\int 5\sin 2\theta\,d\theta = (5)\left(-\frac{1}{2}\right)\cos 2\theta + c$

$\displaystyle = -\frac{5}{2}\cos 2\theta + c$

Problem 11.   Determine (a) $\displaystyle\int 7\sec^2 4t\,dt$

(b) $3\displaystyle\int \mathrm{cosec}^2\,2\theta\,d\theta$

(a)   From Table 23.1(iv),

$$\int 7 \sec^2 4t \, dt = (7) \left(\frac{1}{4}\right) \tan 4t + c$$

$$= \frac{7}{4} \tan 4t + c$$

(b)   From Table 23.1(v),

$$3 \int \mathrm{cosec}^2 \, 2\theta \, d\theta = (3) \left(-\frac{1}{2}\right) \cot 2\theta + c$$

$$= -\frac{3}{2} \cot 2\theta + c$$

Problem 12.   Determine

(a) $\int 4 \, \mathrm{cosec} \, 3x \cot 3x \, dx$

(b) $\int 7 \sec 5t \tan 5t \, dt$

(a)   From Table 23.1(vi),

$$\int 4 \, \mathrm{cosec} \, 3x \cot 3x \, dx = (4) \left(-\frac{1}{3}\right) \mathrm{cosec} \, 3x + c$$

$$= -\frac{4}{3} \mathrm{cosec} \, 3x + c$$

(b)   From Table 23.1(vii),

$$\int 7 \sec 5t \tan 5t \, dt = (7) \left(\frac{1}{5}\right) \sec 5t + c$$

$$= \frac{7}{5} \sec 5t + c$$

Problem 13.   Determine

(a) $\int 5e^{3x} \, dx$ (b) $\int \frac{2}{3e^{4t}} \, dt$

(a)   From Table 23.1(viii),

$$\int 5e^{3x} \, dx = (5) \left(\frac{1}{3}\right) e^{3x} + c = \frac{5}{3} e^{3x} + c$$

(b) $\int \frac{2}{3e^{4t}} \, dt = \int \frac{2}{3} e^{-4t} \, dt$

$$= \left(\frac{2}{3}\right) \left(-\frac{1}{4}\right) e^{-4t} + c$$

$$= -\frac{1}{6} e^{-4t} = -\frac{1}{6e^{4t}} + c$$

Problem 14.   Determine

(a) $\int \frac{3}{5x} \, dx$ (b) $\int \left(\frac{2m^2 + 1}{m}\right) dm$

(a) $\int \frac{3}{5x} \, dx = \int \left(\frac{3}{5}\right) \left(\frac{1}{x}\right) dx$

$$= \frac{3}{5} \ln x + c \text{ (from Table 23.1(ix))}$$

(b) $\int \left(\frac{2m^2 + 1}{m}\right) dm = \int \left(\frac{2m^2}{m} + \frac{1}{m}\right) dm$

$$= \int \left(2m + \frac{1}{m}\right) dm$$

$$= \frac{2m^2}{2} + \ln m + c$$

$$= m^2 + \ln m + c$$

Further problems on standard integrals may be found in Section 23.5, Problems 1 to 14, pages 240 and 241.

23.4   Definite integrals

Integrals containing an arbitrary constant c in their results are called **indefinite integrals** since their precise value cannot be determined without further information. **Definite integrals** are those in which limits are applied.

If an expression is written as $[x]_a^b$, 'b' is called the upper limit and 'a' the lower limit.

The operation of applying the limits is defined as $[x]_a^b = (b) - (a)$.

The increase in the value of the integral x^2 as x increases from 1 to 3 is written as $\int_1^3 x^2 \, dx$.

Applying the limits gives:

$$\int_1^3 x^2 \, dx = \left[\frac{x^3}{3} + c\right]_1^3$$

$$= \left(\frac{(3)^3}{3} + c\right) - \left(\frac{(1)^3}{3} + c\right)$$

$$= (9 + c) - \left(\frac{1}{3} + c\right) = 8\frac{2}{3}$$

Note that the 'c' term always cancels out when limits are applied and it need not be shown with definite integrals.

Problem 15.   Evaluate (a) $\int_1^2 3x\,dx$

(b) $\int_{-2}^3 (4 - x^2)\,dx$

(a) $\int_1^2 3x\,dx = \left[\dfrac{3x^2}{2}\right]_1^2 = \left\{\dfrac{3}{2}(2)^2\right\} - \left\{\dfrac{3}{2}(1)^2\right\}$

$$= 6 - 1\dfrac{1}{2} = 4\dfrac{1}{2}$$

(b) $\int_{-2}^3 (4 - x^2)\,dx = \left[4x - \dfrac{x^3}{3}\right]_{-2}^3$

$$= \left\{4(3) - \dfrac{(3)^3}{3}\right\} - \left\{4(-2) - \dfrac{(-2)^3}{3}\right\}$$

$$= \{12 - 9\} - \left\{-8 - \left(\dfrac{-8}{3}\right)\right\}$$

$$= (3) - \left(-5\dfrac{1}{3}\right) = 8\dfrac{1}{3}$$

Problem 16.   Evaluate $\int_1^4 \left(\dfrac{\theta + 2}{\sqrt{\theta}}\right) d\theta$, taking positive square roots only

$\int_1^4 \left(\dfrac{\theta + 2}{\sqrt{\theta}}\right) d\theta = \int_1^4 \left(\dfrac{\theta}{\theta^{1/2}} + \dfrac{2}{\theta^{1/2}}\right) d\theta$

$$= \int_1^4 (\theta^{1/2} + 2\theta^{-1/2})\,d\theta$$

$$= \left[\dfrac{\theta^{(1/2)+1}}{\left(\dfrac{1}{2}+1\right)} + \dfrac{2\theta^{(-1/2)+1}}{\left(-\dfrac{1}{2}+1\right)}\right]_1^4$$

$$= \left[\dfrac{\theta^{3/2}}{\dfrac{3}{2}} + \dfrac{2\theta^{1/2}}{\dfrac{1}{2}}\right]_1^4 = \left[\dfrac{2}{3}\sqrt{\theta^3} + 4\sqrt{\theta}\right]_1^4$$

$$= \left\{\dfrac{2}{3}\sqrt{(4)^3} + 4\sqrt{4}\right\} - \left\{\dfrac{2}{3}\sqrt{(1)^3} + 4\sqrt{(1)}\right\}$$

$$= \left\{\dfrac{16}{3} + 8\right\} - \left\{\dfrac{2}{3} + 4\right\}$$

$$= 5\dfrac{1}{3} + 8 - \dfrac{2}{3} - 4 = 8\dfrac{2}{3}$$

Problem 17.   Evaluate $\int_0^{\pi/2} 3\sin 2x\,dx$

$\int_0^{\pi/2} 3\sin 2x\,dx = \left[(3)\left(-\dfrac{1}{2}\right)\cos 2x\right]_0^{\pi/2}$

$$= \left[-\dfrac{3}{2}\cos 2x\right]_0^{\pi/2}$$

$$= \left\{-\dfrac{3}{2}\cos 2\left(\dfrac{\pi}{2}\right)\right\} - \left\{-\dfrac{3}{2}\cos 2(0)\right\}$$

$$= \left\{-\dfrac{3}{2}\cos \pi\right\} - \left\{-\dfrac{3}{2}\cos 0\right\}$$

$$= \left\{-\dfrac{3}{2}(-1)\right\} - \left\{-\dfrac{3}{2}(1)\right\}$$

$$= \dfrac{3}{2} + \dfrac{3}{2} = 3$$

Problem 18.   Evaluate $\int_1^2 4\cos 3t\,dt$

$\int_1^2 4\cos 3t\,dt = \left[(4)\left(\dfrac{1}{3}\right)\sin 3t\right]_1^2 = \left[\dfrac{4}{3}\sin 3t\right]_1^2$

$$= \left\{\dfrac{4}{3}\sin 6\right\} - \left\{\dfrac{4}{3}\sin 3\right\}$$

Limits of trigonometric functions are expressed in radians. Thus $\sin 6$ means the sine of 6 radians,

$6\text{ rads} = \left(6 \times \dfrac{180}{\pi}\right)^\circ = 343.77^\circ$

Similarly, $3\text{ rads} = \left(3 \times \dfrac{180}{\pi}\right)^\circ = 171.89^\circ$

Hence $\int_1^2 4\cos 3t\,dt$

$$= \left\{\dfrac{4}{3}\sin 343.77^\circ\right\} - \left\{\dfrac{4}{3}\sin 171.89^\circ\right\}$$

$$= (-0.3727) - (0.1881) = -0.5608$$

Problem 19.   Show that $\int_{\pi/6}^{\pi/3} (2\sec^2 \theta - 3\cos 2\theta)\,d\theta = \dfrac{4}{\sqrt{3}}$

$$\int_{\pi/6}^{\pi/3} (2\sec^2\theta - 3\cos 2\theta)\,d\theta$$

$$= \left[2\tan\theta - \frac{3}{2}\sin 2\theta\right]_{\pi/6}^{\pi/3}$$

$$= \left\{2\tan\frac{\pi}{3} - \frac{3}{2}\sin 2\left(\frac{\pi}{3}\right)\right\}$$

$$- \left\{2\tan\frac{\pi}{6} - \frac{3}{2}\sin 2\left(\frac{\pi}{6}\right)\right\}$$

$$= \left\{2(\sqrt{3}) - \frac{3}{2}\left(\frac{\sqrt{3}}{2}\right)\right\}$$

$$- \left\{2\left(\frac{1}{\sqrt{3}}\right) - \frac{3}{2}\left(\frac{\sqrt{3}}{2}\right)\right\}$$

$$= 2\sqrt{3} - \frac{3}{4}\sqrt{3} - 2\left(\frac{\sqrt{3}}{3}\right) + \frac{3}{4}\sqrt{3}$$

$$\left(\text{since } \frac{1}{\sqrt{3}} = \frac{\sqrt{3}}{3}\right)$$

$$= \frac{4}{3}\sqrt{3} = \frac{4}{\sqrt{3}}$$

Problem 20. Evaluate (a) $\int_{1}^{2} 4e^{2x}\,dx$

(b) $\int_{1}^{4} \frac{3}{4u}\,du$, each correct to 4 significant figures

(a) $\int_{1}^{2} 4e^{2x}\,dx = \left[\frac{4}{2}e^{2x}\right]_{1}^{2} = 2[e^{2x}]_{1}^{2} = 2[e^4 - e^2]$

$$= 2[54.5982 - 7.3891] = \mathbf{94.42}$$

(b) $\int_{1}^{4} \frac{3}{4u}\,du = \left[\frac{3}{4}\ln u\right]_{1}^{4}$

$$= \frac{3}{4}[\ln 4 - \ln 1] = \frac{3}{4}[1.3863 - 0]$$

$$= \mathbf{1.040}$$

Further problems on definite integrals may be found in the following Section 23.5, Problems 15 to 25, page 241.

23.5 Further problems on methods of integration

Standard integrals

In Problems 1 to 14, determine the indefinite integrals:

1 (a) $\int 4\,dx$ (b) $\int 7x\,dx$

$$\left[\text{(a) } 4x + c \text{ (b) } \frac{7x^2}{2} + c\right]$$

2 (a) $\int \frac{2}{5}x^2\,dx$ (b) $\int \frac{5}{6}x^3\,dx$

$$\left[\text{(a) } \frac{2}{15}x^3 + c \text{ (b) } \frac{5}{24}x^4 + c\right]$$

3 (a) $\int (2 + 3x - 4x^2)\,dx$ (b) $2\int (x - 4x^2)\,dx$

$$\left[\begin{array}{l}\text{(a) } 2x + \dfrac{3x^2}{2} - \dfrac{4x^3}{3} + c \\[2mm] \text{(b) } x^2 - \dfrac{8x^3}{3} + c\end{array}\right]$$

4 (a) $\int \left(\frac{3x^2 - 5x}{x}\right)dx$ (b) $\int (2 + \theta)^2\,d\theta$

$$\left[\begin{array}{l}\text{(a) } \dfrac{3x^2}{2} - 5x + c \\[2mm] \text{(b) } 4\theta + 2\theta^2 + \dfrac{\theta^3}{3} + c\end{array}\right]$$

5 (a) $\int \frac{4}{3x^2}\,dx$ (b) $\int \frac{3}{4x^4}\,dx$

$$\left[\text{(a) } \frac{-4}{3x} + c \text{ (b) } \frac{-1}{4x^3} + c\right]$$

6 (a) $2\int \sqrt{x^3}\,dx$ (b) $\int \frac{1}{4}\sqrt[4]{x^5}\,dx$

$$\left[\text{(a) } \frac{4}{5}\sqrt{x^5} + c \text{ (b) } \frac{1}{9}\sqrt[4]{x^9} + c\right]$$

7 (a) $\int \frac{-5}{\sqrt{t^3}}\,dt$ (b) $\int \frac{3}{7\sqrt[5]{x^4}}\,dx$

$$\left[\text{(a) } \frac{10}{\sqrt{t}} + c \text{ (b) } \frac{15}{7}\sqrt[5]{x} + c\right]$$

8 (a) $\int \frac{5}{2}x^{1.8}\,dx$ (b) $\int \frac{3}{2t^{1.4}}\,dt$

$$\left[\text{(a) } \frac{x^{2.8}}{1.12} + c \text{ (b)} \frac{-3.75}{t^{0.4}} + c\right]$$

9   (a) $\int 3\cos 2x\,dx$ (b) $\int 7\sin 3\theta\,d\theta$

$\left[\text{(a) } \dfrac{3}{2}\sin 2x + c \text{ (b) } -\dfrac{7}{3}\cos 3\theta + c\right]$

10   (a) $\int \dfrac{3}{4}\sec^2 3x\,dx$ (b) $\int 2\operatorname{cosec}^2 4\theta\,d\theta$

$\left[\text{(a) } \dfrac{1}{4}\tan 3x + c \text{ (b) } -\dfrac{1}{2}\cot 4\theta + c\right]$

11   (a) $5\int \cot 2t\operatorname{cosec} 2t\,dt$ (b) $\int_3^4 \sec 4t\tan 4t\,dt$

$\left[\text{(a) } -\dfrac{5}{2}\operatorname{cosec} 2t + c \text{ (b) } \dfrac{1}{3}\sec 4t + c\right]$

12   (a) $\int \dfrac{3}{4}e^{2x}\,dx$ (b) $\dfrac{2}{3}\int \dfrac{dx}{e^{5x}}$

$\left[\text{(a) } \dfrac{3}{8}e^{2x} + c \text{ (b) } \dfrac{-2}{15e^{5x}} + c\right]$

13   (a) $\int \dfrac{2}{3x}\,dx$ (b) $\int \left(\dfrac{u^2-1}{u}\right)du$

$\left[\text{(a) } \dfrac{2}{3}\ln x + c \text{ (b) } \dfrac{u^2}{2} - \ln u + c\right]$

14   (a) $\int \dfrac{(2+3x)^2}{\sqrt{x}}\,dx$ (b) $\int \left(\dfrac{1}{t} + 2t\right)^2 dt$

$\left[\begin{array}{l}\text{(a) } 8\sqrt{x} + 8\sqrt{x^3} + \dfrac{18}{5}\sqrt{x^5} + c \\[2mm] \text{(b) } -\dfrac{1}{t} + 4t + \dfrac{4t^3}{3} + c\end{array}\right]$

Definite integrals

In Problems 15 to 23, evaluate the definite integrals (where necessary, correct to 4 significant figures):

15   (a) $\int_2^3 2\,dx$ (b) $\int_{-1}^2 2x\,dx$ \qquad [(a) 2 (b) 3]

16   (a) $\int_1^4 5x^2\,dx$ (b) $\int_{-1}^1 -\dfrac{3}{4}t^2\,dt$

$\left[\text{(a) } 105 \text{ (b) } -\dfrac{1}{2}\right]$

17   (a) $\int_{-1}^2 (3 - x^2)\,dx$ (b) $\int_1^3 (x^2 - 4x + 3)\,dx$

$\left[\text{(a) } 6 \text{ (b) } -1\dfrac{1}{3}\right]$

18   (a) $\int_0^\pi \dfrac{3}{2}\cos\theta\,d\theta$ (b) $\int_0^{\pi/2} 4\cos\theta\,d\theta$

[(a) 0 (b) 4]

19   (a) $\int_{\pi/6}^{\pi/3} 2\sin 2\theta\,d\theta$ (b) $\int_0^2 3\sin t\,dt$

[(a) 1 (b) 4.248]

20   (a) $\int_0^1 5\cos 3x\,dx$ (b) $\int_0^{\pi/6} 3\sec^2 2x\,dx$

[(a) 0.2352 (b) 2.598]

21   (a) $\int_1^2 \operatorname{cosec}^2 4t\,dt$

(b) $\int_{\pi/4}^{\pi/2} (3\sin 2x - 2\cos 3x)\,dx$

[(a) 0.2527 (b) 2.638]

22   (a) $\int_0^1 3e^{3t}\,dt$ (b) $\int_{-1}^2 \dfrac{2}{3e^{2x}}\,dx$

[(a) 19.09 (b) 2.457]

23   (a) $\int_2^3 \dfrac{2}{3x}\,dx$ (b) $\int_1^3 \dfrac{2x^2+1}{x}\,dx$

[(a) 0.2703 (b) 9.099]

24   Show that
$$\int_1^2 \dfrac{(2x-1)(3x+4)}{x}\,dx = 2(7 - 2\ln 2)$$

25   The entropy change ΔS for an ideal gas is given by:

$$\Delta S = \int_{T_1}^{T_2} C_v \dfrac{dT}{T} - R\int_{V_1}^{V_2} \dfrac{dV}{V}$$

where T is the thermodynamic temperature, V is the volume and $R = 8.314$. Determine the entropy change when a gas expands from 1 litre to 3 litres for a temperature rise from 100 K to 400 K given that:

$$C_v = 45 + 6 \times 10^{-3}T + 8 \times 10^{-6}T^2 \quad [55.65]$$

24

Areas under and between curves

24.1 Area under a curve

The area shown shaded in Fig. 24.1 may be determined using approximate methods (such as the trapezoidal rule, the mid-ordinate rule or Simpson's rule) or, more precisely, by using integration.

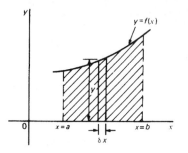

Figure 24.1

(i) Let A be the area shown shaded in Fig. 24.1 and let this area be divided into a number of strips each of width δx. One such strip is shown and let the area of this strip be δA.
Then: $\delta A \approx y\,\delta x$ (1)
The accuracy of statement (1) increases when the width of each strip is reduced, i.e. area A is divided into a greater number of strips

(ii) Area A is equal to the sum of all the strips from $x = a$ to $x = b$,

i.e. $A = \displaystyle\lim_{\delta x \to 0} \sum_{x=a}^{x=b} y\,\delta x$ (2)

(iii) From statement (1), $\dfrac{\delta A}{\delta x} \approx y$ (3)

In the limit, as δx approaches zero, $\dfrac{\delta A}{\delta x}$ becomes the differential coefficient $\dfrac{dA}{dx}$.

Hence $\displaystyle\lim_{\delta x \to 0}\left(\dfrac{\delta A}{\delta x}\right) = \dfrac{dA}{dx} = y$, from statement (3). By integration,

$$\int \dfrac{dA}{dx}\,dx = \int y\,dx, \text{ i.e. } A = \int y\,dx$$

The ordinates $x = a$ and $x = b$ limit the area and such ordinate values are shown as limits. Hence

$$A = \int_a^b y\,dx \qquad (4)$$

(iv) Equating statements (2) and (4) gives:

$$\text{Area } A = \lim_{\delta x \to 0} \sum_{x=a}^{x=b} y\,\delta x = \int_a^b y\,dx$$

$$= \int_a^b f(x)\,dx$$

(v) If the area between a curve $x = f(y)$, the y-axis and ordinates $y = p$ and $y = q$ is required then area $= \int_p^q x\,dy$

Thus determining the area under a curve by integration merely involves evaluating a definite integral.

There are several instances in engineering and science where the area beneath a curve needs to be accurately determined. For example, the areas between limits of a:

    velocity/time graph gives distance travelled,
    force/distance graph gives work done,
    voltage/current graph gives power, and so on

Should a curve drop below the x-axis, then $y(= f(x))$ becomes negative and $\int f(x)\,dx$ is negative. When determining such areas by integration, a negative sign is placed before the integral. For the curve shown in Fig. 24.2, the total shaded area is given by (area E + area F + area G). By integration, **total shaded area**

$$= \int_a^b f(x)\,dx - \int_b^c f(x)\,dx + \int_c^d f(x)\,dx$$

(Note that this is **not** the same as $\int_a^d f(x)\,dx$.)
It is usually necessary to sketch a curve in order to check whether it crosses the x-axis.

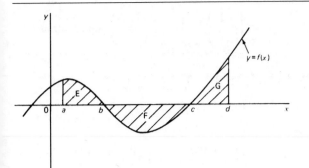

Figure 24.2

Problem 1.   Determine the area enclosed by $y = 2x + 3$, the x-axis and ordinates $x = 1$ and $x = 4$

$y = 2x + 3$ is a straight line graph as shown in Fig. 24.3, where the required area is shown shaded.

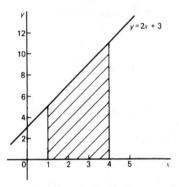

Figure 24.3

By integration, shaded area $= \int_{1}^{4} y \, dx$

$$= \int_{1}^{4} (2x + 3) \, dx$$

$$= \left[\frac{2x^2}{2} + 3x \right]_{1}^{4}$$

$$= [(16 + 12) - (1 + 3)]$$

$$= \textbf{24 square units}$$

(This answer may be checked since the shaded area is a trapezium.

Area of trapezium $= \dfrac{1}{2}$ (sum of parallel sides)
$\times$ (perpendicular distance between parallel sides)

$$= \frac{1}{2}(5 + 11)(3)$$

$$= \textbf{24 square units})$$

Problem 2.   The velocity v of a body t seconds after a certain instant is $(2t^2 + 5)$ m/s. Find by integration how far it moves in the interval from $t = 0$ to $t = 4$ s

Since $2t^2 + 5$ is a quadratic expression, the curve $v = 2t^2 + 5$ is a parabola cutting the v-axis at $v = 5$, as shown in Fig. 24.4.

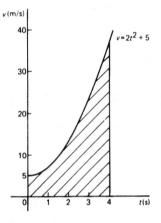

Figure 24.4

The distance travelled is given by the area under the v/t curve (shown shaded in Fig. 24.4).

By integration, shaded area $= \int_{0}^{4} v \, dt$

$$= \int_{0}^{4} (2t^2 + 5) \, dt$$

$$= \left[\frac{2t^3}{3} + 5t \right]_{0}^{4}$$

i.e. **distance travelled** $= 62\dfrac{2}{3}$ **m.**

Problem 3.   Sketch the graph $y = x^3 + 2x^2 - 5x - 6$ between $x = -3$ and $x = 2$ and determine the area enclosed by the curve and the x-axis

A table of values is produced and the graph sketched as shown in Fig. 24.5 where the area enclosed by the curve and the x-axis is shown shaded.

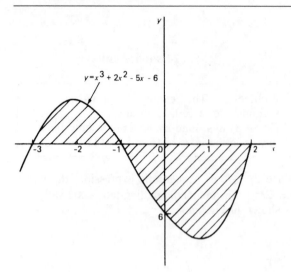

$y = x^3 + 2x^2 - 5x - 6$

Figure 24.5

x	-3	-2	-1	0	1	2
x^3	-27	-8	-1	0	1	8
$2x^2$	18	8	2	0	2	8
$-5x$	15	10	5	0	-5	-10
-6	-6	-6	-6	-6	-6	-6
y	0	4	0	-6	-8	0

Shaded area $= \int_{-3}^{-1} y\,dx - \int_{-1}^{2} y\,dx$, the minus sign before the second integral being necessary since the enclosed area is below the x-axis.
Hence shaded area

$$= \int_{-3}^{-1} (x^3 + 2x^2 - 5x - 6)\,dx$$

$$- \int_{-1}^{2} (x^3 + 2x^2 - 5x - 6)\,dx$$

$$\int_{-3}^{-1} (x^3 + 2x^2 - 5x - 6)\,dx$$

$$= \left[\frac{x^4}{4} + \frac{2x^3}{3} - \frac{5x^2}{2} - 6x \right]_{-3}^{-1}$$

$$= \left\{ \frac{1}{4} - \frac{2}{3} - \frac{5}{2} + 6 \right\}$$

$$- \left\{ \frac{81}{4} - 18 - \frac{45}{2} + 18 \right\}$$

$$= \left\{ 3\frac{1}{12} \right\} - \left\{ -2\frac{1}{4} \right\} = 5\frac{1}{3} \text{ square units}$$

$$\int_{-1}^{2} (x^3 + 2x^2 - 5x - 6)\,dx$$

$$= \left[\frac{x^4}{4} + \frac{2x^3}{3} - \frac{5x^2}{2} - 6x \right]_{-1}^{2}$$

$$= \left\{ 4 + \frac{16}{3} - 10 - 12 \right\} - \left\{ 3\frac{1}{12} \right\}$$

$$= \left\{ -12\frac{2}{3} \right\} - \left\{ 3\frac{1}{12} \right\} = -15\frac{3}{4} \text{ square units}$$

Hence shaded area $= \left(5\frac{1}{3} \right) - \left(-15\frac{3}{4} \right)$

$$= 21\frac{1}{12} \text{ square units}$$

Problem 4. Determine the area enclosed by the curve $y = 3x^2 + 4$, the x-axis and ordinates $x = 1$ and $x = 4$ by (a) the trapezoidal rule, (b) the mid-ordinate rule, (c) Simpson's rule, and (d) integration

---

The curve $y = 3x^2 + 4$ is shown plotted in Fig. 24.6.

(a) **By the trapezoidal rule**

area $=$ (width of interval)$\times$

$$\left[\frac{1}{2} \left(\begin{matrix} \text{first + last} \\ \text{ordinate} \end{matrix} \right) + \left(\begin{matrix} \text{sum of remaining} \\ \text{ordinates} \end{matrix} \right) \right]$$

Selecting 6 intervals each of width 0.5 gives:

$$\text{Area} = (0.5) \left[\frac{1}{2}(7 + 52) + 10.75 + 16 \right.$$

$$\left. + 22.75 + 31 + 40.75 \right]$$

$$= 75.375 \text{ square units}$$

(b) **By the mid-ordinate rule**, area $=$ (width of interval)(sum of mid-ordinates). Selecting 6 intervals, each of width 0.5 gives the mid-ordinates as shown by the broken lines in Fig. 24.6.
Thus, area $= (0.5)(8.5 + 13 + 19 + 26.5$

$$+ 35.5 + 46)$$

$$= 74.25 \text{ square units}$$

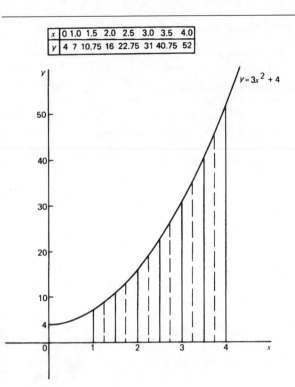

x	0	1.0	1.5	2.0	2.5	3.0	3.5	4.0
y	4	7	10.75	16	22.75	31	40.75	52

Figure 24.6

(c) **By Simpon's rule**,

$$\text{area} = \frac{1}{3}\begin{pmatrix}\text{width of}\\\text{interval}\end{pmatrix}\left[\begin{pmatrix}\text{first + last}\\\text{ordinates}\end{pmatrix}\right.$$

$$+\,4\begin{pmatrix}\text{sum of even}\\\text{ordinates}\end{pmatrix}$$

$$\left.+\,2\begin{pmatrix}\text{sum of remaining}\\\text{odd ordinates}\end{pmatrix}\right]$$

Selecting 6 intervals, each of width 0.5, gives:

$$\text{area} = \frac{1}{3}(0.5)[(7 + 52) + 4(10.75 + 22.75$$

$$+\,40.75) + 2(16 + 31)]$$

$$= \textbf{75 square units}$$

(d) **By integration**, shaded area

$$= \int_1^4 y\,dx$$

$$= \int_1^4 (3x^2 + 4)\,dx$$

$$= [x^3 + 4x]_1^4$$

$$= \textbf{75 square units}$$

Integration gives the precise value for the area under a curve. In this case Simpson's rule is seen to be the most accurate of the three approximate methods.

> **Problem 5.**   Find the area enclosed by the curve $y = \sin 2x$, the x-axis and the ordinates $x = 0$ and $x = \pi/3$.

A sketch of $y = \sin 2x$ is shown in Fig. 24.7. (Note that $y = \sin 2x$ has a period of $\dfrac{2\pi}{2}$, i.e. π radians.)

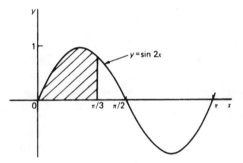

Figure 24.7

$$\text{Shaded area} = \int_0^{\pi/3} y\,dx$$

$$= \int_0^{\pi/3} \sin 2x\,dx$$

$$= \left[-\frac{1}{2}\cos 2x\right]_0^{\pi/3}$$

$$= \left\{-\frac{1}{2}\cos\frac{2\pi}{3}\right\} - \left\{-\frac{1}{2}\cos 0\right\}$$

$$= \left\{-\frac{1}{2}\left(-\frac{1}{2}\right)\right\} - \left\{-\frac{1}{2}(1)\right\}$$

$$= \frac{1}{4} + \frac{1}{2} = \frac{3}{4}\ \textbf{square units}$$

> **Problem 6.**   A gas expands according to the law $pv = $ constant. When the volume is 3 m³ the pressure is 150 kPa. Given that work done $= \int_{v_1}^{v_2} p\,dv$, determine the work done as the gas expands from 2 m³ to a volume of 6 m³

$pv = $ constant. When $v = 3$ m^3 and $p = 150$ kPa the constant is given by $(3 \times 150) = 450$ kPa m^3 or 450 kJ.

Hence $pv = 450$, or $p = \dfrac{450}{v}$

$$\text{Work done} = \int_2^6 \frac{450}{v}\, dv$$

$$= [450\ \ln v]_2^6 = 450[\ln 6 - \ln 2]$$

$$= 450\ \ln \frac{6}{2} = 450\ \ln 3 = \textbf{494.4 kJ}$$

Problem 7. Determine the area enclosed by the curve $y = 4\cos(\theta/2)$, the θ-axis and ordinates $\theta = 0$ and $\theta = (\pi/2)$

The curve $y = 4\cos(\theta/2)$ is shown in Fig. 24.8. (Note that $y = 4\cos(\theta/2)$ has a maximum value of 4 and period $2\pi/(1/2)$, i.e. 4π rads.)

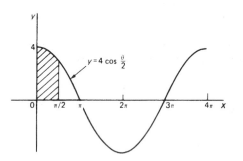

Figure 24.8

$$\text{Shaded area} = \int_0^{\pi/2} y\, d\theta = \int_0^{\pi/2} 4\cos\frac{\theta}{2}\, d\theta$$

$$= \left[4\left(\frac{1}{\frac{1}{2}}\right)\sin\frac{\theta}{2}\right]_0^{\pi/2}$$

$$= \left(8\sin\frac{\pi}{4}\right) - (8\sin 0)$$

$$= \frac{8\sqrt{2}}{2} = 4\sqrt{2} \text{ or } \textbf{5.657 square units}$$

Problem 8. Determine the area bounded by the curve $y = 3e^{t/4}$, the t-axis and ordinates $t = -1$ and $t = 4$, correct to 4 significant figures

---

A table of values is produced as shown.

t	-1	0	1	2	3	4
$y = 3e^{t/4}$	2.34	3.0	3.85	4.95	6.35	8.15

Since all the values of y are positive the area required is wholly above the t-axis.

$$\text{Hence area} = \int_1^4 y\, dt$$

$$= \int_1^4 3e^{t/4}\, dt = \left[\frac{3}{\left(\frac{1}{4}\right)}e^{t/4}\right]_{-1}^4$$

$$= 12[e^{t/4}]_{-1}^4 = 12(e^1 - e^{-1/4})$$

$$= 12(2.7183 - 0.7788)$$

$$= 12(1.9395) = \textbf{23.27 square units}$$

Problem 9. Sketch the curve $y = x^2 + 5$ between $x = -1$ and $x = 4$. Find the area enclosed by the curve, the x-axis and the ordinates $x = 0$ and $x = 3$. Determine also by integration the area enclosed by the curve and the y-axis, between the same limits

---

A table of values is produced and the curve $y = x^2 + 5$ plotted as shown in Fig. 24.9.

x	-1	0	1	2	3
y	6	5	6	9	14

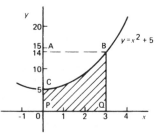

Figure 24.9

$$\text{Shaded area} = \int_0^3 y\, dx = \int_0^3 (x^2 + 5)\, dx$$

$$= \left[\frac{x^3}{3} + 5x\right]_0^3$$

$$= \textbf{24 square units}$$

When $x = 3$, $y = 3^2 + 5 = 14$ and when $x = 0$, $y = 5$.

Since $y = x^2 + 5$ then $x^2 = y - 5$ and $x = \sqrt{(y-5)}$. The area enclosed by the curve $y = x^2 + 5$ (i.e. $x = \sqrt{(y-5)}$), the y-axis and the ordinates $y = 5$ and $y = 14$ (i.e. area ABC of Fig. 24.9) is given by:

$$\text{Area} = \int_{y=5}^{y=14} x\,dy = \int_5^{14} \sqrt{(y-5)}\,dy$$

$$= \int_5^{14} (y-5)^{1/2}\,dy$$

Let $u = y - 5$, then $\dfrac{du}{dy} = 1$ and $dy = du$

Hence $\displaystyle\int (y-5)^{1/2}\,dy = \int u^{1/2}\,du = \frac{2}{3}u^{3/2}$ (for algebraic substitutions, see Chapter 41)

Since $u = y - 5$ then

$$\int_5^{14} \sqrt{(y-5)}\,dy = \frac{2}{3}[(y-5)^{3/2}]_5^{14}$$

$$= \frac{2}{3}[\sqrt{9^3} - 0]$$

$$= 18 \text{ square units}$$

[Check: From Fig. 24.9, area $BCPQ$ + area ABC = 24 + 18 = 42 square units, which is the area of rectangle $ABQP$.]

Problem 10. Determine the area between the curve $y = x^3 - 2x^2 - 8x$ and the x-axis

---

$$y = x^3 - 2x^2 - 8x = x(x^2 - 2x - 8)$$
$$= x(x+2)(x-4)$$

When $y = 0$, then $x = 0$ or $(x + 2) = 0$ or $(x - 4) = 0$, i.e. when $y = 0$, $x = 0$ or -2 or 4, which means that the curve cuts the x-axis at 0, -2 and 4. Since the curve is a continuous function, only one other coordinate value needs to be calculated before a sketch of the curve can be produced. When $x = 1$, $y = -9$, showing that the part of the curve between $x = 0$ and $x = 4$ is negative. A sketch of $y = x^3 - 2x^2 - 8x$ is shown in Fig. 24.10.

$$\text{Shaded area} = \int_{-2}^0 (x^3 - 2x^2 - 8x)\,dx$$

$$- \int_0^4 (x^3 - 2x^2 - 8x)\,dx$$

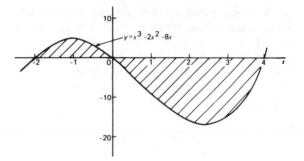

Figure 24.10

$$= \left[\frac{x^4}{4} - \frac{2x^3}{3} - \frac{8x^2}{2}\right]_{-2}^0$$

$$- \left[\frac{x^4}{4} - \frac{2x^3}{3} - \frac{8x^2}{2}\right]_0^4$$

$$= \left(6\frac{2}{3}\right) - \left(-42\frac{2}{3}\right)$$

$$= 49\frac{1}{3} \text{ square units}$$

24.2 The area between curves

The area enclosed between curves $y = f_1(x)$ and $y = f_2(x)$ (shown shaded in Fig. 24.11) is given by:

$$\text{Shaded area} = \int_a^b f_2(x)\,dx - \int_a^b f_1(x)\,dx$$

$$= \int_a^b [f_2(x) - f_1(x)]\,dx$$

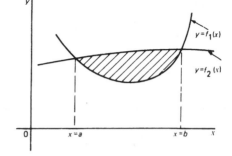

Figure 24.11

Problem 11. Determine the area enclosed between the curves $y = x^2 + 1$ and $y = 7 - x$

At the points of intersection, the curves are equal. Thus, equating the y-values of each curve gives: $x^2 + 1 = 7 - x$, from which $x^2 + x - 6 = 0$. Factorizing gives $(x - 2)(x + 3) = 0$, from which $x = 2$ and $x = -3$. By firstly determining the points of intersection the range of x-values has been found. Tables of values are produced as shown below.

x	-3	-2	-1	0	1	2
$y = x^2 + 1$	10	5	2	1	2	5

x	-3	0	2
$y = 7 - x$	10	7	5

$y = 7 - x$ is a straight line thus only two points are needed, plus one more as a check. A sketch of the two curves is shown in Fig. 24.12.

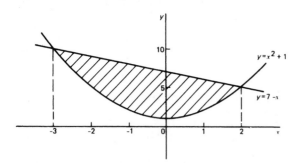

Figure 24.12

Shaded area $= \displaystyle\int_{-3}^{2} (7 - x)\, dx - \int_{-3}^{2} (x^2 + 1)\, dx$

$= \displaystyle\int_{-3}^{2} [(7 - x) - (x^2 + 1)]\, dx$

$= \displaystyle\int_{-3}^{2} (6 - x - x^2)\, dx$

$= \left[6x - \dfrac{x^2}{2} - \dfrac{x^3}{3} \right]_{-3}^{2}$

$= \left(12 - 2 - \dfrac{8}{3} \right) - \left(-18 - \dfrac{9}{2} + 9 \right)$

$= \left(7\dfrac{1}{3} \right) - \left(-13\dfrac{1}{2} \right)$

$= 20\dfrac{5}{6}$ **square units**

Problem 12. (a) Determine the coordinates of the points of intersection of the curves $y = x^2$ and $y^2 = 8x$. (b) Sketch the curves $y = x^2$ and $y^2 = 8x$ on the same axes. (c) Calculate the area enclosed by the two curves

(a) At the points of intersection the coordinates of the curves are equal. When $y = x^2$ then $y^2 = x^4$.
Hence at the points of intersection $x^4 = 8x$, by equating the y^2 values.
Thus $x^4 - 8x = 0$, from which $x(x^3 - 8) = 0$, i.e. $x = 0$ or $(x^3 - 8) = 0$.
Hence at the points of intersection $x = 0$ or $x = 2$.
When $x = 0$, $y = 0$ and when $x = 2$, $y = 2^2 = 4$.
Hence the points of intersection of the curves $y = x^2$ and $y^2 = 8x$ are (0, 0) and (2, 4)

(b) A sketch of $y = x^2$ and $y^2 = 8x$ is shown in Fig. 24.13

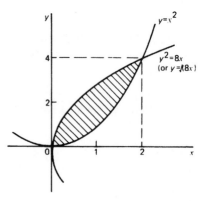

Figure 24.13

(c) Shaded area $= \displaystyle\int_{0}^{2} \{ \sqrt{(8x)} - x^2 \}\, dx$

$= \displaystyle\int_{0}^{2} \{ (\sqrt{8})x^{1/2} - x^2 \}\, dx$

$= \left[(\sqrt{8}) \dfrac{x^{3/2}}{\left(\frac{3}{2}\right)} - \dfrac{x^3}{3} \right]_{0}^{2}$

$= \left\{ \dfrac{\sqrt{8}\sqrt{8}}{\left(\frac{3}{2}\right)} - \dfrac{8}{3} \right\} - \{0\}$

$$= \frac{16}{3} - \frac{8}{3} = \frac{8}{3}$$

$$= 2\frac{2}{3} \text{ square units}$$

Problem 13.   Determine by integration the area bounded by the three straight lines $y = 4 - x$, $y = 3x$ and $3y = x$

Each of the straight lines is shown sketched in Fig. 24.14.

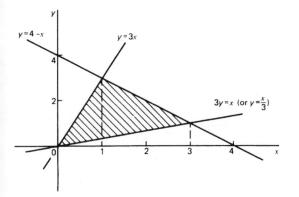

Figure 24.14

Shaded area $= \displaystyle\int_0^1 \left(3x - \frac{x}{3}\right) dx$

$$+ \int_1^3 \left[(4 - x) - \frac{x}{3}\right] dx$$

$$= \left[\frac{3x^2}{2} - \frac{x^2}{6}\right]_0^1 + \left[4x - \frac{x^2}{2} - \frac{x^2}{6}\right]_1^3$$

$$= \left(1\frac{1}{3}\right) + \left(6 - 3\frac{1}{3}\right)$$

$$= 4 \text{ square units}$$

Further problems on areas under and between curves may be found in the following Section 24.3, Problems 1 to 20.

24.3   Further problems on areas under and between curves

Unless otherwise stated all answers are in square units.

1   Show by integration that the area of the triangle formed by the line $y = 2x$, the ordinates $x = 0$ and $x = 4$ and the x-axis is 16 square units.

2   Sketch the curve $y = 3x^2 + 1$ between $x = -2$ and $x = 4$. Determine by integration the area enclosed by the curve, the x-axis and ordinates $x = -1$ and $x = 3$. Use an approximate method to find the area and compare the result with that obtained by integration.   [32]

In Problems 3 to 10, find the area enclosed between the given curves, the horizontal axis and the given ordinates:

3   $y = 5x$; $x = 1$, $x = 4$ $\left[37\frac{1}{2}\right]$

4   $y = 2x^2 - x + 1$; $x = -1$, $x = 2$ $\left[7\frac{1}{2}\right]$

5   $y = 2 \sin 2\theta$; $\theta = 0$, $\theta = \dfrac{\pi}{4}$ [1]

6   $\theta = t + e^t$; $t = 0$, $t = 2$ [8.389]

7   $y = 5 \cos 3t$; $t = 0$, $t = \dfrac{\pi}{6}$ $\left[1\frac{2}{3}\right]$

8   $y = (x - 1)(x - 3)$; $x = 0$, $x = 3$ $\left[2\frac{2}{3}\right]$

9   $y = 2x^3$; $x = -2$, $x = 2$ [16]

10   $xy = 4$; $x = 1$, $x = 4$ [5.545]

11   The force F newtons acting on a body at a distance x metres from a fixed point is given by $F = 3x + 2x^2$. If work done $= \int_{x_1}^{x_2} F \, dx$, determine the work done when the body moves from the position where $x = 1$ m to that when $x = 3$ m. $\left[29\frac{1}{3} \text{ Nm}\right]$

12   Find the area between the curve $y = 4x - x^2$ and the x-axis. $\left[10\frac{2}{3}\right]$

13   Sketch the curves $y = x^2 + 3$ and $y = 7 - 3x$ and determine the area enclosed by them. $\left[20\frac{5}{6}\right]$

14   Determine the area enclosed by the curves $y = \sin x$ and $y = \cos x$ and the y-axis.   [0.4142]

15   The velocity v of a vehicle t seconds after a certain instant is given by $v = (3t^2 + 4)$ m/s. Determine how far it moves in the interval from $t = 1$ s to $t = 5$ s.   [140 m]

16 Determine the coordinates of the points of intersection and the area enclosed between the parabolas $y^2 = 3x$ and $x^2 = 3y$.

[(0, 0), (3, 3); 3]

17 Determine the area enclosed by the curve $y = 5x^2 + 2$, the x-axis and the ordinates $x = 0$ and $x = 3$. Find also the area enclosed by the curve and the y-axis between the same limits.

[51; 90]

18 Calculate the area enclosed between $y = x^3 - 4x^2 - 5x$ and the x-axis using an approximate method and compare your result with the true area obtained by integration.

$$\left[73\frac{5}{6} \text{ or } 73.83\right]$$

19 A gas expands according to the law $pv =$ constant. When the volume is 2 m^3 the pressure is 250 kPa. Find the work done as the gas expands from 1 m^3 to a volume of 4 m^3 given that work done $= \int_{v_1}^{v_2} p \, dv$. [693.1 kJ]

20 Determine the area enclosed by the three straight lines $y = 3x$, $2y = x$ and $y + 2x = 5$.

$$\left[2\frac{1}{2}\right]$$

25

Mean and root mean square values

25.1 Mean or average values

(i) The mean or average value of the curve shown in Fig. 25.1, between $x = a$ and $x = b$, is given by:

mean or average value,

$$\bar{y} = \frac{\text{area under curve}}{\text{length of base}}$$

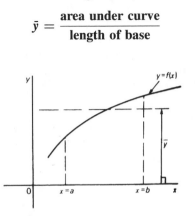

Figure 25.1

(ii) When the area under a curve may be obtained by integration then:

mean or average value,

$$\bar{y} = \frac{\int_a^b y\,dx}{b - a}$$

i.e.

$$\boxed{\bar{y} = \frac{1}{b - a} \int_a^b f(x)\,dx}$$

(iii) For a periodic function, such as a sine wave, the mean value is assumed to be 'the mean value over half a cycle', since the mean value over a complete cycle is zero

Problem 1.   Determine, using integration, the mean value of $y = 5x^2$ between $x = 1$ and $x = 4$

Mean value, $\bar{y} = \dfrac{1}{4 - 1} \displaystyle\int_1^4 y\,dx = \dfrac{1}{3} \int_1^4 5x^2\,dx$

$$= \frac{1}{3}\left[\frac{5x^3}{3}\right]_1^4 = \frac{5}{9}[x^3]_1^4$$

$$= \frac{5}{9}(64 - 1) = \mathbf{35}$$

Problem 2.   A sinusoidal voltage is given by $v = 100 \sin\theta$ volts. Determine the mean value of the voltage over half a cycle using integration

Half a cycle indicates that the limits are 0 and π rads.
Hence mean value,

$$\bar{y} = \frac{1}{\pi - 0}\int_0^\pi v\,d\theta = \frac{1}{\pi}\int_0^\pi 100 \sin\theta\,d\theta$$

$$= \frac{100}{\pi}[-\cos\theta]_0^\pi$$

$$= \frac{100}{\pi}\{(-\cos\pi) - (-\cos 0)\}$$

$$= \frac{100}{\pi}\{(+1) - (-1)\}$$

$$= \frac{200}{\pi} = \mathbf{63.66 \text{ volts}}$$

Note that for a sine wave, the

$$\mathbf{mean\ value} = \frac{\mathbf{2}}{\boldsymbol{\pi}} \times \mathbf{maximum\ value}$$

In this case, mean value $= 2/\pi \times 100 = 63.66$ V.

Problem 3.   Calculate the mean value of $y = 3x^2 + 2$ in the range $x = 0$ to $x = 3$ by (a) the mid-ordinate rule, and (b) integration

(a) A graph of $y = 3x^2$ over the required range is shown in Fig. 25.2 using the following table:

x	0	0.5	1.0	1.5	2.0	2.5	3.0
y	2.0	2.75	5.0	8.75	14.0	20.75	29.0

Using the mid-ordinate rule, mean value

$$= \frac{\text{area under curve}}{\text{length of base}}$$

$$= \frac{\text{sum of mid-ordinates}}{\text{number of mid-ordinates}}$$

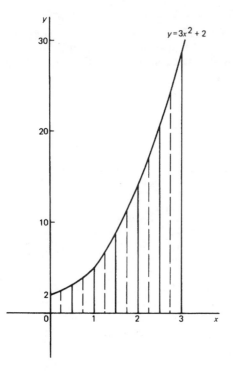

Figure 25.2

Selecting 6 intervals, each of width 0.5, the mid-ordinates are erected as shown by the broken lines in Fig. 25.2

$$\text{Mean value} = \frac{\begin{array}{c} 2.2 + 3.7 + 6.7 + 11.2 \\ + 17.2 + 24.7 \end{array}}{6}$$

$$= \frac{65.7}{6} = \mathbf{10.95}$$

(b) By integration, mean value

$$= \frac{1}{3-0} \int_0^3 y \, dx$$

$$= \frac{1}{3} \int_0^3 (3x^2 + 2) \, dx = \frac{1}{3}[x^3 + 2x]_0^3$$

$$= \frac{1}{3}\{(27 + 6) - (0)\} = \mathbf{11}$$

The answer obtained by integration is exact; greater accuracy may be obtained by the mid-ordinate rule if a larger number of intervals are selected.

Problem 4. The number of atoms, N, remaining in a mass of material during radioactive decay after time t seconds is given by $N = N_0 e^{-\lambda t}$, where N_0 and λ are constants. Determine the mean number of atoms in the mass of material for the time period $t = 0$ and $t = 1/\lambda$

$$\text{Mean number of atoms} = \frac{1}{\frac{1}{\lambda} - 0} \int_0^{1/\lambda} N \, dt$$

$$= \frac{1}{\frac{1}{\lambda}} \int_0^{1/\lambda} N_0 e^{-\lambda t} \, dt$$

$$= \lambda N_0 \int_0^{1/\lambda} e^{-\lambda t} \, dt$$

$$= \lambda N_0 \left[\frac{e^{-\lambda t}}{-\lambda} \right]_0^{1/\lambda}$$

$$= -N_0 [e^{-\lambda(1/\lambda)} - e^0]$$

$$= -N_0 [e^{-1} - e^0]$$

$$= +N_0 [e^0 - e^{-1}]$$

$$= N_0 [1 - e^{-1}]$$

$$= \mathbf{0.632 N_0}$$

25.2 Root mean square values

The **root mean square value** of a quantity is 'the square root of the mean value of the squared values of the quantity' taken over an interval. With reference to Fig. 25.1, the r.m.s. value of $y = f(x)$ over the range $x = a$ to $x = b$ is given by:

$$\boxed{\text{r.m.s. value} = \sqrt{\left\{ \frac{1}{b-a} \int_a^b y^2 dx \right\}}}$$

One of the principal applications of r.m.s. values is with alternating currents and voltages. The r.m.s. value of an alternating current is defined as that current which will give the same heating effect as the equivalent direct current.

Problem 5.   Determine the r.m.s. value of $y = 2x^2$ between $x = 1$ and $x = 4$

$$\text{r.m.s. value} = \sqrt{\left\{ \frac{1}{4-1} \int_1^4 y^2 dx \right\}}$$

$$= \sqrt{\left\{ \frac{1}{3} \int_1^4 (2x^2)^2 dx \right\}}$$

$$= \sqrt{\left\{ \frac{1}{3} \int_1^4 4x^4 dx \right\}}$$

$$= \sqrt{\left\{ \frac{4}{3} \left[\frac{x^5}{5} \right]_1^4 \right\}}$$

$$= \sqrt{\left\{ \frac{4}{15} (1024 - 1) \right\}}$$

$$= \sqrt{(272.8)} = \textbf{16.5}$$

Problem 6.   A sinusoidal voltage has a maximum value of 10 V. Calculate its r.m.s. value

---

A sinusoidal voltage v having a maximum value of 10 V may be written as $v = 10 \sin \theta$ V.
Over the range $\theta = 0$ to $\theta = \pi$,

$$\text{r.m.s. value} = \sqrt{\left\{ \frac{1}{\pi - 0} \int_0^\pi v^2 d\theta \right\}}$$

$$= \sqrt{\left\{ \frac{1}{\pi} \int_0^\pi (10 \sin \theta)^2 d\theta \right\}}$$

$$= \sqrt{\left\{ \frac{100}{\pi} \int_0^\pi \sin^2 \theta \, d\theta \right\}},$$

which is not a 'standard' integral

It is shown in Chapter 38 that $\cos 2\theta = 1 - 2\sin^2 \theta$ and this formula is used whenever $\sin^2 \theta$ needs to be integrated.
Rearranging $\cos 2\theta = 1 - 2\sin^2 \theta$ gives

$$\sin^2 \theta = \frac{1}{2}(1 - \cos 2\theta)$$

Hence $\sqrt{\left\{ \frac{100}{\pi} \int_0^\pi \sin^2 \theta \, d\theta \right\}}$

$$= \sqrt{\left\{ \frac{100}{\pi} \int_0^\pi \frac{1}{2}(1 - \cos 2\theta) \, d\theta \right\}}$$

$$= \sqrt{\left\{ \frac{100}{\pi} \frac{1}{2} \left[\theta - \frac{\sin 2\theta}{2} \right]_0^\pi \right\}}$$

$$= \sqrt{\left\{ \frac{100}{\pi} \frac{1}{2} \left[\left(\pi - \frac{\sin 2\pi}{2} \right) - \left(0 - \frac{\sin 0}{2} \right) \right] \right\}}$$

$$= \sqrt{\left\{ \frac{100}{\pi} \frac{1}{2} [\pi] \right\}} = \sqrt{\left\{ \frac{100}{2} \right\}}$$

$$= \frac{10}{\sqrt{2}} = \textbf{7.071 volts}$$

Note that for a sine wave, the

$$\textbf{r.m.s. value} = \frac{1}{\sqrt{2}} \times \textbf{maximum value.}$$

In this case, r.m.s. value $= \dfrac{1}{\sqrt{2}} \times 10 = 7.071$ V

Problem 7.   In a frequency distribution the average distance from the mean, y, is related to the variable, x, by the equation $y = 2x^2 - 1$. Determine, correct to 3 significant figures, the r.m.s. deviation from the mean for values of x from -1 to $+4$

---

r.m.s. deviation

$$= \sqrt{\left\{ \frac{1}{4 - -1} \int_{-1}^4 y^2 dx \right\}}$$

$$= \sqrt{\left\{ \frac{1}{5} \int_{-1}^4 (2x^2 - 1)^2 dx \right\}}$$

$$= \sqrt{\left\{ \frac{1}{5} \int_{-1}^4 (4x^4 - 4x^2 + 1) \, dx \right\}}$$

$$= \sqrt{\left\{\frac{1}{5}\left[\frac{4x^5}{5} - \frac{4x^3}{3} + x\right]_{-1}^{4}\right\}}$$

$$= \sqrt{\left\{\frac{1}{5}\left[\left(\frac{4}{5}(4)^5 - \frac{4}{3}(4)^3 + 4\right)\right.\right.}$$

$$\left.\left. - \left(\frac{4}{5}(-1)^5 - \frac{4}{3}(-1)^3 + (-1)\right)\right]\right\}$$

$$= \sqrt{\left\{\frac{1}{5}[(737.87) - (-0.467)]\right\}}$$

$$= \sqrt{\left\{\frac{1}{5}[738.34]\right\}}$$

$$= \sqrt{(147.67)} = 12.152 = \mathbf{12.2},$$

correct to 3 significant figures

25.3 Further problems on mean and root mean square values

1   Determine the mean value of (a) $y = 3\sqrt{x}$ from $x = 0$ to $x = 4$, (b) $y = \sin 2\theta$ from $\theta = 0$ to $\theta = \pi/4$, and (c) $y = 4e^t$ from $t = 1$ to $t = 4$.

$$\left[\text{(a) 4 (b) } \frac{2}{\pi} \text{ or 0.637 (c) 69.17}\right]$$

2   Calculate the mean value of $y = 2x^2 + 5$ in the range $x = 1$ to $x = 4$ by (a) the mid-ordinate rule, and (b) integration.   [19]

3   The speed v of a vehicle is given by $v = (4t + 3)$ m/s, where t is the time in seconds. Determine the average value of the speed from $t = 0$ to $t = 3$ s.   [9 m/s]

4   Find the mean value of the curve $y = 6 + x - x^2$ which lies above the x-axis by using an approximate method. Check the result using integration. $[4\frac{1}{6}]$

5   The vertical height h km of a missile varies with the horizontal distance d km, and is given by

$h = 4d - d^2$. Determine the mean height of the missile from $d = 0$ to $d = 4$ km. $[2\frac{2}{3}\text{km}]$

6   The velocity v of a piston moving with simple harmonic motion at any time t is given by $v = c \sin \omega t$, where c is a constant. Determine the mean velocity between $t = 0$ and $t = \pi/\omega$.

$$\left[\frac{2c}{\pi}\right]$$

7   Determine the r.m.s. values of
(a) $y = 3x$ from $x = 0$ to $x = 4$
(b) $y = t^2$ from $t = 1$ to $t = 3$
(c) $y = 25 \sin \theta$ from $\theta = 0$ to $\theta = 2\pi$

$$\left[\text{(a) 6.928 (b) 4.919 (c) } \frac{25}{\sqrt{2}} \text{ or 17.68}\right]$$

8   Calculate the r.m.s. values of
(a) $y = \sin 2\theta$ from $\theta = 0$ to $\theta = \frac{\pi}{4}$
(b) $y = 1 + \sin t$ from $t = 0$ to $t = 2\pi$
(c) $y = 3 \cos 2x$ from $x = 0$ to $x = \pi$
(Note that $\cos^2 t = \frac{1}{2}(1 + \cos 2t)$, from Chapter 38.)

$$\left[\text{(a) } \frac{1}{\sqrt{2}} \text{ or 0.707 (b) 1.225 (c) 2.121}\right]$$

9   The distance, p, of points from the mean value of a frequency distribution are related to the variable, q, by the equation $p = (1/q) + q$. Determine the standard deviation (i.e. the r.m.s. value), correct to 3 significant figures, for values from $q = 1$ to $q = 3$.   [2.58]

10   A current, $i = 30 \sin 100\pi t$ amperes is applied across an electric circuit. Determine its mean and r.m.s. values, each correct to 4 significant figures, over the range $t = 0$ to $t = 10$ ms.   [19.10 A, 21.21 A]

11   A sinusoidal voltage has a peak value of 340 V. Calculate its mean and r.m.s. values, correct to 3 significant figures.   [216 V, 240 V]

12   Determine the form factor, correct to 3 significant figures, of a sinusoidal voltage of maximum value 100 volts, given that

$$\text{form factor} = \frac{\text{r.m.s. value}}{\text{average value}} \qquad [1.11]$$

26

Volumes of solids of revolution

26.1 Introduction

If the area under the curve $y = f(x)$ (shown in Fig. 26.1(a)), between $x = a$ and $x = b$ is rotated 360° about the x-axis, then a volume known as a **solid of revolution** is produced as shown in Fig. 26.1(b).

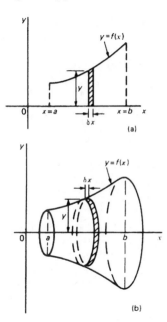

Figure 26.1

The volume of such a solid may be determined precisely using integration.

(i) Let the area shown in Fig. 26.1(a) be divided into a number of strips each of width δx. One such strip is shown shaded.

(ii) When the area is rotated 360° about the x-axis, each strip produces a solid of revolution approximating to a circular disc of radius y and thickness δx.

Volume of disc

$= $ (circular cross-sectional area)(thickness)

$= (\pi y^2)(\delta x)$

(iii) Total volume, V, between ordinates $x = a$ and $x = b$ is given by:

$$\textbf{Volume, } V = \lim_{\delta x \to 0} \sum_{x=a}^{x=b} \pi y^2 \, \delta x$$

$$= \int_a^b \pi y^2 \, dx$$

If a curve $x = f(y)$ is rotated about the y-axis 360° between the limits $y = c$ and $y = d$, as shown in Fig. 26.2, then the volume generated is given by:

$$\textbf{Volume} = \lim_{\delta y \to 0} \sum_{y=c}^{y=d} \pi x^2 \, \delta y = \int_c^d \pi x^2 \, dy$$

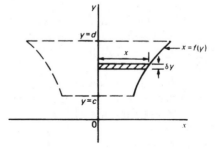

Figure 26.2

26.2 Worked problems on volumes of solids of revolution

> **Problem 1.** Determine the volume of the solid of revolution formed when the curve $y = 2$ is rotated 360° about the x-axis between the limits $x = 0$ to $x = 3$

When $y = 2$ is rotated 360° about the x-axis between $x = 0$ and $x = 3$ (see Fig. 26.3):

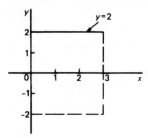

Figure 26.3

volume generated $= \int_0^3 \pi y^2 \, dx = \int_0^3 \pi (2)^2 \, dx$

$$= \int_0^3 4\pi \, dx = 4\pi [x]_0^3$$

$$= 12\pi \text{ cubic units}$$

(Check: The volume generated is a cylinder of radius 2 and height 3. Volume of cylinder $= \pi r^2 h = \pi (2)^2 (3) = \mathbf{12\pi}$ **cubic units.**)

Problem 2.   Find the volume of the solid of revolution when the curve $y = 2x$ is rotated one revolution about the x-axis between the limits $x = 0$ and $x = 5$

When $y = 2x$ is revolved one revolution about the x-axis between $x = 0$ and $x = 5$ (see Fig. 26.4) then:

volume generated $= \int_0^5 \pi y^2 \, dx = \int_0^5 \pi (2x)^2 \, dx$

$$= \int_0^5 4\pi x^2 \, dx = 4\pi \left[\frac{x^3}{3} \right]_0^5$$

$$= \frac{500\pi}{3} = \mathbf{166\frac{2}{3}\pi} \text{ cubic units}$$

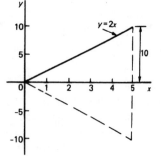

Figure 26.4

(Check: The volume generated is a cone of radius 10 and height 5. Volume of cone $= \frac{1}{3}\pi r^2 h =$

$\frac{1}{3}\pi (10)^2 5 = \frac{500\pi}{3} = \mathbf{166\frac{2}{3}\pi}$ **cubic units.**)

Problem 3.   The curve $y = x^2 + 4$ is rotated one revolution about the x-axis between the limits $x = 1$ and $x = 4$. Determine the volume of the solid of revolution produced

---

Revolving the shaded area shown in Fig. 26.5 360° about the x-axis produces a solid of revolution given by:

$$\text{Volume} = \int_1^4 \pi y^2 \, dx = \int_1^4 \pi (x^2 + 4)^2 \, dx$$

$$= \pi \int_1^4 (x^4 + 8x^2 + 16) \, dx$$

$$= \pi \left[\frac{x^5}{5} + \frac{8x^3}{3} + 16x \right]_1^4$$

$$= \pi [(204.8 + 170.67 + 64)$$

$$- (0.2 + 2.67 + 16)]$$

$$= \mathbf{420.6\pi} \text{ cubic units}$$

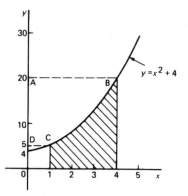

Figure 26.5

Problem 4.   If the curve in Problem 3 is revolved about the y-axis between the same limits, determine the volume of the solid of revolution produced

---

The volume produced when the curve $y = x^2 + 4$ is rotated about the y-axis between $y = 5$ (when

$x = 1$) and $y = 20$ (when $x = 4$)–i.e. rotating area $ABCD$ of Fig. 26.5 about the y-axis is given by:

$$\text{volume} = \int_5^{20} \pi x^2 \, dy$$

Since $y = x^2 + 4$, then $x^2 = y - 4$

Hence volume $= \int_5^{20} \pi(y-4)\, dy = \pi\left[\frac{y^2}{2} - 4y\right]_5^{20}$

$$= \pi\left[(120) - \left(-7\frac{1}{2}\right)\right]$$

$$= 127\frac{1}{2}\pi \text{ cubic units}$$

Problem 5. The area enclosed by the curve $y = 3e^{x/3}$, the x-axis and ordinates $x = -1$ and $x = 3$ is rotated $360°$ about the x-axis. Determine the volume generated

---

A sketch of $y = 3e^{x/3}$ is shown in Fig. 26.6.

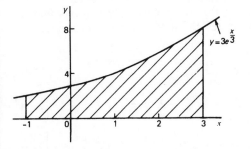

Figure 26.6

When the shaded area is rotated $360°$ about the x-axis then:

volume generated $= \int_{-1}^{3} \pi y^2 \, dx = \int_{-1}^{3} \pi(3e^{x/3})^2 \, dx$

$$= 9\pi \int_{-1}^{3} e^{2x/3} \, dx$$

$$= 9\pi\left[\frac{e^{2x/3}}{\left(\frac{2}{3}\right)}\right]_{-1}^{3}$$

$$= \frac{27\pi}{2}(e^2 - e^{-2/3})$$

$$= 92.82\pi \text{ cubic units}$$

Problem 6. Determine the volume generated when the area above the x-axis bounded by the curve $x^2 + y^2 = 9$ and the ordinates $x = 3$ and $x = -3$ is rotated one revolution about the x-axis

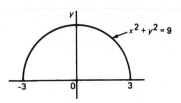

Figure 26.7

Figure 26.7 shows the part of the curve $x^2 + y^2 = 9$ lying above the x-axis. Since, in general, $x^2 + y^2 = r^2$ represents a circle, centre 0 and radius r, then $x^2 + y^2 = 9$ represents a circle, centre 0 and radius 3. When the semicircular area of Fig. 26.7 is rotated one revolution about the x-axis then:

volume generated $= \int_{-3}^{3} \pi y^2 \, dx = \int_{-3}^{3} \pi(9 - x^2)\, dx$

$$= \pi\left[9x - \frac{x^3}{3}\right]_{-3}^{3}$$

$$= \pi[(18) - (-18)]$$

$$= 36\pi \text{ cubic units}$$

(Check: The volume generated is a sphere of radius 3. Volume of sphere $= \frac{4}{3}\pi r^3 = \frac{4}{3}\pi(3)^3 = 36\pi$ cubic units.)

Problem 7. Calculate the volume of a frustum of a sphere of radius 4 cm which lies between two parallel planes at 1 cm and 3 cm from the centre and on the same side of it

---

The volume of a frustum of a sphere may be determined by integration by rotating the curve $x^2 + y^2 = 4^2$ (i.e. a circle, centre 0, radius 4) one revolution about the x-axis, between the limits $x = 1$ and $x = 3$ (i.e. rotating the shaded area of Fig. 26.8).

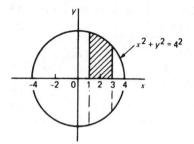

Figure 26.8

Volume of frustum $= \displaystyle\int_1^3 \pi y^2 \, dx = \int_1^3 \pi(4^2 - x^2) \, dx$

$$= \pi \left[16x - \frac{x^3}{3} \right]_1^3$$

$$= \pi \left[(39) - \left(15\frac{2}{3} \right) \right]$$

$$= 23\frac{1}{3} \pi \text{ cubic units}$$

Problem 8.   The area enclosed between the two parabolas $y = x^2$ and $y^2 = 8x$ of Problem 12, Chapter 24, page 248, is rotated $360°$ about the x-axis. Determine the volume of the solid produced

---

The area enclosed by the two curves is shown in Fig. 24.13. The volume produced by revolving the shaded area $360°$ about the x-axis is given by {(volume produced by revolving $y^2 = 8x$) − (volume produced by revolving $y = x^2$)}

i.e. volume $= \displaystyle\int_0^2 \pi(8x) \, dx - \int_0^2 \pi(x^4) \, dx$

$$= \pi \int_0^2 (8x - x^4) \, dx = \pi \left[\frac{8x^2}{2} - \frac{x^5}{5} \right]_0^2$$

$$= \pi \left[\left(16 - \frac{32}{5} \right) - (0) \right]$$

$$= 9.6\pi \text{ cubic units}$$

Further problems on volumes of solids of revolution may be found in the following Section 26.3, Problems 1 to 18.

26.3   Further problems on volumes of solids of revolution

Answers to Problems 1 to 17 are in cubic units and in terms of π. In Problems 1 to 7, determine the volume of the solid of revolution formed by revolving the areas enclosed by the given curve, the x-axis and the given ordinates through one revolution about the x-axis:

1   $y = 5x;\ x = 1,\ x = 4$ $[525\pi]$

2   $y = x^2;\ x = -2,\ x = 3$ $[55\pi]$

3   $y = 2x^2 + 3;\ x = 0,\ x = 2$ $[75.6\pi]$

4   $\dfrac{y^2}{4} = x;\ x = 1,\ x = 5$ $[48\pi]$

5   $xy = 3;\ x = 2,\ x = 3$ $[1.5\pi]$

6   $y = 4e^x;\ x = 0,\ x = 2$ $[428.8\ \pi]$

7   $y = \sec x;\ x = 0,\ x = \dfrac{\pi}{4}$ $[\pi]$

In Problems 8 to 12, determine the volume of the solid of revolution formed by revolving the areas enclosed by the given curves, the y-axis and the given ordinates through one revolution about the y-axis:

8   $y = x^2;\ y = 1,\ y = 3$ $[4\pi]$

9   $y = 3x^2 - 1;\ y = 2,\ y = 4$ $\left[2\frac{2}{3}\pi \right]$

10   $y = \dfrac{2}{x};\ y = 1,\ y = 3$ $\left[2\frac{2}{3}\pi \right]$

11   $x^2 + y^2 = 16;\ y = 0,\ y = 4$ $\left[42\frac{2}{3}\pi \right]$

12   $x\sqrt{y} = 2;\ y = 2,\ y = 3$ $[1.622\pi]$

13   The curve $y = 2x^2 + 3$ is rotated about (a) the x-axis between the limits $x = 0$ and $x = 3$, and (b) the y-axis, between the same limits. Determine the volume generated in each case.

    $[(a)\ 329.4\pi\ (b)\ 81\pi]$

14   Determine the volume of a plug formed by the frustum of a sphere of radius 6 cm which lies between two parallel planes at 2 cm and 4 cm from the centre and on the same side of it. (The equation of a circle, centre 0, radius 4 is $x^2 + y^2 = r^2$.)

    $\left[53\frac{1}{3}\pi \right]$

15 The area enclosed between the two curves $x^2 = 3y$ and $y^2 = 3x$ is rotated about the x-axis. Determine the volume of the solid formed.

$$[8.1\pi]$$

16 The portion of the curve $y = x^2 + \dfrac{1}{x}$ lying between $x = 1$ and $x = 3$ is revolved 360° about the x-axis. Determine the volume of the solid formed. $\left[57\frac{1}{15}\pi\right]$

17 Calculate the volume of the frustum of a sphere of radius 5 cm which lies between two parallel planes at 3 cm and 2 cm from the centre and on opposite sides of it. $\left[113\frac{1}{3}\pi\right]$

18 Sketch the curves $y = x^2 + 2$ and $y - 12 = 3x$ from $x = -3$ to $x = 6$. Determine (a) the coordinates of the points of intersection of the two curves, and (b) the area enclosed by the two curves. (c) If the enclosed area is rotated 360° about the x-axis, calculate the volume of the solid produced.

$$\left[\begin{array}{l} \text{(a) } (-2, 6) \text{ and } (5, 27) \text{ (b) } 57\dfrac{1}{6} \text{ square units} \\ \text{(c) } 1326\pi \text{ cubic units} \end{array}\right]$$

27

Centroids of simple shapes

27.1 Centroids

A **lamina** is a thin flat sheet having uniform thickness. The **centre of gravity** of a lamina is the point where it balances perfectly, i.e. the lamina's **centre of mass.** When dealing with an area (i.e. a lamina of negligible thickness and mass) the term **centre of area** or **centroid** is used for the point where the centre of gravity of a lamina of that shape would lie.

27.2 The first moment of area

The **first moment of area** is defined as the product of the area and the perpendicular distance of its centroid from a given axis in the plane of the area. In Fig. 27.1, the first moment of area A about axis XX is given by (Ay) cubic units.

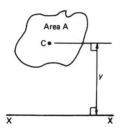

Figure 27.1

27.3 Centroid of area between a curve and the x-axis

(i) Figure 27.2 shows an area $PQRS$ bounded by the curve $y = f(x)$, the x-axis and ordinates $x = a$ and $x = b$. Let this area be divided into a large number of strips, each of width δx. A typical strip is shown shaded drawn at point (x, y) on $f(x)$. The area of the strip is approximately rectangular and is given by $y\,\delta x$.

The centroid, C, has coordinates $\left(x, \dfrac{y}{2}\right)$

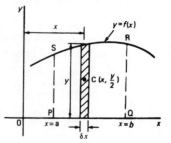

Figure 27.2

(ii) First moment of area of shaded strip about axis $OY = (y\,\delta x)(x) = xy\,\delta x$
Total first moment of area $PQRS$ about axis OY

$$= \lim_{\delta x \to 0} \sum_{x=a}^{x=b} xy\,\delta x = \int_a^b xy\,dx$$

(iii) First moment of area of shaded strip about axis OX

$$= (y\,\delta x)\left(\frac{y}{2}\right) = \frac{1}{2}y^2\,\delta x$$

Total first moment of area $PQRS$ about axis OX

$$= \lim_{\delta x \to 0} \sum_{x=a}^{x=b} \frac{1}{2}y^2\,\delta x = \frac{1}{2}\int_a^b y^2\,dx$$

(iv) Area of $PQRS$, $A = \int_a^b y\,dx$
(from Chapter 24)

(v) Let $\bar{x}$ and $\bar{y}$ be the distances of the centroid of area A about OY and OX, respectively, then:

$(\bar{x})(A) =$ total first moment of area A about axis $OY = \int_a^b xy\,dx$

from which,
$$\bar{x} = \frac{\int_a^b xy\,dx}{\int_a^b y\,dx}$$

and $(\bar{y})(A) =$ total moment of area A about

axis $OX = \dfrac{1}{2}\displaystyle\int_a^b y^2\,dx$

from which, $\quad \bar{y} = \dfrac{\dfrac{1}{2}\displaystyle\int_a^b y^2\,dx}{\displaystyle\int_a^b y\,dx}$

27.4 Centroid of area between a curve and the y-axis

If $\bar{x}$ and $\bar{y}$ are the distances of the centroid of area $EFGH$ in Fig. 27.3 from OY and OX, respectively, then, by similar reasoning as above:

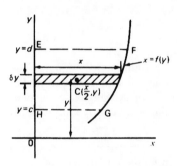

Figure 27.3

$(\bar{x})(\text{total area}) = \displaystyle\lim_{\delta y \to 0} \sum_{y=c}^{y=d} (x\,\delta y)\left(\dfrac{x}{2}\right)$

$= \dfrac{1}{2}\displaystyle\int_c^d x^2\,dy$

from which, $\quad \bar{x} = \dfrac{\dfrac{1}{2}\displaystyle\int_c^d x^2\,dy}{\displaystyle\int_c^d x\,dy}$

and $(\bar{y})(\text{total area}) = \displaystyle\lim_{\delta y \to 0} \sum_{y=c}^{y=d} (x\,\delta y)y$

$= \displaystyle\int_c^d xy\,dy$

from which, $\quad \bar{y} = \dfrac{\displaystyle\int_c^d xy\,dy}{\displaystyle\int_c^d x\,dy}$

27.5 Worked problems on centroids of simple shapes

Problem 1. Show, by integration, that the centroid of a rectangle lies at the intersection of the diagonals

Let a rectangle be formed by the line $y = b$, the x-axis and ordinates $x = 0$ and $x = l$ as shown in Fig. 27.4. Let the coordinates of the centroid C of this area be $(\bar{x}, \bar{y})$.

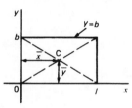

Figure 27.4

By integration, $\bar{x} = \dfrac{\displaystyle\int_0^l xy\,dx}{\displaystyle\int_0^l y\,dx} = \dfrac{\displaystyle\int_0^l (x)(b)\,dx}{\displaystyle\int_0^l b\,dx}$

$= \dfrac{\left[b\dfrac{x^2}{2}\right]_0^l}{[bx]_0^l} = \dfrac{\dfrac{bl^2}{2}}{bl} = \dfrac{l}{2}$

and $\bar{y} = \dfrac{\dfrac{1}{2}\displaystyle\int_0^l y^2\,dx}{\displaystyle\int_0^l y\,dx} = \dfrac{\dfrac{1}{2}\displaystyle\int_0^l b^2\,dx}{bl}$

$= \dfrac{\dfrac{1}{2}[b^2 x]_0^l}{bl} = \dfrac{\dfrac{1}{2}b^2 l}{bl} = \dfrac{b}{2}$

i.e. **the centroid lies at** $\left(\dfrac{l}{2}, \dfrac{b}{2}\right)$ **which is at the intersection of the diagonals.**

Problem 2.   Find the position of the centroid of the area bounded by the curve $y = 3x^2$, the x-axis and the ordinates $x = 0$ and $x = 2$

If $(\bar{x}, \bar{y})$ are the coordinates of the centroid of the given area then:

$$\bar{x} = \frac{\int_0^2 xy\,dx}{\int_0^2 y\,dx} = \frac{\int_0^2 x(3x^2)\,dx}{\int_0^2 3x^2\,dx}$$

$$= \frac{\int_0^2 3x^3\,dx}{\int_0^2 3x^2\,dx} = \frac{\left[\frac{3x^4}{4}\right]_0^2}{\left[x^3\right]_0^2} = \frac{12}{8} = 1.5$$

$$\bar{y} = \frac{\frac{1}{2}\int_0^2 y^2\,dx}{\int_0^2 y\,dx} = \frac{\frac{1}{2}\int_0^2 (3x^2)^2\,dx}{8}$$

$$= \frac{\frac{1}{2}\int_0^2 9x^4\,dx}{8} = \frac{\frac{9}{2}\left[\frac{x^5}{5}\right]_0^2}{8}$$

$$= \frac{\frac{9}{2}\left(\frac{32}{5}\right)}{8} = \frac{18}{5} = 3.6$$

Hence the centroid lies at (1.5, 3.6).

Problem 3.   Determine by integration the position of the centroid of the area enclosed by the line $y = 4x$, the x-axis and ordinates $x = 0$ and $x = 3$

Let the coordinates of the area be $(\bar{x}, \bar{y})$ as shown in Fig. 27.5.

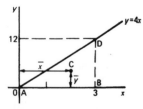

Figure 27.5

Then $\bar{x} = \dfrac{\int_0^3 xy\,dx}{\int_0^3 y\,dx} = \dfrac{\int_0^3 (x)(4x)\,dx}{\int_0^3 4x\,dx}$

$$= \frac{\int_0^3 4x^2\,dx}{\int_0^3 4x\,dx} = \frac{\left[4\frac{x^3}{3}\right]_0^3}{\left[2x^2\right]_0^3} = \frac{36}{18} = 2$$

$$\bar{y} = \frac{\frac{1}{2}\int_0^3 y^2\,dx}{\int_0^3 y\,dx} = \frac{\frac{1}{2}\int_0^3 (4x)^2\,dx}{18}$$

$$= \frac{\frac{1}{2}\int_0^3 16x^2\,dx}{18} = \frac{\frac{1}{2}\left[16\frac{x^3}{3}\right]_0^3}{18}$$

$$= \frac{72}{18} = 4$$

Hence the centroid lies at (2, 4).

In Fig. 27.5, ABD is a right-angled triangle. The centroid lies 4 units from AB and 1 unit from BD showing that the centroid of a triangle lies at one-third of the perpendicular height above any side as base.

Problem 4.   Determine the coordinates of the centroid of the area lying between the curve $y = 5x - x^2$ and the x-axis

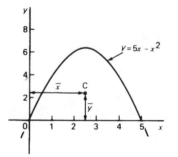

Figure 27.6

$y = 5x - x^2 = x(5 - x)$. When $y = 0$, $x = 0$ or $x = 5$. Hence the curve cuts the x-axis at 0 and 5 as shown in Fig. 27.6. Let the coordinates of the centroid be $(\bar{x}, \bar{y})$ then, by integration,

$$\bar{x} = \frac{\int_0^5 xy\,dx}{\int_0^5 y\,dx} = \frac{\int_0^5 x(5x - x^2)\,dx}{\int_0^5 (5x - x^2)\,dx}$$

$$= \frac{\int_0^5 (5x^2 - x^3)\,dx}{\int_0^5 (5x - x^2)\,dx} = \frac{\left[5\dfrac{x^3}{3} - \dfrac{x^4}{4}\right]_0^5}{\left[\dfrac{5x^2}{2} - \dfrac{x^3}{3}\right]_0^5}$$

$$= \frac{\dfrac{625}{3} - \dfrac{625}{4}}{\dfrac{125}{2} - \dfrac{125}{3}} = \frac{\dfrac{625}{12}}{\dfrac{125}{6}}$$

$$= \left(\frac{625}{12}\right)\left(\frac{6}{125}\right) = \frac{5}{2} = 2.5$$

$$\bar{y} = \frac{\dfrac{1}{2}\int_0^5 y^2\,dx}{\int_0^5 y\,dx} = \frac{\dfrac{1}{2}\int_0^5 (5x - x^2)^2\,dx}{\int_0^5 (5x - x^2)\,dx}$$

$$= \frac{\dfrac{1}{2}\int_0^5 (25x^2 - 10x^3 + x^4)\,dx}{\dfrac{125}{6}}$$

$$= \frac{\dfrac{1}{2}\left[\dfrac{25x^3}{3} - \dfrac{10x^4}{4} + \dfrac{x^5}{5}\right]_0^5}{\dfrac{125}{6}}$$

$$= \frac{\dfrac{1}{2}\left(\dfrac{25(125)}{3} - \dfrac{6250}{4} + 625\right)}{\dfrac{125}{6}} = 2.5$$

Hence the centroid of the area lies at (2.5, 2.5).
(Note from Fig. 27.6 that the curve is symmetrical about $x = 2.5$ and thus x could have been determined 'on sight'.)

Problem 5. Locate the centroid of the area enclosed by the curve $y = 2x^2$, the y-axis and ordinates $y = 1$ and $y = 4$, correct to 3 decimal places

---

From Section 27.4,

$$\bar{x} = \frac{\dfrac{1}{2}\int_1^4 x^2\,dy}{\int_1^4 x\,dy} = \frac{\dfrac{1}{2}\int_1^4 \dfrac{y}{2}\,dy}{\int_1^4 \sqrt{\left(\dfrac{y}{2}\right)}\,dy}$$

$$= \frac{\dfrac{1}{2}\left[\dfrac{y^2}{4}\right]_1^4}{\left[\dfrac{2y^{3/2}}{3\sqrt{2}}\right]_1^4} = \frac{\dfrac{15}{8}}{\dfrac{14}{3\sqrt{2}}} = 0.568$$

$$\bar{y} = \frac{\int_1^4 xy\,dy}{\int_1^4 x\,dy} = \frac{\int_1^4 \sqrt{\left(\dfrac{y}{2}\right)}(y)\,dy}{\dfrac{14}{3\sqrt{2}}} = \frac{\int_1^4 \dfrac{y^{3/2}}{\sqrt{2}}\,dy}{\dfrac{14}{3\sqrt{2}}}$$

$$= \frac{\dfrac{1}{\sqrt{2}}\left[\dfrac{y^{5/2}}{\left(\dfrac{5}{2}\right)}\right]_1^4}{\dfrac{14}{3\sqrt{2}}} = \frac{\dfrac{2}{5\sqrt{2}}(31)}{\dfrac{14}{3\sqrt{2}}} = 2.657$$

Hence the position of the centroid is at (0.568, 2.657).

Problem 6. Locate the position of the centroid enclosed by the curves $y = x^2$ and $y^2 = 8x$

Figure 27.7 shows the two curves intersecting at $(0, 0)$ and $(2, 4)$. These are the same curves as used in Problem 12, Chapter 24, where the shaded area was calculated as $2\frac{2}{3}$ square units. Let the coordinates of centroid C be $\bar{x}$ and $\bar{y}$.

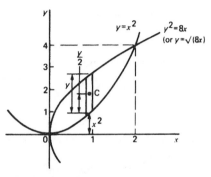

Figure 27.7

By integration, $\bar{x} = \dfrac{\displaystyle\int_0^2 xy\,dx}{\displaystyle\int_0^2 y\,dx}$

The value of y is given by the height of the typical strip shown in Fig. 27.7, i.e. $y = [\sqrt{(8x)} - x^2]$

Hence, $\bar{x} = \dfrac{\displaystyle\int_0^2 x[\sqrt{(8x)} - x^2]\,dx}{2\frac{2}{3}}$

$= \dfrac{\displaystyle\int_0^2 (\sqrt{(8)}x^{3/2} - x^3)\,dx}{2\frac{2}{3}}$

$= \dfrac{\left[\sqrt{(8)}\dfrac{x^{5/2}}{\left(\frac{5}{2}\right)} - \dfrac{x^4}{4}\right]_0^2}{2\frac{2}{3}}$

$= \dfrac{\left(\sqrt{(8)}\dfrac{\sqrt{2^5}}{\frac{5}{2}} - 4\right)}{2\frac{2}{3}} = \dfrac{2\frac{2}{5}}{2\frac{2}{3}} = 0.9$

Care needs to be taken when finding $\bar{y}$ in such examples as this.

From Fig. 27.7, $y = \sqrt{[(8x) - x^2]}$ and $\dfrac{y}{2} = \dfrac{1}{2}[\sqrt{(8x)} - x^2]$

The perpendicular distance from centroid C of the strip to OX is $\dfrac{1}{2}[\sqrt{(8x)} - x^2] + x^2$

Taking moments about OX gives:

(total area) $(\bar{y})$

$= \displaystyle\sum_{x=0}^{x=2} \text{(area of strip)(perpendicular distance of centroid of strip to } OX)$

Hence (area) $(\bar{y})$

$= \displaystyle\int_0^2 [\sqrt{(8x)} - x^2]\left[\dfrac{1}{2}\{\sqrt{(8x)} - x^2\} + x^2\right]\,dx$

$\left(2\dfrac{2}{3}\right)(\bar{y}) = \displaystyle\int_0^2 [\sqrt{(8x)} - x^2]\left(\dfrac{\sqrt{(8x)}}{2} + \dfrac{x^2}{2}\right)\,dx$

$= \displaystyle\int_0^2 \left(\dfrac{8x}{2} - \dfrac{x^4}{2}\right)\,dx = \left[\dfrac{8x^2}{4} - \dfrac{x^5}{10}\right]_0^2$

$= \left(8 - 3\dfrac{1}{5}\right) - (0) = 4\dfrac{4}{5}$

Hence $\bar{y} = \dfrac{4\frac{4}{5}}{2\frac{2}{3}} = 1.8$

Thus the position of the centroid of the shaded area in Fig. 27.7 is at (0.9, 1.8).

Further problems on centroids of simple shapes may be found in Section 27.7, Problems 1 to 14, page 267.

27.6   Theorem of Pappus

The theorem of Pappus states:

'If a plane area is rotated about an axis in its own plane but not intersecting it, the volume of the solid formed is given by the product of the area and the distance moved by the centroid of the area.'

With reference to Fig. 27.8, when the curve $y = f(x)$ is rotated one revolution about the x-axis between the limits $x = a$ and $x = b$, the volume V generated is given by:

volume $V = (A)(2\pi\bar{y})$, from which, $\boxed{\bar{y} = \dfrac{V}{2\pi A}}$

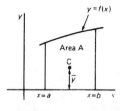

Figure 27.8

Problem 7.   Determine the position of the centroid of a semicircle of radius r by using the theorem of Pappus. Check the answer by using integration (given that the equation of a circle, centre 0, radius r is $x^2 + y^2 = r^2$)

A semicircle is shown in Fig. 27.9 with its diameter lying on the x-axis and its centre at the origin.

Area of semicircle $= \dfrac{\pi r^2}{2}$

When the area is rotated about the x-axis one revolution a sphere is generated of volume $\dfrac{4}{3}\pi r^3$

Let centroid C be at a distance $\bar{y}$ from the origin as shown in Fig. 27.9.

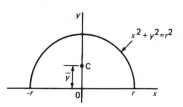

Figure 27.9

From the theorem of Pappus, volume generated $=$ area $\times$ distance moved through by centroid

i.e. $\dfrac{4}{3}\pi r^3 = \left(\dfrac{\pi r^2}{2}\right)(2\pi\bar{y})$

Hence $\qquad \bar{y} = \dfrac{\dfrac{4}{3}\pi r^3}{\pi^2 r^2} = \dfrac{4r}{3\pi}$

By integration, $\bar{y} = \dfrac{\dfrac{1}{2}\displaystyle\int_{-r}^{r} y^2\,dx}{\text{area}}$

$= \dfrac{\dfrac{1}{2}\displaystyle\int_{-r}^{r}(r^2 - x^2)\,dx}{\dfrac{\pi r^2}{2}}$

$= \dfrac{\dfrac{1}{2}\left[r^2 x - \dfrac{x^3}{3}\right]_{-r}^{r}}{\dfrac{\pi r^2}{2}}$

$= \dfrac{\dfrac{1}{2}\left[\left(r^3 - \dfrac{r^3}{3}\right) - \left(-r^3 + \dfrac{r^3}{3}\right)\right]}{\dfrac{\pi r^2}{2}}$

$= \dfrac{4r}{3\pi}$

Hence the centroid of a semicircle lies on the axis of symmetry, distance $\dfrac{4r}{3\pi}$ (or $0.424r$) from its diameter.

Problem 8.   Calculate the area bounded by the curve $y = 2x^2$, the x-axis and ordinates $x = 0$ and $x = 3$. (b) If this area is revolved (i) about the x-axis, and (ii) about the y-axis, find the volumes of the solids produced. (c) Locate the position of the centroid using (i) integration, and (ii) the theorem of Pappus

(a) The required area is shown shaded in Fig. 27.10.

$$\text{Area} = \int_0^3 y\,dx = \int_0^3 2x^2\,dx$$

$$= \left[\dfrac{2x^3}{3}\right]_0^3 = \textbf{18 square units}$$

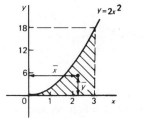

Figure 27.10

(b) (i)   When the shaded area of Fig. 27.10 is revolved $360°$ about the x-axis, the volume generated

$$= \int_0^3 \pi y^2\,dx = \int_0^3 \pi(2x^2)^2\,dx$$

$$= \int_0^3 4\pi x^4\,dx = 4\pi\left[\dfrac{x^5}{5}\right]_0^3 = 4\pi\left(\dfrac{243}{5}\right)$$

$$= \textbf{194.4}\pi \text{ cubic units}$$

(ii)   When the shaded area of Fig. 27.10 is revolved $360°$ about the y-axis, the volume generated

$= $ (volume generated by $x = 3$)

$\quad - $ (volume generated by $y = 2x^2$)

$$= \int_0^{18} \pi(3)^2\,dy - \int_0^{18} \pi\left(\dfrac{y}{2}\right)dy$$

$$= \pi\int_0^{18}\left(9 - \dfrac{y}{2}\right)dy = \pi\left[9y - \dfrac{y^2}{4}\right]_0^{18}$$

$$= \textbf{81}\pi \text{ cubic units}$$

(c) If the coordinates of the centroid of the shaded area in Fig. 27.10 are $(\bar{x}, \bar{y})$ then:

(i) by integration,

$$\bar{x} = \frac{\displaystyle\int_0^3 xy\,dx}{\displaystyle\int_0^3 y\,dx} = \frac{\displaystyle\int_0^3 x(2x^2)\,dx}{18}$$

$$= \frac{\displaystyle\int_0^3 2x^3\,dx}{18} = \frac{\left[\dfrac{2x^4}{4}\right]_0^3}{18} = \frac{81}{36} = 2.25$$

$$\bar{y} = \frac{\dfrac{1}{2}\displaystyle\int_0^3 y^2\,dx}{\displaystyle\int_0^3 y\,dx} = \frac{\dfrac{1}{2}\displaystyle\int_0^3 (2x^2)^2\,dx}{18}$$

$$= \frac{\dfrac{1}{2}\displaystyle\int_0^3 4x^4\,dx}{18} = \frac{\dfrac{1}{2}\left[\dfrac{4x^5}{5}\right]_0^3}{18} = 5.4$$

(ii) Using the theorem of Pappus:

Volume generated when shaded area is revolved about OY = (area)$(2\pi\bar{x})$

i.e. $81\pi = (18)(2\pi\bar{x})$, from which,

$$\bar{x} = \frac{81\pi}{36\pi} = 2.25$$

Volume generated when shaded area is revolved about OX = (area)$(2\pi\bar{y})$

i.e. $194.4\pi = (18)(2\pi\bar{y})$, from which,

$$\bar{y} = \frac{194.4\pi}{36\pi} = 5.4$$

Hence the centroid of the shaded area in Fig. 27.10 is at (2.25, 5.4)

Problem 9. A cylindrical pillar of diameter 400 mm has a groove cut round its circumference. The section of the groove is a semicircle of diameter 50 mm. Determine the volume of material removed, in cubic centimetres, correct to 4 significant figures

---

A part of the pillar showing the groove is shown in Fig. 27.11.

The distance of the centroid of the semicircle from its base is $\dfrac{4r}{3\pi}$ (see Problem 7) $= \dfrac{4(25)}{3\pi} = \dfrac{100}{3\pi}$ mm.

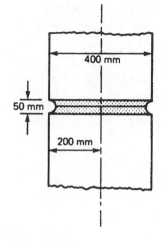

Figure 27.11

The distance of the centroid from the centre of the pillar

$$= \left(200 - \frac{100}{3\pi}\right) \text{ mm}$$

The distance moved by the centroid in one revolution

$$= 2\pi\left(200 - \frac{100}{3\pi}\right)$$

$$= \left(400\pi - \frac{200}{3}\right) \text{ mm}$$

From the theorem of Pappus,

volume = area × distance moved by centroid

$$= \left(\frac{1}{2}\pi 25^2\right)\left(400\pi - \frac{200}{3}\right)$$

$$= 1\,168\,250 \text{ mm}^3$$

Hence the volume of material removed = 1168 cm³, correct to 4 significant figures.

Problem 10. A metal disc has a radius of 5.0 cm and is of thickness 2.0 cm. A semicircular groove of diameter 2.0 cm is machined centrally around the rim to form a pulley. Determine, using Pappus' theorem, the volume and mass of metal removed and the volume and mass of the pulley if the density of the metal is 8000 kg m⁻³

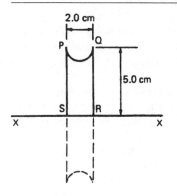

2.0 cm

Figure 27.12

A side view of the rim of the disc is shown in Fig. 27.12.

When area $PQRS$ is rotated about axis XX the volume generated is that of the pulley. The centroid of the semicircular area removed is at a distance of $\dfrac{4r}{3\pi}$ from its diameter (see Problem 7), i.e. $\dfrac{4(1.0)}{3\pi}$, i.e. 0.424 cm from PQ. Thus the distance of the centroid from XX is $5.0 - 0.424$, i.e. 4.576 cm. The distance moved through in one revolution by the centroid is $2\pi(4.576)$ cm.

Area of semicircle $= \dfrac{\pi r^2}{2} = \dfrac{\pi(1.0)^2}{2} = \dfrac{\pi}{2}$ cm^2

By the theorem of Pappus,

volume generated = area × distance moved by centroid $= \left(\dfrac{\pi}{2}\right)(2\pi 4.576)$

i.e. **volume of metal removed = 45.16 cm^3**

Mass of metal removed = density × volume

$$= 8000 \text{ kg m}^{-3} \times \frac{45.16}{10^6} \text{ m}^3$$

$$= 0.3613 \text{ kg or } 361.3 \text{ g}$$

Volume of pulley = volume of cylindrical disc

$$- \text{ volume of metal removed}$$

$$= \pi(5.0)^2(2.0) - 45.16$$

$$= \textbf{111.9 cm}^3$$

Mass of pulley = density × volume

$$= 8000 \text{ kg m}^{-3} \times \frac{111.9}{10^6} \text{ m}^3$$

$$= \textbf{0.8952 kg or 895.2 g}$$

Further problems on the theorem of Pappus may be found in the following Section 27.7, Problems 15 to 19, page 268.

27.7   Further problems on centroids of simple shapes

Centroids of simple shapes

In Problems 1 to 5, find the position of the centroids of the areas bounded by the given curves, the x-axis and the given ordinates:

1 $y = 2x;$ $x = 0, x = 3$ \hfill $[(2, 2)]$

2 $y = 3x + 2;$ $x = 0, x = 4$ \hfill $\left[\left(2\frac{1}{2}, 4\frac{3}{4}\right)\right]$

3 $y = 5x^2;$ $x = 1, x = 4$ \hfill $[(3.036, 24.36)]$

4 $y = 2x^3;$ $x = 0, x = 2$ \hfill $\left[\left(1\frac{3}{5}, 4\frac{4}{7}\right)\right]$

5 $y = x(3x + 1);$ $x = -1, x = 0$ \hfill $\left[\left(-\frac{5}{6}, \frac{19}{30}\right)\right]$

6 Determine the position of the centroid of a sheet of metal formed by the curve $y = 4x - x^2$ which lies above the x-axis. \hfill $[(2, 1.6)]$

7 Find the coordinates of the centroid of the area which lies between the curve $y/x = x - 2$ and the x-axis. \hfill $[(1, -0.4)]$

8 Determine the coordinates of the centroid of the area formed between the curve $y = 9 - x^2$ and the x-axis. \hfill $[(0, 3.6)]$

9 Determine the centroid of the area lying between $y = 4x^2$, the y-axis and the ordinates $y = 0$ and $y = 4$. \hfill $\left[\left(\frac{3}{8}, 2\frac{2}{5}\right)\right]$

10 Find the position of the centroid of the area enclosed by the curve $y = \sqrt{(5x)}$, the x-axis and the ordinate $x = 5$. \hfill $\left[\left(3, 1\frac{7}{8}\right)\right]$

11 Sketch the curve $y^2 = 9x$ between the limits $x = 0$ and $x = 4$. Determine the position of the centroid of this area. \hfill $[(2.4, 0)]$

12 Calculate the points of intersection of the curves $x^2 = 4y$ and $\dfrac{y^2}{4} = x$, and determine the position of the centroid of the area enclosed by them. \hfill $[(0, 0) \text{ and } (4, 4), (1.8, 1.8)]$

13 Determine the position of the centroid of the sector of a circle of radius 3 cm whose angle subtended at the centre is 40°. \hfill [On the centre line, 1.96 cm from the centre]

14 Sketch the curves $y = 2x^2 + 5$ and $y - 8 = x(x + 2)$ on the same axes and determine their points of intersection. Calculate the coordinates of the centroid of the area enclosed by the two curves.   [(−1, 7) and (3, 23), (1, 10.20)]

Theorem of Pappus

15 A right-angled isosceles triangle having a hypotenuse of 8 cm is revolved one revolution about one of its equal sides as axis. Determine the volume of the solid generated using Pappus' theorem.   [189.6 cm³]

16 A rectangle measuring 10.0 cm by 6.0 cm rotates one revolution about one of its longest sides as axis. Determine the volume of the resulting cylinder by using the theorem of Pappus.   [1131 cm²]

17 Using (a) the theorem of Pappus, and (b) integration, determine the position of the centroid of a metal template in the form of a quadrant of a circle of radius 4 cm. (The equation of a circle, centre 0, radius r is $x^2 + y^2 = r^2$.)

[On the centre line, distance 2.40 cm from the centre, i.e. at coordinates (1.70, 1.70)]

18 (a) Determine the area bounded by the curve $y = 5x^2$, the x-axis and the ordinates $x = 0$ and $x = 3$

  (b) If this area is revolved 360° about (i) the x-axis, and (ii) the y-axis, find the volumes of the solids of revolution produced in each case

  (c) Determine the coordinates of the centroid of the area using (i) integral calculus, and (ii) the theorem of Pappus
        [(a) 45 square units
        (b) (i) 1215π cubic units
        (ii) 202.5π cubic units
        (c) (2.25, 13.5)]

19 A metal disc has a radius of 7.0 cm and is of thickness 2.5 cm. A semicircular groove of diameter 2.0 cm is machined centrally around the rim to form a pulley. Determine the volume of metal removed using Pappus' theorem and express this as a percentage of the original volume of the disc. Find also the mass of metal removed if the density of the metal is 7800 kg m⁻³.   [64.90 cm³, 16.86%, 506.2 g]

28

Numerical integration

28.1 Introduction

Even with advanced methods of integration there are many mathematical functions which cannot be integrated by analytical methods and thus approximate methods have then to be used. Approximate methods of definite integrals may be determined by what is termed **numerical integration**.

It is shown in Chapter 24 that determining the value of a definite integral is, in fact, finding the area between a curve, the horizontal axis and the specified ordinates. Three methods of finding approximate areas under curves are the trapezoidal rule, the mid-ordinate rule and Simpson's rule (as explained in Chapter 18), and these rules are used as a basis for numerical integration.

28.2 The trapezoidal rule

Let a required definite integral be denoted by $\int_a^b y\,dx$ and be represented by the area under the graph of $y = f(x)$ between the limits $x = a$ and $x = b$ as shown in Fig. 28.1.

Let the range of integration be divided into n equal intervals each of width d, such that $nd = b-a$,

i.e. $d = \dfrac{b - a}{n}$

The ordinates are labelled $y_1, y_2, y_3, \ldots y_{n+1}$ as shown.

An approximation to the area under the curve may be determined by joining the tops of the ordinates by straight lines. Each interval is thus a trapezium, and since the area of a trapezium is given by:

area $= \dfrac{1}{2}$ (sum of parallel sides)(perpendicular distance between them) then

$$\int_a^b y\,dx \approx \frac{1}{2}(y_1 + y_2)d + \frac{1}{2}(y_2 + y_3)d$$

$$+ \frac{1}{2}(y_3 + y_4)d + \ldots \frac{1}{2}(y_n + y_{n+1})d$$

$$\approx d\left[\frac{1}{2}y_1 + y_2 + y_3 + y_4 + \ldots + y_n\right.$$

$$\left. + \frac{1}{2}y_{n+1}\right]$$

i.e. the trapezoidal rule states:

$$\int_a^b y\,dx \approx \begin{pmatrix}\text{width of} \\ \text{interval}\end{pmatrix}\left\{\frac{1}{2}\begin{pmatrix}\text{first + last} \\ \text{ordinate}\end{pmatrix}\right.$$
$$\left. + \begin{pmatrix}\text{sum of remaining} \\ \text{ordinates}\end{pmatrix}\right\} \quad (1)$$

Problem 1. (a) Use integration to evaluate, correct to 3 decimal places, $\displaystyle\int_1^3 \frac{2}{\sqrt{x}}\,dx$

(b) Use the trapezoidal rule with 4 intervals to evaluate the integral in part (a), correct to 3 decimal places

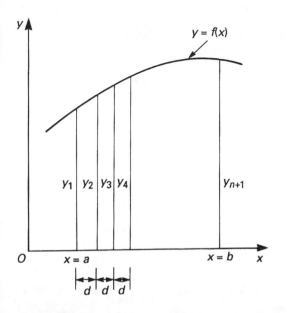

Figure 28.1

(a) $\int_1^3 \dfrac{2}{\sqrt{x}} dx = \int_1^3 2x^{-1/2} dx$

$$= \left[\dfrac{2x^{(-1/2)+1}}{-\dfrac{1}{2}+1}\right]_1^3 = [4x^{1/2}]_1^3$$

$$= 4[\sqrt{x}]_1^3 = 4[\sqrt{3} - \sqrt{1}]$$

$= \mathbf{2.928}$, correct to 3 decimal places

(b) The range of integration is the difference between the upper and lower limits, i.e. $3 - 1 = 2$. Using the trapezoidal rule with 4 intervals gives an interval width $d = \dfrac{3-1}{4} = 0.5$ and ordinates situated at 1.0, 1.5, 2.0, 2.5 and 3.0.

Corresponding values of $\dfrac{2}{\sqrt{x}}$ are shown in the table below, each correct to 4 decimal places (which is one more decimal place than required in the problem).

x	1.0	1.5	2.0	2.5	3.0
$\dfrac{2}{\sqrt{x}}$	2.0000	1.6330	1.4142	1.2649	1.1547

From equation (1):

$$\int_1^3 \dfrac{2}{\sqrt{x}} dx \approx (0.5)\left\{\dfrac{1}{2}(2.000 + 1.1547)\right.$$

$$\left. + 1.6330 + 1.4142 + 1.2649\right\}$$

$= \mathbf{2.945}$, correct to 3 decimal places

This problem demonstrates that even with just 4 intervals a close approximation to the true value of 2.928 (correct to 3 decimal places) is obtained using the trapezoidal rule.

Problem 2.   Use the trapezoidal rule with 8 intervals to evaluate $\int_1^3 \dfrac{2}{\sqrt{x}} dx$, correct to 3 decimal places

---

With 8 intervals, the width of each is $\dfrac{3-1}{8}$, i.e. 0.25 giving ordinates at 1.00, 1.25, 1.50, 1.75, 2.00, 2.25, 2.50, 2.75 and 3.00.

Corresponding values of $\dfrac{2}{\sqrt{x}}$ are shown in the table below.

x	1.00	1.25	1.50	1.75	2.00
$\dfrac{2}{\sqrt{x}}$	2.0000	1.7889	1.6330	1.5119	1.4142

x	2.25	2.50	2.75	3.00
$\dfrac{2}{\sqrt{x}}$	1.3333	1.2649	1.2060	1.1547

From equation (1):

$$\int_1^3 \dfrac{2}{\sqrt{x}} dx \approx (0.25)\left\{\dfrac{1}{2}(2.000 + 1.1547) + 1.7889\right.$$

$$+ 1.6330 + 1.5119 + 1.4142$$

$$\left. + 1.3333 + 1.2649 + 1.2060\right\}$$

$= \mathbf{2.932}$, correct to 3 decimal places

This problem demonstrates that the greater the number of intervals chosen (i.e. the smaller the interval width) the more accurate will be the value of the definite integral. The exact value is found when the number of intervals is infinite, which is, of course, what the process of integration is based upon.

Problem 3.   Use the trapezoidal rule to evaluate $\int_0^{\pi/2} \dfrac{1}{1 + \sin x} dx$, using 6 intervals. Give the answer correct to 4 significant figures

---

With 6 intervals, each will have a width of $\dfrac{\dfrac{\pi}{2} - 0}{6}$, i.e. $\dfrac{\pi}{12}$ rad (or 15°) and the ordinates occur at 0, $\dfrac{\pi}{12}, \dfrac{\pi}{6}, \dfrac{\pi}{4}, \dfrac{\pi}{3}, \dfrac{5\pi}{12}$ and $\dfrac{\pi}{2}$.

Corresponding values of $\dfrac{1}{1 + \sin x}$ are shown in the table below.

x	0	$\dfrac{\pi}{12}$	$\dfrac{\pi}{6}$	$\dfrac{\pi}{4}$
		(or 15°)	(or 30°)	(or 45°)
$\dfrac{1}{1+\sin x}$	1.0000	0.79440	0.66667	0.58579

x	$\dfrac{\pi}{3}$	$\dfrac{5\pi}{12}$	$\dfrac{\pi}{2}$
	(or 60°)	(or 75°)	(or 90°)
$\dfrac{1}{1+\sin x}$	0.53590	0.50867	0.50000

From equation (1):

$$\int_0^{\pi/2} \frac{1}{1+\sin x}\,dx \approx \left(\frac{\pi}{12}\right)\left\{\frac{1}{2}(1.00000+0.50000)\right.$$

$$+\,0.79440+0.66667$$

$$+\,0.58579+0.53590$$

$$\left.+\,0.50867\right\}$$

$$= \mathbf{1.006}, \text{ correct to 4 significant}$$
$$\text{figures}$$

Further problems on the trapezoidal rule may be found in Section 28.5, Problems 1 to 5, page 275, and 17 to 21, page 276.

28.3 The mid-ordinate rule

Let a required definite integral be denoted again by $\int_a^b y\,dx$ and represented by the area under the graph of $y = f(x)$ between the limits $x = a$ and $x = b$, as shown in Fig. 28.2.

With the mid-ordinate rule each interval of width d is assumed to be replaced by a rectangle of height equal to the ordinate at the middle point of each interval, shown as $y_1, y_2, y_3, \ldots y_n$ in Fig. 28.2.

Thus $\displaystyle\int_a^b y\,dx \approx dy_1 + dy_2 + dy_3 + \cdots + dy_n$

$$\approx d(y_1 + y_2 + y_3 + \cdots + y_n)$$

i.e. the mid-ordinate rule states:

$$\int_a^b y\,dx \approx \left(\begin{array}{c}\text{width of}\\\text{interval}\end{array}\right)\left(\begin{array}{c}\text{sum of}\\\text{mid-ordinates}\end{array}\right) \quad (2)$$

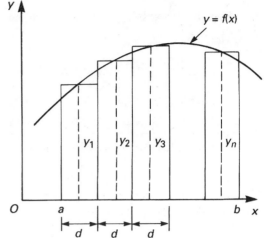

Figure 28.2

Problem 4. Use the mid-ordinate rule with (a) 4 intervals, (b) 8 intervals, to evaluate $\displaystyle\int_1^3 \frac{2}{\sqrt{x}}\,dx$, correct to 3 decimal places

(a) With 4 intervals, each will have a width of $\dfrac{3-1}{4}$, i.e. 0.5 and the ordinates will occur at 1.0, 1.5, 2.0, 2.5 and 3.0. Hence the mid-ordinates y_1, y_2, y_3 and y_4 occur at 1.25, 1.75, 2.25 and 2.75. Corresponding values of $\dfrac{2}{\sqrt{x}}$ are shown in the following table.

x	1.25	1.75	2.25	2.75
$\dfrac{2}{\sqrt{x}}$	1.7889	1.5119	1.3333	1.2060

From equation (2):

$$\int_1^3 \frac{2}{\sqrt{x}}\,dx \approx (0.5)[1.7889 + 1.5119$$

$$+\,1.3333 + 1.2060]$$

$$= \mathbf{2.920}, \text{ correct to 3 decimal}$$
$$\text{places}$$

(b) With 8 intervals, each will have a width of 0.25 and the ordinates will occur at 1.00, 1.25, 1.50, 1.75, ... and thus mid-ordinates at 1.125, 1.375, 1.625, 1.875 ...

Corresponding values of $\dfrac{2}{\sqrt{x}}$ are shown in the following table.

x	1.125	1.375	1.625	1.875
$\dfrac{2}{\sqrt{x}}$	1.8856	1.7056	1.5689	1.4606

x	2.125	2.375	2.625	2.875
$\dfrac{2}{\sqrt{x}}$	1.3720	1.2978	1.2344	1.1795

From equation (2):

$$\int_1^3 \frac{2}{\sqrt{x}}\,dx \approx (0.25)[1.8856 + 1.7056$$

$$+ 1.5689 + 1.4606 + 1.3720$$

$$+ 1.2978 + 1.2344 + 1.1795]$$

$$= \mathbf{2.926}, \text{ correct to 3 decimal}$$
$$\text{places}$$

As previously, the greater the number of intervals the nearer the result is to the true value (of 2.928, correct to 3 decimal places).

Problem 5. Evaluate $\displaystyle\int_0^{2.4} e^{-x^2/3}\,dx$, correct to 4 significant figures, using the mid-ordinate rule with 6 intervals

---

With 6 intervals each will have a width of $\dfrac{2.4 - 0}{6}$, i.e. 0.40 and the ordinates will occur at 0, 0.40, 0.80, 1.20, 1.60, 2.00 and 2.40 and thus mid-ordinates at 0.20, 0.60, 1.00, 1.40, 1.80 and 2.20. Corresponding values of $e^{-x^2/3}$ are shown in the following table.

x	0.20	0.60	1.00	1.40	1.80	2.20
$e^{-x^2/3}$	0.9868	0.8869	0.7165	0.5203	0.3396	0.1992

From equation (2):

$$\int_0^{2.4} e^{-x^2/3}\,dx \approx (0.40)[0.9868 + 0.8869 + 0.7165$$

$$+ 0.5203 + 0.3396 + 0.1992]$$

$$= \mathbf{1.460}, \text{ correct to 4 significant}$$
$$\text{figures}$$

Further problems on the mid-ordinate rule may be found in Section 28.5, Problems 6 to 10, pages 275 and 276, and 17 to 21, page 276.

28.4   Simpson's rule

The approximation made with the trapezoidal rule is to join the top of two successive ordinates by a straight line, i.e. by using a linear approximation of the form $a + bx$. With Simpson's rule, the approximation made is to join the tops of three successive ordinates by a parabola, i.e. by using a quadratic approximation of the form $a + bx + cx^2$.

Figure 28.3 shows a parabola $y = a + bx + cx^2$ with ordinates y_1, y_2 and y_3 at $x = -d$, $x = 0$ and $x = d$, respectively.

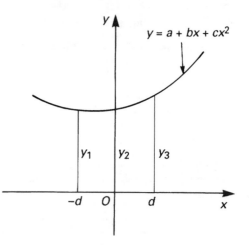

Figure 28.3

Thus the width of each of the two intervals is d. The area enclosed by the parabola, the x-axis and ordinates $x = -d$ and $x = d$ is given by:

$$\int_{-d}^{d} (a + bx + cx^2)\,dx = \left[ax + \frac{bx^2}{2} + \frac{cx^3}{3} \right]_{-d}^{d}$$

$$= \left(ad + \frac{bd^2}{2} + \frac{cd^3}{3} \right)$$

$$- \left(-ad + \frac{bd^2}{2} - \frac{cd^3}{3} \right)$$

$$= 2ad + \frac{2}{3}cd^3 \text{ or}$$

$$\frac{1}{3}d(6a + 2cd^2) \qquad (3)$$

Since $y = a + bx + cx^2$,

at $x = -d$, $y_1 = a - bd + cd^2$

at $x = 0$, $y_2 = a$

and at $x = d$, $y_3 = a + bd + cd^2$

Hence $y_1 + y_3 = 2a + 2cd^2$

And $y_1 + 4y_2 + y_3 = 6a + 2cd^2$ (4)

Thus the area under the parabola between $x = -d$ and $x = d$ in Fig. 28.3 may be expressed as $\frac{1}{3}d(y_1 + 4y_2 + y_3)$, from equations (3) and (4), and the result can be seen to be independent of the position of the origin.

Let a definite integral be denoted by $\int_a^b y\,dx$ and represented by the area under the graph of $y = f(x)$ between the limits $x = a$ and $x = b$, as shown in Fig. 28.4. The range of integration, $b - a$, is divided into an **even** number of intervals, say $2n$, each of width d.

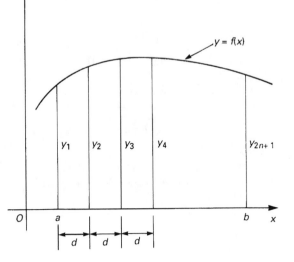

Y

$y = f(x)$

y_1 y_2 y_3 y_4 y_{2n+1}

O   a   b   x

d   d   d

Figure 28.4

Since an even number of intervals is specified, an odd number of ordinates, $2n + 1$, exists. Let an approximation to the curve over the first two intervals be a parabola of the form $y = a + bx + cx^2$ which passes through the tops of the three ordinates y_1, y_2 and y_3. Similarly, let an approximation to the curve over the next two intervals be the parabola which passes through the tops of the ordinates y_3, y_4 and y_5, and so on.

Then

$$\int_a^b y\,dx \approx \frac{1}{3}d(y_1 + 4y_2 + y_3) + \frac{1}{3}d(y_3 + 4y_4 + y_5)$$

$$+ \frac{1}{3}d(y_{2n-1} + 4y_{2n} + y_{2n+1})$$

$$\approx \frac{1}{3}d[(y_1 + y_{2n+1}) + 4(y_2 + y_4 + \cdots + y_{2n})$$

$$+ 2(y_3 + y_5 + \cdots + y_{2n-1})]$$

i.e. Simpson's rule states:

$$\int_a^b y\,dx \approx \frac{1}{3}\left(\begin{array}{c}\text{width of}\\\text{interval}\end{array}\right)\left\{\left(\begin{array}{c}\text{first and last}\\\text{ordinate}\end{array}\right)\right.$$

$$+ 4\left(\begin{array}{c}\text{sum of even}\\\text{ordinates}\end{array}\right)$$

$$\left.+ 2\left(\begin{array}{c}\text{sum of remaining}\\\text{odd ordinates}\end{array}\right)\right\}$$

    (5)

Note that Simpson's rule can only be applied when an even number of intervals is chosen, i.e. an odd number of ordinates.

Problem 6. Use Simpson's rule with (a) 4 intervals, (b) 8 intervals, to evaluate $\int_1^3 \frac{2}{\sqrt{x}}\,dx$, correct to 3 decimal places

(a) With 4 intervals, each will have a width of $\frac{3-1}{4}$, i.e. 0.5 and the ordinates will occur at 1.0, 1.5, 2.0, 2.5 and 3.0. The values of the ordinates are as shown in the table of Problem 1(b).

Thus, from equation (5):

$$\int_1^3 \frac{2}{\sqrt{x}}\,dx \approx \frac{1}{3}(0.5)\{(2.0000 + 1.1547)$$

$$+ 4(1.6330 + 1.2649)$$

$$+ 2(1.4142)\}$$

$$= \frac{1}{3}(0.5)\{3.1547 + 11.5916$$

$$+ 2.8284\}$$

$$= \mathbf{2.929}, \text{ correct to 3 decimal places}$$

(b)   With 8 intervals, each will have a width of $\dfrac{3-1}{8}$, i.e. 0.25 and the ordinates occur at 1.00, 1.25, 1.50, 1.75, . . . , 3.0. The values of the ordinates are as shown in the table in Problem 2.

Thus, from equation (5):

$$\int_1^3 \frac{2}{\sqrt{x}}\,dx \approx \frac{1}{3}(0.25)\{(2.0000 + 1.1547)$$
$$+ 4(1.7889 + 1.5119 + 1.3333$$
$$+ 1.2060) + 2(1.6330 + 1.4142$$
$$+ 1.2649)\}$$
$$= \frac{1}{3}(0.25)\{3.1547 + 23.3604$$
$$+ 8.6242\}$$
$$= \mathbf{2.928}, \text{ correct to 3 decimal}$$
$$\text{places}$$

It is noted that the latter answer is exactly the same as that obtained by integration. In general, Simpson's rule is regarded as the most accurate of the three approximate methods used in numerical integration.

Problem 7.   Evaluate

$$\int_0^{\pi/3} \sqrt{\left(1 - \frac{1}{3}\sin^2\theta\right)}\,d\theta,$$

correct to 3 decimal places, using Simpson's rule with 6 intervals

---

With 6 intervals, each will have a width of $\dfrac{\dfrac{\pi}{3} - 0}{6}$, i.e. $\dfrac{\pi}{18}$ rad (or 10°), and the ordinates will occur at

$0, \dfrac{\pi}{18}, \dfrac{\pi}{9}, \dfrac{\pi}{6}, \dfrac{2\pi}{9}, \dfrac{5\pi}{18}$ and $\dfrac{\pi}{3}$.

Corresponding values of $\sqrt{\left(1 - \dfrac{1}{3}\sin^2\theta\right)}$ are shown in the table below.

θ	0	$\dfrac{\pi}{18}$	$\dfrac{\pi}{9}$	$\dfrac{\pi}{6}$
		(or 10°)	(or 20°)	(or 30°)
$\sqrt{\left(1 - \frac{1}{3}\sin^2\theta\right)}$	1.0000	0.9950	0.9803	0.9574

θ	$\dfrac{2\pi}{9}$	$\dfrac{5\pi}{18}$	$\dfrac{\pi}{3}$
	(or 40°)	(or 50°)	(or 60°)
$\sqrt{\left(1 - \frac{1}{3}\sin^2\theta\right)}$	0.9286	0.8969	0.8660

From equation (5):

$$\int_0^{\pi/3} \sqrt{\left(1 - \frac{1}{3}\sin^2\theta\right)}\,d\theta$$
$$\approx \frac{1}{3}\left(\frac{\pi}{18}\right)\{(1.0000 + 0.8660)$$
$$+ 4(0.9950 + 0.9574 + 0.8969)$$
$$+ 2(0.9803 + 0.9286)\}$$
$$= \frac{1}{3}\left(\frac{\pi}{18}\right)\{1.8660 + 11.3972 + 3.8178\}$$
$$= \mathbf{0.994}, \text{ correct to 3 decimal places}$$

Problem 8.   An alternating current i has the following values at equal intervals of 2.0 milliseconds.

Time (ms)	0	2.0	4.0	6.0	8.0	10.0	12.0
Current i (A)	0	3.5	8.2	10.0	7.3	2.0	0

Charge, q, in millicoulombs, is given by $q = \int_0^{12.0} i\,dt$.

Use Simpson's rule to determine the approximate charge in the 12 millisecond period

---

From equation (5):

$$\text{Charge, } q = \int_0^{12.0} i\,dt \approx \frac{1}{3}(2.0)\{(0 + 0)$$
$$+ 4(3.5 + 10.0 + 2.0) + 2(8.2 + 7.3)\}$$
$$= \mathbf{62\ mC}$$

Problem 9.   Evaluate $\int_1^4 \ln x\,dx$, correct to 4 significant figures, using (i) the trapezoidal rule, (ii) the mid-ordinate rule, (iii) Simpson's rule, each using 6 intervals

(i) Trapezoidal rule

With 6 intervals, each will have a width of $\dfrac{4-1}{6}$, i.e. 0.5, and the ordinates will occur at 1.0, 1.5, 2.0, 2.5, 3.0, 3.5 and 4.0. Corresponding values of $\ln x$ are shown in the table below.

x	1.0	1.5	2.0	2.5
$\ln x$	0	0.4055	0.6931	0.9163

x	3.0	3.5	4.0
$\ln x$	1.0986	1.2528	1.3863

From equation (1):

$$\int_1^4 \ln x\, dx \approx (0.5)\left\{\frac{1}{2}(0 + 1.3863) + 0.4055\right.$$
$$+ 0.6931 + 0.9163 + 1.0986$$
$$\left. + 1.2528\right\}$$
$$= \textbf{2.530}, \text{ correct to 4 significant figures}$$

(ii) Mid-ordinate rule

Each mid-ordinate will occur at 1.25, 1.75, 2.25, 2.75, 3.25 and 3.75. Corresponding values of $\ln x$ are shown in the following table.

x	1.25	1.75	2.25	2.75
$\ln x$	0.2231	0.5596	0.8109	1.0116

x	3.25	3.75
$\ln x$	1.1787	1.3218

From equation (2):

$$\int_1^4 \ln x\, dx \approx (0.5)(0.2331 + 0.5596 + 0.8109$$
$$+ 1.0116 + 1.1787 + 1.3218)$$
$$= \textbf{2.553}, \text{ correct to 4 significant figures}$$

(iii) Simpson's rule

Using the table for part (i) above and from equation (5):

$$\int_1^4 \ln x\, dx \approx \frac{1}{3}(0.5)\{(0 + 1 + 1.3863)$$

$$+ 4(0.4055 + 0.9163 + 1.2528)$$
$$+ 2(0.6931 + 1.0986)\}$$
$$= \frac{1}{3}(0.5)\{1.3863 + 10.2984$$
$$+ 3.5834\}$$
$$= \textbf{2.545}, \text{ correct to 4 significant figures}$$

Further problems on Simpson's rule may be found in the following Section 28.5, Problems 11 to 24, page 276.

28.5 Further problems on numerical integration

In Problems 1 to 5, evaluate the definite integrals using the **trapezoidal rule**, giving the answers correct to 3 decimal places:

1   $\displaystyle\int_0^1 \frac{2}{1+x^2}\, dx$ (Use 8 intervals)   [1.569]

2   $\displaystyle\int_1^3 2\ln 3x\, dx$ (Use 8 intervals)   [6.979]

3   $\displaystyle\int_{\pi/6}^{\pi/3} \tan\theta\, d\theta$ (Use 6 intervals)   [0.551]

4   $\displaystyle\int_0^{\pi/3} \sqrt{(\sin\theta)}\, d\theta$ (Use 6 intervals)   [0.672]

5   $\displaystyle\int_0^{1.4} e^{-x^2}\, dx$ (Use 7 intervals)   [0.843]

In Problems 6 to 10, evaluate the definite integrals using the **mid-ordinate rule**, giving the answers correct to 3 decimal places:

6   $\displaystyle\int_0^2 \frac{3}{1+t^2}\, dt$ (Use 8 intervals)   [3.323]

7   $\displaystyle\int_0^{\pi/2} \frac{1}{1+\sin\theta}\, d\theta$ (Use 6 intervals)   [0.997]

8   $\displaystyle\int_1^3 \frac{\ln x}{x}\, dx$ (Use 10 intervals)   [0.605]

9   $\displaystyle\int_0^{\pi/3} \sqrt{(\cos^3 x)}\, dx$ (Use 6 intervals)   [0.799]

10 $\int_0^{1.6} e^{\frac{2x^3}{5}} \, dx$ (Use 8 intervals)   [2.703]

In Problems 11 to 16, evaluate the definite integrals using **Simpson's rule**, giving the answers correct to 3 decimal places

11 $\int_0^{\pi/2} \sqrt{(\sin x)} \, dx$ (Use 6 intervals)   [1.187]

12 $\int_0^{1.6} \frac{1}{1+\theta^4} \, d\theta$ (Use 8 intervals)   [1.034]

13 $\int_{\pi/6}^{2\pi/3} \frac{1}{1+\cos t} \, dt$ (Use 6 intervals)   [1.465]

14 $\int_{0.2}^{1.0} \frac{\sin \theta}{\theta} \, d\theta$ (Use 8 intervals)   [0.747]

15 $\int_0^{\pi/2} x \cos x \, dx$ (Use 6 intervals)   [0.571]

16 $\int_0^{\pi/3} e^{x^2} \sin 2x \, dx$ (Use 10 intervals)   [1.260]

In Problems 17 and 18 evaluate the definite integrals using (a) integration, (b) the trapezoidal rule, (c) the mid-ordinate rule, (d) Simpson's rule. Give answers correct to 3 decimal places:

17 $\int_1^4 \frac{4}{x^3} \, dx$ (Use 6 intervals)

[(a) 1.875 (b) 2.107 (c) 1.765 (d) 1.916]

18 $\int_2^6 \frac{1}{\sqrt{(2x-1)}} \, dx$ (Use 8 intervals)

[(a) 1.585 (b) 1.588 (c) 1.583 (d) 1.585]

In Problems 19 to 21 evaluate the definite integrals using (a) the trapezoidal rule, (b) the mid-ordinate rule, (c) Simpson's rule. Use 6 intervals in each case and give answers correct to 3 decimal places:

19 $\int_0^{\pi/3} \sqrt{\left(1 - \frac{1}{5}\sin^2 \theta\right)} \, d\theta$

[(a) 1.015 (b) 1.016 (c) 1.016]

20 $\int_0^3 \sqrt{(1+x^4)} \, dx$

[(a) 10.194 (b) 10.007 (c) 10.070]

21 $\int_{0.1}^{0.7} \frac{1}{\sqrt{(1-y^2)}} \, dy$

[(a) 0.677 (b) 0.674 (c) 0.675]

22 A curve is given by the following values:

x	0	1.0	2.0	3.0	4.0	5.0	6.0
y	2	19	40	44	50	10	0

The area under the curve between $x = 0$ and $x = 6.0$ is given by $\int_0^{6.0} y \, dx$. Determine the approximate value of this definite integral, using Simpson's rule.   [158]

23 A vehicle starts from rest and its velocity is measured every second for 8 seconds, with values as follows:

time t (s)	0	1.0	2.0	3.0	4.0	5.0	6.0	7.0	8.0
velocity v (ms^{-1})	0	0.4	1.0	1.7	2.9	4.1	6.2	8.0	9.4

The distance travelled in 8.0 seconds is given by $\int_0^{8.0} v \, dt$.

Estimate this distance using Simpson's rule, giving the answer correct to 3 significant figures.
[28.8 m]

24 A pin moves along a straight guide so that its velocity v (m/s) when it is a distance x (m) from the beginning of the guide at time t (s) is given in the table below.

t (s)	0	0.5	1.0	1.5	2.0
v (m/s)	0	0.052	0.082	0.125	0.162

t (s)	2.5	3.0	3.5	4.0
v (m/s)	0.175	0.186	0.160	0

Use Simpson's rule with 8 intervals to determine the approximate total distance travelled by the pin in the 4.0 second period.   [0.485 m]

Solving equations by iterative methods

29.1 Introduction to iterative methods

Many equations can only be solved graphically or by methods or successive approximations to the roots, called **iterative methods**. Two methods of successive approximations are (i) an algebraic method, introduction in Section 29.2, and (ii) by using the Newton-Raphson formula, given in Section 29.3. Both successive approximation methods rely on a reasonably good first estimate of the value of a root being made. One way of doing this is to sketch a graph of the function, say $y = f(x)$, and determine the approximate values of roots from the points where the graph cuts the x-axis. Another way is by using a functional notation method. This method uses the property that the value of the graph of $f(x) = 0$ changes sign for values of x just before and just after the value of a root. For example, one root of the equation $x^2 - x - 6 = 0$ is $x = 3$. Using functional notation:

$$f(x) = x^2 - x - 6$$
$$f(2) = 2^2 - 2 - 6 = -4$$
$$f(4) = 4^2 - 4 - 6 = +6$$

It can be seen from these results that the value of $f(x)$ changes from -4 at $f(2)$ to $+6$ at $f(4)$, indicating that a root lies between 2 and 4.

29.2 An algebraic method of successive approximations

This method can be used to solve equations of the form:
$a + bx + cx^2 + dx^3 + \ldots = 0$, where $a, b, c, d, \ldots$ are constants.

Procedure:

First approximation

(a) Using a graphical or the functional notation method (see Section 29.1) determine an approximate value of the root required, say x_1

Second approximation

(b) Let the true value of the root be $(x_1 + \delta_1)$
(c) Determine x_2 the approximate value of $(x_1 + \delta_1)$ by determining the value of $f(x_1 + \delta_1) = 0$, but neglecting terms containing products of δ_1

Third approximation

(d) Let the true value of the root be $(x_2 + \delta_2)$
(e) Determine x_3, the approximate value of $(x_2 + \delta_2)$ by determining the value of $f(x_2 + \delta_2) = 0$, but neglecting terms containing products of δ_2
(f) The fourth and higher approximations are obtained in a similar way.

Using the techniques given in paragraphs (b) to (f), it is possible to continue getting values nearer and nearer to the required root. The procedure is repeated until the value of the required root does not change on two consecutive approximations, when expressed to the required degree of accuracy

---

Problem 1. Use an algebraic method of successive approximations to determine the value of the negative root of the quadratic equation: $4x^2 - 6x - 7 = 0$ correct to 3 significant figures. Check the value of the root by using the quadratic formula

---

A first estimate of the values of the roots is made by using the functional notation method

$$f(x) = 4x^2 - 6x - 7$$
$$f(0) = 4(0)^2 - 6(0) - 7 = -7$$
$$f(-1) = 4(-1)^2 - 6(-1) - 7 = 3$$

These results show that the negative root lies between 0 and -1, since the values of $f(x)$ change sign between $f(0)$ and $f(-1)$ (see Section 29.1). The procedure given above for the root lying between 0 and -1 is followed.

First approximation

(a) Let a first approximation be such that it divides the interval 0 to -1 in the ratio of -7 to 3, i.e. let $x_1 = -0.7$

Second approximation

(b) Let the true value of the root, x_2, be $(x_1 + \delta_1)$
(c) Let $f(x_1 + \delta_1) = 0$, then, since $x_1 = -0.7$,

$$4(-0.7 + \delta_1)^2 - 6(-0.7 + \delta_1) - 7 = 0$$

Hence, $4[(-0.7)^2 + (2)(-0.7)(\delta_1) + \delta_1^2] - (6)$
$(-0.7) - 6\delta_1 - 7 = 0$. Neglecting terms containing products of δ_1 gives:
$$1.96 - 5.6\delta_1 + 4.2 - 6\delta_1 - 7 \approx 0$$

i.e. $\delta_1 \approx \dfrac{-1.96 - 4.2 + 7}{-5.6 - 6}$

$$\approx \frac{0.84}{-11.6} \approx -0.072$$

Thus, x_2, a second approximation to the root is $[-0.7 + (-0.072)]$, i.e. $x_2 = -0.772$, correct to 3 significant figures.
The procedure given in (b) and (c) is now repeated for $x_2 = -0.772$

Third approximation

(d) Let the true value of the root, x_3, be $(x_2 + \delta_2)$
(e) Let $f(x_2 + \delta_2) = 0$, then, since $x_2 = -0.772$,
$$4(-0.772 + \delta_2)^2 - 6(-0.772 + \delta_2) - 7 = 0$$

$$4[(-0.772)^2 + (2)(-0.772)(\delta_2) + \delta_2^2]$$
$$- (6)(-0.772) - 6\delta_2 - 7 = 0$$

Neglecting terms containing products of δ_2 gives:
$$2.384 - 6.176\delta_2 + 4.632 - 6\delta_2 - 7 \approx 0$$

i.e. $\delta_2 \approx \dfrac{-2.384 - 4.632 + 7}{-6.176 - 6}$

$$\approx \frac{-0.016}{-12.176} \approx +0.0013$$

Thus x_3, the third approximation to the root is $(-0.772 + 0.0013)$, i.e $x_3 = -0.771$, correct to 3 significant figures

Fourth approximation

(f) The procedure given for the second and third approximations is now repeated for $x_3 = -0.771$

Let the true value of the root, x_4, be $(x_3 + \delta_3)$
Let $f(x_3 + \delta_3) = 0$, then since $x_3 = -0.771$,

$$4(-0.771 + \delta_3)^2 - 6(-0.771 + \delta_3) - 7 = 0$$
$$4[(-0.771)^2 + (2)(-0.771)\delta_3 + \delta_3^2]$$
$$- 6(-0.771) - 6\delta_3 - 7 = 0$$

Neglecting terms containing products of δ_3 gives:
$$2.3778 - 6.168\delta_3 + 4.626 - 6\delta_3 - 7 \approx 0$$

i.e. $\delta_3 \approx \dfrac{-2.3778 - 4.626 + 7}{-6.168 - 6}$

$$\approx \frac{-0.0038}{-12.168} \approx +0.0003$$

Thus, x_4, the fourth approximation to the root is $(-0.771 + 0.0003)$, i.e. $x_4 = -0.771$, correct to 3 significant figures. Since the values of the roots are the same on two consecutive approximations, when stated to the required degree of accuracy, then the negative root of $4x^2 - 6x - 7 = 0$ is $-\mathbf{0.771}$, correct to 3 significant figures. Checking, using the quadratic formula:

$$x = \frac{-(-6) \pm \sqrt{[(-6)^2 - (4)(4)(-7)]}}{(2)(4)}$$

$$= \frac{6 \pm 12.166}{8} = -0.771 \text{ and } 2.27, \text{ correct}$$
$$\text{to 3 significant figures}$$

Problem 2. Determine the value of the smallest positive root of the equation $3x^3 - 10x^2 + 4x + 7 = 0$, correct to 3 significant figures, using an algebraic method of successive approximations

---

The functional notation method is used to find the value of the first approximation.

$$f(x) = 3x^3 - 10x^2 + 4x + 7$$

$$f(0) = 3(0)^3 - 10(0)^2 + 4(0) + 7 = 7$$

$$f(1) = 3(1)^3 - 10(1)^2 + 4(1) + 7 = 4$$

$$f(2) = 3(2)^3 - 10(2)^2 + 4(2) + 7 = -1$$

Following the above procedure:

First approximation

(a) Let the first approximation be such that it divides the interval 1 to 2 in the ratio of 4 to -1, i.e. let x_1 be 1.8

Second approximation

(b) Let the true value of the root, x_2, be $(x_1 + \delta_1)$

(c) Let $f(x_1 + \delta_1) = 0$, then since $x_1 = 1.8$,
$3(1.8+\delta_1)^3 - 10(1.8+\delta_1)^2 + 4(1.8+\delta_1)+7 = 0$
Neglecting terms containing products of δ_1 and using the binomial series gives:

$$3[1.8^3 + 3(1.8)^2\,\delta_1] - 10[1.8^2 + (2)(1.8)\,\delta_1]$$
$$+ 4(1.8 + \delta_1) + 7 \approx 0$$

$$3(5.832 + 9.72\,\delta_1) - 32.4 - 36\,\delta_1 + 7.2$$
$$+ 4\,\delta_1 + 7 \approx 0$$

$$17.496 + 29.16\,\delta_1 - 32.4 - 36\,\delta_1 + 7.2$$
$$+ 4\,\delta_1 + 7 \approx 0$$

$$\delta_1 \approx \frac{-17.496 + 32.4 - 7.2 - 7}{29.16 - 36 + 4} \approx -\frac{0.704}{2.84}$$
$$\approx -0.25$$

Thus $x_2 \approx 1.8 - 0.25 = 1.55$

Third approximation

(d) Let the true value of the root, x_3, be $(x_2 + \delta_2)$

(e) Let $f(x_2 + \delta_2) = 0$, then since $x_2 = 1.55$,
$3(1.55 + \delta_2)^3 - 10(1.55 + \delta_2)^2 + 4(1.55 + \delta_2) + 7 = 0$
Neglecting terms containing products of δ_2 gives:

$$11.17 + 21.62\,\delta_2 - 24.03 - 31\,\delta_2 + 6.2 + 4\,\delta_2 + 7 \approx 0$$

$$\delta_2 \approx \frac{-11.17 + 24.03 - 6.2 - 7}{21.62 - 31 + 4}$$
$$\approx \frac{-0.34}{-5.38} \approx 0.063$$

Thus $x_3 \approx 1.55 + 0.063 \approx 1.613$

(f) Values of x_4 and x_5 are found in a similar way.
$f(x_3 + \delta_3) = 3(1.613 + \delta_3)^3 - 10(1.613 + \delta_3)^2 + 4(1.613 + \delta_3) + 7 = 0$ giving $\delta_3 \approx 0.005$ and $x_4 \approx 1.618$, i.e. 1.62 correct to 3 significant figures.

$f(x_4 + \delta_4) = 3(1.618 + \delta_4)^3 - 10(1.618 + \delta_4)^2 + 4(1.618 + \delta_4) + 7 = 0$ giving $\delta_4 \approx 0$, correct to 4 significant figures and $x_5 \approx 1.62$, correct to 3 significant figures.

Since x_4 and x_5 are the same when expressed to the required degree of accuracy, then the required root is **1.62**, correct to 3 significant figures

(**Note on accuracy and errors**. Depending on the accuracy of evaluating the $f(x+\delta)$ terms, one or two iterations (i.e. successive approximations) might be saved. However, it is not usual to work to more than about 4 significant figures accuracy in this type of calculation. If a small error is made in calculations, the only likely effect is to increase the number of iterations.)

Further problems on solving equations by an algebraic method of successive approximations may be found in Section 29.4, Problems 1 to 5, page 281.

29.3 The Newton-Raphson method

The Newton-Raphson formula, often just referred to as **Newton's method**, may be stated as follows:

if r_1 is the approximate value of a real root of the equation $f(x) = 0$, then a closer approximation to the root r_2 is given by:

$$r_2 = r_1 - \frac{f(r_1)}{f'(r_1)}$$

If, as occasionally happens, the successive approximations of a root do not converge towards the value of the root, a new value of r_1 should be selected so that $f(r_1)$ has the same sign as $f''(r_1)$. The advantages of Newton's method over the algebraic method of successive approximations is that it can be used for any type of mathematical equation (i.e. ones containing trigonometric, exponential, logarithmic, hyperbolic and algebraic functions), and it is usually easier to apply than the algebraic method.

> Problem 3. Use Newton's method to determine the positive root of the quadratic equation $5x^2 + 11x - 17 = 0$, correct to 3 significant figures. Check the value of the root by using the quadratic formula

The functional notation method is used to determine the first approximation to the root.

$$f(x) = 5x^2 + 11x - 17$$
$$f(0) = 5(0)^2 + 11(0) - 17 = -17$$
$$f(1) = 5(1)^2 + 11(1) - 17 = -1$$
$$f(2) = 5(2)^2 + 11(2) - 17 = 25$$

This shows that the value of the root is close to $x = 1$.

Let the first approximation to the root, r_1, be 1. Newton's formula states that:

a closer approximation, $\quad r_2 = r_1 - \dfrac{f(r_1)}{f'(r_1)}$

$f(x) = 5x^2 + 11x - 17$, thus,

$f(r_1) = 5(r_1)^2 + 11(r_1) - 17$

$\qquad = 5(1)^2 + 11(1) - 17 = -1$

$f'(x)$ is the differential coefficient of $f(x)$, i.e. $f'(x) = 10x + 11$. Thus $f'(r_1) = 10(r_1) + 11 = 10(1) + 11 = 21$.

By Newton's formula, a better approximation to the root is:

$r_2 = 1 - \dfrac{-1}{21} = 1 - (-0.048) = 1.05$, correct to 3 significant figures

A still better approximation to the root, r_3, is given by:

$r_3 = r_2 - \dfrac{f(r_2)}{f'(r_2)}$

$\quad = 1.05 - \dfrac{[5(1.05)^2 + 11(1.05) - 17]}{[10(1.05) + 11]}$

$\quad = 1.05 - \dfrac{0.063}{21.5} = 1.05 - 0.003$

$\quad = 1.047$

i.e. 1.05, correct to 3 significant figures.

Since the values of r_2 and r_3 are the same when expressed to the required degree of accuracy, the required root is **1.05**, correct to 3 significant figures. Checking, using the quadratic equation formula,

$x = \dfrac{-11 \pm \sqrt{[121 - 4(5)(-17)]}}{(2)(5)}$

$\quad = \dfrac{-11 \pm 21.47}{10}$

The positive root is 1.047, i.e. **1.05**, correct to 3 significant figures.

Problem 4. Taking the first approximation as 2, determine the root of the equation $x^2 - 3\sin x + 2\ln(x+1) = 3.5$, correct to 3 significant figures, by using Newton's method

---

Newton's formula states that $r_2 = r_1 - (f(r_1)/f'(r_1))$, where r_1 is a first approximation to the root and r_2 is a better approximation to the root.

Since $f(x) = x^2 - 3\sin x + 2\ln(x+1) - 3.5$

$f(r_1) = f(2) = 2^2 - 3\sin 2 + 2\ln 3 - 3.5$

where $\sin 2$ means the sine of 2 radians

$\qquad = 4 - 2.7279 + 2.1972 - 3.5$

$\qquad = -0.0307$

$f'(x) = 2x - 3\cos x + \dfrac{2}{x+1}$

$f'(r_1) = f'(2) = 2(2) - 3\cos 2 + \dfrac{2}{3}$

$\qquad = 4 + 1.2484 + 0.6667$

$\qquad = 5.9151$

Hence, $r_2 = r_1 - \dfrac{f(r_1)}{f'(r_1)} = 2 - \dfrac{-0.0307}{5.9151}$

$\qquad = 2.005$ or 2.01, correct to 3 significant figures

A still better approximation to the root, r_3, is given by:

$r_3 = r_2 - \dfrac{f(r_2)}{f'(r_2)}$

$\quad = 2.005 - \dfrac{[(2.005)^2 - 3\sin 2.005 + 2\ln 3.005 - 3.5]}{\left[2(2.005) - 3\cos 2.005 + \dfrac{2}{2.005+1}\right]}$

$\quad = 2.005 - \dfrac{(-0.0010)}{5.938} = 2.005 + 0.00017$

i.e. $r_3 = 2.01$, correct to 3 significant figures.

Since the values of r_2 and r_3 are the same when expressed to the required degree of accuracy, then the required root is **2.01**, correct to 3 significant figures.

Problem 5. Use Newton's method to find the positive root of:
$(x+4)^3 - e^{1.92x} + 5\cos\dfrac{x}{3} = 9$, correct to 3 significant figures

The functional notational method is used to determine the approximate value of the root.

$$f(x) = (x + 4)^3 - e^{1.92x} + 5 \cos \frac{x}{3} - 9$$

$$f(0) = (0 + 4)^3 - e^0 + 5 \cos 0 - 9 = 59$$

$$f(1) = 5^3 - e^{1.92} + 5 \cos \frac{1}{3} - 9 \approx 114$$

$$f(2) = 6^3 - e^{3.84} + 5 \cos \frac{2}{3} - 9 \approx 164$$

$$f(3) = 7^3 - e^{5.76} + 5 \cos 1 - 9 \approx 19$$

$$f(4) = 8^3 - e^{7.68} + 5 \cos \frac{4}{3} - 9 \approx -1660$$

From these results, let a first approximation to the root be $r_1 = 3$. Newton's formula states that a better approximation to the root,

$$r_2 = r_1 - \frac{f(r_1)}{f'(r_1)}$$

$$f(r_1) = f(3) = 7^3 - e^{5.76} + 5 \cos 1 - 9$$

$$= 19.35$$

$$f'(x) = 3(x + 4)^2 - 1.92e^{1.92x} - \frac{5}{3} \sin \frac{x}{3}$$

$$f'(r_1) = f'(3) = 3(7)^2 - 1.92e^{5.76} - \frac{5}{3} \sin 1$$

$$= -463.7$$

Thus, $r_3 = 3 - \dfrac{19.35}{-463.7} = 3 + 0.042 = 3.042 = 3.04$, correct to 3 significant figures.

Similarly, $r_3 = 3.042 - \dfrac{f(3.042)}{f'(3.042)}$

$$= 3.042 - \frac{(-1.146)}{(-513.1)}$$

$$= 3.042 - 0.0022$$

$$= 3.0398 = 3.04,$$

correct to 3 significant figures.

Since r_2 and r_3 are the same when expressed to the required degree of accuracy, then the required root is **3.04**, correct to 3 significant figures.

Further problems on Newton's method may be found in the following Section 29.4, Problems 6 to 15.

29.4 Further problems on solving equations by iterative methods

In Problems 1 to 5, use an **algebraic method of successive approximation** to solve the equations given to the accuracy stated:

1 $3x^2 + 5x - 17 = 0$, correct to 3 significant figures.   [−3.36, 1.69]

2 $x^3 - 2x + 14 = 0$, correct to 3 decimal places.   [−2.686]

3 $2x^3 - 10x + 4 = 0$, correct to 3 significant figures.   [−2.41, 0.410, 2.00]

4 $x^4 - 3x^3 + 7x - 5.5 = 0$, correct to 3 significant figures.   [−1.53, 1.68]

5 $x^4 + 12x^3 - 13 = 0$, correct to 4 significant figures.   [−12.01, 1.000]

In Problems 6 to 12, use **Newton's method** to solve the equations given to the accuracy stated:

6 $x^2 - 2x - 13 = 0$, correct to 3 decimal places.   [−2.742, 4.742]

7 $3x^3 - 10x = 14$, correct to 4 significant figures.   [2.313]

8 $4x^3 - 16x^2 - 2x + 7 = 0$, correct to 3 significant places.   [−0.668, 0.652, 4.02]

9 $x^4 - 3x^3 + 7x = 12$, correct to 3 decimal places.   [−1.721, 2.648]

10 $3x^4 - 4x^3 + 7x - 12 = 0$, correct to 3 decimal places.   [−1.386, 1.491]

11 $3 \ln x + 4x = 5$, correct to 3 decimal places.   [1.147]

12 $x^3 = 5 \cos 2x$, correct to 3 significant figures.   [−1.693, −0.846, 0.744]

13 $300e^{-2\theta} + \dfrac{\theta}{2} = 6$, correct to 3 significant figures.   [2.05]

14 A Fourier analysis of the instantaneous value of a waveform can be represented by

$$y = \left(t + \frac{\pi}{4}\right) + \sin t + \frac{1}{8} \sin 3t$$

Use Newton's method to determine the value of t near to 0.04, correct to 4 decimal places, when the amplitude, y, is 0.880.   [0.0399]

15 A damped oscillation of a system is given by the equation: $y = -7.4e^{0.5t} \sin 3t$. Determine the value of t near to 4.2, correct to 3 significant figures, when the magnitude y of the oscillation is zero.   [4.19]

Arithmetic and geometric progressions

30.1 Arithmetic progressions

When a sequence has a constant difference between successive terms it is called an **arithmetic progression** (often abbreviated to AP).

Examples include:

(i) 1, 4, 7, 10, 13, ...

where the **common difference** is 3

and (ii) $a, a + d, a + 2d, a + 3d, ...$

where the common difference is d

If the first term of an AP is 'a' and the common difference is 'd' then the nth term is $a + (n - 1)d$.

In example (i) above, the 7th term is given by $1 + (7 - 1)3 = 19$, which may be readily checked.

The sum S of an AP can be obtained by multiplying the average of all the terms by the number of terms.

The average of all the terms $= (a + l)/2$, where 'a' is the first term and l is the last term, i.e. $l = a + (n - 1)d$, for n terms.

Hence the sum of n terms,

$$S_n = n\left(\frac{a+l}{2}\right) = \frac{n}{2}\{a + [a + (n - 1)d]\}$$

i.e. $S_n = \frac{n}{2}[2a + (n - 1)d]$

For example, the sum of the first 7 terms of the series 1, 4, 7, 10, 13, ... is given by

$$S_7 = \frac{7}{2}[2(1) + (7 - 1)3], \text{ since } a = 1 \text{ and } d = 3$$

$$= \frac{7}{2}[2 + 18] = \frac{7}{2}[20] = 70$$

Problem 1. Determine (a) the ninth, and (b) the sixteenth term of the series 2, 7, 12, 17, ...

2, 7, 12, 17, ... is an arithmetic progression with a common difference, d, of 5.

(a) The nth term of an AP is given by $a + (n - 1)d$

Since the first term $a = 2$, $d = 5$ and $n = 9$ then the 9th term is $2 + (9 - 1)5 = 2 + (8)(5) = 2 + 40 = 42$

(b) The 16th term is $2 + (16 - 1)5 = 2 + (15)(5) = 2 + 75 = 77$

Problem 2. The 6th term of an AP is 17 and the 13th term is 38. Determine the 19th term

---

The nth term of an AP is $a + (n - 1)d$.

The 6th term is $\qquad a + 5d = 17 \qquad$ (1)

The 13th term is $\qquad a + 12d = 38 \qquad$ (2)

Equation (2) − equation (1) gives $7d = 21$, from which, $d = 21/7 = 3$.

Substituting in equation (1) gives: $a + 15 = 17$, from which, $a = 2$.

Hence the 19th term is $a + (n - 1)d = 2 + (19 - 1)3 = 2 + (18)(3) = 2 + 54 = 56$.

Problem 3. Determine the number of the term whose value is 22 in the series $2\frac{1}{2}, 4, 5\frac{1}{2}, 7, ...$

$2\frac{1}{2}, 4, 5\frac{1}{2}, 7, ...$ is an AP where $a = 2\frac{1}{2}$ and $d = 1\frac{1}{2}$.

Hence if the nth term is 22 then: $a + (n - 1)d = 22$

i.e. $2\frac{1}{2} + (n - 1)1\frac{1}{2} = 22$; $(n - 1)1\frac{1}{2} = 22 - 2\frac{1}{2} = 19\frac{1}{2}$

$$n - 1 = \frac{19\frac{1}{2}}{1\frac{1}{2}} = 13 \text{ and } n = 13 + 1 = 14$$

i.e. **the 14th term of the AP is 22**.

Problem 4. Find the sum of the first 12 terms of the series 5, 9, 13, 17, ...

5, 9, 13, 17, ... is an AP where $a = 5$ and $d = 4$.
The sum of n terms of an AP,

$$S_n = \frac{n}{2}[2a + (n-1)d]$$

Hence the sum of the first 12 terms,

$$S_{12} = \frac{12}{2}[2(5) + (12-1)4]$$

$$= 6[10 + 44] = 6(54) = \mathbf{324}$$

Problem 5.   Find the sum of the first 21
terms of the series 3.5, 4.1, 4.7, 5.3, ...

3.5, 4.1, 4.7, 5.3, ... is an AP where $a = 3.5$ and
$d = 0.6$.
The sum of the first 21 terms,

$$S_{21} = \frac{n}{2}[2a + (n-1)d]$$

$$= \frac{21}{2}[2(3.5) + (21-1)0.6] = \frac{21}{2}[7 + 12]$$

$$= \frac{21}{2}(19) = \frac{399}{2} = \mathbf{199.5}$$

Problem 6.   The sum of 7 terms of an AP is
35 and the common difference is 1.2.
Determine the first term of the series

$n = 7$, $d = 1.2$ and $S_7 = 35$.
Since the sum of n terms of an AP is given by

$$S_n = \frac{n}{2}[2a + (n-1)d]$$

then

$$35 = \frac{7}{2}[2a + (7-1)1.2]$$

$$= \frac{7}{2}[2a + 7.2]$$

Hence

$$\frac{35 \times 2}{7} = 2a + 7.2$$

$$10 = 2a + 7.2$$

Thus

$$2a = 10 - 7.2 = 2.8$$

from which

$$a = \frac{2.8}{2} = 1.4$$

i.e. **the first term, $a = 1.4$.**

Problem 7.   Three numbers are in arithmetic
progression. Their sum is 15 and their
product is 80. Determine the three numbers

Let the three numbers be $(a - d)$, a and $(a + d)$.
Then $(a-d) + a + (a+d) = 15$, i.e. $3a = 15$, from
which, $a = 5$.
Also, $a(a-d)(a+d) = 80$, i.e. $a(a^2 - d^2) = 80$.

Since $a = 5, 5(5^2 - d^2) = 80$

$$125 - 5d^2 = 80$$

$$125 - 80 = 5d^2$$

$$45 = 5d^2$$

from which, $d^2 = \frac{45}{9} = 9$. Hence $d = \sqrt{9} = \pm 3$

The three numbers are thus $(5 - 3)$, 5 and $(5 + 3)$,
i.e. **2, 5 and 8.**

Problem 8.   Find the sum of all the numbers
between 0 and 207 which are exactly
divisible by 3

The series 3, 6, 9, 12, ..., 207 is an AP whose first
term $a = 3$ and common difference $d = 3$.

The last term is $a + (n-1)d = 207$

i.e. $3 + (n-1)3 = 207$

from which $(n-1) = \frac{207 - 3}{3} = 68$

Hence $n = 68 + 1 = 69$

The sum of all 69 terms is given by

$$S_{69} = \frac{n}{2}[2a + (n-1)d]$$

$$= \frac{69}{2}[2(3) + (69-1)3]$$

$$= \frac{69}{2}[6 + 204] = \frac{69}{2}(210) = \mathbf{7245}$$

Problem 9.   The first, twelfth and last terms
of an arithmetic progression are 4, $31\frac{1}{2}$, and
$376\frac{1}{2}$, respectively. Determine (a) the number
of terms in the series, (b) the sum of all the
terms, and (c) the 80th term

(a) Let the AP be $a, a+d, a+2d, \ldots, a+(n-1)d$, where $a = 4$.

The 12th term is $a + (12 - 1)d = 31\frac{1}{2}$

i.e. $4 + 11d = 31\frac{1}{2}$

from which, $11d = 31\frac{1}{2} - 4 = 27\frac{1}{2}$

Hence $d = \dfrac{27\frac{1}{2}}{11} = 2\frac{1}{2}$

The last term is $a + (n - 1)d$

i.e. $4 + (n - 1)2\frac{1}{2} = 376\frac{1}{2}$

$(n - 1) = \dfrac{376\frac{1}{2} - 4}{2\frac{1}{2}}$

$= \dfrac{372\frac{1}{2}}{2\frac{1}{2}} = 149$

Hence the number of terms in the series, $n = 149 + 1 = 150$

(b) Sum of all the terms,

$$S_{150} = \frac{n}{2}[2a + (n - 1)d]$$

$$= \frac{150}{2}\left[2(4) + (150 - 1)2\frac{1}{2}\right]$$

$$= 75\left[8 + (149)\left(2\frac{1}{2}\right)\right]$$

$$= 85[8 + 372.5] = 75(380.5)$$

$$= \mathbf{28\,537\frac{1}{2}}$$

(c) The 80th term is

$$a + (n - 1)d = 4 + (80 - 1)2\frac{1}{2}$$

$$= 4 + (79)\left(2\frac{1}{2}\right)$$

$$= 4 + 197.5 = \mathbf{201\frac{1}{2}}$$

Further problems on arithmetic progressions may be found in Section 30.3, Problems 1 to 15, page 287.

30.2 Geometric progressions

When a sequence has a constant ratio between successive terms it is called a **geometric progression** (often abbreviated to GP). The constant is called the **common ratio, r**.

Examples include

(i) $1, 2, 4, 8, \ldots$

where the common ratio is 2

and (ii) $a, ar, ar^2, ar^3, \ldots$

where the common ratio is r

If the first term of a GP is 'a' and the common ratio is r, then the nth term is $\boldsymbol{ar^{n-1}}$ which can be readily checked from the above examples.

For example, the 8th term of the GP $1, 2, 4, 8, \ldots$ is $(1)(2)^7 = \mathbf{128}$, since $a = 1$ and $r = 2$.

Let a GP be $a, ar, ar^2, ar^3, \ldots ar^{n-1}$

then the sum of n terms,

$$S_n = a + ar + ar^2 + ar^3 + \ldots + ar^{n-1} \ldots \quad (1)$$

Multiplying throughout by r gives:

$$rS_n = ar + ar^2 + ar^3 + ar^4 + \ldots + ar^{n-1} + ar^n \ldots \quad (2)$$

Subtracting equation (2) from equation (1) gives:

$$S_n - rS_n = a - ar^n$$

i.e. $S_n(1 - r) = a(1 - r^n)$

Thus the sum of n terms, $S_n = \dfrac{a(1 - r^n)}{(1 - r)}$, which is valid when $r < 1$.

Subtracting equation (1) from equation (2) gives $S_n = a(r^n - 1)/(r - 1)$, which is valid when $r > 1$.

For example, the sum of the first 8 terms of the GP $1, 2, 4, 8, 16, \ldots$ is given by $S_8 = 1(2^8 - 1)/(2 - 1)$, since $a = 1$ and $r = 2$

i.e. $S_8 = \dfrac{1(256 - 1)}{1} = \mathbf{255}$

When the common ratio r of a GP is less than unity, the sum of n terms, $S_n = a(1 - r^n)/(1 - r)$, which may be written as $S_n = [a/(1 - r)] - [ar^n/(1 - r)]$.

Since $r < 1$, r^n becomes less as n increases, i.e. $r^n \to 0$ as $n \to \infty$.

Hence $ar^n/(1-r) \to 0$ as $n \to \infty$.

Thus $S_n \to a/(1-r)$ as $n \to \infty$.

The quantity $a/(1-r)$ is called the **sum to infinity**, S_∞, and is the limiting value of the sum of an infinite number of terms,

i.e. $S_\infty = a/(1-r)$, which is valid when $-1 < r < 1$.

For example, the sum to infinity of the GP $1, \frac{1}{2}, \frac{1}{4}, \ldots$ is $S_\infty = 1/(1-\frac{1}{2})$, since $a = 1$ and $r = \frac{1}{2}$, i.e. $S_\infty = 2$.

Problem 10. Determine the tenth term of the series $3, 6, 12, 24, \ldots$

$3, 6, 12, 24, \ldots$ is a geometric progression with a common ratio r of 2. The nth term of a GP is ar^{n-1}, where a is the first term. Hence the 10th term is $(3)(2)^{10-1} = (3)(2)^9 = 3(512) = \mathbf{1536}$.

Problem 11. Find the sum of the first 7 terms of the series, $\frac{1}{2}, 1\frac{1}{2}, 4\frac{1}{2}, 13\frac{1}{2}, \ldots$

$\frac{1}{2}, 1\frac{1}{2}, 4\frac{1}{2}, 13\frac{1}{2}, \ldots$ is a GP with a common ratio $r = 3$.

The sum of n terms, $S_n = \dfrac{a(r^n - 1)}{(r - 1)}$

Hence $S_7 = \dfrac{\frac{1}{2}(3^7 - 1)}{(3 - 1)} = \dfrac{\frac{1}{2}(2187 - 1)}{2} = \mathbf{546\frac{1}{2}}$

Problem 12. The first term of a geometric progression is 12 and the fifth term is 55. Determine the 8th term and the 11th term

---

The 5th term is given by $ar^4 = 55$, where the first term $a = 12$.

Hence $r^4 = 55/a = 55/12$

and $r = \sqrt[4]{\left(\dfrac{55}{12}\right)} = 1.4632$

The 8th term is $ar^7 = (12)(1.4632)^7 = \mathbf{172.3}$.

The 11th term is $ar^{10} = (12)(1.4632)^{10} = \mathbf{539.8}$.

Problem 13. Which term of the series $2187, 729, 243, \ldots$ is $\frac{1}{9}$?

$2187, 729, 243, \ldots$ is a GP with a common ratio $r = \frac{1}{3}$ and first term $a = 2187$.

The nth term of a GP is given by ar^{n-1}.

Hence $\dfrac{1}{9} = (2187)\left(\dfrac{1}{3}\right)^{n-1}$

from which $\left(\dfrac{1}{3}\right)^{n-1} = \dfrac{1}{(9)(2187)} = \dfrac{1}{3^2 3^7}$

$= \dfrac{1}{3^9} = \left(\dfrac{1}{3}\right)^9$

Thus $(n-1) = 9$, from which, $n = 9 + 1 = 10$.

i.e. $\dfrac{1}{9}$ **is the 10th term of the GP**.

Problem 14. Find the sum of the first 9 terms of the series $72.0, 57.6, 46.08, \ldots$

The common ratio, $r = ar/a = 57.6/72.0 = 0.8$ (also $ar^2/ar = 46.08/57.6 = 0.8$).

The sum of 9 terms,

$$S_9 = \dfrac{a(1 - r^n)}{(1 - r)} = \dfrac{72.0(1 - 0.8^9)}{(1 - 0.8)}$$

$$= \dfrac{72.0(1 - 0.1342)}{0.2} = \mathbf{311.7}$$

Problem 15. Find the sum to infinity of the series $3, 1, \frac{1}{3}, \ldots$

$3, 1, \frac{1}{3}, \ldots$ is a GP of common ratio, $r = \frac{1}{3}$.

The sum to infinity,

$$S_\infty = \dfrac{a}{1 - r} = \dfrac{3}{1 - \frac{1}{3}} = \dfrac{3}{\frac{2}{3}} = \dfrac{9}{2} = 4\frac{1}{2}$$

Problem 16. In a geometric progression the sixth term is 8 times the third term and the sum of the seventh and eighth terms is 192. Determine (a) the common ratio, (b) the first term, and (c) the sum of the fifth to eleventh terms, inclusive.

(a) Let the GP be $a, ar, ar^2, ar^3, \ldots, ar^{n-1}$.

The 3rd term $= ar^2$ and the sixth term $= ar^5$.

The 6th term is 8 times the 3rd.

Hence $ar^5 = 8ar^2$ from which, $r^3 = 8, r = \sqrt[3]{8}$

i.e. **the common ratio $r = 2$**

(b)   The sum of the 7th and 8th terms is 192. Hence $ar^6 + ar^7 = 192$.

Since $r = 2$,   then $64a + 128a = 192$

$$192a = 192$$

from which, a, **the first term $= 1$**

(c)   The sum of the 5th to 11th terms (inclusive) is given by:

$$S_{11} - S_4 = \frac{a(r^{11} - 1)}{(r - 1)} - \frac{a(r^4 - 1)}{(r - 1)}$$

$$= \frac{1(2^{11} - 1)}{(2 - 1)} - \frac{1(2^4 - 1)}{(2 - 1)}$$

$$= (2^{11} - 1) - (2^4 - 1)$$

$$= 2^{11} - 2^4 = 2048 - 16 = \mathbf{2032}$$

Problem 17.   A hire tool firm finds that their net return from hiring tools is decreasing by 10% per annum. If their net gain on a certain tool this year is £400, find the possible total of all future profits from this tool (assuming the tool lasts for ever)

The net gain forms a series:

$$£400 + £400 \times 0.9 + £400 \times 0.9^2 + \dots,$$

which is a GP with $a = 400$ and $r = 0.9$.

The sum to infinity,

$$S_\infty = a/(1 - r) = 400/(1 - 0.9)$$

$$= \mathbf{£4000} = \textbf{total future profits}$$

Problem 18.   If £100 is invested at compound interest of 8% per annum, determine (a) the value after 10 years, and (b) the time, correct to the nearest year, it takes to reach more than £300

(a)   Let the GP be $a, ar, ar^2, \dots ar^n$.

The first term $a = £100$.

The common ratio $r = 1.08$.

Hence the second term is

$$ar = (100)(1.08) = £108$$

which is the value after 1 year,

the third term is

$$ar^2 = (100)(1.08)^2 = £116.64,$$

which is the value after 2 years, and so on.

Thus the value after 10 years

$$= ar^{10} = (100)(1.08)^{10} = \mathbf{£215.89}$$

(b)   When £300 has been reached, $300 = ar^n$

i.e. $300 = 100(1.08)^n$

and $3 = (1.08)^n$

Taking logarithms to base 10 of both sides gives:

$$\lg 3 = \lg(1.08)^n$$

$$= n \lg(1.08), \text{ by the laws of logarithms}$$

from which, $n = \dfrac{\lg 3}{\lg 1.08} = 14.3$

Hence it will take 15 years to reach more than £300

Problem 19.   A drilling machine is to have 6 speeds ranging from 50 rev/min to 750 rev/min. If the speeds form a geometric progression determine their values, each correct to the nearest whole number

---

Let the GP of n terms be given by $a, ar, ar^2, \dots ar^{n-1}$.

The first term $a = 50$ rev/min.

The 6th term is given by ar^{6-1}, which is 750 rev/min,

i.e. $ar^5 = 750$

from which, $r^5 = \dfrac{750}{a} = \dfrac{750}{50} = 15$

Thus the common ratio, $r = \sqrt[5]{15} = 1.7188$.

The first term is $a = 50$ rev/min,

the second term is $ar = (50)(1.7188) = 85.94$,

the third term is $ar^2 = (50)(1.7188)^2 = 147.71$,

the fourth term is $ar^3 = (50)(1.7188)^3 = 253.89$,

the fifth term is $ar^4 = (50)(1.7188)^4 = 436.39$,

the sixth term is $ar^5 = (50)(1.7188)^5 = 750.06$

Hence, correct to the nearest whole number, the 6 speeds of the drilling machine are **50, 86, 148, 254, 436** and **750 rev/min.**

Further problems on geometric progressions may be found in the following Section 30.3, Problems 16 to 28.

30.3 Further problems on arithmetic and geometric progressions

Arithmetic progressions

1 Find the 11th term of the series
8, 14, 20, 26, ... [68]

2 Find the 17th term of the series
11, 10.7, 10.4, 10.1, ... [6.2]

3 The seventh term of a series is 29 and the eleventh term is 54. Determine the sixteenth term. [85.25]

4 Find the 15th term of an arithmetic progression of which the first term is $2\frac{1}{2}$ and the tenth term is 16. $\left[23\frac{1}{2}\right]$

5 Determine the number of the term which is 29 in the series 7, 9.2, 11.4, 13.6, ... [11]

6 Find the sum of the first 11 terms of the series 4, 7, 10, 13, ... [209]

7 Determine the sum of the series
6.5, 8.0, 9.5, 11.0, ..., 32. [346.5]

8 The sum of 15 terms of an arithmetic progression is 202.5 and the common difference is 2. Find the first term of the series. $\left[-\frac{1}{2}\right]$

9 Three numbers are in arithmetic progression. Their sum is 9 and their product is $20\frac{1}{4}$. Determine the three numbers. $\left[1\frac{1}{2}, 3, 4\frac{1}{2}\right]$

10 Find the sum of all the numbers between 5 and 250 which are exactly divisible by 4. [7808]

11 Find the number of terms of the series 5, 8, 11, ... of which the sum is 1025. [25]

12 Insert four terms between 5 and $22\frac{1}{2}$ to form an arithmetic progression. $\left[8\frac{1}{2}, 12, 15\frac{1}{2}, 19\right]$

13 The first, tenth and last terms of an arithmetic progression are 9, 40.5, and 425.5, respectively. Find (a) the number of terms, (b) the sum of all the terms, and (c) the 70th term.
[(a) 120 (b) 26 070 (c) 250.5]

14 On commencing employment a man is paid a salary of £7200 per annum and receives annual increments of £350. Determine his salary in the 9th year and calculate the total he will have received in the first 12 years.
[£10 000, £109 500]

15 An oil company bores a hole 80 m deep. Estimate the cost of boring if the cost is £30 for drilling the first metre with an increase in cost of £2 per metre for each succeeding metre.
[£8720]

Geometric progressions

16 Find the 10th term of the series
5, 10, 20, 40, ... [2560]

17 Determine the sum of the first 7 terms of the series $\frac{1}{4}, \frac{3}{4}, 2\frac{1}{4}, 6\frac{3}{4}, ...$ [273.25]

18 The first term of a geometric progression is 4 and the 6th term is 128. Determine the 8th and 11th terms. [512, 4096]

19 Which term of the series 3, 9, 27, ... is 59 049? [10th]

20 Find the sum of the first 7 terms of the series 2, 5, $12\frac{1}{2}$, ... (correct to 4 significant figures). [812.5]

21 Determine the sum to infinity of the series 4, 2, 1, ... [8]

22 Find the sum to infinity of the series $2\frac{1}{2}, -1\frac{1}{4}, \frac{5}{8}, ...$ $\left[1\frac{2}{3}\right]$

23 In a geometric progression the 5th term is 9 times the 3rd term and the sum of the 6th and 7th terms is 1944. Determine (a) the common ratio, (b) the first term, and (c) the sum of the 4th to 10th terms inclusive.
[(a) 3 (b) 2 (c) 59 022]

24 The value of a lathe originally valued at £3000 depreciates 15% per annum. Calculate its value after 4 years. The machine is sold when its value is less than £550. After how many years is the lathe sold? [£1566, 11 years]

25 If the population of Great Britain is 55 million and is decreasing at 2.4% per annum, what will be the population in 5 years time?
[48.71 M]

26 100 grammes of a radioactive substance disintegrates at a rate of 3% per annum. How much of the substance is left after 11 years?
[71.53 g]

27 If £250 is invested at compound interest of 6% per annum determine (a) the value after 15

years, (b) the time, correct to the nearest year, it takes to reach £750.

[(a) £599.14 (b) 19 years]

28 A drilling machine is to have 8 speeds ranging from 100 rev/min to 1000 rev/min. If the speeds form a geometric progression determine their values, each correct to the nearest whole number.

[100, 139, 193, 268, 373, 518, 720, 1000 rev/min]

31

The binomial theorem

31.1 Pascal's triangle

A binomial expression is one which contains two terms connected by a plus or minus sign. Thus $(p+q)$, $(a+x)^2$, $(2x+y)^3$ are examples of binomial expressions. Expanding $(a+x)^n$ for integer values of n from 0 to 6 gives the following results:

1. $(a+x)^0 =$
2. $(a+x)^1 =$
3. $(a+x)^2 = (a+x)(a+x) =$
4. $(a+x)^3 = (a+x)^2(a+x) =$
5. $(a+x)^4 = (a+x)^3(a+x) =$
6. $(a+x)^5 = (a+x)^4(a+x) =$
7. $(a+x)^6 = (a+x)^5(a+x) =$

1. $\qquad 1$
2. $\qquad a+x$
3. $\qquad a^2 + 2ax + x^2$
4. $\qquad a^3 + 3a^2x + 3ax^2 + x^3$
5. $\qquad a^4 + 4a^3x + 6a^2x^2 + 4ax^3 + x^4$
6. $\qquad a^5 + 5a^4x + 10a^3x^2 + 10a^2x^3 + 5ax^4 + x^5$
7. $a^6 + 6a^5x + 15a^4x^2 + 20a^3x^3 + 15a^2x^4 + 6ax^5 + x^6$

From the above results the following patterns emerge:

(i)   '*a*' decreases in power moving from left to right
(ii) '*x*' increases in power moving from left to right
(iii) The coefficients of each term of the expansions are symmetrical about the middle coefficient when n is even and symmetrical about the two middle coefficients when n is odd
(iv) The coefficients are shown separately in Table 31.1 and this arrangement is known as **Pascal's triangle**. A coefficient of a term may be obtained by adding the two adjacent coefficients immediately above in the previous row. This is shown by the triangles in Table 31.1, where, for example, $1 + 3 = 4$, $10 + 5 = 15$, and so on
(v)   Pascal's triangle method is used for expansions of the form $(a+x)^n$ for integer values of n less than about 8

Table 31.1

$(a+x)^0$					1				
$(a+x)^1$				1		1			
$(a+x)^2$				1	2	1			
$(a+x)^3$			1	3	3	1			
$(a+x)^4$		1	4	6	4	1			
$(a+x)^5$	1	5	10	10	5	1			
$(a+x)^6$	1	6	15	20	15	6	1		

> **Problem 1.** Use the Pascal's triangle method to determine the expansion of $(a+x)^7$

From Table 31.1, the row of Pascal's triangle corresponding to $(a+x)^6$ is as shown in (1) below. Adding adjacent coefficients gives the coefficients of $(a+x)^7$ as shown in (2) below.

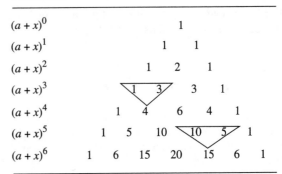

$$1 \quad 6 \quad 15 \quad 20 \quad 15 \quad 6 \quad 1 \qquad (1)$$
$$1 \quad 7 \quad 21 \quad 35 \quad 35 \quad 21 \quad 7 \quad 1 \qquad (2)$$

The first and last terms of the expansion of $(a+x)^7$ are a^7 and x^7, respectively. The powers of '*a*' decrease and the powers of '*x*' increase moving from left to right.

Hence $(a+x)^7 = a^7 + 7a^6x + 21a^5x^2 + 35a^4x^3$
$$+ 35a^3x^4 + 21a^2x^5 + 7ax^6 + x^7$$

> **Problem 2.** Determine, using Pascal's triangle method, the expansion of $(2p - 3q)^5$

Comparing $(2p - 3q)^5$ with $(a+x)^5$ shows that $a = 2p$ and $x = -3q$. Using Pascal's triangle method:
$(a+x)^5 = a^5 + 5a^4x + 10a^3x^2 + 10a^2x^3 + \cdots$

Hence $(2p - 3q)^5 = (2p)^5 + 5(2p)^4(-3q)$

$$+ 10(2p)^3(-3q)^2$$
$$+ 10(2p)^2(-3q)^3$$
$$+ 5(2p)(-3q)^4 + (-3q)^5$$

i.e. $(2p - 3q)^5 = 32p^5 - 240p^4q + 720p^3q^2$
$$- 1080p^2q^3 + 810pq^4 - 243q^5$$

Further problems on Pascal's triangle may be found in Section 31.4, Problems 1 and 2, page 295.

31.2   The binomial theorem

The **binomial theorem** is a formula for raising a binomial expression to any power without lengthy multiplication. The general binomial expansion of $(a + x)^n$ is given by:

$$\boxed{\begin{aligned} (a + x)^n &= a^n + na^{n-1}x + \frac{n(n-1)}{2!}a^{n-2}x^2 \\ &+ \frac{n(n-1)(n-2)}{3!}a^{n-3}x^3 + \cdots + x^n \end{aligned}}$$

where 3! denotes $3 \times 2 \times 1$ and is termed 'factorial 3'. With the binomial theorem n may be a fraction, a decimal fraction or a positive or negative integer.

In the general expansion of $(a + x)^n$ it is noted that the 4th term is: $\frac{n(n-1)(n-2)}{3!}a^{n-3}x^3$. The number 3 is very evident in this expression.

For any term in a binomial expansion, say the rth term, $(r - 1)$ is very evident. It may therefore be reasoned that **the rth term of the expansion $(a + x)^n$** is:

$$\frac{n(n-1)(n-2) \ldots \text{ to } (r-1) \text{ terms}}{(r-1)!}a^{n-(r-1)}x^{r-1}$$

If $a = 1$ in the binomial expansion of $(a + x)^n$ then:

$$\boxed{\begin{aligned} (1 + x)^n &= 1 + nx + \frac{n(n-1)}{2!}x^2 \\ &+ \frac{n(n-1)(n-2)}{3!}x^3 + \ldots, \end{aligned}}$$

which is valid for $-1 < x < 1$.

When x is small compared with 1 then: $(1 + x)^n \approx 1 + nx$.

Problem 3.   Use the binomial theorem to determine the expansion of $(2 + x)^7$

The binomial expansion of

$$(a + x)^n = a^n + na^{n-1}x + \frac{n(n-1)}{2!}a^{n-2}x^2$$
$$+ \frac{n(n-1)(n-2)}{3!}a^{n-3}x^3 + \ldots$$

Where $a = 2$ and $n = 7$:

$$(2 + x)^7 = 2^7 + 7(2)^6x + \frac{(7)(6)}{(2)(1)}(2)^5x^2$$
$$+ \frac{(7)(6)(5)}{(3)(2)(1)}(2)^4x^3 + \frac{(7)(6)(5)(4)}{(4)(3)(2)(1)}(2)^3x^4$$
$$+ \frac{(7)(6)(5)(4)(3)}{(5)(4)(3)(2)(1)}(2)^2x^5$$
$$+ \frac{(7)(6)(5)(4)(3)(2)}{(6)(5)(4)(3)(2)(1)}(2)x^6$$
$$+ \frac{(7)(6)(5)(4)(3)(2)(1)}{(7)(6)(5)(4)(3)(2)(1)}x^7$$

i.e. $(2 + x)^7 = 128 + 448x + 672x^2 + 560x^3$
$$+ 280x^4 + 84x^5 + 14x^6 + x^7$$

Problem 4.   Expand $\left(c - \dfrac{1}{c}\right)^5$ using the binomial theorem

---

$$\left(c - \frac{1}{c}\right)^5 = c^5 + 5c^4\left(-\frac{1}{c}\right) + \frac{(5)(4)}{(2)(1)}c^3\left(-\frac{1}{c}\right)^2$$
$$+ \frac{(5)(4)(3)}{(3)(2)(1)}c^2\left(-\frac{1}{c}\right)^3$$
$$+ \frac{(5)(4)(3)(2)}{(4)(3)(2)(1)}c\left(-\frac{1}{c}\right)^4$$
$$+ \frac{(5)(4)(3)(2)(1)}{(5)(4)(3)(2)(1)}\left(-\frac{1}{c}\right)^5$$

i.e. $\left(c - \dfrac{1}{c}\right)^5 = c^5 - 5c^3 + 10c - \dfrac{10}{c} + \dfrac{5}{c^3} - \dfrac{1}{c^5}$

Problem 5. Without fully expanding $(3 + x)^7$, determine the fifth term

The rth term of the expansion $(a + x)^n$ is given by:

$$\frac{n(n-1)(n-2)\ldots \text{ to } (r-1) \text{ terms}}{(r-1)!} a^{n-(r-1)} x^{r-1}$$

Substituting $n = 7$, $a = 3$ and $r - 1 = 5 - 1 = 4$ gives:

$$\frac{(7)(6)(5)(4)}{(4)(3)(2)(1)}(3)^{7-4} x^4$$

i.e. the fifth term of $(3 + x)^7 = 35(3)^3 x^4 = \mathbf{945x^4}$.

Problem 6. Find the middle term of $$\left(2p - \frac{1}{2q}\right)^{10}$$

In the expansion of $(a+x)^{10}$ there are $10+1$, i.e. 11 terms. Hence the middle term is the sixth. Using the general expression for the rth term where $a = 2p$, $x = -\frac{1}{2q}$, $n = 10$ and $r - 1 = 5$ gives:

$$\frac{(10)(9)(8)(7)(6)}{(5)(4)(3)(2)(1)}(2p)^{10-5}\left(-\frac{1}{2q}\right)^5$$

$$= 252(32p^5)\left(-\frac{1}{32q^5}\right)$$

Hence the middle term of $\left(2p - \frac{1}{2q}\right)^{10}$ is

$$-252\frac{p^5}{q^5}$$

Problem 7. Evaluate $(1.002)^9$ using the binomial theorem correct to (a) 3 decimal places, and (b) 7 significant figures

$$(1+x)^n = 1 + nx + \frac{n(n-1)}{2!}x^2$$

$$+ \frac{n(n-1)(n-2)}{3!}x^3 + \ldots$$

$$(1.002)^9 = (1 + 0.002)^9$$

Substituting $x = 0.002$ and $n = 9$ in the general expansion for $(1+x)^n$ gives:

$$(1 + 0.002)^9 = 1 + 9(0.002) + \frac{(9)(8)}{(2)(1)}(0.002)^2$$

$$+ \frac{(9)(8)(7)}{(3)(2)(1)}(0.002)^3 + \ldots$$

$$= 1 + 0.018 + 0.000144$$

$$+ 0.000000672 + \ldots$$

$$= 1.018144672\ldots$$

Hence $(1.002)^9 = \mathbf{1.018}$,

correct to 3 decimal places

$$= \mathbf{1.018145},$$

correct to 7 significant figures

Problem 8. Evaluate $(0.97)^6$ correct to 4 significant figures using the binomial expansion

$(0.97)^6$ is written as $(1 - 0.03)^6$.
Using the expansion of $(1 + x)^n$ where $n = 6$ and $x = -0.03$ gives:

$$(1 - 0.03)^6 = 1 + 6(-0.03) + \frac{(6)(5)}{(2)(1)}(-0.03)^2$$

$$+ \frac{(6)(5)(4)}{(3)(2)(1)}(-0.03)^3$$

$$+ \frac{(6)(5)(4)(3)}{(4)(3)(2)(1)}(-0.03)^4 + \ldots$$

$$= 1 - 0.18 + 0.0135 - 0.00054$$

$$+ 0.00001215 - \ldots$$

$$\approx 0.83297215$$

i.e. $(\mathbf{0.97})^6 = \mathbf{0.8330}$,

correct to 4 significant figures

Problem 9. Determine the value of $(3.039)^4$, correct to 6 significant figures using the binomial theorem

$(3.039)^4$ may be written in the form $(1 + x)^n$ as:

$$(3.039)^4 = (3 + 0.039)^4 = \left[3\left(1 + \frac{0.039}{3}\right)\right]^4$$

$$= 3^4(1 + 0.013)^4$$

$$(1 + 0.013)^4 = 1 + 4(0.013) + \frac{(4)(3)}{(2)(1)}(0.013)^2$$

$$+ \frac{(4)(3)(2)}{(3)(2)(1)}(0.013)^3 + \ldots$$

$$= 1 + 0.052 + 0.001014$$

$$+ 0.000008788 + \ldots$$

$$= 1.0530228$$

correct to 8 significant figures

Hence $(3.039)^4 = 3^4(1.0530228) = \mathbf{85.2948}$,

correct to 6 significant figures

Problem 10.

(a) Expand $\dfrac{1}{(1 + 2x)^3}$ in ascending powers of x as far as the term in x^3, using the binomial theorem

(b) State the limits of x for which the expansion is valid

(a) Using the binomial expansion of $(1 + x)^n$, where $n = -3$ and x is replaced by $2x$ gives:

$$\frac{1}{(1 + 2x)^3} = (1 + 2x)^{-3} = 1 + (-3)(2x)$$

$$+ \frac{(-3)(-4)}{2!}(2x)^2$$

$$+ \frac{(-3)(-4)(-5)}{3!}(2x)^3 +$$

$$= 1 - 6x + 24x^2 - 80x^3 +$$

(b) The expansion is valid provided $|2x| < 1$, i.e. $|x| < \frac{1}{2}$ or $-\frac{1}{2} < x < \frac{1}{2}$

Problem 11.

(a) Expand $\dfrac{1}{(4 - x)^2}$ in ascending powers of x as far as the term in x^3, using the binomial theorem

(b) What are the limits of x for which the expansion in (a) is true?

(a) $\dfrac{1}{(4 - x)^2} = \dfrac{1}{\left[4\left(1 - \dfrac{x}{4}\right)\right]^2} = \dfrac{1}{4^2\left(1 - \dfrac{x}{4}\right)^2}$

$$= \frac{1}{16}\left(1 - \frac{x}{4}\right)^{-2}$$

Using the expansion of $(1 + x)^n$

$$\frac{1}{(4 - x)^2} = \frac{1}{16}\left(1 - \frac{x}{4}\right)^{-2}$$

$$= \frac{1}{16}\left[1 + (-2)\left(-\frac{x}{4}\right)\right.$$

$$+ \frac{(-2)(-3)}{2!}\left(-\frac{x}{4}\right)^2$$

$$\left. + \frac{(-2)(-3)(-4)}{3!}\left(-\frac{x}{4}\right)^3 + \ldots\right]$$

$$= \frac{1}{16}\left(1 + \frac{x}{2} + \frac{3x^2}{16} + \frac{x^3}{16} + \ldots\right)$$

(b) The expansion in (a) is true provided $\left|\dfrac{x}{4}\right| < 1$, i.e. $|x| < 4$ or $-4 < x < 4$

Problem 12.   Use the binomial theorem to expand $\sqrt{(4 + x)}$ in ascending powers of x to four terms. Give the limits of x for which the expansion is valid

---

$$\sqrt{(4 + x)} = \sqrt{\left[4\left(1 + \frac{x}{4}\right)\right]} = \sqrt{4}\sqrt{\left(1 + \frac{x}{4}\right)}$$

$$= 2\left(1 + \frac{x}{4}\right)^{1/2}$$

Using the expansion of $(1 + x)^n$,

$$2\left(1 + \frac{x}{4}\right)^{1/2} = 2\left[1 + (1/2)\left(\frac{x}{4}\right)\right.$$

$$+ \frac{(1/2)(-1/2)}{2!}\left(\frac{x}{4}\right)^2$$

$$\left. + \frac{(1/2)(-1/2)(-3/2)}{3!}\left(\frac{x}{4}\right)^3 + \ldots\right]$$

$$= 2\left(1 + \frac{x}{8} - \frac{x^2}{128} + \frac{x^3}{1024} - \ldots\right)$$

$$= 2 + \frac{x}{4} - \frac{x^2}{64} + \frac{x^3}{512} - \ldots$$

This is valid when $\left|\dfrac{x}{4}\right| < 1$, i.e. $|x| < 4$ or $-4 < x < 4$

Problem 13.   Expand $\dfrac{1}{\sqrt{(1 - 2t)}}$ in ascending powers of t as far as the term in t^3. State the limits of t for which the expression is valid

$$\frac{1}{\sqrt{(1-2t)}} = (1-2t)^{-1/2} = 1 + (-1/2)(-2t)$$

$$+ \frac{(-1/2)(-3/2)}{2!}(-2t)^2$$

$$+ \frac{(-1/2)(-3/2)(-5/2)}{3!}(-2t)^3 + \dots,$$

using the expansion for $(1+x)^n$

$$= 1 + t + \frac{3}{2}t^2 + \frac{5}{2}t^3 + \dots$$

The expression is valid when $|2t| < 1$, i.e. $|t| < \frac{1}{2}$ or $-\frac{1}{2} < t < \frac{1}{2}$

Problem 14. Simplify $\dfrac{\sqrt[3]{(1-3x)}\sqrt{(1+x)}}{\left(1+\dfrac{x}{2}\right)^3}$

given that powers of x above the first may be neglected

$$\frac{\sqrt[3]{(1-3x)}\sqrt{(1+x)}}{\left(1+\dfrac{x}{2}\right)^3}$$

$$= (1-3x)^{1/3}(1+x)^{1/2}\left(1+\frac{x}{2}\right)^{-3}$$

$$\approx \left[1 + \frac{1}{3}(-3x)\right]\left[1 + \frac{1}{2}(x)\right]\left[1 + (-3)\left(\frac{x}{2}\right)\right],$$

when expanded by the binomial theorem as far as the x term only,

$$= (1-x)\left(1+\frac{x}{2}\right)\left(1-\frac{3x}{2}\right)$$

$$= \left(1 - x + \frac{x}{2} - \frac{3x}{2}\right) \text{ when powers of } x \text{ higher than unity are neglected}$$

$$= (1 - 2x)$$

Problem 15. Express $\dfrac{\sqrt{(1+2x)}}{\sqrt[3]{(1-3x)}}$ as a power series as far as the term in x^2. State the range of values of x for which the series is convergent

---

$$\frac{\sqrt{(1+2x)}}{\sqrt[3]{(1-3x)}} = (1+2x)^{1/2}(1-3x)^{-1/3}$$

$$(1+2x)^{1/2} = 1 + (1/2)(2x)$$

$$+ \frac{(1/2)(-1/2)}{2!}(2x)^2 + \dots$$

$$= 1 + x - \frac{x^2}{2} + \dots \text{ which is valid for } |2x| < 1, \text{ i.e. } |x| < \frac{1}{2}$$

$$(1-3x)^{-1/3} = 1 + (-1/3)(-3x)$$

$$+ \frac{(-1/3)(-4/3)}{2!}(-3x)^2 + \dots$$

$$= 1 + x + 2x^2 + \dots \text{ which is valid for } |3x| < 1, \text{ i.e. } |x| < \frac{1}{3}$$

Hence $\dfrac{\sqrt{(1+2x)}}{\sqrt[3]{(1-3x)}} = (1+2x)^{1/2}(1-3x)^{-1/3}$

$$= \left(1 + x - \frac{x^2}{2} + \dots\right)$$

$$\times (1 + x + 2x^2 + \dots)$$

$$= 1 + x + 2x^2 + x + x^2 - \frac{x^2}{2},$$

neglecting terms of higher power than 2,

$$= 1 + 2x + \frac{5}{2}x^2$$

The series is convergent if $-\dfrac{1}{3} < x < \dfrac{1}{3}$.

Further problems on the binomial theorem may be found in Section 31.4, Problems 3 to 20, pages 295 and 296.

31.3   Practical problems involving the binomial theorem

Binomial expansions may be used for numerical approximations, for calculations with small variations and in probability theory.

Problem 16. The radius of a cylinder is reduced by 4% and its height is increased by 2%. Determine the approximate percentage change in (a) its volume, and (b) its curved surface area (neglecting the products of small quantities)

Volume of cylinder $= \pi r^2 h$

Let r and h be the original values of radius and height.

The new values are $0.96r$ or $(1-0.04)r$ and $1.02h$ or $(1+0.02)h$.

(a) New volume $= \pi[(1-0.04)r]^2[(1+0.02)h]$

$$= \pi r^2 h(1-0.04)^2(1+0.02)$$

Now $(1-0.04)^2 = 1-(2\times 0.04)+(0.04)^2 = (1-0.08)$, neglecting powers of small terms.

Hence new volume

$$\approx \pi r^2 h(1-0.08)(1+0.02)$$

$$\approx \pi r^2 h(1-0.08+0.02), \text{ neglecting products of small terms}$$

$$\approx \pi r^2 h(1-0.06) \text{ or } 0.94\pi r^2 h, \text{ i.e. 94\% of the original volume}$$

Hence the volume is reduced by approximately 6%

(b) Curved surface area of cylinder $= 2\pi rh$.

New surface area

$$= 2\pi[(1-0.04)r][(1+0.02)h]$$

$$= 2\pi rh(1-0.04)(1+0.02)$$

$$\approx 2\pi rh(1-0.04+0.02), \text{ neglecting products of small terms}$$

$$\approx 2\pi rh(1-0.02) \text{ or } 0.98(2\pi rh), \text{ i.e. 98\% of the original surface area}$$

Hence the curved surface area is reduced by approximately 2%

Problem 17.   The second moment of area of a rectangle through its centroid is given by $\dfrac{bl^3}{12}$. Determine the approximate change in the second moment of area if b is increased by 3.5% and l is reduced by 2.5%

New values of b and l are $(1+0.035)b$ and $(1-0.025)l$, respectively.

New second moment of area

$$= \frac{1}{12}[(1+0.035)b][(1-0.025)l]^3$$

$$= \frac{bl^3}{12}(1+0.035)(1-0.025)^3$$

$$\approx \frac{bl}{12}(1+0.035)(1-0.075), \text{ neglecting powers of small terms}$$

$$\approx \frac{bl^3}{12}(1+0.035-0.075), \text{ neglecting products of small terms}$$

$$\approx \frac{bl^3}{12}(1-0.040) \text{ or } (0.96)\frac{bl^3}{12}, \text{ i.e. 96\% of the original second moment of area}$$

Hence the second moment of area is reduced by approximately 4%.

Problem 18.   The resonant frequency of a vibrating shaft is given by: $f = \dfrac{1}{2\pi}\sqrt{\left(\dfrac{k}{I}\right)}$, where k is the stiffness and I is the inertia of the shaft. Use the binomial theorem to determine the approximate percentage error in determining the frequency using the measured values of k and I when the measured value of k is 4% too large and the measured value of I is 2% too small

---

Let f, k and I be the true values of frequency, stiffness and inertia, respectively. Since the measured value of stiffness, k_1, is 4% too large, then

$$k_1 = \frac{104}{100}k = (1+0.04)k$$

The measured value of inertia, I_1, is 2% too small, hence

$$I_1 = \frac{98}{100}I = (1-0.02)I$$

The measured value of frequency,

$$f_1 = \frac{1}{2\pi}\sqrt{\left(\frac{k_1}{I_1}\right)} = \frac{1}{2\pi}k_1^{1/2}I_1^{-1/2}$$

$$= \frac{1}{2\pi}[(1+0.04)k]^{1/2}[(1-0.02)I]^{-1/2}$$

$$= \frac{1}{2\pi}(1+0.04)^{1/2}k^{1/2}(1-0.02)^{-1/2}I^{-1/2}$$

$$= \frac{1}{2\pi}k^{1/2}I^{-1/2}(1+0.04)^{1/2}(1-0.02)^{-1/2}$$

i.e. $f_1 = f(1+0.04)^{1/2}(1-0.02)^{-1/2}$

$$\approx f[1+(1/2)(0.04)][(1+(-1/2)(-0.02)]$$

$$\approx f(1+0.02)(1+0.01)$$

Neglecting the products of small terms,

$$f_1 \approx (1 + 0.02 + 0.01)f \approx 1.03f$$

Thus the percentage error in f based on the measured values of k and I is approximately $[(1.03)(100) - 100]$, i.e. **3% too large**.

Further problems on practical binomial theorem examples may be found in the following Section 31.4, Problems 21 to 30, page 296.

31.4   Further problems on the binomial theorem

Pascal's triangle

1 Use Pascal's triangle to expand $(x - y)^7$.

$$\left[\begin{array}{l} x^7 - 7x^6y + 21x^5y^2 - 35x^4y^3 \\ +35x^3y^4 - 21x^2y^5 + 7xy^6 - y^7 \end{array}\right]$$

2 Expand $(2a + 3b)^5$ using Pascal's triangle.

$$\left[\begin{array}{l} 32a^5 + 240a^4b + 720a^3b^2 \\ +1080a^2b^3 + 810ab^4 + 243b^5 \end{array}\right]$$

Binomial theorem

3 Use the binomial theorem to expand
(a) $(a + 2x)^4$ (b) $(2 - x)^6$.

$$\left[\begin{array}{l} \text{(a) } a^4 + 8a^3x + 24a^2x^2 + 32ax^3 + 16x^4 \\ \text{(b) } 64 - 192x + 240x^2 - 160x^3 \\ \quad +60x^4 - 12x^5 + x^6 \end{array}\right]$$

4 Expand (a) $(2x - 3y)^4$ (b) $\left(2x + \dfrac{2}{x}\right)^5$.

$$\left[\begin{array}{l} \text{(a) } 16x^4 - 96x^3y + 216x^2y^2 \\ \quad -216xy^3 + 81y^4 \\ \text{(b) } 32x^5 + 160x^3 + 320x + \dfrac{320}{x} \\ \quad +\dfrac{160}{x^3} + \dfrac{32}{x^5} \end{array}\right]$$

5 Expand $(p + 2q)^{11}$ as far as the fifth term.

$$\left[\begin{array}{l} p^{11} + 22p^{10}q + 220p^9q^2 \\ +1320p^8q^3 + 5280p^7q^4 \end{array}\right]$$

6 Determine the sixth term of $\left(3p + \dfrac{q}{3}\right)^{13}$.

$$\left[34\,749\,p^8q^5\right]$$

7 Determine the middle term of $(2a - 5b)^8$.

$$\left[700\,000a^4b^4\right]$$

8 Use the binomial theorem to determine, correct to 4 decimal places:

(a) $(1.003)^8$ (b) $(1.042)^7$

$$[\text{(a) } 1.0243 \text{ (b) } 1.3337]$$

9 Use the binomial theorem to determine, correct to 5 significant figures:

(a) $(0.98)^7$ (b) $(2.01)^9$

$$[\text{(a) } 0.86813 \text{ (b) } 535.51]$$

10 Evaluate $(4.044)^6$ correct to 3 decimal places.

$$[4373.880]$$

11 Expand $\left(1 - \dfrac{1}{2}x\right)^5$ using the binomial theorem.

$$\left[1 - \dfrac{5x}{2} + \dfrac{5}{2}x^2 - \dfrac{5}{4}x^3 + \dfrac{5}{16}x^4 - \dfrac{x^5}{32}\right]$$

In Problems 12 to 16 expand in ascending powers of x as far as the term in x^3, using the binomial theorem. State in each case the limits of x for which the series is valid:

12 $\dfrac{1}{(1 - x)}$ $\left[(1 + x + x^2 + x^3 + \ldots),\ |x| < 1\right]$

13 $\dfrac{1}{(1 + x)^2}$

$$\left[1 - 2x + 3x^2 - 4x^3 + \ldots),\ |x| < 1\right]$$

14 $\dfrac{1}{(2 + x)^3}$

$$\left[\dfrac{1}{8}\left(1 - \dfrac{3x}{2} + \dfrac{3x^2}{2} - \dfrac{5}{4}x^3 + \ldots\right),\ |x| < 2\right]$$

15 $\sqrt{(2 + x)}$

$$\left[\sqrt{2}\left(1 + \dfrac{x}{4} - \dfrac{x^2}{32} + \dfrac{x^3}{128} - \ldots\right),\ |x| < 2\right]$$

16 $\dfrac{1}{(\sqrt{(1 + 3x)})}$

$$\left[\left(1 - \dfrac{3x}{2} + \dfrac{27}{8}x^2 - \dfrac{135}{16}x^3 + \ldots\right),\ |x| < \dfrac{1}{3}\right]$$

17 Expand $(2 + 3x)^{-6}$ to three terms. For what values of x is the expansion valid?

$$\left[\dfrac{1}{64}\left(1 - 9x + \dfrac{189}{4}x^2\right),\ |x| < \dfrac{2}{3}\right]$$

18 When x is very small show that:

(a) $\dfrac{1}{(1 - x)^2\sqrt{(1 - x)}} \approx 1 + \dfrac{5}{2}x$

(b) $\dfrac{(1 - 2x)}{(1 - 3x)^4} \approx 1 + 10x$

(c) $\dfrac{\sqrt{(1 + 5x)}}{\sqrt[3]{(1 - 2x)}} \approx 1 + \dfrac{19}{6}x$

19 If x is very small such that x^2 and higher powers may be neglected, determine the power series for $\dfrac{\sqrt{(x+4)}\sqrt[3]{(8-x)}}{\sqrt[5]{(1+x)^3}}$. $\left[4 - \dfrac{31}{15}x\right]$

20 Express the following as power series in ascending powers of x as far as the term in x^2. State in each case the range of x for which the series is valid.

(a) $\sqrt{\left(\dfrac{1-x}{1+x}\right)}$ (b) $\dfrac{(1+x)\sqrt[3]{(1-3x)^2}}{\sqrt{(1+x^2)}}$

$$\left[\begin{array}{l} \text{(a) } 1 - x + \dfrac{1}{2}x^2, \ |x| < 1 \\[2mm] \text{(b) } 1 - x - \dfrac{3}{2}x^2, \ |x| < \dfrac{1}{3} \end{array}\right]$$

Practical problems involving the binomial theorem

21 Pressure p and volume v are related by $pv^3 = c$, where c is a constant. Determine the approximate percentage change in c when p is increased by 3% and v decreased by 1.2%.
[0.6% decrease]

22 Kinetic energy is given by $\frac{1}{2}mv^2$. Determine the approximate change in the kinetic energy when mass m is increased by 2.5% and the velocity v is reduced by 3%.   [3.5% decrease]

23 An error of +1.5% was made when measuring the radius of a sphere. Ignoring the products of small quantities determine the approximate error in calculating (a) the volume, and (b) the surface area.
[(a) 4.5% increase (b) 3.0% increase]

24 The power developed by an engine is given by $I = k$ PLAN, where k is a constant. Determine the approximate percentage change in the power when P and A are each increased by 2.5% and L and N are each decreased by 1.4%.
[2.2% increase]

25 The radius of a cone is increased by 2.7% and its height reduced by 0.9%. Determine the approximate percentage change in its volume, neglecting the products of small terms.
[4.5% increase]

26 The electric field strength H due to a magnet of length $2l$ and moment M at a point on its axis distance x from the centre is given by

$$H = \dfrac{M}{2l}\left\{\dfrac{1}{(x-l)^2} - \dfrac{1}{(x+l)^2}\right\}$$

Show that if l is very small compared with x, then $H \approx \dfrac{2M}{x^3}$.

27 The shear stress τ in a shaft of diameter D under a torque T is given by: $\tau = \dfrac{kT}{\pi D^3}$. Determine the approximate percentage error in calculating τ if T is measured 3% too small and D 1.5% too large.   [7.5% decrease]

28 The energy W stored in a flywheel is given by: $W = kr^5N^2$, where k is a constant, r is the radius and N the number of revolutions. Determine the approximate percentage change in W when r is increased by 1.3% and N is decreased by 2%.   [2.5% increase]

29 In a series electrical circuit containing inductance L and capacitance C the resonant frequency is given by: $f_r = \dfrac{1}{2\pi\sqrt{(LC)}}$. If the values of L and C used in the calculation are 2.6% too large and 0.8% too small, respectively, determine the approximate percentage error in the frequency.   [0.9% too small]

30 The viscosity η of a liquid is given by: $\eta = \dfrac{kr^4}{vl}$, where k is a constant. If there is an error in r of +2%, in v of +4% and l of −3%, what is the resultant error in η?   [+7%]

Multiple choice questions on Part 2 Further Mathematics for Engineering

All questions have only one answer

Suggested time allowed: 1 hour

1 The vertical height y kilometres of a rocket fired from a launcher varies with the horizontal distance x kilometres and is given by $y = 6x - x^2$. The mean height of the rocket from $x = 0$ to $x = 6$ km is

A 3 km   B 6 km
C 9 km   D 36 km

2 A vehicle starts from rest and its velocity is measured every second for 6 seconds, with values as follows:

time t (s)	0	1.0	2.0	3.0	4.0	5.0	6.0
velocity v (m/s)	0	5	11	19	28	41	55

The distance covered in 6 seconds, using Simpson's rule, is

A 131.5 m   B 159 m
C 131 m   D 393 m

3 The equation $x^4 - 3x^2 - 3x + 1 = 0$ has

A 1 real root   B 2 real roots
C 3 real roots   D 4 real roots

4 If $y = 4x^2 - \ln 3x$, then $\dfrac{d^2 y}{dx^2}$ is equal to

A $8 + \dfrac{1}{x^2}$ B $8x - \dfrac{1}{x}$

C $8 - \dfrac{1}{3x}$ D $8 + \dfrac{1}{3x^2}$

5 $(a + x)^4 = a^4 + 4a^3 x + 6a^2 x^2 + 4ax^3 + x^4$
Using Pascal's triangle, the third term of $(a + x)^5$ is

A $10a^2 x^3$ B $5a^4 x$
C $5a^3 x^2$ D $10a^3 x^2$

6 $\displaystyle \int \left(1 + \frac{3}{e^{4x}} \right) dx$ is equal to

A $\dfrac{12}{e^{4x}} + c$ B $x + \dfrac{12}{e^{4x}} + c$

C $x - \dfrac{3}{4e^{4x}} + c$ D $1 - \dfrac{3}{4e^{4x}} + c$

7 The second moment of area of a rectangle through its centroid is given by $\dfrac{bl^3}{12}$. Using the binomial theorem, the approximate percentage change in the second moment of area if b is increased by 3% and l is reduced by 2% is

A -6% B $+1\%$ C $+3\%$ D -3%

8 The volume of the solid of revolution when the curve $y = 2x$ is rotated one revolution about the x-axis between the limits $x = 0$ and $x = 4$ cm is

A $85\dfrac{1}{3}\pi$ cm^3 B 8 cm^3

C $85\dfrac{1}{3}$ cm^3 D 64π cm^3

9 The resistance to motion R_m of a moving vehicle is given by $R_m = \dfrac{4}{x} + 100x$.
The minimum value of resistance is

A $\dfrac{2}{9}$ B -36

C $-\dfrac{2}{9}$ D 36

10 The area enclosed by the curve $y = 5 \cos 2\theta$, the ordinates $\theta = 0$ and $\theta = \dfrac{\pi}{4}$ and the θ-axis is

A -5.0 B 2.5 C 10.0 D 5.0

11 The motion of a particle in an electrostatic field is described by the equation

$y = x^3 + 3x^2 + 5x - 28$. When $x = 2$, y is approximately zero. Using one iteration of the Newton-Raphson method, a better approximation (correct to 2 decimal places) is

A   1.89   B   2.07
C   2.11   D   1.93

12   If $f(t) = e^{2t} \ln 2t$, $f'(t)$ is equal to

A $\dfrac{2e^{2t}}{t}$ B $e^{2t}\left(\dfrac{1}{t} + 2\ln 2t\right)$

C $\dfrac{e^{2t}}{2t}$ D $\dfrac{e^{2t}}{2t} + 2e^{2t}\ln 2t$

13   The area under a force/distance graph gives the work done. The shaded area shown between a and b is

A $k\ln\dfrac{b}{a}$ B $k(\ln a - \ln b)$

C $-\dfrac{k}{2}\left(\dfrac{1}{b^2} - \dfrac{1}{a^2}\right)$ D $\dfrac{k}{2}(\ln b - \ln a)$

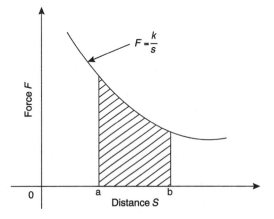

Figure 1

14   A metal template is bounded by the curve $y = x^2$, the x-axis and ordinates $x = 0$ and $x = 2$. The x-coordinate of the centroid of the area is

A   1.0   B   2.0   C   1.5   D   2.5

15   The fifth term of an arithmetic progression is 18 and the twelth term is 46. The eighteenth term is:

A   72   B   74   C   68   D   70

16   The length l metres of a certain metal rod at temperature $t\,°C$ is given by $l = 1 + 3 \times 10^{-5}t + 5 \times 10^{-7}t^2$. The rate of change of length, in mm/°C, when the temperature is $500\,°C$ is

A   0.53   B   1.14
C 5.3×10^{-4} D   1.00053

17   The turning point on the curve $y = x^2 - 6x$ is at

A   (3, 0)   B   (0, 6)
C   (3, −9)   D   (−3, 27)

18   An alternating current i has the following values at equal intervals of 2 ms:

Time ms	0	2.0	4.0	6.0	8.0	10.0	12.0
Current i (A)	0	4.6	7.4	10.8	8.5	3.7	0

Charge q (in millicoulombs) is given by $q = \int_0^{12.0} i\,dt$

Using the trapezoidal rule, the approximate charge in the 12 ms period is

A   70 mC   B   72.1 mC
C   35 mC   D   216.4 mC

19   The vertical displacement, s, of a prototype model in a tank is given by $s = 50\sin 0.2t$ mm, where t is the time in seconds. The vertical velocity of the model, in mm/s, is

A $250\cos 0.2t$ B $10\cos 0.2t$
C $-250\cos 0.2t$ D $-10\cos 0.2t$

20   The value of $\int_0^1 (4\sin 2\theta - 3\cos\theta)\,d\theta$, correct to 4 significant figures, is

A   −1.692   B   −0.05114
C   0.3079   D   −1.855

(Answers on page 454)

Part 3

Additional Mathematics for Engineering

32

Solution of first order differential equations by separation of variables

32.1 Family of curves

Integrating both sides of the derivative $dy/dx = 3$ with respect to x gives $y = \int 3\,dx$, i.e. $y = 3x + c$, where c is an arbitrary constant. $y = 3x + c$ represents a **family of curves**, each of the curves in the family depending on the value of c. Examples include $y = 3x + 8$, $y = 3x + 3$, $y = 3x$ and $y = 3x - 10$ and these are shown in Fig. 32.1.

Each is a straight line of gradient 3. A particular curve of a family may be determined when a point on the curve is specified. Thus, if $y = 3x + c$ passes through the point $(1, 2)$ then $2 = 3(1) + c$, from which, $c = -1$. The equation of the curve passing through $(1, 2)$ is therefore $y = 3x - 1$.

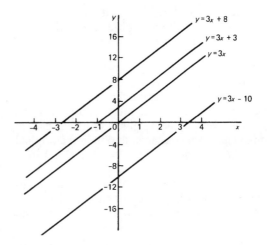

Figure 32.1

Problem 1. Sketch the family of curves given by the equation $\dfrac{dy}{dx} = 4x$ and determine the equation of one of these curves which passes through the point $(2, 3)$

Integrating both sides of $\dfrac{dy}{dx} = 4x$ with respect to x gives:

$$\int \frac{dy}{dx}\,dx = \int 4x\,dx, \ \text{ i.e. } y = 2x^2 + c$$

Some members of the family of curves having an equation $y = 2x^2 + c$ include $y = 2x^2 + 15$, $y = 2x^2 + 8$, $y = 2x^2$ and $y = 2x^2 - 6$, and these are shown in Fig. 32.2. To determine the equation of the curve passing through the point $(2, 3)$, $x = 2$ and $y = 3$ are substituted into the equation $y = 2x^2 + c$. Thus $3 = 2(2)^2 + c$, from which $c = 3 - 8 = -5$.

Hence the equation of the curve passing through the point $(2, 3)$ is $y = 2x^2 - 5$.

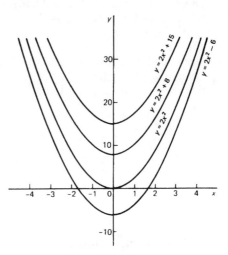

Figure 32.2

Further problems on families of curves may be found in Section 32.6, Problems 1 and 2, pages 307 and 308.

32.2   Differential equations

A **differential equation** is one that contains differential coefficients.

Examples include

(i) $\dfrac{dy}{dx} = 7x$ and (ii) $\dfrac{d^2y}{dx^2} + 5\dfrac{dy}{dx} + 2y = 0$

Differential equations are classified according to the highest derivative which occurs in them. Thus example (i) above is a **first order differential equation**, and example (ii) is a **second order differential equation**.

The **degree** of a differential equation is that of the highest power of the highest differential which the equation contains after simplification.

Thus $\left(\dfrac{d^2x}{dt^2}\right)^3 + 2\left(\dfrac{dx}{dt}\right)^5 = 7$ is a second order differential equation of degree three.

Starting with a differential equation it is possible, by integration and by being given sufficient data to determine unknown constants, to obtain the original function. This process is called '**solving the differential equation**'.

A solution to a differential equation which contains one or more arbitrary constants of integration is called the **general solution** of the differential equation.

When additional information is given so that constants may be calculated the **particular solution** of the differential equation is obtained. The additional information is called **boundary conditions**. It was shown in Section 32.1 that $y = 3x + c$ is the general solution of the differential equation $dy/dx = 3$. Given the boundary conditions $x = 1$ and $y = 2$, produces the particular solution of $y = 3x - 1$.

Equations which can be written in the form

$$\frac{dy}{dx} = f(x),\ \frac{dy}{dx} = f(y)\ \text{and}\ \frac{dy}{dx} = f(x) \cdot g(y)$$

can all be solved by integration. In each case it is possible to separate 'the ys' to one side of the equation and 'the xs' to the other. Solving such equations is therefore known as solution by **separation of variables**.

32.3   The solution of equations of the form $\dfrac{dy}{dx} = f(x)$

A differential equation of the form $\dfrac{dy}{dx} = f(x)$ is solved by direct integration,

i.e.

$$y = \int f(x)\,dx$$

Problem 2.   Find the general solution of the differential equation $\dfrac{dy}{dx} = 3x^2 - \sin 2x$

Integrating both sides of $\dfrac{dy}{dx} = 3x^2 - \sin 2x$ gives:

$$\int \frac{dy}{dx}\,dx = \int (3x^2 - \sin 2x)\,dx$$

i.e. $y = x^3 + \dfrac{1}{2}\cos 2x + c$, which is the general solution.

Problem 3.   Determine the general solution of $x\dfrac{dy}{dx} = 2 - 4x^3$

Rearranging $x\dfrac{dy}{dx} = 2 - 4x^3$ gives:

$$\frac{dy}{dx} = \frac{2 - 4x^3}{x} = \frac{2}{x} - \frac{4x^3}{x} = \frac{2}{x} - 4x^2$$

Integrating both sides gives:

$$y = \int \left(\frac{2}{x} - 4x^2\right) dx$$

$$y = 2\ln x - \frac{4}{3}x^3 + c,$$

which is the general solution

Problem 4.   Find the particular solution of the differential equation $5\dfrac{dy}{dx} + 2x = 3$, given the boundary conditions $y = 1\dfrac{2}{5}$ when $x = 2$

Since $5\dfrac{dy}{dx} + 2x = 3$ then $\dfrac{dy}{dx} = \dfrac{3 - 2x}{5} = \dfrac{3}{5} - \dfrac{2x}{5}$

Hence $y = \displaystyle\int \left(\frac{3}{5} - \frac{2x}{5}\right) dx$

i.e. $y = \dfrac{3x}{5} - \dfrac{x^2}{5} + c$, which is the general solution

Substituting the boundary conditions $y = 1\frac{2}{5}$ and $x = 2$ to evaluate c gives:

$$1\frac{2}{5} = \frac{6}{5} - \frac{4}{5} + c, \text{ from which, } c = 1$$

Hence the particular solution is $y = \dfrac{3x}{5} - \dfrac{x^2}{5} + 1$

Problem 5.   Solve the equation
$$2t\left(t - \frac{d\theta}{dt}\right) = 5, \text{ given } \theta = 2 \text{ when } t = 1$$

Rearranging gives:

$$t - \frac{d\theta}{dt} = \frac{5}{2t} \text{ and } \frac{d\theta}{dt} = t - \frac{5}{2t}$$

Integrating gives:

$$\theta = \int \left(t - \frac{5}{2t}\right) dt$$

i.e. $\theta = \dfrac{t^2}{2} - \dfrac{5}{2}\ln t + c,$

which is the general solution

When $\theta = 2$, $t = 1$, thus $2 = \dfrac{1}{2} - \dfrac{5}{2}\ln 1 + c$ from which, $c = \dfrac{3}{2}$

Hence the particular solution is:

$$\theta = \frac{t^2}{2} - \frac{5}{2}\ln t + \frac{3}{2}$$

i.e. $\theta = \dfrac{1}{2}(t^2 - 5\ln t + 3)$

Problem 6.   The angular velocity ω of a flywheel of moment of inertia I is given by $I(d\omega/dt) + N = 0$, where N is a constant. Determine ω in terms of t given that $\omega = \omega_0$ when $t = 0$

Rearranging $I\dfrac{d\omega}{dt} + N = 0$ gives $\dfrac{d\omega}{dt} = \dfrac{-N}{I}$

Integrating gives: $\omega = \displaystyle\int \frac{-N}{I} dt$, i.e. $\omega = \dfrac{-Nt}{I} + c$

When $t = 0$, $\omega = \omega_0$ thus $\omega_0 = \dfrac{-N(0)}{I} + c$, from which, $c = \omega_0$

Hence $\omega = \dfrac{-Nt}{I} + \omega_0$ or $\omega = \omega_0 - \dfrac{Nt}{I}$

Problem 7.   The bending moment M of the beam is given by $\dfrac{dM}{dx} = -w(l - x)$, where w and x are constants. Determine M in terms of x given: $M = \dfrac{1}{2}wl^2$ when $x = 0$

$$\frac{dM}{dx} = -w(l - x) = -wl + wx$$

Integrating with respect to x gives:

$$M = -wlx + \frac{wx^2}{2} + c,$$

which is the general solution

When $M = \dfrac{1}{2}wl^2$, $x = 0$

Thus $\dfrac{1}{2}wl^2 = -wl(0) + \dfrac{w(0)^2}{2} + c$ from which, $c = \dfrac{1}{2}wl^2$

Hence the particular solution is:

$$M = -wlx + \frac{wx^2}{2} + \frac{1}{2}wl^2$$

i.e. $M = \dfrac{1}{2}w(l^2 - 2lx + x^2)$

or $M = \dfrac{1}{2}w(l - x)^2$

Further problems on equations of the form $dy/dx = f(x)$ may be found in Section 32.6, Problems 3 to 14, page 308.

32.4   The solution of equations of the form $\dfrac{dy}{dx} = f(y)$

A differential equation of the form $dy/dx = f(y)$ is initially rearranged to give $dx = dy/f(y)$ and then the solution is obtained by direct integration,

i.e.
$$\int dx = \int \frac{dy}{f(y)}$$

i.e.
$$x = 5\left(-\frac{1}{3}\cot 3y\right) + c,$$

from standard integrals

Hence the general solution is $x = c - \dfrac{5}{3}\cot 3y$.

Problem 8. Find the general solution of
$$\frac{dy}{dx} = 3 + 2y$$

Rearranging $\dfrac{dy}{dx} = 3 + 2y$ gives: $dx = \dfrac{dy}{3 + 2y}$

Integrating both sides gives: $\int dx = \int \dfrac{dy}{(3 + 2y)}$

Thus, by using the substitution

$$u = (3 + 2y),$$

$$x = \frac{1}{2}\ln(3 + 2y) + c \qquad (1)$$

(see Chapter 41 for algebraic substitutions).
It is possible to give the general solution of a differential equation in a different form. For example, if $c = \ln k$, where k is a constant, then:

$$x = \frac{1}{2}\ln(3 + 2y) + \ln k,$$

i.e. $x = \ln(3 + 2y)^{1/2} + \ln k$

$$x = \ln[k\sqrt{(3 + 2y)}] \qquad (2)$$

by the laws of logarithms, from which,

$$e^x = k\sqrt{(3 + 2y)} \qquad (3)$$

Equations (1), (2) and (3) are all acceptable general solutions of the differential equation $\dfrac{dy}{dx} = 3 + 2y$.

Problem 9. Solve the differential equation
$$5\left(\frac{dy}{dx}\right) = \sin^2 3y$$

Rearranging $5\dfrac{dy}{dx} = \sin^2 3y$ gives:

$$dx = \frac{5}{\sin^2 3y}\,dy = 5\cosec^2 3y\,dy$$

Integrating both sides gives:

$$\int dx = \int 5\cosec^2 3y\,dy$$

Problem 10. Determine the particular solution of $(y^2 - 1)\dfrac{dy}{dx} = 3y$ given that $y = 1$ when $x = 2\dfrac{1}{6}$

Rearranging gives:

$$dx = \left(\frac{y^2 - 1}{3y}\right)dy = \left(\frac{y}{3} - \frac{1}{3y}\right)dy$$

Integrating gives:

$$\int dx = \int \left(\frac{y}{3} - \frac{1}{3y}\right)dy$$

i.e.
$$x = \frac{y^2}{6} - \frac{1}{3}\ln y + c,$$

which is the general solution

When $y = 1$, $x = 2\dfrac{1}{6}$, thus $2\dfrac{1}{6} = \dfrac{1}{6} - \dfrac{1}{3}\ln 1 + c$,
from which, $c = 2$

Hence the particular solution is: $x = \dfrac{y^2}{6} - \dfrac{1}{3}\ln y + 2$

Problem 11. (a) The variation of resistance, R ohms, of an aluminium conductor with temperature $\theta\,°C$ is given by $dR/d\theta = \alpha R$, where α is the temperature coefficient of resistance of aluminium. If $R = R_0$ when $\theta = 0\,°C$, solve the equation for R. (b) If $\alpha = 38 \times 10^{-4}/°C$, determine the resistance of an aluminium conductor at $50\,°C$, correct to 3 significant figures, when its resistance at $0\,°C$ is $24.0\,\Omega$

(a) $\dfrac{dR}{d\theta} = \alpha R$ is of the form $\dfrac{dy}{dx} = f(y)$

Rearranging gives: $d\theta = \dfrac{dR}{\alpha R}$

Integrating both sides gives:

$$\int d\theta = \int \frac{dR}{\alpha R}$$

i.e. $\theta = \dfrac{1}{\alpha}\ln R + c,$

which is the general solution

Substituting the boundary conditions $R = R_0$ when $\theta = 0$ gives:

$$0 = \dfrac{1}{\alpha}\ln R_0 + c$$

from which $c = -\dfrac{1}{\alpha}\ln R_0$

Hence the particular solution is

$$\theta = \dfrac{1}{\alpha}\ln R - \dfrac{1}{\alpha}\ln R_0 = \dfrac{1}{\alpha}(\ln R - \ln R_0)$$

i.e. $\theta = \dfrac{1}{\alpha}\ln\left(\dfrac{R}{R_0}\right)$ or $\alpha\theta = \ln\left(\dfrac{R}{R_0}\right)$

Hence $e^{\alpha\theta} = \dfrac{R}{R_0}$ from which, $R = R_0 e^{\alpha\theta}$

(b)   Substituting $\alpha = 38 \times 10^{-4}$, $R_0 = 24.0$ and $\theta = 50$ into $R = R_0 e^{\alpha\theta}$ gives the resistance at $50\,^\circ$C, i.e. $R_{50} = 24.0e^{(38\times10^{-4}\times50)} = \mathbf{29.0\ ohms}$

Further problems on equations of the form $dy/dx = f(y)$ may be found in Section 32.6, Problems 15 to 26, pages 308 and 309.

32.5   The solution of equations of the form $\dfrac{dy}{dx} = f(x)\cdot g(y)$

A differential equation of the form $\dfrac{dy}{dx} = f(x)\cdot g(y)$, where $f(x)$ is a function of x only and $g(y)$ is a function of y only, may be rearranged as $dy/g(y) = f(x)\,dx$, and then the solution is obtained by direct integration, i.e.

$$\int \dfrac{dy}{g(y)} = \int f(x)\,dx$$

Problem 12.   Solve: $\dfrac{dy}{dx} = \dfrac{(2x^3 - 1)}{(3 - 2y)}$

Separating the variables gives:

$$(3 - 2y)\,dy = (2x^3 - 1)\,dx$$

Integrating both sides gives:

$$\int (3 - 2y)\,dy = \int (2x^3 - 1)\,dx$$

Hence the general solution is:

$$3y - y^2 = \dfrac{x^4}{2} - x + c$$

Problem 13.   Solve the equation $4xy\dfrac{dy}{dx} = y^2 - 1$

Separating variables gives: $\left(\dfrac{4y}{y^2 - 1}\right)dy = \dfrac{1}{x}dx$

Integrating both sides gives:

$$\int \left(\dfrac{4y}{y^2 - 1}\right)dy = \int \left(\dfrac{1}{x}\right)dx$$

Hence the general solution is:

$$\mathbf{2\ln(y^2 - 1) = \ln x + c}\qquad(1)$$

or $\ln(y^2 - 1)^2 - \ln x = c$

from which, $\ln\left\{\dfrac{(y^2 - 1)^2}{x}\right\} = c$

and $\dfrac{(y^2 - 1)^2}{x} = e^c \qquad(2)$

If, in equation (1), $c = \ln A$, where A is a different constant,

then $\ln(y^2 - 1)^2 = \ln x + \ln A$

i.e. $\ln(y^2 - 1)^2 = \ln Ax$

i.e. $(y^2 - 1)^2 = Ax \qquad(3)$

Equations (1) to (3) are thus three valid solutions of the differential equations $4xy\,dy/dx = y^2 - 1$.

Problem 14.   Determine the particular solution of $\dfrac{d\theta}{dt} = 2e^{3t-2\theta}$, given that $t = 0$ when $\theta = 0$

$\dfrac{d\theta}{dt} = 2e^{3t-2\theta} = 2(e^{3t})(e^{-2\theta})$, by the laws of indices

Separating the variables gives:

$$\dfrac{d\theta}{e^{-2\theta}} = 2e^{3t}\, dt, \text{ i.e. } e^{2\theta}\, d\theta = 2e^{3t}\, dt$$

Integrating both sides gives: $\displaystyle\int e^{2\theta}\, d\theta = \int 2e^{3t}\, dt$

Thus the general solution is: $\dfrac{1}{2}e^{2\theta} = \dfrac{2}{3}e^{3t} + c$

When $t = 0$, $\theta = 0$, thus: $\dfrac{1}{2}e^{0} = \dfrac{2}{3}e^{0} + c$

from which, $c = \dfrac{1}{2} - \dfrac{2}{3} = -\dfrac{1}{6}$

Hence the particular solution is:

$$\dfrac{1}{2}e^{2\theta} = \dfrac{2}{3}e^{3t} - \dfrac{1}{6} \text{ or } 3e^{2\theta} = 4e^{3t} - 1$$

Problem 15. Find the curve which satisfies the equation $xy = (1 + x^2)\dfrac{dy}{dx}$ and passes through the point $(0, 1)$

Separating the variables gives: $\dfrac{x}{(1 + x^2)}\, dx = \dfrac{dy}{y}$

Integrating both sides gives: $\dfrac{1}{2}\ln(1 + x^2) = \ln y + c$

When $x = 0$, $y = 1$ thus $\dfrac{1}{2}\ln 1 = \ln 1 + c$, from which, $c = 0$

Hence the particular solution is $\dfrac{1}{2}\ln(1 + x^2) = \ln y$

i.e. $\ln(1 + x^2)^{1/2} = \ln y$, from which, $(1 + x^2)^{1/2} = y$

Hence the equation of the curve is $y = \sqrt{(1 + x^2)}$.

Problem 16. By considering a particle moving with constant acceleration, a, in a straight line, and that $\dfrac{d^2x}{dt^2} = a$, at time t, derive the equations of motion: (i) $v = u + at$ where v is the final velocity,

(ii) $x = ut + \dfrac{1}{2}at^2$ where u is the initial velocity, and (iii) $v^2 = u^2 + 2ax$. Assume boundary conditions of initial velocity u at $t = 0$ and also displacement $x = 0$ at $t = 0$

The basic relationships between distance, velocity and acceleration for a particle moving in a straight line with constant acceleration is that at time t and for a displacement x on the line from the origin, dx/dt is the velocity and d^2x/dt^2 is the acceleration.

(i)   Acceleration, $a = \dfrac{d^2x}{dt^2}$

Integrating once with respect to time t gives,

$v = \dfrac{dx}{dt} = \int a\, dt$ that is, $v = at + A$, where A is an arbitrary constant

Since the velocity at $t = 0$ is u, $A = u$ and $v = u + at$

(ii) Integrating $v = \dfrac{dx}{dt} = u + at$ once gives

$x = ut + \dfrac{1}{2}at^2 + B$, where B is an arbitrary constant

Since, $x = 0$ at $t = 0, B = 0$

i.e. $x = ut + \dfrac{1}{2}at^2$

(iii) The acceleration, $\dfrac{d^2x}{dt^2} = \dfrac{dv}{dt}$

But $\dfrac{dv}{dt} = \dfrac{dv}{dx} \cdot \dfrac{dx}{dt}$ by the chain rule of differentiation $= v\dfrac{dv}{dx}$

Thus $a = v\dfrac{dv}{dx}$. Rearranging and integrating gives $\displaystyle\int v\, dv = \int a\, dx$

i.e. $\dfrac{1}{2}v^2 = ax + C$, where C is an arbitrary constant

But at $x = 0$, the initial velocity is u, i.e. $v = u$, thus $C = \dfrac{u^2}{2}$

It follows that $v^2 = u^2 + 2ax$

Problem 17. The current i in an electric circuit containing resistance R and inductance L in series with a constant voltage source E is given by the differential equation $E - L\left(\dfrac{di}{dt}\right) = Ri$. Solve the equation and find i in terms of time t given that when $t = 0$, $i = 0$

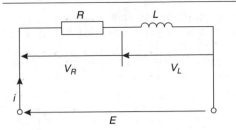

Figure 32.3

In the R–L series circuit shown in Fig. 32.3, the supply p.d., E, is given by

$$E = V_R + V_L$$

$$V_R = iR \text{ and } V_L = L\frac{di}{dt}$$

Hence $E = iR + L\dfrac{di}{dt}$

from which $E - L\dfrac{di}{dt} = Ri$

Most electrical circuits can be reduced to a differential equation.

Rearranging $E - L\dfrac{di}{dt} = Ri$ gives $\dfrac{di}{dt} = \dfrac{E - Ri}{L}$

and separating the variables gives: $\dfrac{di}{E - Ri} = \dfrac{dt}{L}$

Integrating both sides gives: $\displaystyle\int \frac{di}{E - Ri} = \int \frac{dt}{L}$

Hence the general solution is: $-\dfrac{1}{R}\ln(E - Ri) =$

$\dfrac{t}{L} + c$ (by making a substitution $u = E - Ri$)

When $t = 0$, $i = 0$, thus $-\dfrac{1}{R}\ln E = c$

Thus the particular solution is:

$$-\frac{1}{R}\ln(E - Ri) = \frac{t}{L} - \frac{1}{R}\ln E$$

Transposing gives:

$$-\frac{1}{R}\ln(E - Ri) + \frac{1}{R}\ln E = \frac{t}{L}$$

$$\frac{1}{R}[\ln E - \ln(E - Ri)] = \frac{t}{L}$$

$$\ln\left(\frac{E}{E - Ri}\right) = \frac{Rt}{L} \text{ from which } \frac{E}{E - Ri} = e^{Rt/L}$$

Hence $\dfrac{E - Ri}{E} = e^{-Rt/L}$ and $Ri = E - Ee^{-Rt/L}$

Hence current, $i = E/R(1 - e^{-Rt/L})$, which represents the law of growth of current in an inductive circuit.

Problem 18. For an adiabatic expansion of a gas $C_v\dfrac{dp}{p} + C_p\dfrac{dV}{V} = 0$, where C_p and C_v are constants. Given $n = \dfrac{C_p}{C_v}$, show that

$$pV^n = \text{constant}$$

Separating the variables gives:

$$C_v\frac{dp}{p} = -C_p\frac{dV}{V}$$

Integrating both sides gives:

$$C_v\int \frac{dp}{p} = -C_p\int \frac{dV}{V}$$

i.e. $C_v \ln p = -C_p \ln V + k$

Dividing throughout by constant C_v gives:

$$\ln p = -\frac{C_p}{C_v}\ln V + \frac{k}{C_v}$$

Since $\dfrac{C_p}{C_v} = n$, then $\ln p + n \ln V = K$,

where $K = \dfrac{k}{C_v}$

i.e. $\ln p + \ln V^n = K$ or $\ln pV^n = K$, by the laws of logarithms

Hence $pV^n = e^K$, i.e. $pV^n = \textbf{constant}$.

Further problems on variable-separable types of differential equations may be found in the following Section 32.6, Problems 27 to 38, pages 309 and 310.

32.6 Further problems on the solution of first order differential equations by separation of the variables

Family of curves

1 Sketch a family of curves represented by each of the following differential equations:

(a) $\dfrac{dy}{dx} = 6$ (b) $\dfrac{dy}{dx} = 3x$ (c) $\dfrac{dy}{dx} = x + 2$

2   Sketch the family of curves given by the equation $dy/dx = 2x+3$ and determine the equation of one of these curves which passes through the point $(1, 3)$. $[y = x^2 + 3x - 1]$

Differential equations of the form $\dfrac{dy}{dx} = f(x)$

In Problems 3 to 9, solve the differential equations:

3   $\dfrac{dy}{dx} = 3x^5$ $\left[y = \dfrac{1}{2}x^6 + c\right]$

4   $\dfrac{dy}{dx} = \cos 4x - 2x$ $\left[y = \dfrac{\sin 4x}{4} - x^2 + c\right]$

5   $2x\dfrac{dy}{dx} = 3 - x^3$ $\left[y = \dfrac{3}{2}\ln x - \dfrac{x^3}{6} + c\right]$

6   $3\dfrac{dy}{dx} - 2x^2 = e^{4x}$ $\left[y = \dfrac{2}{9}x^3 + \dfrac{e^{4x}}{12} + c\right]$

7   $\dfrac{dy}{dx} + x = 3$, given $y = 2$ when $x = 1$.

$\left[y = 3x - \dfrac{x^2}{2} - \dfrac{1}{2}\right]$

8   $3\dfrac{dy}{d\theta} + \sin\theta = 0$, given $y = \dfrac{2}{3}$ when $\theta = \dfrac{\pi}{3}$.

$\left[y = \dfrac{1}{3}\cos\theta + \dfrac{1}{2}\right]$

9   $\dfrac{1}{e^x} + 2 = x - 3\dfrac{dy}{dx}$, given $y = 1$ when $x = 0$.

$\left[y = \dfrac{1}{6}(x^2 - 4x + \dfrac{2}{e^x} + 4)\right]$

10   The gradient of a curve is given by $\dfrac{dy}{dx} + \dfrac{x^2}{2} = 3x$. Find the equation of the curve if it passes through the point $\left(1, \dfrac{1}{3}\right)$.

$\left[y = \dfrac{3}{2}x^2 - \dfrac{x^3}{6} - 1\right]$

11   The acceleration, a, of a body is equal to its rate of change of velocity, $\dfrac{dv}{dt}$. Find an equation for v in terms of t, given that when $t = 0$, velocity $v = u$. $[v = u + at]$

12   The bending moment, M, of a beam is given by the equation $\dfrac{dM}{dx} = w(x - 1)$, where x is

the distance from one end of a beam of length l and w is a constant. Solve the equation and show that $M = \dfrac{1}{2}w(l - x)^2$ given $M = \dfrac{wl^2}{2}$ when $x = 0$.

13   The velocity, v, of a body is equal to its rate of change of distance $\dfrac{dx}{dt}$. Find an equation for x in terms of t given $v = u + at$, where u and a are constants and $x = 0$ when $t = 0$.

$\left[x = ut + \dfrac{1}{2}at^2\right]$

14   An object is thrown vertically upwards with an initial velocity, u, of 20 m/s. The motion of the object follows the differential equation $\dfrac{ds}{dt} = u - gt$, where s is the height of the object in metres at time t seconds and $g = 9.8$ m/s^2. Determine the height of the object after 3 seconds if $s = 0$ when $t = 0$.   [15.9 m]

Differential equations of the form $\dfrac{dy}{dx} = f(y)$

In Problems 15 to 20, solve the differential equations:

15   $\dfrac{dy}{dx} = 2 + 3y$ $\left[x = \dfrac{1}{3}\ln(2 + 3y) + c\right]$

16   $2\dfrac{dy}{dx} = \cot 4y$ $\left[\dfrac{1}{4}\ln(\sec 4y) = \dfrac{x}{2} + c\right]$

17   $3y\dfrac{dy}{dt} = 2 - y^2$ $\left[-\dfrac{1}{2}\ln(2 - y^2) = \dfrac{t}{3} + c\right]$

18   $\dfrac{dy}{dx} = 2\cos^2 y$ $[\tan y = 2x + c]$

19   $\left(\dfrac{y}{2}\right)\dfrac{dy}{dx} = 3 - y$, given $y = 2$ when $x = -1$

$[2x + y + 3\ln(3 - y) = 0]$

20   $(y^2 + 2)\dfrac{dy}{dx} = 5y$, given $y = 1$ when $x = \dfrac{1}{2}$

$\left[\dfrac{y^2}{2} + 2\ln y = 5x - 2\right]$

21   The current in an electric circuit is given by the equation $Ri + L\dfrac{di}{dt} = 0$, where L and R are constants. Show that $i = Ie^{-Rt/L}$, given that $i = I$ when $t = 0$.

22   The velocity of a chemical reaction is given by $\dfrac{dx}{dt} = k(a - x)$, where x is the amount transferred in time t, k is a constant and a is the concentration at time $t = 0$ when $x = 0$. Solve the equation and determine x in terms of t.

$$[x = a(1 - e^{-kt})]$$

23 (a)   Charge Q coulombs at time t seconds is given by the differential equation $R\dfrac{dQ}{dt} + \dfrac{Q}{C} = 0$, where C is the capacitance in farads and R the resistance in ohms. Solve the equation for Q given that $Q = Q_0$ when $t = 0$

(b)   A circuit possesses a resistance of 250×10^3 ohms and a capacitance of 8.5×10^{-6} farads, and after 0.32 seconds the charge falls to 8.0 C. Determine the initial charge and the charge after 1 second, each correct to 3 significant figures

$$[\text{(a) } Q = Q_0 e^{-t/CR} \text{ (b) } 9.30 \text{ C, } 5.81 \text{ C}]$$

24   The rate of decay of a radioactive substance is given by $\dfrac{dN}{dt} = -\lambda N$, where λ is the decay constant and λN the number of radioactive atoms disintegrating per second. Determine the half-life of a mercury isotope, in hours (i.e. the time for N to become one-half of its original value), assuming the decay constant for mercury to be 2.917×10^{-6} atoms per second.

$$[66 \text{ hours}]$$

25   A differential equation relating the difference in tension T, pulley contact angle θ and coefficient of friction μ is $\dfrac{dT}{d\theta} = \mu T$. When $\theta = 0$, $T = 150$ N, and $\mu = 0.30$ as slipping starts. Determine the tension at the point of slipping when $\theta = 2$ radians. Determine also the value of θ when T is 300 N.   [273 N, 2.31 rads]

26   The rate of cooling of a body is given by $\dfrac{d\theta}{dt} = k\theta$, where k is a constant. If $\theta = 60\,^{\circ}\text{C}$ when $t = 2$ minutes and $\theta = 50\,^{\circ}\text{C}$ when $t = 5$ minutes, determine the time taken for θ to fall to $40\,^{\circ}\text{C}$, correct to the nearest second.

$$[8 \text{ minutes } 40 \text{ seconds}]$$

Variable-separable types of differential equations

In Problems 27 to 33, solve the differential equations:

27   $\dfrac{dy}{dx} = yx^3$ $\left[\ln y = \dfrac{x^4}{4} + c\right]$

28   $\dfrac{dy}{dx} = \dfrac{3x^2 - 2}{2y + 1}$ $[x^3 - 2x = y^2 + y + c]$

29   $\dfrac{dy}{dx} = 2y\cos x$ $[\ln y = 2\sin x + c]$

30   $2xy\dfrac{dy}{dx} = y^2 + 3$

$$[\ln(y^2 + 3) = \ln x + c \text{ or } y^2 + 3 = Ax]$$

31   $(2y - 1)\dfrac{dy}{dx} = (3x^2 + 1)$, given $x = 1$ when $y = 2$ $[y^2 - y = x^3 + x]$

32   $\dfrac{dy}{dx} = e^{2x-y}$, given $x = 0$ when $y = 0$

$$\left[e^y = \dfrac{1}{2}e^{2x} + \dfrac{1}{2}\right]$$

33   $2y(1 - x) + x(1 + y)\dfrac{dy}{dx} = 0$, given $x = 1$ when $y = 1$

$$\left[\ln(x\sqrt{y}) = x - \dfrac{y}{2} - \dfrac{1}{2}\right]$$

34   Show that the solution of the equation $\dfrac{y^2 + 1}{x^2 + 1} = \dfrac{y}{x}\dfrac{dy}{dx}$ is of the form $\sqrt{\left(\dfrac{y^2 + 1}{x^2 + 1}\right)} = $ constant.

35   Solve $xy = (1 - x^2)\dfrac{dy}{dx}$ for y, given $x = 0$ when $y = 1$.

$$\left[y = \dfrac{1}{\sqrt{(1 - x^2)}}\right]$$

36   Determine the equation of the curve which satisfies the equation $xy\dfrac{dy}{dx} = x^2 - 1$, and which passes through the point $(1, 2)$.

$$[x^2 - 2\ln x = y^2 - 3]$$

37   Solve the equation $y\cos^2\theta\dfrac{dy}{d\theta} = 3 + \sin\theta$, given that $y = \sqrt{6}$ when $\theta = \dfrac{\pi}{4}$.

$$[y^2 = 2(3\tan\theta + \sec\theta - \sqrt{2})]$$

38 The p.d., V, between the plates of a capacitor C charged by a steady voltage E through a resistor R is given by the equation $CR\dfrac{dV}{dt} + V = E$.

(a) Solve the equation for V given that at $t = 0$, $V = 0$

(b) Calculate V, correct to 3 significant figures, when $E = 25$ volts, $C = 20 \times 10^{-6}$ farads, $R = 200 \times 10^{3}$ ohms and $t = 3.0$ seconds

$[$(a) $V = E(1 - e^{-t/CR})$ (b) 13.2 volts$]$

33

Introduction to second order differential equations

33.1 Introduction

An equation of the form $a\dfrac{d^2y}{dx^2} + b\dfrac{dy}{dx} + cy = 0$, where a, b and c are constants, is called a **linear second order differential equation with constant coefficients**. If D represents d/dx and D^2 represents d^2/dx^2, then the above equation may be stated as $(aD^2 + bD + c)y = 0$. This equation is said to be in 'D-operator' form.

If $y = Ae^{mx}$,

then $\dfrac{dy}{dx} = Ame^{mx}$ and $\dfrac{d^2y}{dx^2} = Am^2e^{mx}$

Substituting these values into $a\dfrac{d^2y}{dx^2} + b\dfrac{dy}{dx} + cy = 0$ gives:

$$a(Am^2e^{mx}) + b(Ame^{mx}) + cy = 0$$

i.e. $Ae^{mx}(am^2 + bm + c) = 0$

Thus $y = Ae^{mx}$ is a solution of the given equation provided that

$$(am^2 + bm + c) = 0$$

$am^2 + bm + c = 0$ is called the **auxiliary equation**, and since the equation is a quadratic, m may be obtained either by factorizing or by using the quadratic formula. Since, in the auxiliary equation, a, b and c are real values, then the equation may have either

    (i) two different real roots   (when $b^2 > 4ac$)
or (ii) two real equal roots   (when $b^2 = 4ac$)
or (iii) two complex roots   (when $b^2 < 4ac$)

33.2 Procedure to solve differential equations of the form $a\dfrac{d^2y}{dx^2} + b\dfrac{dy}{dx} + cy = 0$

(a) Rewrite the differential equation

$$a\dfrac{d^2y}{dx^2} + b\dfrac{dy}{dx} + cy = 0$$

as $(aD^2 + bD + c)y = 0$

(b) Substitute m for D and solve the auxiliary equation $am^2 + bm + c = 0$ for m

(c) If the roots of the auxiliary equation are:

   (i) **real and different**, say $m = \alpha$ and $m = \beta$, then the general solution is $y = Ae^{\alpha x} + Be^{\beta x}$

   (ii) **real and equal**, say $m = \alpha$ twice, then the general solution is $y = (Ax + B)e^{\alpha x}$

   (iii) **complex**, say $m = \alpha \pm j\beta$, then the general solution is

$$y = e^{\alpha x}\{A\,\cos\beta x + B\,\sin\beta x\}$$

(d) Given boundary conditions, constants A and B may be determined, and the **particular solution** of the differential equation obtained

The particular solutions obtained in the worked problems of Section 33.3 may each be **verified** by substituting expressions for y, dy/dx and d^2y/dx^2 into the original equation.

33.3 Worked problems on differential equations of the form

$$a\frac{d^2y}{dx} + b\frac{dy}{dx} + cy = 0$$

Problem 1. Determine the general solution of $2\frac{d^2y}{dx^2} + 5\frac{dy}{dx} - 3y = 0$. Find also the particular solution given that when $x = 0$, $y = 4$ and $dy/dx = 9$

Using the above procedure:

(a) $2\frac{d^2y}{dx^2} + 5\frac{dy}{dx} - 3y = 0$ in D-operator form is $(2D^2 + 5D - 3)y = 0$, where $D \equiv d/dx$

(b) Substituting m for D gives the auxiliary equation $2m^2 + 5m - 3 = 0$
   Factorizing gives: $(2m - 1)(m + 3) = 0$, from which, $m = \frac{1}{2}$ or $m = -3$

(c) Since the roots are real and different **the general solution is** $y = Ae^{(1/2)x} + Be^{-3x}$

(d) When $x = 0$, $y = 4$, hence

$$4 = A + B \qquad (1)$$

Since $y = Ae^{(1/2)x} + Be^{-3x}$

then $\frac{dy}{dx} = \frac{1}{2}Ae^{(1/2)x} - 3Be^{-3x}$

When $x = 0$, $\frac{dy}{dx} = 9$, thus

$$9 = \frac{1}{2}A - 3B \qquad (2)$$

Solving the simultaneous equations (1) and (2) gives $A = 6$ and $B = -2$.

Hence the particular solution is

$$y = 6e^{(1/2)x} - 2e^{-3x}$$

Problem 2. Find the general solution of $9\frac{d^2y}{dt^2} - 24\frac{dy}{dt} + 16y = 0$ and also the particular solution given the boundary conditions that when $t = 0$, $y = dy/dt = 3$

Using the above procedure:

(a) $9\frac{d^2y}{dt^2} - 24\frac{dy}{dt} + 16y = 0$ in D-operator form is $(9D^2 - 24D + 16)y = 0$ where $D \equiv d/dt$

(b) Substituting m for D gives the auxiliary equation $9m^2 - 24m + 16 = 0$
   Factorizing gives: $(3m - 4)(3m - 4) = 0$,
   i.e. $m = \frac{4}{3}$ twice

(c) Since the roots are real and equal, **the general solution is** $y = (At + B)e^{(4/3)t}$

(d) When $y = 0$, $y = 3$ hence $3 = (0 + B)e^0$,
   i.e. $B = 3$
   Since $y = (At + B)e^{(4/3)t}$

   then $\frac{dy}{dt} = (At + B)\left(\frac{4}{3}e^{(4/3)t}\right) + Ae^{(4/3)t}$, by the product rule

   When $t = 0$, $\frac{dy}{dt} = 3$

   thus $3 = (0 + B)\frac{4}{3}e^0 + Ae^0$,

   i.e. $3 = \frac{4}{3}B + A$ from which, $A = -1$, since $B = 3$.

Hence the particular solution is

$$y = (-t + 3)e^{(4/3)t} \text{ or}$$

$$y = (3 - t)e^{(4/3)t}$$

Problem 3. Solve the differential equation $\frac{d^2y}{dx^2} + 6\frac{dy}{dx} + 13y = 0$, given that when $x = 0$, $y = 3$ and $dy/dx = 7$

Using the above procedure:

(a) $\frac{d^2y}{dx^2} + 6\frac{dy}{dx} + 13y = 0$ in D-operator form is $(D^2 + 6D + 13)y = 0$, where $D \equiv d/dx$

(b) Substituting m for D gives the auxiliary equation $m^2 + 6m + 13 = 0$

Using the quadratic formula:

$$m = \frac{-6 \pm \sqrt{[(6)^2 - 4(1)(13)]}}{2(1)}$$

$$= \frac{-6 \pm \sqrt{(-16)}}{2}$$

i.e. $m = \dfrac{-6 \pm j4}{2} = -3 \pm j2$

(see Chapter 43 for more on j notation)

(c) Since the roots are complex, **the general solution is**

$$y = e^{-3x}(A \cos 2x + B \sin 2x)$$

(d) When $x = 0$, $y = 3$ hence

$3 = e^0(A \cos 0 + B \sin 0)$, i.e. $A = 3$

Since $y = e^{-3x}(A \cos 2x + B \sin 2x)$

then $\dfrac{dy}{dx}$

$= e^{-3x}(-2A \sin 2x + 2B \cos 2x) - 3e^{-3x}$

$\times (A \cos 2x + B \sin 2x)$ by the product rule

$= e^{-3x}[(2B - 3A) \cos 2x - (2A + 3B) \sin 2x]$

When $x = 0$, $\dfrac{dy}{dx} = 7$,

hence $7 = e^0[(2B - 3A) \cos 0 - (2A + 3B) \sin 0]$

i.e. $7 = 2B - 3A$, from which, $B = 8$, since $A = 3$.

Hence the particular solution is

$$y = e^{-3x}(3 \cos 2x + 8 \sin 2x)$$

Problem 4. The equation of motion of a body oscillating on the end of a spring is

$$\frac{d^2x}{dt^2} + 100x = 0$$

where x is the displacement in metres of the body from its equilibrium position after time t seconds. Determine x in terms of t given that at time $t = 0$, $x = 2m$ and $dx/dt = 0$

An equation of the form $\dfrac{d^2x}{dt^2} + m^2x = 0$ is a differential equation representing simple harmonic motion (S.H.M.). Using the procedure of Section 33.2:

(a) $\dfrac{d^2x}{dt^2} + 100x = 0$ in D-operator form is

$(D^2 + 100)x = 0$

(b) The auxiliary equation is $m^2 + 100 = 0$, i.e. $m^2 = -100$ and $m = \sqrt{(-100)}$, i.e. $m = \pm j10$

(c) Since the roots are complex, the general solution is $x = e^0(A \cos 10t + B \sin 10t)$, i.e. $x = (A \cos 10t + B \sin 10t)$ metres

(d) When $t = 0$, $x = 2$ thus $2 = A$

$$\frac{dx}{dt} = -10A \sin 10t + 10B \cos 10t$$

When $t = 0$, $\dfrac{dx}{dt} = 0$

thus $0 = -10A \sin 0 + 10B \cos 0$, i.e. $B = 0$.

Hence the particular solution is

$$x = 2 \cos 10t \text{ metres}$$

Problem 5. The equation

$$\frac{d^2i}{dt^2} + \frac{R}{L}\frac{di}{dt} + \frac{1}{LC}i = 0$$

represents a current i flowing in an electrical circuit containing resistance R, inductance L and capacitance C connected in series. If $R = 200$ ohms, $L = 0.20$ henry and $C = 20 \times 10^{-6}$ farads, solve the equation for i given the boundary conditions that when $t = 0$, $i = 0$ and $di/dt = 100$

Using the procedure of Section 33.2:

(a) $\dfrac{d^2i}{dt^2} + \dfrac{R}{L}\dfrac{di}{dt} + \dfrac{1}{LC}i = 0$ in D-operator form is

$\left(D^2 + \dfrac{R}{L}D + \dfrac{1}{LC}\right)i = 0$ where $D \equiv \dfrac{d}{dt}$

(b) The auxiliary equation is $m^2 + \dfrac{R}{L}m + \dfrac{1}{LC} = 0$

Hence $m = \dfrac{-\dfrac{R}{L} \pm \sqrt{\left[\left(\dfrac{R}{L}\right)^2 - 4(1)\left(\dfrac{1}{LC}\right)\right]}}{2}$

When $R = 200$, $L = 0.2$ and $C = 20 \times 10^{-6}$, then

$m = \dfrac{-\dfrac{200}{0.2} \pm \sqrt{\left[\left(\dfrac{200}{0.2}\right)^2 - \dfrac{4}{(0.2)(20 \times 10^{-6})}\right]}}{2}$

$= \dfrac{-1000 \pm \sqrt{0}}{2} = -500$

(c) Since the two roots are real and equal (i.e. -500 twice, since for a second order differential equation there must be two solutions), the general solution is $i = (At + B)e^{-500t}$

(d)   When $t = 0$, $i = 0$ hence $B = 0$

$$\frac{di}{dt} = (At + B)(-500e^{-500t}) + (e^{-500t})(A),$$

by the product rule

When $t = 0$, $\dfrac{di}{dt} = 100$, thus $100 = -500B + A$

i.e. $A = 100$, since $B = 0$.

Hence the particular solution is

$$i = 100te^{-500t}$$

Problem 6.   The oscillations of a heavily damped pendulum satisfy the differential equation $\dfrac{d^2x}{dt^2} + 6\dfrac{dx}{dt} + 8x = 0$, where x cm is the displacement of the bob at time t seconds. The initial displacement is equal to $+4$ cm and the initial velocity (i.e. dx/dt) is 8 cm/s. Solve the equation for x

---

Using the procedure of Section 33.2:

(a) $\dfrac{d^2x}{dt^2} + 6\dfrac{dx}{dt} + 8x = 0$ in D-operator form is
$(D^2 + 6D + 8)x = 0$, where $D \equiv d/dt$

(b)   The auxiliary equation is $m^2 + 6m + 8 = 0$. Factorizing gives: $(m + 2)(m + 4) = 0$, from which, $m = -2$ or $m = -4$

(c)   Since the roots are real and different, the general solution is $x = Ae^{-2t} + Be^{-4t}$

(d)   Initial displacement means that time $t = 0$. At this instant, $x = 4$.
Thus $4 = A + B$ (1)

Velocity, $\dfrac{dx}{dt} = -2Ae^{-2t} - 4Be^{-4t}$

$\dfrac{dx}{dt} = 8$ cm/s when $t = 0$,

thus $8 = -2A - 4B$ (2)
From equations (1) and (2),

$A = 12$ and $B = -8$

Hence the particular solution is

$$x = 12e^{-2t} - 8e^{-4t}$$

i.e. **displacement**, $x = 4(3e^{-2t} - 2e^{-4t})$ **cm**

33.4   Further problems on differential equations of the form $a\dfrac{d^2y}{dx^2} + b\dfrac{dy}{dx} + cy = 0$

In Problems 1 to 6, determine the general solution of the given differential equations:

1   $\dfrac{d^2y}{dx^2} + 3\dfrac{dy}{dx} - 4y = 0$ $[y = Ae^x + Be^{-4x}]$

2   $6\dfrac{d^2y}{dt^2} - \dfrac{dy}{dx} - 2y = 0$

$$[y = Ae^{(2/3)t} + Be^{(-1/2)t}]$$

3   $\dfrac{d^2y}{dx^2} - 4\dfrac{dy}{dx} + 4y = 0$ $[y = (Ax + B)e^{2x}]$

4   $4\dfrac{d^2\theta}{dt^2} + 4\dfrac{d\theta}{dt} + \theta = 0$ $[\theta = (At + B)e^{(-1/2)t}]$

5   $\dfrac{d^2y}{dx^2} + 2\dfrac{dy}{dx} + 5y = 0$

$$[y = e^{-x}(A\cos 2x + B\sin 2x)]$$

6   $\dfrac{d^2y}{dx^2} + y = 0$ $[y = A\cos x + B\sin x]$

In Problems 7 to 12, find the particular solution of the given differential equations for the stated boundary conditions:

7   $6\dfrac{d^2y}{dx^2} + 5\dfrac{dy}{dx} - 6y = 0$,

when $x = 0$, $y = 5$ and $\dfrac{dy}{dx} = -1$
$$[y = 3e^{(2/3)x} + 2e^{(-3/2)x}]$$

8   $4\dfrac{d^2y}{dt^2} - 5\dfrac{dy}{dt} + y = 0$, when $t = 0$, $y = 1$ and

$\dfrac{dy}{dt} = -2$ $[y = 4e^{(1/4)t} - 3e^t]$

9   $(9D^2 + 30D + 25)y = 0$, where $D \equiv \dfrac{d}{dx}$, when

$x = 0$, $y = 0$ and $\dfrac{dy}{dx} = 2$ $[y = 2xe^{(-5/3)x}]$

10   $\dfrac{d^2x}{dt^2} - 6\dfrac{dx}{dt} + 9x = 0$, where $t = 0$, $x = 2$ and

$\dfrac{dx}{dt} = 0$ $[x = 2(1 - 3t)e^{3t}]$

11 $\dfrac{d^2y}{dx^2} + 6\dfrac{dy}{dx} + 13y = 0$, where $x = 0$, $y = 4$

and $\dfrac{dy}{dx} = 0$ $[y = 2e^{-3x}(2\cos 2x + 3\sin 2x)]$

12 $(4D^2 + 20D + 125)\theta = 0$, where $D \equiv \dfrac{d}{dt}$, when

$t = 0$, $\theta = 3$ and $\dfrac{d\theta}{dt} = 2.5$

$[\theta = e^{-2.5t}(3\cos 5t + 2\sin 5t)]$

13 A body moves in a straight line so that its distance s metres from the origin after time t seconds is given by $\dfrac{d^2s}{dt^2} + a^2s = 0$, where a is a constant. Solve the equation for s given that $s = c$ and $ds/dt = 0$ when $t = 2\pi/a$.

$[s = c\cos at]$

14 Determine an expression for x for a differential equation $\dfrac{d^2x}{dt^2} + 2n\dfrac{dx}{dt} + n^2x = 0$ which represents a critically damped oscillator, given that at time $t = 0$, $x = s$ and $dx/dt = u$.

$[x = \{s + (u + ns)t\}e^{-nt}]$

15 The motion of the pointer of a galvanometer about its position of equilibrium is represented by the equation $I\dfrac{d^2\theta}{dt^2} + K\dfrac{d\theta}{dt} + F\theta = 0$. If I, the moment of inertia of the pointer about its pivot, is 5×10^{-3}, K, the resistance due to friction at unit angular velocity, is 2×10^{-2} and F, the force on the spring necessary to produce unit displacement, is 0.20, solve the equation for θ in terms of t given that when $t = 0$, $\theta = 0.3$ and $d\theta/dt = 0$.

$[\theta = e^{-2t}(0.3\cos 6t + 0.1\sin 6t)]$

16 $L\dfrac{d^2i}{dt^2} + R\dfrac{di}{dt} + \dfrac{1}{c}i = 0$ is an equation representing current i in an electric circuit. If inductance L is 0.25 henry, capacitance C is 29.76×10^{-6} farads and R is 250 ohms, solve the equation for i given the boundary conditions that when $t = 0$, $i = 0$ and $di/dt = 34$.

$\left[i = \dfrac{1}{20}(e^{-160t} - e^{-840t})\right]$

34

A numerical solution of differential equations

34.1 Introduction

Not all differential equations may be solved by separating the variables (as in Chapter 32) or by the method used for second order differential equations (as in Chapter 33). A number of other analytical methods of solving differential equations exist (and some are covered in *Higher Engineering Mathematics*). However, the differential equations that can be solved by such analytical methods are fairly restricted.

Where a differential equation and known boundary conditions are given, an approximate solution may be obtained by applying a **numerical method**. There are a number of such numerical methods available and the simplest of these is called **Euler's method**.

34.2 Euler's method

If h is the interval between two near ordinates y_0 and y_1, as shown in Fig. 34.1, and if $f(a) = y_0$ and $y_1 = f(a+h)$, then Euler's method states:

$$f(a+h) = f(a) + hf'(a)$$

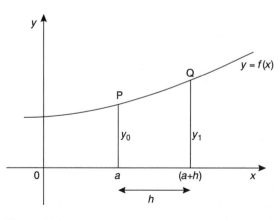

Figure 34.1

i.e.
$$y_1 = y_0 + h(y')_0 \tag{1}$$

Hence if y_0, h and $(y')_0$ are known, y_1, which is an approximate value for the function at Q in Fig. 34.1, can be calculated.

Euler's method is demonstrated in the following worked problems.

34.3 Worked problems on Euler's method

Problem 1. Obtain a numerical solution of the differential equation

$$\frac{dy}{dx} = 3(1+x) - y$$

given the initial conditions that $x = 1$ when $y = 4$, for the range $x = 1.0$ to $x = 2.0$ with intervals of 0.2. Draw the graph of the solution

$$\frac{dy}{dx} = y' = 3(1+x) - y$$

With $x_0 = 1$ and $y_0 = 4$, $(y')_0 = 3(1+1) - 4 = 2$
By Euler's method:

$$y_1 = y_0 + h(y')_0, \text{ from equation (1)}$$

Hence $y_1 = 4 + (0.2)(2) = \mathbf{4.4}$, since $h = 0.2$
At point Q in Fig. 34.2, $x_1 = 1.2$, $y_1 = 4.4$

and $(y')_1 = 3(1 + x_1) - y_1$

i.e. $(y')_1 = 3(1 + 1.2) - 4.4 = \mathbf{2.2}$

If the values of x, y and y' found for point Q are regarded as new starting values of x_0, y_0 and $(y')_0$, the above process can be repeated and values found for the point R shown in Fig. 34.3.

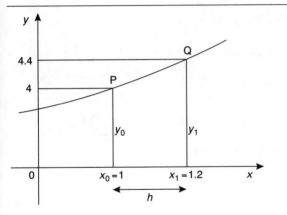

Figure 34.2

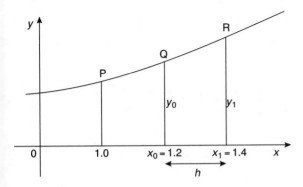

Figure 34.3

Thus at point R,

$$y_1 = y_0 + h(y')_0, \text{ from equation (1)}$$
$$= 4.4 + (0.2)(2.2) = \mathbf{4.84}$$

When $x_1 = 1.4$ and $y_1 = 4.84$,

$$(y')_1 = 3(1 + 1.4) - 4.84 = \mathbf{2.36}$$

This step-by-step Euler's method can be continued and it is easiest to list the results in a table, as shown in Table 34.1. The results for lines 1 to 3 have been produced above.

Table 34.1

	x_0	y_0	$(y')_0$
1.	1	4	2
2.	1.2	4.4	2.2
3.	1.4	4.84	2.36
4.	1.6	5.312	2.488
5.	1.8	5.8096	2.5904
6.	2.0	6.32768	

For line 4, where $x_0 = 1.6$:

$$y_1 = y_0 + h(y')_0$$
$$= 4.84 + (0.2)(2.36) = \mathbf{5.312}$$
and $(y')_0 = 3(1 + 1.6) - 5.312 = \mathbf{2.488}$

For line 5, where $x_0 = 1.8$:

$$y_1 = y_0 + h(y')_0$$
$$= 5.312 + (0.2)(2.488) = \mathbf{5.8096}$$
and $(y')_0 = 3(1 + 1.8) - 5.8096 = \mathbf{2.5904}$

For line 6, where $x_0 = 2.0$:

$$y_1 = y_0 + h(y')_0$$
$$= 5.8096 + (0.2)(2.5904) = \mathbf{6.32768}$$

(As the range is 1.0 to 2.0 there is no need to calculate $(y')_0$ in line 6.) The particular solution is given by the value of y against x.

A graph of the solution of $\frac{dy}{dx} = 3(1 + x) - y$ with initial conditions $x = 1$ and $y = 4$ is shown in Fig. 34.4.

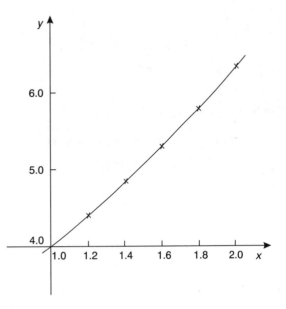

Figure 34.4

In practice it is probably best to plot the graph as each calculation is made, which checks that there is a smooth progression and that no calculation errors have occurred.

Problem 2. Use Euler's method to obtain a numerical solution of the differential equation $\dfrac{dy}{dx} + y = 2x$, given the initial conditions that at $x = 0$, $y = 1$, for the range $x = 0(0.2)1.0$. Draw the graph of the solution in this range.

$x = 0(0.2)1.0$ means that x ranges from 0 to 1.0 in equal intervals of 0.2 (i.e. $h = 0.2$ in Euler's method).

$$\frac{dy}{dx} + y = 2x, \text{ hence } \frac{dy}{dx} = 2x - y,$$

i.e. $y' = 2x - y$

If initially $x_0 = 0$ and $y_0 = 1$, then $(y')_0 = 2(0) - 1 = -1$.
Hence line 1 in Table 34.2 can be completed with $x = 0$, $y = 1$ and $y'(0) = -1$.

Table 34.2

	x_0	y_0	$(y')_0$
1.	0	1	−1
2.	0.2	0.8	−0.4
3.	0.4	0.72	0.08
4.	0.6	0.736	0.464
5.	0.8	0.8288	0.7712
6.	1.0	0.98304	

For line 2, where $x_0 = 0.2$ and $h = 0.2$:

$$y_1 = y_0 + h(y'), \text{ from equation (1)}$$
$$= 1 + (0.2)(-1) = \mathbf{0.8}$$

and $(y')_0 = 2x_0 - y_0 = 2(0.2) - 0.8 = \mathbf{-0.4}$

For line 3, where $x_0 = 0.4$:

$$y_1 = y_0 + h(y')_0$$
$$= 0.8 + (0.2)(-0.4) = \mathbf{0.72}$$

and $(y')_0 = 2x_0 - y_0 = 2(0.4) - 0.72 = \mathbf{0.08}$

For line 4, where $x_0 = 0.6$:

$$y_1 = y_0 + h(y')_0$$
$$= 0.72 + (0.2)(0.08) = \mathbf{0.736}$$

and $(y')_0 = 2x_0 - y_0 = 2(0.6) - 0.736 = \mathbf{0.464}$

For line 5, where $x_0 = 0.8$:

$$y_1 = y_0 + h(y')_0$$
$$= 0.736 + (0.2)(0.464) = \mathbf{0.8288}$$

and $(y')_0 = 2x_0 - y_0 = 2(0.8) - 0.8288$
$$= \mathbf{0.7712}$$

For line 6, where $x_0 = 1.0$:

$$y_1 = y_0 + h(y')_0$$
$$= 0.8288 + (0.2)(0.7712) = 0.98304$$

As the range is 0 to 1.0, $(y')_0$ in line 6 is not needed. A graph of the solution of $\dfrac{dy}{dx} + y = 2x$, with initial conditions $x = 0$ and $y = 1$ is shown in Fig. 34.5.

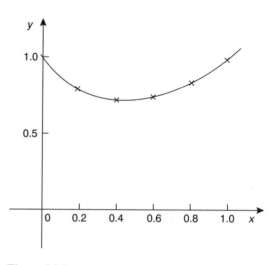

Figure 34.5

Problem 3.

(a) Obtain a numerical solution, using Euler's method, of the differential equation $\dfrac{dy}{dx} = y - x$, with the initial conditions that at $x = 0$, $y = 2$, for the range $x = 0(0.1)0.5$. Draw the graph of the solution

(b) By an analytical method (not done in this text), the solution of the above differential equation is given by $y = x + 1 + e^x$.
Determine the percentage error at $x = 0.3$

(a) $\dfrac{dy}{dx} = y' = y - x$

If initially $x_0 = 0$ and $y_0 = 2$, then $(y')_0 = y_0 - x_0 = 2 - 0 = 2$. Hence line 1 of Table 34.3 is completed.

Table 34.3

	x_0	y_0	$(y')_0$
1.	0	2	2
2.	0.1	2.2	2.1
3.	0.2	2.41	2.21
4.	0.3	2.631	2.331
5.	0.4	2.8641	2.4641
6.	0.5	3.11051	

For line 2, where $x_0 = 0.1$:

$$y_1 = y_0 + h(y')_0, \text{ from equation (1)},$$

$$= 2 + (0.1)(2) = \mathbf{2.2}$$

and $(y')_0 = y_0 - x_0 = 2.2 - 0.1 = \mathbf{2.1}$

For line 3, where $x_0 = 0.2$:

$$y_1 = y_0 + h(y')_0$$

$$= 2.2 + (0.1)(2.1) = \mathbf{2.41}$$

and $(y')_0 = y_0 - x_0 = 2.41 - 0.2 = \mathbf{2.21}$

For line 4, where $x_0 = 0.3$:

$$y_1 = y_0 + h(y')_0$$

$$= 2.41 + (0.1)(2.21) = \mathbf{2.631}$$

and $(y')_0 = y_0 - x_0 = 2.631 - 0.3 = \mathbf{2.331}$

For line 5, where $x_0 = 0.4$:

$$y_1 = y_0 + h(y')_0$$

$$= 2.631 + (0.1)(2.331) = \mathbf{2.8641}$$

and $(y')_0 = y_0 - x_0 = 2.8641 - 0.4 = \mathbf{2.4641}$

For line 6, where $x_0 = 0.5$:

$$y_1 = y_0 + h(y')_0$$

$$= 2.8641 + (0.1)(2.4641) = \mathbf{3.11051}$$

A graph of the solution of $\dfrac{dy}{dx} = y - x$ with $x = 0$, $y = 2$ is shown in Fig. 34.6

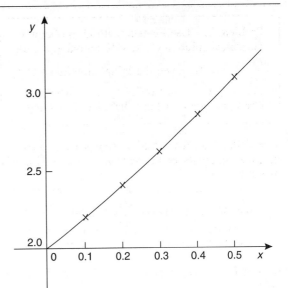

Figure 34.6

(b) If the solution of the differential equation $\dfrac{dy}{dx} = y - x$ is given by $y = x + 1 + e^x$, then when $x = 0.3$, $y = 0.3 + 1 + e^{0.3} = 2.649859$. By Euler's method, when $x = 0.3$ (i.e. line 4 in Table 34.3), $y = 2.631$

Percentage error

$$= \left(\frac{\text{actual} - \text{estimated}}{\text{actual}} \right) \times 100\%$$

$$= \left(\frac{2.649859 - 2.631}{2.649859} \right) \times 100\%$$

$$= \mathbf{0.712\%}$$

Euler's method of numerical solution of differential equations is simple, but approximate. The method is most useful when the interval h is small.

34.4 Further problems on Euler's method

1 Use Euler's method to obtain a numerical solution of the differential equation $\dfrac{dy}{dx} = 3 - \dfrac{y}{x}$, with the initial conditions that $x = 1$ when $y = 2$, for the range $x = 1.0$ to $x = 1.5$ with intervals of 0.1. Draw the graph of the solution in this range.   [See Table 34.4]

Table 34.4

x	y
1.0	2
1.1	2.1
1.2	2.209091
1.3	2.3250
1.4	2.446154
1.5	2.571429

2 Obtain a numerical solution of the differential equation $\dfrac{1}{x}\dfrac{dy}{dx} + 2y = 1$, given the initial conditions that $x = 0$ when $y = 1$, in the range $x = 0(0.2)1.0$.   [See Table 34.5]

Table 34.5

x	y
0	1
0.2	1
0.4	0.96
0.6	0.8864
0.8	0.793664
1.0	0.699692

3 (a) The differential equation $\dfrac{dy}{dx} + 1 = -\dfrac{y}{x}$ has the initial conditions that $y = 1$ at $x = 2$. Produce a numerical solution of the differential equation in the range $x = 2.0(0.1)2.5$

  (b) If the solution of the differential equation by an analytical method is given by $y = \dfrac{4}{x} - \dfrac{x}{2}$,

determine the percentage error at $x = 2.2$
          [(a) see Table 34.6 (b) 1.206%]

Table 34.6

x	y
2.0	1
2.1	0.85
2.2	0.709524
2.3	0.577273
2.4	0.452174
2.5	0.333334

4 Use Euler's method to obtain a numerical solution of the differential equation $\dfrac{dy}{dx} = x - \dfrac{2y}{x}$, given the initial conditions that $y = 1$ when $x = 2$, in the range $x = 2.0(0.2)3.0$.

If the solution of the differential equation is given by $y = \dfrac{x^2}{4}$, determine the percentage error by using Euler's method when $x = 2.8$.
          [See Table 34.7, 1.596%]

Table 34.7

x	y
2.0	1
2.2	1.2
2.4	1.421818
2.6	1.664849
2.8	1.928718
3.0	2.213187

35

Vectors and phasors

35.1 Introduction

Some physical quantities are entirely defined by a numerical value and are called **scalar quantities** or **scalars**. Examples of scalars include time, mass, temperature, energy and volume. Other physical quantities are defined by both a numerical value and a direction in space and these are called **vector quantities** or **vectors**. Examples of vectors include force, velocity, moment and displacement.

35.2 Vector addition

A vector may be represented by a straight line, the length of line being directly proportional to the magnitude of the quantity and the direction of the line being in the same direction as the line of action of the quantity. An arrow is used to denote the sense of the vector, that is, for a horizontal vector, say, whether it acts from left to right or vice versa. The arrow is positioned at the end of the vector and this position is called the 'nose' of the vector. Figure 35.1 shows a velocity of 20 m/s at an angle of 45° to the horizontal and may be depicted by $oa = 20$ m/s at 45° to the horizontal.

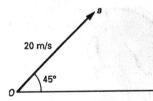

Figure 35.1

To distinguish between vector and scalar quantities, various ways are used. These include

(i) bold print
(ii) two capital letters with an arrow above them to denote the sense of direction, e.g. $\vec{AB}$, where A is the starting point and B the end point of the vector
(iii) a line over the top of letters, e.g. $\overline{AB}$ or $\bar{a}$

(iv) letters with an arrow above, e.g. $\vec{a}$, $\vec{A}$
(v) underlined letters, e.g. $\underline{a}$

The one adopted in this text is to denote vector quantities in bold print.

Thus, **oa** represents a vector quantity, but oa is the magnitude of the vector **oa**. Also, positive angles are measured in an anticlockwise direction from a horizontal, right facing line and negative angles in a clockwise direction from this line. Thus 90° is a line vertically upwards and −90° is a line vertically downwards.

The resultant of adding two vectors together, say V_1 at an angle θ_1 and V_2 at angle $(-\theta_2)$, as shown in Fig. 35.2(a), can be obtained by drawing **oa** to represent V_1 and then drawing **ar** to represent V_2. The resultant of $V_1 + V_2$ is given by **or**. This is shown in Fig. 35.2(b), the vector equation being **oa** + **ar** = **or**. This is called the **'nose-to-tail' method** of vector addition.

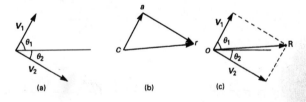

Figure 35.2

Alternatively, by drawing lines parallel to V_1 and V_2 from the noses of V_2 and V_1, respectively, and letting the point of intersection of these parallel lines be R, gives **OR** as the magnitude and direction of the resultant of adding V_1 and V_2, as shown in Fig. 35.2(c). This is called the **'parallelogram' method** of vector addition.

Problem 1. A force of 4 N is inclined at an angle of 45° to a second force of 7 N, both forces acting at a point. Find the magnitude of the resultant of these two forces and the direction of the resultant with respect to the 7 N force by both the 'triangle' and the 'parallelogram' methods

The forces are shown in Fig. 35.3(a). Although the 7 N force is shown as a horizontal line, it could have been drawn in any direction. Using the 'nose-to-tail' method, a line 7 units long is drawn horizontally to give vector **oa** in Fig. 35.3(b). To the nose of this vector **ar** is drawn 4 units long at an angle of 45° to **oa**. The resultant of vector addition is **or** and by measurement is **10.2 units long** and at an angle of **16°** to the 7 N force.

Figure 35.3(c) uses the 'parallelogram' method in which lines are drawn parallel to the 7 N and 4 N forces from the noses of the 4 N and 7 N forces, respectively. These intersect at R. Vector **OR** gives the magnitude and direction of the resultant of vector addition and as obtained by the 'nose-to-tail' method is 10.2 units long at an angle of 16° to the 7 N force. Thus by both methods, the resultant of vector addition is **a force of 10.2 N at an angle of 16° to the 7 N force**.

Often it is easier to use the 'nose-to-tail' method when more than two vectors are being added. The order in which the vectors are added is immaterial. In this case the order taken is v_1, then v_2, then v_3 but just the same result would have been obtained if the order had been, say, v_1, v_3 and finally v_2. v_1 is drawn 10 units long at an angle of 20° to the horizontal, shown by **oa** in Fig. 35.5. v_2 is added to v_1 by drawing a line 15 units long vertically upwards from a, shown as **ab**. Finally, v_3 is added to $v_1 + v_2$ by drawing a line 7 units long at an angle at 190° from b, shown as **br**. The resultant of vector addition is **or** and by measurement is 17.5 units long at an angle of 82° to the horizontal. Thus

$$v_1 + v_2 + v_3 = \textbf{17.5 m/s at 82° to the horizontal}$$

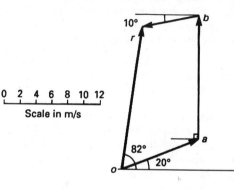

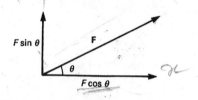

Figure 35.5

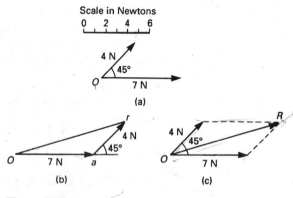

Scale in Newtons

(a)

(b)   (c)

Figure 35.3

35.3 Resolution of vectors

A vector can be resolved into two component parts such that the vector addition of the component parts is equal to the original vector. The two components usually taken are a horizontal component and a vertical component. For the vector shown as **F** in Fig. 35.6, the horizontal component is $F\cos\theta$ and the vertical component is $F\sin\theta$.

Figure 35.6

For the vectors F_1 and F_2 shown in Fig. 35.7, the horizontal component of vector addition is

$$H = F_1\cos\theta_1 + F_2\cos\theta_2$$

Problem 2. Use a graphical method to determine the magnitude and direction of the resultant of the three velocities shown in Fig. 35.4

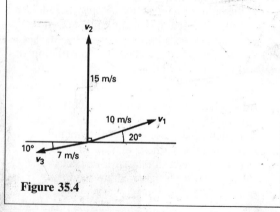

Figure 35.4

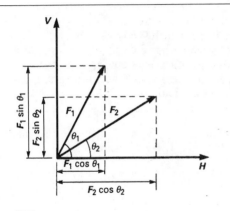

Figure 35.7

and the vertical component of vector addition is

$$V = F_1 \sin \theta_1 + F_2 \sin \theta_2$$

Having obtained H and V, the magnitude of the resultant vector R is given by $\sqrt{(H^2 + V^2)}$ and its angle to the horizontal is given by arctan (V/H).

Problem 3.   Resolve the acceleration vector of 17 m/s² at an angle of 120° to the horizontal into a horizontal and a vertical component

---

For a vector A at angle θ to the horizontal, the horizontal component is given by $A \cos \theta$ and the vertical component by $A \sin \theta$. Any convention of signs may be adopted, in this case horizontally from left to right is taken as positive and vertically upwards is taken as positive.

Horizontal   component H = $17 \cos 120°$ =   **−8.5 m/s²**, acting from left to right.

Vertical component $V = 17 \sin 120° = $ **14.72 m/s²**, acting vertically upwards.

These component vectors are shown in Fig. 35.8.

Problem 4.   Calculate the resultant force of the two forces given in Problem 1

Reference to Fig. 35.3 shows that there are horizontal components due to both the 7 N and the 4 N forces. Horizontal component of force,

$$H = 7 \cos 0° + 4 \cos 45°$$
$$= 7 + 2.828 = 9.828 \text{ N}$$

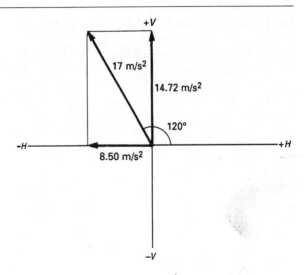

Figure 35.8

Vertical component of force,
$$V = 7 \sin 0° + 4 \sin 45°$$
$$= 0 + 2.828 = 2.828 \text{ N}$$

The magnitude of the resultant of vector addition

$$= \sqrt{(H^2 + V^2)} = \sqrt{(9.828^2 + 2.828^2)}$$
$$= \sqrt{(104.59)} = 10.23 \text{ N}$$

The direction of the resultant of vector addition

$$= \arctan \left(\frac{V}{H} \right)$$

$$= \arctan \left(\frac{2.828}{9.828} \right) = 16.05°$$

Thus, the resultant of the two forces is a single vector of 10.23 N at 16.05° to the 7 N vector.

Problem 5.   Calculate the resultant velocity of the three velocities given in Problem 2

With reference to Fig. 35.4:
Horizontal component of the velocity,

$$H = 10 \cos 20° + 15 \cos 90° + 7 \cos 190°$$
$$= 9.397 + 0 + (-6.894) = 2.503 \text{ m/s}$$

Vertical component of the velocity,

$$V = 10 \sin 20° + 15 \sin 90° + 7 \sin 190°$$
$$= 3.420 + 15 + (-1.216) = 17.204 \text{ m/s}$$

Magnitude of the resultant of vector addition

$$= \sqrt{(H^2 + V^2)} = \sqrt{(2.503^2 + 17.204^2)}$$

$$= \sqrt{302.24} = 17.39 \text{ m/s}$$

Direction of the resultant of vector addition

$$= \arctan\left(\frac{V}{H}\right) = \arctan\left(\frac{17.204}{2.503}\right)$$

$$= \arctan 6.8734 = 81.72°$$

Thus, the resultant of the three velocities is a single vector of 17.39 m/s at 81.72° to the horizontal.

35.4   Vector subtraction

In Fig. 35.9, a force vector F is represented by oa. The vector $(-oa)$ can be obtained by drawing a vector from o in the opposite sense to oa but having the same magnitude, shown as ob in Fig. 35.9, i.e. $ob = (-oa)$.

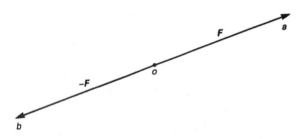

Figure 35.9

For two vectors acting at a point, as shown in Fig. 35.10(a), the resultant of vector addition is $os = oa + ob$. Figure 35.10(b) shows vectors $ob + (-oa)$, that is, $ob - oa$ and the vector equation is $ob - oa = od$. Comparing od in Fig. 35.10(b) with the broken line ab in Fig. 35.10(a) shows that the second diagonal of the 'parallelogram' method of vector addition gives the magnitude and direction of vector subtraction of oa from ob.

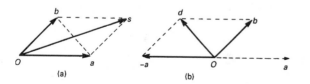

Figure 35.10

Problem 6. Accelerations of $a_1 = 1.5$ m/s^2 at 90° and $a_2 = 2.6$ m/s^2 at 145° act at a point. Find $a_1 + a_2$ and $a_1 - a_2$ by
(i) drawing a scale vector diagram, and
(ii) by calculation

(i) The scale vector diagram is shown in Fig. 35.11. By measurement,

$$a_1 + a_2 = \textbf{3.7 m/s}^2 \textbf{ at 126°}$$

$$a_1 - a_2 = \textbf{2.1 m/s}^2 \textbf{ at 0°}$$

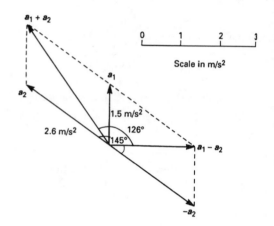

Figure 35.11

(ii) Resolving horizontally and vertically gives:

Horizontal component of $a_1 + a_2$

$$H = 0 + 2.6 \cos 145° = -2.13$$

Vertical component of $a_1 + a_2$

$$V = 1.5 + 2.6 \sin 145° = 2.99$$

Magnitude of $a_1 + a_2 = \sqrt{(-2.13^2 + 2.99^2)}$

$$= 3.67 \text{ m/s}^2$$

Direction of $a_1 + a_2 = \arctan\left(\frac{2.99}{-2.13}\right)$

and must lie in the second quadrant since H is negative and V is positive.

Arctan $(2.99/-2.13) = -54.5°$, and for this to be in the second quadrant, the true angle is 180° displaced, i.e. 180° − 54.5° or 125.5°. Thus

$$a_1 + a_2 = \textbf{3.67 m/s}^2 \textbf{ at 125.5°}$$

Horizontal component of $a_1 - a_2$, that is

$$a_1 + (-a_2) = 0 + 2.6\cos(145° - 180°)$$
$$= 2.6\cos(-35°) = 2.13$$

Vertical component of $a_1 - a_2$, that is

$$a_1 + (-a_2) = 1.5 + 2.6\sin(-35°) = 0$$

Magnitude of $a_1 - a_2 = \sqrt{(2.13^2 + 0_2)}$
$$= 2.13 \text{ m/s}^2$$

Direction of $a_1 - a_2 = \arctan\left(\dfrac{0}{2.13}\right) = 0°$

Thus $a_1 - a_2 = \textbf{2.13 m/s}^2 \textbf{ at } \textbf{0}°$

Problem 7.   Calculate the resultant of
(i) $v_1 - v_2 + v_3$ and (ii) $v_2 - v_1 - v_3$ when
$v_1 = 22$ units at $140°$, $v_2 = 40$ units at $190°$
and $v_3 = 15$ units at $290°$

(i)   The vectors are shown in Fig. 35.12 with their
angles referred to the horizontal axis. When
this is done, the sign is determined by reference to the horizontal and vertical axes. Thus
for horizontal resolution:

$$v_1 = -22\cos 40°, v_2 = -40\cos 10° \text{ and}$$
$$v_3 = 15\cos 70°$$

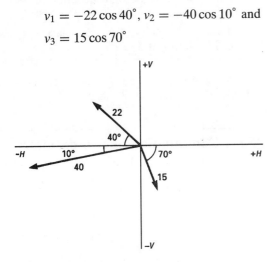

Figure 35.12

The horizontal component of $v_1 - v_2 + v_3$

$$= (-22\cos 40°) - (-40\cos 10°)$$
$$+ (15\cos 70°)$$
$$= -16.85 + 39.39 + 5.13 = 27.67 \text{ units}$$

For vertical resolution:

$$v_1 = 22\sin 40°, v_2 = -40\sin 10° \text{ and}$$
$$v_3 = -15\sin 70°$$

The vertical component of $v_1 - v_2 + v_3$

$$= (22\sin 40°) - (-40\sin 10°)$$
$$+ (-15\sin 70°)$$
$$= 14.14 + 6.95 - 14.10 = 6.99 \text{ units}$$

The magnitude of the resultant, R, which can
be represented by the mathematical symbol for
'the **modulus** of' as $|v_1 - v_2 + v_3|$ is given by:

$$|R| = \sqrt{(27.67^2 + 6.99^2)} = 28.5 \text{ units}$$

The direction of the resultant, R, which can
be represented by the mathematical symbol for
'the **argument** of' as $\arg(v_1 - v_2 + v_3)$ is given
by:

$$\textbf{arg } R = \arctan\left(\dfrac{6.99}{27.67}\right) = 14.2°$$

Thus $v_1 - v_2 + v_3 = \textbf{28.5 units at } \textbf{14.2}°$

(ii)   The horizontal component of $v_2 - v_1 - v_3$

$$= (-40\cos 10°) - (-22\cos 40°)$$
$$- (15\cos 70°)$$
$$= -39.39 + 16.85 - 5.13$$
$$= -27.67 \text{ units}$$

The vertical component of $v_2 - v_1 - v_3$

$$= (-40\sin 10°) - (22\sin 40°)$$
$$- (-15\sin 70°)$$
$$= -6.95 - 14.14 + 14.10$$
$$= -6.99 \text{ units}$$

Let $R = v_2 - v_1 - v_3$.

Then $|R| = \sqrt{[(-27.67)^2 + (-6.99)^2]} = 28.5 \text{ units}$

and $\textbf{arg } R = \arctan\left(\dfrac{-6.99}{-27.67}\right)$

and must lie in the third quadrant since both H and
V are negative quantities. Arctan $(6.99/27.67) =$
$14.2°$, hence the required angle is $180° + 14.2° =$
$194.2°$.

Thus $v_2 - v_1 - v_3 = \textbf{28.5 units at } \textbf{194.2}°$

This result is as expected, since $v_2 - v_1 - v_3 = -(v_1 - v_2 + v_3)$ and the vector 28.5 units at 194.2° is minus times the vector 28.5 units at 14.2°.

Problem 8. Two cars, P and Q, are travelling towards the junction of two roads which are at right angles to one another. Car P has a velocity of 45 km/h due east and car Q a velocity of 55 km/h due south. Calculate (i) the velocity of car P relative to car Q, and (ii) the velocity of car Q relative to car P

For relative velocity problems, some fixed datum point should be selected. This is often a fixed point on the earth's surface. In any vector equation, only the start and finish points affect the resultant vector of a system. Two different systems are shown in Fig. 35.13, but in each of the systems, the resultant vector is *ad*.

The vector equation of the system shown in Fig. 35.13(a) is:

$$ad = ab + bd$$

and that for the system shown in Fig. 35.13(b) is:

$$ad = ab + bc + cd$$

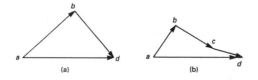

Figure 35.13

Thus in vector equations of this form, only the first and last letters, a and d, respectively, fix the magnitude and direction of the resultant vector. This principle is used in relative velocity problems.

(i) The directions of the cars are shown in Fig. 35.14(a), called a **space diagram**. The velocity diagram is shown in Fig. 35.14(b), in which *pe* is taken as the velocity of car P relative to point e on the earth's surface. The velocity of P relative to Q is vector *pq* and the vector equation is $pq = pe + eq$. Hence the vector directions are as shown, *eq* being in the opposite direction to *qe*. From the geometry of the vector triangle,

$$|pq| = \sqrt{(45^2 + 55^2)} = 71.1 \text{ km/h}$$

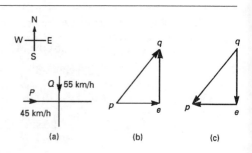

Figure 35.14

$$\arg pq = \arctan\left(\frac{55}{45}\right) = 50.7°$$

That is, the velocity of car P relative to car Q is 71.1 km/h at 50.7°

(ii) The velocity of car Q relative to car P is given by the vector equation $qp = qe + ep$ and the vector diagram is as shown in Fig. 35.14(c), having *ep* opposite in direction to *pe*. From the geometry of this vector triangle:

$$|qp| = \sqrt{(45^2 + 55^2)} = 71.1 \text{ m/s}$$

$$\arg qp = \arctan\left(\frac{55}{45}\right) = 50.7°$$

but must lie in the third quadrant, that is, the required angle is $180° + 50.7° = 230.7°$.

Thus the velocity of car Q relative to car P is 71.1 km/h at 230.7°

Further problems on vectors may be found in Section 35.6, Problems 1 to 8, pages 330 and 331.

35.5 Combination of two periodic functions of the same frequency

There are a number of instances in engineering and science where waveforms combine and where it is required to determine the single phasor (called the resultant) which could replace two or more separate phasors. (From Chapter 17, Section 4, a phasor is a rotating vector.) Uses are found in electrical alternating current theory, in mechanical vibrations, in the addition of forces and with sound waves. There are several methods of determining the resultant and two such methods are shown below.

(i) **Plotting the periodic functions graphically**
This may be achieved by sketching the separate functions on the same axes and then

adding (or subtracting) ordinates at regular intervals (see Problems 9 to 11). Alternatively, a table of values may be drawn up before plotting the resultant waveforms (see Problem 12)

(ii) **Resolution of phasors by drawing or calculation**

The resultant of two periodic functions may be found from their relative positions when the time is zero. For example, if $y_1 = 4 \sin \omega t$ and $y_2 = 3 \sin(\omega t - \pi/3)$ then each may be represented as phasors as shown in Fig. 35.15, y_1 being 4 units long and drawn horizontally and y_2 being 3 units long, lagging y_1 by $\pi/3$ radians or 60°. To determine the resultant of $y_1 + y_2$, y_1 is drawn horizontally as shown in Fig. 35.16 and y_2 is joined to the end of y_1 at 60° to the horizontal. The resultant is given by y_R. This is the same as the diagonal of a parallelogram which is shown completed in Fig. 35.17. Resultant y_R, in Figs 35.16 and 35.17, is determined either by:

(a) scaled drawing and measurement, or
(b) by use of the cosine rule (and then sine rule to calculate angle ϕ), or
(c) by determining horizontal and vertical components of lengths oa and ab in Fig. 35.16, and then using Pythagoras' theorem to calculate ob

Figure 35.15

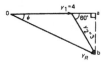

Figure 35.16

Figure 35.17

In this case, by calculation, $y_R = 6.083$ and angle $\phi = 25.28°$ or 0.441 rad. Thus the resultant may be

expressed in sinusoidal form as $y_R = 6.083 \sin(\omega t - 0.441)$. If the resultant phasor $y_R = y_1 - y_2$ is required, then y_2 is still 3 units long but is drawn in the opposite direction, as shown in Fig. 35.18, and y_R is determined by measurement or calculation. (See Problems 13 to 15.)

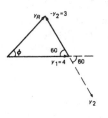

Figure 35.18

> Problem 9. Plot the graph of $y_1 = 3 \sin A$ from $A = 0°$ to $A = 360°$. On the same axes plot $y_2 = 2 \cos A$. By adding ordinates plot $y_R = 3 \sin A + 2 \cos A$ and obtain a sinusoidal expression for this resultant waveform

$y_1 = 3 \sin A$ and $y_2 = 2 \cos A$ are shown plotted in Fig. 35.19. Ordinates may be added at, say, 15° intervals. For example,

$$\text{at} \quad 0°, \ y_1 + y_2 = 0 + 2 = 2$$
$$\text{at} \quad 15°, \ y_1 + y_2 = 0.78 + 1.93 = 2.71$$
$$\text{at} \quad 120°, \ y_1 + y_2 = 2.60 + -1 = 1.6$$
$$\text{at} \quad 210°, \ y_1 + y_2 = -1.50 - 1.73$$
$$= -3.23, \text{ and so on}$$

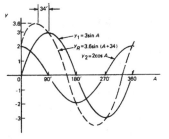

Figure 35.19

The resultant waveform, shown by the broken line, has the same period, i.e. 360°, and thus the same frequency as the single phasors. The maximum value, or amplitude, of the resultant is 3.6. The resultant waveform leads $y_1 = 3 \sin A$ by 34° or 0.593 rad.

The sinusoidal expression for the resultant waveform is:

$$y_R = 3.6\sin(A + 34°) \text{ or}$$

$$y_R = 3.6\sin(A + 0.593)$$

Problem 10. Plot the graphs of $y_1 = 4\sin\omega t$ and $y_2 = 3\sin(\omega t - \pi/3)$ on the same axis, over one cycle. By adding ordinates at intervals plot $y_R = y_1 + y_2$ and obtain a sinusoidal expression for the resultant waveform

---

$y_1 = 4\sin\omega t$ and $y_2 = 3\sin(\omega t - \pi/3)$ are shown plotted in Fig. 35.20.

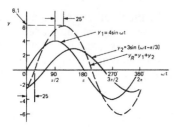

Figure 35.20

Ordinates are added at 15° intervals and the resultant is shown by the broken line. The amplitude of the resultant is 6.1 and it lags y_1 by 25° or 0.436 rad. Hence the sinusoidal expression for the resultant waveform is

$$y_R = 6.1\sin(\omega t - 0.436)$$

Problem 11. Determine a sinusoidal expression for $y_1 - y_2$ when $y_1 = 4\sin\omega t$ and $y_2 = 3\sin(\omega t - \pi/3)$

---

y_1 and y_2 are shown plotted in Fig. 35.21. At 15° intervals y_2 is subtracted from y_1. For example:

at   0°, $y_1 - y_2 = 0 - (-2.6) = +2.6$

at   30°, $y_1 - y_2 = 2 - (-1.5) = +3.5$

at   150°, $y_1 - y_2 = 2 - 3 = -1$,   and so on

The amplitude, or peak value of the resultant (shown by the broken line), is 3.6 and it leads y_1 by 45° or 0.79 rad. Hence

$$y_1 - y_2 = 3.6\sin(\omega t + 0.79)$$

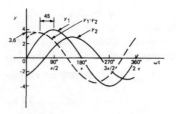

Figure 35.21

Problem 12. Draw a graph to represent current $i = 2.4\sin t + 3.2\sin(t + 40°)$ amperes and express i in the general form $i = A\sin(t \pm \alpha)$

A table of values may be drawn up as shown below.

$t°$	0	30	60
$\sin t$	0	0.5	0.866
$2.4\sin t$	0	1.2	2.08
$(t + 40°)$	40	70	100
$\sin(t + 40°)$	0.64	0.94	0.98
$3.2\sin(t + 40°)$	2.05	3.01	3.14
$i = 2.4\sin t +$ $3.2\sin(t + 40°)$	2.05	4.21	5.22

$t°$	90	120	150
$\sin t$	1.0	0.866	0.5
$2.4\sin t$	2.4	2.08	1.2
$(t + 40°)$	130	160	190
$\sin(t + 40°)$	0.77	0.34	-0.17
$3.2\sin(t + 40°)$	2.46	1.09	-0.54
$i = 2.4\sin t +$ $3.2\sin(t + 40°)$	4.86	3.17	0.66

$t°$	180	210	240
$\sin t$	0	-0.5	-0.866
$2.4\sin t$	0	-1.2	-2.08
$(t + 40°)$	220	250	280
$\sin(t + 40°)$	-0.64	-0.94	-0.98
$3.2\sin(t + 40°)$	-2.05	-3.01	-3.14
$i = 2.4\sin t +$ $3.2\sin(t + 40°)$	-2.05	-4.21	-5.22

$t°$		270	300	330	360
$\sin t$		−1.0	−0.866	−0.5	0
$2.4 \sin t$		−2.4	−2.08	−1.2	0
$(t + 40°)$		310	340	370	400
$\sin(t + 40°)$		−0.77	−0.34	0.17	0.64
$3.2 \sin(t + 40°)$		−2.46	−1.09	0.54	2.05
$i = 2.4 \sin t +$					
$3.2 \sin(t + 40°)$		−4.86	−3.17	−0.66	2.05

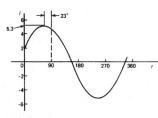

Figure 35.22

A graph of $i = 2.4 \sin t + 3.2 \sin(t + 40°)$ is shown in Fig. 35.22. The amplitude is 5.3 amperes and i is 23°, i.e. 0.40 rad, ahead of a sine wave starting at 0°. Hence

$$i = 5.3 \sin(t + 0.40) \text{ amperes}$$

Problem 13. Given $y_1 = 2 \sin \omega t$ and $y_2 = 3 \sin(\omega t + \pi/4)$, obtain an expression for the resultant $y_R = y_1 + y_2$, (a) by drawing, and (b) by calculation

---

(a) When time $t = 0$ the position of phasors y_1 and y_2 are as shown in Fig. 35.23(a). To obtain the resultant, y_1 is drawn horizontally, 2 units long, y_2 is drawn 3 units long at an angle of $\pi/4$ rads or 45° and joined to the end of y_1 as shown in Fig. 35.23(b). y_R is measured as 4.6 units long and angle ϕ is measured as 27° or 0.47 rad.
Alternatively, y_R is the diagonal of the parallelogram formed as shown in Fig. 35.23(c).
Hence, by drawing, $y_R = 4.6 \sin(\omega t + 0.47)$

(b) From Fig. 35.23(b), and using the cosine rule:

$$y_R^2 = 2^2 + 3^2 - [2(2)(3) \cos 135°]$$

$$= 4 + 9 - [-8.485] = 21.49$$

Hence $y_R = \sqrt{(21.49)} = 4.64$

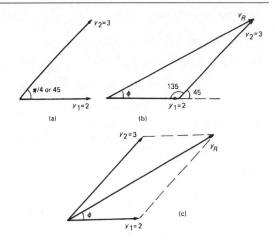

Figure 35.23

Using the sine rule:

$$\frac{3}{\sin \phi} = \frac{4.64}{\sin 135°} \quad \text{from which}$$

$$\sin \phi = \frac{3 \sin 135°}{4.64} = 0.4572$$

Hence $\phi = \arcsin 0.4572 = 27°12'$ or 0.475 rad
By calculation $y_R = 4.64 \sin(\omega t + 0.475)$

Problem 14. Two alternating voltages are given by $v_1 = 15 \sin \omega t$ volts and $v_2 = 25 \sin(\omega t - \pi/6)$ volts. Determine a sinusoidal expression for the resultant $v_R = v_1 + v_2$ by finding horizontal and vertical components

---

The relative positions of v_1 and v_2 at time $t = 0$ are shown in Fig. 35.24(a) and the phasor diagram is shown in Fig. 35.24(b).
The horizontal component of $v_R = oa + ab$

$$= 15 + 25 \cos 30°$$

$$= 36.65 \text{ V}$$

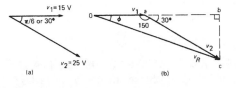

Figure 35.24

The vertical component of v_R

$$= bc = 25\sin 30° = 12.50 \text{ V}$$

Hence $\quad v_R(= oc) = \sqrt{[(36.65)^2 + (12.50)^2]}$

by Pythagoras' theorem

$$= 38.72 \text{ volts}$$

$$\tan\phi = \frac{bc}{ob} = \frac{12.50}{36.65} = 0.3411$$

from which,

$$\phi = \arctan 0.3411 = 18°50' \text{ or } 0.329 \text{ radians}$$

Hence

$$v_R = v_1 + v_2 = 38.72\sin(\omega t - 0.329) \text{ V}$$

Problem 15. For the voltages in Problem 14, determine the resultant $v_R = v_1 - v_2$

To find the resultant $v_R = v_1 - v_2$, the phasor v_2 of Fig. 35.24(b) is reversed in direction as shown in Fig. 35.25. Using the cosine rule:

$$v_R^2 = 15^2 + 25^2 - 2(15)(25)\cos 30°$$

$$= 225 + 625 - 649.5 = 200.5$$

$$v_R = \sqrt{(200.5)} = 14.16 \text{ volts}$$

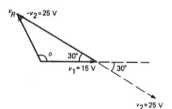

Figure 35.25

Using the sine rule:

$$\frac{25}{\sin\phi} = \frac{14.16}{\sin 30°}, \text{ from which}$$

$$\sin\phi = \frac{25\sin 30°}{14.16} = 0.8828$$

Hence $\phi = \arcsin 0.8828 = 61.98°$ or $118.02°$. From Fig. 35.25, ϕ is obtuse, hence $\phi = 118.02°$ or 2.06 radians. Hence

$$v_R = v_1 - v_2 = 14.16\sin(\omega t + 2.06) \text{ V}$$

Further problems on the combination of two periodic functions of the same frequency may be found in the following Section 35.6, Problems 9 to 13, page 331.

35.6 Further problems on vectors and phasors

Vectors

1 Forces of 23 N and 41 N act at a point and are inclined at 90° to each other. Find, by drawing, the resultant force and its direction relative to the 41 N force. [47 N at 29°]

2 Forces A, B and C are coplanar and act at a point. Force A is 12 kN at 90°, B is 5 kN at 180° and C is 13 kN at 293°. Determine graphically the resultant force. [Zero]

3 Calculate the magnitude and direction of velocities of 3 m/s at 18° and 7 m/s at 115° when acting simultaneously on a point. [7.27 m/s at 90.8°]

4 Three forces of 2 N, 3 N and 4 N act as shown in Fig. 35.26. Calculate the magnitude of the resultant force and its direction relative to the 2 N force. [6.24 N at 76.10°]

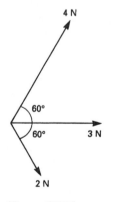

Figure 35.26

5 A load of 5.89 N is lifted by two strings, making angles of 20° and 35° with the vertical. Calculate the tensions in the strings. (For a system such as this, the vectors representing the forces form a closed triangle when the system is in equilibrium.) [2.46 N, 4.12 N]

6 The acceleration of a body is due to four component, coplanar accelerations. These are

2 m/s^2 due north, 3 m/s^2 due east, 4 m/s^2 to the south-west and 5 m/s^2 to the south-east. Calculate the resultant acceleration and its direction.   [5.7 m/s^2 at 310°]

7   A car is moving along a straight horizontal road at 79.2 km/h and rain is falling vertically downwards at 26.4 km/h. Find the velocity of the rain relative to the driver of the car.
[83.5 km/h at 71.6° to the vertical]

8   Calculate the time needed to swim across a river 142 m wide when the swimmer can swim at 2 km/h in still water and the river is flowing at 1 km/h.   [4 minutes 55 seconds]

Combination of two periodic functions of the same frequency

9   Plot the graph of $y = 2\sin A$ from $A = 0°$ to $A = 360°$. On the same axes plot $y = 4\cos A$. By adding ordinates at intervals plot $y = 2\sin A + 4\cos A$ and obtain a sinusoidal expression for the waveform.
$$[4.5\sin(A + 63°26')]$$

10   Two alternating voltages are given by $v_1 = 10\sin\omega t$ volts and $v_2 = 14\sin(\omega t + \pi/3)$ volts. By plotting v_1 and v_2 on the same axes over one cycle obtain a sinusoidal expression for (a) $v_1 + v_2$ (b) $v_1 - v_2$.
$$\begin{bmatrix} \text{(a) } 20.9\sin(\omega t + 0.62) \text{ volts} \\ \text{(b) } 12.5\sin(\omega t - 1.33) \text{ volts} \end{bmatrix}$$

11   Draw up a table of values for the waveform $y = 5\sin(\omega t + \pi/4) + 3\sin(\omega t - \pi/3)$ and plot a graph of y against ωt. Express y in the form $y = A\sin(\omega t \pm \alpha)$.   [$5.1\sin(\omega t + 0.184)$]

In Problems 12 and 13, express the combination of periodic functions in the form $A\sin(\omega t \pm \alpha)$ using phasors, either by drawing or by calculation:

12   (a) $12\sin\omega t + 5\cos\omega t$

(b) $7\sin\omega t + 5\sin\left(\omega t + \dfrac{\pi}{4}\right)$

(c) $6\sin\omega t + 3\sin\left(\omega t - \dfrac{\pi}{6}\right)$

$$\begin{bmatrix} \text{(a) } 13\sin(\omega t + 0.395) \\ \text{(b) } 11.11\sin(\omega t + 0.324) \\ \text{(c) } 8.73\sin(\omega t - 0.173) \end{bmatrix}$$

13   (a) $i = 25\sin\omega t - 15\sin\left(\omega t + \dfrac{\pi}{3}\right)$

(b) $v = 8\sin\omega t - 5\sin\left(\omega t - \dfrac{\pi}{4}\right)$

(c) $x = 9\sin\left(\omega t + \dfrac{\pi}{3}\right) - 7\sin\left(\omega t - \dfrac{3\pi}{8}\right)$

$$\begin{bmatrix} \text{(a) } i = 21.79\sin(\omega t - 0.639) \\ \text{(b) } v = 5.695\sin(\omega t + 0.670) \\ \text{(c) } x = 14.38\sin(\omega t + 1.444) \end{bmatrix}$$

36

Scalar and vector products

36.1 The unit triad

When a vector x of magnitude x units and direction $\theta°$ is divided by the magnitude of the vector, the result is a vector of unit length at angle $\theta°$. The unit vector for a velocity of 10 m/s at 50° is (10 m/s at 50°)/(10 m/s), i.e. 1 at 50°. In general, the unit vector for oa is $oa/|oa|$, the oa being a vector and having both magnitude and direction and $|oa|$ being the magnitude of the vector only. One method of completely specifying the direction of a vector in space relative to some reference point is to use three unit vectors, mutually at right angles to each other, as shown in Fig. 36.1. Such a system is called a **unit triad**.

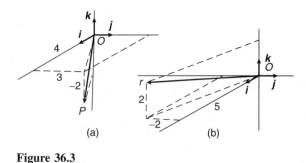

Figure 36.2

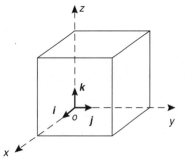

Figure 36.1

In Fig. 36.2, one way to get from o to r is to move x units along i to point a, then y units in direction j to get to b and finally z units in direction k to get to r. The vector or is specified as

$$or = xi + yj + zk$$

Problem 1.   With reference to three axes drawn mutually at right angles, depict the vectors (i) $op = 4i + 3j - 2k$, and (ii) $or = 5i - 2j + 2k$

The required vectors are depicted in Fig. 36.3, op being shown in Fig. 36.3(a) and or in Fig. 36.3(b).

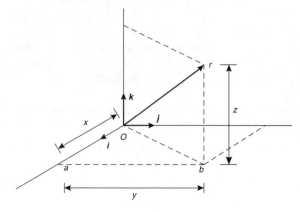

Figure 36.3

36.2 The scalar product of two vectors

When vector oa is multiplied by a scalar quantity, say k, the magnitude of the resultant vector will be k times the magnitude of oa and its direction will remain the same. Thus $2 \times (5$ N at $20°)$ results in a vector of magnitude 10 N at 20°.

One of the products of two vector quantities is called the **scalar** or **dot product** of two vectors and is defined as the product of their magnitudes multiplied by the cosine of the angle between them. The scalar product of oa and ob is shown as $oa \cdot ob$. For vectors $oa = oa$ at $\theta_1°$, and $ob = ob$ at θ_2 where $\theta_2 > \theta_1$, the scalar product is:

$$oa \cdot ob = oa\,ob\cos(\theta_2 - \theta_1)$$

For vectors v_1 and v_2 shown in Fig. 36.4, the scalar product is

$$v_1 \cdot v_2 = v_1 v_2 \cos \theta$$

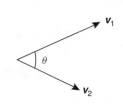

Figure 36.4

The commutative law of algebra, $a \times b = b \times a$ applies to scalar products. This is demonstrated in Fig. 36.5. Let oa represent vector v_1 and ob represent vector v_2. Then:

$$oa \cdot ob = v_1 v_2 \cos \theta$$

(by definition of a scalar product)

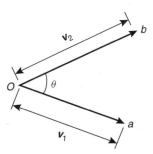

Figure 36.5

Similarly, $ob \cdot oa = v_2 v_1 \cos \theta = v_1 v_2 \cos \theta$ by the commutative law of algebra. Thus $oa \cdot ob = ob \cdot oa$. The projection of ob on oa is shown in Fig. 36.6(a) and by the geometry of triangle obc, it can be seen that the projection is $v_2 \cos \theta$. Since, by definition $oa \cdot ob = v_1(v_2 \cos \theta)$, it follows that

$$oa \cdot ob = v_1 \text{ (the projection of } v_2 \text{ on } v_1)$$

Similarly the projection of oa on ob is shown in Fig. 36.6(b) and is $v_1 \cos \theta$. Since by definition $ob \cdot oa = v_2(v_1 \cos \theta)$, it follows that

$$ob \cdot oa = v_2 \text{ (the projection of } v_1 \text{ on } v_2)$$

This shows that the scalar product of two vectors is the product of the magnitude of one vector and the magnitude of the projection of the other vector on it. The angle between two vectors can be expressed in terms of the vector constants as follows:

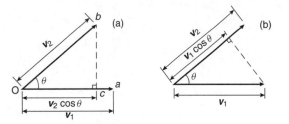

Figure 36.6

Because $a \cdot b = ab \cos \theta$,

then
$$\boxed{\cos \theta = \frac{a \cdot b}{ab}} \quad (1)$$

Let $a = a_1 i + a_2 j + a_3 k$ and $b = b_1 i + b_2 j + b_3 k$

$$a \cdot b = (a_1 i + a_2 j + a_3 k) \cdot (b_1 i + b_2 j + b_3 k)$$

Multiplying out the brackets gives

$$a \cdot b = a_1 b_1 i \cdot i + a_1 b_2 i \cdot j + a_1 b_3 i \cdot k + a_2 b_1 j \cdot i$$
$$+ a_2 b_2 j \cdot j + a_2 b_3 j \cdot k + a_3 b_1 k \cdot i$$
$$+ a_3 b_2 k \cdot j + a_3 b_3 k \cdot k$$

However, the unit vectors i, j and k all have a magnitude of 1 and $i \cdot i = (1)(1) \cos 0° = 1$, $i \cdot j = (1)(1) \cos 90° = 0$, $i \cdot k = (1)(1) \cos 90° = 0$ and similarly $j \cdot j = 1, j \cdot k = 0$ and $k \cdot k = 1$. Thus, only terms containing $i \cdot i, j \cdot j$ or $k \cdot k$ in the expansion above will not be zero.

Thus,
$$\boxed{a \cdot b = a_1 b_1 + a_2 b_2 + a_3 b_3} \quad (2)$$

Both a and b in equation (1) can be expressed in terms of a_1, b_1, a_2, b_2, a_3 and b_3.
From the geometry of Fig. 36.7, the length of diagonal OP in terms of side lengths a, b and c can be obtained as follows.

$$OP^2 = OB^2 + BP^2 \text{ and } OB^2 = OA^2 + AB^2$$

Thus, $OP^2 = OA^2 + AB^2 + BP^2$

$$= a^2 + b^2 + c^2, \text{ in terms of side lengths}$$

i.e.
$$\boxed{OP = \sqrt{(a^2 + b^2 + c^2)}} \quad (3)$$

Relating this result to the two vectors $a_1 i + a_2 j + a_3 k$ and $b_1 i + b_2 j + b_3 k$, gives $a = \sqrt{(a_1^2 + a_2^2 + a_3^2)}$ and $b = \sqrt{(b_1^2 + b_2^2 + b_3^2)}$.

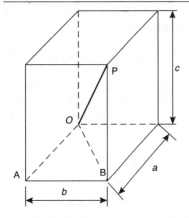

Figure 36.7

That is, from equation (1),

$$\cos\theta = \frac{a_1 b_1 + a_2 b_2 + a_3 b_3}{\sqrt{(a_1^2 + a_2^2 + a_3^2)}\sqrt{(b_1^2 + b_2^2 + b_3^2)}}$$

(4)

Problem 2. Find vector a joining points P and Q where point P has coordinates $(4, -1, 3)$ and point Q has coordinates $(2, 5, 0)$. Also find $|a|$, the magnitude of a

---

Let 0 be the origin, i.e. its coordinates are $(0, 0, 0)$. The position vector of P and Q are given by

$$OP = 4i - j + 3k \text{ and } OQ = 2i + 5j$$

By the addition law of vectors $OP + PQ = OQ$

Hence $a = PQ = OQ - OP$

i.e. $a = PQ = (2i + 5j) - (4i - j + 3k)$

$$= -2i + 6j - 3k$$

From equation (3), the magnitude of a,

$$|a| = \sqrt{(a^2 + b^2 + c^2)}$$

$$= \sqrt{[(-2)^2 + (6)^2 + (-3)^2]}$$

$$= \sqrt{49} = 7$$

Problem 3. If $p = 2i + j - k$ and $q = i - 3j + 2k$ determine:

   (i) $p \cdot q$ (ii) $p + q$

  (iii) $|p + q|$ (iv) $|p| + |q|$

(i)   From equation (2), if $p = a_1 i + a_2 j + a_3 k$ and $q = b_1 i + b_2 j + b_3 k$ then $p \cdot q = a_1 b_1 + a_2 b_2 + a_3 b_3$

When $p = 2i + j - k$, $a_1 = 2$, $a_2 = 1$ and $a_3 = -1$

and when $q = i - 3j + 2k$, $b_1 = 1$, $b_2 = -3$ and $b_3 = 2$

Hence $p \cdot q = (2)(1) + (1)(-3) + (-1)(2)$

i.e. $p \cdot q = -3$

(ii) $p + q = (2i + j - k) + (i - 3j + 2k)$

$$= 3i - 2j + k$$

(iii) $|p + q| = |3i - 2j + k|$

From equation (3),

$$|p + q| = \sqrt{[(3)^2 + (-2)^2 + (1)^2]} = \sqrt{14}$$

(iv) From equation (3),

$$|p| = |2i + j - k| = \sqrt{[(2)^2 + (1)^2 + (-1)^2]}$$

$$= \sqrt{6}$$

Similarly,

$$|q| = |i - 3j + 2k| = \sqrt{[(1)^2 + (-3)^2 + (2)^2]}$$

$$= \sqrt{14}$$

Hence $|p| + |q| = \sqrt{6} + \sqrt{14} = \mathbf{6.191}$, correct to 3 decimal places

---

Problem 4. Determine the angle between vectors oa and ob when $oa = i + 2j - 3k$ and $ob = 2i - j + 4k$

An equation for $\cos\theta$ is given in equation (4)

$$\cos\theta = \frac{a_1 b_1 + a_2 b_2 + a_3 b_3}{\sqrt{(a_1^2 + a_2^2 + a_3^2)}\sqrt{(b_1^2 + b_2^2 + b_3^2)}}$$

Since $oa = i + 2j - 3k$, $a_1 = 1$, $a_2 = 2$ and $a_3 = -3$
Since $ob = 2i - j + 4k$, $b_1 = 2$, $b_2 = -1$ and $b_3 = 4$

Thus,

$$\cos\theta = \frac{(1 \times 2) + (2 \times (-1)) + ((-3) \times 4)}{\sqrt{(1^2 + 2^2 + (-3)^2)}\sqrt{(2^2 + (-1)^2 + 4^2)}}$$

$$= \frac{-12}{\sqrt{14}\sqrt{21}} = -0.6999$$

$$\theta = 134.4° \text{ or } 225.6°$$

By sketching the position of the two vectors as shown in Problem 1, it will be seen that 225.6° is not an acceptable answer. **Thus $\theta = 134.4°$.**

Practical application of scalar product

Problem 5.   A constant force of $F = 10i + 2j - k$ newtons displaces an object from $A = i + j + k$ to $B = 2i - j + 3k$ (in metres). Find the work done in newton metres

---

One of the applications of scalar products is to the work done by a constant force when moving a body. The work done is the product of the applied force and the distance moved in the direction of the force, i.e. **work done $= F \cdot d$**

The principles developed in Problem 8, Chapter 35, apply equally to this problem when determining the displacement. From the sketch shown in Fig. 36.8,

$$AB = AO + OB = OB - OA$$

that is $AB = (2i - j + 3k) - (i + j + k)$

$$= i - 2j + 2k$$

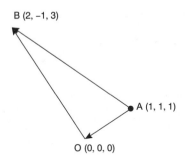

Figure 36.8

The work done is $F \cdot d$, that is $F \cdot AB$ in this case, i.e. work done $= (10i + 2j - k) \cdot (i - 2j + 2k)$
But from equation (2), $a \cdot b = a_1b_1 + a_2b_2 + a_3b_3$
Hence **work done** $= (10 \times 1) + (2 \times (-2))$
$$+ (-1) \times 2) = \textbf{4 Nm}$$
(Theoretically, it is quite possible to get a negative answer to a 'work done' problem. This indicates that the force must be in the opposite sense to that given, in order to give the displacement stated.)

Further problems on scalar products may be found in Section 36.4, Problems 1 to 10, page 338.

36.3   Vector products

A second product of two vectors is called the **vector** or **cross product** and is defined in terms of its modulus and the magnitudes of the two vectors and the sine of the angle between them. The vector product of vectors oa and ob is written as $oa \times ob$ and is defined by:

$$|oa \times ob| = oa\, ob \sin \theta$$

where θ is the angle between the two vectors.
    The direction of $oa \times ob$ is perpendicular to both oa and ob, as shown in Fig. 36.9.

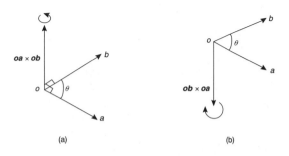

(a)   (b)

Figure 36.9

The direction is obtained by considering that a right-handed screw is screwed along $oa \times ob$ with its head at the origin and if the direction of $oa \times ob$ is correct, the head should rotate from oa to ob, as shown in Fig. 36.9(a). It follows that the direction of $ob \times oa$ is as shown in Fig. 36.9(b). Thus $oa \times ob$ is not equal to $ob \times oa$. The magnitudes of $oa\, ob \sin \theta$ are the same but their directions are 180° displaced, i.e.

$$oa \times ob = -ob \times oa$$

The vector product of two vectors may be expressed in terms of the unit vectors. Let two vectors, a and b, be such that:

$$a = a_1i + a_2j + a_3k \text{ and } b = b_1i + b_2j + b_3k$$

Then,

$$a \times b = (a_1i + a_2j + a_3k) \times (b_1i + b_2j + b_3k)$$
$$= a_1b_1i \times i + a_1b_2i \times j + a_1b_3i \times k$$
$$+ a_2b_1j \times i + a_2b_2j \times j + a_2b_3j \times k$$
$$+ a_3b_1k \times i + a_3b_2k \times j + a_3b_3k \times k$$

But by the definition of a vector product,

$$i \times j = k, j \times k = i \text{ and } k \times i = j$$

Also $i \times i = j \times j = k \times k = (1)(1) \sin 0° = 0$

Remembering that $a \times b = -b \times a$ gives:

$$a \times b = a_1 b_2 k - a_1 b_3 j - a_2 b_1 k + a_2 b_3 i$$
$$+ a_3 b_1 j - a_3 b_2 i$$

Grouping the i, j and k terms together, gives

$$a \times b = (a_2 b_3 - a_3 b_2)i + (a_3 b_1 - a_1 b_3)j$$
$$+ (a_1 b_2 - a_2 b_1)k$$

The vector product can be written in determinant form as:

$$a \times b = \begin{vmatrix} i & j & k \\ a_1 & a_2 & a_3 \\ b_1 & b_2 & b_3 \end{vmatrix} \quad (1)$$

The 3×3 determinant $\begin{vmatrix} i & j & k \\ a_1 & a_2 & a_3 \\ b_1 & b_2 & b_3 \end{vmatrix}$ is evaluated as:

$$i \begin{vmatrix} a_2 & a_3 \\ b_2 & b_3 \end{vmatrix} - j \begin{vmatrix} a_1 & a_3 \\ b_1 & b_3 \end{vmatrix} + k \begin{vmatrix} a_1 & a_2 \\ b_1 & b_2 \end{vmatrix}$$

where $\begin{vmatrix} a_2 & a_3 \\ b_2 & b_3 \end{vmatrix} = a_2 b_3 - a_3 b_2$

$\begin{vmatrix} a_1 & a_3 \\ b_1 & b_3 \end{vmatrix} = a_1 b_3 - a_3 b_1$

and $\begin{vmatrix} a_1 & a_2 \\ b_1 & b_2 \end{vmatrix} = a_1 b_2 - a_2 b_1$ (see page 440)

The magnitude of the vector product of two vectors can be found by expressing it in scalar product form and then using the relationship

$$a \cdot b = a_1 b_1 + a_2 b_2 + a_3 b_3$$

Squaring both sides of a vector product equation gives:

$$(|a \times b|)^2 = a^2 b^2 \sin^2 \theta = a^2 b^2 (1 - \cos^2 \theta)$$
$$= a^2 b^2 - a^2 b^2 \cos^2 \theta \dots \quad (2)$$

It is stated in Section 36.2 that $a \cdot b = ab \cos \theta$, hence

$$a \cdot a = a^2 \cos \theta. \text{ But } \theta = 0°, \text{ thus}$$
$$a \cdot a = a^2$$

Also, $\cos \theta = \dfrac{a \cdot b}{ab}$

Multiplying both sides of this equation by $a^2 b^2$ and squaring gives

$$a^2 b^2 \cos^2 \theta = \frac{a^2 b^2 (a \cdot b)^2}{a^2 b^2} = (a \cdot b)^2$$

Substituting in equation (2) above for $a^2 = a \cdot a$, $b^2 = b \cdot b$ and $a^2 b^2 \cos^2 \theta = (a \cdot b)^2$ gives

$$(|a \times b|)^2 = (a \cdot a)(b \cdot b) - (a \cdot b)^2$$

That is, $|a \times b| = \sqrt{[(a \cdot a)(b \cdot b) - (a \cdot b)^2]}$ (3)

Problem 6. For the vectors $a = i + 4j - 2k$ and $b = 2i - j + 3k$ find (i) $a \times b$ and (ii) $|a \times b|$

(i) From equation (1),

$$a \times b = \begin{vmatrix} i & j & k \\ a_1 & a_2 & a_3 \\ b_1 & b_2 & b_3 \end{vmatrix} = i \begin{vmatrix} a_2 & a_3 \\ b_2 & b_3 \end{vmatrix}$$
$$- j \begin{vmatrix} a_1 & a_3 \\ b_1 & b_3 \end{vmatrix} + k \begin{vmatrix} a_1 & a_2 \\ b_1 & b_2 \end{vmatrix}$$

$$= i \begin{vmatrix} 4 & -2 \\ -1 & 3 \end{vmatrix} - j \begin{vmatrix} 1 & -2 \\ 2 & 3 \end{vmatrix}$$
$$+ k \begin{vmatrix} 1 & 4 \\ 2 & -1 \end{vmatrix}$$

$$= i(12 - 2) - j(3 + 4) + k(-1 - 8)$$
$$= 10i - 7j - 9k$$

(ii) From equation (3)

$$|a \times b| = \sqrt{[(a \cdot a)(b \cdot b) - (a \cdot b)^2]}$$

Now $a \cdot a = (1)(1) + (4 \times 4) +$
$$(-2)(-2)) = 21$$

$b \cdot b = (2)(2) + (-1)(-1) + (3)(3)$
$$= 14$$

and $a \cdot b = (1)(2) + (4)(-1) + (-2)(3)$
$$= -8$$

Thus $|a \times b| = \sqrt{(21 \times 14 - 64)} = \sqrt{230}$
$$= \mathbf{15.17}$$

Problem 7. If $p = 4i + j - 2k$, $q = 3i - 2j + k$ and $r = i - 2k$ find (a) $(p - 2q) \times r$ (b) $p \times (2r \times 3q)$

(a) $(p - 2q) \times r = [4i + j - 2k - 2(3i - 2j$

$+ k)] \times (i - 2k)$

$= (-2i + 5j - 4k) \times (i - 2k)$

$$= \begin{vmatrix} i & j & k \\ -2 & 5 & -4 \\ 1 & 0 & -2 \end{vmatrix}$$

(from equation (1)

$$= i \begin{vmatrix} 5 & -4 \\ 0 & -2 \end{vmatrix} - j \begin{vmatrix} -2 & -4 \\ 1 & -2 \end{vmatrix}$$

$$+ k \begin{vmatrix} -2 & 5 \\ 1 & 0 \end{vmatrix}$$

$= i(-10 - 0) - j(4 + 4)$

$+ k(0 - 5)$

i.e. $(p - 2q) \times r = -10i - 8j - 5k$

(b) $(2r \times 3q) = (2i - 4k) \times (9i - 6j + 3k)$

$$= \begin{vmatrix} i & j & k \\ 2 & 0 & -4 \\ 9 & -6 & 3 \end{vmatrix}$$

$= i(0 - 24) - j(6 + 36)$

$+ k(-12 - 0)$

$= -24i - 42j - 12k$

Hence $p \times (2r \times 3q)$

$= (4i + j - 2k) \times (-24i - 42j - 12k)$

$$= \begin{vmatrix} i & j & k \\ 4 & 1 & -2 \\ -24 & -42 & -12 \end{vmatrix}$$

$= i(-12 - 84) - j(-48 - 48) + k(-168 + 24)$

$= -96i + 96j - 144k$ or $-48(2i - 2j + 3k)$

Practical applications of vector products

Problem 8.   Find the moment and the magnitude of the moment of a force of $(i + 2j - 3k)$ newtons about point B having coordinates $(0, 1, 1)$, when the force acts on a line through A whose coordinates are $(1, 3, 4)$

The moment M about point B of a force vector F which has a position vector of r from A is given by

$M = r \times F$

r is the vector from B to A, i.e. $r = BA$

But $BA = BO + OA = OA - OB$ (see Problem 8, Chapter 35), that is, $r = (i + 3j + 4k) - (j + k) = i + 2j + 3k$

Moment, $M = r \times F$

$= (i + 2j + 3k) \times (i + 2j - 3k)$

$$= \begin{vmatrix} i & j & k \\ 1 & 2 & 3 \\ 1 & 2 & -3 \end{vmatrix}$$

$= i(-6 - 6) - j(-3 - 3) + k(2 - 2)$

$= -12i + 6j$ Nm

The magnitude of M,

$|M| = |r \times F|$

$= \sqrt{[(r \cdot r)(F \cdot F) - (r \cdot F)^2]}$

$r \cdot r = (1)(1) + (2)(2) + (3)(3) = 14$

$F \cdot F = (1)(1) + (2)(2) + (-3)(-3) = 14$

$r \cdot F = (1)(1) + (2)(2) + (3)(-3) = -4$

$|M| = \sqrt{(14 \times 14 - (-4)^2)} = \sqrt{180}$ Nm

$= \textbf{13.42 Nm}$

Problem 9.   The axis of a circular cylinder coincides with the z-axis and it rotates with an angular velocity of $(2i - 5j + 7k)$ rad/s. Determine the tangential velocity at a point P on the cylinder, whose coordinates are $(j + 3k)$ metres, and also determine the magnitude of the tangential velocity

---

The velocity v of point P on a body rotating with angular velocity ω about a fixed axis is given by:

$v = \omega \times r,$

where r is the point on vector P

Thus $v = (2i - 5j + 7k) \times (j + 3k)$

$$= \begin{vmatrix} i & j & k \\ 2 & -5 & 7 \\ 0 & 1 & 3 \end{vmatrix}$$

$= i(-15 - 7) - j(6 - 0) + k(2 - 0)$

$= (-22i - 6j + 2k)$ m/s

The magnitude of v,

$|v| = \sqrt{\{(\omega \cdot \omega)(r \cdot r) - (r \cdot \omega)^2\}}$

$$\omega \cdot \omega = (2)(2) + (-5)(-5) + (7)(7) = 78$$
$$r \cdot r = (0)(0) + (1)(1) + (3)(3) = 10$$
$$\omega \cdot r = (2)(0) + (-5)(1) + (7)(3) = 16$$

Hence, $v = \sqrt{(78 \times 10 - 16^2)} = \sqrt{524}$ m/s
$$= 22.89 \text{ m/s}$$

Further problems on vector products may be found in the following Section 36.4, Problems 11 to 20.

36.4   Further problems on scalar and vector products

Scalar products

1   Find the scalar product $a \cdot b$ when
   (i) $a = i + 2j - k$, $b = 2i + 3j + k$
   (ii) $a = i - 3j + k$, $b = 2i + j + k$

   [(i) 7, (ii) 0]

Given $p = 2i - 3j$, $q = 4j - k$ and $r = i + 2j - 3k$, determine the quantities stated in Problems 2 to 7:

2   (a) $p \cdot q$ (b) $p \cdot r$ [(a) −12 (b) −4]
3   (a) $q \cdot r$ (b) $r \cdot q$ [(a) 11 (b) 11]
4   (a) $|p|$ (b) $|r|$ [(a) $\sqrt{13}$ (b) $\sqrt{14}$]
5   (a) $p \cdot (q + r)$ (b) $2r \cdot (q - 2p)$

   [(a) −16 (b) 38]
6   (a) $|p + r|$ (b) $|p| + |r|$ [(a) $\sqrt{19}$ (b) 7.347]
7   Find the angle between (a) p and q (b) q and r.

   [(a) 143.82° (b) 44.52°]
8   Determine the angle between the forces:
   $F_1 = 3i + 4j + 5k$ and $F_2 = i + j + k$

   [11.5°]
9   Find the angle between the velocity vectors $v_1 = 5i + 2j + 7k$ and $v_2 = 4i + j - k$.

   [66.4°]
10   Calculate the work done by a force $F = (-5i + j + 7k)$ N when its point of application moves from point $(-2i - 6j + k)$ m to the point $(i - j + 10k)$ m.   [53 Nm]

Vector products

In Problems 11 to 14, determine the quantities stated when $p = 3i + 2k$, $q = i - 2j + 3k$ and $r = -4i + 3j - k$.

11   (a) $p \times q$ (b) $q \times p$

   [(a) $4i - 7j - 6k$ (b) $-4i + 7j + 6k$]
12   (a) $|p \times r|$ (b) $|r \times q|$ [(a) 11.92 (b) 13.96]
13   (a) $2p \times 3r$ (b) $(p + r) \times q$

   [(a) $-36i - 30j + 54k$ (b) $11i + 4j - k$]
14   (a) $p \times (r \times q)$ (b) $(3p \times 2r) \times q$

   [(a) $-22i - j + 33k$ (b) $18i + 162j + 102k$]
15   For vectors $p = 4i - j + 2k$ and $q = -2i + 3j - 2k$ determine

   (i) $p \cdot q$ (ii) $p \times q$ (iii) $|p \times q|$ (iv) $q \times p$ and (v) the angle between the vectors

   $$\begin{bmatrix} \text{(i) } -15 \text{ (ii) } -4i + 4j + 10k \text{ (iii) } 11.49 \\ \text{(iv) } 4i - 4j - 10k \text{ (v) } 142.5° \end{bmatrix}$$

16   For vectors $a = -7i + 4j + \dfrac{k}{2}$ and $b = 6i - 5j - k$ find (i) $a \cdot b$ (ii) $a \times b$ (iii) $|a \times b|$ (iv) $b \times a$ and (v) the angle between the vectors

   $$\begin{bmatrix} \text{(i) } -62\frac{1}{2} \text{ (ii) } -1\frac{1}{2}i - 4j + 11k \\ \text{(iii) } 11.8 \\ \text{(iv) } 1\frac{1}{2}i + 4j - 11k \qquad \text{(v) } 169.3° \end{bmatrix}$$

17   Forces of $(i + 3j)$, $(-2i - j)$, $(i - 2j)$ newtons act at three points having position vectors of $(2i + 5j)$, $4j$ and $(-i + j)$ metres, respectively. Calculate the magnitude of the moment

   [10 Nm]

18   A force of $(2i - j + k)$ newtons acts on a line through point P having coordinates $(0, 3, 1)$ metres. Determine the moment vector and its magnitude about point Q having coordinates $(4, 0, -1)$ metres.

   [$M = (5i + 8j - 2k)$ Nm, $M = 9.64$ Nm]

19   A sphere is rotating with angular velocity ω about the z-axis of a system, the axis coinciding with the axis of the sphere. Determine the velocity vector and its magnitude at position $(-5i + 2j - 7k)$ m, when the angular velocity is $(i + 2j)$ rad/s.

   [$v = -14i + 7j + 12k$, $v = 19.7$ m/s]

20   Calculate the velocity vector and its magnitude for a particle rotating about the z-axis at an angular velocity of $(3i - j + 2k)$ rad/s when the position vector of the particle is at $(i - 5j + 4k)$ m.   [$6i - 10j - 14k$, 18.2 m/s]

37

Trigonometric ratios, graphs, identities and equations

37.1 Trigonometric ratios of acute angles

(a) With reference to the right-angled triangle shown in Fig. 37.1:

(i) sine $\theta = \dfrac{\text{opposite side}}{\text{hypotenuse}}$,

i.e. $\sin \theta = \dfrac{b}{c}$

(ii) cosine $\theta = \dfrac{\text{adjacent side}}{\text{hypotenuse}}$,

i.e. $\cos \theta = \dfrac{a}{c}$

(iii) tangent $\theta = \dfrac{\text{opposite side}}{\text{adjacent side}}$,

i.e. $\tan \theta = \dfrac{b}{a}$

Figure 37.1

(iv) secant $\theta = \dfrac{\text{hypotenuse}}{\text{adjacent side}}$,

i.e. $\sec \theta = \dfrac{c}{a}$

(v) cosecant $\theta = \dfrac{\text{hypotenuse}}{\text{opposite side}}$,

i.e. $\csc \theta = \dfrac{c}{b}$

(vi) cotangent $\theta = \dfrac{\text{adjacent side}}{\text{opposite side}}$,

i.e. $\cot \theta = \dfrac{a}{b}$

(b) From above,

(i) $\dfrac{\sin \theta}{\cos \theta} = \dfrac{\frac{b}{c}}{\frac{a}{c}} = \dfrac{b}{a} = \tan \theta$,

i.e. $\tan \theta = \dfrac{\sin \theta}{\cos \theta}$

(ii) $\dfrac{\cos \theta}{\sin \theta} = \dfrac{\frac{a}{c}}{\frac{b}{c}} = \dfrac{a}{b} = \cot \theta$,

i.e. $\cot \theta = \dfrac{\cos \theta}{\sin \theta}$

(iii) $\sec \theta = \dfrac{1}{\cos \theta}$

(iv) $\csc \theta = \dfrac{1}{\sin \theta}$

(Note 's' and 'c' go together)

(v) $\cot \theta = \dfrac{1}{\tan \theta}$

Secants, cosecants and cotangents are called the **reciprocal ratios**.

Problem 1. Determine the values of the six trigonometric ratios for angle θ shown in the right-angled triangle ABC in Fig. 37.2

Figure 37.2

By definition:

$$\sin \theta = \frac{\text{opposite side}}{\text{hypotenuse}} = \frac{5}{13} = \textbf{0.3846}$$

Hence

$$\text{cosec}\,\theta = \frac{13}{5} = \textbf{2.6000}$$

$$\cos \theta = \frac{\text{adjacent side}}{\text{hypotenuse}} = \frac{12}{13} = \textbf{0.9231}$$

Hence

$$\sec \theta = \frac{13}{12} = \textbf{1.0833}$$

$$\tan \theta = \frac{\text{opposite side}}{\text{adjacent side}} = \frac{5}{12} = \textbf{0.4167}$$

Hence

$$\cot \theta = \frac{12}{5} = \textbf{2.4000}$$

Problem 2. If $\cos X = \dfrac{9}{41}$ determine the values of the other five trigonometric ratios

---

Figure 37.3 shows a right-angled triangle.

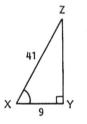

Figure 37.3

Since $\cos X = \dfrac{9}{41}$, then $XY = 9$ units and $XZ = 41$ units.

Using Pythagoras' theorem: $41^2 = 9^2 + YZ^2$ from which $YZ = \sqrt{(41^2 - 9^2)} = 40$ units.
Thus

$$\sin X = \frac{40}{41}, \quad \tan X = \frac{40}{9} = 4\frac{4}{9},$$

$$\textbf{cosec}\,X = \frac{41}{40} = 1\frac{1}{40}, \quad \sec X = \frac{41}{9} = 4\frac{5}{9}$$

and $\cot X = \dfrac{9}{40}$

Problem 3. If $\sin \theta = 0.625$ and $\cos \theta = 0.500$ determine, without using trigonometric tables or calculators, the values of $\text{cosec}\,\theta$, $\sec \theta$, $\tan \theta$ and $\cot \theta$

$$\text{cosec}\,\theta = \frac{1}{\sin \theta} = \frac{1}{0.625} = \textbf{1.60}$$

$$\sec \theta = \frac{1}{\cos \theta} = \frac{1}{0.500} = \textbf{2.00}$$

$$\tan \theta = \frac{\sin \theta}{\cos \theta} = \frac{0.625}{0.500} = \textbf{1.25}$$

$$\cot \theta = \frac{\cos \theta}{\sin \theta} = \frac{0.500}{0.625} = \textbf{0.80}$$

Further problems on trigonometric ratios of acute angles may be found in Section 37.8, Problems 1 to 3, page 356.

37.2   Evaluating trigonometric ratios

As stated in Chapter 10, the easiest method of evaluating trigonometric functions of any angle is by using a calculator.

The following values, correct to 4 decimal places, may be checked:

$$\sin 72° = 0.9511,$$

$$\cos 191° = -0.9816, \text{ and}$$

$$\tan 273.61° = -15.8504$$

(For more examples on sine, cosine and tangent, see Chapter 10.)

Most calculators contain only sine, cosine and tangent functions. Thus to evaluate secants, cosecants and cotangents, reciprocals need to be used.

The following values, correct to 4 decimal places, may be checked:

$$\sec 32° = \frac{1}{\cos 32°} = 1.1792$$

$$\sec 215.12° = \frac{1}{\cos 215.12°} = -1.2226$$

$$\text{cosec } 75° = \frac{1}{\sin 75°} = 1.0353$$

$$\text{cosecant } 321.62° = \frac{1}{\text{sine } 321.62°} = -1.6106$$

$$\text{cotangent } 41° = \frac{1}{\text{tangent } 41°} = 1.1504$$

$$\text{cotangent } 263.59° = \frac{1}{\text{tangent } 263.59°} = 0.1123$$

Problem 4. Evaluate, correct to 4 decimal places: (a) sine 168°14′ (b) cosine 271.41° (c) tangent 98°4′

(a) sine $168°14' = \text{sine } 168\frac{14°}{60} = \textbf{0.2039}$

(b) cosine $271.41° = \textbf{0.0246}$

(c) tangent $98°4' = \tan 98\frac{4°}{60} = \textbf{−7.0558}$

Problem 5. Evaluate, correct to 4 decimal places: (a) secant 161° (b) secant 22.45° (c) secant 302°29′

(a) $\sec 161° = \dfrac{1}{\cos 161°} = \textbf{−1.0576}$

(b) $\sec 22.45° = \dfrac{1}{\cos 22.45°} = \textbf{1.0820}$

(c) $\sec 302°29' = \dfrac{1}{\cos 302°29'}$

$$= \frac{1}{\cos 302\frac{29°}{60}} = \textbf{1.8620}$$

Problem 6. Evaluate, correct to 4 significant figures: (a) cosecant 97° (b) cosecant 279.16° (c) cosecant 49°7′

(a) $\operatorname{cosec} 97° = \dfrac{1}{\sin 97°} = \textbf{1.008}$

(b) $\operatorname{cosec} 279.16° = \dfrac{1}{\sin 279.16°} = \textbf{−1.013}$

(c) $\operatorname{cosec} 49°7' = \dfrac{1}{\sin 49°7'} = \dfrac{1}{\sin 49\frac{7°}{60}} = \textbf{1.323}$

Problem 7. Evaluate, correct to 4 decimal places: (a) cotangent 341° (b) cotangent 17.49° (c) cotangent 163°52′

(a) $\cot 341° = \dfrac{1}{\tan 341°} = \textbf{−2.9042}$

(b) $\cot 17.49° = \dfrac{1}{\tan 17.49°} = \textbf{3.1735}$

(c) $\cot 163°52' = \dfrac{1}{\tan 163°52'}$

$$= \frac{1}{\tan 163\frac{52°}{60}} = \textbf{−3.4570}$$

Problem 8. Evaluate, correct to 4 significant figures: (a) sin 1.481 (b) cos(3π/5) (c) tan 2.93

(a) sin 1.481 means the sine of 1.481 radians. Hence a calculator needs to be on the radian function.
Hence sin 1.481 = **0.9960**

(b) $\cos(3\pi/5) = \cos 1.884955... = \textbf{−0.3090}$

(c) $\tan 2.93 = \textbf{−0.2148}$

Problem 9. Evaluate, correct to 4 decimal places: (a) secant 5.37 (b) cosecant π/4 (c) cotangent π/24

(a) Again, with no degrees sign, it is assumed that 5.37 means 5.37 radians.

Hence $\sec 5.37 = \dfrac{1}{\cos 5.37} = \textbf{1.6361}$

(b) $\operatorname{cosec}(\pi/4) = \dfrac{1}{\sin(\pi/4)}$

$$= \frac{1}{\sin 0.785398...} = \textbf{1.4142}$$

(c) $\cot(5\pi/24) = \dfrac{1}{\tan(5\pi/24)}$

$$= \frac{1}{\tan 0.654498...} = \textbf{1.3032}$$

Problem 10. Determine the acute angles: (a) arcsec 2.3164 (b) arccosec 1.1784 (c) arccot 2.1273

(a) $\arccos 2.3164 = \arccos\left(\dfrac{1}{2.3164}\right)$

$= \arccos 0.4317\ldots$

$= \mathbf{64.42°}$ or $\mathbf{64°25'}$

or $\mathbf{1.124}$ **radians**

(b) $\text{arccosec}\, 1.1784 = \arcsin\left(\dfrac{1}{1.1784}\right)$

$= \arcsin 0.8486\ldots$

$= \mathbf{58.06°}$ or $\mathbf{58°4'}$

or $\mathbf{1.013}$ **radians**

(c) $\arccos 2.1273 = \arctan\left(\dfrac{1}{2.1273}\right)$

$= \arctan 0.4700\ldots$

$= \mathbf{25.18°}$ or $\mathbf{25°11'}$

or $\mathbf{0.439}$ **radians**

Problem 11.   Evaluate the following expression, correct to 4 significant figures:

$$\frac{4\sec 32°10' - 2\cot 15°19'}{3\,\text{cosec}\,63°8'\tan 14°57'}$$

By calculator:

$\sec 32°10' = 1.1813,\ \ \cot 15°19' = 3.6512$

$\text{cosec}\,63°8' = 1.1210,\ \ \tan 14°57' = 0.2670$

Hence

$\dfrac{4\sec 32°10' - 2\cot 15°19'}{3\,\text{cosec}\,63°8'\tan 14°57'}$

$= \dfrac{4(1.1813) - 2(3.6512)}{3(1.1210)(0.2670)}$

$= \dfrac{4.7252 - 7.3024}{0.8979} = \dfrac{-2.5772}{0.8979}$

$= \mathbf{-2.870},$ correct to 4 significant figures

Problem 12.   Evaluate, correct to 4 decimal places: (a) $\sec(-115°)$ (b) $\cot(-189.31°)$ (c) $\text{cosec}(-95°47')$

(a)   Positive angles are considered by convention to be anticlockwise and negative angles as clockwise.

Hence $-115°$ is actually the same as $245°$ (i.e. $360° - 115°$)

Hence $\sec(-115°) = \sec 245°$

$$= \frac{1}{\cos 245°} = -2.3662$$

(b). $\cot(-189.31°) = \dfrac{1}{\tan(-189.31°)} = \mathbf{-6.1000}$

(c) $\text{cosec}(-95°47') = \dfrac{1}{\sin\left(-95\dfrac{47°}{60}\right)} = \mathbf{-1.0051}$

Further problems on evaluating trigonometric ratios may be found in Section 37.8, Problems 4 to 15, pages 356 and 357.

37.3   Graphs of trigonometric functions

By drawing up tables of values from $0°$ to $360°$, graphs of $y = \sin A$, $y = \cos A$ and $y = \tan A$ may be plotted. Values obtained with a calculator (correct to 3 decimal places – which is more than sufficient for plotting graphs), using $30°$ intervals, are shown below, with the respective graphs shown in Fig. 37.4.

(a) $y = \sin A$

A	0	30°	60°	90°	120°
$\sin A$	0	0.500	0.866	1.000	0.866

A	150°	180°	210°	240°	270°
$\sin A$	0.500	0	−0.500	−0.866	−1.000

A	300°	330°	360°
$\sin A$	−0.866	−0.500	0

(b) $y = \cos A$

A	0	30°	60°	90°	120°
$\cos A$	1.000	0.866	0.500	0	−0.500

A	150°	180°	210°	240°	270°
$\cos A$	−0.866	−1.000	−0.866	−0.500	0

A	300°	330°	360°
$\cos A$	0.500	0.866	1.000

(c) $y = \tan A$

A	0	30°	60°	90°	120°
$\tan A$	0	0.577	1.732	∞	-1.732

A	150°	180°	210°	240°	270°
$\tan A$	-0.577	0	0.577	1.732	∞

A	300°	330°	360°
$\tan A$	-1.732	-0.577	0

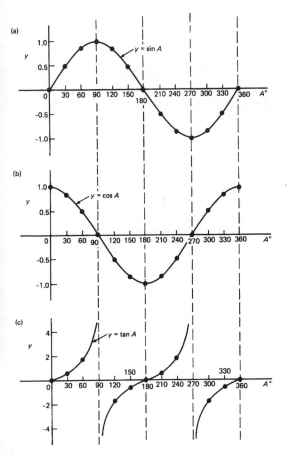

(a)

(b)

(c)

Figure 37.4

From Fig. 37.4 it is seen that:

(i)   Sine and cosine graphs oscillate between peak values of ± 1

(ii) The cosine curve is the same shape as the sine curve but displaced by 90°

(iii) The sine and cosine curves are continuous and they repeat at intervals of 360°; the tangent curve appears to be discontinuous and repeats at intervals of 180°

37.4   Angles of any magnitude

(i)   Figure 37.5 shows rectangular axes XX' and YY' intersecting at origin 0. As with graphical work, measurements made to the right and above 0 are positive while those to the left and downwards are negative. Let OA be free to rotate about 0. By convention, when OA moves anticlockwise angular measurement is considered positive, and vice versa.

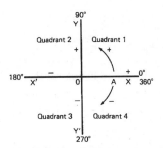

Figure 37.5

(ii) Let OA be rotated anticlockwise so that θ_1 is any angle in the first quadrant and let perpendicular AB be constructed to form the right-angled triangle OAB (see Fig. 37.6). Since all three sides of the triangle are positive, all six trigonometric ratios are positive in the first quadrant. (Note: OA is always positive since it is the radius of a circle.)

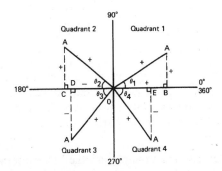

Figure 37.6

(iii)   Let OA be further rotated so that θ_2 is any angle in the second quadrant and let AC be constructed to form the right-angled triangle OAC. Then:

$$\sin\theta_2 = \frac{+}{+} = +, \cos\theta_2 = \frac{-}{+} = -,$$

$$\tan\theta_2 = \frac{+}{-} = -$$

$$\csc\theta_2 = \frac{+}{+} = +, \sec\theta_2 = \frac{+}{-} = -,$$

$$\cot\theta_2 = \frac{-}{+} = -$$

(iv)   Let OA be further rotated so that θ_3 is any angle in the third quadrant and let AD be constructed to form the right-angled triangle OAD. Then:

$$\sin\theta_3 = \frac{-}{+} = -\text{(and hence } \csc\theta_3 \text{ is } -)$$

$$\cos\theta_3 = \frac{-}{+} = -\text{(and hence } \sec\theta_3 \text{ is } -)$$

$$\tan\theta_3 = \frac{-}{-} = +\text{(and hence } \cot\theta_3 \text{ is } +)$$

(v)   Let OA be further rotated so that θ_4 is any angle in the fourth quadrant and let AE be constructed to form the right-angled triangle OAE. Then:

$$\sin\theta_4 = \frac{-}{+} = -\text{(and hence } \csc\theta_4 \text{ is } -)$$

$$\cos\theta_4 = \frac{+}{+} = +\text{(and hence } \sec\theta_4 \text{ is } +)$$

$$\tan\theta_4 = \frac{-}{+} = -\text{(and hence } \cot\theta_4 \text{ is } -)$$

(vi)   The results obtained in (ii) to (v) are summarized in Fig. 37.7. The letters underlined spell the word CAST when starting in the fourth quadrant and moving in an anticlockwise direction

(vii)   In the first quadrant of Fig. 37.4 all the curves have positive values; in the second only sine is positive; in the third only tangent is positive; in the fourth only cosine is positive (exactly as summarized in Fig. 37.7)

A knowledge of angles of any magnitude is needed when finding, for example, all the angles between $0°$ and $360°$ whose sine is, say, 0.3261. If 0.3261 is entered into a calculator and then the inverse

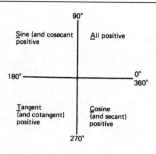

Figure 37.7

sine key pressed (or $\sin^{-1}$ key) the answer $19.03°$ appears. However, there is a second angle between $0°$ and $360°$ which the calculator does not give. Sine is also positive in the second quadrant (either from CAST or from Fig. 37.4(a)). The other angle is shown in Fig. 37.8 as angle θ where $\theta = 180° - 19.03° = 160.97°$. Thus $19.03°$ **and** $160.97°$ are the angles between $0°$ and $360°$ whose sine is 0.3261 (check that $\sin 160.97° = 0.3261$ on your calculator).

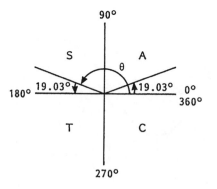

Figure 37.8

Be careful! Your calculator only gives you one of these answers. The second answer needs to be deduced from a knowledge of angles of any magnitude, as shown in the following problems.

Problem 13.   Determine all the angles between $0°$ and $360°$ (a) whose sine is -0.4638 and (b) whose tangent is 1.7629

(a)   The angles whose sine is -0.4638 occurs in the third and fourth quadrants since sine is negative in these quadrants (see Fig. 37.9(a)). From Fig. 37.9(b), $\theta = \arcsin 0.4638 = 27°38'$.

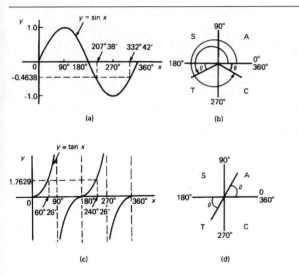

Figure 37.9

Measured from $0°$, the two angles between $0°$ and $360°$ whose sine is -0.4638 are $180° + 27°38'$, i.e. **207°38'** and $360° - 27°38'$, i.e. **332°22'**.

(Note that a calculator generally only gives one answer, i.e. $-27.632588°$)

(b) A tangent is positive in the first and third quadrants (see Fig. 37.9(c)). From Fig. 37.9(d), $\theta =$ arctan $1.7629 = 60°26'$. Measured from $0°$, the two angles between $0°$ and $360°$ whose tangent is 1.7629 are **60°26'** and $180° + 60°26'$, i.e. **240°26'**

Problem 14.   Solve for angles of α between $0°$ and $360°$: (a) $\operatorname{arcsec}(-2.1499) = \alpha$ (b) $\operatorname{arccot} 1.3111 = \alpha$

(a) Secant is negative in the second and third quadrants (i.e. the same as for cosine). From Fig. 37.10(a), $\theta =$ arcsec $2.1499 = 62°17'$. Measured from $0°$, the two angles between $0°$ and $360°$ whose secant is -2.1499 are

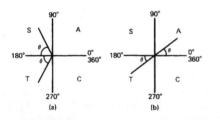

Figure 37.10

$$\alpha = 180° - 62°17' = \mathbf{117°43'}$$

and $\alpha = 180° + 62°17' = \mathbf{242°17'}$

(b) Cotangent is positive in the first and third quadrants (i.e. same as for tangent). From Fig. 37.10(b), $\theta =$ arccot $1.3111 = 37°20'$. Hence

$$\alpha = \mathbf{37°20'} \text{ and } 180° + 37°20' = \mathbf{217°20'}$$

Further problems on evaluating trigonometric ratios of any magnitude may be found in Section 37.8, Problems 16 and 17, page 357.

37.5   Sine and cosine curves

Graphs of sine and cosine waveforms

(i)   A graph of $y = \sin A$ is shown by the broken line in Fig. 37.11 and is obtained by drawing up a table of values as in Section 37.3. A similar table may be produced for $y = \sin 2A$.

$A°$	0	30	45	60	90	120
$2A$	0	60	90	120	180	240
$\sin A$	0	0.866	1.0	0.866	0	-0.866

$A°$	135	150	180	210	225	240
$2A$	270	300	360	420	450	480
$\sin A$	-1.0	-0.866	0	0.866	1.0	0.866

$A°$	270	300	315	330	360
$2A$	540	600	630	660	720
$\sin A$	0	-0.866	-1.0	-0.866	0

A graph of $y = \sin 2A$ is shown in Fig. 37.11

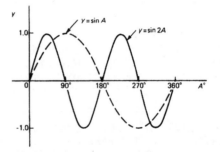

Figure 37.11

(ii)   A graph of $y = \sin \frac{1}{2}A$ is shown in Fig. 37.12 using the following table of values.

$A°$	0	30	60	90	120
$\frac{1}{2}A$	0	15	30	45	60
$\sin \frac{1}{2}A$	0	0.259	0.500	0.707	0.866

$A°$	150	180	210	240	270
$\frac{1}{2}A$	75	90	105	120	135
$\sin \frac{1}{2}A$	0.966	1.00	0.966	0.866	0.707

$A°$	300	330	360
$\frac{1}{2}A$	150	165	180
$\sin \frac{1}{2}A$	0.500	0.259	0

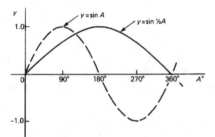

Figure 37.12

(iii)   A graph of $y = \cos A$ is shown by the broken line in Fig. 37.13 and is obtained by drawing up a table of values. A similar table may be produced for $y = \cos 2A$.

$A°$	0	30	45	60	90	120
$2A$	0	60	90	120	180	240
$\cos A$	1.0	0.50	0	−0.50	−1.0	−0.50

$A°$	135	150	180	210	225	240
$2A$	270	300	360	420	450	480
$\cos A$	0	0.50	1.0	0.50	0	−0.50

$A°$	270	300	315	330	360
$2A$	540	600	630	660	720
$\cos A$	−1.0	−0.50	0	0.50	1.0

A graph of $y = \cos 2A$ is shown in Fig. 37.13

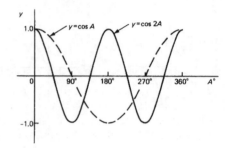

Figure 37.13

(iv)   A graph of $y = \cos \frac{1}{2}A$ is shown in Fig. 37.14 using the following table of values.

$A°$	0	30	60	90	120
$\frac{1}{2}A$	0	15	30	45	60
$\cos \frac{1}{2}A$	1.0	0.966	0.866	0.707	0.50

$A°$	150	180	210	240	270
$\frac{1}{2}A$	75	90	105	120	135
$\sin \frac{1}{2}A$	0.259	0	−0.259	−0.50	−0.707

$A°$	300	330	360
$\frac{1}{2}A$	150	165	180
$\sin \frac{1}{2}A$	−0.866	−0.966	−1.0

Periodic functions and period

(i)   Each of the graphs shown in Figs 37.11 to 37.14 will repeat itself as angle A increases – these graphs are thus called **periodic functions**.

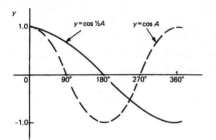

Figure 37.14

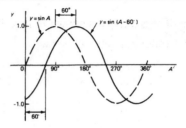

Figure 37.15

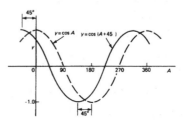

Figure 37.16

(ii) $y = \sin A$ and $y = \cos A$ repeat themselves every 360° (or 2π radians); thus 360° is called the **period** of these waveforms. $y = \sin 2A$ and $y = \cos 2A$ repeat themselves every 180° (or π radians); thus 180° is the period of these waveforms

(iii) In general, if $y = \sin pA$ or $y = \cos pA$ (where p is a constant) then the period of the waveform is 360°/p (or $2\pi/p$ rad). Hence if $y = \sin 3A$ then the period is 360/3, i.e. 120°, and if $y = \cos 4A$ then the period is 360/4, i.e. 90°

Amplitude

Amplitude is the name given to the maximum or peak value of a sine wave. Each of the graphs shown in Figs 37.11 to 37.14 has an amplitude of +1 (i.e. they oscillate between +1 and −1). However, if $y = 4\sin A$, each of the values in the table is multiplied by 4 and the maximum value, and thus amplitude, is 4. Similarly, if $y = 5\cos 2A$, the amplitude is 5 and the period is 360°/2, i.e. 180°.

Lagging and leading angles

(i)   A sine or cosine curve may not always start at 0°. To show this a periodic function is represented by $y = \sin(A \pm \alpha)$ or $y = \cos(A \pm \alpha)$ where α is a phase displacement compared with $y = \sin A$ or $y = \cos A$

(ii) By drawing up a table of values, a graph of $y = \sin(A - 60°)$ may be plotted as shown in Fig. 37.15. If $y = \sin A$ is assumed to start at 0° then $y = \sin(A - 60°)$ starts 60° later (i.e. has a zero value 60° later). Thus $y = \sin(A - 60°)$ is said to **lag** $y = \sin A$ by 60°

(iii) By drawing up a table of values, a graph of $y = \cos(A + 45°)$ may be plotted as shown in Fig. 37.16. If $y = \cos A$ is assumed to start at 0° then $y = \cos(A + 45°)$ starts 45°

earlier (i.e. has a zero value 45° earlier). Thus $y = \cos(A + 45°)$ is said to **lead** $y = \cos A$ by 45°

(iv) Generally, a graph of $y = \sin(A - \alpha)$ lags $y = \sin A$ by angle α, and a graph of $y = \sin(A + \alpha)$ leads $y = \sin A$ by angle α

(v)   A cosine curve is the same shape as a sine curve but starts 90° earlier, i.e. leads by 90°. Hence $\cos A = \sin(A + 90°)$

Graphs of $\sin^2 A$ and $\cos^2 A$

(i)   A graph of $y = \sin^2 A$ is shown in Fig. 37.17 using the following table of values.

$A°$	0	30	60	90	120
$\sin A$	0	0.50	0.866	1.0	0.866
$(\sin A)^2 = \sin^2 A$	0	0.25	0.75	1.0	0.75

$A°$	150	180	210	240
$\sin A$	0.50	0	−0.50	−0.866
$(\sin A)^2 = \sin^2 A$	0.25	0	0.25	0.75

$A°$	270	300	330	360
$\sin A$	−1.0	−0.866	−0.50	0
$(\sin A)^2 = \sin^2 A$	1.0	0.75	0.25	0

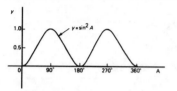

Figure 37.17

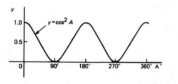

Figure 37.18

(ii) A graph of $y = \cos^2 A$ is shown in Fig. 37.18 using the following table of values.

$A°$	0	30	60	90
$\cos A$	1.0	0.866	0.50	0
$(\cos A)^2 = \cos^2 A$	1.0	0.75	0.25	0

$A°$	120	150	180
$\cos A$	−0.50	−0.866	−1.0
$(\cos A)^2 = \cos^2 A$	0.25	0.75	1.0

$A°$	210	240	270
$\cos A$	−0.866	−0.50	0
$(\cos A)^2 = \cos^2 A$	0.75	0.25	0

$A°$	300	330	360
$\cos A$	0.50	0.866	1.0
$(\cos A)^2 = \cos^2 A$	0.25	0.75	1.0

(iii) $y = \sin^2 A$ and $y = \cos^2 A$ are both periodic functions of period 180° (or π rad) and both contain only positive values. Thus a graph of $y = \sin^2 2A$ has a period 180°/2, i.e. 90°. Similarly, a graph of $y = 4\cos^2 3A$ has a maximum value of 4 and a period of 180°/3, i.e. 60°

Problem 15. Sketch $y = \sin 3A$ between $A = 0°$ and $A = 360°$

Amplitude = 1; period = 360°/3 = 120°.

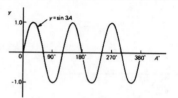

Figure 37.19

A sketch of $y = \sin 3A$ is shown in Fig. 37.19.

Problem 16. Sketch $y = 3\sin 2A$ from $A = 0$ to $A = 2\pi$ radians

---

Amplitude = 3, period = $2\pi/2 = \pi$ rads (or 180°). A sketch of $y = 3\sin 2A$ is shown in Fig. 37.20.

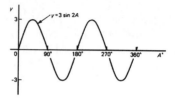

Figure 37.20

Problem 17. Sketch $y = 4\cos 2x$ from $x = 0°$ to $x = 360°$

Amplitude = 4; period = 360°/2 = 180°. A sketch of $y = 4\cos 2x$ is shown in Fig. 37.21.

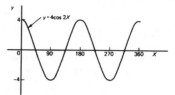

Figure 37.21

Problem 18. Sketch $y = 2\sin \frac{3}{5}A$ over one cycle

Amplitude = 2; period = $\dfrac{360°}{\dfrac{3}{5}} = \dfrac{360° \times 5}{3}$

$$= 600°$$

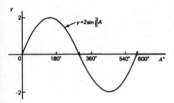

Figure 37.22

A sketch of $y = 2 \sin \dfrac{3}{5} A$ is shown in Fig. 37.22.

Problem 19.   Sketch $y = 5 \sin(A + 30°)$
from $A = 0°$ to $A = 360°$

Amplitude $= 5$; period $= 360°/1 = 360°$.
$5 \sin(A + 30°)$ leads $5 \sin A$ by $30°$ (i.e. starts $30°$
earlier).
A sketch of $y = 5 \sin(A + 30°)$ is shown in
Fig. 37.23.

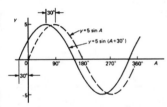

Figure 37.23

Problem 20.   Sketch $y = 7 \sin(2A - \pi/3)$
over one cycle

Amplitude $= 7$; period $= 2\pi/2 = \pi$ radians.
In general, $y = \sin(pt - \alpha)$ **lags** $y = \sin py$ by
α/p, hence $7 \sin(2A - \pi/3)$ lags $7 \sin 2A$ by $(\pi/3)/2$,
i.e. $\pi/6$ rad or $30°$.
A sketch of $y = 7 \sin(2A - \pi/3)$ is shown in
Fig. 37.24.

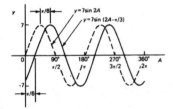

Figure 37.24

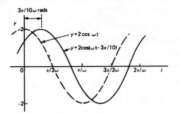

Figure 37.25

Problem 21.   Sketch $y = 2 \cos(\omega t - 3\pi/10)$
over one cycle

Amplitude $= 2$; period $= 2\pi/\omega$ rad.
$2 \cos(\omega t - 3\pi/10)$ lags $2 \cos \omega t$ by $3\pi/10\omega$ radians.
A sketch of $y = 2 \cos(\omega t - 3\pi/10)$ is shown in
Fig. 37.25.

Problem 22.   Sketch $y = 3 \sin^2 \dfrac{1}{2} A$ in the
range $0 < A < 360°$

Maximum value $= 3$; period $= 180°/(1/2) = 360°$.
A sketch of $3 \sin^2 \dfrac{1}{2} A$ is shown in Fig. 37.26.

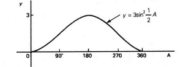

Figure 37.26

Problem 23.   Sketch $y = 7 \cos^2 2A$ between
$A = 0°$ and $A = 360°$

Maximum value $= 7$, period $= 180°/2 = 90°$.
A sketch of $y = 7 \cos^2 2A$ is shown in Fig. 37.27.

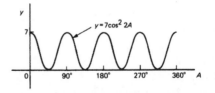

Figure 37.27

Problem 24. The instantaneous value of voltage in an a.c. circuit at any time t seconds is given by $v = 340\sin(50\pi t - 0.541)$ volts. Determine:

(a) the amplitude, periodic time, frequency and phase angle (in degrees)
(b) the value of the voltage when $t = 0$
(c) the value of the voltage when $t = 10$ ms
(d) the time when the voltage first reaches 200 V, and
(e) the time when the voltage is a maximum

Sketch one cycle of the waveform

(a) Amplitude = **340 V**
Angular velocity, $\omega = 50\pi$
Hence periodic time,

$$T = \frac{2\pi}{\omega} = \frac{2\pi}{50\pi} = \frac{1}{25} = \textbf{0.04 s or 40 ms}$$

Frequency $f = \dfrac{1}{T} = \dfrac{1}{0.04} = \textbf{25 Hz}$

Phase angle $= 0.541$ rad $= \left(0.541 \times \dfrac{180}{\pi}\right)^{\circ}$

$= \textbf{31}^{\circ}$ **lagging** $v = 340\sin(50\pi t)$

(b) When $t = 0$, $v = 340\sin(0 - 0.541)$

$= 340\sin(-31^{\circ}) = \textbf{-175.1 V}$

(c) When $t = 10$ ms then

$$v = 340\sin\left(50\pi\frac{10}{10^3} - 0.541\right)$$

$$= 340\sin(1.0298)$$

$$= 340\sin 59^{\circ} = \textbf{291.4 volts}$$

(d) When $v = 200$ volts then

$$200 = 340\sin(50\pi t - 0.541)$$

$$\frac{200}{340} = \sin(50\pi t - 0.541)$$

Hence $(50\pi t - 0.541)$

$$= \arcsin\frac{200}{340}$$

$$= 36.03^{\circ} \text{ or } 0.6288 \text{ rad}$$

$$50\pi t = 0.6288 + 0.541 = 1.1698$$

Hence time, $t = \dfrac{1.1698}{50\pi} = \textbf{7.447 ms}$

(e) When the voltage is a maximum, $v = 340$ V. Hence

$$340 = 340\sin(50\pi t - 0.541)$$

$$1 = \sin(50\pi t - 0.541)$$

$$50\pi t - 0.541 = \arcsin 1 = 90^{\circ} \text{ or } 1.5708 \text{ rad}$$

$$50\pi t = 1.5708 + 0.541 = 2.1118$$

Hence time, $t = \dfrac{2.1118}{50\pi} = \textbf{13.44 ms}$

A sketch of $v = 340\sin(50\pi t - 0.541)$ volts is shown in Fig. 37.28.

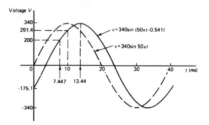

Figure 37.28

Further problems on sine and cosine curves may be found in Section 37.8, Problems 18 to 24, page 357.

37.6   Trigonometric identities

A **trigonometric identity** is an expression that is true for all values of the unknown variable.

$$\tan\theta = \frac{\sin\theta}{\cos\theta}, \quad \cot\theta = \frac{\cos\theta}{\sin\theta}, \quad \sec\theta = \frac{1}{\cos\theta}$$

$$\operatorname{cosec}\theta = \frac{1}{\sin\theta} \text{ and } \cot\theta = \frac{1}{\tan\theta}$$

are examples of trigonometric identities from Section 37.1.

Applying Pythagoras' theorem to the right-angled triangle shown in Fig. 37.29 gives:

$$a^2 + b^2 = c^2 \tag{1}$$

Dividing each term of equation (1) by c^2 gives:

$$\frac{a^2}{c^2} + \frac{b^2}{c^2} = \frac{c^2}{c^2}, \text{ i.e. } \left(\frac{a}{c}\right)^2 + \left(\frac{b}{c}\right)^2 = 1$$

Figure 37.29

$$(\cos\theta)^2 + (\sin\theta)^2 = 1$$

Hence $\cos^2\theta + \sin^2\theta = 1$ (2)

Dividing each term of equation (1) by a^2 gives:

$$\frac{a^2}{a^2} + \frac{b^2}{a^2} = \frac{c^2}{a^2}, \text{ i.e. } 1 + \left(\frac{b}{a}\right)^2 = \left(\frac{c}{a}\right)^2$$

Hence $1 + \tan^2\theta = \sec^2\theta$ (3)

Dividing each term of equation (1) by b^2 gives:

$$\frac{a^2}{b^2} + \frac{b^2}{b^2} = \frac{c^2}{b^2}, \text{ i.e. } \left(\frac{a}{b}\right)^2 + 1 = \left(\frac{c}{b}\right)^2$$

Hence $\cot^2\theta + 1 = \cosec^2\theta$ (4)

Equations (2), (3) and (4) are three further examples of trigonometric identities. For the proof of further trigonometric identities, see the following worked problem.

Problem 25. Prove the identity
$\sin^2\theta \cot\theta \sec\theta = \sin\theta$

With trigonometric identities it is necessary to start with the left-hand side (LHS) and attempt to make it equal to the right-hand side (RHS) or vice versa. It is often useful to change all of the trigonometric ratios into sines and cosines where possible. Thus

$$\text{LHS} = \sin^2\theta \cot\theta \sec\theta = \sin^2\theta \left(\frac{\cos\theta}{\sin\theta}\right)\left(\frac{1}{\cos\theta}\right)$$

$$= \sin\theta \text{ (by cancelling)} = \text{RHS}$$

Problem 26. Prove that
$$\frac{\tan x + \sec x}{\sec x \left(1 + \dfrac{\tan x}{\sec x}\right)} = 1$$

$$\text{LHS} = \frac{\tan x + \sec x}{\sec x \left(1 + \dfrac{\tan x}{\sec x}\right)}$$

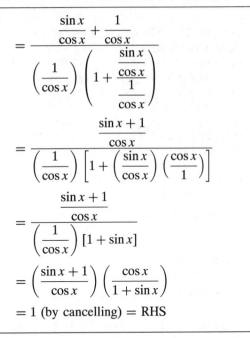

$$= \frac{\dfrac{\sin x}{\cos x} + \dfrac{1}{\cos x}}{\left(\dfrac{1}{\cos x}\right)\left(1 + \dfrac{\dfrac{\sin x}{\cos x}}{\dfrac{1}{\cos x}}\right)}$$

$$= \frac{\dfrac{\sin x + 1}{\cos x}}{\left(\dfrac{1}{\cos x}\right)\left[1 + \left(\dfrac{\sin x}{\cos x}\right)\left(\dfrac{\cos x}{1}\right)\right]}$$

$$= \frac{\dfrac{\sin x + 1}{\cos x}}{\left(\dfrac{1}{\cos x}\right)[1 + \sin x]}$$

$$= \left(\frac{\sin x + 1}{\cos x}\right)\left(\frac{\cos x}{1 + \sin x}\right)$$

$$= 1 \text{ (by cancelling)} = \text{RHS}$$

Problem 27. Prove that $\dfrac{1 + \cot\theta}{1 + \tan\theta} = \cot\theta$

$$\text{LHS} = \frac{1 + \cot\theta}{1 + \tan\theta} = \frac{1 + \dfrac{\cos\theta}{\sin\theta}}{1 + \dfrac{\sin\theta}{\cos\theta}} = \frac{\dfrac{\sin\theta + \cos\theta}{\sin\theta}}{\dfrac{\cos\theta + \sin\theta}{\cos\theta}}$$

$$= \left(\frac{\sin\theta + \cos\theta}{\sin\theta}\right)\left(\frac{\cos\theta}{\cos\theta + \sin\theta}\right) = \frac{\cos\theta}{\sin\theta}$$

$$= \cot\theta = \text{RHS}$$

Problem 28. Show that
$\cos^2\theta - \sin^2\theta = 1 - 2\sin^2\theta$

From equation (2), $\cos^2\theta + \sin^2\theta = 1$, from which, $\cos^2\theta = 1 - \sin^2\theta$.

Hence, $\text{LHS} = \cos^2\theta - \sin^2\theta$

$$= (1 - \sin^2\theta) - \sin^2\theta$$

$$= 1 - \sin^2\theta - \sin^2\theta$$

$$= 1 - 2\sin^2\theta = \text{RHS}$$

Problem 29. Prove that

$$\sqrt{\left(\frac{1 - \sin x}{1 + \sin x}\right)} = \sec x - \tan x$$

$$\text{LHS} = \sqrt{\left(\frac{1 - \sin x}{1 + \sin x}\right)}$$

$$= \sqrt{\left\{\frac{(1 - \sin x)(1 - \sin x)}{(1 + \sin x)(1 - \sin x)}\right\}}$$

$$= \sqrt{\left\{\frac{(1 - \sin x)^2}{(1 - \sin^2 x)}\right\}}$$

Since $\cos^2 x + \sin^2 x = 1$ then $1 - \sin^2 x = \cos^2 x$

$$\text{LHS} = \sqrt{\left\{\frac{(1 - \sin x)^2}{(1 - \sin^2 x)}\right\}} = \sqrt{\left\{\frac{(1 - \sin x)^2}{\cos^2 x}\right\}}$$

$$= \frac{1 - \sin x}{\cos x} = \frac{1}{\cos x} - \frac{\sin x}{\cos x}$$

$$= \sec x - \tan x = \text{RHS}$$

Problem 30. Use a calculator to show that the identity $1 + \cot^2 \theta = \operatorname{cosec}^2 \theta$ is true when $\theta = 223°$

$$\cot 223° = \frac{1}{\tan 223°} = 1.0723687\ldots$$

$$\cot^2 223° = (\cot 223°)^2 = (1.0723687\ldots)^2$$

$$= 1.1499746\ldots$$

Hence

$$\text{LHS} = 1 + \cot^2 223° = 2.1499746\ldots$$

$$\operatorname{cosec} 223° = \frac{1}{\sin 223°} = -1.46627918\ldots$$

$$\text{RHS} = \operatorname{cosec}^2 223° = (\operatorname{cosec} 223°)^2$$

$$= (-1.46627918\ldots)^2 = 2.1499746\ldots$$

Since LHS = RHS the identity is shown to be true when $\theta = 223°$.

Further problems on trigonometric identities may be found in Section 37.8, Problems 25 to 31, page 357.

37.7   Trigonometric equations

Equations which contain trigonometric ratios are called **trigonometric equations**. There are usually an infinite number of solutions to such equations: however, solutions are often restricted to those between $0°$ and $360°$. A knowledge of angles of any

magnitude is essential in the solution of trigonometric equations and calculators cannot be relied upon to give all the solutions (as shown earlier). Figure 37.30 shows a summary for angles of any magnitude.

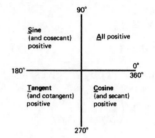

Figure 37.30

Equations of the type $a \sin^2 A + b \sin A + c = 0$

(i)   **When $a = 0$,** $b \sin A + c = 0$, hence

$$\sin A = -\frac{c}{b} \text{ and } A = \arcsin\left(-\frac{c}{b}\right)$$

There are two values of A between $0°$ and $360°$ which satisfy such an equation, provided $-1 \le c/b \le 1$ (see Problems 31 to 34).

(ii)   **When $b = 0$,** $a \sin^2 A + c = 0$, hence

$$\sin^2 A = -\frac{c}{a}$$

$$\sin A = \sqrt{\left(-\frac{c}{a}\right)}$$

$$\text{and } A = \arcsin \sqrt{\left(-\frac{c}{a}\right)}$$

If either a or c is a negative number, then the value within the square root sign is positive. Since when a square root is taken there is a positive and negative answer there are four values of A between $0°$ and $360°$ which satisfy such an equation, provided $-1 \le c/a \le 1$ (see Problems 35 to 37).

(iii)   **When a, b and c are all non-zero:**
$a \sin^2 A + b \sin A + c = 0$ is a quadratic equation in which the unknown is $\sin A$. The solution of a quadratic equation is obtained either by factorizing (if possible) or by using the quadratic formula:

$$\sin A = \frac{-b \pm \sqrt{(b^2 - 4ac)}}{2a}$$

(see Problems 38 to 40)

(iv) Often the trigonometric identities $\cos^2 A + \sin^2 A = 1$, $1 + \tan^2 A = \sec^2 A$ and $\cot^2 A + 1 = \csc^2 A$ need to be used to reduce equations to one of the above forms (see Problems 41 to 44)

Problem 31. Solve the trigonometric equation $5 \sin \theta + 3 = 0$ for values of θ from $0°$ to $360°$

$5 \sin \theta + 3 = 0$, from which $\sin \theta = -3/5 = -0.6000$.

Hence $\theta = \arcsin(-0.6000)$. Sine is negative in the third and fourth quadrants (see Fig. 37.31). The acute angle $\arcsin(0.6000) = 36°52'$ (shown as α in Fig. 37.31(b)). Hence

$$\theta = 180° + 36°52', \text{ i.e. } \mathbf{216°52'}$$

or $\theta = 360° - 36°52', \text{ i.e. } \mathbf{323°8'}$

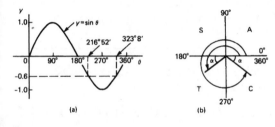

(a)   (b)

Figure 37.31

Problem 32. Solve $1.5 \tan x - 1.8 = 0$ for $0° \le x \le 360°$

$1.5 \tan x - 1.8 = 0$, from which $\tan x = \dfrac{1.8}{1.5} = 1.2000$.

Hence $x = \arctan 1.2000$.

Tangent is positive in the first and third quadrants (see Fig. 37.32). The acute angle $\arctan 1.2000 = 50°12'$. Hence

$$x = \mathbf{50°12'} \text{ or } 180° + 50°12' = \mathbf{230°12'}$$

Problem 33. Solve $4 \sec t = 5$ for values of t between $0°$ and $360°$

$4 \sec t = 5$, from which $\sec t = \dfrac{5}{4} = 1.2500$.

Hence $t = \text{arcsec } 1.2500$.

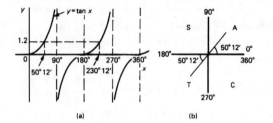

(a)   (b)

Figure 37.32

Secant $= (1/\text{cosine})$ is positive in the first and fourth quadrants (see Fig. 37.33). The acute angle arcsec $1.2500 = 36°52'$. Hence

$$t = \mathbf{36°52'} \text{ or } 360° - 36°52' = \mathbf{323°8'}$$

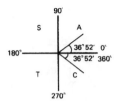

Figure 37.33

Problem 34. Solve $3.2(\cot \theta - 1) = -12$ for values of θ between $0°$ and $360°$

$3.2(\cot \theta - 1) = -12$, hence

$$\cot \theta - 1 = \frac{-12}{3.2} = -3.7500$$

$$\cot \theta = -3.7500 + 1 = -2.7500$$

$$\theta = \text{arccot}(-2.7500)$$

Cotangent $= (1/\text{tangent})$ is negative in the second and fourth quadrants (see Fig. 37.34). The acute angle arccot$(2.7500) = 19°59'$. Hence

$$\theta = 180° - 19°59' = \mathbf{160°1'}$$

or $\theta = 360° - 19°59' = \mathbf{340°1'}$

Figure 37.34

Problem 35. Solve $2 - 4\cos^2 A = 0$ for values of A in the range $0° < A < 360°$

$2 - 4\cos^2 A = 0$, from which $\cos^2 A = \dfrac{2}{4} = 0.5000$.

Hence $\cos A = \sqrt{(0.5000)} = \pm 0.7071$ and $A = \arccos(\pm 0.7071)$.
Cosine is positive in quadrants one and four and negative in quadrants two and three. Thus in this case there are four solutions, one in each quadrant (see Fig. 37.35). The acute angle $\arccos 0.7071 = 45°$. Hence

$$A = 45°, 135°, 225° \text{ or } 315°$$

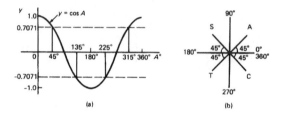

Figure 37.35

Problem 36. Solve $0.75\sec^2 x - 1.2 = 0$ for values of x between $0°$ and $360°$

$0.75\sec^2 x - 1.2 = 0$, from which, $\sec^2 x = \dfrac{1.2}{0.75} = 1.6000$.
Hence $\sec x = \sqrt{(1.6000)} = \pm 1.2649$ and $x = \text{arcsec}(\pm 1.2649)$.
There are four solutions, one in each quadrant. The acute angle arcsec $1.2649 = 37°46'$. Hence

$$x = 37°46', 142°14', 217°46' \text{ or } 322°14'$$

Problem 37. Solve $\dfrac{1}{2}\cot^2 y = 1.3$ for $0° \le y \le 360°$

$\dfrac{1}{2}\cot^2 y = 1.3$, from which, $\cot^2 y = 2(1.3) = 2.6$.

Hence $\cot y = \sqrt{2.6} = \pm 1.6125$, and $y = \text{arccot}(\pm 1.6125)$. There are four solutions, one in each quadrant. The acute angle arccot $1.6125 = 31°48'$. Hence

$$y = 31°48', 148°12', 211°48' \text{ or } 328°12'$$

Problem 38. Solve the equation $8\sin^2 \theta + 2\sin \theta - 1 = 0$, for all values of θ between $0°$ and $360°$

Factorizing $8\sin^2 \theta + 2\sin \theta - 1 = 0$ gives $(4\sin \theta - 1)(2\sin \theta + 1) = 0$.

Hence $4\sin \theta - 1 = 0$, from which, $\sin \theta = \dfrac{1}{4} = 0.2500$, or $2\sin \theta + 1 = 0$, from which $\sin \theta = -\dfrac{1}{2} = -0.5000$.

$\theta = \arcsin 0.250 = 14°29'$ or $165°31'$, since sine is positive in the first and second quadrants, or $\theta = \arcsin(-0.5000) = 210°$ or $330°$, since sine is negative in the third and fourth quadrants. Hence

$$\theta = 14°29', 165°31', 210° \text{ or } 330°$$

Problem 39. Solve $6\cos^2 \theta + 5\cos \theta - 6 = 0$ for values of θ from $0°$ to $360°$

Factorizing $6\cos^2 \theta + 5\cos \theta - 6 = 0$ gives $(3\cos \theta - 2)(2\cos \theta + 3) = 0$.

Hence $3\cos \theta - 2 = 0$, from which, $\cos \theta = \dfrac{2}{3} = 0.6667$, or $2\cos \theta + 3 = 0$, from which, $\cos \theta = -\dfrac{3}{2} = -1.5000$.
The minimum value of a cosine is -1, hence the latter expression has no solution and is thus neglected. Hence

$$\theta = \arccos 0.6667 = 48°11' \text{ or } 311°49'$$

since cosine is positive in the first and fourth quadrants.

Problem 40. Solve $9\tan^2 x + 16 = 24\tan x$ in the range $0° \le x \le 360°$

Rearranging gives $9\tan^2 x - 24\tan x + 16 = 0$ and factorizing gives $(3\tan x - 4)(3\tan x - 4) = 0$, i.e. $(3\tan x - 4)^2 = 0$.
Hence $\tan x = \dfrac{4}{3} = 1.3333$ and $x = \arctan 1.3333$ $= 53°8'$ or $233°8'$ since tangent is positive in the first and third quadrants.

Problem 41.   Solve $5\cos^2 t + 3\sin t - 3 = 0$ for values of t from $0°$ to $360°$

Since $\cos^2 t + \sin^2 t = 1$, $\cos^2 t = 1 - \sin^2 t$. Substituting for $\cos^2 t$ in $5\cos^2 t + 3\sin t - 3 = 0$ gives

$$5(1 - \sin^2 t) + 3\sin t - 3 = 0$$

$$5 - 5\sin^2 t + 3\sin t - 3 = 0$$

$$-5\sin^2 t + 3\sin t + 2 = 0$$

$$5\sin^2 t - 3\sin t - 2 = 0$$

Factorizing gives $(5\sin t + 2)(\sin t - 1) = 0$. Hence $5\sin t + 2 = 0$, from which, $\sin t = -\dfrac{2}{5} = -0.4000$, or $\sin t - 1 = 0$, from which, $\sin t = 1$. $t = \arcsin(-0.4000) = 203°35'$ or $336°25'$, since sine is negative in the third and fourth quadrants, or $t = \arcsin 1 = 90°$. Hence

$$t = 90°, 203°35' \text{ or } 336°25'$$

as shown in Fig. 37.36.

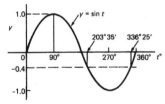

Figure 37.36

Problem 42.   Solve $18\sec^2 A - 3\tan A = 21$ for values of A between $0°$ and $360°$

$1 + \tan^2 A = \sec^2 A$. Substituting for $\sec^2 A$ in $18\sec^2 A - 3\tan A = 21$ gives $18(1 + \tan^2 A) - 3\tan A = 21$, i.e.

$$18 + 18\tan^2 A - 3\tan A - 21 = 0$$

$$18\tan^2 A - 3\tan A - 3 = 0$$

Factorizing gives $(6\tan A - 3)(3\tan A + 1) = 0$.

Hence $6\tan A - 3 = 0$, from which, $\tan A = \dfrac{3}{6} = 0.5000$ or $3\tan A + 1 = 0$, from which, $\tan A = -\dfrac{1}{3} = -0.3333$. Thus $A = \arctan(0.5000) = 26°34'$

or $206°34'$, since tangent is positive in the first and third quadrants, or $A = \arctan(-0.3333) = 161°34'$ or $341°34'$, since tangent is negative in the second and fourth quadrants. Hence

$$A = 26°34', 161°34', 206°34' \text{ or } 341°34'$$

Problem 43.   Solve $3\operatorname{cosec}^2 \theta - 5 = 4\cot\theta$ in the range $0 < \theta < 360°$

$\cot^2\theta + 1 = \operatorname{cosec}^2\theta$. Substituting for $\operatorname{cosec}^2\theta$ in $3\operatorname{cosec}^2\theta - 5 = 4\cot\theta$ gives

$$3(\cot^2\theta + 1) - 5 = 4\cot\theta$$

$$3\cot^2\theta + 3 - 5 = 4\cot\theta$$

$$3\cot^2\theta - 4\cot\theta - 2 = 0$$

Since the left-hand side does not factorize the quadratic formula is used. Thus

$$\cot\theta = \frac{-(-4) \pm \sqrt{[(-4)^2 - 4(3)(-2)]}}{2(3)}$$

$$= \frac{4 \pm \sqrt{(16 + 24)}}{6} = \frac{4 \pm \sqrt{40}}{6}$$

$$= \frac{10.3246}{6} \text{ or } -\frac{2.3246}{6}$$

Hence $\cot\theta = 1.7208$ or -0.3874, $\theta = \operatorname{arccot} 1.7208 = 30°10'$ or $210°10'$, since cotangent is positive in the first and third quadrants, or $\theta = \operatorname{arccot}(-0.3874) = 111°11'$ or $291°11'$, since cotangent is negative in the second and fourth quadrants. Hence

$$\theta = 30°10', 111°11', 210°10' \text{ or } 291°11'$$

Problem 44.   Solve $7\sin^2\theta - 4\cos\theta = 5$ for values of θ between $0°$ and $360°$

Since $\cos^2\theta + \sin^2\theta = 1$, $\sin^2\theta = 1 - \cos^2\theta$. Substituting for $\sin^2\theta$ in $7\sin^2\theta - 4\cos\theta = 5$ gives $7(1 - \cos^2\theta) - 4\cos\theta = 5$, i.e.

$$7 - 7\cos^2\theta - 4\cos\theta = 5$$

$$-7\cos^2\theta - 4\cos\theta + 2 = 0$$

$$7\cos^2\theta + 4\cos\theta - 2 = 0$$

Since the left-hand side does not factorize the quadratic formula is used. Thus

$$\cos\theta = \frac{-4 \pm \sqrt{\{(4)^2 - 4(7)(-2)\}}}{2(7)}$$

$$= \frac{-4 \pm \sqrt{(16 + 56)}}{14}$$

$$= \frac{-4 \pm \sqrt{72}}{14} = \frac{4.4853}{14} \text{ or } \frac{-12.4853}{14}$$

$$= 0.3204 \text{ or } -0.8918$$

$$\theta = \arccos 0.3204 = 71°19' \text{ or } 288°41'$$

or $\qquad \theta = \arccos(-0.8918)$

$$= 153°6' \text{ or } 206°54'$$

Hence $\quad \theta = \mathbf{71°19', 153°6', 206°54' \text{ or } 288°41'}$

Further problems on trigonometric equations may be found in the following Section 37.8, Problems 32 to 41, page 358.

37.8 Further problems on trigonometric ratios, graphs, identities and equations

Trigonometric ratios of acute angles

1 For the right-angled triangle shown in Fig. 37.37, find (a) $\sin\alpha$ (b) $\cos\theta$ (c) $\sec\theta$ (d) $\csc\alpha$ (e) $\tan\theta$ (f) $\cot\theta$ (g) $\cot\alpha$.

$$\left[\begin{array}{l} \text{(a) } \dfrac{15}{17} \text{ (b) } \dfrac{15}{17} \text{ (c) } 1\dfrac{2}{15} \text{ (d) } 1\dfrac{2}{15} \\[2mm] \text{(e) } \dfrac{8}{15} \text{ (f) } 1\dfrac{7}{8} \text{ (g) } \dfrac{8}{15} \end{array} \right]$$

Figure 37.37

2 If $\tan\theta = \dfrac{7}{24}$, find the other five trigonometric ratios in fraction form.

$$\left[\begin{array}{l} \sin\theta = \dfrac{7}{25}, \ \cos\theta = \dfrac{24}{25}, \ \sec\theta = 1\dfrac{1}{24} \\[2mm] \csc\theta = 3\dfrac{4}{7}, \ \cot\theta = 3\dfrac{3}{7} \end{array} \right]$$

3 In a right-angled triangle PQR, $PQ = 11.6$ cm, $QR = 5.3$ cm and $R = 90°$. Evaluate, correct to 4 significant figures, (a) $\sec Q$ (b) $\cot P$

(c) $\tan Q$ (d) $\csc Q$ (e) $\cot Q$.

$$\left[\begin{array}{l} \text{(a) } 2.189 \text{ (b) } 1.947 \text{ (c) } 1.947 \\ \text{(d) } 1.124 \text{ (e) } 0.5137 \end{array} \right]$$

Evaluating trigonometric ratios

In Problems 4 to 9, evaluate correct to 4 decimal places:

4 (a) sine 93°21' (b) cosine 151°10' (c) tangent 301°43'

$$[\text{(a) } 0.9983 \text{ (b) } -0.8760 \text{ (c) } -1.6181]$$

5 (a) secant 73° (b) secant 286.45° (c) secant 155°41'

$$[\text{(a) } 3.4203 \text{ (b) } 3.5313 \text{ (c) } -1.0974]$$

6 (a) cosecant 213° (b) cosecant 15.62° (c) cosecant 311°50'

$$[\text{(a) } -1.8361 \text{ (b) } 3.7139 \text{ (c) } -1.3421]$$

7 (a) cotangent 71° (b) cotangent 151.62° (c) cotangent 321°23'

$$[\text{(a) } 0.3443 \text{ (b) } -1.8510 \text{ (c) } -1.2519]$$

8 (a) sine $\dfrac{\pi}{3}$ (b) cos 1.127 (c) tan 2.962

$$[\text{(a) } 0.8660 \text{ (b) } 0.4294 \text{ (c) } -0.1815]$$

9 (a) sec $\dfrac{\pi}{8}$ (b) cosec 2.961 (c) cot 2.612

$$[\text{(a) } 1.0824 \text{ (b) } 5.5675 \text{ (c) } -1.7083]$$

10 Determine the acute angle in degrees (correct to 2 decimal places), degrees and minutes, and in radians (correct to 3 decimal places) for the following:
(a) arcsec 1.6214 (b) arccosec 2.4891 (c) arccot 1.9614

$$\left[\begin{array}{l} \text{(a) } 51.92°, \ 51°55', \ 0.906 \text{ rad} \\ \text{(b) } 23.69°, \ 23°41', \ 0.413 \text{ rad} \\ \text{(c) } 27.01°, \ 27°1', \ 0.471 \text{ rad} \end{array} \right]$$

11 Evaluate the following, each correct to 4 significant figures:

(a) $3\sec 27°10' - 2\cot 15°19'$

(b) $\dfrac{6.4 \csc 29°5' - \sec 81°}{2\cot 12°}$

(c) $\dfrac{4\cot 127°15'}{5\csc 77° - 3\sec 4°11'}$

$$[\text{(a) } -3.930 \text{ (b) } 0.7199 \text{ (c) } -1.432]$$

12 Determine the acute angle, in degrees and minutes, correct to the nearest minute, given by

$$\text{arcsec} \left(\frac{7.29 \sin 42°16'}{3.76} \right) \qquad [39°56']$$

13   If $\tan x = 1.5276$ determine $\sec x$, $\operatorname{cosec} x$ and $\cot x$. (Assume x is an acute angle.)

$$[1.8258, 1.1952, 0.6546]$$

14   Evaluate, correct to 4 significant figures:

(a) $3 \cot 14°15' \sec 23°9'$

(b) $\dfrac{\operatorname{cosec} 27°19' + \sec 45°29'}{1 - \operatorname{cosec} 27°19' \sec 45°29'}$

(c) $\dfrac{30 \tan 61° \sec 54° - 15 \cot 14°}{2 \operatorname{cosec} 24°}$

$$[\text{(a) } 12.85 \text{ (b) } -1.710 \text{ (c) } 6.491]$$

15   Evaluate (a) $\operatorname{cosec}(-125°)$ (b) $\cot(-241°)$ (c) $\sec(-49°15')$

$$[\text{(a) } -1.2208 \text{ (b) } -0.5543 \text{ (c) } 1.5320]$$

Evaluating trigonometric ratios of any magnitude

16   Find all the angles between $0°$ and $360°$:

(a)   whose sine is -0.7321

(b)   whose cosecant is 2.5317

(c)   whose cotangent is -0.6312

$$\left[\begin{array}{l}\text{(a) } 227°4' \text{ or } 312°56' \text{ (b) } 23°16' \text{ or } 156°44' \\ \text{(c) } 122°16' \text{ or } 302°16' \end{array}\right]$$

17   Solve for all values of θ between $0°$ and $360°$:
(a) $\arccos(-0.5316) = \theta$ (b) $\operatorname{arcsec} 2.3162 = \theta$ (c) $\arctan 0.8314 = \theta$

$$\left[\begin{array}{l}\text{(a) } 122°7' \text{ or } 237°53' \text{ (b) } 64°25' \text{ or } 295°35' \\ \text{(c) } 39°44' \text{ or } 219°44' \end{array}\right]$$

Sine and cosine curves

In Problems 18 to 20 state the amplitude and period of the waveform and sketch the curve between $0°$ and $360°$:

18   (a) $y = \cos 3A$ (b) $y = 2 \sin \dfrac{5x}{2}$

(c) $y = 3 \sin 4t$

$$[\text{(a) } 1, 120° \text{ (b) } 2, 144° \text{ (c) } 3, 90°]$$

19   (a) $y = 3 \cos \dfrac{\theta}{2}$ (b) $y = \dfrac{7}{2} \sin \dfrac{3x}{8}$

(c) $y = 6 \sin(t - 45°)$

$$\left[\text{(a) } 3, 720° \text{ (b) } \dfrac{7}{2}, 960° \text{ (c) } 6, 360°\right]$$

20   (a) $y = 4 \cos(2\theta + 30°)$ (b) $y = 2 \sin^2 2t$

(c) $y = 5 \cos^2 \dfrac{3}{2}\theta$

$$[\text{(a) } 4, 180° \text{ (b) } 2, 90° \text{ (c) } 5, 120°]$$

In Problems 21 to 23 state the amplitude, periodic time, frequency and phase angle (in degrees):

21   $v = 200 \sin(200\pi t + 0.29)$ V

$$[200 \text{ V}, 10 \text{ ms}, 100 \text{ Hz}, 16°37' \text{ leading}]$$

22   $i = 32 \sin(400\pi t - 0.42)$ A

$$[32 \text{ A}, 5 \text{ ms}, 200 \text{ Hz}, 24°4' \text{ lagging}]$$

23   $x = 5 \sin(314.2t + 0.33)$ cm

$$[5 \text{ cm}, 20 \text{ ms}, 50 \text{ Hz}, 18°54' \text{ leading}]$$

24   The current in an a.c. circuit at any time t seconds is given by:

$$i = 5 \sin(100\pi t - 0.432) \text{ amperes}$$

Determine (a) the amplitude, periodic time, frequency and phase angle (in degrees), (b) the value of current at $t = 0$, (c) the value of current at $t = 8$ ms, (d) the time when the current is first a maximum, and (e) the time when the current first reaches 3 A. Sketch one cycle of the waveform showing relevant points.

$$\left[\begin{array}{l}\text{(a) } 5 \text{ A}, 20 \text{ ms}, 50 \text{ Hz}, 24°45' \text{ lagging} \\ \text{(b) } -2.093 \text{ A (c) } 4.363 \text{ A} \\ \text{(d) } 6.375 \text{ ms (e) } 3.423 \text{ ms} \end{array}\right]$$

Trigonometric identities

In Problems 25 to 30 prove the trigonometric identities:

25   $\sin x \cot x = \cos x$

26   $\dfrac{1}{\sqrt{(1 - \cos^2 \theta)}} = \operatorname{cosec} \theta$

27   $2 \cos^2 A - 1 = \cos^2 A - \sin^2 A$

28   $\dfrac{\cos x - \cos^3 x}{\sin x} = \sin x \cos x$

29   $(1 + \cot \theta)^2 + (1 - \cot \theta)^2 = 2 \operatorname{cosec}^2 \theta$

30   $\dfrac{\sin^2 x(\sec x + \operatorname{cosec} x)}{\cos x \tan x} = 1 + \tan x$

31   Show that the trigonometric identities $\cos^2 \theta + \sin^2 \theta = 1$, $1 + \tan^2 \theta = \sec^2 \theta$ and $\cot^2 \theta + 1 = \operatorname{cosec}^2 \theta$ are valid when θ is (a) $124°$, (b) $231°$, and (c) $312°46'$.

Trigonometric equations

In Problems 32 to 41 solve the equations for angles between $0°$ and $360°$:

32   (a) $4 - 7\sin\theta = 0$ (b) $2.5\cos x + 1.75 = 0$

$$\begin{bmatrix} \text{(a) } \theta = 34°51' \text{ or } 145°9' \\ \text{(b) } x = 134°26' \text{ or } 225°34' \end{bmatrix}$$

33   (a) $3\operatorname{cosec} A + 5.5 = 0$
(b) $4(2.32 - 5.4\cot t) = 0$

$$\begin{bmatrix} \text{(a) } A = 213°3' \text{ or } 326°57' \\ \text{(b) } t = 66°45' \text{ or } 246°45' \end{bmatrix}$$

34   (a) $5\sin^2 y = 3$ (b) $3\tan^2\phi - 2 = 0$

$$\begin{bmatrix} \text{(a) } y = 50°46', 129°14', 230°46' \text{ or } 309°14' \\ \text{(b) } \phi = 39°14', 140°46', 219°14' \text{ or } 320°46' \end{bmatrix}$$

35   (a) $5 + 3\operatorname{cosec}^2 D = 8$ (b) $2\cot^2\theta = 5$

$$\begin{bmatrix} \text{(a) } D = 90° \text{ or } 270° \\ \text{(b) } \theta = 32°19', 147°41', 212°19' \text{ or } 327°41' \end{bmatrix}$$

36   (a) $15\sin^2 A + \sin A - 2 = 0$
(b) $8\tan^2\theta + 2\tan\theta = 15$

$$\begin{bmatrix} \text{(a) } A = 19°28', 160°32', 203°35' \text{ or } 336°25' \\ \text{(b) } \theta = 51°20', 123°41', 231°20' \text{ or } 303°41' \end{bmatrix}$$

37   (a) $\sec x + 6 = \sec^2 x$

(b) $2\operatorname{cosec}^2 t - 5\operatorname{cosec} t = 12$

$$\begin{bmatrix} \text{(a) } x = 70°32', 120°, 240° \text{ or } 289°28' \\ \text{(b) } t = 14°29', 165°31', 221°49' \text{ or } 318°11' \end{bmatrix}$$

38   (a) $12\sin^2\theta - 6 = \cos\theta$
(b) $16\sec x - 2 = 14\tan^2 x$

$$\begin{bmatrix} \text{(a) } \theta = 48°11', 138°35', 221°25' \text{ or } 311°49' \\ \text{(b) } x = 52°56' \text{ or } 307°4' \end{bmatrix}$$

39   (a) $4\cot^2 A - 6\operatorname{cosec} A + 6 = 0$
(b) $2\cos^2 y + 3\sin y = 3$

$$[\text{(a) } 90° \text{ (b) } 30°, 90° \text{ or } 150°]$$

40   (a) $3.2\sin^2 x + 2.5\cos x - 1.8 = 0$
(b) $5\sec t + 2\tan^2 t = 3$

$$\begin{bmatrix} \text{(a) } 112°11' \text{ or } 247°49' \\ \text{(b) } 107°50' \text{ or } 252°10' \end{bmatrix}$$

41   (a) $2.9\cos^2\alpha - 7\sin\alpha + 1 = 0$
(b) $3\operatorname{cosec}^2\beta = 8 - 7\cot\beta$

$$\begin{bmatrix} \text{(a) } \alpha = 27°50' \text{ or } 152°10' \\ \text{(b) } \beta = 60°10', 161°1', 240°10' \text{ or } 341°1' \end{bmatrix}$$

38

Compound angles

38.1 Compound-angle formulae

An electric current i may be expressed as $i = 5\sin(\omega t - 0.33)$ amperes. Similarly, the displacement x of a body from a fixed point can be expressed as $x = 10\sin(2t + 0.67)$ metres. The angles $(\omega t - 0.33)$ and $(2t + 0.67)$ are called **compound angles** because they are the sum or difference of two angles. The **compound-angle formulae** for sines and cosines of the sum and difference of two angles A and B are:

$$\sin(A + B) = \sin A \cos B + \cos A \sin B$$

$$\sin(A - B) = \sin A \cos B - \cos A \sin B$$

$$\cos(A + B) = \cos A \cos B - \sin A \sin B$$

$$\cos(A - B) = \cos A \cos B + \sin A \sin B$$

(Note, $\sin(A + B)$ is **not** equal to $(\sin A + \sin B)$, and so on.)

The formulae stated above may be used to derive two further compound-angle formulae:

$$\tan(A + B) = \frac{\tan A + \tan B}{1 - \tan A \tan B}$$

$$\tan(A - B) = \frac{\tan A - \tan B}{1 + \tan A \tan B}$$

The compound-angle formulae are true for all values of A and B, and by substituting values of A and B into the formulae they may be shown to be true.

Problem 1. Show that the compound-angle formula for (a) $\sin(A + B)$ and (b) $\cos(A + B)$ are true, correct to 3 significant figures, when $A = 45°$ and $B = 60°$

(a) $\sin(A + B) = \sin A \cos B + \cos A \sin B$

When $A = 45°$ and $B = 60°$,

$\sin(45° + 60°)$

$= \sin 45° \cos 60° + \cos 45° \sin 60°$

$= (0.7071)(0.5000) + (0.7071)(0.8660)$

$= 0.3536 + 0.6123$

i.e. $\sin 105° = 0.9659$

$\sin 105°(\equiv \sin 75°) = 0.9659$ by calculator.

Thus the formula for $\sin(A + B)$ is true when $A = 45°$ and $B = 60°$

(b) $\cos(A + B) = \cos A \cos B - \sin A \sin B$

When $A = 45°$ and $B = 60°$,

$\cos(45° + 60°)$

$= \cos 45° \cos 60° - \sin 45° \sin 60°$

$= (0.7071)(0.5000) - (0.7071)(0.8660)$

$= 0.3536 - 0.6123$

i.e. $\cos 105° = -0.2587$

$\cos 105°(\equiv \cos 75°) = -0.2588$ by calculator.

Thus the formula for $\cos(A + B)$ is true when $A = 45°$ and $B = 60°$, the small error being due to 'rounding-off'

Problem 2. Show that the compound-angle formula for (a) $\sin(A - B)$ and (b) $\cos(A - B)$ is true when $A = 211°$ and $B = 124°$

(a) $\sin(A - B) = \sin A \cos B - \cos A \sin B$

When $A = 211°$ and $B = 124°$,

$\sin(211° - 124°)$

$= \sin 211° \cos 124° - \cos 211° \sin 124°$

$= (-0.5150)(-0.5592) - (-0.8572)(0.8290)$

$= 0.2880 + 0.7106$

i.e. $\sin 87° = 0.9986$. By calculator $\sin 87° = 0.9986$.

Thus the formula for $\sin(A - B)$ is true when $A = 211°$ and $B = 124°$

(b) $\cos(A - B) = \cos A \cos B + \sin A \sin B$

When $A = 211°$ and $B = 124°$,

$\cos(211° - 124°)$

$= \cos 211° \cos 124° + \sin 211° \sin 124°$

$= (-0.8572)(-0.5592) + (-0.5150)(0.8290)$

$= 0.4793 - 0.4269$

i.e. $\cos 87° = 0.0524$
By calculator, $\cos 87° = 0.0523$.
Thus the formula for $\cos(A - B)$ is true when
$A = 211°$ and $B = 124°$

Problem 3. Verify that when $A = 162°$ and $B = 55°$ the compound-angle formulae for
(a) $\tan(A + B)$ and
(b) $\tan(A - B)$
are true, correct to 3 significant figures

(a) $\tan(A + B) = \dfrac{\tan A + \tan B}{1 - \tan A \tan B}$

When $A = 162°$ and $B = 55°$,

$$\tan(162° + 55°) = \dfrac{\tan 162° + \tan 55°}{1 - \tan 162° \tan 55°}$$

$$= \dfrac{(-0.3249) + (1.4281)}{1 - (-0.3249)(1.4281)}$$

i.e. $\tan 217° = \dfrac{1.1032}{1.4640} = 0.7536$

By calculator, $\tan 217° (= +\tan 37°) = 0.7536$.
Thus the formula for $\tan(A + B)$ is true when
$A = 162°$ and $B = 55°$

(b) $\tan(A - B) = \dfrac{\tan A - \tan B}{1 + \tan A \tan B}$

When $A = 162°$ and $B = 55°$,

$$\tan(162° - 55°) = \dfrac{\tan 162° - \tan 55°}{1 + \tan 162° \tan 55°}$$

$$= \dfrac{(-0.3249) - (1.4281)}{1 + (-0.3249)(1.4281)}$$

i.e. $\tan 107° = \dfrac{-1.7530}{0.5360} = -3.2705 =$
-3.271, correct to 4 significant figures
By calculator,
$\tan 107° (= -\tan 73°) = -3.2709 = -3.271$,
correct to 3 significant figures.
Thus the formula for $\tan(A - B)$ is true when
$A = 162°$ and $B = 55°$

Problem 4. Expand and simplify the following expressions:
(a) $\sin(\pi + \alpha)$
(b) $-\cos(90° + \beta)$
(c) $\sin(A - B) - \sin(A + B)$

(a) $\sin(\pi + \alpha)$

$= \sin \pi \cos \alpha + \cos \pi \sin \alpha$

(from the formula for $\sin(A + B)$)

$= (0)(\cos \alpha) + (-1)\sin \alpha = -\sin \alpha$

(b) $-\cos(90° + \beta)$

$= -[\cos 90° \cos \beta - \sin 90° \sin \beta]$

$= -[(0)(\cos \beta) - (1)\sin \beta] = \sin \beta$

(c) $\sin(A - B) - \sin(A + B)$

$= [\sin A \cos B - \cos A \sin B]$

$- [\sin A \cos B + \cos A \sin B]$

$= -2 \cos A \sin B$

Problem 5. Prove that
$$\cos(y - \pi) + \sin \left(y + \dfrac{\pi}{2} \right) = 0$$

$\cos(y - \pi) = \cos y \cos \pi + \sin y \sin \pi$

$= (\cos y)(-1) + (\sin y)(0) = -\cos y$

$\sin \left(y + \dfrac{\pi}{2} \right) = \sin y \cos \dfrac{\pi}{2} + \cos y \sin \dfrac{\pi}{2}$

$= (\sin y)(0) + (\cos y)(1) = \cos y$

Hence

$\cos(y - \pi) + \sin \left(y + \dfrac{\pi}{2} \right) = (-\cos y) + (\cos y) = 0$

Problem 6. Show that
$$\tan \left(x + \dfrac{\pi}{4} \right) \tan \left(x - \dfrac{\pi}{4} \right) = -1$$

$\tan \left(x + \dfrac{\pi}{4} \right) = \dfrac{\tan x + \tan \dfrac{\pi}{4}}{1 - \tan x \tan \dfrac{\pi}{4}}$

(from the formula for $\tan(A + B)$)

$= \dfrac{\tan x + 1}{1 - (\tan x)(1)} = \left(\dfrac{1 + \tan x}{1 - \tan x} \right)$,

since $\tan \dfrac{\pi}{4} = 1$

$\tan \left(x - \dfrac{\pi}{4} \right) = \dfrac{\tan x - \tan \dfrac{\pi}{4}}{1 + \tan x \tan \dfrac{\pi}{4}} = \left(\dfrac{\tan x - 1}{1 + \tan x} \right)$

Hence $\tan\left(x+\dfrac{\pi}{4}\right)\tan\left(x-\dfrac{\pi}{4}\right)$

$$= \left(\dfrac{1+\tan x}{1-\tan x}\right)\left(\dfrac{\tan x-1}{1+\tan x}\right) = \dfrac{\tan x-1}{1-\tan x}$$

$$= \dfrac{-(1-\tan x)}{1-\tan x} = -1$$

Problem 7. Given that $\sin A = 12/13$ and $\cos B = 8/17$, where A and B are acute angles, determine without using trigonometric tables the values of (a) $\sin(A+B)$, (b) $\cos(A-B)$ and (c) $\tan(A-B)$ each correct to 3 significant figures

---

$\sin A = \dfrac{12}{13} = \dfrac{\text{opposite side}}{\text{hypotenuse}}$,

as shown in Fig. 38.1(a).

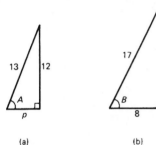

(a)   (b)

Figure 38.1

From Fig. 38.1(a), side $p = \sqrt{(13^2 - 12^2)} = 5$, by Pythagoras' theorem.

Hence $\cos A = \dfrac{5}{13}$ and $\tan A = \dfrac{12}{5}$

Similarly, $\cos B = \dfrac{8}{17} = \dfrac{\text{adjacent side}}{\text{hypotenuse}}$ as shown in

Fig. 38.1(b).

From Fig. 38.1(b), side $q = \sqrt{(17^2 - 8^2)} = 15$, by Pythagoras' theorem.

Hence $\sin B = \dfrac{15}{17}$ and $\tan B = \dfrac{15}{8}$

(a) $\sin(A+B) = \sin A \cos B + \cos A \sin B$

$$= \left(\dfrac{12}{13}\right)\left(\dfrac{8}{17}\right) + \left(\dfrac{5}{13}\right)\left(\dfrac{15}{17}\right)$$

$$= \dfrac{96}{221} + \dfrac{75}{221} = \dfrac{171}{221} \text{ or } \mathbf{0.774}$$

(b) $\cos(A-B) = \cos A \cos B + \sin A \sin B$

$$= \left(\dfrac{5}{13}\right)\left(\dfrac{8}{17}\right) + \left(\dfrac{12}{13}\right)\left(\dfrac{15}{17}\right)$$

$$= \dfrac{40}{221} + \dfrac{180}{221} = \dfrac{220}{221} \text{ or } \mathbf{0.995}$$

(c) $\tan(A-B) = \dfrac{\tan A - \tan B}{1 + \tan A \tan B}$

$$= \dfrac{\dfrac{12}{5} - \dfrac{15}{8}}{1 + \left(\dfrac{12}{5}\right)\left(\dfrac{15}{8}\right)}$$

$$= \dfrac{0.525}{5.5} = \mathbf{0.095}$$

Problem 8. If $\sin P = 0.8142$ and $\cos Q = 0.4432$ evaluate, correct to 3 decimal places (a) $\sin(P-Q)$, (b) $\cos(P+Q)$, and (c) $\tan(P+Q)$, using the compound-angle formulae

---

Since $\sin P = 0.8142$ then $P = \arcsin 0.8142 = 54.51°$.

Thus $\cos P = \cos 54.51° = 0.5806$ and $\tan P = \tan 54.51° = 1.4025$.

Since $\cos Q = 0.4432$, $Q = \arccos 0.4432 = 63.69°$.

Thus $\sin Q = \sin 63.69° = 0.8964$ and $\tan Q = \tan 63.69° = 2.0225$.

(a) $\sin(P-Q)$

$$= \sin P \cos Q - \cos P \sin Q$$

$$= (0.8142)(0.4432) - (0.5806)(0.8964)$$

$$= 0.3609 - 0.5204 = \mathbf{-0.160}$$

(b) $\cos(P+Q)$

$$= \cos P \cos Q - \sin P \sin Q$$

$$= (0.5806)(0.4432) - (0.8142)(0.8964)$$

$$= 0.2573 - 0.7298 = \mathbf{-0.473}$$

(c) $\tan(P+Q) = \dfrac{\tan P + \tan Q}{1 - \tan P \tan Q}$

$$= \dfrac{(1.4025) + (2.0225)}{1 - (1.4025)(2.0225)}$$

$$= \dfrac{3.4250}{-1.8366} = \mathbf{-1.865}$$

Problem 9. Determine, in surd form, the value of sin 75° without using trigonometric tables

$$\sin 75° = \sin(45° + 30°)$$

$$= \sin 45° \cos 30° + \cos 45° \sin 30°,$$
$$\text{from the formula for } \sin(A + B)$$

$$= \left(\frac{1}{\sqrt{2}}\right)\left(\frac{\sqrt{3}}{2}\right) + \left(\frac{1}{\sqrt{2}}\right)\left(\frac{1}{2}\right)$$

$$= \frac{\sqrt{3}}{2\sqrt{2}} + \frac{1}{2\sqrt{2}} = \frac{\sqrt{3} + 1}{2\sqrt{2}}$$

(see page 112 for surd form)

Problem 10. Solve the equation $4 \sin(x - 20°) = 5 \cos x$ for values of x between 0° and 90°

$$4 \sin(x - 20°) = 4[\sin x \cos 20° - \cos x \sin 20°],$$
$$\text{from the formula for } \sin(A - B)$$

$$= 4[\sin x(0.9397) - \cos x(0.3420)]$$

$$= 3.7588 \sin x - 1.3680 \cos x$$

Since $4 \sin(x - 20°) = 5 \cos x$ then
$3.7588 \sin x - 1.3680 \cos x = 5 \cos x$
Rearranging gives:
$$3.7588 \sin x = 5 \cos x + 1.3680 \cos x$$

$$= 6.3680 \cos x$$

$$\text{and } \frac{\sin x}{\cos x} = \frac{6.3680}{3.7588} = 1.6942$$

i.e. $\tan x = 1.6942$, and $x = \arctan 1.6942 = 59.449°$ or **59°27′**

(Check: LHS $= 4 \sin(59.449° - 20°)$
$= 4 \sin 39.449° = 2.542$
RHS $= 5 \cos x = 5 \cos 59.449°$
$= 2.542)$

Further problems on compound-angle formulae may be found in Section 38.6, Problems 1 to 11, page 369.

38.2   Conversion of $a \sin \omega t + b \cos \omega t$ into $R \sin(\omega t + \alpha)$

(i) $R \sin(\omega t + \alpha)$ represents a sine wave of maximum value R, periodic time $2\pi/\omega$, frequency $\omega/2\pi$ and leading $R \sin \omega t$ by angle α. (See Chapter 37)

(ii) $R \sin(\omega t + \alpha)$ may be expanded using the compound-angle formula for $\sin(A+B)$, where $A = \omega t$ and $B = \alpha$

Hence $R \sin(\omega t + \alpha)$
$$= R[\sin \omega t \cos \alpha + \cos \omega t \sin \alpha]$$
$$= R \sin \omega t \cos \alpha + R \cos \omega t \sin \alpha$$
$$= (R \cos \alpha) \sin \omega t + (R \sin \alpha) \cos \omega t$$

(iii)   If $a = R \cos \alpha$ and $b = R \sin \alpha$, where a and b are constants, then $R \sin(\omega t + \alpha) = a \sin \omega t + b \cos \omega t$, i.e. a sine and cosine function of the same frequency when added produce a sine wave of the same frequency

(iv)   Since $a = R \cos \alpha$, then $\cos \alpha = a/R$, and since $b = R \sin \alpha$, then $\sin \alpha = b/R$

If the values of a and b are known then the values of R and α may be calculated. The relationship between constants a, b, R and α are shown in Fig. 38.2.

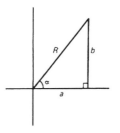

Figure 38.2

From Fig. 38.2, by Pythagoras' theorem:

$$\boldsymbol{R = \sqrt{(a^2 + b^2)}}$$

and from trigonometric ratios:

$$\boldsymbol{\alpha = \arctan b/a}$$

Problem 11. Find an expression for $3 \sin \omega t + 4 \cos \omega t$ in the form $R \sin(\omega t + \alpha)$ and sketch graphs of $3 \sin \omega t$, $4 \cos \omega t$ and $R \sin(\omega t + \alpha)$ on the same axes

Let $3 \sin \omega t + 4 \cos \omega t = R \sin(\omega t + \alpha)$

then $3 \sin \omega t + 4 \cos \omega t$
$$= R[\sin \omega t \cos \alpha + \cos \omega t \sin \alpha]$$
$$= (R \cos \alpha) \sin \omega t + (R \sin \alpha) \cos \omega t$$

Equating coefficients of $\sin \omega t$ gives:

$$3 = R\cos\alpha, \text{ from which, } \cos\alpha = \frac{3}{R}$$

Equating coefficients of $\cos \omega t$ gives:

$$4 = R\sin\alpha, \text{ from which, } \sin\alpha = \frac{4}{R}$$

There is only one quadrant where both $\sin\alpha$ and $\cos\alpha$ are positive, and this is the first, as shown in Fig. 38.3. From Fig. 38.3, by Pythagoras' theorem:

$$R = \sqrt{(3^2 + 4^2)} = 5$$

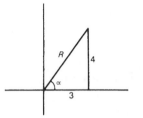

Figure 38.3

From trigonometric ratios: $\alpha = \arctan \dfrac{4}{3} = 53°8'$ or 0.927 radians.
Hence $3\sin\omega t + 4\cos\omega t = 5\sin(\omega t + 0.927)$.
A sketch of $3\sin\omega t$, $4\cos\omega t$ and $5\sin(\omega t + 0.927)$ is shown in Fig. 38.4.

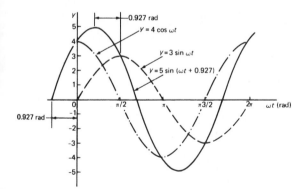

Figure 38.4

Two periodic functions of the same frequency may be combined by

(a) plotting the functions graphically and combining ordinates at intervals, or

(b) by resolution of phasors by drawing or calculation

This problem, together with the following, demonstrate a third method of combining waveforms.

Problem 12.   Express $4.6\sin\omega t - 7.3\cos\omega t$ in the form $R\sin(\omega t + \alpha)$

Let $\quad 4.6\sin\omega t - 7.3\cos\omega t = R\sin(\omega t + \alpha)$

then $\quad 4.6\sin\omega t - 7.3\cos\omega t$

$$= R[\sin\omega t\cos\alpha + \cos\omega t\sin\alpha]$$
$$= (R\cos\alpha)\sin\omega t + (R\sin\alpha)\cos\omega t$$

Equating coefficients of $\sin\omega t$ gives: $4.6 = R\cos\alpha$, from which, $\cos\alpha = \dfrac{4.6}{R}$

Equating coefficients of $\cos\omega t$ gives: $-7.3 = R\sin\alpha$, from which $\sin\alpha = \dfrac{-7.3}{R}$

There is only one quadrant where cosine is positive and sine is negative, i.e. the fourth quadrant, as shown in Fig. 38.5. By Pythagoras' theorem:

$$R = \sqrt{[(4.6)^2 + (-7.3)^2]} = 8.628$$

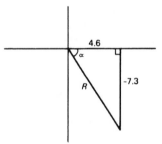

Figure 38.5

By trigonometric ratios:

$$\alpha = \arctan\left(\frac{-7.3}{4.6}\right) = -57.78° \text{ or } -1.008 \text{ radians}$$

Hence

$$4.6\sin\omega t - 7.3\cos\omega t = 8.628\sin(\omega t - 1.008)$$

Problem 13.   Express
$-2.7\sin\omega t - 4.1\cos\omega t$ in the form
$R\sin(\omega t + \alpha)$

Let $-2.7 \sin \omega t - 4.1 \cos \omega t = R \sin(\omega t + \alpha)$

$$= R\{\sin \omega t \cos \alpha + \cos \omega t \sin \alpha\}$$

$$= (R \cos \alpha) \sin \omega t + (R \sin \alpha) \cos \omega t$$

Equating coefficients gives:

$$-2.7 = R \cos \alpha, \text{ from which, } \cos \alpha = \frac{-2.7}{R}$$

and $-4.1 = R \sin \alpha, \text{ from which, } \sin \alpha = \frac{-4.1}{R}$

There is only one quadrant in which both cosine and sine are negative, i.e. the third quadrant, as shown in Fig. 38.6. From Fig. 38.6,

$$R = \sqrt{[(-2.7)^2 + (-4.1)^2]} = 4.909$$

and $\theta = \arctan \dfrac{4.1}{2.7} = 56.63°$

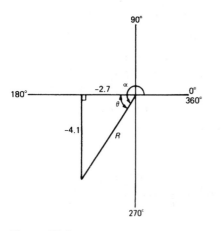

Figure 38.6

Hence $\alpha = 180° + 56.63° = 236.63°$ or 4.130 radians.
Thus

$$-2.7 \sin \omega t - 4.1 \cos \omega t = 4.909 \sin(\omega t + 4.130)$$

An angle of 236.63° is the same as $-123.37°$ or -2.153 radians.
Hence $-2.7 \sin \omega t - 4.1 \cos \omega t$ may be expressed also as **4.909 sin(ωt − 2.153)**, which is preferred since it is the **principal value** (i.e. $-\pi \le \alpha \le \pi$).

Problem 14.   In Problem 13, state when the maximum value first occurs

---

$4.909 \sin(\omega t - 2.153)$ has a maximum value of 4.909 which first occurs when $(\omega t - 2.153)$ is equal to 90°

or $\pi/2$ radians (since a sine wave is at its maximum at 90°).
Hence $\omega t - 2.153 = \pi/2 = 1.571$, from which, $\omega t = 1.571 + 2.153 = 3.724$ radians or 213.37°.
Thus the maximum value of 4.909 occurs first when angle $\omega t = 213.37°$.

Problem 15.   Express $3 \sin \theta + 5 \cos \theta$ in the form $R \sin(\theta + \alpha)$, and hence solve the equation $3 \sin \theta + 5 \cos \theta = 4$, for values of θ between 0° and 360°

Let $3 \sin \theta + 5 \cos \theta = R \sin(\theta + \alpha)$

$$= R[\sin \theta \cos \alpha + \cos \theta \sin \alpha]$$

$$= (R \cos \alpha) \sin \theta + (R \sin \alpha) \cos \theta$$

Equating coefficients gives:

$$3 = R \cos \alpha, \text{ from which, } \cos \alpha = \frac{3}{R}$$

and $5 = R \sin \alpha, \text{ from which, } \sin \alpha = \dfrac{5}{R}$

Since both $\sin \alpha$ and $\cos \alpha$ are positive, R lies in the first quadrant, as shown in Fig. 38.7.

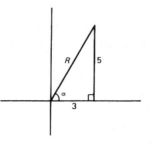

Figure 38.7

From Fig. 38.7, $R = \sqrt{(3^2 + 5^2)} = 5.831$ and
$\alpha = \arctan \dfrac{5}{3} = 59°2'$
Hence $3 \sin \theta + 5 \cos \theta = 5.831 \sin(\theta + 59°2')$
However $3 \sin \theta + 5 \cos \theta = 4$
Thus $5.831 \sin(\theta + 59°2') = 4$, from which

$$(\theta + 59°2') = \arcsin\left(\frac{4}{5.831}\right)$$

i.e. $\theta + 59°2' = 43°19'$ or $136°41'$

Hence $\theta = 43°19' - 59°2' = -15°43'$

or $\theta = 136°41' - 59°2' = 77°39'$

Since $-15°43'$ is the same as $-15°43' + 360°$, i.e. $344°17'$, then the solutions are $\theta = \mathbf{77°39'}$ **or** $\mathbf{344°17'}$, which may be checked by substituting into the original equation.

Problem 16. Solve the equation $3.5\cos A - 5.8\sin A = 6.5$ for $0° \le A \le 360°$.

Let $\quad 3.5\cos A - 5.8\sin A = R\sin(A + \alpha)$

$$= R[\sin A\cos\alpha + \cos A\sin\alpha]$$

$$= (R\cos\alpha)\sin A + (R\sin\alpha)\cos A$$

Equating coefficients gives:

$$3.5 = R\sin\alpha, \text{ from which, } \sin\alpha = \frac{3.5}{R}$$

and $\quad -5.8 = R\cos\alpha$, from which, $\cos\alpha = \dfrac{-5.8}{R}$

There is only one quadrant in which both sine is positive and cosine is negative, i.e. the second, as shown in Fig. 38.8.

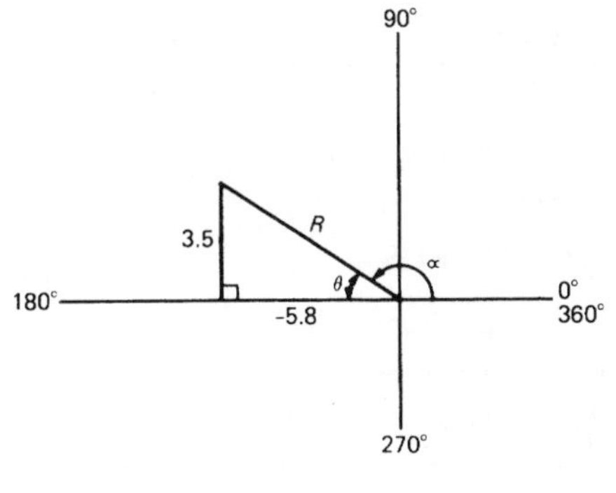

Figure 38.8

From Fig. 38.8, $R = \sqrt{[(3.5)^2 + (-5.8)^2]} = 6.774$ and $\theta = \arctan\dfrac{3.5}{5.8} = 31°7'$

Hence $\alpha = 180° - 31°7' = 148°53'$
Thus

$$3.5\cos A - 5.8\sin A = 6.774\sin(A + 148°53') = 6.5$$

Hence $\quad \sin(A + 148°53') = \dfrac{6.5}{6.774}$, from which,

$$(A + 148°53') = \arcsin\frac{6.5}{6.774}$$

$$= 73°39' \text{ or } 106°21'$$

Thus $\quad A = 73°39' - 148°53' = -75°14'$

$$\equiv (-75°14' + 360°) = 284°46'$$

or $\quad A = 106°21' - 148°53' = -42°32'$

$$\equiv (-42°32' + 360°) = 317°28'$$

The solutions are thus $A = \mathbf{284°46'}$ **or** $\mathbf{317°28'}$, which may be checked in the original equation.

Further problems on the conversion of $a\sin\omega t + b\cos\omega t$ into $R\sin(\omega t + \alpha)$ may be found in Section 38.6, Problems 12 to 22, pages 369 and 370.

38.3 Double angles

(i) If, in the compound-angle formula for $\sin(A + B)$, we let $B = A$ then
$$\sin 2A = 2\sin A\cos A$$

Also, for example,

$$\sin 4A = 2\sin 2A\cos 2A$$

and $\quad \sin 8A = 2\sin 4A\cos 4A$, and so on

(ii) If, in the compound-angle formula for $\cos(A + B)$, we let $B = A$ then
$$\cos 2A = \cos^2 A - \sin^2 A$$

Since $\cos^2 A + \sin^2 A = 1$, then $\cos^2 A = 1 - \sin^2 A$, and $\sin^2 A = 1 - \cos^2 A$, and two further formulae for $\cos 2A$ can be produced.

Thus $\quad \cos 2A = \cos^2 A - \sin^2 A$

$$= (1 - \sin^2 A) - \sin^2 A$$

i.e. $\quad \cos 2A = 1 - 2\sin^2 A$

and $\quad \cos 2A = \cos^2 A - \sin^2 A$

$$= \cos^2 A - (1 - \cos^2 A)$$

i.e. $\quad \cos 2A = 2\cos^2 A - 1$

Also, for example,

$$\cos 4A = \cos^2 2A - \sin^2 2A \text{ or}$$

$$1 - 2\sin^2 2A \text{ or}$$

$$2\cos^2 2A - 1$$

and $\quad \cos 6A = \cos^2 3A - \sin^2 3A \text{ or}$

$$1 - 2\sin^2 3A \text{ or}$$

$$2\cos^2 3A - 1,$$

and so on

(iii)   If, in the compound-angle formula for $\tan(A + B)$, we let $B = A$ then

$$\tan 2A = \frac{2 \tan A}{1 - \tan^2 A}$$

Also, for example,

$$\tan 4A = \frac{2 \tan 2A}{1 - \tan^2 2A}$$

and $$\tan 5A = \frac{2 \tan \frac{5}{2}A}{1 - \tan^2 \frac{5}{2}A} \quad \text{and so on}$$

Problem 17.   If $\sin \theta = 4/5$, where θ is an acute angle, evaluate $\sin 2\theta$, $\cos 2\theta$ and $\tan 2\theta$, without evaluating angle θ

$$\sin \theta = \frac{4}{5} = \frac{\text{opposite side}}{\text{hypotenuse}}$$

By Pythagoras' theorem,

$$\text{the adjacent side} = \sqrt{(5^2 - 4^2)} = 3$$

Hence $\cos \theta = \dfrac{3}{5}$ and $\tan \theta = \dfrac{4}{3}$

$$\sin 2\theta = 2 \sin \theta \cos \theta = 2 \left(\frac{4}{5}\right)\left(\frac{3}{5}\right) = \frac{24}{25} \text{ or } \mathbf{0.96}$$

$$\cos 2\theta = \cos^2 \theta - \sin^2 \theta = \left(\frac{3}{5}\right)^2 - \left(\frac{4}{5}\right)^2$$

$$= \frac{-7}{25} \text{ or } \mathbf{-0.28}$$

$$\left(\text{or } \cos 2\theta = \cos^2 \theta - \sin^2 \theta = 1 - 2\left(\frac{4}{5}\right)^2\right.$$

$$= 1 - \frac{32}{25} = \frac{-7}{25}$$

$$\text{or } \cos 2\theta = 2 \cos^2 \theta - 1 = 2\left(\frac{3}{5}\right)^2 - 1$$

$$\left. = \frac{18}{25} - 1 = \frac{-7}{25}\right)$$

$$\tan 2\theta = \frac{2 \tan \theta}{1 - \tan^2 \theta} = \frac{2(4/3)}{1 - (4/3)^2} = \frac{8/3}{-7/9}$$

$$= \left(\frac{8}{3}\right)\left(-\frac{9}{7}\right) = \frac{-24}{7} \text{ or } \mathbf{-3.4286}$$

Problem 18. $I_3 \sin 3\theta$ is the third harmonic of a waveform. Express the third harmonic in terms of the first harmonic $\sin \theta$, when $I_3 = 1$

When $I_3 = 1$, $I_3 \sin 3\theta$

$$= \sin 3\theta = \sin(2\theta + \theta)$$

$$= \sin 2\theta \cos \theta + \cos 2\theta \sin \theta,$$

from the $\sin(A + B)$ formula

$$= (2 \sin \theta \cos \theta) \cos \theta + (1 - 2 \sin^2 \theta) \sin \theta,$$

from the double angle expansions

$$= 2 \sin \theta \cos^2 \theta + \sin \theta - 2 \sin^3 \theta$$

$$= 2 \sin \theta (1 - \sin^2 \theta) + \sin \theta - 2 \sin^3 \theta$$

$$(\text{since } \cos^2 \theta = 1 - \sin^2 \theta)$$

$$= 2 \sin \theta - 2 \sin^3 \theta + \sin \theta - 2 \sin^3 \theta$$

i.e. $\mathbf{\sin 3\theta = 3 \sin \theta - 4 \sin^3 \theta}$

Problem 19.   Prove that $\dfrac{1 - \cos 2\theta}{\sin 2\theta} = \tan \theta$

$$\text{LHS} = \frac{1 - \cos 2\theta}{\sin 2\theta} = \frac{1 - (1 - 2 \sin^2 \theta)}{2 \sin \theta \cos \theta}$$

$$= \frac{2 \sin^2 \theta}{2 \sin \theta \cos \theta} = \frac{\sin \theta}{\cos \theta}$$

$$= \tan \theta = \text{RHS}$$

Problem 20.   Prove that $\cot 2x + \text{cosec } 2x = \cot x$

$$\text{LHS} = \cot 2x + \text{cosec } 2x = \frac{\cos 2x}{\sin 2x} + \frac{1}{\sin 2x}$$

$$= \frac{\cos 2x + 1}{\sin 2x}$$

$$= \frac{(2 \cos^2 x - 1) + 1}{\sin 2x}$$

$$= \frac{2 \cos^2 x}{\sin 2x} = \frac{2 \cos^2 x}{2 \sin x \cos x}$$

$$= \frac{\cos x}{\sin x} = \cot x = \text{RHS}$$

Further problems on double angles may be found in Section 38.6, Problems 23 to 26, page 370.

38.4   Changing products of sines and cosines into sums or differences

(i) $\sin(A + B) + \sin(A - B) = 2 \sin A \cos B$ (from the formulae in Section 38.1)

i.e. **$\sin A \cos B$**

$$= \frac{1}{2}[\sin(A + B) + \sin(A - B)] \qquad (1)$$

(ii) $\sin(A + B) - \sin(A - B) = 2 \cos A \sin B$

i.e. **$\cos A \sin B$**

$$= \frac{1}{2}[\sin(A + B) - \sin(A - B)] \qquad (2)$$

(iii) $\cos(A + B) + \cos(A - B) = 2 \cos A \cos B$

i.e. **$\cos A \cos B$**

$$= \frac{1}{2}[\cos(A + B) + \cos(A - B)] \qquad (3)$$

(iv) $\cos(A + B) - \cos(A - B) = -2 \sin A \sin B$

i.e. **$\sin A \sin B$**

$$= -\frac{1}{2}[\cos(A + B) - \cos(A - B)] \qquad (4)$$

Problem 21.   Express $\sin 4x \cos 3x$ as a sum or difference of sines and cosines

---

From equation (1),

$$\sin 4x \cos 3x = \frac{1}{2}[\sin(4x + 3x) + \sin(4x - 3x)]$$

$$= \frac{1}{2}(\sin 7x + \sin x)$$

Problem 22.   Express $2 \cos 5\theta \sin 2\theta$ as a sum or difference of sines or cosines

---

From equation (2),

$$2 \cos 5\theta \sin 2\theta = 2\left\{\frac{1}{2}[\sin(5\theta + 2\theta) - \sin(5\theta - 2\theta)]\right\}$$

$$= \sin 7\theta - \sin 3\theta$$

Problem 23.   Express $3 \cos 4t \cos t$ as a sum or difference of sines or cosines

---

From equation (3),

$$3 \cos 4t \cos t = 3\left\{\frac{1}{2}[\cos(4t + t) + \cos(4t - t)]\right\}$$

$$= \frac{3}{2}(\cos 5t + \cos 3t)$$

Thus, if the integral $\int 3 \cos 4t \cos t \, dt$ was required, then

$$\int 3 \cos 4t \cos t \, dt = \int \frac{3}{2}(\cos 5t + \cos 3t) \, dt$$

$$= \frac{3}{2}\left[\frac{\sin 5t}{5} + \frac{\sin 3t}{3}\right] + c$$

Problem 24.   Express $\sin \dfrac{\pi}{3} \sin \dfrac{\pi}{4}$ as a sum or difference of sines or cosines

---

From equation (4),

$$\sin \frac{\pi}{3} \sin \frac{\pi}{4} = -\frac{1}{2}\left[\cos\left(\frac{\pi}{3} + \frac{\pi}{4}\right) - \cos\left(\frac{\pi}{3} - \frac{\pi}{4}\right)\right]$$

$$= -\frac{1}{2}\left[\cos \frac{7\pi}{12} - \cos \frac{\pi}{12}\right]$$

$$= \frac{1}{2}\left[\cos \frac{\pi}{12} - \cos \frac{7\pi}{12}\right]$$

Problem 25.   In an alternating current circuit, voltage $v = 5 \sin \omega t$ and current $i = 10 \sin(\omega t - \pi/6)$. Find an expression for the instantaneous power p at time t given that $p = vi$, expressing the answer as a sum or difference of sines and cosines

---

$$p = vi = (5 \sin \omega t)\left[10 \sin\left(\omega t - \frac{\pi}{6}\right)\right]$$

$$= 50 \sin \omega t \sin\left(\omega t - \frac{\pi}{6}\right)$$

From equation (4),

$$50 \sin \omega t \sin\left(\omega t - \frac{\pi}{6}\right)$$

$$= 50\left\{-\frac{1}{2}\cos\left(\omega t + \omega t - \frac{\pi}{6}\right)\right.$$

$$\left. - \cos\left[\omega t - \left(\omega t - \frac{\pi}{6}\right)\right]\right\}$$

$$= -25\left\{\cos\left(2\omega t - \frac{\pi}{6}\right) - \cos \frac{\pi}{6}\right\}$$

i.e. instantaneous power,

$$p = 25 \left[\cos \frac{\pi}{6} - \cos \left(2\omega t - \frac{\pi}{6} \right) \right]$$

Further problems on changing products of sines and cosines into sums or differences may be found in Section 38.6, Problems 27 to 35, page 370.

38.5   Changing sums or differences of sines and cosines into products

In the compound-angle formula let $(A+B) = X$ and $(A - B) = Y$.
Solving the simultaneous equations gives $A = \dfrac{X+Y}{2}$ and $B = \dfrac{X-Y}{2}$
Thus $\sin(A + B) + \sin(A - B) = 2 \sin A \cos B$

becomes

$$\sin X + \sin Y$$

$$= 2 \sin \left(\frac{X+Y}{2} \right) \cos \left(\frac{X-Y}{2} \right) \qquad (5)$$

Similarly,

$$\sin X - \sin Y$$

$$= 2 \cos \left(\frac{X+Y}{2} \right) \sin \left(\frac{X-Y}{2} \right) \qquad (6)$$

$$\cos X + \cos Y$$

$$= 2 \cos \left(\frac{X+Y}{2} \right) \cos \left(\frac{X-Y}{2} \right) \qquad (7)$$

$$\cos X - \cos Y$$

$$= -2 \sin \left(\frac{X+Y}{2} \right) \sin \left(\frac{X-Y}{2} \right) \qquad (8)$$

Problem 26.   Express $\sin 5\theta + \sin 3\theta$ as a product

---

From equation (5),

$$\sin 5\theta + \sin 3\theta = 2 \sin \left(\frac{5\theta + 3\theta}{2} \right) \cos \left(\frac{5\theta - 3\theta}{2} \right)$$

$$= 2 \sin 4\theta \cos \theta$$

Problem 27.   Express $\sin 7x - \sin x$ as a product

From equation (6),

$$\sin 7x - \sin x = 2 \cos \left(\frac{7x + x}{2} \right) \sin \left(\frac{7x - x}{2} \right)$$

$$= 2 \cos 4x \sin 3x$$

Problem 28.   Express $\cos 8x + \cos 4x$ as a product

From equation (7),

$$\cos 8x + \cos 4x = 2 \cos \left(\frac{8x + 4x}{2} \right) \cos \left(\frac{8x - 4x}{2} \right)$$

$$= 2 \cos 6x \cos 2x$$

Problem 29.   Express $\cos 2t - \cos 5t$ as a product

From equation (8),

$$\cos 2t - \cos 5t = -2 \sin \left(\frac{2t + 5t}{2} \right) \sin \left(\frac{2t - 5t}{2} \right)$$

$$= -2 \sin \frac{7}{2}t \sin \left(-\frac{3}{2}t \right)$$

$$= 2 \sin \frac{7}{2}t \sin \frac{3}{2}t$$

$$\left[\text{since } \sin \left(-\frac{3t}{2} \right) = -\sin \frac{3}{2}t \right]$$

Problem 30.   Show that
$$\frac{\cos 6x + \cos 2x}{\sin 6x + \sin 2x} = \cot 4x$$

From equation (7),

$$\cos 6x + \cos 2x = 2 \cos 4x \cos 2x$$

From equation (5),

$$\sin 6x + \sin 2x = 2 \sin 4x \cos 2x$$

Hence

$$\frac{\cos 6x + \cos 2x}{\sin 6x + \sin 2x} = \frac{2 \cos 4x \cos 2x}{2 \sin 4x \cos 2x}$$

$$= \frac{\cos 4x}{\sin 4x} = \cot 4x$$

Further problems on changing sums or differences of sines and cosines into products may be found in the following Section 38.6, Problems 36 to 42, page 370.

38.6 Further problems on compound angles

Compound-angle formulae

1 Show that the compound-angle formulae for $\sin(A+B)$ and $\cos(A+B)$ are true when $A = 32°$ and $B = 59°$.

2 Verify that the compound-angle formulae for $\sin(A - B)$ and $\tan(A + B)$ are true when $A = 115°$ and $B = 51°$.

3 Prove that the compound-angle formulae for $\tan(A - B)$ and $\cos(A - B)$ are true when $A = 278°$ and $B = 141°$.

4 Reduce the following to the sine of one angle:
(a) $\sin 37° \cos 21° + \cos 37° \sin 21°$
(b) $\sin 2x \cos 5x + \cos 2x \sin 5x$
(c) $\sin 7t \cos 3t - \cos 7t \sin 3t$

$$[\text{(a) } \sin 58° \text{ (b) } \sin 7x \text{ (c) } \sin 4t]$$

5 Reduce the following to the cosine of one angle:
(a) $\cos 71° \cos 33° - \sin 71° \sin 33°$
(b) $\cos 5\theta \cos 2\theta - \sin 5\theta \sin 2\theta$
(c) $\cos \dfrac{\pi}{3} \cos \dfrac{\pi}{4} + \sin \dfrac{\pi}{3} \sin \dfrac{\pi}{4}$

$$\left[\text{(a) } \cos 104° \equiv -\cos 76° \text{ (b) } \cos 7\theta \text{ (c) } \cos \dfrac{\pi}{12}\right]$$

6 Show that:
(a) $\sin\left(x + \dfrac{\pi}{3}\right) + \sin\left(x + \dfrac{2\pi}{3}\right) = \sqrt{3}\cos x$

and (b) $-\sin\left(\dfrac{3\pi}{2} - \phi\right) = \cos\phi$

7 Prove that:
(a) $\sin\left(\theta + \dfrac{\pi}{4}\right) - \sin\left(\theta - \dfrac{3\pi}{4}\right)$
$= \sqrt{2}(\sin\theta + \cos\theta)$

and (b) $\dfrac{\cos(270° + \theta)}{\cos(360° - \theta)} = \tan\theta$

8 If $\sin A = \dfrac{5}{13}$ and $\cos B = \dfrac{9}{41}$ evaluate
(a) $\sin(A + B)$, (b) $\cos(A + B)$ (c) $\tan(A - B)$, correct to 4 decimal places.

$$[\text{(a) } 0.9850 \text{ (b) } -0.1726 \text{ (c) } -1.4123]$$

9 Given $\cos A = 0.42$ and $\sin B = 0.73$ evaluate (a) $\sin(A - B)$, (b) $\cos(A - B)$, (c) $\tan(A + B)$, correct to 4 decimal places.

$$[\text{(a) } 0.3136 \text{ (b) } 0.9495 \text{ (c) } -2.4687]$$

10 Determine, in surd form, the value of $\cos 15°$ without using trigonometric tables.

$$\left[\dfrac{\sqrt{3} + 1}{2\sqrt{2}}\right]$$

11 Solve the following equations for values of θ between 0° and 360°: (a) $3\sin(\theta + 30°) = 7\cos\theta$ (b) $4\sin(\theta - 40°) = 2\sin\theta$.

$$[\text{(a) } 64°43' \text{ or } 244°43' \text{ (b) } 67°31' \text{ or } 247°31']$$

Conversion of $a \sin \omega t + b \cos \omega t$ into $R \sin(\omega t + \alpha)$

In Problems 12 to 16, change the functions into the form $R \sin(\omega t \pm \alpha)$:

12 $5 \sin \omega t + 8 \cos \omega t$ $[9.434 \sin(\omega t + 1.012)]$

13 $9 \sin \omega t + 5 \cos \omega t$ $[10.30 \sin(\omega t + 0.507)]$

14 $4 \sin \omega t - 3 \cos \omega t$ $[5 \sin(\omega t - 0.644)]$

15 $-7 \sin \omega t + 4 \cos \omega t$ $[8.062 \sin(\omega t + 2.622)]$

16 $-3 \sin \omega t - 6 \cos \omega t$ $[6.708 \sin(\omega t - 2.034)]$

17 Solve the following equations for values of θ between 0° and 360°: (a) $2\sin\theta + 4\cos\theta = 3$ (b) $12\sin\theta - 9\cos\theta = 7$.

$$[\text{(a) } 74°26' \text{ or } 338°42' \text{ (b) } 64°41' \text{ or } 189°3']$$

18 Solve the following equations for $0° \le A \le 360°$: (a) $3\cos A + 2\sin A = 2.8$ (b) $12\cos A - 4\sin A = 11$.

$$[\text{(a) } 72°44' \text{ or } 354°38' \text{ (b) } 11°9' \text{ or } 311°59']$$

19 The third harmonic of a wave motion is given by $4.3\cos 3\theta - 6.9\sin 3\theta$. Express this in the form $R\sin(3\theta \pm \alpha)$. $[8.13 \sin(3\theta + 2.584)]$

20 The displacement x metres of a mass from a fixed point about which it is oscillating is given by $x = 2.4\sin \omega t + 3.2\cos \omega t$, where t is the time in seconds. Express x in the form $R\sin(\omega t + \alpha)$. $[x = 4.0\sin(\omega t + 0.927)]$

21 Alternating currents are given by $i_1 = 6\sin \omega t$ amperes and $i_2 = 10\cos \omega t$ amperes. Calculate the maximum value of the resultant $(i_1 + i_2)$ and its phase displacement relative to i_1.

$$[11.6 \text{ A}, 59°2' \text{ leading}]$$

22 Two voltages, $v_1 = 5\cos \omega t$ and $v_2 = -8\sin \omega t$ are inputs to an analogue circuit. Determine an

expression for the output voltage if this is given by $(v_1 + v_2)$.   [$9.434 \sin(\omega t + 2.583)$]

Double angles

23 Given $\cos \theta = \dfrac{5}{13}$, where θ is an acute angle, evaluate (a) $\sin 2\theta$ (b) $\cos 2\theta$ (c) $\tan 2\theta$, without evaluating angle θ, each correct to 4 decimal places.
        [(a) 0.7101 (b) $-$0.7041 (c) $-$1.0084]

24 The power p in an electrical circuit is given by $p = \dfrac{v^2}{R}$. Determine the power in terms of V, R and $\cos 2t$ when $v = V \cos t$.
        $\left[\dfrac{V^2}{2R}(1 + \cos 2t)\right]$

25 Prove the following identities:

(a) $1 - \dfrac{\cos 2\phi}{\cos^2 \phi} = \tan^2 \phi$

(b) $\dfrac{1 + \cos 2t}{\sin^2 t} = 2 \cot^2 t$

(c) $\dfrac{(\tan 2x)(1 + \tan x)}{\tan x} = \dfrac{2}{1 - \tan x}$

(d) $2 \operatorname{cosec} 2\theta \cos 2\theta = \cot \theta - \tan \theta$

26 If the third harmonic of a waveform is given by $V_3 \cos 3\theta$, express the third harmonic in terms of the first harmonic $\cos \theta$, when $V_3 = 1$.
        [$\cos 3\theta = 4\cos^3 \theta - 3\cos \theta$]

Products into sums or differences

In Problems 27 to 32, express as sums or differences:

27 $\sin 7t \cos 2t$ $\left[\dfrac{1}{2}(\sin 9t + \sin 5t)\right]$

28 $\cos 8x \sin 2x$ $\left[\dfrac{1}{2}(\sin 10x - \sin 6x)\right]$

29 $2 \sin 7t \sin 3t$ [$\cos 4t - \cos 10t$]

30 $4 \cos 3\theta \cos \theta$ [$2(\cos 4\theta + \cos 2\theta)$]

31 $3 \sin \dfrac{\pi}{3} \cos \dfrac{\pi}{6}$ $\left[\dfrac{3}{2}\left(\sin \dfrac{\pi}{2} + \sin \dfrac{\pi}{6}\right)\right]$

32 $\sin 23° \sin 68°$ $\left[\dfrac{1}{2}(\cos 45° - \cos 91°)\right]$

33 Determine $\displaystyle\int 2 \sin 3t \cos t \, dt$
        $\left[-\dfrac{\cos 4t}{4} - \dfrac{\cos 2t}{2} + c\right]$

34 Evaluate $\displaystyle\int_0^{\pi/2} 4 \cos 5x \cos 2x \, dx$ $\left[-\dfrac{20}{21}\right]$

35 Solve the equation: $2 \sin 2\phi \sin \phi = \cos \phi$ in the range $\phi = 0$ to $\phi = 180°$.   [$30°$, $90°$ or $150°$]

Sums or differences into products

In Problems 36 to 41, express as products:

36 $\sin 3x + \sin x$ [$2 \sin 2x \cos x$]

37 $\dfrac{1}{2}(\sin 9\theta - \sin 7\theta)$ [$\cos 8\theta \sin \theta$]

38 $\cos 5t + \cos 3t$ [$2 \cos 4t \cos t$]

39 $\dfrac{1}{8}(\cos 5t - \cos t)$ $\left[-\dfrac{1}{4}\sin 3t \sin 2t\right]$

40 $\dfrac{1}{2}\left(\cos \dfrac{\pi}{3} + \cos \dfrac{\pi}{4}\right)$ $\left[\cos \dfrac{7\pi}{24} \cos \dfrac{\pi}{24}\right]$

41 $3(\sin 7\phi + \sin 2\phi)$ $\left[6 \sin \dfrac{9}{2}\phi \cos \dfrac{5}{2}\phi\right]$

42 Show that: (a) $\dfrac{\sin 4x - \sin 2x}{\cos 4x + \cos 2x} = \tan x$

(b) $\frac{1}{2}\{\sin(5x - \alpha) - \sin(x + \alpha)\}$
        $= \cos 3x \sin(2x - \alpha)$

Angular properties of circles

39.1 Introduction

A **circle** is a plain figure enclosed by a curved line, every point on which is equidistant from a point within, called the **centre**.

39.2 Properties of circles

(i) The distance from the centre to the curve is called the **radius**, r, of the circle (see OP in Fig. 39.1)

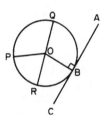

Figure 39.1

(ii) The boundary of a circle is called the **circumference**, c

(iii) Any straight line passing through the centre and touching the circumference at each end is called the **diameter**, d (see QR in Fig. 39.1). Thus $d = 2r$

(iv) The ratio

$$\frac{\text{circumference}}{\text{diameter}} = \text{a constant for any circle}$$

This constant is denoted by the Greek letter π (pronounced 'pie'),

where $\pi = 3.14159$, correct to 5 decimal places.

Hence $c/d = \pi$ or $c = \pi d$ or $c = 2\pi r$

(v) A **semicircle** is one-half of the whole circle

(vi) A **quadrant** is one-quarter of a whole circle

(vii) A **tangent** to a circle is a straight line which meets the circle in one point only and does not cut the circle when produced. AC in Fig. 39.1 is a tangent to the circle since it touches the curve at point B only. If radius OB is drawn, then angle ABO is a right angle

(viii) A **sector** of a circle is the part of a circle between radii (for example, the portion OXY of Fig. 39.2 is a sector). If a sector is less than a semicircle it is called a **minor sector**, if greater than a semicircle it is called a **major sector**

Figure 39.2

(ix) A **chord** of a circle is any straight line which divides the circle into two parts and is terminated at each end by the circumference. ST in Fig. 39.2 is a chord

(x) A **segment** is the name given to the parts into which a circle is divided by a chord. If the segment is less than a semicircle it is called a **minor segment** (see shaded area in Fig. 39.2). If the segment is greater than a semicircle it is called a **major segment** (see the unshaded area in Fig. 39.2)

(xi) An **arc** is a portion of the circumference of a circle. The distance SRT in Fig. 39.2 is called a **minor arc** and the distance $SXYT$ is called a **major arc**

(xii) The angle at the centre of a circle, subtended by an arc, is double the angle at the circumference subtended by the same arc. With reference to Fig. 39.3, **angle AOC = 2 × angle ABC**

(xiii) The angle in a semicircle is a right angle (see angle BQP in Fig. 39.3)

Figure 39.3

39.3   Radian measure

One **radian** is defined as the angle subtended at the centre of a circle by an arc equal in length to the radius. With reference to Fig. 39.4, for arc length l, θ radians $= l/r$ or $\boxed{l = r\theta}$, where θ is in radians. When l = whole circumference ($= 2\pi r$) then $\theta = l/r = 2\pi r/r = 2\pi$

i.e. 2π radians $= 360°$ or $\boxed{\pi \text{ radians} = 180°}$

Thus 1 rad $= 180°/\pi = 57.30°$, correct to 2 decimal places.
Since π rad $= 180°$, then $\pi/2 = 90°$, $\pi/3 = 60°$, $\pi/4 = 45°$, and so on.

Figure 39.4

39.4   Worked problems on circles

> **Problem 1.**   Find the circumference of a circle of radius 12.0 cm

Circumference, $c = 2 \times \pi \times \text{radius}$
$$= 2\pi r = 2\pi(12.0) = \textbf{75.40 cm}$$

> **Problem 2.**   If the diameter of a circle is 75 mm, find its circumference

Circumference, $c = \pi \times$ diameter
$$= \pi d = \pi(75) = \textbf{235.6 mm}$$

> **Problem 3.**   Determine the radius of a circle if its perimeter is 112 cm

Perimeter = circumference, $c = 2\pi r$
Hence $r = \dfrac{c}{2\pi} = \dfrac{112}{2\pi} = \textbf{17.83 cm}$

> **Problem 4.**   In Fig. 39.5, AB is a tangent to the circle at B. If the circle radius is 40 mm and $AB = 150$ mm, calculate the length AO

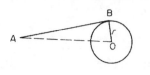

Figure 39.5

A tangent to a circle is at right angles to a radius drawn from the point of contact, i.e. $\angle ABO = 90°$. Hence, using Pythagoras' theorem:

$$AO^2 = AB^2 + OB^2$$
$$AO = \sqrt{(AB^2 + OB^2)} = \sqrt{[(150)^2 + (40)^2]}$$
$$= \textbf{155.2 mm}$$

> **Problem 5.**   Convert to radians: (a) 125° (b) 69°47′

(a)   Since $180° = \pi$ rad then $1° = \pi/180$ rad, therefore

$$125° = 125\left(\frac{\pi}{180}\right)^c = \textbf{2.182 radians}$$

(Note that ^{c} means 'circular measure' and indicates radian measure)

(b) $69°47′ = 69\dfrac{47°}{60} = 69.783°$

$$69.783° = 69.783\left(\frac{\pi}{180}\right)^c$$
$$= \textbf{1.218 radians}$$

> **Problem 6.**   Convert to degrees and minutes: (a) 0.749 radians (b) $3\pi/4$ radians

(a) Since π rad $= 180°$ then 1 rad $= 180°/\pi$, therefore

$$0.749 = 0.749\left(\frac{180}{\pi}\right)° = 42.915°$$

$0.915° = (0.915 \times 60)' = 55'$, correct to the nearest minute, hence

0.749 radians $= 42°55'$

(b) Since 1 rad $= \left(\dfrac{180}{\pi}\right)°$ then $\dfrac{3\pi}{4}$ rad $=$

$$\frac{3\pi}{4}\left(\frac{180}{\pi}\right)° = \frac{3}{4}(180)° = \mathbf{135°}$$

Problem 7.   Express in radians, in terms of π, (a) 45° (b) 60° (c) 90° (d) 150° (e) 270° (f) 37.5°

Since $180° = \pi$ rad then $1° = 180/\pi$, hence

(a) $45° = 45\left(\dfrac{\pi}{180}\right)$ rad $= \dfrac{\pi}{4}$ **rad**

(b) $60° = 60\left(\dfrac{\pi}{180}\right)$ rad $= \dfrac{\pi}{3}$ **rad**

(c) $90° = 90\left(\dfrac{\pi}{180}\right)$ rad $= \dfrac{\pi}{2}$ **rad**

(d) $150° = 150\left(\dfrac{\pi}{180}\right)$ rad $= \dfrac{5\pi}{6}$ **rad**

(e) $270° = 270\left(\dfrac{\pi}{180}\right)$ rad $= \dfrac{3\pi}{2}$ **rad**

(f) $37.5° = 37.5\left(\dfrac{\pi}{180}\right)$ rad $= \dfrac{75\pi}{360}$ rad

$$= \frac{5\pi}{24}\ \mathbf{rad}$$

Problem 8.   Find the length of arc of a circle of radius 5.5 cm when the angle subtended at the centre is 1.20 radians

---

Length of arc, $l = r\theta$, where θ is in radians, hence

$$l = (5.5)(1.20) = \mathbf{6.60\ cm}$$

Problem 9.   Determine the diameter and circumference of a circle if an arc of length 4.75 cm subtends an angle of 0.91 radians

---

Since $l = r\theta$ then $r = \dfrac{l}{\theta} = \dfrac{4.75}{0.91} = 5.22$ cm

Diameter $= 2 \times$ radius $= 2 \times 5.22 = \mathbf{10.44\ cm}$

Circumference, $c = \pi d = \pi(10.44) = \mathbf{32.80\ cm}$

Problem 10.   If an angle of 125° is subtended by an arc of a circle of radius 8.4 cm, find the length of (a) the minor arc, and (b) the major arc, correct to 3 significant figures

---

Since $180° = \pi$ rad then $1° = \left(\dfrac{\pi}{180}\right)$ rad

and $125° = 125\left(\dfrac{\pi}{180}\right)$ rad

Length of minor arc,

$$l = r\theta = (8.4)(125)\left(\frac{\pi}{180}\right)$$

$$= \mathbf{18.3\ cm},$$

correct to 3 significant figures

Length of major arc

$$= (\text{circumference} - \text{minor arc})$$

$$= 2\pi(8.4) - 18.3 = \mathbf{34.5\ cm},$$

correct to 3 significant figures

(Alternatively, major arc $= r\theta$

$$= 8.4(360 - 125)(\pi/180)$$

$$= 34.5\ \text{cm})$$

Problem 11.   Determine the angle, in degrees and minutes, subtended at the centre of a circle of diameter 42 mm by an arc of length 36 mm

---

Since length of arc, $l = r\theta$ then $\theta = l/r$

Radius, $r = \dfrac{\text{diameter}}{2} = \dfrac{42}{2} = 21$ mm

hence $\theta = \dfrac{l}{r} = \dfrac{36}{21} = 1.7143$ radians

1.7143 rad $= 1.7143 \times (180/\pi)° = 98.22° = \mathbf{98°13'}$

$= $ angle subtended at centre of circle

Problem 12. If an arc of length 11.48 cm subtends an angle of 168°27′ at the centre of a circle, find its radius correct to the nearest millimetre

$$168°27' = 168\frac{27°}{60} = 168.45°$$

$$= 168.45(\pi/180) \text{ radians},$$

hence $\theta = 2.94$ radians

Since arc length $l = r\theta$

then $r = \dfrac{l}{r} = \dfrac{11.48}{2.94} = 3.905$ cm $= 39.05$ mm

$= \mathbf{39\ mm}$ to the nearest millimetre

Further problems on circles may be found in the following Section 39.5, Problems 1 to 15.

39.5 Further problems on angular properties of circles

1 Calculate the length of the circumference of a circle of radius 7.2 cm. [45.24 cm]

2 If the diameter of a circle is 82.6 mm, calculate the circumference of the circle. [259.5 mm]

3 Determine the radius of a circle whose circumference is 16.52 cm. [2.629 cm]

4 Find the diameter of a circle whose perimeter is 149.8 cm. [47.68 cm]

5 Convert to radians in terms of π: (a) 30° (b) 75° (c) 225°

$$\left[\text{(a) } \frac{\pi}{6} \text{ (b) } \frac{5\pi}{12} \text{ (c) } \frac{5\pi}{4} \right]$$

6 Convert to radians: (a) 48° (b) 84°51′ (c) 232°15′

[(a) 0.838 (b) 1.481 (c) 4.054]

7 Convert to degrees: (a) $\dfrac{5\pi}{6}$ rad (b) $\dfrac{4\pi}{9}$ rad (c) $\dfrac{7\pi}{12}$ rad [(a) 150° (b) 80° (c) 105°]

8 Convert to degrees and minutes: (a) 0.0125 rad (b) 2.69 rad (c) 7.241 rad

[(a) 0°43′ (b) 154°8′ (c) 414°53′]

9 Find the length of an arc of a circle of radius 8.32 cm when the angle subtended at the centre is 2.14 radians. [17.80 cm]

10 If the angle subtended at the centre of a circle of diameter 82 mm is 1.46 rad, find the lengths of the (a) minor arc (b) major arc.

[(a) 59.86 mm (b) 197.8 mm]

11 A pendulum of length 1.5 m swings through an angle of 10° in a single swing. Find, in centimetres, the length of the arc traced by the pendulum bob. [26.2 cm]

12 Determine the length of the radius and circumference of a circle if an arc length of 32.6 cm subtends an angle of 3.76 radians.

[8.67 cm, 54.48 cm]

13 An arc subtends an angle of 96° at the centre of a circle of radius 125 mm. Find the length of the arc. [209.4 mm]

14 Determine the angle of lap, in degrees and minutes, if 180 mm of a belt drive are in contact with a pulley of diameter 250 mm.

[82°30′]

15 Determine the number of complete revolutions a motorcycle wheel will make in travelling 2 km, if the wheel's diameter is 85.1 cm.

[748]

Multiple choice questions on Part 3 Additional Mathematics for Engineering

All questions have only one correct answer

Suggested time allowed: 1 hour

1 A sinusoidal voltage is given by $v = R\sin(\omega t + \alpha)$ volts. Which of the following statements is incorrect?

A R is the average value of the voltage

B periodic time $= \dfrac{2\pi}{\omega}$ s

C frequency $= \dfrac{\omega}{2\pi}$ Hz

D $\omega =$ angular velocity

2 The solution of the differential equation $4\dfrac{dy}{dx} + x = 2$, given $y = 3$ when $x = 2$ is

A $y = 2x^2 + \dfrac{x^2}{2} + 1$

B $y = \dfrac{1}{2}x - \dfrac{x^2}{8} - \dfrac{5}{2}$

C $y = 2x - \dfrac{x^2}{2} - \dfrac{5}{2}$

D $y = \dfrac{1}{2}\left(x - \dfrac{x^2}{4} + 5\right)$

3 A pendulum of length 1.2 m swings through an angle of $12°$ in a single swing. The length of arc traced by the pendulum bob is

A 14.40 cm   B 25.13 cm

C 10.00 cm   D 45.24 cm

4 Two current phasors are shown in Fig. 1. If $i_1 = 5$ A and $i_2 = 12$ A, the resultant (i.e. length OP) is

A 15.93 A at $12.82°$ to i_1

B 9.17 A at $22.68°$ to i_1

C 15.93 A at $32.18°$ to i_1

D 9.17 A at $67.72°$ to i_1

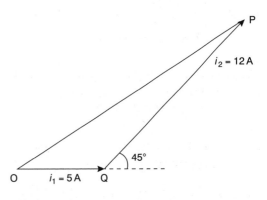

Figure 1

5 The angles between $0°$ and $360°$ whose tangent is -1.6341 are

A $121.46°$ and $301.46°$

B $58.54°$ and $238.54°$

C $121.46°$ and $238.54°$

D $211.46°$ and $301.46°$

6 The gradient of a curve is given by $3\dfrac{dy}{dx} = 6x - x^3$. The equation of the curve which passes through the point $\left(2, \dfrac{2}{3}\right)$ is

A $y = 2x - \dfrac{x^3}{3} - \dfrac{2}{3}$

B $y = x^2 - \dfrac{x^4}{12} - 2$

C $y = 2 - x^2 + \dfrac{8}{3}$

D $y = \dfrac{1}{12}(12x^2 - x^4 + 24)$

Questions 7 and 8 relate to the following information.

Accelerations of $a_1 = 3$ m/s at 90° and $a_2 = 4$ m/s at 150° act at a point.

7   $a_1 + a_2$ is given by
   A   2.05 m/s² at 76.94°
   B   6.08 m/s² at 55.28°
   C   2.05 m/s² at 13.06°
   D   6.08 m/s² at 124.72°

8   $a_1 - a_2$ is given by
   A   2.05 m/s² at 76.94°
   B   6.08 m/s² at 55.28°
   C   2.05 m/s² at 13.06°
   D   6.08 m/s² at 124.72°

9   Which of the following trigonometric identities is incorrect?
   A $\cos 2\theta = \cos^2 \theta + \sin^2 \theta$
   B $\cos 4\theta = 2\cos^2 2\theta - 1$
   C $\cos 4\theta = \cos^2 2\theta - \sin^2 2\theta$
   D $\cos 2\theta = 1 - 2\sin^2 \theta$

10   A wheel on a car has a diameter of 800 mm. If the car travels 5 miles, the number of complete revolutions the wheel makes $\left(\text{given } 1 \text{ km} = \dfrac{5}{8} \text{ mile}\right)$ is

   A   1989   B   1591
   C   3183   D   10 000

11   An alternating voltage v is given by $v = 150\sin\left(200\pi t + \dfrac{\pi}{3}\right)$ volts. When $v = 75$ volts, the time t is equal to

   A   0.046 s   B   −0.83 ms
   C   −1.046 s   D   2.50 ms

12   The solution of the differential equation $\dfrac{dy}{dx} = 2e^{x-y}$, given $x = 0$ when $y = 0$, is

   A $-\dfrac{1}{e^y} = 2e^x - 1$ B $e^y - 2e^x = 1$

   C $2e^x - \dfrac{1}{e^y} + 1 = 0$ D $e^y = 2e^x - 1$

13   Three forces of 2 N, 3 N and 4 N act as shown in Fig. 2. The magnitude of the resultant force is

   A   1 N   B   6.88 N
   C   8.26 N   D   9 N

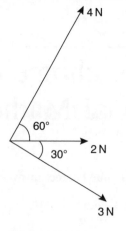

Figure 2

14   The acute angle arccot 2.468 is equal to
   A   22.06°   B   66.10°
   C   67.94°   D   23.90°

15   The current in an electric circuit is given by $E\dfrac{di}{dt} + Ri = 0$ where inductance L and resistance R are constants.

   Solving for i, given that $i = I$ when $t = 0$, gives

   A $i = I + e^{-Rt/L}$ B $i = Ie^{Lt/R}$

   C $i = Ie^{Rt/L}$ D $i = Ie^{-Rt/L}$

16   The magnitude of the resultant of velocities of 5.2 m/s at 20° and 8.5 m/s at 120° when acting simultaneously at a point is
   A   13.70 m/s   B   9.16 m/s
   C   3.30 m/s   D   44.20 m/s

17   Correct to 4 significant figures, the value of $\operatorname{cosec}(-130°)$ is
   A   −0.7660   B   −1.556
   C   −1.305   D   −0.6428

18   The displacement x metres of a mass from a fixed point about which it is oscillating is given by $x = 3\cos \omega t - 4\sin \omega t$, where t is the time in seconds. x may be expressed as

   A $5\sin(\omega t + 2.50)$ metres

   B $7\sin(\omega t - 36.87°)$ metres

   C $5\sin \omega t$ metres

   D $-\sin(\omega t - 2.50)$ metres

19   A force of 5 N is inclined at an angle of 50° to a second force of 12 N, both forces acting at a

Figure 3

point, as shown in Fig. 3. The magnitude of the resultant of these two forces and the direction of the resultant with respect to the 12 N force is

A   7 N at 50°   B   9.6 N at 23.6°

C   17 N at 130°   D   15.7 N at 14.1°

20   If $\tan\theta = 1.5291$, $\sec\theta$ is equal (correct to 4 decimal places) to

A   0.8369   B   1.1949

C   1.8271   D   0.6540

(Answers on page 454.)

Part 4

Extended Mathematics for Engineering

40

Functions and their curves

40.1 Standard curves

When a mathematical equation is known, coordinates may be calculated for a limited range of values, and the equation may be represented pictorially as a graph, within this range of calculated values. Sometimes it is useful to show all the characteristic features of an equation, and in this case a sketch depicting the equation can be drawn, in which all the important features are shown, but the accurate plotting of points is less important. This technique is called 'curve sketching' and can involve the use of differential calculus, with, for example, calculations involving turning points.

If, say, y depends on, say, x, then y is said to be a function of x and the relationship is expressed as $y = f(x)$; x is called the independent variable and y is the dependent variable.

In engineering and science, corresponding values are obtained as a result of tests or experiments.

Here is a brief resumé of standard curves, some of which have been met earlier in this text.

(i) **Straight line** (see Chapter 13, page 133)

The general equation of a straight line is $y = mx + c$, where m is the gradient $\left(\text{i.e. } \dfrac{dy}{dx}\right)$ and c is the y-axis intercept.

Two examples are shown in Fig. 40.1

(ii) **Quadratic graphs** (see Chapter 16, page 159)

The general equation of a quadratic graph is $y = ax^2 + bx + c$, and its shape is that of a parabola.

The simplest example of a quadratic graph, $y = x^2$, is shown in Fig. 40.2

(iii) **Cubic equations** (see Chapter 16, page 163)

The general equation of a cubic graph is $y = ax^3 + bx^2 + cx + d$.

The simplest example of a cubic graph, $y = x^3$, is shown in Fig. 40.3

(iv) **Trigonometric functions** (see Chapter 37, page 342)

Graphs of $y = \sin\theta$, $y = \cos\theta$ and $y = \tan\theta$ are shown in Fig. 40.4

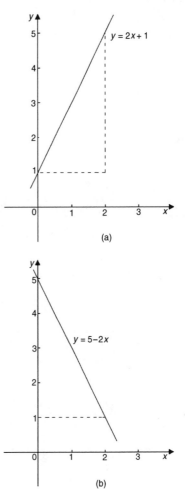

Figure 40.1

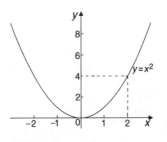

Figure 40.2

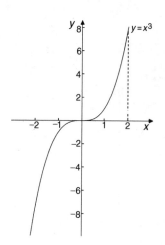

Figure 40.3

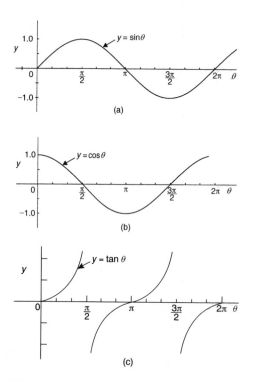

(a)

(b)

(c)

Figure 40.4

(v) **Circle**

The simplest equation of a circle is $x^2 + y^2 = r^2$, with centre at the origin and radius r, as shown in Fig. 40.5.

More generally, the equation of a circle, centre (a, b), radius r, is given by:

$$(x - a)^2 + (y - b)^2 = r^2 \qquad (1)$$

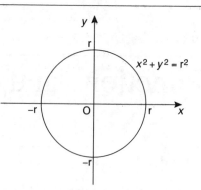

Figure 40.5

Figure 40.6 shows a circle

$$(x - 2)^2 + (y - 3)^2 = 4$$

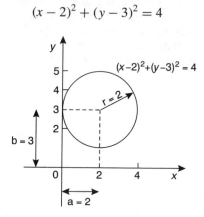

Figure 40.6

The general equation of a circle is:

$$x^2 + y^2 + 2ex + 2fy + c = 0 \qquad (2)$$

Multiplying out the bracketed terms in equation (1) gives:

$$x^2 - 2ax + a^2 + y^2 - 2by + b^2 = r^2$$

Comparing this with equation (2) gives:

$$2e = -2a, \text{ i.e. } a = -\frac{2e}{2}$$

$$2f = -2b, \text{ i.e. } b = -\frac{2f}{2}$$

and $\quad c = a^2 + b^2 - r^2,$

i.e. $\quad r = \sqrt{(a^2 + b^2 - c)}$

Thus, for example, the equation $x^2 + y^2 - 4x - 6y + 9 = 0$ represents a circle with centre

$a = -(-4/2)$, $b = -(-6/2)$, i.e. at $(2, 3)$
and radius $r = \sqrt{(2^2 + 3^2 - 9)} = 2$.
Hence $x^2 + y^2 - 4x - 6y + 9 = 0$ is the
circle shown in Fig. 40.6 (which may be
checked by multiplying out the brackets in
the equation $(x - 2)^2 + (y - 3)^2 = 4$)

(vi) **Ellipse**
The equation of an ellipse is

$$\frac{x^2}{a^2} + \frac{y^2}{b^2} = 1$$

and the general shape is as shown in
Fig. 40.7.

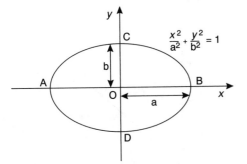

Figure 40.7

The length AB is called the major axis and
CD the minor axis. In the above equation,
'a' is the semi-major axis and 'b' is the semi-
minor axis.
(Note that if $b = a$, the equation becomes

$$\frac{x^2}{a^2} + \frac{y^2}{a^2} = 1$$

i.e. $x^2 + y^2 = a^2$, which is a circle of radius a)

(vii) **Hyperbola**
The equation of a hyperbola is

$$\frac{x^2}{a^2} - \frac{y^2}{b^2} = 1$$

and the general shape is shown in Fig. 40.8.
The curve is seen to be symmetrical about
both the x- and y-axes. The distance AB in
Fig. 40.8 is given by $2a$

(viii) **Rectangular hyperbola**
The equation of a rectangular hyperbola is

$$xy = c \text{ or } y = \frac{c}{x}$$ and the general shape is
shown in Fig. 40.9

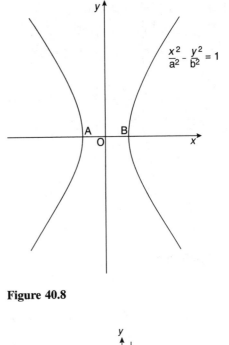

Figure 40.8

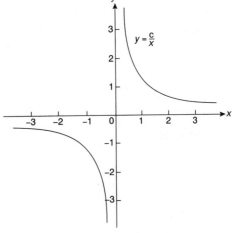

Figure 40.9

(ix) **Logarithmic function** (see Chapter 17,
page 168)
$y = \ln x$ and $y = \lg x$ are both of the general
shape shown in Fig. 40.10

(x) **Exponential functions** (see Chapter 17,
page 166)
$y = e^x$ is of the general shape shown in
Fig. 40.11

(xi) **Polar curves** (see Chapter 17, page 171)
The equation of a polar curve is of the form
$r = f(\theta)$. An example of a polar curve,
$r = a \sin \theta$, is shown in Fig. 40.12

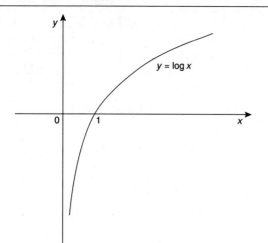

Figure 40.10

transformations of $y = f(x)$. For example, knowing the graph of $y = f(x)$, can help us draw the graphs of $y = af(x)$, $y = f(x) + a$, $y = f(x + a)$, $y = f(ax)$, $y = -f(x)$ and $y = f(-x)$.

(i) $y = af(x)$

For each point (x_1, y_1) on the graph of $y = f(x)$ there exists a point (x_1, ay_1) on the graph of $y = af(x)$. Thus the graph of $y = af(x)$ can be obtained by stretching $y = f(x)$ parallel to the y-axis by a scale factor 'a'.
Graphs of $y = x + 1$ and $y = 3(x + 1)$ are shown in Fig. 40.13(a) and graphs of $y = \sin\theta$ and $y = 2\sin\theta$ are shown in Fig. 40.13(b)

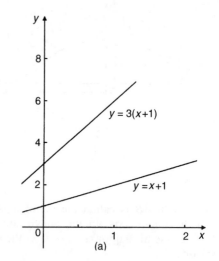

(a)

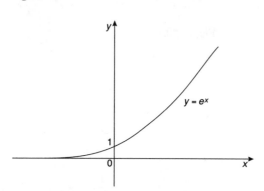

Figure 40.11

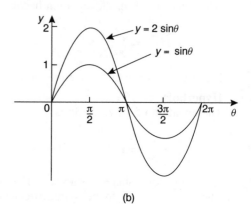

(b)

Figure 40.13

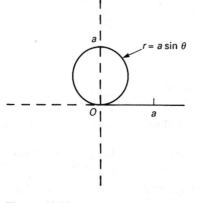

Figure 40.12

40.2   Simple transformations

From the graph of $y = f(x)$ it is possible to deduce the graphs of other functions which are

(ii) $y = f(x) + a$

The graph of $y = f(x)$ is translated by 'a' units parallel to the y-axis to obtain $y = f(x) + a$.

For example, if $f(x) = x$, $y = f(x) + 3$ becomes $y = x + 3$, as shown in Fig. 40.14(a). Similarly, if $f(\theta) = \cos\theta$, then $y = f(\theta) + 2$ becomes $y = \cos\theta + 2$, as shown in Fig. 40.14(b). Also, if $f(x) = x^2$, then $y = f(x) + 3$ becomes $y = x^2 + 3$, as shown in Fig. 40.14(c)

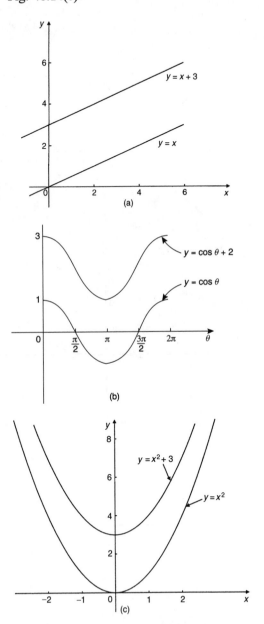

(a)

(b)

(c)

Figure 40.14

(iii) $y = f(x + a)$
The graph of $y = f(x)$ is translated by 'a' units parallel to the x-axis to obtain $y =$

$f(x + a)$. If 'a' > 0 it moves $y = f(x)$ in the negative direction on the x-axis (i.e. to the left), and if 'a' < 0 it moves $y = f(x)$ in the positive direction on the x-axis (i.e. to the right).

For example, if $f(x) = \sin x$, $y = f\left(x - \dfrac{\pi}{3}\right)$ becomes $y = \sin\left(x - \dfrac{\pi}{3}\right)$ as shown in Fig. 40.15(a) and $y = \sin\left(x + \dfrac{\pi}{4}\right)$ is shown in Fig. 40.15(b).

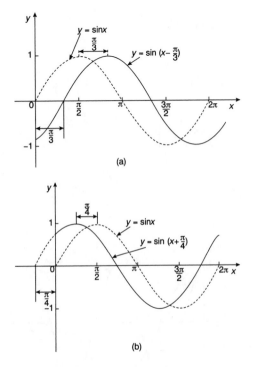

(a)

(b)

Figure 40.15

Similarly graphs of $y = x^2$, $y = (x - 1)^2$ and $y = (x + 2)^2$ are shown in Fig. 40.16

(iv) $y = f(ax)$
For each point (x_1, y_1) on the graph of $y = f(x)$, there exists a point $\left(\dfrac{x_1}{a}, y_1\right)$ on the graph of $y = f(ax)$. Thus the graph of $y = f(ax)$ can be obtained by stretching $y = f(x)$ parallel to the x-axis by a scale factor $\dfrac{1}{a}$.

For example, if $f(x) = (x - 1)^2$, and $a = \frac{1}{2}$, then $f(ax) = \left(\dfrac{x}{2} - 1\right)^2$.

Both of these curves are shown in Fig. 40.17(a).

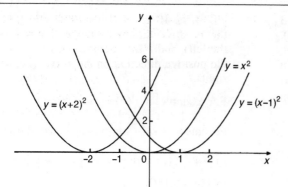

Figure 40.16

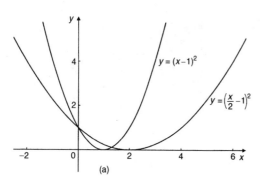

(a)

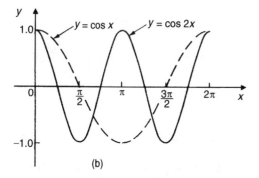

(b)

Figure 40.17

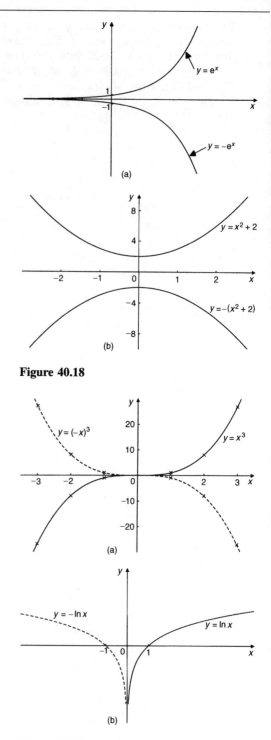

Figure 40.18

(a)

(b)

Figure 40.19

Similarly, $y = \cos x$ and $y = \cos 2x$ are shown in Fig. 40.17(b)

(v) $y = -f(x)$

The graph of $y = -f(x)$ is obtained by reflecting $y = f(x)$ in the x-axis. For example, graphs of $y = e^x$ and $y = -e^x$ are shown in Fig. 40.18(a), and graphs of $y = x^2 + 2$ and $y = -(x^2 + 2)$ are shown in Fig. 40.18(b)

(vi) $y = f(-x)$

The graph of $y = f(-x)$ is obtained by reflecting $y = f(x)$ in the y-axis. For example,

graphs of $y = x^3$ and $y = (-x)^3 = -x^3$ are shown in Fig. 40.19(a) and graphs of $y = \ln x$ and $y = -\ln x$ are shown in Fig. 40.19(b)

Problem 1.   Sketch the following graphs, showing relevant points:
(a) $y = (x - 4)^2$ (b) $y = x^3 - 8$

(a)   In Fig. 40.20 a graph of $y = x^2$ is shown by the broken line. The graph of $y = (x - 4)^2$ is of the form $y = f(x + a)$. Since $a = -4$, then $y = (x - 4)^2$ is translated 4 units to the right of $y = x^2$, parallel to the x-axis.
(See (iii) above)

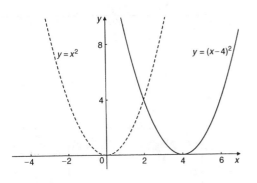

Figure 40.20

(b)   In Fig. 40.21 a graph of $y = x^3$ is shown by the broken line. The graph of $y = x^3 - 8$ is of the form $y = f(x) + a$. Since $a = -8$, then $y = x^3 - 8$ is translated 8 units down from $y = x^3$, parallel to the y-axis.
(See (ii) above)

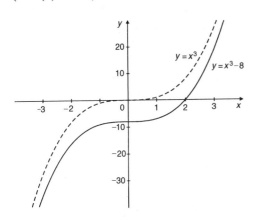

Figure 40.21

Problem 2.   Sketch the following graphs, showing relevant points:
(a) $y = 5 - (x + 2)^3$ (b) $y = 1 + 3 \sin 2x$

(a)   Figure 40.22(a) shows a graph of $y = x^3$. Figure 40.22(b) shows a graph of $y = (x + 2)^3$ (see $f(x + a)$, (iii) above). Figure 40.22(c) shows a graph of $y = -(x + 2)^3$ (see $-f(x)$, (v) above). Figure 40.22(d) shows the graph of $y = 5 - (x + 2)^3$ (see $f(x) + a$, (ii) above)

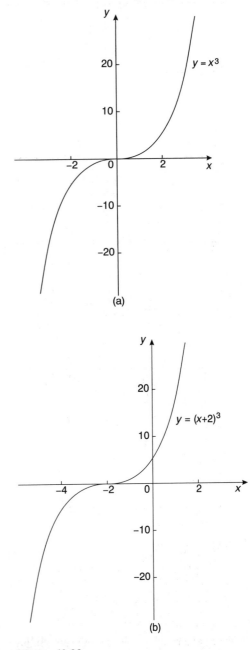

Figure 40.22

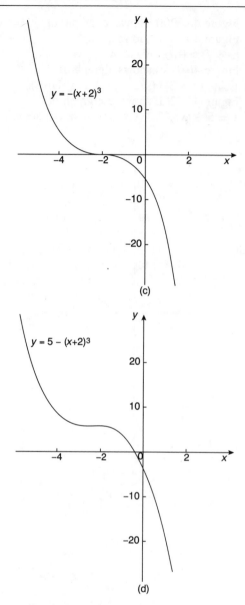

(c)

(d)

Figure 40.22 (*continued*)

(b) Figure 40.23(a) shows a graph of $y = \sin x$.
Figure 40.23(b) shows a graph of $y = \sin 2x$
(see $f(ax)$, (iv) above).
Figure 40.23(c) shows a graph of $y = 3 \sin 2x$
(see $af(x)$, (i) above).
Figure 40.23(d) shows a graph of $y = 1 + 3 \sin 2x$ (see $f(x) + a$, (ii) above)

Further problems on simple transformations with curve sketching may be found in Section 40.9, Problems 1 to 10, page 395.

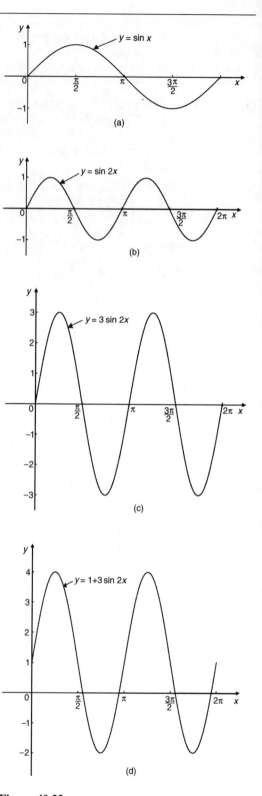

Figure 40.23

40.3 Periodic functions

A function $f(x)$ is said to be **periodic** if $f(x+T) = f(x)$ for all values of x, where T is some positive number. T is the interval between two successive repetitions and is called the **period** of the function $f(x)$. For example, $y = \sin x$ is periodic in x with period 2π since $\sin x = \sin(x + 2\pi) = \sin(x + 4\pi)$, and so on. Similarly, $y = \cos x$ is a periodic function with period 2π since $\cos x = \cos(x+2\pi) = \cos(x + 4\pi)$, and so on. In general, if $y = \sin \omega t$ or $y = \cos \omega t$ then the period of the waveform is $2\pi/\omega$. The function shown in Fig. 40.24 is also periodic of period 2π and is defined by:

$$f(x) = \begin{cases} -1, & \text{when } -\pi < x < 0 \\ 1, & \text{when } 0 < x < \pi \end{cases}$$

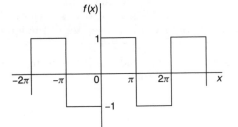

Figure 40.24

40.4 Continuous and discontinuous functions

If a graph of a function has no sudden jumps or breaks it is called a **continuous function**, examples being the graphs of sine and cosine functions. However, other graphs make finite jumps at a point or points in the interval. The square wave shown in Fig. 40.24 has **finite discontinuities** as $x = \pi$, 2π, 3π, and so on, and is therefore a discontinuous function. $y = \tan x$ is another example of a discontinuous function.

When integrating discontinuous functions over a range in which it is discontinuous, integration is done a part at a time.

For example, a function $f(x)$ is defined as

$$f(x) = \begin{cases} -1 \text{ when } -\pi < 0 < 0 \\ 3 \text{ when } 0 < x < \pi \end{cases}$$

To determine $1/2\pi \int_{-\pi}^{\pi} f(x)\,dx$, which is involved in Fourier series, the integration is determined in two parts, i.e.

$$\frac{1}{2\pi} \int_{-\pi}^{\pi} f(x)\,dx = \frac{1}{2\pi} \left\{ \int_{-\pi}^{0} -1\,dx + \int_{0}^{\pi} 3\,dx \right\}$$

$$= \frac{1}{2\pi} \left\{ [-x]_{-\pi}^{0} + [3x]_{0}^{\pi} \right\}$$

$$= \frac{1}{2\pi} \{(0 - \pi) + (3\pi - 0)\} = 1$$

40.5 Even and odd functions

Even functions

A function $y = f(x)$ is said to be even if $f(-x) = f(x)$ for all values of x. Graphs of even functions are always **symmetrical about the y-axis** (i.e. is a mirror image). Two examples of even functions are $y = x^2$ and $y = \cos x$ as shown in Fig. 40.25.

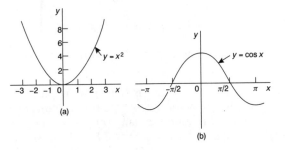

Figure 40.25

Odd functions

A function $y = f(x)$ is said to be **odd** if $f(-x) = -f(x)$ for all values of x. Graphs of odd functions are always **symmetrical about the origin**. Two examples of odd functions are $y = x^3$ and $y = \sin x$ as shown in Fig. 40.26.

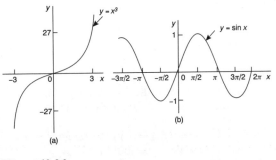

Figure 40.26

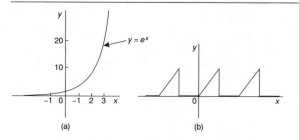

(a)

(b)

Figure 40.27

Many functions are neither even nor odd, two such examples being shown in Fig. 40.27.

Problem 3.   Sketch the following functions and state whether they are even or odd functions: (a) $y = \tan x$

$$\text{(b)} \quad f(x) = \begin{cases} 2, & \text{when } 0 < x < \dfrac{\pi}{2} \\ -2, & \text{when } \dfrac{\pi}{2} < x < \dfrac{3\pi}{2} \\ 2, & \text{when } \dfrac{3\pi}{2} < x < 2\pi \end{cases},$$

and is periodic of period 2π

(a)   A graph of $y = \tan x$ is shown in Fig. 40.28(a) and is symmetrical about the origin and is thus an **odd function** (i.e. $\tan(-x) = -\tan x$)

(b)   A graph of $f(x)$ is shown in Fig. 40.28(b) and is symmetrical about the $f(x)$ axis hence the function is an **even** one, $(f(-x) = f(x))$

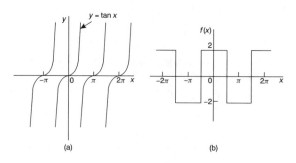

(a)

(b)

Figure 40.28

Problem 4.   Sketch the following graphs and state whether the functions are even, odd or neither even nor odd:
(a) $y = \ln x$ (b) $f(x) = x$ in the range $-\pi$ to π and is periodic of period 2π

(a)   A graph of $y = \ln x$ is shown in Fig. 40.29(a) and the curve is neither symmetrical about the y-axis nor symmetrical about the origin and is thus **neither even nor odd**

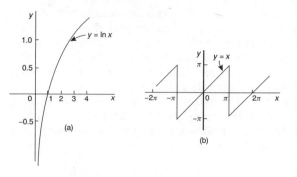

Figure 40.29

(b)   A graph of $y = x$ in the range $-\pi$ to π is shown in Fig. 40.29(b) and is symmetrical about the origin and is thus an **odd function**

Further problems on even and odd functions may be found in Section 40.9, Problems 11 to 13, pages 395 and 396.

40.6   Inverse functions

If y is a function of x, the graph of y against x can be used to find x when any value of y is given. Thus the graph also expresses that x is a function of y. Two such functions are called **inverse functions**.

In general, given a function $y = f(x)$, its inverse may be obtained by interchanging the roles of x and y and then transposing for y. The inverse function is denoted by $y = f^{-1}(x)$.

For example, if $y = 2x + 1$, the inverse is obtained by

(i)   transposing for x, i.e. $x = \dfrac{y-1}{2} = \dfrac{y}{2} - \dfrac{1}{2}$

and (ii)   interchanging x and y, giving the inverse

as $y = \dfrac{x}{2} - \dfrac{1}{2}$

Thus if $f(x) = 2x + 1$, then $f^{-1}(x) = \dfrac{x}{2} - \dfrac{1}{2}$.

A graph of $f(x) = 2x+1$ and its inverse $f^{-1}(x) = \dfrac{x}{2} - \dfrac{1}{2}$ is shown in Fig. 40.30 and $f^{-1}(x)$ is seen to be a reflection of $f(x)$ in the line $y = x$.

Similarly, if $y = x^2$, the inverse is obtained by

(i)   transposing for x, i.e. $x = \pm\sqrt{y}$

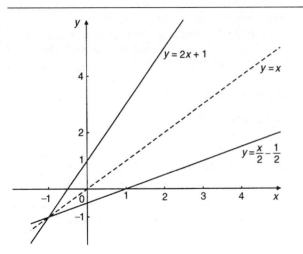

Figure 40.30

and (ii)   interchanging x and y, giving the inverse
$$y = \pm\sqrt{x}$$

Hence the inverse has two values for every value of x. Thus $f(x) = x^2$ does not have an inverse. In such a case the domain of the original function may be restricted to $y = x^2$ for $x \geq 0$. Thus the inverse is then $y = +\sqrt{x}$. A graph of $f(x) = x^2$ and its inverse $f^{-1}(x) = \sqrt{x}$ for $x > 0$ is shown in Fig. 40.31 and, again, $f^{-1}(x)$ is seen to be a reflection of $f(x)$ in the line $y = x$.

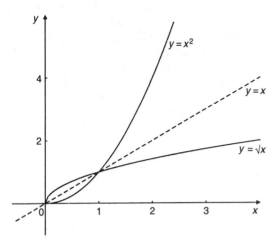

Figure 40.31

It is noted from the latter example, that not all functions have an inverse. An inverse, however, can be determined if the range is restricted.

Problem 5.   Determine the inverse for each of the following functions:
(a) $f(x) = x - 1$ (b) $f(x) = x^2 - 4$ $(x \geq 0)$
(c) $f(x) = x^2 + 1$

(a)   If $y = f(x)$, then $y = x - 1$
Transposing for x gives $x = y + 1$
Interchanging x and y gives $y = x + 1$
Hence if $f(x) = x - 1$, then
$$f^{-1}(x) = x + 1$$

(b)   If $y = f(x)$, then $y = x^2 - 4$ $(x \geq 0)$
Transposing for x gives $x = \sqrt{(y + 4)}$
Interchanging x and y gives $y = \sqrt{(x + 4)}$
Hence if $f(x) = x^2 - 4(x \geq 0)$ then
$$f^{-1}(x) = \sqrt{(x + 4)} \text{ if } x \geq -4$$

(c)   If $y = f(x)$, then $y = x^2 + 1$
Transposing for x gives $x = \sqrt{(y - 1)}$
Interchanging x and y gives $y = \sqrt{(x - 1)}$,
which has two values. **Hence there is no inverse of $f(x) = x^2 + 1$**, since the domain of $f(x)$ is not restricted

Inverse trigonometric functions

If $y = \sin x$, then x is the angle whose sine is y. Inverse trigonometric functions are denoted by prefixing the function with 'arc'. Hence transposing $y = \sin x$ for x gives $x = \arcsin y$. Interchanging x and y gives the inverse $y = \arcsin x$.

Similarly, $y = \arccos x$, $y = \arctan x$, $y = \text{arcsec } x$, $y = \text{arccosec } x$ and $y = \text{arccot } x$ are all inverse trigonometric functions. The angle is always expressed in radians.

Inverse trigonometric functions are periodic so it is necessary to specify the smallest or **principal** value of the angle. For $\arcsin x$, $\arctan x$, $\text{arccosec } x$ and $\text{arccot } x$, the principal value is in the range $-\dfrac{\pi}{2} \leq y \leq \dfrac{\pi}{2}$. For $\arccos x$ and $\text{arcsec } x$ the principal value is in the range $0 \leq y \leq \pi$.

Graphs of the six inverse trigonometric functions are shown in Fig. 40.32.

Problem 6.   Determine the principal values of
(a) $\arcsin 0.5$ (b) $\arctan(-1)$

(c) $\arccos\left(-\dfrac{\sqrt{3}}{2}\right)$ (d) $\text{arccosec}(\sqrt{2})$

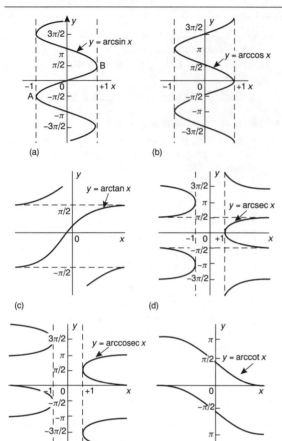

Figure 40.32

Using a calculator,

(a) $\arcsin 0.5 \equiv \sin^{-1} 0.5 = 30°$

$$= \frac{\pi}{6} \text{ rad or } \mathbf{0.5236 \text{ rad}}$$

(b) $\arctan(-1) \equiv \tan^{-1}(-1) = -45°$

$$= -\frac{\pi}{4} \text{ rad or } \mathbf{-0.7854 \text{ rad}}$$

(c) $\arccos\left(-\dfrac{\sqrt{3}}{2}\right) \equiv \cos^{-1}\left(-\dfrac{\sqrt{3}}{2}\right) = 150°$

$$= \frac{5\pi}{6} \text{ rad or } \mathbf{2.6180 \text{ rad}}$$

(d) $\text{arccosec}(\sqrt{2}) = \arcsin\left(\dfrac{1}{\sqrt{2}}\right) \equiv \sin^{-1}\left(\dfrac{1}{\sqrt{2}}\right)$

$$= 45° = \frac{\pi}{4} \text{ rad or } \mathbf{0.7854 \text{ rad}}$$

Problem 7. Evaluate (in radians), correct to 3 decimal places: $\arcsin 0.30 + \arccos 0.65$

$\arcsin 0.30 = 17.4576° = 0.3047$ rad

$\arccos 0.65 = 49.4584° = 0.8632$ rad

Hence $\arcsin 0.30 + \arccos 0.65 = 0.3047 + 0.8632 = \mathbf{1.168}$, correct to 3 decimal places.

Further problems on inverse functions may be found in Section 40.9, Problems 14 to 26, page 396.

40.7   Brief guide to curve sketching

The following steps will give information from which the graphs of many types of functions $y = f(x)$ can be sketched.

(i)   Use calculus to determine the location and nature of maximum and minimum points (see Chapter 22)

(ii)   Determine where the curve cuts the x- and y-axes

(iii)   Inspect the equation for symmetry.
    (a)   If the equation is unchanged when $-x$ is substituted for x, the graph will be symmetrical about the y-axis (i.e. it is an **even function**)
    (b)   If the equation is unchanged when $-y$ is substituted for y, the graph will be symmetrical about the x-axis
    (c)   If $f(-x) = -f(x)$, the graph is symmetrical about the origin (i.e. it is an **odd function**)

40.8   Worked problems on curve sketching

Problem 8.   Sketch the graphs of
(a) $y = 2x^2 + 12x + 20$
(b) $y = -3x^2 + 12x - 15$

(a) $y = 2x^2 + 12x + 20$ is a parabola since the equation is a quadratic. To determine the turning point:

Gradient $= \dfrac{dy}{dx} = 4x + 12 = 0$ for a turning point

Hence $4x = -12$ and $x = -3$

When $x = -3$, $y = 2(-3)^2 + 12(-3) + 20 = 2$

Hence $(-3, 2)$ are the coordinates of the turning point

$\dfrac{d^2y}{dx^2} = 4$, which is positive, hence $(-3, 2)$ is a minimum point

When $x = 0$, $y = 20$, hence the curve cuts the y-axis at $y = 20$

Thus knowing the curve passes through $(-3, 2)$ and $(0, 20)$ and appreciating the general shape of a parabola results in the sketch given in Fig. 40.33

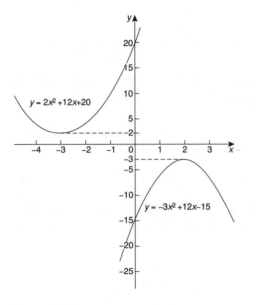

Figure 40.33

(b) $y = -3x^2 + 12x - 15$ is also a parabola (but 'upside down' due to the minus sign in front of the x^2 term)

Gradient $= \dfrac{dy}{dx} = -6x + 12 = 0$ for a turning point

Hence $6x = 12$ and $x = 2$

When $x = 2$, $y = -3(2)^2 + 12(2) - 15 = -3$

Hence $(2, -3)$ are the coordinates of the turning point

$\dfrac{d^2y}{dx^2} = -6$, which is negative, hence $(2, -3)$ is a maximum point

When $x = 0$, $y = -15$, hence the curve cuts the axis at $y = -15$

The curve is shown sketched in Fig. 40.33

Problem 9.   Sketch the curves depicting the following equations:
(a) $x = \sqrt{(9 - y^2)}$ (b) $y^2 = 16x$ (c) $xy = 5$

(a)   Squaring both sides of the equation and transposing gives $x^2 + y^2 = 9$. Comparing this with the standard equation of a circle, centre origin and radius a, i.e. $x^2 + y^2 = a^2$, shows that $x^2 + y^2 = 9$ represents a circle, centre origin and radius 3. A sketch of this circle is shown in Fig. 40.34(a)

(b)   The equation $y^2 = 16x$ is symmetrical about the x-axis and has its vertex at the origin $(0, 0)$. Also, when $x = 1$, $y = \pm 4$. A sketch of this parabola is shown in Fig. 40.34(b)

(c)   The equation $y = \dfrac{a}{x}$ represents a rectangular hyperbola lying entirely within the first and third quadrants. Transposing $xy = 5$ gives $y = \dfrac{5}{x}$, and therefore represents the rectangular hyperbola shown in Fig. 40.34(c)

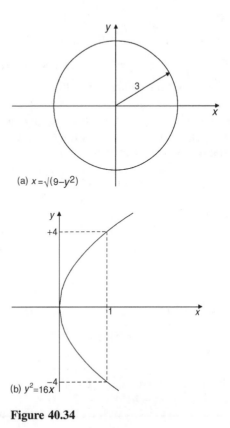

(a) $x = \sqrt{(9 - y^2)}$

(b) $y^2 = 16x$

Figure 40.34

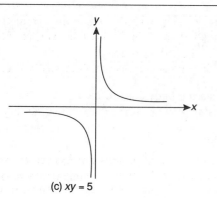

(c) $xy = 5$

Figure 40.34   (*continued*)

Problem 10.   Sketch the curves depicting the following equations:
(a) $4x^2 = 36 - 9y^2$ (b) $3y^2 + 15 = 5x^2$

(a)   By dividing throughout by 36 and transposing the equation $4x^2 = 36 - 9y^2$ can be written as $\dfrac{x^2}{9} + \dfrac{y^2}{4} = 1$. The equation of an ellipse is of the form $\dfrac{x^2}{a^2} + \dfrac{y^2}{b^2} = 1$, where $2a$ and $2b$ represent the length of the axes of the ellipse. Thus $\dfrac{x^2}{(3)^2} + \dfrac{y^2}{(2)^2} = 1$ represents an ellipse, having its axes coinciding with the x- and y-axes of a rectangular coordinate system, the major axis being 2(3), i.e. 6 units long and the minor axis 2(2), i.e. 4 units long, as shown in Fig. 40.35(a)

(b)   Dividing $3y^2 + 15 = 5x^2$ throughout by 15 and transposing gives $\dfrac{x^2}{3} - \dfrac{y^2}{5} = 1$. The equation $\dfrac{x^2}{a^2} - \dfrac{y^2}{b^2} = 1$ represents a hyperbola which is symmetrical about both the x- and y-axes, the distance between the vertices being given by $2a$.

Thus a sketch of $\dfrac{x^2}{3} - \dfrac{y^2}{5} = 1$ is as shown in Fig. 40.35(b), having a distance of $2\sqrt{3}$ between its vertices

Problem 11.   Sketch the circle given by the equation $x^2 + y^2 - 4x + 6y - 3 = 0$

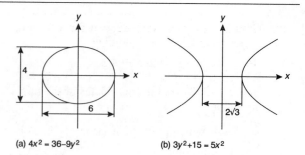

(a) $4x^2 = 36 - 9y^2$ (b) $3y^2 + 15 = 5x^2$

Figure 40.35

The equation of a circle, centre (a, b), radius r is given by

$$(x - a)^2 + (y - b)^2 = r^2$$

The general equation of a circle is $x^2 + y^2 + 2ex + 2fy + c = 0$ and from Section 40.1(v), $a = -\dfrac{2e}{2}, b = -\dfrac{2f}{2}$ and $r = \sqrt{(a^2 + b^2 - c)}$

Hence if $x^2 + y^2 - 4x + 6y - 3 = 0$

then $a = -\left(\dfrac{-4}{2}\right) = 2, b = -\left(\dfrac{6}{2}\right) = -3$

and $r = \sqrt{[(2)^2 + (-3)^2 - (-3)]} = \sqrt{16} = 4$

Thus the circle has centre $(2, -3)$ and radius 4, as shown in Fig. 40.36.

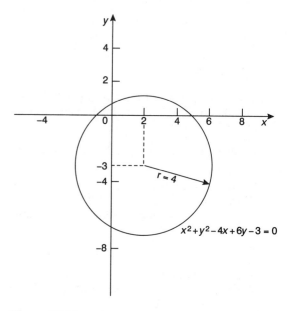

$x^2 + y^2 - 4x + 6y - 3 = 0$

Figure 40.36

Problem 12.   Describe the shape of the curves represented by the following

equations: (a) $x = 2\sqrt{\left[1 - \left(\dfrac{y}{2}\right)^2\right]}$

(b) $\dfrac{y^2}{8} = 2x$ (c) $y = 6\left(1 - \dfrac{x^2}{16}\right)^{1/2}$

(a)   Squaring the equation gives $x^2 = 4\left[1 - \left(\dfrac{y}{2}\right)^2\right]$

and transposing gives $x^2 = 4 - y^2$, i.e. $x^2 + y^2 = 4$. Comparing this equation with $x^2 + y^2 = a^2$ shows that $x^2 + y^2 = 4$ is the equation of a **circle** having centre at the origin $(0, 0)$ and of radius 2 units

(b)   Transposing $\dfrac{y^2}{8} = 2x$ gives $y = 4\sqrt{x}$. Thus

$\dfrac{y^2}{8} = 2x$ is the equation of a **parabola** having its axis of symmetry coinciding with the x-axis and its vertex at the origin of a rectangular coordinate system

(c) $y = 6\left(1 - \dfrac{x^2}{16}\right)^{1/2}$ can be transposed to $\dfrac{y}{6} = \left(1 - \dfrac{x}{16}\right)^{1/2}$ and squaring both sides gives

$\dfrac{y^2}{36} = 1 - \dfrac{x^2}{16}$, i.e. $\dfrac{x^2}{16} + \dfrac{y^2}{36} = 1$.

This is the equation of an **ellipse**, centre at the origin of a rectangular coordinate system, the major axis coinciding with the y-axis and being $2\sqrt{36}$, i.e. 12 units long. The minor axis coincides with the x-axis and is $2\sqrt{16}$, i.e. 8 units long

Problem 13.   Describe the shape of the curves represented by the following equations:

(a) $\dfrac{x}{5} = \sqrt{\left[1 + \left(\dfrac{y}{2}\right)^2\right]}$ (b) $\dfrac{y}{4} = \dfrac{15}{2x}$

(a)   Since $\dfrac{x}{5} = \sqrt{\left[1 + \left(\dfrac{y}{2}\right)^2\right]}$

$\dfrac{x^2}{25} = 1 + \left(\dfrac{y}{2}\right)^2$

i.e. $\dfrac{x^2}{25} - \dfrac{y^2}{4} = 1$

This is a **hyperbola** which is symmetrical about both the x- and y-axes, the vertices being $2\sqrt{25}$, i.e. 10 units apart.
(With reference to Section 40.1(vii), a is equal to ± 5)

(b)   The equation $\dfrac{y}{4} = \dfrac{15}{2x}$ is of the form $y = \dfrac{a}{x}$,

where $a = \dfrac{60}{2} = 30$.

This represents a **rectangular hyperbola**, symmetrical about both the x- and y-axes, and lying entirely in the first and third quadrants, similar in shape to the curves shown in Fig. 40.9

Further problems on curve sketching may be found in the following Section 40.9, Problems 27 to 45, pages 396 and 397.

40.9   Further problems on functions and their curves

Simple transformations

In Problems 1 to 10, sketch the graphs, showing relevant points: (Answers on pages 397 and 398)

1   $y = 3x - 5$
2   $y = -3x + 4$
3   $y = x^2 + 3$
4   $y = (x - 3)^2$
5   $y = (x - 4)^2 + 2$
6   $y = x - x^2$
7   $y = x^3 + 2$
8   $y = 1 + 2\cos 3x$
9   $y = 3 - 2\sin\left(x + \dfrac{\pi}{4}\right)$
10   $y = 2\ln x$

Even and odd functions

In Problems 11 and 12 determine whether the given functions are even, odd or neither even nor odd:

11   (a) x^4 (b) $\tan 3x$ (c) $2e^{3t}$ (d) $\sin^2 x$

[(a) even (b) odd (c) neither (d) even]

12   (a) $5t^3$ (b) $e^x + e^{-x}$ (c) $\dfrac{\cos\theta}{\theta}$ (d) e^{x^2}

[(a) odd (b) even (c) odd (d) even]

13   State whether the following functions which are periodic of period 2π are even or odd:

(a) $f(\theta) = \begin{cases} \theta, & \text{when } -\pi < \theta < 0 \\ -\theta, & \text{when } 0 < \theta < \pi \end{cases}$

(b) $f(x) = \begin{cases} x, & \text{when } -\dfrac{\pi}{2} < x < \dfrac{\pi}{2} \\ 0, & \text{when } \dfrac{\pi}{2} < x < \dfrac{3\pi}{2} \end{cases}$

[(a) even (b) odd]

Inverse functions

Determine the inverse of the functions given in Problems 14 to 17:

14    $f(x) = x + 1$ $[f^{-1}(x) = x - 1]$

15    $f(x) = 5x - 1$ $\left[f^{-1}(x) = \dfrac{1}{5}(x + 1)\right]$

16    $f(x) = x^3 + 1$ $[f^{-1}(x) = \sqrt[3]{(x - 1)}]$

17    $f(x) = \dfrac{1}{x} + 2$ $\left[f^{-1}(x) = \dfrac{1}{x - 2}\right]$

Determine the principal value of the inverse functions in Problems 18 to 24:

18   arcsin(-1) $\left[-\dfrac{\pi}{2} \text{ or } -1.5708 \text{ rad}\right]$

19   arccos 0.5 $\left[\dfrac{\pi}{3} \text{ or } 1.0472 \text{ rad}\right]$

20   arctan 1 $\left[\dfrac{\pi}{4} \text{ or } 0.7854 \text{ rad}\right]$

21   arccot 2   [0.4636 rad]

22   arccosec 2.5   [0.4115 rad]

23   arcsec 1.5   [0.8411 rad]

24   arcsin $\left(\dfrac{1}{\sqrt{2}}\right)$ $\left[\dfrac{\pi}{4} \text{ or } 0.7854 \text{ rad}\right]$

25   Evaluate x, correct to 3 decimal places:

$x = \arcsin \dfrac{1}{3} + \arccos \dfrac{4}{5} - \arctan \dfrac{8}{9}$ [0.257]

26   Evaluate y, correct to 4 significant figures:
$y = 3 \operatorname{arcsec} \sqrt{2} - 4 \operatorname{arccosec} \sqrt{2} + 5 \operatorname{arccot} 2$
[1.533]

Curve sketching

27   Sketch the graphs of

(a) $y = 3x^2 + 9x + \dfrac{7}{4}$

(b) $y = -5x^2 + 20x + 50$

$\left[\begin{array}{l}\text{(a) Parabola with minimum value at} \\ \left(-\dfrac{3}{2}, -5\right) \text{ and passing through } \left(0, 1\dfrac{3}{4}\right) \\ \text{(b) Parabola with minimum value at} \\ (2, 70) \text{ and passing through } (0, 50)\end{array}\right]$

28   Sketch the graphs of
(a) $y + 3x^2 = 1 - 6x$
(b) $y - 19 = 3x^2 - 12x$

$\left[\begin{array}{l}\text{(a) Parabola with maximum value at} \\ (-1, 4) \text{ and passing through } (0, 1) \\ \text{(b) Parabola with minimum value at} \\ (2, 7) \text{ and passing through } (0, 19)\end{array}\right]$

In Problems 29 to 36, sketch the curves depicting the equations given:

29    $x = 6\sqrt{\left[1 - \left(\dfrac{y}{6}\right)^2\right]}$

[circle, centre $(0, 0)$, radius 6 units]

30    $\dfrac{y^2}{3x} = 5$ $\left[\begin{array}{l}\text{parabola, symmetrical about} \\ \text{x-axis, vertex at } (0, 0)\end{array}\right]$

31    $\sqrt{x} = \dfrac{y}{9}$ $\left[\begin{array}{l}\text{parabola, symmetrical about} \\ \text{x-axis, vertex at } (0, 0)\end{array}\right]$

32    $y^2 = \dfrac{x^2 - 16}{4}$

$\left[\begin{array}{l}\text{hyperbola, symmetrical about x- and} \\ \text{y-axes, distance between vertices} \\ \text{8 units along x-axis}\end{array}\right]$

33    $\dfrac{y^2}{5} = 5 - \dfrac{x^2}{2}$

$\left[\begin{array}{l}\text{ellipse, centre } (0, 0), \text{ major axis} \\ \text{10 units along y-axis, minor} \\ \text{axis } 2\sqrt{10} \text{ units along x-axis}\end{array}\right]$

34    $x = \dfrac{1}{3}\sqrt{(36 - 18y^2)}$

$\left[\begin{array}{l}\text{ellipse, centre } (0, 0), \text{ major axis} \\ \text{4 units along x-axis, minor axis} \\ 2\sqrt{2} \text{ units along y-axis}\end{array}\right]$

35    $x = 3\sqrt{(1 + y^2)}$

$\left[\begin{array}{l}\text{hyperbola, symmetrical about} \\ \text{x- and y-axes, distance between} \\ \text{vertices 6 units along x-axis}\end{array}\right]$

36    $x^2 y^2 = 9$

$\left[\begin{array}{l}\text{rectangular hyperbola, lying in} \\ \text{first and third quadrants only}\end{array}\right]$

37   Determine (a) the radius, and (b) the coordinates of the centre of the circle given by the

equation $x^2 + y^2 + 8x - 2y + 8 = 0$.

[(a) 3 (b) (−4, 1)]

38 Sketch the circle given by the equation
$x^2 + y^2 - 6x + 4y + 3 = 0$.

[Centre at (3, −2), radius 4]

In Problems 39 to 45 describe the shape of the curves represented by the equations given:

39 $y = \sqrt{[3(1 - x^2)]}$

⎡ellipse, centre (0, 0), major axis⎤
⎢$2\sqrt{3}$ units along y-axis, minor⎥
⎣axis 2 units along x-axis   ⎦

40 $y = \sqrt{[3(x^2 - 1)]}$

⎡hyperbola, symmetrical about x-⎤
⎢and y-axes, vertices 2 units apart⎥
⎣along x-axis   ⎦

41 $y = \sqrt{(9 - x^2)}$

[circle, centre (0, 0), radius 3 units]

42 $y = 7x^{-1}$

⎡rectangular hyperbola, lying in first⎤
⎢and third quadrants, symmetrical   ⎥
⎣about x- and y-axes   ⎦

43 $y = (3x)^{1/2}$

⎡parabola, vertex at (0, 0),   ⎤
⎣symmetrical about the x-axis⎦

44 $y^2 - 8 = -2x^2$

⎡ellipse, centre (0, 0), major axis   ⎤
⎢$2\sqrt{8}$ units along the y-axis, minor⎥
⎣axis 4 units along the x-axis   ⎦

45 $4x^2 - 3y^2 = 12$

⎡hyperbola, symmetrical about x-⎤
⎢and y-axes, vertices on x-axis ⎥
⎣distance $2\sqrt{3}$ apart   ⎦

Answers to Problems 1 to 10

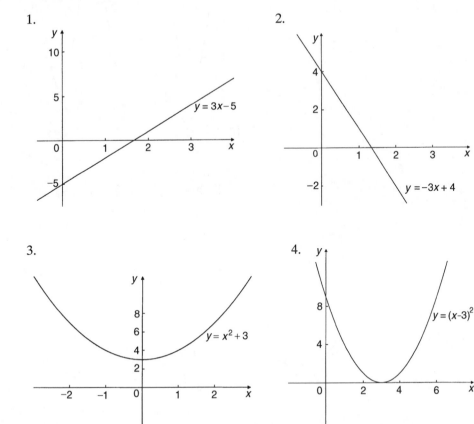

Figure 40.37

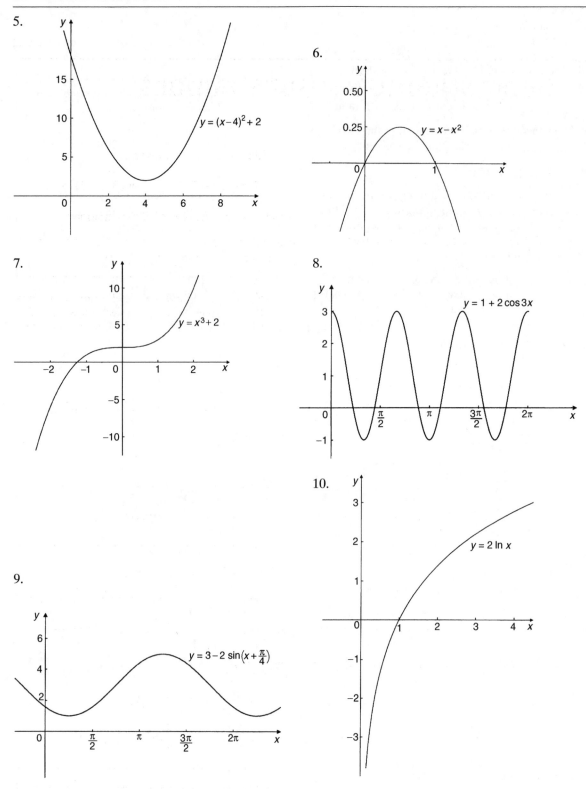

Figure 40.37   *(continued)*

Integration using substitutions

41.1 Introduction

Functions which require integrating are not always in the 'standard form' shown in Chapter 23. However, it is often possible to change a function into a form which can be integrated by using either:

(i) an algebraic substitution,
(ii) a trigonometric substitution,
(iii) partial fractions, or
(iv) integration by parts (see Chapter 42)

41.2 Algebraic substitutions

With **algebraic substitutions**, the substitution usually made is to let u be equal to $f(x)$ such that $f(u)\,du$ is a standard integral. It is found that integrals of the forms

$$k\int [f(x)]^n f'(x)\,dx \text{ and } k\int \frac{f'(x)}{[f(x)]^n}\,dx$$

(where k and n are constants) can both be integrated by substituting u for $f(x)$.

Problem 1.   Determine $\int \cos(3x + 7)\,dx$

$\int \cos(3x+7)\,dx$ is not a standard integral of the form shown in Table 23.1, page 235, thus an algebraic substitution is made.

Let $u = 3x + 7$ then $\dfrac{du}{dx} = 3$ and rearranging gives

$dx = \dfrac{du}{3}$

Hence $\displaystyle\int \cos(3x + 7)\,dx$

$$= \int (\cos u)\frac{du}{3} = \frac{1}{3}\int \cos u\,du,$$

which is a standard integral $= \dfrac{1}{3}\sin u + c$

Rewriting u as $(3x + 7)$ gives:

$$\int \cos(3x + 7)\,dx = \frac{1}{3}\sin(3x + 7) + c,$$

which may be checked by differentiation.

Problem 2.   Find $\int (2x - 5)^7\,dx$

$(2x-5)$ may be multiplied by itself 7 times and then each term of the result integrated. However, this would be a lengthy process and thus an algebraic substitution is made.

Let $u = (2x - 5)$ then $\dfrac{du}{dx} = 2$ and $dx = \dfrac{du}{2}$

Hence

$$\int (2x - 5)^7\,dx = \int u^7 \frac{du}{2} = \frac{1}{2}\int u^7\,du$$

$$= \frac{1}{2}\left(\frac{u^8}{8}\right) + c = \frac{1}{16}u^8 + c$$

Rewriting u as $(2x - 5)$ gives:

$$\int (2x - 5)^7\,dx = \frac{1}{16}(2x - 5)^8 + c$$

Problem 3.   Find $\displaystyle\int \frac{4}{(5x - 3)}\,dx$

Let $u = (5x - 3)$ then $\dfrac{du}{dx} = 5$ and $dx = \dfrac{du}{5}$

Hence $\displaystyle\int \frac{4}{(5x - 3)}\,dx$

$$= \int \frac{4}{u}\frac{du}{5} = \frac{4}{5}\int \frac{1}{u}\,du$$

$$= \frac{4}{5}\ln u + c = \frac{4}{5}\ln(5x - 3) + c$$

Problem 4.   Evaluate $\int_0^1 2e^{6x-1}\,dx$, correct to 4 significant figures

---

Let $u = 6x - 1$ then $\dfrac{du}{dx} = 6$ and $dx = \dfrac{du}{6}$

Hence $\displaystyle\int 2e^{6x-1}\,dx = \int 2e^u \dfrac{du}{6} = \dfrac{1}{3}\int e^u\,du$

$$= \dfrac{1}{3}e^u + c = \dfrac{1}{3}e^{6x-1} + c$$

Thus $\displaystyle\int_0^1 2e^{6x-1}\,dx = \dfrac{1}{3}[e^{6x-1}]_0^1$

$$= \dfrac{1}{3}[e^5 - e^{-1}] = \mathbf{49.35},$$

correct to 4 significant figures.

Problem 5. Determine $\displaystyle\int 3x(4x^2 + 3)^5\,dx$

Let $u = (4x^2 + 3)$ then $\dfrac{du}{dx} = 8x$ and $dx = \dfrac{du}{8x}$

Hence $\displaystyle\int 3x(4x^2 + 3)^5\,dx$

$$= \int 3x(u)^5 \dfrac{du}{8x} = \dfrac{3}{8}\int u^5\,du, \text{ by cancelling}$$

The original variable 'x' has been completely removed and the integral is now only in terms of u and is a standard integral.

Hence $\dfrac{3}{8}\displaystyle\int u^5\,du = \dfrac{3}{8}\left(\dfrac{u^6}{6}\right) + c = \dfrac{1}{16}u^6 + c$

$$= \dfrac{1}{16}(4x^2 + 3)^6 + c$$

Problem 6. Evaluate $\int_0^{\pi/6} 24\sin^5\theta\cos\theta\,d\theta$

Let $u = \sin\theta$ then $\dfrac{du}{d\theta} = \cos\theta$ and $d\theta = \dfrac{du}{\cos\theta}$

Hence $\displaystyle\int 24\sin^5\theta\cos\theta\,d\theta$

$$= \int 24u^5\cos\theta\dfrac{du}{\cos\theta}$$

$$= 24\int u^5\,du, \text{ by cancelling}$$

$$= 24\dfrac{u^6}{6} + c = 4u^6 + c = 4(\sin\theta)^6 + c$$

$$= 4\sin^6\theta + c$$

Thus $\displaystyle\int_0^{\pi/6} 24\sin^5\theta\cos\theta\,d\theta$

$$= [4\sin^6\theta]_0^{\pi/6} = 4\left[\left(\sin\dfrac{\pi}{6}\right)^6 - (\sin 0)^6\right]$$

$$= 4\left[\left(\dfrac{1}{2}\right)^6 - 0\right] = \dfrac{1}{16} \text{ or } \mathbf{0.0625}$$

Problem 7. Find $\displaystyle\int \dfrac{x}{2 + 3x^2}\,dx$

Let $u = 2 + 3x^2$ then $\dfrac{du}{dx} = 6x$ and $dx = \dfrac{du}{6x}$

Hence $\displaystyle\int \dfrac{x}{2 + 3x^2}\,dx$

$$= \int \dfrac{x}{u}\dfrac{du}{6x} = \dfrac{1}{6}\int \dfrac{1}{u}\,du, \text{ by cancelling,}$$

$$= \dfrac{1}{6}\ln u + x$$

$$= \dfrac{1}{6}\ln(2 + 3x^2) + c$$

Problem 8. Determine $\displaystyle\int \dfrac{2x}{\sqrt{(4x^2 - 1)}}\,dx$

Let $u = 4x^2 - 1$ then $\dfrac{du}{dx} = 8x$ and $dx = \dfrac{du}{8x}$

Hence $\displaystyle\int \dfrac{2x}{\sqrt{(4x^2 - 1)}}\,dx$

$$= \int \dfrac{2x}{\sqrt{u}}\dfrac{du}{8x} = \dfrac{1}{4}\int \dfrac{1}{\sqrt{u}}\,du, \text{ by cancelling}$$

$$= \dfrac{1}{4}\int u^{-1/2}\,du$$

$$= \dfrac{1}{4}\left[\dfrac{u^{(-1/2)+1}}{-\frac{1}{2} + 1}\right] + c = \dfrac{1}{4}\left[\dfrac{u^{1/2}}{\frac{1}{2}}\right] + c$$

$$= \dfrac{1}{2}\sqrt{u} + c = \dfrac{1}{2}\sqrt{(4x^2 - 1)} + c$$

Problem 9. Show that
$\int \tan\theta\,d\theta = \ln(\sec\theta) + c$

$$\int \tan\theta \, d\theta = \int \frac{\sin\theta}{\cos\theta} \, d\theta. \text{ Let } u = \cos\theta \text{ then}$$

$$\frac{du}{d\theta} = -\sin\theta \text{ and } d\theta = \frac{-du}{\sin\theta}$$

Hence $\displaystyle\int \frac{\sin\theta}{\cos\theta} \, d\theta$

$$= \int \frac{\sin\theta}{u}\left(\frac{-du}{\sin\theta}\right) = -\int \frac{1}{u} \, du = -\ln u + c$$

$$= -\ln(\cos\theta) + c = \ln(\cos\theta)^{-1} + c,$$

by the laws of logarithms

Hence $\tan\theta \, d\theta = \ln(\sec\theta) + c$, since

$$(\cos\theta)^{-1} = \frac{1}{\cos\theta} = \sec\theta$$

41.3   Change of limits

When evaluating definite integrals involving substitutions it is sometimes more convenient to **change the limits** of the integral as shown in Problems 10 and 11.

Problem 10.   Evaluate $\int_1^3 5x\sqrt{(2x^2 + 7)} \, dx$, taking positive values of square roots only

---

Let $u = 2x^2 + 7$, then $\dfrac{du}{dx} = 4x$ and $dx = \dfrac{du}{4x}$

It is possible in this case to change the limits of integration. Thus when $x = 3$, $u = 2(3)^2 + 7 = 25$ and when $x = 1$, $u = 2(1)^2 + 7 = 9$.

Hence $\displaystyle\int_{x=1}^{x=3} 5x\sqrt{(2x^2 + 7)} \, dx$

$$= \int_{u=9}^{u=25} 5x\sqrt{u}\frac{du}{4x} = \frac{5}{4}\int_9^{25} \sqrt{u} \, du$$

$$= \frac{5}{4}\int_9^{25} u^{1/2} \, du$$

Thus the limits have been changed, and it is unnecessary to change the integral back in terms of x

Thus $\displaystyle\int_{x=1}^{x=3} 5x\sqrt{(2x^2 + 7)} \, dx$

$$= \frac{5}{4}\left[\frac{u^{3/2}}{3/2}\right]_9^{25} = \frac{5}{6}[\sqrt{u^3}]_9^{25}$$

$$= \frac{5}{6}[\sqrt{25^3} - \sqrt{9^3}] = \frac{5}{6}(125 - 27)$$

$$= 81\frac{2}{3}$$

Problem 11.   Evaluate $\displaystyle\int_0^2 \frac{3x}{\sqrt{(2x^2 + 1)}} \, dx$, taking positive values of square roots only

---

Let $u = 2x^2 + 1$ then $\dfrac{du}{dx} = 4x$ and $dx = \dfrac{du}{4x}$

Hence $\displaystyle\int_0^2 \frac{3x}{\sqrt{(2x^2 + 1)}} \, dx = \int_{x=0}^{x=2} \frac{3x}{\sqrt{u}}\frac{du}{4x}$

$$= \frac{3}{4}\int_{x=0}^{x=2} u^{-1/2} \, du$$

Since $u = 2x^2 + 1$, when $x = 2$, $u = 9$ and when $x = 0$, $u = 1$.

Thus $\dfrac{3}{4}\displaystyle\int_{x=0}^{x=2} u^{-1/2} \, du = \frac{3}{4}\int_{u=1}^{u=9} u^{-1/2} \, du$,

i.e. the limits have been changed

$$= \frac{3}{4}\left[\frac{u^{1/2}}{\frac{1}{2}}\right]_1^9 = \frac{3}{2}[\sqrt{9} - \sqrt{1}] = 3,$$

taking positive values of square roots only.

Further problems on algebraic substitutions may be found in Section 41.7, Problems 1 to 22, pages 412 and 413.

41.4   Trigonometric substitutions

Table 41.1 gives a summary of the integrals that require the use of **trigonometric substitutions**.

Problem 12.   Evaluate $\int_0^{\pi/4} 2\cos^2 4t \, dt$

Since $\cos 2t = 2\cos^2 t - 1$ (from Chapter 38) then

$\cos^2 t = \dfrac{1}{2}(1 + \cos 2t)$ and $\cos^2 4t = \dfrac{1}{2}(1 + \cos 8t)$

Table 41.1

$f(x)$	$\int f(x)\,dx$	Method	See Problem
1 $\cos^2 x$	$\dfrac{1}{2}\left(x+\dfrac{\sin 2x}{2}\right)+c$	Use $\cos 2x = 2\cos^2 x - 1$	12
2 $\sin^2 x$	$\dfrac{1}{2}\left(x-\dfrac{\sin 2x}{2}\right)+c$	Use $\cos 2x = 1 - 2\sin^2 x$	13
3 $\tan^2 x$	$\tan x - x + c$	Use $1 + \tan^2 x = \sec^2 x$	14
4 $\cot^2 x$	$-\cot x - x + c$	Use $\cot^2 x + 1 = \operatorname{cosec}^2 x$	15
5 $\cos^m x \sin^n x$	(a) If either m or n is odd (but not both), use $\cos^2 x + \sin^2 x = 1$ (b) If both m and n are even, use either $\cos 2x = 2\cos^2 x - 1$ or $\cos 2x = 1 - 2\sin^2 x$		16, 17 18, 19
6 $\sin A \cos B$		Use $\dfrac{1}{2}[\sin(A+B)+\sin(A-B)]$	20
7 $\cos A \sin B$		Use $\dfrac{1}{2}[\sin(A+B)-\sin(A-B)]$	21
8 $\cos A \cos B$		Use $\dfrac{1}{2}[\cos(A+B)+\cos(A-B)]$	22
9 $\sin A \sin B$		Use $-\dfrac{1}{2}[\cos(A+B)-\cos(A-B)]$	23
10 $\dfrac{1}{\sqrt{(a^2-x^2)}}$	$\arcsin\dfrac{x}{a}+c$	Use $x = a\sin\theta$ substitution	24, 25
11 $\sqrt{(a^2-x^2)}$	$\dfrac{a^2}{2}\arcsin\dfrac{x}{a}+\dfrac{x}{2}\sqrt{(a^2-x^2)}+c$		26, 27
12 $\dfrac{1}{a^2+x^2}$	$\dfrac{1}{a}\arctan\dfrac{x}{a}+c$	Use $x = a\tan\theta$ substitution	28–30

Hence $\displaystyle\int_0^{\pi/4} 2\cos^2 4t\,dt$

$$= 2\int_0^{\pi/4}\frac{1}{2}(1+\cos 8t)\,dt = \left[t+\frac{\sin 8t}{8}\right]_0^{\pi/4}$$

$$= \left[\frac{\pi}{4}+\frac{\sin 8\left(\frac{\pi}{4}\right)}{8}\right]-\left[0+\frac{\sin 0}{8}\right]=\frac{\pi}{4}$$

Problem 13. Determine $\int \sin^2 3x\,dx$

Since $\cos 2x = 1 - 2\sin^2 x$, then
$\sin^2 x = \frac{1}{2}(1-\cos 2x)$ and $\sin^2 3x = \frac{1}{2}(1-\cos 6x)$.

Hence $\displaystyle\int \sin^2 3x\,dx = \int \frac{1}{2}(1-\cos 6x)\,dx$

$$= \frac{1}{2}\left(x-\frac{\sin 6x}{6}\right)+c$$

Problem 14. Find $3\int \tan^2 4x\,dx$

Since $1 + \tan^2 x = \sec^2 x$, then $\tan^2 x = \sec^2 x - 1$ and $\tan^2 4x = \sec^2 4x - 1$.

Hence $\displaystyle 3\int \tan^2 4x\,dx = 3\int (\sec^2 4x - 1)\,dx$

$$= 3\left(\frac{\tan 4x}{4}-x\right)+c$$

Problem 15. Evaluate $\displaystyle\int_{\pi/6}^{\pi/3}\frac{1}{2}\cot^2 2\theta\,d\theta$

Since $\cot^2 \theta + 1 = \operatorname{cosec}^2 \theta$, then
$\cot^2 \theta = \operatorname{cosec}^2 \theta - 1$ and $\cot^2 2\theta = \operatorname{cosec}^2 2\theta - 1$.

Hence $\displaystyle\int_{\pi/6}^{\pi/3} \frac{1}{2}\cot^2 2\theta\, d\theta$

$= \dfrac{1}{2}\displaystyle\int_{\pi/6}^{\pi/3}(\operatorname{cosec}^2 2\theta - 1)\, d\theta = \dfrac{1}{2}\left[\dfrac{-\cot 2\theta}{2} - \theta\right]_{\pi/6}^{\pi/3}$

$= \dfrac{1}{2}\left[\left(\dfrac{-\cot 2\left(\frac{\pi}{3}\right)}{2} - \dfrac{\pi}{3}\right) - \left(\dfrac{-\cot 2\left(\frac{\pi}{6}\right)}{2} - \dfrac{\pi}{6}\right)\right]$

$= \dfrac{1}{2}[(-(-0.2887) - 1.0472) - (-0.2887 - 0.5236)]$

$= \mathbf{0.0269}$

Problem 16.   Determine $\int \sin^5\theta\, d\theta$

Since $\cos^2\theta + \sin^2\theta = 1$ then $\sin^2\theta = (1 - \cos^2\theta)$

Hence $\int \sin^5\theta\, d\theta$

$\quad = \int \sin\theta(\sin^2\theta)^2\, d\theta = \int \sin\theta(1 - \cos^2\theta)^2\, d\theta$

$\quad = \int \sin\theta(1 - 2\cos^2\theta + \cos^4\theta)\, d\theta$

$\quad = \int(\sin\theta - 2\sin\theta\cos^2\theta + \sin\theta\cos^4\theta)\, d\theta$

$\quad = -\cos\theta + \dfrac{2\cos^3\theta}{3} - \dfrac{\cos^5\theta}{5} + c$

(Whenever a power of a cosine is multiplied by a sine of power 1, or vice versa, the integral may be determined by inspection as shown.

In general, $\displaystyle\int \cos^n\theta\sin\theta\, d\theta = \dfrac{-\cos^{n+1}\theta}{(n+1)}$

and $\displaystyle\int \sin^n\theta\cos\theta\, d\theta = \dfrac{\sin^{n+1}\theta}{(n+1)} + c)$

Problem 17.   Evaluate $\int_0^{\pi/2}\sin^2 x\cos^3 x\, dx$

$\displaystyle\int_0^{\pi/2}\sin^2 x\cos^3 x\, dx$

$\quad = \displaystyle\int_0^{\pi/2}\sin^2 x\cos^2 x\cos x\, dx$

$\quad = \displaystyle\int_0^{\pi/2}(\sin^2 x)(1 - \sin^2 x)(\cos x)\, dx$

$\quad = \displaystyle\int_0^{\pi/2}(\sin^2 x\cos x - \sin^4 x\cos x)\, dx$

$= \left[\dfrac{\sin^3 x}{3} - \dfrac{\sin^5 x}{5}\right]_0^{\pi/2}$

$= \left[\dfrac{\left(\sin\frac{\pi}{2}\right)^3}{3} - \dfrac{\left(\sin\frac{\pi}{2}\right)^5}{5}\right] - [0 - 0]$

$= \dfrac{1}{3} - \dfrac{1}{5} = \dfrac{\mathbf{2}}{\mathbf{15}}$

Problem 18.   Evaluate $\int_0^{\pi/4} 4\cos^4\theta\, d\theta$, correct to 4 significant figures

---

$\displaystyle\int_0^{\pi/4} 4\cos^4\theta\, d\theta$

$= 4\displaystyle\int_0^{\pi/4}(\cos^2\theta)^2\, d\theta$

$= 4\displaystyle\int_0^{\pi/4}\left[\dfrac{1}{2}(1 + \cos 2\theta)\right]^2\, d\theta$

$= \displaystyle\int_0^{\pi/4}(1 + 2\cos 2\theta + \cos^2 2\theta)\, d\theta$

$= \displaystyle\int_0^{\pi/4}\left[1 + 2\cos 2\theta + \dfrac{1}{2}(1 + \cos 4\theta)\right] d\theta$

$= \displaystyle\int_0^{\pi/4}\left(\dfrac{3}{2} + 2\cos 2\theta + \dfrac{1}{2}\cos 4\theta\right) d\theta$

$= \left[\dfrac{3\theta}{2} + \sin 2\theta + \dfrac{\sin 4\theta}{8}\right]_0^{\pi/4}$

$= \left[\dfrac{3}{2}\left(\dfrac{\pi}{4}\right) + \sin\dfrac{2\pi}{4} + \dfrac{\sin 4(\pi/4)}{8}\right] - [0]$

$= \dfrac{3\pi}{8} + 1 = \mathbf{2.178},$

correct to 4 significant figures

Problem 19.   Find $\int \sin^2 t\cos^4 t\, dt$

$\displaystyle\int \sin^2 t\cos^4 t\, dt$

$\quad = \displaystyle\int \sin^2 t(\cos^2 t)^2\, dt$

$\quad = \displaystyle\int\left(\dfrac{1 - \cos 2t}{2}\right)\left(\dfrac{1 + \cos 2t}{2}\right)^2 dt$

$\quad = \dfrac{1}{8}\displaystyle\int(1 - \cos 2t)(1 + 2\cos 2t + \cos^2 2t)\, dt$

$$= \frac{1}{8} \int (1 + 2\cos 2t + \cos^2 2t - \cos 2t$$

$$- 2\cos^2 2t - \cos^3 2t)\, dt$$

$$= \frac{1}{8} \int (1 + \cos 2t - \cos^2 2t - \cos^3 2t)\, dt$$

$$= \frac{1}{8} \int \left[1 + \cos 2t - \left(\frac{1 + \cos 4t}{2} \right) \right.$$

$$\left. - \cos 2t(1 - \sin^2 2t) \right] dt$$

$$= \frac{1}{8} \int \left(\frac{1}{2} - \frac{\cos 4t}{2} + \cos 2t \sin^2 2t \right) dt$$

$$= \frac{1}{8} \left(\frac{t}{2} - \frac{\sin 4t}{8} + \frac{\sin^3 2t}{6} \right) + c$$

Problem 20. Determine $\int \sin 3t \cos 2t \, dt$

$$\int \sin 3t \cos 2t \, dt$$

$$= \int \frac{1}{2}[\sin(3t + 2t) + \sin(3t - 2t)]\, dt,$$

from 6 of Table 41.1

$$= \frac{1}{2} \int (\sin 5t + \sin t)\, dt$$

$$= \frac{1}{2} \left(\frac{-\cos 5t}{5} - \cos t \right) + c$$

Problem 21. Find $\int \frac{1}{3} \cos 5x \sin 2x \, dx$

$$\int \frac{1}{3} \cos 5x \sin 2x \, dx$$

$$= \frac{1}{3} \int \frac{1}{2}[\sin(5x + 2x) - \sin(5x - 2x)]\, dx,$$

from 7 of Table 41.1

$$= \frac{1}{6} \int (\sin 7x - \sin 3x)\, dx$$

$$= \frac{1}{6} \left(\frac{-\cos 7x}{7} + \frac{\cos 3x}{3} \right) + c$$

Problem 22. Evaluate $\int_0^1 2\cos 6\theta \cos \theta \, d\theta$, correct to 4 decimal places

$$\int_0^1 2\cos 6\theta \cos \theta \, d\theta$$

$$= 2 \int_0^1 \frac{1}{2}[\cos(6\theta + \theta) + \cos(6\theta - \theta)]\, d\theta,$$

from 8 of Table 41.1

$$= \int_0^1 (\cos 7\theta + \cos 5\theta)\, d\theta$$

$$= \left[\frac{\sin 7\theta}{7} + \frac{\sin 5\theta}{5} \right]_0^1$$

$$= \left(\frac{\sin 7}{7} + \frac{\sin 5}{5} \right) - \left(\frac{\sin 0}{7} + \frac{\sin 0}{5} \right)$$

'sin 7' means 'the sine of 7 radians' ($\equiv 401°4'$) and $\sin 5 \equiv 286°29'$.

Hence $\int_0^1 2\cos 6\theta \cos \theta \, d\theta$

$$= (0.09386 + -0.19178) - (0)$$

$$= -\mathbf{0.0979}, \text{ correct to 4 decimal places}$$

Problem 23. Find $3 \int \sin 5x \sin 3x \, dx$

$$3 \int \sin 5x \sin 3x \, dx$$

$$= 3 \int -\frac{1}{2}[\cos(5x + 3x) - \cos(5x - 3x)]\, dx,$$

from 9 of Table 41.1

$$= -\frac{3}{2} \int (\cos 8x - \cos 2x)\, dx$$

$$= -\frac{3}{2} \left(\frac{\sin 8x}{8} - \frac{\sin 2x}{2} \right) + c$$

$$\text{or } \frac{3}{16}(4\sin 2x - \sin 8x) + c$$

Problem 24. Determine $\int \frac{1}{\sqrt{(a^2 - x^2)}} \, dx$

Let $x = a \sin \theta$, then $\dfrac{dx}{d\theta} = a \cos \theta$ and $dx = a \cos \theta \, d\theta$.

Hence $\int \dfrac{1}{\sqrt{(a^2 - x^2)}}\, dx$

$= \int \dfrac{1}{\sqrt{(a^2 - a^2 \sin^2 \theta)}}\, a \cos \theta\, d\theta$

$= \int \dfrac{a \cos \theta\, d\theta}{\sqrt{[a^2(1 - \sin^2 \theta)]}}$

$= \int \dfrac{a \cos \theta\, d\theta}{\sqrt{(a^2 \cos^2 \theta)}},$

since $\sin^2 \theta + \cos^2 \theta = 1$

$= \int \dfrac{a \cos \theta\, d\theta}{a \cos \theta} = \int d\theta = \theta + c$

Since $x = a \sin \theta$, then $\sin \theta = \dfrac{x}{a}$ and $\theta = \arcsin \dfrac{x}{a}$

Hence $\int \dfrac{1}{\sqrt{(a^2 - x^2)}}\, dx = \arcsin \dfrac{x}{a} + c$

Problem 25. Evaluate $\displaystyle\int_0^3 \dfrac{1}{\sqrt{(9 - x^2)}}\, dx$

From Problem 24,

$\displaystyle\int_0^3 \dfrac{1}{\sqrt{(9 - x^2)}}\, dx = \left[\arcsin \dfrac{x}{3}\right]_0^3$, since $a = 3$

$= (\arcsin 1 - \arcsin 0)$

$= \dfrac{\pi}{2}$ or $\mathbf{1.5708}$

Problem 26. Find $\displaystyle\int \sqrt{(a^2 - x^2)}\, dx$

Let $x = a \sin \theta$ then $\dfrac{dx}{d\theta} = a \cos \theta$ and $dx = a \cos \theta\, d\theta$.

Hence $\displaystyle\int \sqrt{(a^2 - x^2)}\, dx$

$= \displaystyle\int \sqrt{(a^2 - a^2 \sin^2 \theta)}\,(a \cos \theta\, d\theta)$

$= \displaystyle\int \sqrt{[a^2(1 - \sin^2 \theta)]}\,(a \cos \theta\, d\theta)$

$= \displaystyle\int \sqrt{(a^2 \cos^2 \theta)}\,(a \cos \theta\, d\theta)$

$= \displaystyle\int (a \cos \theta)(a \cos \theta\, d\theta)$

$= a^2 \displaystyle\int \cos^2 \theta\, d\theta = a^2 \displaystyle\int \left(\dfrac{1 + \cos 2\theta}{2}\right) d\theta$

(since $\cos 2\theta = 2 \cos^2 \theta - 1$)

$= \dfrac{a^2}{2}\left(\theta + \dfrac{\sin 2\theta}{2}\right) + c$

In the compound-angle addition formula:
$\sin(A+B) = \sin A \cos B + \cos A \sin B$, let $B = A$, then $\sin 2A = 2 \sin A \cos A$ (as shown in Chapter 38).

Hence $\displaystyle\int \sqrt{(a^2 - x^2)}\, dx$

$= \dfrac{a^2}{2}\left(\theta + \dfrac{2 \sin \theta \cos \theta}{2}\right) + c$

$= \dfrac{a^2}{2}[\theta + \sin \theta \cos \theta] + c$

Since $x = a \sin \theta$, then $\sin \theta = \dfrac{x}{a}$ and $\theta = \arcsin \dfrac{x}{a}$

Also, $\cos^2 \theta + \sin^2 \theta = 1$, from which,

$\cos \theta = \sqrt{(1 - \sin^2 \theta)} = \sqrt{\left[1 - \left(\dfrac{x}{a}\right)^2\right]}$

$= \sqrt{\left(\dfrac{a^2 - x^2}{a^2}\right)} = \dfrac{\sqrt{(a^2 - x^2)}}{a}$

Thus $\displaystyle\int \sqrt{(a^2 - x^2)}\, dx$

$= \dfrac{a^2}{2}[\theta + \sin \theta \cos \theta]$

$= \dfrac{a^2}{2}\left[\arcsin \dfrac{x}{a} + \left(\dfrac{x}{a}\right)\dfrac{\sqrt{(a^2 - x^2)}}{a}\right] + c$

$= \dfrac{a^2}{2} \arcsin \dfrac{x}{a} + \dfrac{x}{2}\sqrt{(a^2 - x^2)} + c$

Problem 27. Evaluate $\displaystyle\int_0^4 \sqrt{(16 - x^2)}\, dx$

From Problem 26, $\displaystyle\int_0^4 \sqrt{(16 - x^2)}\, dx$

$= \left[\dfrac{16}{2} \arcsin \dfrac{x}{4} + \dfrac{x}{2}\sqrt{(16 - x^2)}\right]_0^4$

$= [8 \arcsin 1 + 2\sqrt{(0)}] - [8 \arcsin 0 + 0]$

$$= 8 \arcsin 1 = 8 \left(\frac{\pi}{2} \right)$$

$$= 4\pi \text{ or } 12.57$$

Problem 28. Determine $\displaystyle \int \frac{1}{(a^2 + x^2)} \, dx$

Let $x = a \tan \theta$ then $\dfrac{dx}{d\theta} = a \sec^2 \theta$ and $dx = a \sec^2 \theta \, d\theta$.

Hence $\displaystyle \int \frac{1}{(a^2 + x^2)} \, dx$

$$= \int \frac{1}{(a^2 + a^2 \tan^2 \theta)} (a \sec^2 \theta \, d\theta)$$

$$= \int \frac{a \sec^2 \theta \, d\theta}{a^2 (1 + \tan^2 \theta)}$$

$$= \int \frac{a \sec^2 \theta \, d\theta}{a^2 \sec^2 \theta}, \text{ since } 1 + \tan^2 \theta = \sec^2 \theta$$

$$= \int \frac{1}{a} \, d\theta = \frac{1}{a} (\theta) + c$$

Since $x = a \tan \theta$, $\theta = \arctan \dfrac{x}{a}$. Hence

$$\int \frac{1}{(a^2 + x^2)} dx = \frac{1}{a} \arctan \frac{x}{a} + c$$

Problem 29. Evaluate $\displaystyle \int_0^2 \frac{1}{(4 + x^2)} \, dx$

From Problem 28, $\displaystyle \int_0^2 \frac{1}{(4 + x^2)} \, dx$

$$= \frac{1}{2} \left[\arctan \frac{x}{2} \right]_0^2 \text{ since } a = 2$$

$$= \frac{1}{2} (\arctan 1 - \arctan 0) = \frac{1}{2} \left(\frac{\pi}{4} - 0 \right)$$

$$= \frac{\pi}{8} \text{ or } 0.3927$$

Problem 30. Evaluate $\displaystyle \int_0^1 \frac{5}{(3 + 2x^2)} \, dx$,
correct to 4 decimal places

$$\int_0^1 \frac{5}{(3 + 2x^2)} \, dx$$

$$= \int_0^1 \frac{5}{2[(3/2) + x^2]} \, dx$$

$$= \frac{5}{2} \int_0^1 \frac{1}{[\sqrt{(3/2)}]^2 + x^2} \, dx$$

$$= \frac{5}{2} \left[\frac{1}{\sqrt{(3/2)}} \arctan \frac{x}{\sqrt{(3/2)}} \right]_0^1$$

$$= \frac{5}{2} \sqrt{\left(\frac{2}{3} \right)} \left[\arctan \sqrt{\left(\frac{2}{3} \right)} - \arctan 0 \right]$$

$$= (2.0412)[0.6847 - 0]$$

$$= \mathbf{1.3976}, \text{ correct to 4 decimal places}$$

Further problems on trigonometric substitutions may be found in Section 41.7, Problems 23 to 39, pages 413 and 414.

41.5   Integration using partial fractions

The process of expressing a fraction in terms of simpler fractions–called **partial fractions**–is discussed in Chapter 3. The forms of partial fractions used are summarized on page 31.

Certain functions have to be resolved into partial fractions before they can be integrated as demonstrated in the following worked problems.

Problem 31. Determine $\displaystyle \int \frac{11 - 3x}{x^2 + 2x - 3} \, dx$

As shown in Problem 50, page 32:

$$\frac{11 - 3x}{x^2 + 2x - 3} \equiv \frac{11 - 3x}{(x - 1)(x + 3)}$$

$$\equiv \frac{A}{(x - 1)} + \frac{B}{(x + 3)}$$

$$\equiv \frac{A(x + 3) + B(x - 1)}{(x - 1)(x + 3)}$$

Hence $11 - 3x \equiv A(x + 3) + B(x - 1)$
Let $x = 1$, then $8 = 4A$, from which, $A = 2$
Let $x = -3$, then $20 = -4B$, from which $B = -5$

Hence $\displaystyle\int \frac{11 - 3x}{x^2 + 2x - 3}\, dx$

$$= \int \left\{ \frac{2}{(x - 1)} - \frac{5}{(x + 3)} \right\} dx$$

$$= 2\ln(x - 1) - 5\ln(x + 3) + c$$

or $\ln\left\{ \dfrac{(x - 1)^2}{(x + 3)^5} \right\} + c$ by the laws of logarithms

---

Problem 32. Find
$$\int \frac{2x^2 - 9x - 35}{(x + 1)(x - 2)(x + 3)}\, dx$$

It was shown in Problem 51, page 32, that:

$$\frac{2x^2 - 9x - 35}{(x + 1)(x - 2)(x + 3)}$$

$$\equiv \frac{4}{(x + 1)} - \frac{3}{(x - 2)} + \frac{1}{(x + 3)}$$

Hence $\displaystyle\int \frac{2x^2 - 9x - 35}{(x + 1)(x - 2)(x + 3)}\, dx$

$$\equiv \int \left\{ \frac{4}{(x + 1)} - \frac{3}{(x - 2)} + \frac{1}{(x + 3)} \right\} dx$$

$$= 4\ln(x + 1) - 3\ln(x - 2) + \ln(x + 3) + c$$

$$= \ln\left\{ \frac{(x + 1)^4(x + 3)}{(x - 2)^3} \right\} + c$$

Problem 33. Determine $\displaystyle\int \frac{x^2 + 1}{x^2 - 3x + 2}\, dx$

By dividing out (since the numerator and denominator are of the same degree) and resolving into partial fractions it was shown in Problem 52, page 33, that

$$\frac{x^2 + 1}{x^2 - 3x + 2} \equiv 1 - \frac{2}{(x - 1)} + \frac{5}{(x - 2)}$$

Hence $\displaystyle\int \frac{x^2 + 1}{(x^2 - 3x + 2)}\, dx$

$$\equiv \int \left\{ 1 - \frac{2}{(x - 1)} + \frac{5}{(x - 2)} \right\} dx$$

$$= x - 2\ln(x - 1) + 5\ln(x - 2) + c$$

or $x + \ln\left\{ \dfrac{(x - 2)^5}{(x - 1)^2} \right\} + c$

Problem 34. Evaluate
$$\int_2^3 \frac{x^3 - 2x^2 - 4x - 4}{x^2 + x - 2}\, dx,\ \text{correct to 4}$$
significant figures

By dividing out and resolving into partial fractions it was shown in Problem 53, page 33, that

$$\frac{x^3 - 2x^2 - 4x - 4}{x^2 + x - 2} \equiv x - 3 + \frac{4}{(x + 2)} - \frac{3}{(x - 1)}$$

Hence $\displaystyle\int_2^3 \left(\frac{x^3 - 2x^2 - 4x - 4}{x^2 + x - 2} \right) dx$

$$\equiv \int_2^3 \left\{ x - 3 + \frac{4}{(x + 2)} - \frac{3}{(x - 1)} \right\} dx$$

$$= \left[\frac{x^2}{2} - 3x + 4\ln(x + 2) - 3\ln(x - 1) \right]_2^3$$

$$= \left(\frac{9}{2} - 9 + 4\ln 5 - 3\ln 2 \right)$$

$$- (2 - 6 + 4\ln 4 - 3\ln 1)$$

$$= -1.687,\ \text{correct to 4 significant figures}$$

Problem 35. Determine $\displaystyle\int \frac{2x + 3}{(x - 2)^2}\, dx$

It was shown in Problem 54, page 33, that:

$$\frac{2x + 3}{(x - 2)^2} \equiv \frac{A}{(x - 2)} + \frac{B}{(x - 2)^2}$$

$$\equiv \frac{A(x - 2) + B}{(x - 2)^2}$$

Hence $2x + 3 \equiv A(x - 2) + B$, from which, $A = 2$ and $B = 7$.

Thus

$$\int \frac{2x + 3}{(x - 2)^2}\, dx \equiv \int \left\{ \frac{2}{(x - 2)} + \frac{7}{(x - 2)^2} \right\} dx$$

$$= 2\ln(x - 2) - \frac{7}{(x - 2)} + c$$

$\int \dfrac{7}{(x-2)^2} dx$ is determined using the algebraic substitution $u = (x - 2)$.

Problem 36. Find $\int \dfrac{5x^2 - 2x - 19}{(x+3)(x-1)^2} dx$

It was shown in Problem 55, page 34, that:

$$\frac{5x^2 - 2x - 19}{(x+3)(x-1)^2} \equiv \frac{2}{(x+3)} + \frac{3}{(x-1)} - \frac{4}{(x-1)^2}$$

Hence $\int \dfrac{5x^2 - 2x - 19}{(x+3)(x-1)^2} dx$

$$\equiv \int \left\{ \frac{2}{(x+3)} + \frac{3}{(x-1)} - \frac{4}{(x-1)^2} \right\} dx$$

$$= 2\ln(x+3) + 3\ln(x-1) + \frac{4}{(x-1)} + c$$

or $\ln\{(x+3)^2 (x-1)^3\} + \dfrac{4}{(x-1)} + c$

Problem 37. Evaluate $\displaystyle\int_{-2}^{1} \dfrac{3x^2 + 16x + 15}{(x+3)^3} dx$, correct to 4 significant figures

---

It was shown in Problem 56, page 34, that:

$$\frac{3x^2 + 16x + 15}{(x+3)^3}$$

$$\equiv \frac{3}{(x+3)} - \frac{2}{(x+3)^2} - \frac{6}{(x+3)^3}$$

Hence $\displaystyle\int_{-2}^{1} \dfrac{3x^2 + 16x + 15}{(x+3)^3} dx$

$$\equiv \int_{-2}^{1} \left\{ \frac{3}{(x+3)} - \frac{2}{(x+3)^2} - \frac{6}{(x+3)^3} \right\} dx$$

$$= \left[3\ln(x+3) + \frac{2}{(x+3)} + \frac{3}{(x+3)^2} \right]_{-2}^{1}$$

$$= \left(3\ln 4 + \frac{2}{4} + \frac{3}{16} \right) - \left(3\ln 1 + \frac{2}{1} + \frac{3}{1} \right)$$

$$= -0.1536, \text{ correct to 4 significant figures}$$

Problem 38. Find $\int \dfrac{3 + 6x + 4x^2 - 2x^3}{x^2(x^2 + 3)} dx$

It was shown in Problem 58, page 35, that:

$$\frac{3 + 6x + 4x^2 - 2x^3}{x^2(x^2 + 3)}$$

$$\equiv \frac{A}{x} + \frac{B}{x^2} + \frac{Cx + D}{(x^2 + 3)}$$

$$\equiv \frac{Ax(x^2 + 3) + B(x^2 + 3) + (Cx + D)x^2}{x^2(x^2 + 3)}$$

Hence $3 + 6x + 4x^2 - 2x^3 \equiv Ax(x^2 + 3) + B(x^2 + 3) + (Cx + D)x^2$ from which $A = 2$, $B = 1$, $C = -4$ and $D = 3$.

Thus $\int \dfrac{3 + 6x + 4x^2 - 2x^3}{x^2(x^2 + 3)} dx$

$$\equiv \int \left(\frac{2}{x} + \frac{1}{x^2} + \frac{3 - 4x}{(x^2 + 3)} \right) dx$$

$$= \int \left\{ \frac{2}{x} + \frac{1}{x^2} + \frac{3}{(x^2 + 3)} - \frac{4x}{(x^2 + 3)} \right\} dx$$

$$\int \frac{3}{(x^2 + 3)} dx = 3 \int \frac{1}{x^2 + (\sqrt{3})^2} dx$$

$$= \frac{3}{\sqrt{3}} \arctan \frac{x}{\sqrt{3}},$$

from 12, Table 41.1

$\int \dfrac{4x}{x^2 + 3} dx$ is determined using the algebraic substitution $u = (x^2 + 3)$.

Hence $\int \left\{ \dfrac{2}{x} + \dfrac{1}{x^2} + \dfrac{3}{(x^2 + 3)} - \dfrac{4x}{x^2 + 3} \right\} dx$

$$= 2\ln x - \frac{1}{x} + \frac{3}{\sqrt{3}} \arctan \frac{x}{\sqrt{3}}$$

$$- 2\ln(x^2 + 3) + c$$

$$= \ln \left(\frac{x}{x^2 + 3} \right)^2 - \frac{1}{x} + \sqrt{3} \arctan \frac{x}{\sqrt{3}} + c$$

Problem 39. Determine $\int \dfrac{1}{(x^2 - a^2)} dx$

Let $\dfrac{1}{(x^2 - a^2)} \equiv \dfrac{A}{(x - a)} + \dfrac{B}{(x + a)}$

$$\equiv \frac{A(x + a) + B(x - a)}{(x + a)(x - a)}$$

Equating the numerators gives:

$$1 \equiv A(x + a) + B(x - a)$$

Let $x = a$, then $A = \dfrac{1}{2a}$, and let $x = -a$,

then $B = -\dfrac{1}{2a}$

Hence $\displaystyle\int \frac{1}{(x^2 - a^2)}\, dx$

$$= \int \frac{1}{2a}\left[\frac{1}{(x - a)} - \frac{1}{(x + a)}\right] dx$$

$$= \frac{1}{2a}[\ln(x - a) - \ln(x + a)] + c$$

$$= \frac{1}{2a}\ln\left(\frac{x - a}{x + a}\right) + c$$

Problem 40.   Evaluate $\displaystyle\int_3^4 \frac{3}{(x^2 - 4)}\, dx$, correct to 3 significant figures

---

From Problem 39,

$$\int_3^4 \frac{3}{(x^2 - 4)}\, dx = 3\left[\frac{1}{2(2)}\ln\left(\frac{x - 2}{x + 2}\right)\right]_3^4$$

$$= \frac{3}{4}\left[\ln\frac{2}{6} - \ln\frac{1}{5}\right]$$

$$= \frac{3}{4}\ln\frac{5}{3} = 0.383,$$

correct to 3 significant figures

Problem 41.   Determine $\displaystyle\int \frac{1}{(a^2 - x^2)}\, dx$

Using partial fractions, let

$$\frac{1}{(a^2 - x^2)} \equiv \frac{1}{(a - x)(a + x)}$$

$$\equiv \frac{A}{(a - x)} + \frac{B}{(a + x)}$$

$$\equiv \frac{A(a + x) + B(a - x)}{(a - x)(a + x)}$$

Then $1 \equiv A(a + x) + B(a - x)$

Let $x = a$ then $A = \dfrac{1}{2a}$. Let $x = -a$ then $B = \dfrac{1}{2a}$.

Hence $\displaystyle\int \frac{1}{(a^2 - x^2)}\, dx$

$$= \int \frac{1}{2a}\left[\frac{1}{(a - x)} + \frac{1}{(a + x)}\right] dx$$

$$= \frac{1}{2a}[-\ln(a - x) + \ln(a + x)] + c$$

$$= \frac{1}{2a}\ln\left(\frac{a + x}{a - x}\right) + c$$

Problem 42.   Evaluate $\displaystyle\int_0^2 \frac{5}{(9 - x^2)}\, dx$, correct to 4 decimal places

---

From Problem 41,

$$\int_0^2 \frac{5}{(9 - x^2)}\, dx = 5\left[\frac{1}{2(3)}\ln\left(\frac{3 + x}{3 - x}\right)\right]_0^2$$

$$= \frac{5}{6}\left[\ln\frac{5}{1} - \ln 1\right] = 1.3412,$$

correct to 4 decimal places

Further problems on partial fractions may be found in Section 41.7, Problems 40 to 55, pages 414 and 415.

41.6   A further trigonometric substitution, the $t = \tan\dfrac{\theta}{2}$ substitution

Integrals of the form $\displaystyle\int \frac{d\theta}{a\cos\theta + b\sin\theta + c}$, where a, b and c are constants, may be determined by using the substitution $t = \tan\dfrac{\theta}{2}$

This is explained as follows:

If angle A in the right-angled triangle ABC shown in Fig. 41.1 is made equal to $\dfrac{\theta}{2}$ then, since tangent $= \dfrac{\text{opposite}}{\text{adjacent}}$, if $BC = t$ and $AB = 1$

then $\tan \dfrac{\theta}{2} = t$

By Pythagoras' theorem, $AC = \sqrt{(1 + t^2)}$.

Therefore $\sin \dfrac{\theta}{2} = \dfrac{t}{\sqrt{(1 + t^2)}}$ and

$\cos \dfrac{\theta}{2} = \dfrac{1}{\sqrt{(1 + t^2)}}$

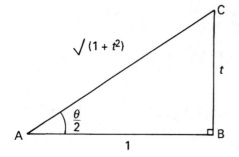

Figure 41.1

Since $\sin 2x = 2 \sin x \cos x$ (from double angle formulae), then

$$\sin \theta = 2 \sin \dfrac{\theta}{2} \cos \dfrac{\theta}{2}$$

$$= 2 \left(\dfrac{t}{\sqrt{(1 + t^2)}} \right) \left(\dfrac{1}{\sqrt{(1 + t^2)}} \right)$$

i.e. $\boldsymbol{\sin \theta = \dfrac{2t}{(1 + t^2)}}$ \hfill (1)

Since $\cos 2x = \cos^2 x - \sin^2 x$ (from double angle formulae), then

$$\cos \theta = \cos^2 \dfrac{\theta}{2} - \sin^2 \dfrac{\theta}{2}$$

$$= \left(\dfrac{1}{\sqrt{(1 + t^2)}} \right)^2 - \left(\dfrac{t}{\sqrt{(1 + t^2)}} \right)^2$$

i.e. $\boldsymbol{\cos \theta = \dfrac{1 - t^2}{1 + t^2}}$ \hfill (2)

Also, since $t = \tan \dfrac{\theta}{2}, \dfrac{dt}{d\theta} = \dfrac{1}{2} \sec^2 \dfrac{\theta}{2}$

$$= \dfrac{1}{2} \left(1 + \tan^2 \dfrac{\theta}{2} \right),$$

from trigonometric identities,

i.e. $\dfrac{dt}{d\theta} = \dfrac{1}{2}(1 + t^2)$

from which, $\boldsymbol{d\theta = \dfrac{2 \, dt}{1 + t^2}}$ \hfill (3)

Equations (1), (2) and (3) are used to determine integrals of the form

$$\int \dfrac{d\theta}{a \cos \theta + b \sin \theta + c},$$

where a, b or c may be zero. This is demonstrated in the following worked problems.

Problem 43. Determine $\displaystyle\int \dfrac{d\theta}{\sin \theta}$

If $t = \tan \dfrac{\theta}{2}$ then $\sin \theta = \dfrac{2t}{1 + t^2}$ and $d\theta = \dfrac{2 \, dt}{1 + t^2}$, from equations (1) and (3).

Thus $\displaystyle\int \dfrac{d\theta}{\sin \theta} = \int \dfrac{1}{2t/(1 + t^2)} \left(\dfrac{2 \, dt}{1 + t^2} \right)$

$$= \int \dfrac{1}{t} \, dt = \ln t + c$$

Hence $\displaystyle\int \dfrac{d\theta}{\sin \theta} = \ln \left(\tan \dfrac{\theta}{2} \right) + c$

Problem 44. Determine $\displaystyle\int \dfrac{dx}{\cos x}$

If $t = \tan \dfrac{x}{2}$ then $\cos x = \dfrac{1 - t^2}{1 + t^2}$ and $dx = \dfrac{2 \, dt}{1 + t^2}$, from equations (2) and (3).

Thus $\displaystyle\int \dfrac{dx}{\cos x} = \int \dfrac{1}{(1 - t^2)/(1 + t^2)} \left(\dfrac{2 \, dt}{1 + t^2} \right)$

$$= \int \dfrac{2}{1 - t^2} \, dt$$

$\dfrac{2}{1 - t^2}$ may be resolved into partial fractions.

Let $\dfrac{2}{1 - t^2} = \dfrac{2}{(1 - t)(1 + t)}$

$$= \dfrac{A}{(1 - t)} + \dfrac{B}{(1 + t)}$$

$$= \frac{A(1+t)+B(1-t)}{(1-t)(1+t)}$$

Hence $2 = A(1+t)+B(1-t)$

When $t = 1$, $2 = 2A$, from which $A = 1$

When $t = -1$, $2 = 2B$, from which $B = 1$

Hence
$$\int \frac{2\,dt}{1-t^2} = \int \frac{1}{(1-t)} + \frac{1}{(1+t)}\,dt$$
$$= -\ln(1-t) + \ln(1+t) + c$$
$$= \ln \left\{ \frac{(1+t)}{(1-t)} \right\} + c$$

Thus
$$\int \frac{dx}{\cos x} = \ln \left\{ \frac{1 + \tan \frac{x}{2}}{1 - \tan \frac{x}{2}} \right\} + c$$

Note that since $\tan \dfrac{\pi}{4} = 1$ the above result may be written as

$$\int \frac{dx}{\cos x} = \ln \left\{ \frac{\tan \frac{\pi}{4} + \tan \frac{x}{2}}{1 - \tan \frac{\pi}{4} \tan \frac{x}{2}} \right\} + c$$
$$= \ln \left\{ \tan \left(\frac{\pi}{4} + \frac{x}{2} \right) \right\} + c,$$

from compound angles (see Chapter 38).

Problem 45. Determine $\displaystyle\int \frac{dx}{1 + \cos x}$

If $t = \tan \dfrac{x}{2}$ then $\cos x = \dfrac{1-t^2}{1+t^2}$ and $dx = \dfrac{2\,dt}{1+t^2}$, from equations (2) and (3).

Thus $\displaystyle\int \frac{dx}{1 + \cos x}$
$$= \int \frac{1}{1 + (1-t^2)/(1+t^2)} \left(\frac{2\,dt}{1+t^2} \right)$$
$$= \int \frac{1}{[(1+t^2)+(1-t^2)]/(1+t^2)} \left(\frac{2\,dt}{1+t^2} \right) = \int dt$$

Hence $\displaystyle\int \frac{dx}{1 + \cos x} = t + c = \tan \frac{x}{2} + c$

Problem 46. Determine $\displaystyle\int \frac{d\theta}{5 + 4\cos\theta}$

If $t = \tan \dfrac{\theta}{2}$ then $\cos\theta = \dfrac{1-t^2}{1+t^2}$ and $d\theta = \dfrac{2\,dt}{1+t^2}$, from equations (2) and (3).

Thus $\displaystyle\int \frac{d\theta}{5 + 4\cos\theta}$

$$= \int \frac{1}{5 + 4(1-t^2)/(1+t^2)} \left(\frac{2\,dt}{1+t^2} \right)$$
$$= \int \frac{1}{[5(1+t^2)+4(1-t^2)]/(1+t^2)} \left(\frac{2\,dt}{1+t^2} \right)$$
$$= 2 \int \frac{dt}{t^2 + 9} = 2 \int \frac{dt}{t^2 + 3^2}$$
$$= 2 \left(\frac{1}{3} \arctan \frac{t}{3} \right) + c$$

Hence $\displaystyle\int \frac{d\theta}{5 + 4\cos\theta} = \frac{2}{3} \arctan \left(\frac{1}{3} \tan \frac{\theta}{2} \right) + c$

Problem 47. Determine $\displaystyle\int \frac{dx}{\sin x + \cos x}$

If $t = \tan \dfrac{x}{2}$ then $\sin x = \dfrac{2t}{1+t^2}$, $\cos x = \dfrac{1-t^2}{1+t^2}$ and $dx = \dfrac{2\,dt}{1+t^2}$, from equations (1), (2) and (3).

Thus $\displaystyle\int \frac{dx}{\sin x + \cos x}$

$$= \int \frac{2\,dt/(1+t^2)}{[2t/(1+t^2)] + [(1-t^2)/(1+t^2)]}$$
$$= \int \frac{2\,dt/(1+t^2)}{(2t + 1 - t^2)/(1+t^2)}$$
$$= \int \frac{2\,dt}{1 + 2t - t^2} = \int \frac{-2\,dt}{t^2 - 2t - 1}$$
$$= \int \frac{-2\,dt}{(t-1)^2 - 2} = \int \frac{2\,dt}{(\sqrt{2})^2 - (t-1)^2}$$
$$= 2 \left[\frac{1}{2\sqrt{2}} \ln \left\{ \frac{\sqrt{2} + (t-1)}{\sqrt{2} - (t-1)} \right\} \right] + c,$$

from Problem 41

$$= \frac{1}{\sqrt{2}} \ln \left\{ \frac{\sqrt{2} - 1 + \tan \frac{x}{2}}{\sqrt{2} + 1 - \tan \frac{x}{2}} \right\} + c$$

Problem 48.   Determine

$$\int \frac{dx}{7 - 3\sin x + 6\cos x}$$

From equations (1) to (3), $\displaystyle\int \frac{dx}{7 - 3\sin x + 6\cos x}$

$$= \int \frac{2\,dt/(1+t^2)}{7 - 3(2t/(1+t^2)) + 6[(1-t^2)/(1+t^2)]}$$

$$= \int \frac{2\,dt/(1+t^2)}{[7(1+t^2) - 3(2t) + 6(1-t^2)]/(1+t^2)}$$

$$= \int \frac{2\,dt}{7 + 7t^2 - 6t + 6 - 6t^2}$$

$$= \int \frac{2\,dt}{t^2 - 6t + 13}$$

$$= \int \frac{2\,dt}{(t-3)^2 + 2^2}$$

$$= 2\left[\frac{1}{2}\arctan\left(\frac{t-3}{2}\right)\right] + c,$$

from Problem 28

Hence $\displaystyle\int \frac{dx}{7 - 3\sin x + 6\cos x}$

$$= \arctan\left(\frac{\tan\dfrac{x}{2} - 3}{2}\right) + c$$

Problem 49.   Determine $\displaystyle\int \frac{d\theta}{4\cos\theta + 3\sin\theta}$

From equations (1) to (3), $\displaystyle\int \frac{d\theta}{4\cos\theta + 3\sin\theta}$

$$= \int \frac{2\,dt/(1+t^2)}{4[(1-t^2)/(1+t^2)] + 3(2t/(1+t^2))}$$

$$= \int \frac{2\,dt}{4 - 4t^2 + 6t} = \int \frac{dt}{2 + 3t - 2t^2}$$

$$= -\frac{1}{2}\int \frac{dt}{(t^2 - (3/2)t - 1)}$$

$$= -\frac{1}{2}\int \frac{dt}{(t - 3/4)^2 - 25/16}$$

$$= \frac{1}{2}\int \frac{dt}{(5/4)^2 - (t - 3/4)^2}$$

$$= \frac{1}{2}\left[\frac{1}{2(5/4)}\ln\left\{\frac{(5/4) + (t - 3/4)}{(5/4) - (t - 3/4)}\right\}\right] + c,$$

from Problem 41,

$$= \frac{1}{5}\ln\left\{\frac{(1/2) + t}{2 - t}\right\} + c$$

Hence $\displaystyle\int \frac{d\theta}{4\cos\theta + 3\sin\theta}$

$$= \frac{1}{5}\ln\left\{\frac{(1/2) + \tan\theta/2}{2 - \tan\theta/2}\right\} + c$$

$$\text{or }\frac{1}{5}\ln\left\{\frac{1 + 2\tan\theta/2}{4 - 2\tan\theta/2}\right\} + c$$

Further problems on the $t = \tan\dfrac{\theta}{2}$ substitution may be found in the following Section 41.7, Problems 56 to 66, pages 415 and 416.

41.7   Further problems on integration using substitutions

Algebraic substitutions

In Problems 1 to 15, integrate with respect to the variable:

1   $2\sin(4x + 9)$ $\left[-\dfrac{1}{2}\cos(4x + 9) + c\right]$

2   $3\cos(2\theta - 5)$ $\left[\dfrac{3}{2}\sin(2\theta - 5) + c\right]$

3   $4\sec^2(3t + 1)$ $\left[\dfrac{4}{3}\tan(3t + 1) + c\right]$

4   $\dfrac{1}{2}(5x - 3)^6$ $\left[\dfrac{1}{70}(5x - 3)^7 + c\right]$

5   $\dfrac{-3}{(2x - 1)}$ $\left[-\dfrac{3}{2}\ln(2x - 1) + c\right]$

6   $3e^{3\theta + 5}$ $\left[e^{3\theta + 5} + c\right]$

7   $2x(2x^2 - 3)^5$ $\left[\dfrac{1}{12}(2x^2 - 3)^6 + c\right]$

8   $\dfrac{1}{3}\sin^2\theta\cos\theta$ $\left[\dfrac{1}{9}\sin^3\theta + c\right]$

9   $5\cos^5 t\sin t$ $\left[-\dfrac{5}{6}\cos^6 t + c\right]$

10 $3 \sec^2 3x \tan 3x$

$$\left[\frac{1}{3} \sec^2 3x + c \text{ or } \frac{1}{3} \tan^2 3x + c \right]$$

11 $2t\sqrt{(3t^2 - 1)}$

$$\left[\frac{2}{9} \sqrt{(3t^2 - 1)^3} + c \right]$$

12 $\dfrac{\ln \theta}{\theta}$

$$\left[\frac{1}{2} (\ln \theta)^2 + c \right]$$

13 $\dfrac{x}{\sqrt{(x^2 + 4)}}$

$$\left[\sqrt{(x^2 + 4)} + c \right]$$

14 $3 \tan 2t$

$$\left[\frac{3}{2} \ln(\sec 2t) + c \right]$$

15 $\dfrac{2e^t}{\sqrt{(e^t + 4)}}$

$$\left[4\sqrt{(e^t + 4)} + c \right]$$

In Problems 16 to 22, evaluate the definite integrals correct to 4 significant figures:

16 $\displaystyle\int_0^1 (3x + 1)^5 \, dx$ [227.5]

17 $\displaystyle\int_0^2 x\sqrt{(2x^2 + 1)} \, dx$ [4.333]

18 $\displaystyle\int_0^{\pi/3} 2 \sin\left(3t + \frac{\pi}{4}\right) dt$ [0.9428]

19 $\displaystyle\int_0^1 3 \cos(4x - 3) \, dx$ [0.7369]

20 $\displaystyle\int_0^1 3xe^{(2x^2 - 1)} \, dx$ [1.763]

21 $\displaystyle\int_0^{\pi/2} 3 \sin^4 \theta \cos \theta \, d\theta$ [0.6000]

22 $\displaystyle\int_0^1 \frac{3x}{(4x^2 - 1)^5} \, dx$ [0.09259]

Trigonometric substitutions

In Problems 23 and 24, integrate with respect to the variable:

23 (a) $\sin^2 2x$ (b) $3 \cos^2 t$

$$\left[(a) \ \frac{1}{2}\left(x - \frac{\sin 4x}{4}\right) + c \right]$$

$$\left[(b) \ \frac{3}{2}\left(t + \frac{\sin 2t}{2}\right) + c \right]$$

24 (a) $5 \tan^2 3\theta$ (b) $2 \cot^2 2t$

$$\left[(a) \ 5\left(\frac{1}{3} \tan 3\theta - \theta\right) + c \right]$$

$$\left[(b) \ -(\cot 2t + 2t) + c \right]$$

In Problems 25 and 26, evaluate the definite integrals, correct to 4 significant figures:

25 (a) $\displaystyle\int_0^{\pi/3} 3 \sin^2 3x \, dx$ (b) $\displaystyle\int_0^{\pi/4} \cos^2 4x \, dx$

$$\left[(a) \ \frac{\pi}{2} \text{ or } 1.571 \right]$$

$$\left[(b) \ \frac{\pi}{8} \text{ or } 0.3927 \right]$$

26 (a) $\displaystyle\int_0^1 2 \tan^2 2t \, dt$ (b) $\displaystyle\int_{\pi/6}^{\pi/3} \frac{1}{2} \cot^2 \theta \, d\theta$

$$[(a) \ -4.185, \ (b) \ 0.3156]$$

Powers of sines and cosines

In Problems 27 to 29, integrate with respect to the variable:

27 (a) $\sin^3 \theta$ (b) $2 \cos^3 2x$

$$\left[(a) \ -\cos \theta + \frac{\cos^3 \theta}{3} + c \right]$$

$$\left[(b) \ \sin 2x - \frac{\sin^3 2x}{3} + c \right]$$

28 (a) $2 \sin^3 t \cos^2 t$ (b) $\sin^3 x \cos^4 x$

$$\left[(a) \ \frac{-2}{3} \cos^3 t + \frac{2}{5} \cos^5 t + c \right]$$

$$\left[(b) \ \frac{-\cos^5 x}{5} + \frac{\cos^7 x}{7} + c \right]$$

29 (a) $2 \sin^4 2\theta$ (b) $\sin^2 t \cos^2 t$

$$\left[(a) \ \frac{3\theta}{4} - \frac{1}{4} \sin 4\theta + \frac{1}{32} \sin 8\theta + c \right]$$

$$\left[(b) \ \frac{t}{8} - \frac{1}{32} \sin 4t + c \right]$$

30 Show that $\displaystyle\int_0^{\pi/2} \sin^4 2t \cos^2 2t \, dt = \frac{\pi}{32}$

Products of sines and cosines

In Problems 31 and 32, integrate with respect to the variable:

31   (a) $\sin 5t \cos 2t$ (b) $2 \sin 3x \sin x$

$$\left[\text{(a) } -\frac{1}{2} \left(\frac{\cos 7t}{7} + \frac{\cos 3t}{3} \right) + c \right.$$

$$\left. \text{(b) } \frac{\sin 2x}{2} - \frac{\sin 4x}{4} + c \right]$$

32   (a) $3 \cos 6x \cos x$ (b) $\frac{1}{2} \cos 4\theta \sin 2\theta$

$$\left[\text{(a) } \frac{3}{2} \left(\frac{\sin 7x}{7} + \frac{\sin 5x}{5} \right) + c \right.$$

$$\left. \text{(b) } \frac{1}{4} \left(\frac{\cos 2\theta}{2} - \frac{\cos 6\theta}{6} \right) + c \right]$$

In Problems 33 and 34, evaluate the definite integrals:

33   (a) $\displaystyle\int_0^{\pi/2} \cos 4x \cos 3x \, dx$

(b) $\displaystyle\int_0^1 2 \sin 7t \cos 3t \, dt$

$$\left[\text{(a) } \frac{3}{7} \text{ or } 0.4286 \text{ (b) } 0.5973 \right]$$

34   (a) $-4 \displaystyle\int_0^{\pi/3} \sin 5\theta \sin 2\theta \, d\theta$

(b) $\displaystyle\int_1^2 3 \cos 8t \sin 3t \, dt$

$$[\text{(a) } 0.2474 \text{ (b) } -0.1999]$$

Sine θ substitution

35   Determine (a) $\displaystyle\int \frac{5}{\sqrt{(4 - t^2)}} \, dt$

(b) $\displaystyle\int \frac{3}{\sqrt{(9 - x^2)}} \, dx$

$$\left[\text{(a) } 5 \arcsin \frac{x}{2} + c \right.$$

$$\left. \text{(b) } 3 \arcsin \frac{x}{3} + c \right]$$

36   Determine (a) $\displaystyle\int \sqrt{(4 - x^2)} \, dx$

(b) $\displaystyle\int \sqrt{(16 - 9t^2)} \, dt$

$$\left[\text{(a) } 2 \arcsin \frac{x}{2} + \frac{x}{2}\sqrt{(4 - x^2)} + c \right.$$

$$\left. \text{(b) } \frac{8}{3} \arcsin \frac{3t}{4} + \frac{t}{2}\sqrt{(16 - 9t^2)} + c \right]$$

37   Evaluate (a) $\displaystyle\int_0^4 \frac{1}{\sqrt{(16 - x^2)}} \, dx$

(b) $\displaystyle\int_0^1 \sqrt{(9 - 4x^2)} \, dx$

$$\left[\text{(a) } \frac{\pi}{2} \text{ or } 1.571 \text{ (b) } 2.760 \right]$$

Tan θ substitution

38   Determine (a) $\displaystyle\int \frac{3}{4 + t^2} \, dt$ (b) $\displaystyle\int \frac{5}{16 + 9\theta^2} \, d\theta$

$$\left[\text{(a) } \frac{3}{2} \arctan \frac{x}{2} + c \right.$$

$$\left. \text{(b) } \frac{5}{12} \arctan \frac{3\theta}{4} + c \right]$$

39   Evaluate (a) $\displaystyle\int_0^1 \frac{3}{1 + t^2} \, dt$ (b) $\displaystyle\int_0^3 \frac{5}{4 + x^2} \, dx$

$$[\text{(a) } 2.356 \text{ (b) } 2.457]$$

Integration using partial fractions

In Problems 40 to 47, integrate with respect to x:

40   $\displaystyle\int \frac{12}{(x^2 - 9)} \, dx$

$$\left[\begin{array}{l} 2\ln(x - 3) - 2\ln(x + 3) + c \\ \text{or } \ln\left\{ \dfrac{x - 3}{x + 3} \right\}^2 + c \end{array} \right]$$

41   $\displaystyle\int \frac{4(x - 4)}{(x^2 - 2x - 3)} \, dx$

$$\left[\begin{array}{l} 5\ln(x + 1) - \ln(x - 3) + c \\ \text{or } \ln\left\{ \dfrac{(x + 1)^5}{(x - 3)} \right\} + c \end{array} \right]$$

42   $\displaystyle\int \frac{3(2x^2 - 8x - 1)}{(x + 4)(x + 1)(2x - 1)} \, dx$

$$\left[\begin{array}{l} 7\ln(x + 4) - 3\ln(x + 1) - \ln(2x - 1) + c \\ \text{or } \ln\left\{ \dfrac{(x + 4)^7}{(x + 1)^3 (2x - 1)} \right\} + c \end{array} \right]$$

43   $\displaystyle\int \frac{x^2 + 9x + 8}{x^2 + x - 6} \, dx$

$$\left[\begin{array}{l} x + 2\ln(x + 3) + 6\ln(x - 2) + c \\ \text{or } x + \ln\{(x + 3)^2 (x - 2)^6\} + c \end{array} \right]$$

44 $\int \dfrac{3x^3 - 2x^2 - 16x + 20}{(x - 2)(x + 2)} dx$

$$\left[\dfrac{3x^2}{2} - 2x + \ln(x - 2) - 5\ln(x + 2) + c\right]$$

45 $\int \dfrac{4x - 3}{(x + 1)^2} dx$

$$\left[4\ln(x + 1) + \dfrac{7}{(x + 1)} + c\right]$$

46 $\int \dfrac{5x^2 - 30x + 44}{(x - 2)^3} dx$

$$\left[5\ln(x - 2) + \dfrac{10}{(x - 2)} - \dfrac{2}{(x - 2)^2} + c\right]$$

47 $\int \dfrac{x^2 - x - 13}{(x^2 + 7)(x - 2)} dx$

$$\left[\ln(x^2 + 7) + \dfrac{3}{\sqrt{7}} \arctan \dfrac{x}{\sqrt{7}} - \ln(x - 2) + c\right]$$

In Problems 48 to 54, evaluate the definite integrals correct to 4 significant figures:

48 $\int_3^4 \dfrac{x^2 - 3x + 6}{x(x - 2)(x - 1)} dx$ [0.6275]

49 $\int_4^6 \dfrac{x^2 - x - 14}{x^2 - 2x - 3} dx$ [0.8122]

50 $\int_1^2 \dfrac{x^2 + 7x + 3}{x^2(x + 3)} dx$ [1.663]

51 $\int_6^7 \dfrac{18 + 21x - x^2}{(x - 5)(x + 2)^2} dx$ [1.089]

52 $\int_5^6 \dfrac{6x - 5}{(x - 4)(x^2 + 3)} dx$ [0.5880]

53 $\int_1^2 \dfrac{4}{(16 - x^2)} dx$ [0.2939]

54 $\int_4^5 \dfrac{2}{(x^2 - 9)} dx$ [0.1865]

55 The velocity constant k of a given chemical reaction is given by:

$$kt = \int \dfrac{dx}{(3 - 0.4x)(2 - 0.6x)}$$

where $x = 0$ when $t = 0$. Determine kt.

$$\left[\ln\left\{\dfrac{2(3 - 0.4x)}{3(2 - 0.6x)}\right\}\right]$$

$t = \tan \dfrac{\theta}{2}$ **substitution**

In Problems 56 to 63, integrate with respect to the variable:

56 $\int \dfrac{d\theta}{1 + \sin \theta}$ $\left[\dfrac{-2}{1 + \tan \dfrac{\theta}{2}} + c\right]$

57 $\int \dfrac{dx}{1 - \cos x + \sin x}$ $\left[\ln\left\{\dfrac{\tan \dfrac{x}{2}}{1 + \tan \dfrac{x}{2}}\right\} + c\right]$

58 $\int \dfrac{d\alpha}{3 + 2\cos \alpha}$

$$\left[\dfrac{2}{\sqrt{5}} \arctan\left(\dfrac{1}{\sqrt{5}} \tan \dfrac{\alpha}{2}\right) + c\right]$$

59 $\int \dfrac{dx}{3 \sin x - 4 \cos x}$

$$\left[\dfrac{1}{5} \ln\left\{\dfrac{2\tan \dfrac{x}{2} - 1}{2\tan \dfrac{x}{2} + 4}\right\} + c\right]$$

60 $\int \dfrac{dp}{3 - 4\sin p + 2\cos p}$

$$\left[\dfrac{1}{\sqrt{11}} \ln\left\{\dfrac{\tan \dfrac{p}{2} - 4 - \sqrt{11}}{\tan \dfrac{p}{2} - 4 + \sqrt{11}}\right\} + c\right]$$

61 $\int \dfrac{d\theta}{5 + 4\sin \theta}$

$$\left[\dfrac{2}{3} \arctan\left(\dfrac{5\tan \dfrac{x}{2} + 4}{3}\right) + c\right]$$

62 $\int \dfrac{dx}{1 + 2\sin x}$

$$\left[\dfrac{1}{\sqrt{3}} \ln\left\{\dfrac{\tan \dfrac{x}{2} + 2 - \sqrt{3}}{\tan \dfrac{x}{2} + 2 + \sqrt{3}}\right\} + c\right]$$

63 $\displaystyle\int \frac{d\theta}{3 - 4\sin\theta}$

$$\left[\frac{1}{\sqrt{7}} \ln \left\{ \frac{3\tan\dfrac{\theta}{2} - 4 - \sqrt{7}}{3\tan\dfrac{\theta}{2} - 4 + \sqrt{7}} \right\} + c \right]$$

64 Show that

$$\int \frac{dt}{1 + 3\cos t} = \frac{1}{2\sqrt{2}} \ln \left\{ \frac{\sqrt{2} + \tan\dfrac{t}{2}}{\sqrt{2} - \tan\dfrac{t}{2}} \right\} + c$$

65 Show that $\displaystyle\int_0^{\pi/3} \frac{3\,d\theta}{\cos\theta} = 1.317$, correct to 4 significant figures.

66 Show that $\displaystyle\int_0^{\pi/2} \frac{d\theta}{2 + \cos\theta} = \frac{\pi}{3\sqrt{3}}$

42

Integration by parts

42.1 Introduction

From the product rule of differentiation:

$$\frac{d}{dx}(uv) = v\frac{du}{dx} + u\frac{dv}{dx}$$

where u and v are both functions of x.

Rearranging gives: $u\frac{dv}{dx} = \frac{d}{dx}(uv) - v\frac{du}{dx}$

Integrating both sides with respect to x gives:

$$\int u\frac{dv}{dx}\,dx = \int \frac{d}{dx}(uv)\,dx - \int v\frac{du}{dx}\,dx$$

i.e.

$$\boxed{\int u\frac{dv}{dx}\,dx = uv - \int v\frac{du}{dx}\,dx}$$

or

$$\boxed{\int u\,dv = uv - \int v\,du}$$

This is known as the **integration by parts formula** and provides a method of integrating such products of simple functions as $\int xe^x\,dx$, $\int t\sin t\,dt$, $\int e^\theta \cos\theta\,d\theta$ and $\int x\ln x\,dx$.

Given a product of two terms to integrate the initial choice is: 'which part to make equal to u' and 'which part to make equal to dv'. The choice must be such that the 'u part' becomes a constant after successive differentiation and the 'dv part' can be integrated from standard integrals. Invariably, the following rule holds: 'If a product to be integrated contains an algebraic term (such as x, t^2 or 3θ) then this term is chosen as the 'u part'. The one exception to this rule is when a '$\ln x$' term is involved: in this case $\ln x$ is chosen as the 'u part'.

42.2 Worked problems on integration by parts

Problem 1. Determine $\int x\cos x\,dx$

From the integration by parts formula,

$$\int u\,dv = uv - \int v\,du$$

Let $u = x$, from which $\frac{du}{dx} = 1$, i.e. $du = dx$ and let $dv = \cos x\,dx$, from which $v = \int \cos x\,dx = \sin x$

Expressions for u, du, v and dv are now substituted into the 'by parts' formula as shown below.

$$
\begin{array}{ccccc}
\int u & dv & = & u & v & -\int & v & du \\
\int x & \cos x\,dx & = & (x) & (\sin x) & -\int & (\sin x) & (dx)
\end{array}
$$

i.e. $\int x\cos x\,dx = x\sin x - (-\cos x) + c$

$$= x\sin x + \cos x + c$$

This result may be checked by differentiating the right-hand side.

i.e. $\dfrac{d}{dx}(x\sin x + \cos x + c)$

$= [(x)(\cos x) + (\sin x)(1)] - \sin x + 0 = x\cos x,$

which is the function being integrated

Problem 2. Find $\int 3te^{2t}\,dt$

Let $u = 3t$, from which, $\frac{du}{dt} = 3$, i.e. $du = 3\,dt$ and

let $dv = e^{2t}\,dt$, from which, $v = \int e^{2t}\,dt = \frac{1}{2}e^{2t}$

Substituting into $\int u\,dv = uv - \int v\,du$ gives:

$$\int 3te^{2t}\,dt = (3t)\left(\frac{1}{2}e^{2t}\right) - \int \left(\frac{1}{2}e^{2t}\right)(3\,dt)$$

$$= \frac{3}{2}te^{2t} - \frac{3}{2}\int e^{2t}\,dt$$

$$= \frac{3}{2}te^{2t} - \frac{3}{2}\left(\frac{e^{2t}}{2}\right) + c$$

Hence $\int 3te^{2t}\,dt = \dfrac{3}{2}e^{2t}\left(t - \dfrac{1}{2}\right) + c$, which may be checked by differentiation.

Problem 3.   Evaluate $\displaystyle\int_0^{\pi/2} 2\theta \sin\theta\,d\theta$

Let $u = 2\theta$, from which, $\dfrac{du}{d\theta} = 2$, i.e. $du = 2\,d\theta$ and let $dv = \sin\theta\,d\theta$, from which,

$$v = \int \sin\theta\,d\theta = -\cos\theta$$

Substituting into $\int u\,dv = uv - \int v\,du$ gives:

$$\int 2\theta \sin\theta\,d\theta = (2\theta)(-\cos\theta) - \int(-\cos\theta)(2\,d\theta)$$

$$= -2\theta\cos\theta + 2\int\cos\theta\,d\theta$$

$$= -2\theta\cos\theta + 2\sin\theta + c$$

Hence $\displaystyle\int_0^{\pi/2} 2\theta\sin\theta\,d\theta$

$$= [-2\theta\cos\theta + 2\sin\theta]_0^{\pi/2}$$

$$= \left[-2\left(\dfrac{\pi}{2}\right)\cos\dfrac{\pi}{2} + 2\sin\dfrac{\pi}{2}\right] - [0 + 2\sin 0]$$

$$= (-0 + 2) - (0 + 0),$$

since $\cos\dfrac{\pi}{2} = 0$ and $\sin\dfrac{\pi}{2} = 1$,

$$= 2$$

Problem 4.   Evaluate $\displaystyle\int_0^1 5xe^{4x}\,dx$, **correct to 3 significant figures**

Let $u = 5x$, from which $\dfrac{du}{dx} = 5$, i.e. $du = 5\,dx$ and let $dv = e^{4x}\,dx$, from which, $v = \int e^{4x}\,dx = \dfrac{1}{4}e^{4x}$.

Substituting into $\int u\,dv = uv - \int v\,du$ gives:

$$\int 5xe^{4x}\,dx = (5x)\left(\dfrac{e^{4x}}{4}\right) - \int\left(\dfrac{e^{4x}}{4}\right)(5\,dx)$$

$$= \dfrac{5}{4}xe^{4x} - \dfrac{5}{4}\int e^{4x}\,dx$$

$$= \dfrac{5}{4}xe^{4x} - \dfrac{5}{4}\left(\dfrac{e^{4x}}{4}\right) + c$$

$$= \dfrac{5}{4}e^{4x}\left(x - \dfrac{1}{4}\right) + c$$

Hence $\displaystyle\int_0^1 5xe^{4x}\,dx$

$$= \left[\dfrac{5}{4}e^{4x}\left(x - \dfrac{1}{4}\right)\right]_0^1$$

$$= \left[\dfrac{5}{4}e^4\left(1 - \dfrac{1}{4}\right)\right] - \left[\dfrac{5}{4}e^0\left(0 - \dfrac{1}{4}\right)\right]$$

$$= \left(\dfrac{15}{16}e^4\right) - \left(-\dfrac{5}{16}\right)$$

$$= 51.186 + 0.313 = 51.499 = \mathbf{51.5},$$

correct to 3 significant figures

Problem 5.   Determine $\displaystyle\int x^2 \sin x\,dx$

Let $u = x^2$, from which, $\dfrac{du}{dx} = 2x$, i.e. $du = 2x\,dx$, and let $dv = \sin x\,dx$, from which,

$$v = \int \sin x\,dx = -\cos x$$

Substituting into $\int u\,dv = uv - \int v\,dv$ gives:

$$\int x^2 \sin x\,dx = (x^2)(-\cos x) - \int(-\cos x)(2x\,dx)$$

$$= -x^2\cos x + 2\left[\int x\cos x\,dx\right]$$

The integral, $\int x\cos x\,dx$, is not a 'standard integral' and it can only be determined by using the integration by parts formula again.

From Problem 1, $\int x\cos x\,dx = x\sin x + \cos x$

Hence $\displaystyle\int x^2 \sin x\,dx$

$$= -x^2\cos x + 2\{x\sin x + \cos x\} + c$$

$$= -x^2\cos x + 2x\sin x + 2\cos x + c$$

$$= (2 - x^2)\cos x + 2x\sin x + c$$

In general, if the algebraic term of a product is of power n, then the integration by parts formula is applied n times.

Problem 6. Find $\int x \ln x \, dx$

The logarithmic function is chosen as the 'u part'.

Thus when $u = \ln x$, then $\dfrac{du}{dx} = \dfrac{1}{x}$ i.e. $du = \dfrac{dx}{x}$

Letting $dv = x \, dx$ gives $v = \int x \, dx = \dfrac{x^2}{2}$

Substituting into $\int u \, dv = uv - \int v \, du$ gives:

$$\int x \ln x \, dx = (\ln x)\left(\frac{x^2}{2}\right) - \int \left(\frac{x^2}{2}\right)\frac{dx}{x}$$

$$= \frac{x^2}{2}\ln x - \frac{1}{2}\int x \, dx$$

$$= \frac{x^2}{2}\ln x - \frac{1}{2}\left(\frac{x^2}{2}\right) + c$$

Hence $\int x \ln x \, dx = \dfrac{x^2}{2}\left(\ln x - \dfrac{1}{2}\right) + c$ or

$$\frac{x^2}{4}(2\ln x - 1) + c$$

Problem 7. Determine $\int \ln x \, dx$

$\int \ln x \, dx$ is the same as $\int (1)\ln x \, dx$

Let $u = \ln x$, from which, $\dfrac{du}{dx} = \dfrac{1}{x}$, i.e. $du = \dfrac{dx}{x}$

and let $dv = 1 \, dx$, from which, $v = \int 1 \, dx = x$.

Substituting into $\int u \, dv = uv - \int v \, du$ gives:

$$\int \ln x \, dx = (\ln x)(x) - \int x\frac{dx}{x}$$

$$= x \ln x - \int dx = x \ln x - x + c$$

Hence $\int \ln x \, dx = x(\ln x - 1) + c$

Problem 8. Evaluate $\int_1^9 \sqrt{x}\ln x \, dx$, correct to 3 significant figures

---

Let $u = \ln x$, from which, $du = \dfrac{dx}{x}$ and

let $dv = \sqrt{x}\, dx = x^{1/2}\, dx$, from which,

$$v = \int x^{1/2}\, dx = \frac{2}{3}x^{3/2}$$

Substituting into $\int u \, dv = uv - \int v \, du$ gives:

$$\int \sqrt{x}\ln x \, dx = (\ln x)\left(\frac{2}{3}x^{3/2}\right) - \int\left(\frac{2}{3}x^{3/2}\right)\left(\frac{dx}{x}\right)$$

$$= \frac{2}{3}\sqrt{x^3}\ln x - \frac{2}{3}\int x^{1/2}\, dx$$

$$= \frac{2}{3}\sqrt{x^3}\ln x - \frac{2}{3}\left(\frac{2}{3}x^{3/2}\right) + c$$

$$= \frac{2}{3}\sqrt{x^3}\left[\ln x - \frac{2}{3}\right] + c$$

Hence $\int_1^9 \sqrt{x}\ln x \, dx$

$$= \left[\frac{2}{3}\sqrt{x^3}\left(\ln x - \frac{2}{3}\right)\right]_1^9$$

$$= \left[\frac{2}{3}\sqrt{9^3}\left(\ln 9 - \frac{2}{3}\right)\right] - \left[\frac{2}{3}\sqrt{1^3}\left(\ln 1 - \frac{2}{3}\right)\right]$$

$$= \left[18\left(\ln 9 - \frac{2}{3}\right)\right] - \left[\frac{2}{3}\left(0 - \frac{2}{3}\right)\right]$$

$$= 27.550 + 0.444 = 27.994 = \mathbf{28.0},$$

correct to 3 significant figures

Problem 9. Find $\int e^{ax}\cos bx \, dx$

When integrating a product of an exponential and a sine or cosine function it is immaterial which part is made equal to 'u'.

Let $u = e^{ax}$, from which $\dfrac{du}{dx} = ae^{ax}$,

i.e. $du = ae^{ax}\, dx$ and let $dv = \cos bx \, dx$, from

which, $v = \int \cos bx \, dx = \dfrac{1}{b} \sin bx$.

Substituting into $\int u \, dv = uv - \int v \, du$ gives:

$$\int e^{ax} \cos bx \, dx$$

$$= (e^{ax}) \left(\frac{1}{b} \sin bx \right) - \int \left(\frac{1}{b} \sin bx \right) (ae^{ax} \, dx)$$

$$= \frac{1}{b} e^{ax} \sin bx - \frac{a}{b} \left[\int e^{ax} \sin bx \, dx \right] \qquad (1)$$

$\int e^{ax} \sin bx \, dx$ is now determined separately using integration by parts again:

Let $u = e^{ax}$ then $du = ae^{ax} \, dx$, and let $dv = \sin bx \, dx$, from which $v = \int \sin bx \, dx = -\dfrac{1}{b} \cos bx$.

Substituting into the parts formula gives:

$$\int e^{ax} \sin bx \, dx = (e^{ax}) \left(-\frac{1}{b} \cos bx \right)$$

$$- \int \left(-\frac{1}{b} \cos bx \right) (ae^{ax} \, dx)$$

$$= -\frac{1}{b} e^{ax} \cos bx$$

$$+ \frac{a}{b} \int e^{ax} \cos bx \, dx$$

Substituting this result into equation (1) gives:

$$\int e^{ax} \cos bx \, dx = \frac{1}{b} e^{ax} \sin bx - \frac{a}{b} \left[-\frac{1}{b} e^{ax} \cos bx \right.$$

$$\left. + \frac{a}{b} \int e^{ax} \cos bx \, dx \right]$$

$$= \frac{1}{b} e^{ax} \sin bx + \frac{a}{b^2} e^{ax} \cos bx$$

$$- \frac{a^2}{b^2} \int e^{ax} \cos bx \, dx$$

The integral on the far right of this equation is the same as the integral on the left-hand side and thus they may be combined.

$$\int e^{ax} \cos bx \, dx + \frac{a^2}{b^2} \int e^{ax} \cos bx \, dx$$

$$= \frac{1}{b} e^{ax} \sin bx + \frac{a}{b^2} e^{ax} \cos bx$$

i.e. $\left(1 + \dfrac{a^2}{b^2} \right) \int e^{ax} \cos bx \, dx$

$$= \frac{1}{b} e^{ax} \sin bx + \frac{a}{b^2} e^{ax} \cos bx$$

i.e. $\left(\dfrac{b^2 + a^2}{b^2} \right) \int e^{ax} \cos bx \, dx$

$$= \frac{e^{ax}}{b^2} (b \sin bx + a \cos bx)$$

Hence $\displaystyle\int e^{ax} \cos bx \, dx$

$$= \left(\frac{b^2}{a^2 + b^2} \right) \left(\frac{e^{ax}}{b^2} \right) (b \sin bx + a \cos bx)$$

$$= \frac{e^{ax}}{a^2 + b^2} (b \sin bx + a \cos bx) + c$$

Using a similar method to the above, that is, integrating by parts twice, the following result may be proved:

$$\int e^{ax} \sin bx \, dx$$

$$= \frac{e^{ax}}{a^2 + b^2} (a \sin bx - b \cos bx) + c \qquad (2)$$

Problem 10. Evaluate $\displaystyle\int_0^{\pi/4} e^t \sin 2t \, dt$, correct to 4 decimal places

Comparing $\int e^t \sin 2t \, dt$ with $\int e^{ax} \sin bx \, dx$ shows that $x = t$, $a = 1$ and $b = 2$.

Hence, substituting into equation (2) gives:

$$\int_0^{\pi/4} e^t \sin 2t \, dt$$

$$= \left[\frac{e^t}{1^2 + 2^2} (1 \sin 2t - 2 \cos 2t) \right]_0^{\pi/4}$$

$$= \left[\frac{e^{\pi/4}}{5} \left(\sin 2 \left(\frac{\pi}{4} \right) - 2 \cos 2 \left(\frac{\pi}{4} \right) \right) \right]$$

$$- \left[\frac{e^0}{5} (\sin 0 - 2 \cos 0) \right]$$

$$= \left[\frac{e^{\pi/4}}{5} (1 - 0) \right] - \left[\frac{1}{5} (0 - 2) \right] = \frac{e^{\pi/4}}{5} + \frac{2}{5}$$

$$= \mathbf{0.8387}, \text{ correct to 4 decimal places}$$

42.3   Further problems on integration by parts

Determine the integrals in Problems 1 to 10 using integration by parts:

1   $\int xe^{2x}\,dx$ $\left[\dfrac{e^{2x}}{2}\left(x-\dfrac{1}{2}\right)+c\right]$

2   $\int \dfrac{4x}{e^{3x}}\,dx$ $\left[-\dfrac{4}{3}e^{-3x}\left(x+\dfrac{1}{3}\right)+c\right]$

3   $\int x\sin x\,dx$ $[-x\cos x+\sin x+c]$

4   $\int 2x^2\ln x\,dx$ $\left[\dfrac{2}{3}x^3\left(\ln x-\dfrac{1}{3}\right)+c\right]$

5   $\int 2\ln 3x\,dx$ $[2x(\ln 3x-1)+c]$

6   $\int 5\theta\cos 2\theta\,d\theta$

$\left[\dfrac{5}{2}\left(\theta\sin 2\theta+\dfrac{1}{2}\cos 2\theta\right)+c\right]$

7   $\int 3t^2e^{2t}\,dt$ $\left[\dfrac{3}{2}e^{2t}\left(t^2-t+\dfrac{1}{2}\right)+c\right]$

8   $\int x^2\sin 3x\,dx$

$\left[\dfrac{\cos 3x}{27}(2-9x^2)+\dfrac{2}{9}x\sin 3x+c\right]$

9   $\int 2e^{5x}\cos 2x\,dx$

$\left[\dfrac{2}{29}e^{5x}(2\sin 2x+5\cos 2x)+c\right]$

10   $\int 2\theta\sec^2\theta\,d\theta$ $[2[\theta\tan\theta-\ln(\sec\theta)]+c]$

Evaluate the integrals in Problems 11 to 18, correct to 4 significant figures:

11   $\int_0^2 2xe^x\,dx$ [16.78]

12   $\int_0^{\pi/4} x\sin 2x\,dx$ [0.2500]

13   $\int_1^2 x\ln x\,dx$ [0.6363]

14   $\int_0^{\pi/2} t^2\cos t\,dt$ [0.4674]

15   $\int_1^2 3x^2e^{x/2}\,dx$ [15.78]

16   $\int_0^1 2e^{3x}\sin 2x\,dx$ [11.31]

17   $\int_0^{\pi/2} e^t\cos 3t\,dt$ [−1.543]

18   $\int_1^4 \sqrt{x^3}\ln x\,dx$ [12.78]

19   In determining a Fourier series to represent $f(x)=x$ in the range $-\pi$ to π, Fourier coefficients are given by:

$$a_n=\dfrac{1}{\pi}\int_{-\pi}^{\pi} x\cos nx\,dx \text{ and}$$

$$b_n=\dfrac{1}{\pi}\int_{-\pi}^{\pi} x\sin nx\,dx$$

where n is a positive integer. Show by using integration by parts that $a_n=0$ and

$$b_n=-\dfrac{2}{n}\cos n\pi.$$

20   If a string is plucked at a point $x=\frac{l}{3}$ with an amplitude a and released, the equation of motion is:

$$K=\dfrac{2}{l}\left\{\int_0^{l/2}\dfrac{3a}{l}x\sin\dfrac{n\pi}{l}x\,dx\right.$$

$$\left.+\int_{1/3}^1\dfrac{3a}{2l}(l-x)\sin\dfrac{n\pi}{l}x\,dx\right\}$$

where n is a constant.

Show that $K=\dfrac{9a}{\pi^2 n^2}\sin\left(\dfrac{n\pi}{3}\right)$.

21   The equation $C=\displaystyle\int_0^1 e^{-0.4\theta}\cos 1.2\theta\,d\theta$

and $S=\displaystyle\int_0^1 e^{-0.4\theta}\sin 1.2\theta\,d\theta$

are involved in the study of damped oscillations. Determine the values of C and S.

$[C=0.66,\ S=0.41]$

43

An introduction to complex numbers

43.1 Cartesian complex numbers

(i) If the quadratic equation $x^2 + 2x + 5 = 0$ is solved using the quadratic formula then

$$x = \frac{-2 \pm \sqrt{[(2)^2 - (4)(1)(5)]}}{2(1)}$$

$$= \frac{-2 \pm \sqrt{-16}}{2}$$

$$= \frac{-2 \pm \sqrt{[(16)(-1)]}}{2}$$

$$= \frac{-2 \pm \sqrt{16}\sqrt{-1}}{2} = \frac{-2 \pm 4\sqrt{-1}}{2}$$

$$= -1 \pm 2\sqrt{-1}$$

It is not possible to evaluate $\sqrt{-1}$ in real terms. However, if an operator j is defined as $j = \sqrt{-1}$ then the solution may be expressed as $x = -1 \pm j2$

(ii) $-1 + j2$ and $-1 - j2$ are known as **complex numbers**. Both solutions are of the form $a + jb$, 'a' being termed the **real part** and jb the **imaginary part.** A complex number of the form $a + jb$ is called a **cartesian complex number**

(iii) In pure mathematics the symbol i is used to indicate $\sqrt{-1}$ (i being the first letter of the word imaginary). However, i is the symbol of electric current, and to avoid possible confusion the next letter in the alphabet, j, is used to represent $\sqrt{-1}$

Problem 1. Solve the quadratic equation $x^2 + 4 = 0$

Since $x^2 + 4 = 0$ then $x^2 = -4$ and $x = \sqrt{-4}$
i.e. $x = \sqrt{[(-1)(4)]} = \sqrt{(-1)}\sqrt{4} = j(\pm 2) = \pm j2$
(since $j = \sqrt{-1}$).
(Note that $\pm j2$ may also be written as $\pm 2j$.)

Problem 2. Solve the quadratic equation $2x^2 + 3x + 5 = 0$

Using the quadratic formula,

$$x = \frac{-3 \pm \sqrt{[(3)^2 - 4(2)(5)]}}{2(2)}$$

$$= \frac{-3 \pm \sqrt{-31}}{4} = \frac{-3 \pm \sqrt{(-1)}\sqrt{31}}{4}$$

$$= \frac{-3 \pm j\sqrt{31}}{4}$$

Hence $x = -\dfrac{3}{4} + j\dfrac{\sqrt{31}}{4}$ or $-0.750 \pm j1.392$, correct to 3 decimal places.
(Note, a graph of $y = 2x^2 + 3x + 5$ does not cross the x-axis and hence $2x^2 + 3x + 5 = 0$ has no real roots.)

Problem 3. Evaluate (a) j^3 (b) j^4 (c) j^{23} (d) $\dfrac{-4}{j^9}$

(a) $j^3 = j^2 \times j = (-1) \times j = -j$, since $j^2 = -1$

(b) $j^4 = j^2 \times j^2 = (-1) \times (-1) = 1$

(c) $j^{23} = j \times j^{22} = j \times (j^2)^{11} = j \times (-1)^{11}$
    $= j \times (-1) = -j$

(d) $j^9 = (j^4)^2(j) = (1)^2(j) = j$

Hence $\dfrac{-4}{j^9} = \dfrac{-4}{j} = \dfrac{-4}{j} \times \dfrac{-j}{-j}$

$$= \dfrac{4j}{-j^2} = \dfrac{4j}{-(-1)} = 4j$$

Further problems on the introduction to complex numbers may be found in Section 43.10, Problems 1 to 4, page 432.

43.2   The Argand diagram

A complex number may be represented pictorially on rectangular or cartesian axes. The horizontal (or x) axis is used to represent the real axis and the vertical (or y) axis is used to represent the imaginary axis. Such a diagram is called an **Argand diagram**. In Fig. 43.1, the point A represents the complex number $(3 + j2)$ and is obtained by plotting the co-ordinates $(3, j2)$ as in graphical work. Figure 43.1 also shows the Argand points B, C and D representing the complex numbers $(-2 + j4)$, $(-3 - j5)$ and $(1 - j3)$, respectively.

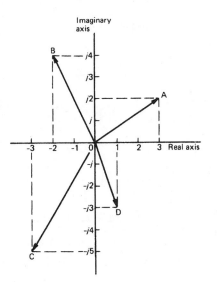

Figure 43.1

43.3   Addition and subtraction of complex numbers

Two complex numbers are added/subtracted by adding/subtracting separately the two real parts and the two imaginary parts.
For example, if $Z_1 = a + jb$ and $Z_2 = c + jd$,

then $Z_1 + Z_2 = (a + jb) + (c + jd)$

$$= (a + c) + j(b + d)$$

and $Z_1 - Z_2 = (a + jb) - (c + jd)$

$$= (a - c) + j(b - d)$$

Thus, for example,

$$(2 + j3) + (3 - j4) = 2 + j3 + 3 - j4$$

$$= 5 - j1$$

and $$(2 + j3) - (3 - j4) = 2 + j3 - 3 + j4$$

$$= -1 + j7$$

The addition and subtraction of complex numbers may be achieved graphically as shown in the Argand diagram of Fig. 43.2. $(2 + j3)$ is represented by vector OP and $(3 - j4)$ by vector OQ. In Fig. 43.2(a) by vector addition (i.e. the diagonal of the parallelogram) $OP + OQ = OR$. R is the point $(5, -j1)$. Hence $(2 + j3) + (3 - j4) = 5 - j1$.

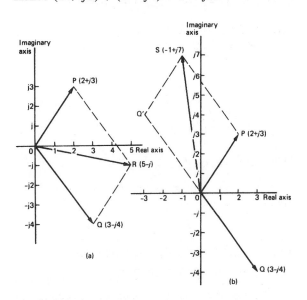

Figure 43.2

In Fig. 43.2(b), vector OQ is reversed (shown as OQ') since it is being subtracted. (Note $OQ = 3 - j4$ and $OQ' = -(3 - j4) = -3 + j4$.) $OP - OQ = OP + OQ' = OS$ is found to be the Argand point $(-1, j7)$.
Hence $(2 + j3) - (3 - j4) = -1 + j7$.

Problem 4.   Given $Z_1 = 2 + j4$ and $Z_2 = 3 - j$ determine (a) $Z_1 + Z_2$, (b) $Z_1 - Z_2$, and (c) $Z_2 - Z_1$, and show the results on an Argand diagram

---

(a) $Z_1 + Z_2 = (2 + j4) + (3 - j)$

$$= (2 + 3) + j(4 - 1) = \mathbf{5 + j3}$$

(b) $Z_1 - Z_2 = (2 + j4) - (3 - j)$

$$= (2 - 3) + j(4 - -1) = \mathbf{-1 + j5}$$

(c) $Z_2 - Z_1 = (3 - j) - (2 + j4)$

$$= (3 - 2) + j(-1 - 4) = \mathbf{1 - j5}$$

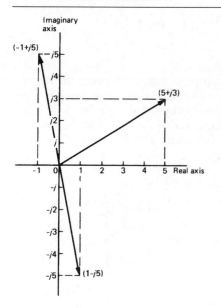

Figure 43.3

Each result is shown in the Argand diagram of Fig. 43.3.

43.4   Multiplication and division of complex numbers

(i)   **Multiplication of complex numbers** is achieved by assuming all quantities involved are real and then using $j^2 = -1$ to simplify.

Hence $(a + jb)(c + jd)$
$$= ac + a(jd) + (jb)c + (jb)(jd)$$
$$= ac + jad + jbc + j^2bd$$
$$= (ac - bd) + j(ad + bc), \text{ since } j^2 = -1$$

Thus $(3 + j2)(4 - j5)$
$$= 12 - j15 + j8 - j^2 10$$
$$= (12 - -10) + j(-15 + 8) = 22 - j7$$

(ii)   The **complex conjugate** of a complex number is obtained by changing the sign of the imaginary part. Hence the complex conjugate of $a + jb$ is $a - jb$. The product of a complex number and its complex conjugate is always a real number.
For example, $(3 + j4)(3 - j4) = 9 - j12 + j12 - j^2 16 = 9 + 16 = 25$. $((a + jb)(a - jb)$ may be evaluated 'on sight' as $a^2 + b^2$)

(iii)   **Division of complex numbers** is achieved by multiplying both numerator and denominator by the complex conjugate of the denominator.

For example,

$$\frac{2 - j5}{3 + j4} = \frac{2 - j5}{3 + j4} \times \frac{(3 - j4)}{(3 - j4)}$$

$$= \frac{6 - j8 - j15 + j^2 20}{3^2 + 4^2}$$

$$= \frac{-14 - j23}{25}$$

$$= \frac{-14}{25} - j\frac{23}{25} \text{ or } -0.56 - j0.92$$

Problem 5.   If $Z_1 = 1 - j3$, $Z_2 = -2 + j5$ and $Z_3 = -3 - j4$, determine in $a + jb$ form:
(a) $Z_1 Z_2$ (b) $\dfrac{Z_1}{Z_3}$ (c) $\dfrac{Z_1 Z_2}{Z_1 + Z_2}$ (d) $Z_1 Z_2 Z_3$

(a) $Z_1 Z_2 = (1 - j3)(-2 + j5)$
$$= -2 + j5 + j6 - j^2 15$$
$$= (-2 + 15) + j(5 + 6), \text{ since } j^2 = -1,$$
$$= \mathbf{13 + j11}$$

(b) $\dfrac{Z_1}{Z_3} = \dfrac{1 - j3}{-3 - j4} = \dfrac{1 - j3}{-3 - j4} \times \dfrac{-3 + j4}{-3 + j4}$

$$= \frac{-3 + j4 + j9 - j^2 12}{3^2 + 4^2} = \frac{9 + j13}{25}$$

$$= \frac{9}{25} + j\frac{13}{25} \text{ or } \mathbf{0.36 + j0.52}$$

(c) $\dfrac{Z_1 Z_2}{Z_1 + Z_2} = \dfrac{(1 - j3)(-2 + j5)}{(1 - j3) + (-2 + j5)}$

$$= \frac{13 + j11}{-1 + j2}, \text{ from part (a)},$$

$$= \frac{13 + j11}{-1 + j2} \times \frac{-1 - j2}{-1 - j2}$$

$$= \frac{-13 - j26 - j11 - j^2 22}{1^2 + 2^2}$$

$$= \frac{9 - j37}{5}$$

$$= \frac{9}{5} - j\frac{37}{5} \text{ or } \mathbf{1.8 - j7.4}$$

(d) $Z_1 Z_2 Z_3 = (13 + j11)(-3 - j4)$, since
$$Z_1 Z_2 = 13 + j11, \text{ from part (a)}$$

i.e. $Z_1Z_2Z_3 = -39 - j52 - j33 - j^2 44$

$$= (-39 + 44) - j(52 + 33)$$

$$= 5 - j85$$

Problem 6. Evaluate: (a) $\dfrac{2}{(1+j)^4}$

(b) $j\left(\dfrac{1+j3}{1-j2}\right)^2$

(a) $(1+j)^2 = (1+j)(1+j) = 1 + j + j + j^2$
$$= 1 + j + j - 1 = j2$$
$(1+j)^4 = [(1+j)^2]^2 = (j2)^2 = j^2 4 = -4$

Hence $\dfrac{2}{(1+j)^4} = \dfrac{2}{-4} = -\dfrac{1}{2}$

(b) $\dfrac{1+j3}{1-j2} = \dfrac{1+j3}{1-j2} \times \dfrac{1+j2}{1+j2}$

$$= \dfrac{1 + j2 + j3 + j^2 6}{1^2 + 2^2} = \dfrac{-5 + j5}{5}$$

$$= -1 + j1 = -1 + j$$

$\left(\dfrac{1+j3}{1-j2}\right)^2 = (-1+j)^2$

$$= (-1+j)(-1+j)$$

$$= 1 - j - j + j^2 = -j2$$

Hence $j\left(\dfrac{1+j3}{1-j2}\right)^2 = j(-j2)$

$$= -j^2 2 = 2,$$
since $j^2 = -1$

Further problems on operations involving Cartesian complex numbers may be found in Section 43.10, Problems 5 to 13, pages 432 and 433.

43.5   Complex equations

If two complex numbers are equal, then their real parts are equal and their imaginary parts are equal. Hence if $a + jb = c + jd$, then $a = c$ and $b = d$.

Problem 7. Solve the complex equations:
(a) $2(x + jy) = 6 - j3$
(b) $(1 + j2)(-2 - j3) = a + jb$

(a) $2(x + jy) = 6 - j3$. Hence $2x + j2y = 6 - j3$
Equating the real parts gives:
$$2x = 6, \text{ i.e. } x = 3$$
Equating the imaginary parts gives:
$$2y = -3, \text{ i.e. } y = -\dfrac{3}{2}$$

(b) $(1 + j2)(-2 - j3) = a + jb$
$-2 - j3 - j4 - j^2 6 = a + jb$
Hence $4 - j7 = a + jb$
Equating real and imaginary terms gives:
$a = 4$ and $b = -7$

Problem 8. Solve the equations:
(a) $(2 - j3) = \sqrt{(a + jb)}$
(b) $(x - j2y) + (y - j3x) = 2 + j3$

(a) $(2 - j3) = \sqrt{(a + b)}$
Hence $(2 - j3)^2 = a + jb,$
i.e. $(2 - j3)(2 - j3) = a + jb$
Hence $4 - j6 - j6 + j^2 9 = a + jb$
and $-5 - j12 = a + jb$
Thus $a = -5$ and $b = -12$

(b) $(x - j2y) + (y - j3x) = 2 + j3$
Hence $(x + y) + j(-2y - 3x) = 2 + j3$
Equating real and imaginary parts gives:
$$x + y = 2 \tag{1}$$
and $-3x - 2y = 3 \tag{2}$

i.e. two simultaneous equations to solve.
Multiplying equation (1) by 2 gives:
$$2x + 2y = 4 \tag{3}$$

Adding equations (2) and (3) gives:
$$-x = 7, \text{ i.e. } x = -7$$

From equation (1), $y = 9$, which may be checked in equation (2)

Further problems on complex equations may be found in Section 43.10, Problems 14 to 20, page 433.

43.6   The polar form of a complex number

(i) Let a complex number Z be $x + jy$ as shown in the Argand diagram of Fig. 43.4. Let distance OZ be r and the angle OZ makes with the positive real axis be θ.
From trigonometry, $x = r\cos\theta$ and $y = 4\sin\theta$
Hence $Z = x + jy = r\cos\theta + jr\sin\theta = r(\cos\theta + j\sin\theta)$
$Z = r(\cos\theta + j\sin\theta)$ is usually abbreviated to $Z = r\angle\theta$ which is known as the **polar form** of a complex number

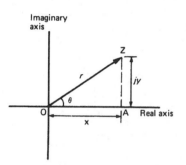

Figure 43.4

(ii) r is called the **modulus** (or magnitude) of Z and is written as mod Z or $|Z|$. r is determined using Pythagoras' theorem on triangle OAZ in Fig. 43.4, i.e. $r = \sqrt{(x^2 + y^2)}$

(iii) θ is called the **argument** (or amplitude) of Z and is written as arg Z. By trigonometry on triangle OAZ, arg $Z = \theta = \arctan y/x$

(iv) Whenever changing from cartesian form to polar form, or vice versa, a sketch is invaluable for determining the quadrant in which the complex number occurs

---

Problem 9.   Determine the modulus and argument of the complex number $Z = 2 + j3$, and express Z in polar form

---

$Z = 2 + j3$ lies in the first quadrant as shown in Fig. 43.5.

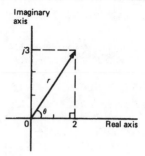

Figure 43.5

Modulus $|Z| = r = \sqrt{(2^2 + 3^2)} = \sqrt{13}$ or **3.606**, correct to 3 decimal places.

Argument, arg $Z = \theta = \arctan\dfrac{3}{2} = 56°19'$

In polar form, $2 + j3$ is written as **3.606∠56°19'**

Problem 10.   Express the following complex numbers in polar form: (a) $3 + j4$ (b) $-3 + j4$ (c) $-3 - j4$ (d) $3 - j4$

(a) $3 + j4$ is shown in Fig. 43.6 and lies in the first quadrant.
Modulus, $r = \sqrt{(3^2 + 4^2)} = 5$
Argument $\theta = \arctan\dfrac{4}{3} = 53.13° = 53°8'$
Hence $3 + j4 = 5\angle53°8'$

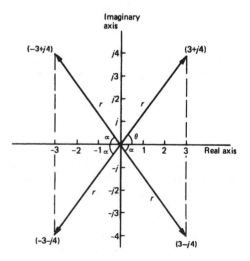

Figure 43.6

(b) $-3 + j4$ is shown in Fig. 43.6 and lies in the second quadrant.

Modulus, $r = 5$ and angle $\alpha = 53°8'$, from part (a)

Argument $= 180° - 53°8' = 126°52'$ (i.e. the argument must be measured from the positive real axis)

Hence $-3 + j4 = 5\angle126°52'$

(c) $-3 - j4$ is shown in Fig. 43.6 and lies in the third quadrant.

Modulus, $r = 5$ and $\alpha = 53°8'$, as above

Hence the argument $= 180° + 53°8' = 233°8'$, which is the same as $-126°52'$

Hence $(-3 - j4) = 5\angle233°8'$ or $5\angle-126°52'$

(By convention the **principal value** is normally used, i.e. the numerically least value, such that $-\pi \le \theta \le \pi$)

(d) $3 - j4$ is shown in Fig. 43.6 and lies in the fourth quadrant.

Modulus, $r = 5$ and angle $\alpha = 53°8'$, as above

Hence $(3 - j4) = 5\angle-53°8'$

Problem 11. Convert (a) $4\angle30°$
(b) $7\angle-145°$ into $a + jb$ form, correct to 4 significant figures

---

(a) $4\angle30°$ is shown in Fig. 43.7(a) and lies in the first quadrant.

Using trigonometric ratios, $x = 4\cos30° = 3.464$ and $y = 4\sin30° = 2.000$

Hence $4\angle30° = 3.464 + j2.000$

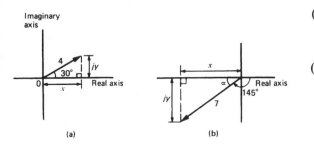

(a)   (b)

Figure 43.7

(b) $7\angle-145°$ is shown in Fig. 43.7(b) and lies in the third quadrant.

Angle $\alpha = 180° - 145° = 35°$

Hence $x = 7\cos35° = 5.734$ and $y = 7\sin35° = 4.015$

Hence $7\angle-145° = -5.734 - j4.015$

Alternatively

$7\angle-145° = 7\cos(-145°) + j7\sin(-145°)$

$= -5.734 - j4.015$

43.7 Multiplication and division in polar form

If $Z_1 = r_1\angle\theta_1$ and $Z_2 = r_2\angle\theta_2$ then:

(i) $Z_1 Z_2 = r_1 r_2 \angle(\theta_1 + \theta_2)$ and

(ii) $\dfrac{Z_1}{Z_2} = \dfrac{r_1}{r_2}\angle(\theta_1 - \theta_2)$

Problem 12. Determine in polar form:
(a) $8\angle25° \times 4\angle60°$
(b) $3\angle16° \times 5\angle-44° \times 2\angle80°$

(a) $8\angle25° \times 4\angle60° = (8 \times 4)\angle(25° + 60°)$

$= 32\angle85°$

(b) $3\angle16° \times 5\angle-44° \times 2\angle80° = (3 \times 5 \times 2)\angle[16° + (-44°) + 80°]$

$= 30\angle52°$

Problem 13. Evaluate in polar form

(a) $\dfrac{16\angle75°}{2\angle15°}$ (b) $\dfrac{10\angle\dfrac{\pi}{4} \times 12\angle\dfrac{\pi}{2}}{6\angle-\dfrac{\pi}{3}}$

(a) $\dfrac{16\angle75°}{2\angle15°} = \dfrac{16}{2}\angle(75° - 15°) = 8\angle60°$

(b) $\dfrac{10\angle\dfrac{\pi}{4} \times 12\angle\dfrac{\pi}{2}}{6\angle-\dfrac{\pi}{3}}$

$= \dfrac{10 \times 12}{6}\angle\left(\dfrac{\pi}{4} + \dfrac{\pi}{2} - \left(-\dfrac{\pi}{3}\right)\right)$

$= 20\angle\dfrac{13\pi}{12}$ or $20\angle195°$ or $20\angle-165°$

Problem 14. Evaluate in polar form
$2\angle30° + 5\angle-45° - 4\angle120°$

Addition and subtraction in polar form are not possible directly. Each complex number has to be converted into Cartesian form first.

$2\angle30° = 2(\cos30° + j\sin30°)$

$= 2\cos30° + j2\sin30° = 1.732 + j1.000$

$5\angle-45° = 5(\cos(-45°) + j\sin(-45°))$

$= 5\cos(-45°) + j5\sin(-45°)$

$= 3.536 - j3.536$

$4\angle120° = 4(\cos 120° + j\sin 120°)$

$= 4\cos 120° + j4\sin 120°$

$= -2.000 + j3.464$

Hence $2\angle30° + 5\angle-45° - 4\angle120°$

$= (1.732 + j1.000) + (3.536 - j3.536)$

$\quad - (-2.000 + j3.464)$

$= 7.268 - j6.000$, which lies in the
fourth quadrant

$= \sqrt{[(7.268)^2 + (6.000)^2]}\angle\arctan\left(\dfrac{-6.000}{7.268}\right)$

$= 9.425\angle-39.54°$ or $9.425\angle-39°32'$

Further problems on polar form may be found in
Section 43.10, Problems 21 to 28, page 433.

43.8 De Moivre's theorem

De Moivre's theorem states: $\boxed{[r\angle\theta]^n = r^n \angle n\theta}$,

which is true for all positive, negative or fractional
values of n, and is thus useful for determining
powers and roots of complex numbers.
For example, $[3\angle20°]^4 = 3^4\angle(4 \times 20°) = 81\angle80°$
The **square root of a complex number** is deter-
mined by letting $n = \frac{1}{2}$ in De Moivre's theorem,
i.e. $\sqrt{[r\angle\theta]} = [r\angle\theta]^{1/2} = r^{1/2}\angle\frac{1}{2}\theta = \sqrt{r}\angle(\theta/2)$
There are two square roots of a real number, equal
in size but opposite in sign.

Problem 15. Determine (a) $[2\angle35°]^5$
(b) $(-2 + j3)^6$ in polar form

(a) $[2\angle35°]^5 = 2^5\angle(5 \times 35°)$, from De Moivre's
theorem

$= 32\angle175°$

(b) $(-2 + j3) = \sqrt{[(-2)^2 + (3)^2]}\angle\arctan\dfrac{3}{-2}$

$= \sqrt{13}\angle123°41'$, since $-2 + j3$
lies in the second quadrant

$(-2 + j3)^6 = [\sqrt{13}\angle123°41']^6$

$= (\sqrt{13})^6\angle(6 \times 123°41')$, by
De Moivre's theorem

$= 2197\angle742°6'$

$= 2197\angle382°6'$(since $742°6' \equiv$
$742°6' - 360° = 382°6'$)

$= \mathbf{2197\angle22°6'}$ (since $382°6' \equiv$
$382°6' - 360° = 22°6'$)

Problem 16. Determine the two square
roots of the complex number $(5 + j12)$ in
polar and Cartesian forms and show the roots
on an Argand diagram

$(5 + j12) = \sqrt{[5^2 + 12^2]}\angle\arctan\dfrac{12}{5} = 13\angle67°23'$
When determining square roots two solutions result.
To obtain the second solution, one way is to
express $13\angle67°23'$ also as $13\angle(67°23' + 360°)$, i.e.
$13\angle427°23'$. When the angle is divided by 2 an angle
less than 360° is obtained.
Hence $\sqrt{(5 + j12)}$

$= \sqrt{[13\angle67°23']}$ and $\sqrt{[13\angle427°23']}$

$= [13\angle67°23']^{1/2}$ and $[13\angle427°23']^{1/2}$

$= 13^{1/2}\angle\left(\dfrac{1}{2} \times 67°23'\right)$ and

$\quad 13^{1/2}\angle\left(\dfrac{1}{2} \times 427°23'\right)$

$= \sqrt{13}\angle33°42'$ and $\sqrt{13}\angle213°42'$

$= 3.61\angle33°42'$ and $3.61\angle213°42'$

**Thus, in polar form, the two roots are
$3.61\angle33°42'$ and $3.61\angle-146°18'$.**

$3.61\angle33°42' = 3.61(\cos 33°42' + j\sin 33°42')$

$= 3.0 + j2.0$

$3.61\angle-146°18' = 3.61(\cos(-146°18')$
$\quad\quad\quad + j\sin(-146°18'))$

$= -3.0 - j2.0$

**Thus, in Cartesian form the two roots are
$\pm(3.0 + j2.0)$.**
From the Argand diagram shown in Fig. 43.8 the
two roots are seen to be 180° apart, which is always
true when finding square roots of complex numbers.

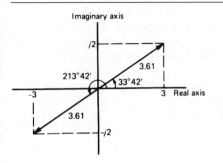

Figure 43.8

Further problems on De Moivre's theorem may be found in Section 43.10, Problems 29 to 35, page 434.

43.9   Applications of complex numbers

There are several applications of complex numbers in science and engineering, in particular in electrical alternating current theory and in mechanical vector analysis.

The effect of multiplying a phasor by j is to rotate it in a positive direction (i.e. anticlockwise) on an Argand diagram through $90°$ without altering its length. Similarly, multiplying a phasor by $-j$ rotates the phasor through $-90°$. These facts are used in a.c. theory since certain quantities in the phasor diagrams lie at $90°$ to each other. For example, in the R-L series circuit shown in Fig. 43.9(a), V_L leads I by $90°$ (i.e. I lags V_L by $90°$) and may be written as jV_L, the vertical axis being regarded as the imaginary axis of an Argand diagram. Thus $V_R + jV_L = V$ and since $V_R = IR$, $V = IX_L$

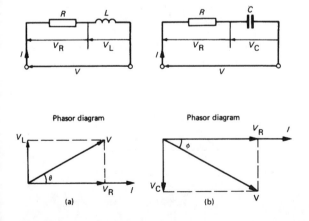

Figure 43.9

(where X_L is the inductive reactance, $2\pi f L$ ohms) and $V = IZ$ (where Z is the impedance) then $R + jX_L = Z$.

Similarly, for the R-C circuit shown in Fig. 43.9(b), V_C lags I by $90°$ (i.e. I leads V_C by $90°$) and $V_R - jV_C = V$, from which $R - jX_C = Z$ (where X_C is the capacitive reactance $1/(2\pi f C)$ ohms.

Problem 17.   Determine the resistance and series inductance (or capacitance) for each of the following impedances, assuming a frequency of 50 Hz:
(a) $(4.0 + j7.0)$ Ω (b) $-j20$ Ω (c) $10\angle30°$ Ω
(d) $15\angle-60°$ Ω

(a)   Impedance, $Z = 4.0 + j7.0$ Ω.
Hence **resistance $= 4.0$ Ω** and reactance $= 7.0$ Ω.
Since the imaginary part is positive, the reactance is inductive, i.e. $X_L = 7.0$ Ω.
Since $X_L = 2\pi f L$ then

$$\text{inductance, } L = \frac{X_L}{2\pi f} = \frac{7.0}{2\pi(50)}$$

$$= \mathbf{0.0223\ H\ or\ 22.3\ mH}$$

(b)   Impedance, $Z = -j20$, i.e. $Z = 0 - j20$. Hence **resistance $= 0$** and reactance $= 20$ Ω. Since the imaginary part is negative, the reactance is capacitive, i.e. $X_C = 20$ Ω.
Since $X_C = \dfrac{1}{2\pi f C}$ then:

$$\text{capacitance, } C = \frac{1}{2\pi f X_C} = \frac{1}{2\pi(50)(20)} F$$

$$= \frac{10^6}{2\pi(50)(20)}\ \mu F = \mathbf{159.2\ \mu F}$$

(c)   Impedance, $Z = 10\angle30°$

$$= 10(\cos 30° + j\sin 30°)$$

$$= 8.66 + j5.0\ \Omega$$

Hence **resistance $= 8.66$ ohms** and inductive reactance, $X_L = 5.0$ Ω.
Since $X_L = 2\pi f L$ then

$$\text{inductance, } L = \frac{X_L}{2\pi f} = \frac{5.0}{2\pi(50)}$$

$$= \mathbf{0.0159\ H\ or\ 15.9\ mH}$$

(d)   Impedance, $Z = 15\angle-60°$

$$= 15[\cos(-60°) + j\sin(-60°)]$$

$$= 7.50 - j12.99 \ \Omega$$

Hence **resistance = 7.50 ohms** and capacitive reactance, $X_C = 12.99 \ \Omega$.

Since $X_C = \dfrac{1}{2\pi fC}$ then

capacitance, $C = \dfrac{1}{2\pi f X_C}$

$$= \dfrac{10^6}{2\pi(50)(12.99)} \ \mu F$$

$$= \textbf{245} \ \boldsymbol{\mu}\textbf{F}$$

Problem 18. An alternating voltage of 240 V, 50 Hz is connected across an impedance of $(60 - j100) \ \Omega$. Determine (a) the resistance, (b) the capacitance, (c) the magnitude of the impedance and its phase angle, and (d) the current flowing

---

(a)   Impedance, $Z = (60 - j100) \ \Omega$.
Hence **resistance = 60 Ω**

(b)   Capacitive reactance, $X_C = 100 \ \Omega$.
Since $X_C = \dfrac{1}{2\pi fC}$ then

capacitance, $C = \dfrac{1}{2\pi f X_C}$

$$= \dfrac{1}{2\pi(50)(100)} \ F$$

$$= \dfrac{10^6}{2\pi(50)(100)} \mu F$$

$$= \textbf{31.83} \ \boldsymbol{\mu}\textbf{F}$$

(c)   Magnitude of impedance

$$|Z| = \sqrt{[(60)^2 + (-100)^2]} = \textbf{116.6} \ \boldsymbol{\Omega}$$

Phase angle, $\arg Z = \arctan\left(\dfrac{-100}{60}\right)$

$$= -\textbf{59°2}'$$

(d)   Current flowing, $I = \dfrac{V}{Z} = \dfrac{240\angle0°}{116.6\angle-59°2'}$

$$= \textbf{2.058}\angle\textbf{59°2}' \ \textbf{A}$$

The circuit and phasor diagrams are as shown in Fig. 43.9(b)

Problem 19. A resistance of 50 Ω is connected in series with an inductance of 0.20 H. If the terminal voltage is 220 V, 50 Hz determine (a) the inductive reactance, (b) the impedance, (c) the current flowing and its phase angle relative to the terminal voltage, (d) the voltage across the resistance, and (e) the voltage across the inductance

---

The circuit and phasor diagrams are as shown in Fig. 43.9(a).

(a)   Inductive reactance, $X_L = 2\pi fL = 2\pi(50)(0.20)$

$$= \textbf{62.83} \ \boldsymbol{\Omega}$$

(b)   Impedance, $Z = R + jX_L = 50 + j62.83$

$$= \sqrt{[(50)^2 + (62.83)^2]}\angle\arctan\left(\dfrac{62.83}{50}\right)$$

$$= \textbf{80.30}\angle\textbf{51°29}' \ \boldsymbol{\Omega}$$

(c)   Current, $I = \dfrac{V}{Z}$. Taking the terminal voltage V as the reference quantity, i.e. $220\angle0°$ V, then

current, $I = \dfrac{220\angle0°}{80.30\angle51°29'} = 2.74\angle-51°29'$ A,

i.e. **the current is 2.74 A lagging the terminal voltage by 51°29'**

(d)   Voltage across resistance

$$V_R = IR = (2.74\angle - 51°29')(50\angle0°)$$

$$= \textbf{137.0}\angle-\textbf{51°29}' \ \textbf{V}$$

(e)   Voltage across inductance

$$V_L = IX_L = (2.74\angle-51°29')(62.83\angle90°)$$

$$= \textbf{172.2}\angle\textbf{38°31}' \ \textbf{V}$$

The phasor sum of V_R and V_L is the terminal voltage V as shown in Fig. 43.10.

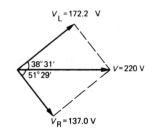

Figure 43.10

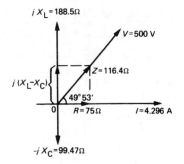

Figure 43.12

Problem 20.   A coil of resistance 75 Ω and inductance 150 mH in series with an 8 μF capacitor is connected to a 500 V, 200 Hz supply. Calculate (a) the current flowing and its phase angle, and (b) the power factor of the circuit

The circuit diagram is shown in Fig. 43.11.

(a)   Inductive reactance,

$$X_L = 2\pi fL = 2\pi(200)(150 \times 10^{-3})$$
$$= 188.5 \ \Omega$$

Capacitive reactance,

$$X_C = \frac{1}{2\pi fC}$$

$$= \frac{1}{2\pi(200)(8 \times 10^{-6})} = 99.47 \ \Omega$$

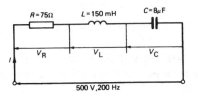

Figure 43.11

Impedance Z

$$= R + j(X_L - X_C)$$
$$= 75 + j(188.5 - 99.47)$$
$$= (75 + j89.03) \ \Omega$$
$$= \sqrt{[(75)^2 + (89.03)^2]} \angle \arctan\left(\frac{89.03}{75}\right)$$
$$= 116.4\angle 49°53' \ \Omega$$

Current, I

$$= \frac{V}{Z} = \frac{500\angle 0°}{116.4\angle 49°53'}$$
$$= 4.296\angle -49°53' \ A$$

i.e. **the current flowing is 4.296 A, lagging the voltage by 49°53'**
The phasor diagram is shown in Fig. 43.12

(b)   Power factor $= \dfrac{R}{Z} = \dfrac{75}{116.4} = $ **0.644 lagging**
(or power factor $= \cos\phi = \cos 49°53' = $ **0.644 lagging**)

Problem 21.   For the parallel circuit shown in Fig. 43.13, determine the value of current I and its phase relative to the 240 V supply, using complex numbers

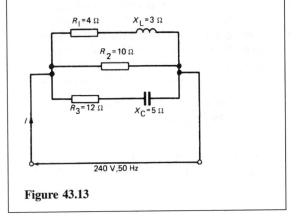

Figure 43.13

Current, $I = \dfrac{V}{Z}$. Impedance Z for the three-branch parallel circuit is given by:

$$\frac{1}{Z} = \frac{1}{Z_1} + \frac{1}{Z_2} + \frac{1}{Z_3},$$

where $Z_1 = 4 + j3$, $Z_2 = 10$ and $Z_3 = 12 - j5$

Admittance, $Y_1 = \dfrac{1}{Z_1} = \dfrac{1}{4 + j3}$

$$= \frac{1}{4 + j3} \times \frac{4 - j3}{4 - j3} = \frac{4 - j3}{4^2 + 3^2}$$
$$= 0.160 - j0.120 \text{ siemens}$$

Admittance, $Y_2 = \dfrac{1}{Z_2} = \dfrac{1}{10} = 0.10$ siemens

Admittance, $Y_3 = \dfrac{1}{Z_3} = \dfrac{1}{12 - j5}$

$$= \frac{1}{12 - j5} \times \frac{12 + j5}{12 + j5} = \frac{12 + j5}{12^2 + 5^2}$$

$$= 0.0710 + j0.0296 \text{ siemens}$$

Total admittance,

$$Y = Y_1 + Y_2 + Y_3$$

$$= (0.160 - j0.120) + (0.10)$$

$$+ (0.0710 + j0.0296)$$

$$= 0.331 - j0.0904 = 0.343\angle -15°17' \text{ siemens}$$

Current, $I = \dfrac{V}{Z} = VY = (240\angle 0°)(0.343\angle -15°17')$

$$= 82.32\angle -15°17' \text{ A}$$

Problem 22. Determine the magnitude and direction of the resultant of the three coplanar forces given below, when they act at a point.
Force A, 10 N acting at 45° from the positive horizontal axis
Force B, 8 N acting at 120° from the positive horizontal axis
Force C, 15 N acting at 210° from the positive horizontal axis

---

The space diagram is shown in Fig. 43.14. The forces may be written as complex numbers.

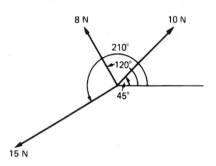

Figure 43.14

Thus force A, $f_A = 10\angle 45°$, force B, $f_B = 8\angle 120°$ and force C, $f_C = 15\angle 210° = 15\angle -150°$

The resultant force

$$= f_A + f_B + f_C$$

$$= 10\angle 45° + 8\angle 120° + 15\angle -150°$$

$$= 10(\cos 45° + j\sin 45°)$$

$$+ 8(\cos 120° + j\sin 120°)$$

$$+ 15[\cos(-150°) + j\sin(-150°)]$$

$$= (7.071 + j7.071) + (-4.00 + j6.928)$$

$$+ (-12.99 - j7.50)$$

$$= -9.919 + j6.499$$

Magnitude of resultant force

$$= \sqrt{[(-9.919)^2 + (6.499)^2]} = \textbf{11.86 N}$$

Direction of resultant force $= \arctan\left(\dfrac{6.499}{-9.919}\right)$

$$= \textbf{146°46'}$$

(since $-9.919 + j6.499$ lies in the second quadrant)

Further problems on applications of complex numbers may be found in the following Section 43.10, Problems 36 to 47, pages 434 and 435.

43.10   Further problems on complex numbers

Introduction to complex numbers

In Problems 1 to 3, solve the quadratic equations:

1   $x^2 + 25 = 0$ $\hspace{2em}$ $[\pm j5]$

2   $2x^2 + 3x + 4 = 0$

$$\left[-\frac{3}{4} \pm j\frac{\sqrt{23}}{4} \text{ or } -0.750 \pm j1.199\right]$$

3   $4t^2 - 5t + 7 = 0$

$$\left[\frac{5}{8} \pm j\frac{\sqrt{87}}{8} \text{ or } 0.625 \pm j1.166\right]$$

4   Evaluate (a) j^8 (b) $-\dfrac{1}{j^7}$ (c) $\dfrac{4}{2j^{13}}$

$$[\text{(a) 1 (b) } -j \text{ (c) } -j2]$$

Operations on Cartesian complex numbers

5   Evaluate (a) $(3 + j2) + (5 - j)$ and
    (b) $(-2 + j6) - (3 - j2)$ and show the results on an Argand diagram.

$$[\text{(a) } 8 + j \text{ (b) } -5 + j8]$$

6   Write down the complex conjugates of
(a) $3 + j4$, (b) $2 - j$

$$[\text{(a) } 3 - j4 \text{ (b) } 2 + j]$$

In Problems 7 to 11 evaluate in $a + jb$ form given $Z_1 = 1 + j2$, $Z_2 = 4 - j3$, $Z_3 = -2 + j3$ and $Z_4 = -5 - j$:

7   (a) $Z_1 + Z_2 - Z_3$ (b) $Z_2 - Z_1 + Z_4$

$$[\text{(a) } 7 - j4 \text{ (b) } -2 - j6]$$

8   (a) $Z_1 Z_2$ (b) $Z_3 Z_4$

$$[\text{(a) } 10 + j5 \text{ (b) } 13 - j13]$$

9   (a) $Z_1 Z_3 + Z_4$ (b) $Z_1 Z_2 Z_3$

$$[\text{(a) } -13 - j2 \text{ (b) } -35 + j20]$$

10 (a) $\dfrac{Z_1}{Z_2}$ (b) $\dfrac{Z_1 + Z_3}{Z_2 - Z_4}$

$$\left[\text{(a) } \frac{-2}{25} + j\frac{11}{25} \text{ (b) } \frac{-19}{85} + j\frac{43}{85}\right]$$

11 (a) $\dfrac{Z_1 Z_3}{Z_1 + Z_3}$ (b) $Z_2 + \dfrac{Z_1}{Z_4} + Z_3$

$$\left[\text{(a) } \frac{3}{26} + j\frac{41}{26} \text{ (b) } \frac{45}{26} - j\frac{9}{26}\right]$$

12 Evaluate (a) $\dfrac{1 - j}{1 + j}$ (b) $\dfrac{1}{1 + j}$

$$\left[\text{(a) } -j \text{ (b) } \frac{1}{2} - j\frac{1}{2}\right]$$

13 Show that

$$\frac{-25}{2}\left(\frac{1 + j2}{3 + j4} - \frac{2 - j5}{-j}\right) = 57 + j24$$

Complex equations

In Problems 14 to 18 solve the complex equations:

14 $5(x + jy) = 10 - j15$ $[x = 2, y = -3]$

15 $(2 + j)(3 - j2) = a + jb$ $[a = 8, b = -1]$

16 $\dfrac{2 + j}{1 - j} = j(x + jy)$ $\left[x = \dfrac{3}{2}, y = -\dfrac{1}{2}\right]$

17 $(2 - j3) = \sqrt{(a + jb)}$ $[a = -5, b = -12]$

18 $(x - j2y) - (y - jx) = 2 + j$

$$[x = 3, y = 1]$$

19 If $Z = R + j\omega L + (1/j\omega C)$, express Z in $(a + jb)$ form when $R = 10$, $L = 5$, $C = 0.04$ and $\omega = 4$. $[Z = 10 + j13.75]$

20 An equation derived from an a.c. bridge network is given by:

$$(R_1)\left(-\frac{j}{\omega C_3}\right) =$$

$$\left(R_x - \frac{j}{\omega C_x}\right)\frac{(R_2)\left(-\dfrac{j}{\omega C_2}\right)}{\left(R_2 - \dfrac{j}{\omega C_2}\right)}$$

Components R_1, R_2, C_2 and C_3 have known values. Determine expressions for R_x and C_x in terms of the known components.

$$\left[R_x = \frac{R_1 C_2}{C_3}, C_x = \frac{C_3 R_2}{R_1}\right]$$

Polar form

21 Determine the modulus and argument of
(a) $2 + j4$ (b) $-5 - j2$ (c) $j(2 - j)$

$$\left[\begin{array}{l}\text{(a) } 4.472, 63°26' \text{ (b) } 5.385, -158°12' \\ \text{(c) } 2.236, 63°26'\end{array}\right]$$

In Problems 22 and 23 express the given Cartesian complex numbers in polar form, leaving answers in surd form:

22 (a) $2 + j3$ (b) -4 (c) $-6 + j$

$$[\text{(a) } \sqrt{13}∠56°19' \text{ (b) } 4∠180° \text{ (c) } \sqrt{37}∠170°32']$$

23 (a) $-j3$ (b) $(-2 + j)^3$ (c) $j^3(1 - j)$

$$[\text{(a) } 3∠-90° \text{ (b) } \sqrt{125}∠100°18' \text{ (c) } \sqrt{2}∠135°]$$

In Problems 24 and 25 convert the given polar complex numbers into $(a + jb)$ form giving answers correct to 4 significant figures:

24 (a) $5∠30°$ (b) $3∠60°$ (c) $7∠45°$

$$\left[\begin{array}{l}\text{(a) } 4.330 + j2.500 \text{ (b) } 1.500 + j2.598 \\ \text{(c) } 4.950 + j4.950\end{array}\right]$$

25 (a) $6∠125°$ (b) $4∠\pi$ (c) $3.5∠-120°$

$$\left[\begin{array}{l}\text{(a) } -3.441 + j4.915 \text{ (b) } -4.000 + j0 \\ \text{(c) } -1.750 - j3.031\end{array}\right]$$

In Problems 26 to 28, evaluate in polar form:

26 (a) $3∠20° \times 15∠45°$ (b) $2.4∠65° \times 4.4∠-21°$

$$[\text{(a) } 45∠65° \text{ (b) } 10.56∠44°]$$

27 (a) $6.4∠27° \div 2∠-15°$

(b) $5∠30° \times 4∠80° \div 10∠-40°$

$$[\text{(a) } 3.2∠42° \text{ (b) } 2∠150°]$$

28 (a) $4∠\dfrac{\pi}{6} + 3∠\dfrac{\pi}{8}$

(b) $2∠120° + 5.2∠58° - 1.6∠-40°$

$$[\text{(a) } 6.986∠26°47' \text{ (b) } 7.190∠85°46']$$

De Moivre's theorem

29   Determine in polar form

(a) $[1.5\angle15°]^5$ (b) $(1 + j2)^6$

$$[(a)\ 7.594\angle75°\ (b)\ 125\angle20°37']$$

30   Determine in polar and Cartesian forms

(a) $[3\angle41°]^4$ (b) $(-2 - j)^5$

$$\begin{bmatrix}(a)\ 81\angle164°,\ -77.85 + j22.33\\(b)\ 55.90\angle-47°10',\ 38.00 - j40.99\end{bmatrix}$$

31   Convert $(3 - j)$ into polar form and hence evaluate $(3 - j)^7$, giving the answer in polar form.

$$[\sqrt{10}\angle-18°26',\ 3162\angle-129°2']$$

In Problems 32 to 34 determine the two square roots of the given complex numbers in Cartesian form and show the results on an Argand diagram:

32   (a) $1 + j$ (b) j

$$[(a)\ \pm(1.099 + j0.455)\ (b)\ \pm(0.707 + j0.707)]$$

33   (a) $3 - j4$ (b) $-1 - j2$

$$[(a)\ \pm(-2 + j)\ (b)\ \pm(-0.786 + j1.272)]$$

34   (a) $7\angle60°$ (b) $12\angle\dfrac{3\pi}{2}$

$$\begin{bmatrix}(a)\ \pm(2.291 + j1.323)\\(b)\ \pm(-2.449 + j2.449)\end{bmatrix}$$

35   Simplify $\dfrac{(\cos \pi/8 - j \sin \pi/8)^3}{(\cos \pi/8 + j \sin \pi/8)^5}$ $[-1]$

Applications of complex numbers

36   Determine the resistance R and series inductance L (or capacitance C) for each of the following impedances assuming the frequency to be 50 Hz

(a) $(3 + j8)$ Ω (b) $(2 - j3)$ Ω (c) $j14$ Ω

(d) $-j150$ Ω (e) $12\left/\dfrac{\pi}{6}\right.$ Ω (f) $8\angle-60°$ Ω

$$\begin{bmatrix}(a)\ R = 3\ \Omega,\ L = 25.5\ \text{mH}\\(b)\ R = 2\ \Omega,\ C = 1061\ \mu\text{F}\\(c)\ R = 0,\ L = 44.56\ \text{mH},\\(d)\ R = 0,\ C = 21.22\ \mu\text{F}\\(e)\ R = 10.39\ \Omega,\ L = 19.10\ \text{mH}\\(f)\ R = 4\ \Omega,\ C = 459.4\ \mu\text{F}\end{bmatrix}$$

37   Two impedances, $Z_1 = (3 + j6)$ Ω and $Z_2 = (4 - j3)$ Ω are connected in series to a supply voltage of 120 V. Determine the magnitude of the current and its phase angle relative to the voltage.   [15.76 A, 23°12' lagging]

38   If the two impedances in Problem 37 are connected in parallel determine the current flowing and its phase relative to the 120 V supply voltage.   [27.25 A, 3°22' lagging]

39   An alternating voltage of 200 V, 50 Hz is applied across an impedance of $(30 - j40)$ Ω. Calculate (a) the resistance, (b) the capacitance, (c) the current, and (d) the phase angle between voltage and current.

$$\begin{bmatrix}(a)\ 30\ \Omega\ (b)\ 79.58\ \mu\text{F}\\(c)\ 4\ \text{A}\ (d)\ 53°8'\ \text{lagging}\end{bmatrix}$$

40   A series circuit consists of a 12 Ω resistor, a coil of inductance 0.10 H and a capacitance of 160 μF. Calculate the current flowing and its phase relative to the supply voltage of 240 V, 50 Hz. Determine also the power factor of the circuit.   [14.42 A, 43°51' lagging, 0.721]

41   For the circuit shown in Fig. 43.15, determine the current I flowing and its phase relative to the applied voltage.   [14.6 A, 2°30' leading]

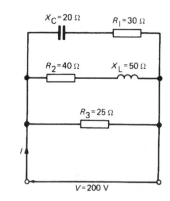

Figure 43.15

42   Determine, using complex numbers, the magnitude and direction of the resultant of the coplanar forces given below, which are acting at a point. Force A, 5 N acting horizontally, force B, 9 N acting at an angle of 135° to force A, force C, 12 N acting at an angle of 240° to force A.   [8.393 N, 208°40' from force A]

43   A delta-connected impedance Z_A is given by:

$$Z_A = \frac{Z_1 Z_2 + Z_2 Z_3 + Z_3 Z_1}{Z_2}$$

Determine Z_A in both Cartesian and polar form given $Z_1 = (10 + j0)$ Ω, $Z_2 = (0 - j10)$ Ω and $Z_3 = (10 + j10)$ Ω.

$$[(10 + j20)\ \Omega,\ 22.36\angle63.43°\ \Omega]$$

44   In the hydrogen atom, the angular momentum, p, of the de Broglie wave is given by:

$$p\psi = -\left(\frac{jh}{2\pi}\right)(\pm jm\psi)$$

Determine an expression for p. $\left[\pm\dfrac{mh}{2\pi}\right]$

45   An aircraft P flying at a constant height has a velocity of $(400 + j300)$ km/h. Another aircraft Q at the same height has a velocity of $(200 - j600)$ km/h. Determine (a) the velocity of P relative to Q, and (b) the velocity of Q relative to P. Express the answers in polar form, correct to the nearest km/h.

$$\left[\begin{array}{l}\text{(a) 922 km/h at } 77.47° \\ \text{(b) 922 km/h at } -102.53°\end{array}\right]$$

46   Three vectors are represented by P, $2\angle30°$, Q, $3\angle90°$ and R, $4\angle-60°$. Determine in polar form the vectors represented by

(a) $P + Q + R$, (b) $P - Q - R$

[(a) $3.770\angle8.17°$ (b) $1.488\angle100.37°$]

47   For a transmission line, the characteristic impedance Z_0 and the propagation coefficient γ are given by:

$$Z_0 = \sqrt{\left(\frac{R + j\omega L}{G + j\omega C}\right)}$$

and $\gamma = \sqrt{(R + j\omega L)(G + j\omega C)}]$

Given $R = 25$ ohms, $L = 5 \times 10^{-3}$ henrys, $G = 80 \times 10^{-6}$ siemens, $C = 0.04 \times 10^{-6}$ farad and $\omega = 2000\pi$ determine, in polar form, Z_0 and γ.

[$Z_0 = 390.2\angle-10.43°$ Ω, $\gamma = 0.1029\angle61.92°$]

44

The theory of matrices and determinants

44.1 Matrix notation

Matrices and determinants are mainly used for the solution of linear simultaneous equations. The coefficients of the variables for linear simultaneous equations may be shown in matrix form. The coefficients of x and y in the simultaneous equations

$$x + 2y = 3$$
$$4x - 5y = 6$$ become

$$\begin{pmatrix} 1 & 2 \\ 4 & -5 \end{pmatrix}$$ in matrix notation

Similarly, the coefficients of p, q and r in the equations

$$1.3p - 2.0q + r = 7$$
$$3.7p + 4.8q - 7r = 3 \quad \text{become}$$
$$4.1p + 3.8q + 12r = -6$$

$$\begin{pmatrix} 1.3 & -2.0 & 1 \\ 3.7 & 4.8 & -7 \\ 4.1 & 3.8 & 12 \end{pmatrix}$$ in matrix form

The numbers within a matrix are called an **array** and the coefficients forming the array are called the **elements** of the matrix. The number of rows in a matrix is usually specified by m and the number of columns by n and a matrix referred to as an 'm by n' matrix. Thus, $\begin{pmatrix} 2 & 3 & 6 \\ 4 & 5 & 7 \end{pmatrix}$ is a '2 by 3' matrix. Matrices cannot be expressed as a single numerical value, but they can often be simplified or combined, and unknown element values can be determined by comparison methods. Just as there are rules for addition, subtraction, multiplication and division of numbers in arithmetic, rules for these operations can be applied to matrices and the rules of matrices are such that they obey most of those governing the algebra of numbers.

44.2 Addition, subtraction and multiplication of matrices

(i) Addition of matrices

Corresponding elements in two matrices may be added to form a single matrix.

> **Problem 1.** Add the matrices
>
> (a) $\begin{pmatrix} 2 & -1 \\ -7 & 4 \end{pmatrix}$ and $\begin{pmatrix} -3 & 0 \\ 7 & -4 \end{pmatrix}$ and
>
> (b) $\begin{pmatrix} 3 & 1 & -4 \\ 4 & 3 & 1 \\ 1 & 4 & -3 \end{pmatrix}$ and $\begin{pmatrix} 2 & 7 & -5 \\ -2 & 1 & 0 \\ 6 & 3 & 4 \end{pmatrix}$

(a) Adding the corresponding elements gives:

$$\begin{pmatrix} 2 & -1 \\ -7 & 4 \end{pmatrix} + \begin{pmatrix} -3 & 0 \\ 7 & -4 \end{pmatrix}$$
$$= \begin{pmatrix} 2+(-3) & -1+0 \\ -7+7 & 4+(-4) \end{pmatrix}$$
$$= \begin{pmatrix} -1 & -1 \\ 0 & 0 \end{pmatrix}$$

(b) Adding the corresponding elements gives:

$$\begin{pmatrix} 3 & 1 & -4 \\ 4 & 3 & 1 \\ 1 & 4 & -3 \end{pmatrix} + \begin{pmatrix} 2 & 7 & -5 \\ -2 & 1 & 0 \\ 6 & 3 & 4 \end{pmatrix}$$
$$= \begin{pmatrix} 3+2 & 1+7 & -4+(-5) \\ 4+(-2) & 3+1 & 1+0 \\ 1+6 & 4+3 & -3+4 \end{pmatrix}$$
$$= \begin{pmatrix} 5 & 8 & -9 \\ 2 & 4 & 1 \\ 7 & 7 & 1 \end{pmatrix}$$

(ii) Subtraction of matrices

If A is a matrix and B is another matrix, then $(A - B)$ is a single matrix formed by subtracting the elements of B from the corresponding elements of A.

Problem 2.   Subtract

(a) $\begin{pmatrix} -3 & 0 \\ 7 & -4 \end{pmatrix}$ from $\begin{pmatrix} 2 & -1 \\ -7 & 4 \end{pmatrix}$ and

(b) $\begin{pmatrix} 2 & 7 & -5 \\ -2 & 1 & 0 \\ 6 & 3 & 4 \end{pmatrix}$ from $\begin{pmatrix} 3 & 1 & -4 \\ 4 & 3 & 1 \\ 1 & 4 & -3 \end{pmatrix}$

(a)   To find matrix A minus matrix B, the elements of B are taken from the corresponding elements of A. Thus:

$$\begin{pmatrix} 2 & -1 \\ -7 & 4 \end{pmatrix} - \begin{pmatrix} -3 & 0 \\ 7 & -4 \end{pmatrix}$$

$$= \begin{pmatrix} 2-(-3) & -1-0 \\ -7-7 & 4-(-4) \end{pmatrix}$$

$$= \begin{pmatrix} 5 & -1 \\ -14 & 8 \end{pmatrix}$$

(b) $\begin{pmatrix} 3 & 1 & -4 \\ 4 & 3 & 1 \\ 1 & 4 & -3 \end{pmatrix} - \begin{pmatrix} 2 & 7 & -5 \\ -2 & 1 & 0 \\ 6 & 3 & 4 \end{pmatrix}$

$$= \begin{pmatrix} 3-2 & 1-7 & -4-(-5) \\ 4-(-2) & 3-1 & 1-0 \\ 1-6 & 4-3 & -3-4 \end{pmatrix}$$

$$= \begin{pmatrix} 1 & -6 & 1 \\ 6 & 2 & 1 \\ -5 & 1 & -7 \end{pmatrix}$$

Problem 3.   If $A = \begin{pmatrix} -3 & 0 \\ 7 & -4 \end{pmatrix}$,

$B = \begin{pmatrix} 2 & -1 \\ -7 & 4 \end{pmatrix}$ and $C = \begin{pmatrix} 1 & 0 \\ -2 & -4 \end{pmatrix}$ find

$A + B - C$

$A + B = \begin{pmatrix} -1 & -1 \\ 0 & 0 \end{pmatrix}$ (from Problem 1).

Hence, $A + B - C = \begin{pmatrix} -1 & -1 \\ 0 & 0 \end{pmatrix} - \begin{pmatrix} 1 & 0 \\ -2 & -4 \end{pmatrix}$

$$= \begin{pmatrix} -1-1 & -1-0 \\ 0-(-2) & 0-(-4) \end{pmatrix}$$

$$= \begin{pmatrix} -2 & -1 \\ 2 & 4 \end{pmatrix}$$

Alternatively $A + B - C$

$$= \begin{pmatrix} -3 & 0 \\ 7 & -4 \end{pmatrix} + \begin{pmatrix} 2 & -1 \\ -7 & 4 \end{pmatrix} - \begin{pmatrix} 1 & 0 \\ -2 & -4 \end{pmatrix}$$

$$= \begin{pmatrix} -3+2-1 & 0+(-1)-0 \\ 7+(-7)-(-2) & -4+4-(-4) \end{pmatrix}$$

$$= \begin{pmatrix} -2 & -1 \\ 2 & 4 \end{pmatrix} \text{ as obtained previously}$$

(iii) Multiplication

When a matrix is multiplied by a number (called scalar multiplication), a single matrix results in which each element of the original matrix has been multiplied by the number.

Problem 4.   If $A = \begin{pmatrix} -3 & 0 \\ 7 & -4 \end{pmatrix}$,

$B = \begin{pmatrix} 2 & -1 \\ -7 & 4 \end{pmatrix}$ and $C = \begin{pmatrix} 1 & 0 \\ -2 & -4 \end{pmatrix}$ find

$2A - 3B + 4C$

For scalar multiplication, each element is multiplied by the scalar quantity, hence

$$2A = 2\begin{pmatrix} -3 & 0 \\ 7 & -4 \end{pmatrix} = \begin{pmatrix} -6 & 0 \\ 14 & -8 \end{pmatrix}$$

$$3B = 3\begin{pmatrix} 2 & -1 \\ -7 & 4 \end{pmatrix} = \begin{pmatrix} 6 & -3 \\ -21 & 12 \end{pmatrix}$$

and $4C = 4\begin{pmatrix} 1 & 0 \\ -2 & -4 \end{pmatrix} = \begin{pmatrix} 4 & 0 \\ -8 & -16 \end{pmatrix}$

Hence $2A - 3B + C$

$$= \begin{pmatrix} -6 & 0 \\ 14 & -8 \end{pmatrix} - \begin{pmatrix} 6 & -3 \\ -21 & 12 \end{pmatrix} + \begin{pmatrix} 4 & 0 \\ -8 & -16 \end{pmatrix}$$

$$= \begin{pmatrix} -6-6+4 & 0-(-3)+0 \\ 14-(-21)+(-8) & -8-12+(-16) \end{pmatrix}$$

$$= \begin{pmatrix} -8 & 3 \\ 27 & -36 \end{pmatrix}$$

When a matrix A is multiplied by another matrix B, a single matrix results in which elements are obtained from the sum of the products of the corresponding rows of A and the corresponding columns of B.

Two matrices A and B may be multiplied together, provided the number of elements in the rows of matrix A are equal to the number of elements in the columns of matrix B. In general terms, when multiplying a matrix of dimensions (m by n) by a matrix of dimensions (n by r), the resulting matrix has dimensions (m by r). Thus a 2 by 3 matrix multiplied by a 3 by 1 matrix gives a matrix of dimensions 2 by 1.

Problem 5. If $A = \begin{pmatrix} 2 & 3 \\ 1 & -4 \end{pmatrix}$ and

$B = \begin{pmatrix} -5 & 7 \\ -3 & 4 \end{pmatrix}$ find $A \times B$

Let $A \times B = C$ where $C = \begin{pmatrix} C_{11} & C_{12} \\ C_{21} & C_{22} \end{pmatrix}$.

C_{11} is the sum of the products of the first row elements of A and the first column elements of B taken one at a time.

Thus $C_{11} = (2 \times (-5)) + (3 \times (-3)) = -19$.

C_{12} is the sum of the products of the first row elements of A and the second column elements of B, taken one at a time,

i.e. $C_{12} = (2 \times 7) + (3 \times 4) = 26$.

C_{21} is the sum of the products of the second row elements of A and the first column elements of B, taken one at a time,

i.e. $C_{21} = (1 \times (-5)) + ((-4) \times (-3)) = 7$.

Finally, C_{22} is the sum of the products of the second row elements of A and the second column elements of B, taken one at a time,

i.e. $C_{22} = (1 \times 7) + ((-4) \times 4) = -9$.

Thus, $A \times B = \begin{pmatrix} -19 & 26 \\ 7 & -9 \end{pmatrix}$.

Problem 6. Simplify

$$\begin{pmatrix} 3 & 4 & 0 \\ -2 & 6 & -3 \\ 7 & -4 & 1 \end{pmatrix} \times \begin{pmatrix} 2 \\ 5 \\ -1 \end{pmatrix}$$

The sum of the products of the elements of each row of the first matrix and the elements of the second matrix (called a column matrix) are taken one at a time. Thus:

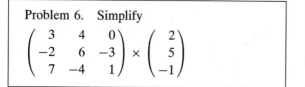

$$= \begin{pmatrix} (3 \times 2) & + & (4 \times 5) & + & (0 \times (-1)) \\ (-2 \times 2) & + & (6 \times 5) & + & (-3 \times (-1)) \\ (7 \times 2) & + & (-4 \times 5) & + & (1 \times (-1)) \end{pmatrix}$$

$$= \begin{pmatrix} 26 \\ 29 \\ -7 \end{pmatrix}$$

Problem 7. If $A = \begin{pmatrix} 3 & 4 & 0 \\ -2 & 6 & -3 \\ 7 & -4 & 1 \end{pmatrix}$ and

$B = \begin{pmatrix} 2 & -5 \\ 5 & -6 \\ -1 & -7 \end{pmatrix}$, find $A \times B$

The sum of the products of the elements of each row of the first matrix and the elements of each column of the second matrix are taken one at a time. Thus:

$$\begin{pmatrix} 3 & 4 & 0 \\ -2 & 6 & -3 \\ 7 & -4 & 1 \end{pmatrix} \times \begin{pmatrix} 2 & -5 \\ 5 & -6 \\ -1 & -7 \end{pmatrix}$$

$$= \begin{pmatrix} [(3 \times 2) & [(3 \times (-5)) \\ + (4 \times 5) & + (4 \times (-6)) \\ + (0 \times (-1))] & + (0 \times (-7))] \\ [(-2 \times 2) & [(-2 \times (-5)) \\ + (6 \times 5) & + (6 \times (-6)) \\ + (-3 \times (-1))] & + (-3 \times (-7))] \\ [(7 \times 2) & [(7 \times (-5)) \\ + (-4 \times 5) & + (-4 \times (-6)) \\ + (1 \times (-1))] & + (1 \times (-7))] \end{pmatrix}$$

$$= \begin{pmatrix} 26 & -39 \\ 29 & -5 \\ -7 & -18 \end{pmatrix}$$

Problem 8. Determine
$$\begin{pmatrix} 1 & 0 & 3 \\ 2 & 1 & 2 \\ 1 & 3 & 1 \end{pmatrix} \times \begin{pmatrix} 2 & 2 & 0 \\ 1 & 3 & 2 \\ 3 & 2 & 0 \end{pmatrix}$$

The sum of the products of the elements of each row of the first matrix and the elements of each column of the second matrix are taken one at a time. Thus:

$$\begin{pmatrix} 1 & 0 & 3 \\ 2 & 1 & 2 \\ 1 & 3 & 1 \end{pmatrix} \times \begin{pmatrix} 2 & 2 & 0 \\ 1 & 3 & 2 \\ 3 & 2 & 0 \end{pmatrix} \text{ is equal to}$$

$$\begin{pmatrix} [(1 \times 2) & [(1 \times 2) & [(1 \times 0) \\ + (0 \times 1) & + (0 \times 3) & + (0 \times 2) \\ + (3 \times 3)] & + (3 \times 2)] & + (3 \times 0)] \\ [(2 \times 2) & [(2 \times 2) & [(2 \times 0) \\ + (1 \times 1) & + (1 \times 3) & + (1 \times 2) \\ + (2 \times 3)] & + (2 \times 2)] & + (2 \times 0)] \\ [(1 \times 2) & [(1 \times 2) & [(1 \times 0) \\ + (3 \times 1) & + (3 \times 3) & + (3 \times 2) \\ + (1 \times 3)] & + (1 \times 2)] & + (1 \times 0)] \end{pmatrix}$$

i.e. $\begin{pmatrix} 11 & 8 & 0 \\ 11 & 11 & 2 \\ 8 & 13 & 6 \end{pmatrix}$

In algebra, the commutative law of multiplication states that $a \times b = b \times a$. For matrices, this law is only true in a few special cases, and in general $A \times B$ is **not** equal to $B \times A$.

Problem 9. If $A = \begin{pmatrix} 2 & 3 \\ 1 & 0 \end{pmatrix}$ and

$B = \begin{pmatrix} 2 & 3 \\ 0 & 1 \end{pmatrix}$ show that $A \times B \neq B \times A$

$A \times B = \begin{pmatrix} 2 & 3 \\ 1 & 0 \end{pmatrix} \times \begin{pmatrix} 2 & 3 \\ 0 & 1 \end{pmatrix}$

$= \begin{pmatrix} [(2 \times 2) + (3 \times 0)] & [(2 \times 3) + (3 \times 1)] \\ [(1 \times 2) + (0 \times 0)] & [(1 \times 3) + (0 \times 1)] \end{pmatrix}$

$= \begin{pmatrix} 4 & 9 \\ 2 & 3 \end{pmatrix}$

$B \times A = \begin{pmatrix} 2 & 3 \\ 0 & 1 \end{pmatrix} \times \begin{pmatrix} 2 & 3 \\ 1 & 0 \end{pmatrix}$

$= \begin{pmatrix} [(2 \times 2) + (3 \times 1)] & [(2 \times 3) + (3 \times 0)] \\ [(0 \times 2) + (1 \times 1)] & [(0 \times 3) + (1 \times 0)] \end{pmatrix}$

$= \begin{pmatrix} 7 & 6 \\ 1 & 0 \end{pmatrix}$

Since $\begin{pmatrix} 4 & 9 \\ 2 & 3 \end{pmatrix} \neq \begin{pmatrix} 7 & 6 \\ 1 & 0 \end{pmatrix}$, then $A \times B \neq B \times A$.

Further problems on addition, subtraction and multiplication of matrices may be found in Section 44.8, Problems 1 to 18, pages 442 and 443.

44.3   The unit matrix

A **unit matrix**, I, is one which all elements of the leading diagonal (\) have a value of 1 and all other elements have a value of 0. Multiplication of a matrix by I is the equivalent of multiplying by 1 in arithmetic.

44.4   The determinant of a 2 by 2 matrix

The **determinant** of a 2 by 2 matrix, $\begin{pmatrix} a & b \\ c & d \end{pmatrix}$ is defined as $(ad - bc)$. The elements of the determinant of a matrix are written between vertical lines. Thus, the determinant of $\begin{pmatrix} 3 & -4 \\ 1 & 6 \end{pmatrix}$ is written as $\begin{vmatrix} 3 & -4 \\ 1 & 6 \end{vmatrix}$ and is equal to $(3 \times 6) - (-4 \times 1)$, i.e. $18 - (-4)$ or 22. Hence the determinant of a matrix can be expressed as a single numerical value, i.e. $\begin{vmatrix} 3 & -4 \\ 1 & 6 \end{vmatrix} = 22$.

Problem 10.   Determine the value of $\begin{vmatrix} 3 & -2 \\ 7 & 4 \end{vmatrix}$

The vertical lines show that this is a determinant and the value of $\begin{vmatrix} p & q \\ r & s \end{vmatrix}$ is defined as $(p \times s) - (q \times r)$.

Thus $\begin{vmatrix} 3 & -2 \\ 7 & 4 \end{vmatrix} = (3 \times 4) - (-2 \times 7)$

$= 12 - (-14) = \mathbf{26}$

Further problems on 2 by 2 determinants may be found in Section 44.8, Problems 19 to 21, page 443.

44.5   The inverse or reciprocal of a 2 by 2 matrix

The inverse of matrix A is A^{-1} such that $A \times A^{-1} = I$, the unit matrix.

Let matrix A be $\begin{pmatrix} 1 & 2 \\ 3 & 4 \end{pmatrix}$ and let the inverse matrix, A^{-1} be $\begin{pmatrix} a & b \\ c & d \end{pmatrix}$. Then, since $A \times A^{-1} = I$,

$$\begin{pmatrix} 1 & 2 \\ 3 & 4 \end{pmatrix} \times \begin{pmatrix} a & b \\ c & d \end{pmatrix} = \begin{pmatrix} 1 & 0 \\ 0 & 1 \end{pmatrix}$$

Multiplying the matrices on the left-hand side, gives

$$\begin{pmatrix} a + 2c & b + 2d \\ 3a + 4c & 3b + 4d \end{pmatrix} = \begin{pmatrix} 1 & 0 \\ 0 & 1 \end{pmatrix}$$

Equating corresponding elements gives:

$b + 2d = 0$, i.e. $b = -2d$

and $\quad 3a + 4c = 0$, i.e. $a = -\dfrac{4}{3}c$

Substituting for a and b gives:

$$\begin{pmatrix} -\dfrac{4}{3}c + 2c & -2d + 2d \\ 3\left(-\dfrac{4}{3}c\right) + 4c & 3(-2d) + 4d \end{pmatrix} = \begin{pmatrix} 1 & 0 \\ 0 & 1 \end{pmatrix}$$

i.e. $\begin{pmatrix} \frac{2}{3}c & 0 \\ 0 & -2d \end{pmatrix} = \begin{pmatrix} 1 & 0 \\ 0 & 1 \end{pmatrix}$

showing that $\frac{2}{3}c = 1$, i.e. $c = \frac{3}{2}$ and $-2d = 1$,

i.e. $d = -\frac{1}{2}$.

Since $b = -2d$, $b = 1$ and since $a = -\frac{4}{3}c$, $a = -2$.

Thus the inverse of matrix $\begin{pmatrix} 1 & 2 \\ 3 & 4 \end{pmatrix}$ is $\begin{pmatrix} a & b \\ c & d \end{pmatrix}$

that is, $\begin{pmatrix} -2 & 1 \\ \frac{3}{2} & -\frac{1}{2} \end{pmatrix}$.

There is, however, a quicker method of obtaining the inverse of a 2 by 2 matrix.

For any matrix $\begin{pmatrix} p & q \\ r & s \end{pmatrix}$ the inverse may be obtained by:

(i)   interchanging the positions of p and s,
(ii)   changing the signs of q and r, and
(iii)   multiplying this new matrix by the reciprocal of the determinant of $\begin{pmatrix} p & q \\ r & s \end{pmatrix}$

Thus the inverse of matrix $\begin{pmatrix} 1 & 2 \\ 3 & 4 \end{pmatrix}$ is

$$\frac{1}{4-6}\begin{pmatrix} 4 & -2 \\ -3 & 1 \end{pmatrix} = \begin{pmatrix} -2 & 1 \\ \frac{3}{2} & -\frac{1}{2} \end{pmatrix}$$

as obtained previously.

Problem 11.   Determine the inverse of $\begin{pmatrix} 3 & -2 \\ 7 & 4 \end{pmatrix}$

The inverse of matrix $\begin{pmatrix} p & q \\ r & s \end{pmatrix}$ is obtained by interchanging the positions of p and s, changing the signs of q and r and multiplying by the reciprocal of the determinant $\begin{vmatrix} p & q \\ r & s \end{vmatrix}$. Thus, the inverse of

$$\begin{pmatrix} 3 & -2 \\ 7 & 4 \end{pmatrix} = \frac{1}{(3 \times 4) - (-2 \times 7)}\begin{pmatrix} 4 & 2 \\ -7 & 3 \end{pmatrix}$$

$$= \frac{1}{26}\begin{pmatrix} 4 & 2 \\ -7 & 3 \end{pmatrix}$$

$$= \begin{pmatrix} \frac{2}{13} & \frac{1}{13} \\ \frac{-7}{26} & \frac{3}{26} \end{pmatrix}$$

Further problems on the inverse of 2 by 2 matrices may be found in Section 44.8, Problems 22 to 24, page 443.

44.6   The determinant of a 3 by 3 matrix

(i)   The **minor** of an element of a 3 by 3 matrix is the value of the 2 by 2 determinant obtained by covering up the row and column containing that element.

Thus for the matrix $\begin{pmatrix} 1 & 2 & 3 \\ 4 & 5 & 6 \\ 7 & 8 & 9 \end{pmatrix}$ the minor of element 4 is obtained by covering the row $(4 \quad 5 \quad 6)$ and the column $\begin{pmatrix} 1 \\ 4 \\ 7 \end{pmatrix}$, leaving the 2 by 2 determinant $\begin{vmatrix} 2 & 3 \\ 8 & 9 \end{vmatrix}$, i.e. the minor of element 4 is $(2 \times 9) - (3 \times 8)$, i.e. -6

(ii)   The sign of a minor depends on its position within the matrix, the sign pattern being $\begin{pmatrix} + & - & + \\ - & + & - \\ + & - & + \end{pmatrix}$. Thus the signed-minor of element 4 in the matrix $\begin{pmatrix} 1 & 2 & 3 \\ 4 & 5 & 6 \\ 7 & 8 & 9 \end{pmatrix}$ is

$$-\begin{vmatrix} 2 & 3 \\ 8 & 9 \end{vmatrix} = -(-6) = 6.$$

The signed-minor of an element is called the **cofactor** of the element

(iii)   **The value of a 3 by 3 determinant is the sum of the products of the elements and their cofactors of any row or any column of the corresponding 3 by 3 matrix**

Problem 12.   Find the value of $\begin{vmatrix} 3 & 4 & -1 \\ 2 & 0 & 7 \\ 1 & -3 & -2 \end{vmatrix}$

The value of this determinant is the sum of the products of the elements and their cofactors, of any row or of any column. If the second row or second column is selected, the element 0 will make the product of the element and its cofactor zero and reduce the amount of arithmetic to be done to a minimum. Supposing a second row expansion is selected.

The minor of 2 is the value of the determinant remaining when the row and column containing the 2 (i.e. the second row and the first column) is covered up. Thus the cofactor of element 2 is $\begin{vmatrix} 4 & -1 \\ -3 & -2 \end{vmatrix}$, i.e. -11. The sign of element 2 is minus (see (ii) above), hence the cofactor of element 2 (the signed-minor) is $+11$. Similarly the minor of element 7 is $\begin{vmatrix} 3 & 4 \\ 1 & -3 \end{vmatrix}$, i.e. -13, and its cofactor is $+13$. Hence the value of the sum of the products of the elements and their cofactors is $2 \times 11 + 7 \times 13$, i.e.

$$\begin{vmatrix} 3 & 4 & -1 \\ 2 & 0 & 7 \\ 1 & -3 & -2 \end{vmatrix} = 113$$

The same result will be obtained whichever row or column is selected. For example, the third column expansion is

$$+(-1)\begin{vmatrix} 2 & 0 \\ 1 & -3 \end{vmatrix} - 7\begin{vmatrix} 3 & 4 \\ 1 & -3 \end{vmatrix} + (-2)\begin{vmatrix} 3 & 4 \\ 2 & 0 \end{vmatrix}$$

i.e. $6 + 91 + 16 = \mathbf{113}$, as obtained previously.

Problem 13. Evaluate $\begin{vmatrix} 1 & 4 & -3 \\ -5 & 2 & 6 \\ -1 & -4 & 2 \end{vmatrix}$

Using the first row: $\begin{vmatrix} 1 & 4 & -3 \\ -5 & 2 & 6 \\ -1 & -4 & 2 \end{vmatrix}$

$$= 1\begin{vmatrix} 2 & 6 \\ -4 & 2 \end{vmatrix} - 4\begin{vmatrix} -5 & 6 \\ -1 & 2 \end{vmatrix} + (-3)\begin{vmatrix} -5 & 2 \\ -1 & -4 \end{vmatrix}$$

$$= (4 + 24) - 4(-10 + 6) - 3(20 + 2)$$

$$= 28 + 16 - 66 = \mathbf{-22}$$

Using the second column: $\begin{vmatrix} 1 & 4 & -3 \\ -5 & 2 & 6 \\ -1 & -4 & 2 \end{vmatrix}$

$$= -4\begin{vmatrix} -5 & 6 \\ -1 & 2 \end{vmatrix} + 2\begin{vmatrix} 1 & -3 \\ -1 & 2 \end{vmatrix} - (-4)\begin{vmatrix} 1 & -3 \\ -5 & 6 \end{vmatrix}$$

$$= -4(-10 + 6) + 2(2 - 3) + 4(6 - 15)$$

$$= 16 - 2 - 36 = \mathbf{-22}, \text{ as previously obtained.}$$

Further problems on 3 by 3 determinants may be found in Section 44.8, Problems 25 to 31, pages 443 and 444.

44.7 The inverse or reciprocal of a 3 by 3 matrix

The **adjoint** of a matrix A is obtained by:

(i) forming a matrix B of the cofactors of A, and
(ii) **transposing** matrix B to give B^T, where B^T is the matrix obtained by writing the rows of B as the columns of B^T. Then **adj** $A = B^T$.

The inverse of matrix A, A^{-1} is given by A^{-1}

$$= \frac{\text{adj } A}{|A|}$$ where adj A is the adjoint of matrix A

and $|A|$ is the determinant of matrix A

Problem 14. Determine the inverse of the

matrix $\begin{pmatrix} 3 & 4 & -1 \\ 2 & 0 & 7 \\ 1 & -3 & -2 \end{pmatrix}$

The inverse of matrix A, $A^{-1} = \dfrac{\text{adj } A}{|A|}$

The adjoint of A is found by:

(i) obtaining the matrix of the cofactors of the elements, and
(ii) transposing this matrix

The cofactor of element 3 is $+\begin{vmatrix} 0 & 7 \\ -3 & -2 \end{vmatrix}$, i.e. 21.

The cofactor of element 4 is $-\begin{vmatrix} 2 & 7 \\ 1 & -2 \end{vmatrix}$, i.e. 11, and so on.

The matrix of cofactors is $\begin{pmatrix} 21 & 11 & -6 \\ 11 & -5 & 13 \\ 28 & -23 & -8 \end{pmatrix}$.

The transpose of the matrix of cofactors, i.e. the adjoint of the matrix, is obtained by writing the rows as columns, and is $\begin{pmatrix} 21 & 11 & 28 \\ 11 & -5 & -23 \\ -6 & 13 & -8 \end{pmatrix}$.

From Problem 12, the determinant of $\begin{vmatrix} 3 & 4 & -1 \\ 2 & 0 & 7 \\ 1 & -3 & -2 \end{vmatrix}$

is 113. Hence the inverse of

$$\begin{pmatrix} 3 & 4 & -1 \\ 2 & 0 & 7 \\ 1 & -3 & -2 \end{pmatrix} \text{ is } \dfrac{\begin{pmatrix} 21 & 11 & 28 \\ 11 & -5 & -23 \\ -6 & 13 & -8 \end{pmatrix}}{113}$$

i.e. $\dfrac{1}{113} \begin{pmatrix} 21 & 11 & 28 \\ 11 & -5 & -23 \\ -6 & 13 & -8 \end{pmatrix}$.

Problem 15. Find the inverse of
$$\begin{pmatrix} 1 & 5 & -2 \\ 3 & -1 & 4 \\ -3 & 6 & -7 \end{pmatrix}$$

$$\text{Inverse} = \frac{\text{adjoint}}{\text{determinant}}$$

The matrix of cofactors is $\begin{pmatrix} -17 & 9 & 15 \\ 23 & -13 & -21 \\ 18 & -10 & -16 \end{pmatrix}$.

The transpose of the matrix of cofactors (i.e. the

adjoint) is $\begin{pmatrix} -17 & 23 & 18 \\ 9 & -13 & -10 \\ 15 & -21 & -16 \end{pmatrix}$.

The determinant of $\begin{pmatrix} 1 & 5 & -2 \\ 3 & -1 & 4 \\ -3 & 6 & -7 \end{pmatrix}$

$$= 1(7 - 24) - 5(-21 + 12) - 2(18 - 3)$$
$$= -17 + 45 - 30 = -2$$

Hence the inverse of

$$\begin{pmatrix} 1 & 5 & -2 \\ 3 & -1 & 4 \\ -3 & 6 & -7 \end{pmatrix} = \dfrac{\begin{pmatrix} -17 & 23 & 18 \\ 9 & -13 & -10 \\ 15 & -21 & -16 \end{pmatrix}}{-2}$$

$$= \begin{pmatrix} 8.5 & -11.5 & -9 \\ -4.5 & 6.5 & 5 \\ -7.5 & 10.5 & 8 \end{pmatrix}$$

Further problems on the inverse of a 3 by 3 matrix may be found in the following Section 44.8, Problems 32 to 40, page 444.

44.8 Further problems on the theory of matrices and determinants

In Problems 1 to 40, the matrices stated are:

$$A = \begin{pmatrix} 3 & -1 \\ -4 & 7 \end{pmatrix}, \quad B = \begin{pmatrix} \frac{1}{2} & \frac{2}{3} \\ -\frac{1}{3} & \frac{3}{5} \end{pmatrix},$$

$$C = \begin{pmatrix} -1.3 & 7.4 \\ 2.5 & -3.9 \end{pmatrix}, \quad D = \begin{pmatrix} 4 & -7 & 6 \\ -2 & 4 & 0 \\ 5 & 7 & -4 \end{pmatrix},$$

$$E = \begin{pmatrix} 3 & 6 & \frac{1}{2} \\ 5 & -\frac{2}{3} & 7 \\ -1 & 0 & \frac{3}{5} \end{pmatrix},$$

$$F = \begin{pmatrix} 3.1 & 2.4 & 6.4 \\ -1.6 & 3.8 & -1.9 \\ 5.3 & 3.4 & -4.8 \end{pmatrix}, \quad G = \begin{pmatrix} \frac{3}{4} \\ 1\frac{2}{5} \end{pmatrix},$$

$$H = \begin{pmatrix} -2 \\ 5 \end{pmatrix}, \quad J = \begin{pmatrix} 4 \\ -11 \\ 7 \end{pmatrix}, \quad K = \begin{pmatrix} 1 & 0 \\ 0 & 1 \\ 1 & 0 \end{pmatrix}$$

Addition, subtraction and multiplication

In Problems 1 to 17, perform the matrix operation stated:

1 $A + B$
$$\left[\begin{pmatrix} 3\frac{1}{2} & -\frac{1}{3} \\ -4\frac{1}{3} & 6\frac{2}{5} \end{pmatrix} \right]$$

2 $D + E$
$$\left[\begin{pmatrix} 7 & -1 & 6\frac{1}{2} \\ 3 & 3\frac{1}{3} & 7 \\ 4 & 7 & -3\frac{2}{5} \end{pmatrix} \right]$$

3 $A - B$
$$\left[\begin{pmatrix} 2\frac{1}{2} & -1\frac{2}{3} \\ -3\frac{2}{3} & 7\frac{3}{5} \end{pmatrix} \right]$$

4 $D - E$

$$\left[\begin{pmatrix} 1 & -13 & 5\frac{1}{2} \\ -7 & 4\frac{2}{3} & -7 \\ 6 & 7 & -4\frac{3}{5} \end{pmatrix} \right]$$

16 $D \times E$

$$\left[\begin{pmatrix} -29 & 28\frac{2}{3} & -43\frac{2}{5} \\ 14 & -14\frac{2}{3} & 27 \\ 54 & 25\frac{1}{3} & 49\frac{1}{10} \end{pmatrix} \right]$$

5 $A + B - C$

$$\left[\begin{pmatrix} 4.8 & -7.7\dot{3} \\ -6.8\dot{3} & 10.3 \end{pmatrix} \right]$$

17 $D \times F$

$$\left[\begin{pmatrix} 55.4 & 3.4 & 10.1 \\ -12.6 & 10.4 & -20.4 \\ -16.9 & 25.0 & 37.9 \end{pmatrix} \right]$$

6 $D - E + F$

$$\left[\begin{pmatrix} 4.1 & -10.6 & 11.9 \\ -8.6 & 8.4\dot{6} & -8.9 \\ 11.3 & 10.4 & -9.4 \end{pmatrix} \right]$$

18 Show that $A \times C \neq C \times A$.

$$\left[\begin{array}{l} A \times C = \begin{pmatrix} -6.4 & 26.1 \\ 22.7 & -56.9 \end{pmatrix}, \\ C \times A = \begin{pmatrix} -33.5 & -53.1 \\ 23.1 & -29.8 \end{pmatrix} \\ \text{Hence they are not equal} \end{array} \right]$$

7 $5A + 6B$

$$\left[\begin{pmatrix} 18.0 & -1.0 \\ -22.0 & 31.4 \end{pmatrix} \right]$$

8 $2D + 3E - 4F$

$$\left[\begin{pmatrix} 4.6 & -5.6 & -12.1 \\ 17.4 & -9.2 & 28.6 \\ -14.2 & 0.4 & 13.0 \end{pmatrix} \right]$$

2 × 2 determinants

19 Calculate the determinant of matrix A.   [17]

20 Calculate the determinant of matrix B.

$$\left[-\frac{7}{90} \right]$$

9 $A \times H$

$$\left[\begin{pmatrix} -11 \\ 43 \end{pmatrix} \right]$$

21 Calculate the determinant of matrix C.

$$[-13.43]$$

10 $B \times G$

$$\left[\begin{pmatrix} 1\frac{37}{120} \\ -1\frac{9}{100} \end{pmatrix} \right]$$

Inverse of 2 by 2 matrix

22 Determine the inverse of matrix A.

$$\left[\begin{pmatrix} \frac{7}{17} & \frac{1}{17} \\ \frac{4}{17} & \frac{3}{17} \end{pmatrix} \right]$$

11 $A \times B$

$$\left[\begin{pmatrix} 1\frac{5}{6} & 2\frac{3}{5} \\ -4\frac{1}{3} & -6\frac{13}{15} \end{pmatrix} \right]$$

12 $A \times C$

$$\left[\begin{pmatrix} -6.4 & 26.1 \\ 22.7 & -56.9 \end{pmatrix} \right]$$

23 Determine the inverse of matrix B.

$$\left[\begin{pmatrix} 7\frac{5}{7} & 8\frac{4}{7} \\ -4\frac{2}{7} & -6\frac{3}{7} \end{pmatrix} \right]$$

13 $D \times J$

$$\left[\begin{pmatrix} 135 \\ -52 \\ -85 \end{pmatrix} \right]$$

14 $F \times J$

$$\left[\begin{pmatrix} 30.8 \\ -61.5 \\ -49.8 \end{pmatrix} \right]$$

24 Determine the inverse of matrix C.

$$\left[\begin{array}{c} \begin{pmatrix} 0.290 & 0.551 \\ 0.186 & 0.097 \end{pmatrix} \\ \text{correct to 3 decimal places} \end{array} \right]$$

15 $E \times K$

$$\left[\begin{pmatrix} 3\frac{1}{2} & 6 \\ 12 & -\frac{2}{3} \\ -\frac{2}{5} & 0 \end{pmatrix} \right]$$

3 by 3 determinants

25 Find the matrix of minors of matrix D.

$$\left[\begin{pmatrix} -16 & 8 & -34 \\ -14 & -46 & 63 \\ -24 & 12 & 2 \end{pmatrix} \right]$$

26 Find the matrix of minors of matrix E.

$$\left[\begin{pmatrix} -\dfrac{2}{5} & 10 & -\dfrac{2}{3} \\ 3\dfrac{3}{5} & 2\dfrac{3}{10} & 6 \\ 42\dfrac{1}{3} & 18\dfrac{1}{2} & -32 \end{pmatrix}\right]$$

27 Find the matrix of cofactors of matrix D.

$$\left[\begin{pmatrix} -16 & -8 & -34 \\ 14 & -46 & -63 \\ -24 & -12 & 2 \end{pmatrix}\right]$$

28 Find the matrix of cofactors of matrix E.

$$\left[\begin{pmatrix} -\dfrac{2}{5} & -10 & -\dfrac{2}{3} \\ -3\dfrac{3}{5} & 2\dfrac{3}{10} & -6 \\ 42\dfrac{1}{3} & -18\dfrac{1}{2} & -32 \end{pmatrix}\right]$$

29 Calculate the determinant of matrix D.

$$[-212]$$

30 Calculate the determinant of matrix E.

$$\left[-61\dfrac{8}{15}\right]$$

31 Calculate the determinant of matrix F.

$$[-242.83]$$

Inverse of 3 by 3 matrix

32 Write down the transpose of matrix D.

$$\left[\begin{pmatrix} 4 & -2 & 5 \\ -7 & 4 & 7 \\ 6 & 0 & -4 \end{pmatrix}\right]$$

33 Write down the transpose of matrix E.

$$\left[\begin{pmatrix} 3 & 5 & -1 \\ 6 & -\dfrac{2}{3} & 0 \\ \dfrac{1}{2} & 7 & \dfrac{3}{5} \end{pmatrix}\right]$$

34 Write down the transpose of matrix F.

$$\left[\begin{pmatrix} 3.1 & -1.6 & 5.3 \\ 2.4 & 3.8 & 3.4 \\ 6.4 & -1.9 & -4.8 \end{pmatrix}\right]$$

35 Determine the adjoint of matrix D.

$$\left[\begin{pmatrix} -16 & 14 & -24 \\ -8 & -46 & -12 \\ -34 & -63 & 2 \end{pmatrix}\right]$$

36 Determine the adjoint of matrix E.

$$\left[\begin{pmatrix} -\dfrac{2}{5} & -3\dfrac{3}{5} & 42\dfrac{1}{3} \\ -10 & 2\dfrac{3}{10} & -18\dfrac{1}{2} \\ -\dfrac{2}{3} & -6 & -32 \end{pmatrix}\right]$$

37 Determine the adjoint of matrix F.

$$\left[\begin{pmatrix} -11.78 & 33.28 & -28.88 \\ -17.75 & -48.80 & -4.35 \\ -25.58 & 2.18 & 15.62 \end{pmatrix}\right]$$

38 Find the inverse of matrix D.

$$\left[-\dfrac{1}{212}\begin{pmatrix} -16 & 14 & -24 \\ -8 & -46 & -12 \\ -34 & -63 & 2 \end{pmatrix}\right]$$

39 Find the inverse of matrix E.

$$\left[-\dfrac{15}{923}\begin{pmatrix} -\dfrac{2}{5} & -3\dfrac{3}{5} & 43\dfrac{1}{3} \\ -10 & 2\dfrac{3}{10} & -18\dfrac{1}{2} \\ -\dfrac{2}{3} & -6 & -32 \end{pmatrix}\right]$$

40 Find the inverse of matrix F.

$$\left[-\dfrac{1}{242.83}\begin{pmatrix} -11.78 & 33.28 & -28.88 \\ -17.75 & -48.80 & -4.35 \\ -25.58 & 2.18 & 15.62 \end{pmatrix}\right]$$

45

The solution of simultaneous equations by matrices and determinants

45.1 Some properties of determinants

There are certain **properties of determinants** which enable the value of a determinant to be found more simply. Some of these are given below.

(i) If all the elements in a row or column are interchanged with the corresponding elements in another row or column, the value of the determinant obtained is -1 times the value of the original determinant. Thus

$$\begin{vmatrix} a & b \\ c & d \end{vmatrix} \equiv (-1) \times \begin{vmatrix} b & a \\ d & c \end{vmatrix}$$

Similarly,

$$\begin{vmatrix} a_1 & b_1 & c_1 \\ a_2 & b_2 & c_2 \\ a_3 & b_3 & c_3 \end{vmatrix} \equiv (-1) \times \begin{vmatrix} a_2 & b_2 & c_2 \\ a_1 & b_1 & c_1 \\ a_3 & b_3 & c_3 \end{vmatrix}$$

To verify this:

$$(-1) \times \begin{vmatrix} a_2 & b_2 & c_2 \\ a_1 & b_1 & c_1 \\ a_3 & b_3 & c_3 \end{vmatrix}$$

$$= (-1) \left[a_2 \begin{vmatrix} b_1 & c_1 \\ b_3 & c_3 \end{vmatrix} - b_2 \begin{vmatrix} a_1 & c_1 \\ a_3 & c_3 \end{vmatrix} \right.$$

$$\left. + c_2 \begin{vmatrix} a_1 & b_1 \\ a_3 & b_3 \end{vmatrix} \right]$$

$$= (-1)[a_2(b_1 c_3 - b_3 c_1) - b_2(a_1 c_3 - a_3 c_1)$$

$$+ c_2(a_1 b_3 - a_3 b_1)]$$

$$= -a_2 b_1 c_3 + a_2 b_3 c_1 + a_1 b_2 c_3 - a_3 b_2 c_1$$

$$- a_1 b_3 c_2 + a_3 b_1 c_2$$

$$= a_1(b_2 c_3 - b_3 c_2) - b_1(a_2 c_3 - a_3 c_2)$$

$$+ c_1(a_2 b_3 - a_3 b_2)$$

$$= \begin{vmatrix} a_1 & b_1 & c_1 \\ a_2 & b_2 & c_2 \\ a_3 & b_3 & c_3 \end{vmatrix}$$

(ii) If two rows or two columns of a determinant are equal, its value is equal to zero. Thus

$$\begin{vmatrix} a & a \\ c & c \end{vmatrix} \equiv 0$$

Similarly,

$$\begin{vmatrix} a_1 & a_1 & c_1 \\ a_2 & a_2 & c_2 \\ a_3 & a_3 & c_3 \end{vmatrix}$$

$$= a_1 \begin{vmatrix} a_2 & c_2 \\ a_3 & c_3 \end{vmatrix} - a_1 \begin{vmatrix} a_2 & c_2 \\ a_3 & c_3 \end{vmatrix} + c_1 \begin{vmatrix} a_2 & a_2 \\ a_3 & a_3 \end{vmatrix}$$

$$= c_1 \begin{vmatrix} a_2 & a_2 \\ a_3 & a_3 \end{vmatrix}$$

$$= c_1 a_2 a_3 - c_1 a_2 a_3 = 0$$

(iii) If all the elements in any row or any column of a determinant have a common factor, the elements in that row or column can be divided by the common factor and the factor becomes a factor of the determinant. Thus

$$\begin{vmatrix} a & b \\ b \times c & b \times d \end{vmatrix} \equiv b \times \begin{vmatrix} a & b \\ c & d \end{vmatrix}$$

Similarly,

$$\begin{vmatrix} a_1 & db_1 & c_1 \\ a_2 & db_2 & c_2 \\ a_3 & db_3 & c_3 \end{vmatrix}$$

$$= a_1 \begin{vmatrix} db_2 & c_2 \\ db_3 & c_3 \end{vmatrix} - db_1 \begin{vmatrix} a_2 & c_2 \\ a_3 & c_3 \end{vmatrix} + c_1 \begin{vmatrix} a_2 & db_2 \\ a_3 & db_3 \end{vmatrix}$$

$$= a_1(db_2 c_3 - db_3 c_2) - db_1(a_2 c_3 - a_3 c_2)$$

$$+ c_1(a_2 db_3 - a_3 db_2)$$

$$= d[a_1(b_2 c_3 - b_3 c_2) - b_1(a_2 c_3 - a_3 c_2)$$

$$+ c_1(a_2 b_3 - a_3 b_2)]$$

$$= d \times \begin{vmatrix} a_1 & b_1 & c_1 \\ a_2 & b_2 & c_2 \\ a_3 & b_3 & c_3 \end{vmatrix}$$

(iv) The value of a determinant remains unaltered if a multiple of the elements in any row or any column are added to the corresponding elements of any other row or column. Thus

$$\begin{vmatrix} a & b \\ c & d \end{vmatrix} \equiv \begin{vmatrix} a & b+ka \\ c & d+kc \end{vmatrix}$$

Similarly when, say, the second row elements are multiplied by d and added to the third row

$$\begin{vmatrix} a_1 & b_1 & c_1 \\ a_2 & b_2 & c_2 \\ a_3 & b_3 & c_3 \end{vmatrix} = \begin{vmatrix} a_1 & b_1 & c_1+db_1 \\ a_2 & b_2 & c_2+db_2 \\ a_3 & b_3 & c_3+db_3 \end{vmatrix} = |A|$$

Then $|A| = a_1 \begin{vmatrix} b_2 & c_2+db_2 \\ b_3 & c_3+db_3 \end{vmatrix} - b_1 \begin{vmatrix} a_2 & c_2+db_2 \\ a_3 & c_3+db_3 \end{vmatrix}$

$\qquad + (c_1+db_1) \begin{vmatrix} a_2 & b_2 \\ a_3 & b_3 \end{vmatrix}$

$= a_1[b_2(c_3+db_3) - b_3(c_2+db_2)]$

$\quad - b_1[a_2(c_3+db_3) - a_3(c_2+db_2)]$

$\quad + c_1(a_2b_3 - a_3b_2) + db_1(a_2b_3 - a_3b_2)$

$= a_1b_2c_3 + a_1b_2db_3 - a_1b_3c_2 - a_1db_2b_3$

$\quad - a_2b_1c_3 - a_2b_1db_3 + a_3b_1c_2 + a_3b_1db_2$

$\quad + a_2b_3c_1 - a_3b_2c_1 + a_2db_1b_3 - a_3db_1b_2$

$= a_1(b_2c_3 - b_3c_2) - b_1(a_2c_3 - a_3c_2)$

$\quad + c_1(a_2b_3 - a_3b_2) + d(a_1b_2b_3 - a_1b_2b_3)$

$\quad - a_2b_1b_3 + a_3b_1b_2 + a_2b_1b_3 - a_3b_1b_2)$

$= \begin{vmatrix} a_1 & b_1 & c_1 \\ a_2 & b_2 & c_2 \\ a_3 & b_3 & c_3 \end{vmatrix}$

The properties of determinants previously listed may be used to:

(i) reduce the size of elements within the determinant, and

(ii) introduce as many zero elements as is practical before evaluating the determinant. Simplification is mainly achieved by using property (iv)

Problem 1.   Simplify and evaluate

$$\begin{vmatrix} 7 & 30 \\ 60 & 252 \end{vmatrix}$$

$\begin{vmatrix} 7 & 30 \\ 60 & 252 \end{vmatrix} = 4 \times \begin{vmatrix} 7 & 30 \\ 15 & 63 \end{vmatrix}$ (property (iii))

Taking 4 times column 1 from column 2 gives:

$$4 \times \begin{vmatrix} 7 & (30-28) \\ 15 & (63-60) \end{vmatrix} = 4 \times \begin{vmatrix} 7 & 2 \\ 15 & 3 \end{vmatrix}$$

Taking 5 times column 2 from column 1 gives:

$$4 \times \begin{vmatrix} (7-10) & 2 \\ (15-15) & 3 \end{vmatrix} = 4 \times \begin{vmatrix} -3 & 2 \\ 0 & 3 \end{vmatrix}$$

Hence, $\begin{vmatrix} 7 & 30 \\ 60 & 252 \end{vmatrix} = 4[(-3 \times 3) - 0]$

$$= -36$$

(If a calculator is available, it is usually easier just to evaluate the original determinant, i.e. $(7 \times 252) - (30 \times 60) = -36$)

Problem 2.   Simplify and evaluate:

$$\begin{vmatrix} 2 & 7 & 26 \\ 1 & 2 & 6 \\ 4 & 11 & 40 \end{vmatrix}$$

Taking 3 times column 2 from column 3 gives

$$\begin{vmatrix} 2 & 7 & (26-21) \\ 1 & 2 & (6-6) \\ 4 & 11 & (40-33) \end{vmatrix} \text{ i.e. } \begin{vmatrix} 2 & 7 & 5 \\ 1 & 2 & 0 \\ 4 & 11 & 7 \end{vmatrix}$$

Taking twice column 1 from column 2 gives

$$\begin{vmatrix} 2 & (7-4) & 5 \\ 1 & (2-2) & 0 \\ 4 & (11-8) & 7 \end{vmatrix} \text{ i.e. } \begin{vmatrix} 2 & 3 & 5 \\ 1 & 0 & 0 \\ 4 & 3 & 7 \end{vmatrix}$$

Since two of the elements in row 2 are zero, the value of this determinant is

$$-1 \times \begin{vmatrix} 3 & 5 \\ 3 & 7 \end{vmatrix} + 0 - 0$$

$$= -3 \times \begin{vmatrix} 1 & 5 \\ 1 & 7 \end{vmatrix} \quad \text{(property (iii))}$$

$$= -3(7-5) = -6$$

(As for second order determinants, it is often easier just to evaluate the original determinant when a calculator is available, rather than simplifying it.)

Further problems on the properties of determinants may be found in Section 45.4, Problems 1 to 5, page 450.

45.2 Solution of simultaneous equations by matrices

(a) The procedure for solving linear simultaneous equations in **two unknowns** using matrices is:

(i) write the equations in the form

$$a_1 x + b_1 y = c_1$$
$$a_2 x + b_2 y = c_2$$

(ii) write the matrix equation corresponding to these equations,

i.e. $\begin{pmatrix} a_1 & b_1 \\ a_2 & b_2 \end{pmatrix} \times \begin{pmatrix} x \\ y \end{pmatrix} = \begin{pmatrix} c_1 \\ c_2 \end{pmatrix}$

(iii) determine the inverse matrix of $\begin{pmatrix} a_1 & b_1 \\ a_2 & b_2 \end{pmatrix}$,

i.e. $\dfrac{1}{a_1 b_2 - b_1 a_2} \begin{pmatrix} b_2 & -b_1 \\ -a_2 & a_1 \end{pmatrix}$

(from Chapter 44)

(iv) multiply each side of (ii) by the inverse matrix, and

(v) solve for x and y by equating corresponding elements

Problem 3. Use matrices to solve the simultaneous equations:

$$3x + 5y - 7 = 0 \tag{1}$$
$$4x - 3y - 19 = 0 \tag{2}$$

(i) Writing the equations in the $a_1 x + b_1 y = c$ form gives:

$$3x + 5y = 7$$
$$4x - 3y = 19$$

(ii) The matrix equation is

$$\begin{pmatrix} 3 & 5 \\ 4 & -3 \end{pmatrix} \times \begin{pmatrix} x \\ y \end{pmatrix} = \begin{pmatrix} 7 \\ 19 \end{pmatrix}$$

(iii) The inverse of matrix $\begin{pmatrix} 3 & 5 \\ 4 & -3 \end{pmatrix}$ is

$$\frac{1}{3 \times (-3) - 5 \times 4} \begin{pmatrix} -3 & -5 \\ -4 & 3 \end{pmatrix}$$

i.e. $\begin{pmatrix} \dfrac{3}{29} & \dfrac{5}{29} \\ \dfrac{4}{29} & \dfrac{-3}{29} \end{pmatrix}$

(iv) Multiplying each side of (ii) by (iii) and remembering that $A \times A^{-1} = I$, the unit matrix gives:

$$\begin{pmatrix} 1 & 0 \\ 0 & 1 \end{pmatrix} \begin{pmatrix} x \\ y \end{pmatrix} = \begin{pmatrix} \dfrac{3}{29} & \dfrac{5}{29} \\ \dfrac{4}{29} & \dfrac{-3}{29} \end{pmatrix} \times \begin{pmatrix} 7 \\ 19 \end{pmatrix}$$

Thus $\begin{pmatrix} x \\ y \end{pmatrix} = \begin{pmatrix} \dfrac{21}{29} + \dfrac{95}{29} \\ \dfrac{28}{29} - \dfrac{57}{29} \end{pmatrix}$

i.e. $\begin{pmatrix} x \\ y \end{pmatrix} = \begin{pmatrix} 4 \\ -1 \end{pmatrix}$

(v) By comparing corresponding elements. $x = \mathbf{4}$ and $y = \mathbf{-1}$

Checking: equation (1),

$$3 \times 4 + 5 \times (-1) - 7 = 0 = \text{RHS}$$

equation (2),

$$4 \times 4 - 3 \times (-1) - 19 = 0 = \text{RHS}$$

(b) The procedure for solving linear simultaneous equations in **three unknowns** by matrices is:

(i) write the equations in the form

$$a_1 x + b_1 y + c_1 z = d_1$$
$$a_2 x + b_2 y + c_2 z = d_2$$
$$a_3 x + b_3 y + c_3 z = d_3$$

(ii) write the matrix equation corresponding to these equations, i.e.

$$\begin{pmatrix} a_1 & b_1 & c_1 \\ a_2 & b_2 & c_2 \\ a_3 & b_3 & c_3 \end{pmatrix} \times \begin{pmatrix} x \\ y \\ z \end{pmatrix} = \begin{pmatrix} d_1 \\ d_2 \\ d_3 \end{pmatrix}$$

(iii) determine the inverse matrix of

$$\begin{pmatrix} a_1 & b_1 & c_1 \\ a_2 & b_2 & c_2 \\ a_3 & b_3 & c_3 \end{pmatrix}$$

(iv) multiply each side of (ii) by the inverse matrix, and

(v) solve for x, y and z by equating the corresponding elements

Problem 4. Use matrices to solve the simultaneous equations

$$x + y + z - 4 = 0 \tag{1}$$

$$2x - 3y + 4z - 33 = 0 \tag{2}$$

$$3x - 2y - 2z - 2 = 0 \tag{3}$$

(i) Writing the equations in the $a_1x + b_1y + c_1z = d_1$ form gives:

$$x + y + z = 4$$

$$2x - 3y + 4z = 33$$

$$3x - 2y - 2z = 2$$

(ii) The matrix equation is

$$\begin{pmatrix} 1 & 1 & 1 \\ 2 & -3 & 4 \\ 3 & -2 & -2 \end{pmatrix} \times \begin{pmatrix} x \\ y \\ z \end{pmatrix} = \begin{pmatrix} 4 \\ 33 \\ 2 \end{pmatrix}$$

(iii) The inverse matrix of $A = \begin{pmatrix} 1 & 1 & 1 \\ 2 & -3 & 4 \\ 3 & -2 & -2 \end{pmatrix}$

is given by $A^{-1} = \dfrac{\text{adj } A}{|A|}$

The adjoint of A is the transpose of the matrix of the cofactors of the elements (see Chapter 44). The matrix of cofactors is

$$\begin{pmatrix} 14 & 16 & 5 \\ 0 & -5 & 5 \\ 7 & -2 & -5 \end{pmatrix}$$

and the transpose of this matrix gives

$$\text{adj } A = \begin{pmatrix} 14 & 0 & 7 \\ 16 & -5 & -2 \\ 5 & 5 & -5 \end{pmatrix}$$

The determinant of A, i.e. the sum of the products of elements and their cofactors, using a first row expansion is

$$1 \begin{vmatrix} -3 & 4 \\ -2 & -2 \end{vmatrix} - 1 \begin{vmatrix} 2 & 4 \\ 3 & -2 \end{vmatrix} + 1 \begin{vmatrix} 2 & -3 \\ 3 & -2 \end{vmatrix}$$

i.e. $(1 \times 14) - (1 \times -16) + (1 \times 5)$, that is, 35.

Hence the inverse of A,

$$A^{-1} = \frac{1}{35} \begin{pmatrix} 14 & 0 & 7 \\ 16 & -5 & -2 \\ 5 & 5 & -5 \end{pmatrix}$$

(iv) Multiplying each side of (ii) by (iii), remembering that $A \times A^{-1} = I$, the unit matrix gives

$$\begin{pmatrix} 1 & 0 & 0 \\ 0 & 1 & 0 \\ 0 & 0 & 1 \end{pmatrix} \times \begin{pmatrix} x \\ y \\ z \end{pmatrix}$$

$$= \frac{1}{35} \begin{pmatrix} 14 & 0 & 7 \\ 16 & -5 & -2 \\ 5 & 5 & -5 \end{pmatrix} \times \begin{pmatrix} 4 \\ 33 \\ 2 \end{pmatrix}$$

$$\begin{pmatrix} x \\ y \\ z \end{pmatrix} = \frac{1}{35} \begin{pmatrix} (14 \times 4) + (0 \times 33) \\ + (7 \times 2) \\ (16 \times 4) + (-5 \times 33) \\ + ((-2) \times 2) \\ (5 \times 4) + (5 \times 33) \\ + ((-5) \times 2) \end{pmatrix}$$

$$= \frac{1}{35} \begin{pmatrix} 70 \\ -105 \\ 175 \end{pmatrix} = \begin{pmatrix} 2 \\ -3 \\ 5 \end{pmatrix}$$

(v) By comparing corresponding elements, $x = 2$, $y = -3$, $z = 5$, which can be checked in the original equations

Further problems on solving simultaneous equations using matrices may be found in Section 45.4, Problems 6 to 14, pages 450 and 451.

45.3 Solution of simultaneous equations by determinants

(a) When solving linear simultaneous equations in **two unknowns** using determinants:

(i) write the equations in the form

$$a_1x + b_1y + c_1 = 0$$

$$a_2x + b_2y + c_2 = 0$$

and then

(ii) the solution is given by

$$\frac{x}{D_x} = \frac{-y}{D_y} = \frac{1}{D}$$

where $D_x = \begin{vmatrix} b_1 & c_1 \\ b_2 & c_2 \end{vmatrix}$

i.e. the determinant of the coefficients left when the x column is covered up

$$D_y = \begin{vmatrix} a_1 & c_1 \\ a_2 & c_2 \end{vmatrix}$$

i.e. the determinant of the coefficients left when the y column is covered up, and

$$D = \begin{vmatrix} a_1 & b_1 \\ a_2 & b_2 \end{vmatrix}$$

i.e. the determinant of the coefficients left when the constants column is covered up

---

Problem 5. The velocity of a car, accelerating at uniform acceleration a between two points, is given by $v = u + at$, where u is its velocity when passing the first point and t is the time taken to pass between the two points. If $v = 21$ m/s when $t = 3.5$ s and $v = 33$ m/s when $t = 6.1$ s, use determinants to find the values of u and a, each correct to 4 significant figures

---

Substituting the given values in $v = u + at$ gives

$$u + 3.5a = 21 \tag{1}$$

$$u + 6.1a = 33 \tag{2}$$

(i) The equations are written in the form $a_1x + b_1y + c_1 = 0$, i.e. $u + 3.5a - 21 = 0$ and $u + 6.1a - 33 = 0$

(ii) The solution is given by $\dfrac{u}{D_u} = \dfrac{-a}{D_a} = \dfrac{1}{D}$,

where D_u is the determinant of coefficients left when the u column is covered up,

i.e. $\quad D_u = \begin{vmatrix} 3.5 & -21 \\ 6.1 & -33 \end{vmatrix} = 12.6$

Similarly, $\quad D_a = \begin{vmatrix} 1 & -21 \\ 1 & -33 \end{vmatrix} = -12$

and $\quad D = \begin{vmatrix} 1 & 3.5 \\ 1 & 6.1 \end{vmatrix} = 2.6$

Thus $\dfrac{u}{12.6} = \dfrac{-a}{-12} = \dfrac{1}{2.6}$

i.e. $u = \dfrac{12.6}{2.6} = \mathbf{4.846}$ **m/s**

and $a = \dfrac{12}{2.6} = \mathbf{4.615}$ **m/s²**, each correct to 4 significant figures

---

(b) When solving simultaneous equations in **three unknowns** using determinants:

(i) Write the equations in the form

$$a_1x + b_1y + c_1z + d_1 = 0$$

$$a_2x + b_2y + c_2z + d_2 = 0$$

$$a_3x + b_3y + c_3z + d_3 = 0$$

and then

(ii) the solution is given by

$$\frac{x}{D_x} = \frac{-y}{D_y} = \frac{z}{D_z} = \frac{-1}{D}$$

where D_x is $\begin{vmatrix} b_1 & c_1 & d_1 \\ b_2 & c_2 & d_2 \\ b_3 & c_3 & d_3 \end{vmatrix}$

i.e. the determinant of the coefficients obtained by covering up the x column

$$D_y \text{ is } \begin{vmatrix} a_1 & c_1 & d_1 \\ a_2 & c_2 & d_2 \\ a_3 & c_3 & d_3 \end{vmatrix}$$

i.e. the determinant of the coefficients obtained by covering up the y column

$$D_z \text{ is } \begin{vmatrix} a_1 & b_1 & d_1 \\ a_2 & b_2 & d_2 \\ a_3 & b_3 & d_3 \end{vmatrix}$$

i.e. the determinant of the coefficients obtained by covering up the z column

and D is $\begin{vmatrix} a_1 & b_1 & c_1 \\ a_2 & b_2 & c_2 \\ a_3 & b_3 & c_3 \end{vmatrix}$

i.e. the determinant of the coefficients obtained by covering up the constants column

---

Problem 6. A d.c. circuit comprises three closed loops. Applying Kirchhoff's laws to the closed loops gives the following equations for current flow in milliamperes:

$$2I_1 + 3I_2 - 4I_3 = 26$$

$$I_1 - 5I_2 - 3I_3 = -87$$

$$-7I_1 + 2I_2 + 6I_3 = 12$$

Use determinants to solve for I_1, I_2 and I_3

(i) Writing the equations in the $a_1x + b_1y + c_1z + d_1 = 0$ form gives:

$$2I_1 + 3I_2 - 4I_3 - 26 = 0$$

$$I_1 - 5I_2 - 3I_3 + 87 = 0$$

$$-7I_1 + 2I_2 + 6I_3 - 12 = 0$$

(ii) The solution is given by

$$\frac{I_1}{D_{I_1}} = \frac{-I_2}{D_{I_2}} = \frac{I_3}{D_{I_3}} = \frac{-1}{D},$$

where D_{I_1} is the determinant of coefficients obtained by covering up the I_1 column, i.e.

$$D_{I_1} = \begin{vmatrix} 3 & -4 & -26 \\ -5 & -3 & 87 \\ 2 & 6 & -12 \end{vmatrix}$$

The evaluation of this determinant may be simplified as follows:
Adding twice column 2 to column 3 gives

$$\begin{vmatrix} 3 & -4 & -34 \\ -5 & -3 & 81 \\ 2 & 6 & 0 \end{vmatrix}$$

Taking three times column 1 from column 2 gives

$$\begin{vmatrix} 3 & -13 & -34 \\ -5 & 12 & 81 \\ 2 & 0 & 0 \end{vmatrix}$$

Hence

$$D_{I_1} = 2[(-13 \times 81) - (12 \times (-34))] = -1290$$

Also, $\quad D_{I_2} = \begin{vmatrix} 2 & -4 & -26 \\ 1 & -3 & 87 \\ -7 & 6 & -12 \end{vmatrix} = 1806$

$$D_{I_3} = \begin{vmatrix} 2 & 3 & -26 \\ 1 & -5 & 87 \\ -7 & 2 & -12 \end{vmatrix} = 1161$$

and $\quad D = \begin{vmatrix} 2 & 3 & -4 \\ 1 & -5 & -3 \\ -7 & 2 & 6 \end{vmatrix} = 129$

Thus $\dfrac{I_1}{-1290} = \dfrac{-I_2}{1806} = \dfrac{I_3}{-1161} = \dfrac{-1}{129}$

giving $I_1 = 10$ mA, $I_2 = 14$ mA and $I_3 = 9$ mA

Further problems on solving simultaneous equations using determinants may be found in the following Section 45.4, Problems 15 to 22, page 451.

45.4 Further problems on the solution of simultaneous equations by matrices and determinants

Properties of determinants

In Problems 1 to 5 simplify and evaluate the determinants given:

1 $\begin{vmatrix} 7 & 5 \\ 21 & 26 \end{vmatrix}$ [77]

2 $\begin{vmatrix} -5 & -4 \\ 10 & 5 \end{vmatrix}$ [15]

3 $\begin{vmatrix} 14 & 14 & 20 \\ 11 & 9 & 13 \\ 15 & 17 & 19 \end{vmatrix}$ [144]

4 $\begin{vmatrix} 8 & -2 & -10 \\ 2 & -3 & -2 \\ 6 & 3 & 8 \end{vmatrix}$ [−328]

5 $\begin{vmatrix} 14 & 42 & 33 \\ 13 & 47 & 36 \\ 17 & 24 & 22 \end{vmatrix}$ [1]

Solving simultaneous equations by matrices

In Problems 6 to 10 use **matrices** to solve the simultaneous equations given:

6 $3x + 4y = 0$
   $2x + 5y + 7 = 0$ $[x = 4, \ y = -3]$

7 $2p + 5q + 14.6 = 0$
   $3.1p + 1.7q + 2.06 = 0$
   $[p = 1.2, \ q = -3.4]$

8 $x + 2y + 3z = 5$
   $2x - 3y - z = 3$
   $-3x + 4y + 5z = 3$
   $[x = 1, \ y = -1, \ z = 2]$

9 $3a + 4b - 3c = 2$
   $-2a + 2b + 2c = 15$
   $7a - 5b + 4c = 26$
   $[a = 2.5, \ b = 3.5, \ c = 6.5]$

10 $p + 2q + 3r + 7.8 = 0$
   $2p + 5q - r - 1.4 = 0$
   $5p - q + 7r - 3.5 = 0$
   $[p = 4.1, \ q = -1.9, \ r = -2.7]$

11 In two closed loops of an electrical circuit, the currents flowing are given by the simultaneous equations

$$I_1 + 2I_2 + 4 = 0$$
$$5I_1 + 3I_2 - 1 = 0$$

Use matrices to solve for I_1 and I_2.
   $[I_1 = 2, \ I_2 = -3]$

12 The relationship between the displacement, s, velocity, v, and acceleration, a, of a piston is given by the equations

$$s + 2v + 2a = 4$$

$$3s - v + 4a = 25$$

$$3s + 2v - a = -4$$

Use matrices to determine the values of s, v and a. $[s = 2, v = -3, a = 4]$

13 In a mechanical system, acceleration $\ddot{x}$, velocity $\dot{x}$ and distance x are related by the simultaneous equations

$$3.4\ddot{x} + 7.0\dot{x} - 13.2x = -11.39$$

$$-6.0\ddot{x} + 4.0\dot{x} + 3.5x = 4.98$$

$$2.7\ddot{x} + 6.0\dot{x} + 7.1x = 15.91$$

Use matrices to find the values of $\ddot{x}$, $\dot{x}$ and x. $[\ddot{x} = 0.5, \dot{x} = 0.77, x = 1.4]$

14 Prove that the inverse matrix,

$$A^{-1} = 0.2 \begin{pmatrix} -1 & 9 & -3 \\ -2 & 3 & -1 \\ 2 & -13 & 6 \end{pmatrix}$$

when $A = \begin{pmatrix} 1 & -3 & 0 \\ 2 & 0 & 1 \\ 4 & 1 & 3 \end{pmatrix}$

Hence solve the equations
$x - 3y = a$, $2x + z = b$, $4x + y + 3z = c$
for x, y and z in terms of a, b and c.

$$\begin{bmatrix} x & = -0.2a + 1.8b - 0.6c \\ y & = -0.4a + 0.6b - 0.2c \\ z & = 0.4a - 2.6b + 1.2c \end{bmatrix}$$

Solving simultaneous equations by determinants

In Problems 15 to 19 use **determinants** to solve the simultaneous equations given:

15 $3x - 5y = -17.6$
$7y - 2x - 22 = 0$ $[x = -1.2, y = 2.8]$

16 $2.3m - 4.4n = 6.84$
$8.5n - 6.7m = 1.23$

$[m = -6.4, n = -4.9]$

17 $3x + 4y + z = 10$
$2x - 3y + 5z + 9 = 0$
$x + 2y - z = 6$ $[x = 1, y = 2, z = -1]$

18 $1.2p - 2.3q - 3.1r + 10.1 = 0$
$4.7p + 3.8q - 5.3r - 21.5 = 0$
$3.7p - 8.3q + 7.4r + 28.1 = 0$
$[p = 1.5, q = 4.5, r = 0.5]$

19 $\dfrac{x}{2} - \dfrac{y}{3} + \dfrac{2z}{5} = -\dfrac{1}{20}$

$\dfrac{x}{4} + \dfrac{2y}{3} - \dfrac{z}{2} = \dfrac{19}{40}$

$x + y - z = \dfrac{59}{60}$

$$\left[x = \frac{7}{20}, \ y = \frac{17}{40}, \ z = -\frac{5}{24} \right]$$

20 In a system of forces, the relationship between two forces F_1 and F_2 is given by

$$5F_1 + 3F_2 + 6 = 0$$

$$3F_1 + 5F_2 + 18 = 0$$

Use determinants to solve for F_1 and F_2.
$[F_1 = 1.5, F_2 = -4.5]$

21 Kirchhoff's laws are used to determine the current equations in an electrical network and show that

$$i_1 + 8i_2 + 3i_3 = -31$$

$$3i_1 - 2i_2 + i_3 = -5$$

$$2i_1 - 3i_2 + 2i_3 = 6$$

Use determinants to solve for i_1, i_2 and i_3.
$[i_1 = -5, i_2 = -4, i_3 = 2]$

22 The forces in three members of a framework are F_1, F_2 and F_3. They are related by the simultaneous equations shown below. Find the values of F_1, F_2 and F_3 using determinants.

$$1.4F_1 + 2.8F_2 + 2.8F_3 = 5.6$$

$$4.2F_1 - 1.4F_2 + 5.6F_3 = 35.0$$

$$4.2F_1 + 2.8F_2 - 1.4F_3 = -5.6$$

$$[F_1 = 2, F_2 = -3, F_3 = 4]$$

Multiple choice questions on Part 4 Extended Mathematics for Engineering

All questions have only one correct answer

Suggested time allowed: $1\dfrac{1}{4}$ hours

1 $\dfrac{5}{j^6}$ is equivalent to

   A $j5$ B -5 C $-j5$ D 5

2 The value of $\displaystyle\int_0^{\pi/6} 2\sin\left(3t+\dfrac{\pi}{2}\right)dt$ is

   A 6   B $-\dfrac{2}{3}$ C -6 D $\dfrac{2}{3}$

3 The value of the determinant $\begin{vmatrix} 2 & -1 & 4 \\ 0 & 1 & 5 \\ 6 & 0 & -1 \end{vmatrix}$ is

   A 4   B 52   C -56 D 8

4 $2\angle\dfrac{\pi}{3}+3\angle\dfrac{\pi}{6}$ in polar form is

   A $5\angle\dfrac{\pi}{2}$ B $4.84\angle0.84$

   C $6\angle0.55$ D $4.84\angle0.73$

5 The matrix product $\begin{pmatrix} 2 & 3 \\ -1 & 4 \end{pmatrix}\begin{pmatrix} 1 & -5 \\ -2 & 6 \end{pmatrix}$ is equal to

   A $\begin{pmatrix} -13 \\ 26 \end{pmatrix}$ B $\begin{pmatrix} 3 & -2 \\ -3 & 10 \end{pmatrix}$

   C $\begin{pmatrix} -4 & 8 \\ -9 & 29 \end{pmatrix}$ D $\begin{pmatrix} 1 & -2 \\ -3 & -2 \end{pmatrix}$

6 $\displaystyle\int \ln x\,dx$ is equal to

   A $x(\ln x - 1)+c$ B $\dfrac{1}{x}+c$

   C $x\ln x - 1 + c$ D $\dfrac{1}{x}+\dfrac{1}{x^2}+c$

7 The value of $\displaystyle\int_0^{\pi/3} 16\cos^4\theta\sin\theta\,d\theta$ is

   A -0.1 B 3.1 C 0.1 D -3.1

8 $(1+j)^4$ is equivalent to

   A 4   B $-j4$ C $j4$ D -4

9 The inverse of the matrix $\begin{pmatrix} 5 & -3 \\ -2 & 1 \end{pmatrix}$ is

   A $\begin{pmatrix} -5 & 3 \\ 2 & -1 \end{pmatrix}$ B $\begin{pmatrix} -1 & -3 \\ -2 & -5 \end{pmatrix}$

   C $\begin{pmatrix} -1 & 3 \\ 2 & -5 \end{pmatrix}$ D $\begin{pmatrix} 1 & 3 \\ 2 & 5 \end{pmatrix}$

10 The graph of $y = 2\tan 3\theta$ is

   A a continuous, periodic, even function
   B a discontinuous, non-periodic, odd function
   C a discontinuous, periodic, odd function
   D a continuous, non-periodic, even function

11 The two square roots of $(-3+j4)$ are

   A $\pm(1+j2)$ B $\pm(0.71+j2.12)$
   C $\pm(1-j2)$ D $\pm(0.71-j2.12)$

12 $\displaystyle\int_2^3 \dfrac{3}{x^2+x-2}\,dx$ is equal to

   A $3\ln 2.5$ B $\dfrac{1}{3}\lg 1.6$

   C $\ln 40$ D $\ln 1.6$

13 The adjoint of the matrix $\begin{pmatrix} 1 & 2 & 3 \\ -4 & 0 & 1 \\ 0 & 5 & -2 \end{pmatrix}$ is

   A $\begin{pmatrix} -5 & 8 & -20 \\ -19 & -2 & 5 \\ 2 & 13 & 8 \end{pmatrix}$

   B $\begin{pmatrix} 1 & -4 & 0 \\ 2 & 0 & 5 \\ 3 & 1 & -2 \end{pmatrix}$

C $\begin{pmatrix} -5 & -16 & -60 \\ 76 & 0 & -5 \\ 0 & -65 & -16 \end{pmatrix}$

D $\begin{pmatrix} -5 & 19 & 2 \\ -8 & -2 & -13 \\ -20 & -5 & 8 \end{pmatrix}$

14 $(-4 - j3)$ in polar form is

A $5\angle -143.13°$ B $5\angle 126.87°$

C $5\angle 143.13°$ D $5\angle -126.87°$

15 $\int xe^{2x}$ is

A $\dfrac{x^2}{4}e^{2x} + c$

B $2e^{2x} + c$

C $\dfrac{e^{2x}}{4}(2x - 1) + c$

D $2e^{2x}(x - 2) + c$

16 The value of $\left| \begin{matrix} j2 & -(1+j) \\ (1-j) & 1 \end{matrix} \right|$ is

A $2(1 + j)$ B 2

C $-j2$ D $-2 + j2$

17 $\int_0^{\pi/2} 2\sin^3 t\, dt$ is equal to

A $1\dfrac{1}{3}$ B $-\dfrac{1}{4}$ C $-1\dfrac{1}{3}$ D $\dfrac{1}{4}$

18 $[2\angle 30°]^4$ in Cartesian form is

A $(0.50 + j0.06)$ B $(-8 + j13.86)$

C $(-4 + j6.93)$ D $(13.86 + j8)$

19 If matrix $P = \begin{pmatrix} 1 & 3 & -2 \\ 0 & 4 & 1 \\ -2 & 5 & 0 \end{pmatrix}$ then the transpose of P, P^T is

A $\begin{pmatrix} -5 & -2 & 8 \\ -10 & -4 & -11 \\ 11 & -1 & 4 \end{pmatrix}$

B $\begin{pmatrix} -5 & 2 & 8 \\ 10 & -4 & 11 \\ 11 & 1 & 4 \end{pmatrix}$

C $\begin{pmatrix} 1 & 0 & -2 \\ 3 & 4 & 5 \\ -2 & 1 & 0 \end{pmatrix}$

D $-\dfrac{1}{27} \begin{pmatrix} -5 & -2 & 8 \\ -10 & -4 & -11 \\ 11 & -1 & 4 \end{pmatrix}$

20 Evaluating $\int_0^4 \dfrac{1}{\sqrt{(16 - t^2)}}\, dt$ gives

A 2   B $\dfrac{\pi}{2}$

C -2 D 90

(Answers on page 454.)

Answers to multiple choice questions

Part 1 (pages 200 to 202)

1	B	2	D	3	A	4	C
5	A	6	D	7	C	8	A
9	D	10	B	11	D	12	C
13	B	14	B	15	A	16	C
17	D	18	C	19	B	20	A

Part 3 (pages 375 to 377)

1	A	2	D	3	B	4	C
5	A	6	B	7	D	8	C
9	A	10	C	11	B	12	D
13	B	14	A	15	D	16	B
17	C	18	A	19	D	20	C

Part 2 (pages 297 and 298)

1	B	2	C	3	B	4	A
5	D	6	C	7	D	8	A
9	D	10	B	11	D	12	B
13	A	14	C	15	D	16	A
17	C	18	A	19	B	20	C

Part 4 (pages 452 and 453)

1	B	2	D	3	C	4	D
5	C	6	A	7	B	8	D
9	B	10	C	11	A	12	D
13	D	14	A	15	C	16	A
17	A	18	B	19	C	20	B

Index